Brains Through Time

Brains Through Time

A Natural History of Vertebrates

GEORG F. STRIEDTER AND R. GLENN NORTHCUTT

OXFORD
UNIVERSITY PRESS

OXFORD
UNIVERSITY PRESS

Oxford University Press is a department of the University of Oxford. It furthers the University's objective of excellence in research, scholarship, and education by publishing worldwide. Oxford is a registered trade mark of Oxford University Press in the UK and certain other countries.

Published in the United States of America by Oxford University Press
198 Madison Avenue, New York, NY 10016, United States of America.

Library of Congress Cataloging-in-Publication Data
Names: Striedter, Georg F., 1962– author. | Northcutt, R. Glenn, author.
Title: Brains through time : a natural history of vertebrates /
Georg F. Striedter, Department of Neurobiology & Behavior,
University of California, Irvine, R. Glenn Northcutt,
Department of Neuroscience, University of California, San Diego.
Description: New York, NY, United States of America : Oxford University Press, [2020] |
Includes bibliographical references and index.
Identifiers: LCCN 2019017476 | ISBN 9780195125689 (hardback) |
ISBN 9780190055462 (updf) | ISBN 9780190079734 (epub) | ISBN 9780190081461 (oso)
Subjects: LCSH: Vertebrates—Evolution. | Brain—Evolution. | Nervous system—Evolution.
Classification: LCC QL607.5 .S77 2020 | DDC 596—dc23
LC record available at https://lccn.loc.gov/2019017476

5 7 9 8 6 4

Printed by Integrated Books International, United States of America

Dedicated to all students of evolutionary neurobiology,
past, present, and future

Contents

Preface

For a professor to write a book with one of his PhD students 20 years after said student left the lab is highly unusual; it suggests a complicated back-story. Indeed, the idea for this book was hatched a long, long time ago. Here is what Glenn, the senior author, wrote in 1996, in his initial proposal to publishers:

> The evolution of vertebrate brains and the behaviors they subserved can best be understood in the broader context of vertebrate evolution. In order to understand why brains exhibit different patterns of organization among existing vertebrates, it is first necessary to understand the origin and subsequent radiation of each vertebrate group. This, in turn, necessitates some understanding of the physical and biological worlds in which these events took place. When vertebrate diversity is viewed in this context, it becomes apparent that the origin of each vertebrate group was based on a few key innovations that allowed an ancestral population to exploit the environment in a novel way; subsequent evolution of that population was channeled (both facilitated and constrained) by these innovations. Once the key innovations of a particular group are identified, it is possible to see how brains were modified and how they reflected and subserved these innovations. While this approach is widely used in most areas of biological research, neurobiology is a notable exception. Most neurobiological research is model driven and results in detailed data that are frequently unintelligible and uninteresting to most integrative and organismal biologists. This book is an attempt to bring neurobiology within the purview of the wider integrative biological audience and, at the same time, to demonstrate the strengths of a comparative approach to the neurobiological community.

This proposal was met with great excitement by Glenn's students and friends, and it resulted in a book contract. However, execution of the plan was far more challenging than expected. As Glenn began to write, the number of planned pages swelled, and many pages filled with jargon and a slew of anatomical details. The number of planned figures also grew, as Glenn developed plans to illustrate the internal organization of many different vertebrate brains, their cranial nerves, and all the major sense organs, not to mention their phylogenetic relationships and ecological contexts. Long fragments of two chapters gradually emerged, but Glenn became increasingly concerned about losing his audience. Who would want to read this book? Who would be able to absorb all of the included detail? As the concerns grew, the book project languished.

When Glenn retired in the summer of 2014, Glenn's friends urged him to pick the project up again. Especially adamant were some of Glenn's former students, among them Georg. Toward the end of 2014, over lunch in San Diego, Georg offered to help in any way he could. Having just finished the writing of an introductory textbook on neurobiology, Georg was also casting about for a new "big project." By the end of lunch, a plan was hatched for Georg to join the project as junior partner, using his book writing skills to complement Glenn's deep knowledge of vertebrate brains, both extant and extinct. Georg's knowledge of avian and mammalian nervous systems would also nicely complement Glenn's expertise on piscine, amphibian, and reptilian brains. As first author, Georg would do the writing and generate most of the figures, but both authors planned to meet regularly to harmonize their vision, hatch specific plans, and discuss revisions to the figures and text. Four years and more than a dozen face-to-face meetings later, the book is here.

Reflecting on the process now, the metaphor of two people paddling a canoe (illustrated in Figure P.1) seems apt. Georg was sitting in the front, paddling furiously and alert to diverse obstacles and possible paths. Glenn was in the back, steering in more subtle ways and delivering occasional power strokes. Together, they managed to get through unharmed (and without major fights!). More importantly, they took a path that neither one of them alone would have been able to pursue. Nor was the path entirely predictable. In several cases, both authors abandoned some of their own long-standing views. Those moments were especially thrilling, perhaps because they were risky.

Of course, Glenn and Georg had lots of help along the way. Mary Sue Northcutt provided strong support, including editorial feedback on the entire manuscript. Anna Striedter was similarly supportive, putting up with Georg disappearing to San Diego on numerous Sundays and being distracted on many other days. Then there is Jeremy Lewis, the editor, who was willing to revise the first contract and sought helpful anonymous reviews. In addition, invaluable feedback came from many colleagues and friends, including Ann Butler, Mark Braford, Chris Braun, Jenny Clack, Barbara Finlay, Bernd Fritzsch, Agustin Gonzalez, Andrew Iwaniuk, Jon

Figure P.1 The Paddler twins. Artwork from Tikal's Temple I (burial 116), showing the two Paddler gods transporting the Mayan maize god, as well as supernatural anthropomorphic iguana, spider monkey, parrot, and dog.

From Schele and Miller (1986, *The Blood of Kings*. London: Thames & Hudson); drawing by Linda Schele, © David Schele, SD-2014, with permission.

Kaas, Shigeru Kuratani, Catherine McCormick, Michael Pritz, Luis Puelles, Walt Wilczynski, Steven Wise, and Mario Wullimann. Several of them provided an enormous number of comments, which were often challenging to address but in the end improved the manuscript substantially. We are lucky to have such selfless friends, and grateful for their contributions to this book. Hopefully you, dear reader, will find the end result informative.

Georg Striedter and Glenn Northcutt—Irvine and Poway, CA

1

Reconstructing History

Aims and Methodology

> There is one purpose to life and one only: to bear witness to and understand as much as possible of the complexity of the world—its beauty, its mysteries, its riddles. The more you understand, the more you look, the greater is your enjoyment of life and your sense of peace.
>
> —Anne Rice, *Servant of the Bones* (1996, p. 135)

> It is interesting to contemplate a tangled bank, clothed with many plants of many kinds, with birds singing on the bushes, with various insects flitting about, and with worms crawling through the damp earth, and to reflect that these elaborately constructed forms, so different from each other, and dependent upon each other in so complex a manner, have all been produced by laws acting around us. . . . There is grandeur in this view of life.
>
> —Charles Darwin, *The Origin of Species*, 6th ed. (1872, pp. 444–445)

Separated by more than 100 years, by the Atlantic Ocean, and by very different lines of work, Anne Rice and Charles Darwin both described the pleasure that can come from observing nature's complexity. One can, of course, appreciate nature without giving it much thought, but Rice and Darwin were describing a deeper sense of joy: one that comes from not just looking at nature, but from understanding how its diverse elements are connected to one another and to us. This deeper sense of enjoyment lies at the heart of natural history and is what we hope to facilitate.

But how does one go from merely looking at nature to understanding it? Two main approaches have been recognized. One is to collect a large number of observations in various places, under a variety of different conditions, and then to look for meaningful patterns in the collected data. This comparative approach can suggest hypotheses that might explain the discovered pattern. These hypotheses can then be tested using the second, complementary approach, which is to manipulate nature in highly specific ways and then observe whether the manipulations have the expected effects; i.e., the experimental method. The crucial difference is that the comparative method relies on naturally occurring variation, whereas the experimental method creates artificial variation that can be tightly controlled.

Brains Through Time. Georg F. Striedter and R. Glenn Northcutt, Oxford University Press (2020). © Oxford University Press.
DOI: 10.1093/oso/9780195125689.001.0001

Although the experimental method is often celebrated as the key to scientific success, the comparative approach has played a crucial role in most scientific fields. In the following sections, we illustrate this role, first in several non-biological domains and then within biology. Our aim is not to elevate comparative approaches over experimental ones, but to highlight the utility of comparative research, especially in disciplines that are not very amenable to experimentation. This sampling of comparative research in diverse disciplines also allows us to introduce a few concepts that are important for later chapters, such as the idea that a layer of rock can be dated to a specific period in the earth's history, that continental plates move slowly across the globe, and that the history of species long extinct can be reconstructed from present-day observations.

1.1. Comparative Approaches Outside of Biology

Comparative approaches are used in virtually all fields of academic inquiry, ranging from literature and art to social sciences and economics (e.g., Lijphart, 1971; Collier, 1993). For our purposes, it is most instructive to examine how comparative research has advanced our knowledge in some of the physical sciences, notably astrophysics, chemistry, and geology.

1.1.1. Stellar Evolution

When you look up at the sky on a clear, moonless night, you may see a few thousand stars. It can be fun and awe inspiring, especially if you are in the mountains or the desert, but most of us will not see much of a pattern in those stars. You might notice that some are brighter than others, that they vary slightly in color, and that they are more common in some parts of the sky (e.g., the Milky Way). You might even recognize some of the constellations you learned about, but the pattern still appears largely random.

Not so for astronomers. Over many years, they diligently collected enormous amounts of data on the position, brightness (luminance), and color of the stars and then looked for patterns. A major breakthrough came in 1911–1913, when Ejnar Hertzsprung and Henry N. Russel plotted luminance versus color for hundreds of stars in the same general region of space, which meant that the included stars were likely similar in age and chemical composition (Gingerich, 2013). The resulting plot, now called the Hertzsprung-Russel diagram, revealed that roughly 90% of all the data points cluster along the diagonal, forming what astronomers call the "main sequence" (Figure 1.1). A second cluster forms a "horizontal branch," which diverges from the main sequence toward the right side of the diagram. A few additional data points aggregate in the diagram's top right corner, and a few are found near the bottom left. How should one explain this surprisingly

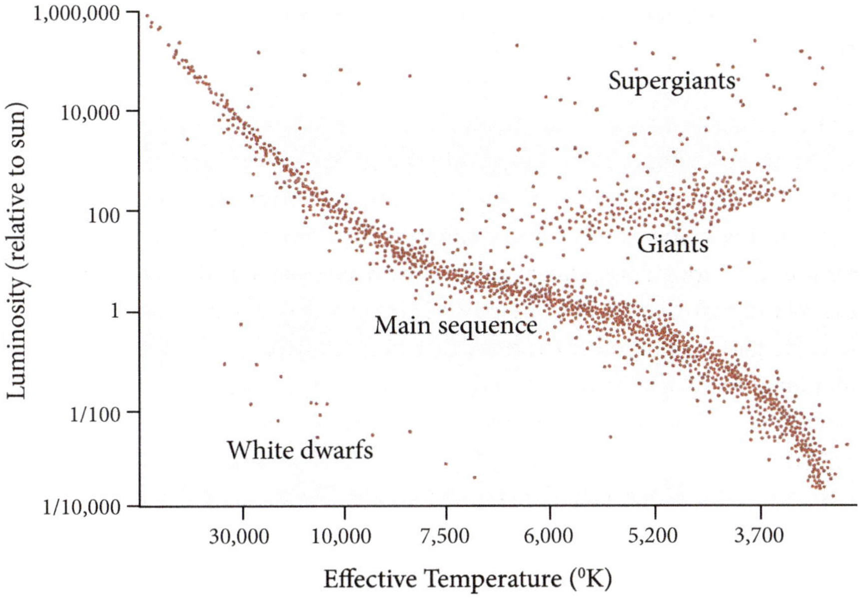

Figure 1.1 Hertzsprung-Russel diagram for nearby stars. Plotted here are the luminance and temperature (and color) values for several hundred stars in the Pleiades cluster near our sun. The stars on the diagonal "main sequence" are thought to be close to equilibrium while burning hydrogen in their core. When the hydrogen in the core runs out, they start burning helium and moving off the main sequence. Depending on their initial size, they expand either into giants or super-giants. When the fuel in these massive stars is depleted, they shrink until, eventually, most of them become "white dwarfs."

Adapted from Wiley and Lieberman (2011), which is based on original data from the Centre de Donnée astronomiques de Strasbourg.

heterogeneous, non-random distribution of stars in the Hertzsprung-Russel diagram?

The answer is that the observed pattern reflects the dynamics of stellar evolution. Astronomers had suspected since the 1880s that stars do not exist for all eternity but, instead, are born from the contraction of cosmic dust clouds and die when they shrink dramatically and cease to give off light. As a star goes through its life cycle, it changes in brightness and in temperature (i.e., color). The importance of the Hertzsprung-Russel diagram lay in its implication that stars spend most of their life cycle in an equilibrium state that places them along the main sequence. Combining these data with additional insights from experimental studies and theoretical considerations, astronomers hypothesized that, during their time on the main sequence, stars are burning hydrogen in their core and remain relatively constant in size. When the hydrogen fuel runs out, the stars begin to burn helium in their core. At that point they move off the main sequence along the horizontal branch (or nearby paths) and swell until they become "giants" or (depending on their initial mass) "supergiants." Eventually, every star becomes completely depleted of fuel and begins to shrink, collapsing into itself. Such depleted stars then become white dwarfs or, if they had been supergiants, black holes.

The Hertzsprung-Russel diagram was not, in and of itself, sufficient to generate all this information about stellar evolution, but it was a major factor, for example, in the realization that stars heat up and expand before they cool and shrink (Arny, 1990). It also influenced Arthur Eddington's theory about what happens inside stars and how this changes over time (Eddington, 1926). Remarkably, these advances in astronomy occurred long before scientists fully understood the chain reactions that happen at the extremely high pressures and temperatures inside a stellar core. That is, they predated our understanding of thermonuclear fusion, which flourished only in the 1940s and 1950s. The new knowledge did force some revision of the original ideas about stellar evolution, but the Hertzsprung-Russel diagram is still widely taught and often used to illustrate the developmental trajectories of individual stars.

1.1.2. The Periodic Table

Chemists labored for many years to identify the 94 different elements of which the natural world is built, as well as a few that can be synthesized. As scientists learned about the characteristics of these elements, they noticed both similarities and differences. Early attempts to make sense of this pattern included the recognition that similar elements sometimes come in groups of three (triads), that elemental properties tend to covary with atomic weight, and that some properties recur at regular intervals (i.e., periodically) along a series of increasing atomic weights (Scerri, 2015). The most successful attempt to recognize a meaningful pattern in the available data was that of Mendeléef, who grouped similar elements by arranging them in columns and rows, as shown in Figure 1.2. This effort allowed him to identify some mistakes in the data of the time and, more importantly, to predict the existence of several elements that had not yet been identified (Mendeléef, 1889). Mendeléef's original "periodic table" had to be amended in light of subsequent discoveries, such as the identification of noble gases, but his insights were based on a comparative approach and clearly drove the field of chemistry forward.

The pattern that Mendeléef identified is not inherently historical, but most chemical elements do have a history. The discovery of this fact owes much to a number of scientists who had collected data on the relative abundance of the various elements. They noticed that hydrogen is by far the most abundant element in the universe, with helium a distant second (Suess and Urey, 1956; Lodders, 2003). All of the other, heavier elements are relatively rare. Moreover, plotting an element's abundance against its atomic number revealed a distinct sawtooth pattern, with the even-numbered elements being more abundant than their odd-numbered neighbors (Figure 1.2). Combined with other knowledge, these relative abundance curves suggested that most elements were not generated at the time of the Big Bang, but rather are generated inside stars as they proceed through

Mendeleev 1869

			Ti = 50	Zr = 90	? = 180
			V = 51	Nb = 94	Ta = 182
			Cr = 52	Mo = 96	W = 186
			Mn = 55	Rh = 104.4	Pt = 197.4
			Fe = 56	Ru = 104.4	Ir = 198
			Ni = Co = 59	Pd = 106.6	Os = 199
H = 1			Cu = 63.4	Ag = 108	Hg = 200
	Be = 9.4	Mg = 24	Zn = 65.2	Cd = 112	
	B = 11	Al = 27.4	? = 68	Ur = 116	An = 197 ?
	C = 12	Si = 28	? = 70	Sn = 118	
	N = 14	P = 31	As = 75	Sb = 122	Bi = 210 ?
	O = 16	S = 32	Se = 79.4	Te = 128	
	F = 19	Cl = 35.5	Br = 80	J = 127	
Li = 1	Na = 23	K = 39	Rb = 85.4	Ce = 133	Tl = 204
		Ca = 40	Sr = 87.6	Ba = 137	Pb =207
		? = 45	Ce = 92		
		? Er = 56	La = 94		
		? Yt = 60	Di = 95		
		? In = 75.6	Th = 118 ?		

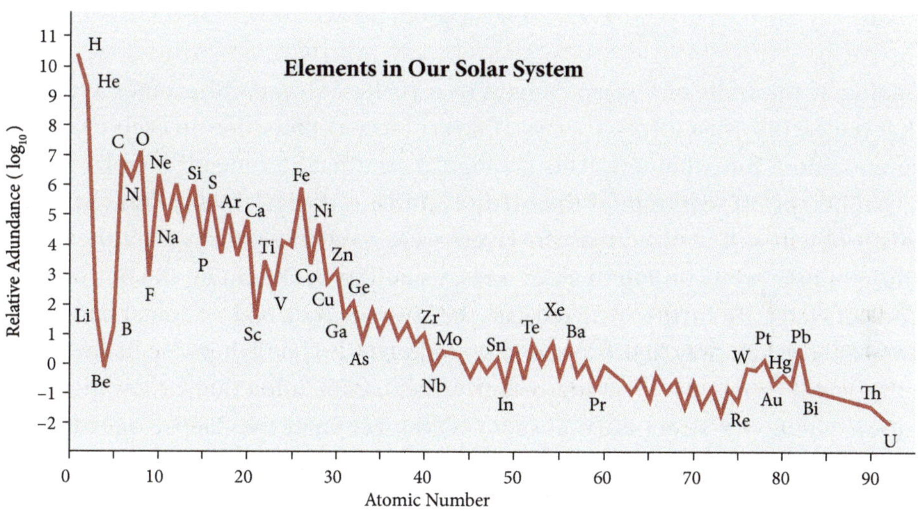

Figure 1.2 Patterns among the elements. Shown at the top is the original periodic table of elements that Dimitri Mendeléef (aka Mendeleev) proposed in 1869. Importantly, his schema allowed him to identify some mistakes in the earlier data and predict some elements that had not been discovered yet (red question marks). The bottom graph depicts the relative abundance of the various elements in our solar system. Hydrogen and helium are by far the most abundant elements (note the log scale!), and elements with even atomic numbers are more common than those with odd atomic numbers. Such graphs helped to inspire theories about how the vast majority of elements are synthesized inside of stars.

Data from Lodders (2003).

their life cycle (Burbidge et al., 1957). The fusion of two hydrogen nuclei inside a star generates helium. Subsequently, the fusion of helium nuclei with one another, and the fusion of their fusion products in various combinations, produces the other elements with even atomic numbers. Fusing some of them with hydrogen atoms creates the odd-numbered elements, but they are less commonly formed and less stable. Hence the bias toward even-numbered elements (aka the Oddo-Harkins rule).

Of course, stellar nucleosynthesis is more complex than we describe, but the comparative approach clearly helped to advance our understanding of how chemical elements are born. We also think that it is beautiful and grand, in Rice's and Darwin's sense, to realize that most of the elements that sustain life on earth originated in the stars. Since roughly 40 tons of cosmic dust enter our atmosphere each day, and many of its elements end up inside our bodies, we heartily assent to Joni Mitchell's lyrics that "we are stardust."

1.1.3. Geological Strata

Gazing at the walls of a steep canyon or a rocky outcrop, it becomes apparent that the earth's crust forms a series of layers (strata) that differ in their chemical composition. But what does this geological stratification mean? Nicolas Steno, a Catholic priest working for the Medici family in the late 1600s, thought about this problem and realized that the layers were most likely formed by the deposition of new layers on top of older ones, usually as sediment on the bottom of a body of water. He further realized that the layers of rock and soil must have been horizontal when they first formed, but could be tilted or otherwise distorted by subsequent processes. Most importantly, Steno concluded that, at any given location, the upper strata must always be younger than the deeper ones (Doyle et al., 2001).

Steno's insights were extended in the 1790s by William Smith, a British geologist who noticed that the ordering of the geological strata was consistent from place to place, at least within large parts of England. Being an avid fossil collector, Smith also realized that each layer contained a unique set of fossils, such that he could discriminate geologically similar layers by their fossil content. Combining this insight with Steno's "principle of superposition," Smith inferred that the fossils in the upper layers must be younger than those in the lower strata. Similar ideas had been developed independently (convergently!) by Georges Cuvier and Alexandre Brongniart in France around the same time. Curiously, none of these authors interpreted the succession of different fossils in the geological strata as evidence of evolution. Cuvier and Brongniart thought the changes were due to local ecological catastrophes, followed by migrations of new species into the formerly devastated areas, and Smith was not very interested in theoretical ideas at all. He was more interested in using his knowledge to find coal deposits or nutrient-rich soil. Indeed, the scientific understanding of geological stratification has frequently helped explorers discover coal, oil, and diverse minerals.

As geologists identified more and more geological strata, they grouped adjacent strata into a hierarchical arrangement of stages, series, systems, and erathems (Figure 1.3). Initially, the ages of the various layers could only be determined relative to one another, using Steno's principle of superposition. However, the development of carbon dating and various other techniques eventually made it possible to assign specific chronological dates to many of the layer boundaries. As a result, it is now possible to think of individual geological stages, series, systems, and erathems as chronological ages, epochs, periods, and eras, respectively. The Mesozoic era, for example, lasted from 252 to 66 million years ago, and the Jurassic period lasted from 201 to 145 years before the present (Figure 1.3). This absolute dating of geological strata is important because it allows us to specify when a particular species first appeared in the fossil record and, if the species became extinct, when it disappeared from that record. Since the fossil record is notoriously incomplete, a species might have existed earlier than the record suggests, and it might well have persisted longer. However, the dated fossils give us minimum estimates for both the first appearance of a species and its longevity. As we discuss shortly, these estimates are very useful for dating the divergence of various groups of animals from one another.

1.1.4. Plate Tectonics

The comparative approach further advanced geology by helping to reveal that the earth's continents have slowly moved across the globe. One of the first steps in this major discovery was the realization that the Western coastline of Africa runs roughly parallel with the Eastern coastline of South America. The fit of these continental boundaries is even better when one includes the shallow seas around the continents, (i.e., the continental shelves; Figure 1.4). This fit of the continental boundaries came to the attention of Alfred Wegener, who cleverly combined it with the knowledge that surprisingly similar fossils were found on opposite sides of the Atlantic (Demhardt, 2006). In 1912, Wegener proposed that the continents of today had once been assembled into a giant supercontinent, which he called Pangea. The flip side of this argument was Wegeners's "continental drift" hypothesis, according to which Pangea broke up when its component continents drifted apart. Unfortunately, the mechanism that Wegener proposed to explain why the continents would move as he proposed was widely deemed implausible and, as a result, his theory was rarely taken seriously.

Despite this setback, evidence in favor of continental movements continued to accumulate (Molnar, 2015). Especially important was the discovery that earthquakes and other seismic events were concentrated in specific lines across the globe (Figure 1.4), which we now recognize as defining the edges of continental plates. Additional evidence came from paleomagnetic studies, which are based on the fact

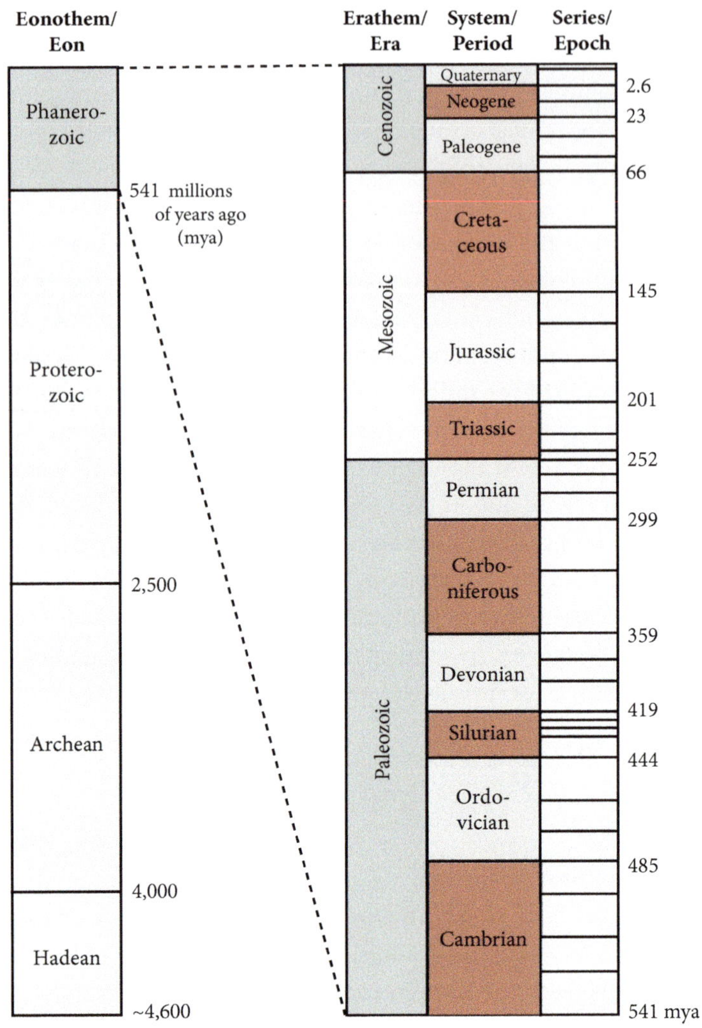

Figure 1.3 The geological timescale. The earth's crust can be divided into numerous layers (or strata), which can be grouped into a hierarchical set of series, systems, erathems, and eonothems. For non-geologists, it is generally simpler to think of these layers as epochs, periods, eras, and eons in the earth's history. Note that unicellular organisms arose around 4,000 mya, multicellular organisms around 800 mya, jawed vertebrates in the Cambrian, tetrapods in the Devonian, and amniotes in the Carboniferous period. Birds, mammals, teleosts, and modern elasmobranchs originated in the Jurassic period.

The illustrated dates for boundaries between the geological layers were taken from www.stratigraphy.com. Adapted from a figure by Alessandro Grippo (http://homepage.smc.edu/grippo_alessandro/gss1.html).

that the magnetic particles in rocks reflect the orientation of the earth's magnetic field at the time and place where the rock was formed. Such studies have shown that the ocean floor adjacent to some of the continental plate boundaries exhibits bands of alternating magnetic polarity that run parallel to the plate boundaries (Mason

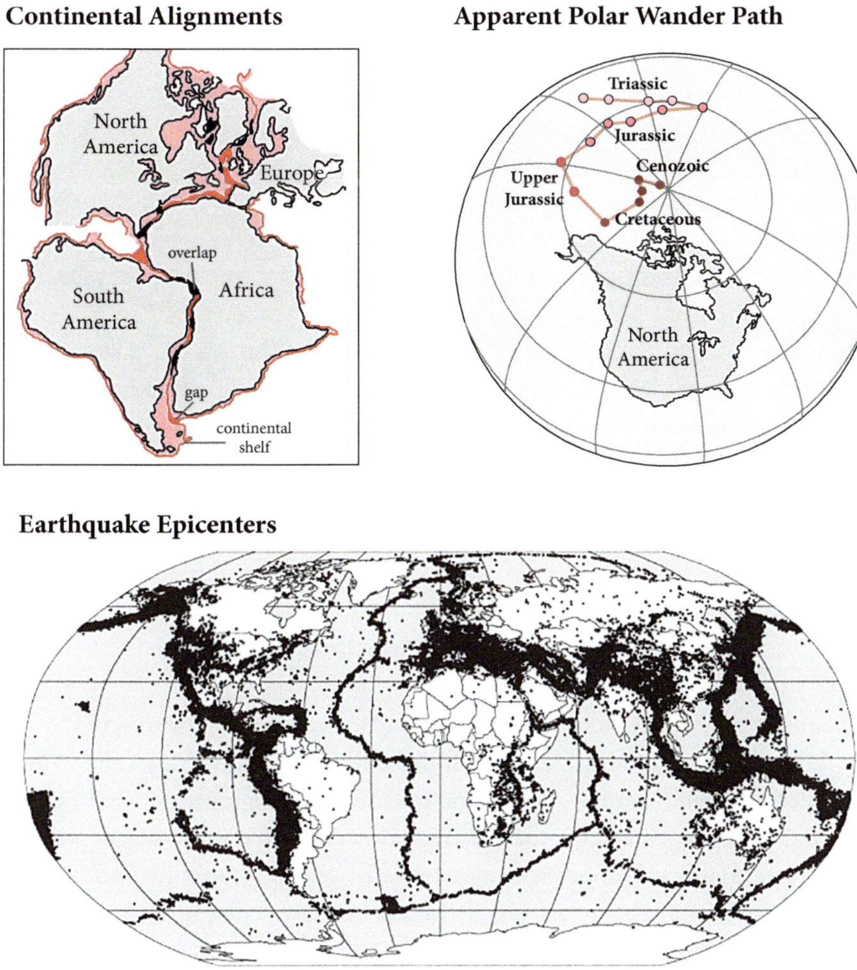

Figure 1.4 Evidence for plate tectonics. As shown at the top left, the Eastern coastlines of North and South America align rather well with the Western coasts of Europe and Africa, especially if one includes the continental shelf (to 900 m below sea level). The diagram at the top right illustrates the apparent polar wander path as determined from sites in North America. The map at the bottom depicts the estimated location of nearly 360,000 earthquake epicenters between 1963 and 1998. It highlights the boundaries between the various continental plates.
Adapted from Bullard et al. (1965) and Butler (1992); epicenter map from NASA, DTAM project team.

and Raff, 1961). Given that the earth's magnetic polarity is thought to reverse every few thousand years, these alternating bands suggest that new rock is created at those plate boundaries by molten rock that ascends from deep within the earth and fills the widening gaps between the continental plates (Hess, 1962). Additional evidence came from studies that measured the magnetic field in various rock strata on a specific continent and used these data to estimate the location of the earth's magnetic poles over millions of years. Such studies showed that the earth's magnetic poles appear to have moved much more than one would expect from a slight wobbling of the earth's rotational axis (Figure 1.4). The simplest explanation for this observation is that it is not, in fact, the poles that wandered, but the continents.

Geologists later combined these diverse forms of evidence with a more plausible mechanism for continental movements, namely a set of convection currents beneath the earth's crust that can force the continents apart or crash them into one another. These efforts resulted in the theory of plate tectonics, which is now widely accepted (Oreskes and LeGrand, 2001). Importantly for our purposes, geologists can use the "apparent polar wander paths," as well as other forms of evidence, to estimate the location of the various continental plates during the various periods of the earth's history. The estimates get more uncertain the further back one goes, but for the period during which vertebrates emerged and thrived (i.e., after the Cambrian), those estimates are reasonably robust. They indicate, for example, that Pangea was assembled roughly 335 million years ago and started to break apart ~160 million years later.

1.2. The Comparative Method in Biology

Comparative approaches have contributed substantially to many different areas of biology. For example, Alexander von Humboldt at the beginning of the 19th century collected vast amounts of information on the geographical distribution of plants and animals, which eventually led him to realize that particular types of species occur in predictable combinations across the globe (Humboldt and Bonpland, 2009). With this insight he invented the concept of ecological zones and, one might say, ecology in general. Other biologists were preoccupied with arranging the multitude of plant and animal species according to some kind of logical framework. First came the notion of arranging species along a linear scale, but Darwin and others later argued that the underlying pattern was actually a huge family tree. Subsequent generations of biologists focused on how one might reconstruct this tree of life. They also wanted to know how best to trace the evolution of individual traits, and how to test hypotheses about adaptation and constraint. In the following, we discuss all these subjects in turn.

1.2.1. Scale of Nature versus Family Tree

Starting at least with Aristotle, scientists have tried to order the different types of animals and plants along a linear scale, called *scala naturae* (Latin for "scale of Nature"; Lovejoy, 1936). The metric for this scale is usually the similarity between a given species and *Homo sapiens* such that the animals most similar to us are placed on the highest rungs, while species progressively more different from us are assigned to successively lower positions along the scale. A major problem with this view is that "similarity to humans" is a highly multidimensional measure, and different dimensions yield different species rankings. For example, rankings based on presumed intelligence might place dolphins high on the scale, while rankings based on mode of locomotion elevate the flightless birds (e.g., kiwis and ostriches) far above dolphins and other aquatic mammals. Regardless of the measure used, adherents of the *scala naturae* view tend to equate increases in similarity to humans with evolutionary "progress."

Biologists in the second half of the 19th century sought alternative ways to make sense of species' similarities and differences. The most successful effort was that of Charles Darwin, who realized in the mid-1800s that the connections between species formed not a scale but a family tree, a genealogy. In the only figure in his *Origin of Species* (Darwin, 1859), Darwin illustrated how two species can diverge over the course of many, many generations into multiple distinct populations and, eventually, into several new species (Figure 1.5). This schema is clearly incompatible with the concept of *scala naturae*. After all, could you rank your own extended family members along a simple linear scale? Therefore, Darwin described the genealogy of species not as a scale but as a "tree of life." In one of his notebooks, he even called it a "coral of life" (Barrett et al., 2009) because, as in healthy corals, only the tips are currently alive. Although Darwin remained conflicted about the relationship between evolution and progress, he was quite clear that any such progress could be reversed and, in any case, evolution did not follow a single, linear path.

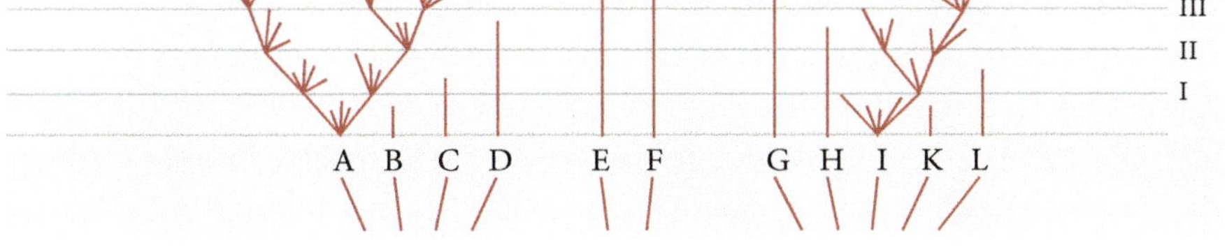

Figure 1.5 Darwin's view of speciation. This diagram is the only figure in Darwin's *The Origin of Species* (1859). Darwin's own description of this figure is lengthy, but letters A–L represent distinct species, two of which go on to form multiple distinct varieties and eventually, over the course of thousands of generations (each horizontal line represents 1,000 generations), two and three distinct species. As Darwin summarized: "Thus, the diagram illustrates the steps by which the small differences distinguishing varieties are increased into the larger differences distinguishing species." (p. 92)

Although Darwin's ideas became widely accepted, *scala naturae* thinking remains widespread. Even expert neurobiologists often resurrect it in the form of a "phylogenetic scale" (Hodos and Campbell, 1969). For example, they often draw phylogenetic trees that look like conifers (e.g., fir trees), with humans perched on top and all the lower branches neatly pruned. Other diagrams may look more deeply branched (i.e., bush-like), but all the living species are arranged in series from low to high, with humans at the top (Figure 1.6). As further evidence that the *scala naturae* concept lives on, the official *Guide for the Care and Use of Laboratory Animals* (8th edition, National Academy of Sciences, 2011, p. 5) discusses "replacing animals such as vertebrates with animals that are lower on the phylogenetic scale." According to the National Academy of Sciences, therefore, an institutional animal care and use committee may be compelled to use the fallacious concept of a phylogenetic scale when deciding whether to approve a research proposal.

Scala naturae thinking may also manifest in phylogenetic diagrams that seem, at first glance, to contain no explicit scale. Consider, for example, the top diagram in Figure 1.7. This "cladogram" depicts the evolutionary branching order of the hypothetical taxonomic groups listed across the top (letters A–F). Taxon E is more closely related to taxon F than any of the other taxa, D is more closely related to E-plus-F than to any other taxa, and so on. Importantly, taxa A–F are arranged in an orderly sequence from left to right, creating the impression that taxa A, B, C, D, E, and F form a phylogenetic scale (at least for people from cultures in which writing is read from left to right). This impression vanishes, however, when the branches are displayed in fully equivalent but graphically different ways. As shown in Figure 1.7, the underlying phylogenetic relationships remain the same when the lines are graphically rotated around one or more of the branch points (aka nodes), even though the apparent "sequence" (e.g., from left to right) is rearranged. Similarly, the original phylogenetic scale disappears when some taxonomic groups are deleted from the diagram while others are expanded (bottom diagram of Figure 1.7). Thus, the letters at the top of the cladogram do not constitute a phylogenetic sequence, let alone a *scale*. Instead, phylogenetic sequences flow from the bottom of each cladogram toward the top, tracing out multiple, diverging trajectories.

Moreover, all cladograms are biased in the sense that they represent only a subset of all species. Comparative biologists often focus on a specific subset of species because they are interested in following the evolution of a specific lineage. Often, this is our own hominid phylogeny. As O'Hara (1992) has pointed out, goldfish would likely focus on a different set of species and tell a very different story; and they would illustrate it with a cladogram that places goldfish on the right and includes far more ray-finned fishes than mammals. That said, we freely acknowledge that the organization of our book also serves an anthropocentric storyline. In our view, there is nothing inherently wrong with emphasizing a particular set of evolutionary transformations, as long as one acknowledges that similar transformations may have occurred in other lineages.

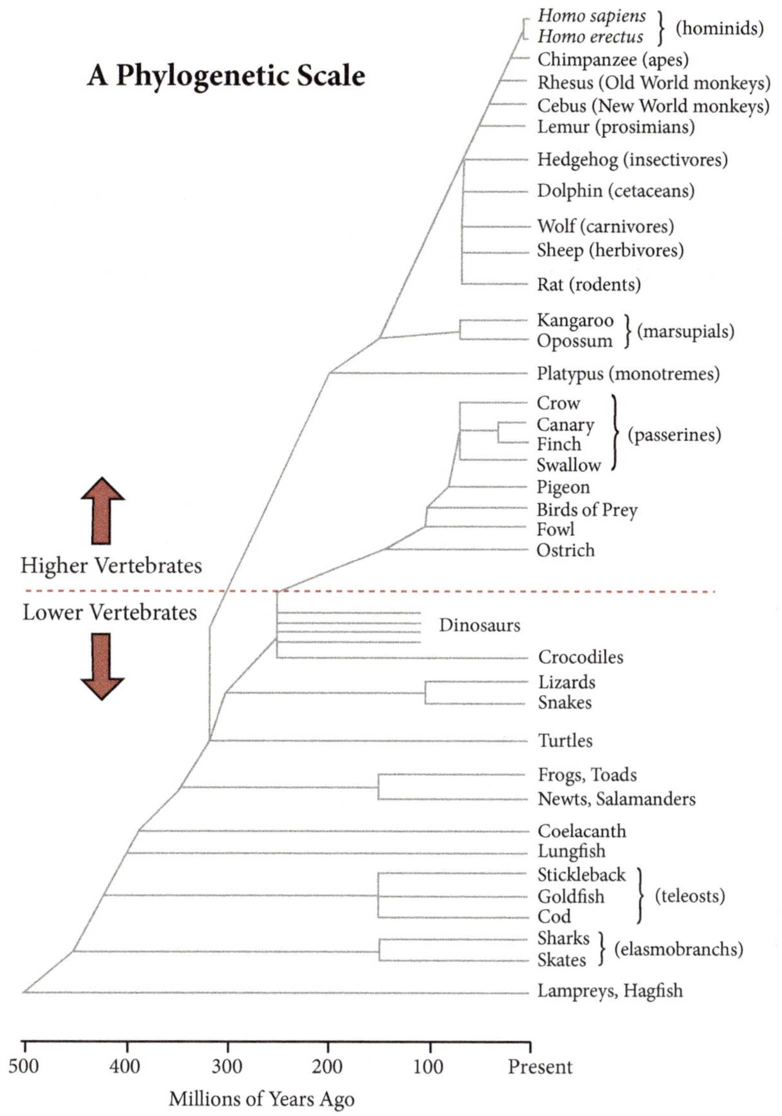

Figure 1.6 One version of the phylogenetic scale. This diagram appeared in a book on "Animal Thought" and is supposed to represent "the presumed sequence of vertebrate evolution," although the author readily admits that the selection of representative species is arbitrary, and the ordering of species on the vertical axis is unsystematic. Nonetheless, the author also concludes that "in general the higher a species is on this axis, the more closely related it is to *Homo sapiens*" (p. 118). It clearly is not the case, however, that humans are more closely related to crows than to lizards or ostriches. Indeed, it seems that all endothermic vertebrates (birds and mammals) are classified as "higher vertebrates" without explicit justification.

From Walker (1983), with the author's permission.

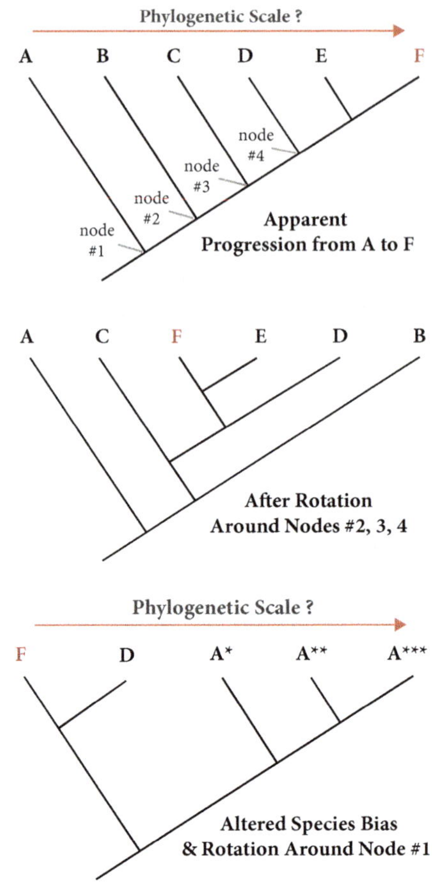

Figure 1.7 Equivalent cladograms. The top cladogram depicts the phylogenetic relationships of six hypothetical taxa (A–F). By putting the deepest branches on the left, and the most recently diverged taxa on the far right, this cladogram makes it appear as if the taxa A–F fall onto a linear phylogenetic scale. This impression vanishes, however, when one rotates the cladogram around some of its nodes (middle) or expands some lineages while pruning others (bottom). Indeed, the bottom diagram makes it appear as if taxon F is at the bottom of a phylogenetic scale, rather than at the top.

1.2.2. Reconstructing Phylogenies

Darwin's insight that species are connected to one another through their genealogy was slow to change the practice of species classification (de Queiroz, 1988), but change did come eventually. Historically, taxonomic classifications had been based on the similarity between species, with the most similar species grouped together into a genus, the most similar genera grouped into families, similar families into orders, etc. This "phenetic" approach to classification worked well for many years, but had the same fundamental flaw as the *scala naturae* view, namely that emphasizing different forms of similarity produced different classifications. Taxonomists tried to solve this problem by quantifying the similarity between species and then subjecting the data to hierarchical clustering algorithms, but the results still depended heavily on the types of similarities that were considered and the algorithms that were used for clustering (Sokal and Sneath, 1963; Ridley, 1986).

In contrast, scientists who took Darwin's discovery to heart argued that species classifications should be based solely on genealogy, rather than similarity. Just as your brother might be more similar to one of your cousins than to you, so a species might be quite different from its "sister species" yet similar to a more distant relative (at least in some respects). But if similarity is not a reliable indicator of phylogenetic relationship, how can scientists infer the phylogenetic relationships between species without having a time machine?

1.2.2.1. Phylogenetic Systematics

The solution to this quandary was articulated most clearly and forcefully by the German entomologist Willi Hennig (1966). He argued that the key to reconstructing phylogeny is to identify features (aka characters) that arose just once in some ancestral species and were then inherited by the descendent species. In contrast, characters that arose multiple times in diverse lineages are unhelpful, if not downright misleading, when it comes to reconstructing phylogenetic relationships. Characters that evolved long before the last common ancestor of the species being examined also offer no real clue about the species' relationships.

To illustrate the latter point, consider how you might determine the phylogenetic relationships between lampreys (a group of eel-shaped, jawless vertebrates), sharks, and cows (Figure 1.8; Gardiner, 1979). At first, you might think that lampreys and sharks must be more closely related to one another than they are to cows, because both have gills, swim in water, and have a cartilaginous (rather than bony) skeleton. However, all of these characters are also found in the nearest invertebrate relatives of all vertebrates (tunicates and amphioxus; see Chapter 2), which means that they probably evolved long before the three taxa diverged from one another. In contrast, jaws and paired appendages are found in sharks and cows (and other "gnathostomes") but not in lampreys or any invertebrates that are closely related to vertebrates (i.e., outgroups to vertebrates). Thus, jaws and paired appendages are "shared derived characters" (aka synapomorphies) for sharks and cows. According to Hennig and his followers, who came to be known as cladists, this means that they support the hypothesis that cows and sharks are more closely related to one another than they are to lampreys.

One potential problem for Hennig's cladistic approach is that some shared derived characters might be lost in some descendants of the ancestor in which they first emerged. For example, dolphins and other toothed whales have lost the body hair that presumably evolved in the last common ancestor of all mammals (see Chapter 6). This problem can be easily avoided, however, if enough other mammals are included in the analysis, because then it becomes abundantly clear that having body hair is the "primitive condition" for mammals. A more serious problem is that an attribute that had been suspected of being a shared derived character for some

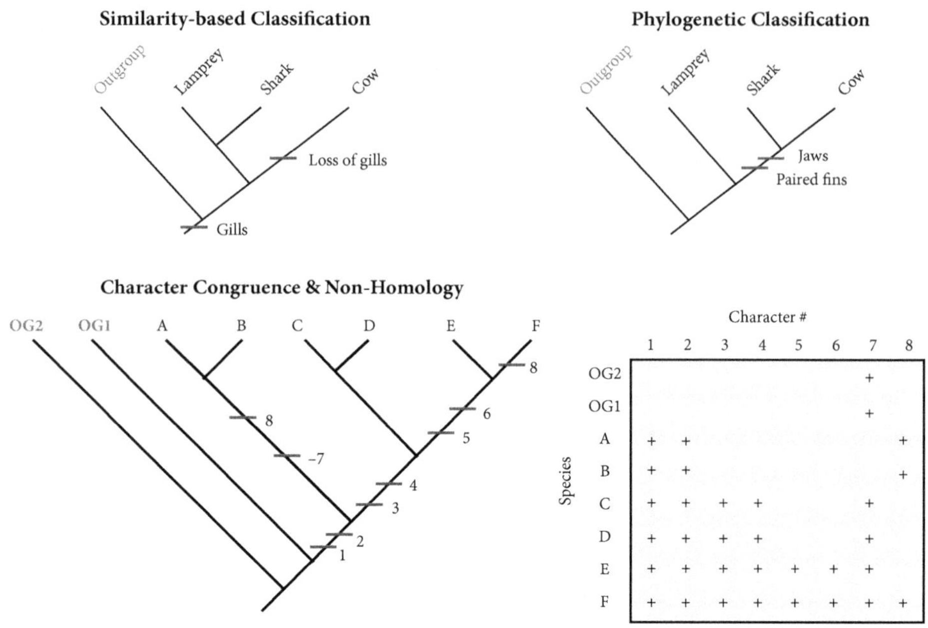

Figure 1.8 Systematics and homology. Based on overall similarity (e.g., the presence of gills), lampreys and sharks should be more closely related to one another than to cows (top left). In contrast, the identification of shared derived characters, such as jaws and paired fins, indicates that sharks and cows are more closely related to one another than to lampreys (top right). In practice, phylogenetic systematics involves the construction of a data matrix listing all characters suspected of being shared and derived for some group (bottom right). This matrix is then used to find the cladogram that best fits the data (bottom left). In this case, only character #8 turns out to be independently derived in different lineages, meaning that it is not a shared derived character for the species in which it is found, and therefore not homologous between species A or B and F. The outgroups (OG1 and OG2) represent taxa that are most closely related to the main species of interest; they help determine which characters are primitive, rather than derived.

Bottom diagram adapted from Wiley and Lieberman (2011), with permission of John Wiley & Sons through PLSclear.

lineage (Rieppel and Kearney, 2002) might actually have evolved independently in several different lineages. The best way to avoid such errors is to consider many different characters when reconstructing any phylogeny. When this is done, the preponderance of characters will frequently provide strong support for a specific phylogeny (or small group of very similar phylogenies) so that the incongruent, repeatedly evolved features can be identified with ease (e.g., character #8 in Figure 1.8).Historically, systematists used mainly morphological characters to build phylogenies, but in the 1980s, they increasingly began to use molecular data. The early attempts at molecular phylogenetics were problematic, because they relied on overall similarity measures derived from DNA-DNA hybridization, rather than specific genomic characters (Sibley and Ahlquist, 1990; O'Hara, 1991). Even studies that looked at variation in specific genes provided uncertain, variable results, because different genes evolve at different rates in different lineages; because genes were lost or duplicated differentially between the studied lineages (i.e., incomplete lineage sorting; e.g., Pollard et al., 2006); and because most studies examined a small number of species (Barker et al., 2004). Over the years, however, the accumulation of vast amounts of comparative molecular data, as well as major improvements in analytical techniques, have mitigated these concerns. As a result, many of the phylogenies we have relied on in this book are based on the analysis of dozens if not thousands of genes, as well as non-coding DNA sequences (e.g., Pyron and Wiens, 2011; Jarvis et al., 2014; Prum et al., 2015). These phylogenies are often consistent with the comparative morphological data, and some of the best phylogenies are based on both molecular and morphological data (e.g., Reeder et al., 2015). An important exception to this rule is that the phylogenies of extinct species must, for obvious reasons, rely on morphological data alone. That said, fossil-based phylogenies also continued to improve as more specimens have come to light and the analysis techniques became more powerful (e.g., Ruta et al., 2003; Anquetin, 2012).

The new generation of phylogenic analyses has supported many traditional ideas about species relationships, but it has also spawned some major revisions. Those revisions have forced biologists to come up with new names for some taxonomic groups, and to abandon other names. In particular, modern biologists have tended to discard old names that refer to paraphyletic groups, defined as taxonomic groupings that include some but not all descendants of their last common ancestor. For example, the term "reptiles" specifically excludes birds, even though birds are more closely related to crocodilians than to any other living vertebrates (Figure 1.9). Instead, modern biologists prefer taxonomic names that refer to monophyletic groups, defined as all the descendants of a common ancestor, as well as that ancestor itself. Thus, "reptiles" and birds (flying reptiles!) together form a monophyletic group that we call the sauropsids (Figure 1.9). Cladists also discourage names that refer to polyphyletic groups, which are defined by similarities that evolved independently. For example, the term "warm-blooded vertebrates" refers to birds and mammals, because these are the only major extant (i.e., living) groups of vertebrates capable of generating their own body heat (see Chapter 6), but birds and mammals do not share a common ancestor that is not also shared by numerous "cold-blooded reptiles." Therefore, the term does not reflect phylogenetic relatedness and should be avoided. We generally concur with these recommendations but plead guilty to sometimes using the old, established names when their meaning is clear.

Extinct species are frequently included in phylogenetic analyses and depicted in cladograms, where they are often marked with a cross (to signify their extinction). When this is done, it is common practice to refer to the monophyletic group that includes all the extant members of a lineage as the "crown group" of that lineage (Figure 1.9). The closest extinct relatives of that crown group are called the "stem groups" for that lineage.

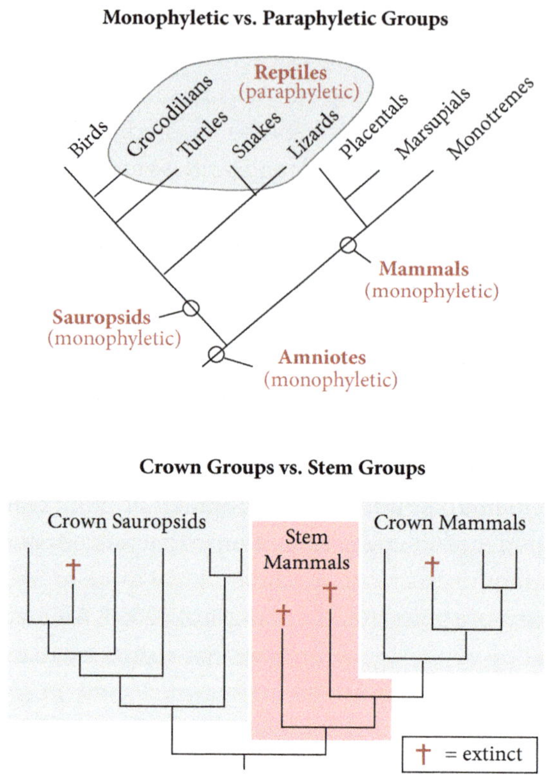

Figure 1.9 Naming conventions in phylogenetic systematics. Practitioners of phylogenetic systematics (aka cladists) prefer to give formal names only to monophyletic groups, defined as an ancestor and all its descendants. They dislike paraphyletic names, such as "reptiles," even though such names are often deeply entrenched. For a given lineage, the monophyletic group that includes all of its extant (living) species is called the "crown group" (bottom). The extinct members of a lineage that fall outside of this crown group are said to form that lineage's "stem."

1.2.2.2. Timetrees

The phylogenetic diagrams produced by cladists (i.e., cladograms) depict the branching order of the investigated species, but they usually do not contain an absolute timescale (Avise, 2009). To provide estimates of species divergence times, fossil specimens of a known age (e.g., from a carbon-dated layer of rock) are certainly useful. However, one can never know whether a given fossil is an ancestor of a specific group of living species, or merely a side-branch that became extinct. Nonetheless, it is usually possible to assign a fossil specimen to a specific lineage, which then establishes the lineage's minimum age. Because the fossil record is generally incomplete, all lineages have a "ghost lineage" component that goes back beyond the known fossil record. As a result of this limitation, even carefully selected fossils can establish only the minimum divergence time between two or more lineages.

An important step toward resolving this problem was taken when comparative molecular biologists realized that some gene and protein sequences have changed at relatively constant rates over evolutionary time (Zuckerkandl and Pauling, 1962, 1965; Margoliash, 1963). If the rate of this "molecular clock" is known, then it should be possible to determine how long ago the genes and proteins in different species diverged from one another. This divergence time, in turn, ought to reflect the divergence time between the species themselves. Although this idea was fascinating, it soon became apparent that the rate of evolutionary change is different for different genes and proteins. It also varies between lineages, if only because species often differ substantially in generation time. To address these issues, some researchers compute average rates of change across multiple genes and species, but this approach leads to very uncertain estimates.

A more productive approach has been to develop "relaxed clock" models that explicitly allow for unequal rates of molecular evolution (Thorne and Kishino, 2002; Drummond et al., 2006). To make these models work, researchers use strategically selected fossils to "calibrate" the molecular clocks in multiple lineages (Hedges and Kumar, 2004; Near and Sanderson, 2004; Benton et al., 2009). Thus, molecular and fossil data are combined to generate "timetrees" (Figure 1.10) that illustrate both phylogenetic relationships and absolute divergence times (Hedges and Kumar, 2009). The divergence estimates produced in these analyses are usually older than those produced by purely morphological (fossil) studies, but this is not surprising, given the universal presence of ghost lineages. Only

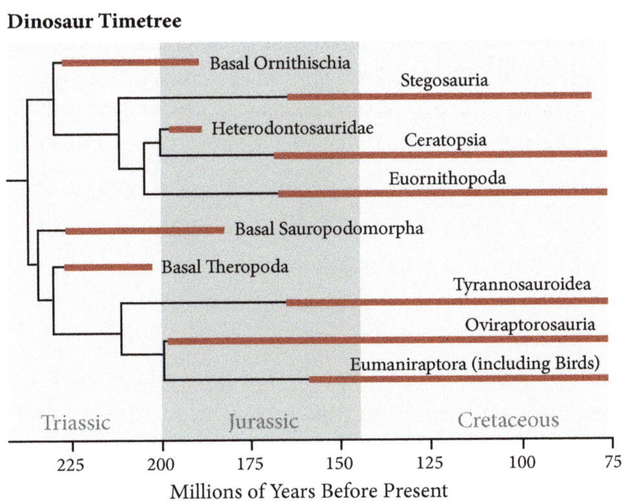

Figure 1.10 Dated phylogeny of dinosaurs. This diagram is a simplified excerpt from a larger timetree for more than 400 dinosaur species (see http://palaeo.gly.bris.ac.uk/macro/supertree). The thick red lines represent the known fossil record; the thin black lines show inferred ancestral nodes and ghost ranges. Note that birds belong to the eumaniraptor clade of dinosaurs, and that all divergence times are estimates.
Adapted from Lloyd et al. (2008) and Benton (2009).

when the molecular data suggest that divergence times are much, much older than the oldest fossils in those lineages do heated controversies ensue. This is not the case for most of the lineages we discuss in this book (Erwin et al., 2011; Goswami, 2012).

1.2.3. Character Reconstruction and Homology

Once a well-supported, robust phylogeny has been obtained, it becomes possible to reconstruct the evolutionary history of individual characters within that phylogeny (Northcutt, 1984). Those characters might be a stretch of DNA, a specific gene, a protein, a morphological structure, a physiological process, or even a species-typical behavior (Striedter and Northcutt, 1991). Regardless of the type of character one is examining, the first step in character reconstruction is to map the distribution of the character across the extant species in the phylogenetic tree (i.e., the tips of the cladogram or timetree). Next, one must reconstruct at which nodes in the phylogeny the character was present, and where it was not. This reconstruction of conditions at the nodes requires testing various possibilities and then selecting the most parsimonious scenario, which is defined as the scenario requiring the smallest number of evolutionary transformations (gains or losses) of the character (Figure 1.11; Wiley and Lieberman, 2011).

Often this kind of character reconstruction reveals that the character in question is confined to a monophyletic group of species, meaning that it arose with the origin of that group and was retained in most of its descendants, but does not appear in other portions of the cladogram. In such cases, the character is said to be homologous among the members of this group (it is also a shared derived character for that group). In other instances, however, parsimony analysis will indicate that the character in question evolved more than once, in disparate lineages. In that case, the independently evolved characters are not homologous to one another. Although the concept of homology is much maligned and often misunderstood, in the context of phylogenetic systematics and parsimony analysis, its definition is clear (Striedter, 1998; Hall, 2000; Cracraft, 2005).

For example, camera-type eyes (i.e., eyes with a single lens) evolved independently in vertebrates and cephalopods (octopus and squid), because it would be much less parsimonious to argue that such eyes were lost repeatedly in all the intervening lineages (Figure 1.11). Therefore, the eyes in these two lineages are not homologous to one another, even though they are remarkably similar. Vertebrate and cephalopod eyes are also not homologous to the compound eyes of insects and other arthropods, which are structurally quite different from camera eyes (e.g., they contain thousands of separate lenses). This is interesting, because eye development in both insects and vertebrates requires the expression of *pax6*, which is found in all metazoan animals and, therefore, is homologous among them (Tomarev et al., 1997; Gehring and Ikeo, 1999). Thus, we here have a case of non-homologous

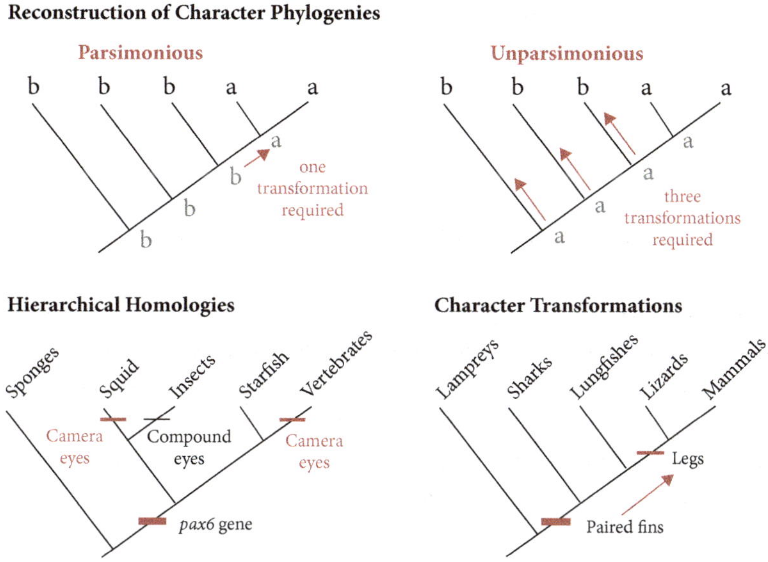

Figure 1.11 Character transformations and homologies. Given a phylogeny, one can map the distribution of individual characters (aka features) on that phylogeny and try to infer when they appeared or were lost. According to the principle of parsimony, the scenario requiring the smallest number of evolutionary transformations is most likely to be correct. The diagram at the bottom left shows that the evolution of both morphological structures (e.g., eyes) and genes (e.g. *pax6*) can be reconstructed in this manner. The transformation of paired fins into legs (bottom right) is complicated because one can think of them as two separate characters (fins and legs) or as a single character (limb) that changed substantially.

structures requiring the expression of homologous genes. Such instances are often cited as examples of "deep homology" (Shubin et al., 2009), but the morphological structures are nonetheless not really homologous. Instead, the most parsimonious conclusion is that the homologous genes were recruited independently, in separate lineages, to build non-homologous structures (Fernald, 2006; Wagner, 2007; McCune and Schimenti, 2012). Indeed, it is difficult to imagine how novel morphological structures could arise in evolution if they did not involve the activation of at least some preexisting, ancestral genes (Wagner and Lynch, 2010). *Pax6*, for example, may have been recruited into a variety of "master regulator" roles because it has two independent DNA-binding domains, which allow it to control an unusually broad spectrum of target genes (Kozmic, 2008).

Most characters vary only in relatively minor ways once they have evolved in a particular lineage, which means that they can easily be recognized as "the same character" in different species (Wagner, 2000). The main olfactory bulb, for example, is readily identifiable in virtually all vertebrates (though it was lost in the toothed whales), because it is remarkably conserved in its position within the brain, in its connection with the olfactory epithelium, and in its histological structure. In contrast, some other characters vary substantially across the members of the lineage in which the character is found. For example, some of the jaw bones in jawed fishes are homologous to the middle ear bones of mammals, meaning that they are considered "the same character," but they have clearly changed their structure, position, and function (see Chapters 5 and 6). Such cases of "transformational homology" (Brower, 2014) are fascinating, because they tend to be associated with major changes in the behavior and ecology of a species, but they are often controversial, precisely because their differences raise doubts about whether they should really be considered "the same character."

Some of these debates are inevitable, but they can be minimized by clearly defining the characters that one is discussing (Campbell and Hodos, 1970; Striedter, 1999). In the case of eye evolution, it is important to be clear whether one is talking about the *pax6* gene or eyes (i.e., the complex morphological structures). Similarly, one can say that paired appendages are homologous across all jawed vertebrates, while also acknowledging that legs are an innovation of tetrapods and wings have evolved independently (as novel characters) in birds and bats. Systematists sometimes solve this issue by distinguishing between characters and character states, with legs and wings being different states of the "paired appendage" character. This terminology can help resolve debates about uncertain homologies insofar as researchers may agree that a particular character is homologous between two or more taxa, even as they disagree about variations in character states. In this book, however, we avoid the term "character state" (see also Patterson, 1988; Briscoe, 2019). Instead, we strive always to be clear about which character is being considered, where it fits in the hierarchy of biological characters (Striedter and Northcutt, 1991), and how it changed across phylogeny.

To reiterate, we define homologous characters as those that are most parsimoniously interpreted as having a continuous history that can be traced back to a single origin in a common ancestral species (Northcutt, 1984; Striedter and Northcutt, 1991; Striedter, 1998, 1999). Thus, we explicitly exclude characters that evolved independently in two or more lineages, no matter how similar those characters may be. This view is grounded in the theory and practice of phylogenetic systematics (Hennig, 1966; Wiley and Lieberman, 2011). It differs from the view of some other authors, who have defined homologous characters as those that share some essential similarity, which then becomes necessary and sufficient for the identification of characters as homologs. Often these essential similarities are given special standing because they relate to developmental mechanisms, such as similarities in the underlying gene networks, embryonic origins, or both (see Butler and Saidel, 2000; Wagner, 2007). Other "essential criteria" may include neuronal connections (Karten, 1969) or a character's position within a general morphological framework (Nieuwenhuys and Puelles, 2016). Indeed, these features are often remarkably conserved across phylogeny and very useful for identifying *potential* homologs. However, in our view, all characters, at any level of organization or stage of development, are capable of changing in some subset of their attributes without necessarily forfeiting their homology (Striedter, 1998). Conversely, it is possible for non-homologous characters to arise through similar developmental mechanisms, to exhibit similar connections to other characters, or to occupy similar positions in a topological framework. If those similarities arose independently in more than one lineage, then they are not homologous.

1.2.4. The Evolution of Development

Traditionally, comparative biologists have compared almost exclusively the characters of adult organisms, and they reconstructed their evolutionary history as if one adult character had transformed directly into another (Bateson, 1892). However, an adult animal never produces another adult organism directly, at least not among most animals. Instead, they produce a zygote, which then develops into the adult form of the next generation (Figure 1.12). As Walter Garstang put it in 1922, "the real phylogeny of Metazoa [multicellular animals] has never been a direct succession of adult forms, but a succession of ontogenies or life cycles" (p. 82). From this perspective, evolutionary change is not accomplished by direct modifications of an adult character, but by changes in the character's developmental trajectory, in its ontogeny. To quote Garstang again: "Ontogeny does not recapitulate phylogeny; it creates it" (p. 82).

If this is true, then comparative biologists should be encouraged to compare not only adult forms, but also entire developmental pathways, i.e., ontogenies. Some comparative biologists had long been interested in the evolution of development, but most of their attention had been focused on the idea that "ontogeny recapitulates phylogeny" (De Beer, 1940; Gould, 1977). Specifically, they usually assumed that organisms in their development pass through the adult stages of their successive ancestors, which would allow researchers to elucidate phylogenetic relationships by studying ontogenetic sequences. This "recapitulation theory" is based on the idea that evolution modifies ontogenies only by adding stages to their end, extending development beyond the adult stages of the immediate ancestor (Figure 1.12). This assumption has always been problematic because it implies that ontogenies would become deleteriously long over evolutionary time, unless some developmental stages were compressed or deleted. Moreover, the recapitulation theory disregarded plenty of evidence for "larval adaptations" that probably never existed in any adult form. For example, chickens develop an "egg tooth" on top of their beak just before they hatch. This tooth helps them escape from the egg shell, but it probably never existed in any adult ancestors of birds.

More recent comparative embryological studies have further clarified that the evolutionary addition of stages to the end of ancestral ontogenies (i.e., terminal addition) is by no means the principal form of evolutionary changes in development (Mabee, 1989). Consider, for example, the development and evolution of the lateral line system, which consists of mechanosensory and electrosensory cells on or just

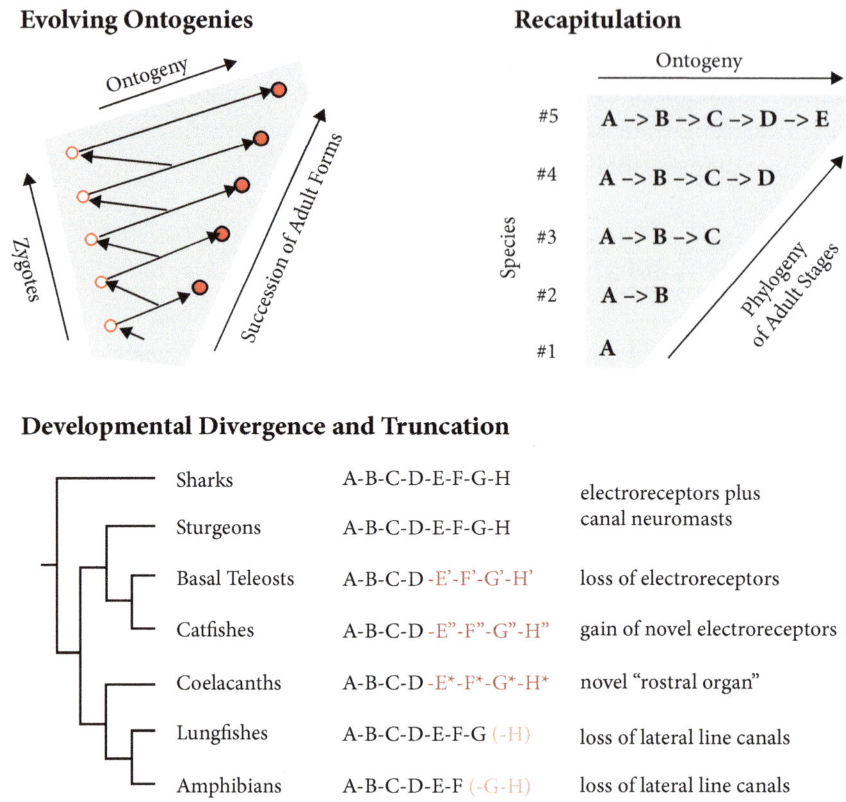

Figure 1.12 The evolution of development. The diagram at the top left shows that phylogeny is not a succession of adult forms, but a succession of entire ontogenies (during which zygotes become adults and produce the next generation of zygotes). The old idea that "ontogeny recapitulates phylogeny" can be true only if phylogeny only adds stages (A–E) to the end of ancestral ontogenies (top right). That this is not the case was shown, for example, by a comparative analysis of the lateral line system in vertebrates (bottom). Lungfishes and aquatic amphibians lost terminal stages of the ancestral ontogeny (red letters in parentheses), coelacanths modified the second half of lateral line development to create a unique "rostral organ," early teleosts lost the electroreceptive component of the lateral line (E'–H'), and a few teleosts (e.g., catfishes) re-evolved electroreceptors through mechanisms that are not well understood but almost certainly involve changes midway through lateral line development.
Adapted from Garstang (1922), Gould (1977), and Northcutt (1992).

below the body surface of most aquatic vertebrates (these sensory cells can detect mechanical and weak electrical stimuli that are transmitted through the water surrounding the animal; see Chapter 2). A parsimony analysis reveals that the primitive developmental pathway for this sensory system includes a series of eight stages (Figure 1.12; Northcutt, 1992; 1997). During the last two stages of this sequence, the sensory receptor primordia sink deeper into the skin and become enclosed in a series of canals, the lateral line canals. Importantly, the comparative analysis also showed that this developmental sequence is cut short in aquatic salamanders and lepidosirenid lungfishes, such that their lateral line receptor cells remain at the body surface, not enclosed by a canal. These "terminal deletions" clearly falsify the recapitulation theory. Even more interesting is that those salamanders and lungfishes have similar terminal deletions in the ontogenies of many different characters, which means that these animals in numerous respects look like the juveniles of their most likely ancestors. We return to this phenomenon of "paedomorphosis" in Chapter 4.

In addition to deleting stages from the end of a developmental sequence, evolution frequently alters the trajectory of ancestral ontogenies before they reach their ancestral endpoint. One example of such a divergence between primitive and derived ontogenies is provided by the evolution of the "rostral organ" in coelacanths. This unique set of electrosensory lateral line canals on the dorsal snout of coelacanths is probably used to detect potential prey at close range (Berquist et al., 2015). It is clearly homologous to part of the lateral line system in other fishes, but its developmental pathway must have diverged from the ancestral developmental sequence long before its end, especially since the rostral organ lacks mechanosensory receptors. Another good example from comparative neurobiology is the divergent development of the telencephalon in mammals and birds (Striedter, 1997). Some early stages of telencephalic development are similar between the two lineages, but the later stages diverge dramatically (see Chapters 5 and 6). Thus, neither of the two ontogenies "recapitulates" the other.

Aside from refuting the idea that "ontogeny recapitulates phylogeny," comparative analysis of developmental pathways can provide important insights into the mechanisms that are responsible for evolutionary change. Traditionally, biologists considered natural selection and changes in gene frequencies to be the principal mechanisms of evolution, but comparative developmental (i.e., evo-devo) studies can provide an important complementary perspective by identifying the molecular and cellular mechanisms that account for species differences in development (Gilbert et al., 1996). For example, one study showed that experimental changes in the expression levels of specific genes can change the shape of a bird's beak in ways that mimic naturally occurring variation (Abzhanov et al., 2004). Manipulating two genes at a time can even make the beak of a chicken look very much like the snout of its ancestors, the dinosaurs (Bhullar et al., 2015). Such work is difficult, because evolutionary changes in development may have required coordinated changes in many different cellular and molecular processes, and because tracking characters across different developmental stages entails both technical and conceptual challenges. Nonetheless, further exploration of the evo-devo approach promises to provide a more complete, synthetic understanding of the mechanisms that have driven evolutionary change (Kuratani, 2009; Brigandt and Love, 2010).

1.2.5. Adaptation and Constraint

In addition to seeking the proximate (or direct) causes of evolutionary changes in biological characters, evolutionary biologists often aim to understand their distal (or ultimate) causes. Specifically, they seek to explain the reasons behind the evolutionary changes. Much work focuses on the benefits they might provide to individuals in terms of their survival and reproduction, although traits may also change for other, non-adaptive reasons. The search for adaptive significance was, of course, Darwin's second profound contribution to biology. Although biologists generally assume that most biological characters provide some adaptive value, many also entail some costs. Therefore, most characters reflect some sort of compromise. In addition, some variants may be impossible for a species to generate, creating developmental or genomic constraints (Alberch, 1989; Roux and Robinson-Rechavi, 2008; Bolstad et al., 2015). In any case, the search for all such explanations usually begins with an examination of how the character in question co-varies with other body parts or processes, as well as the species' life history, behavior, and ecology.

One common type of study examines how the size of a specific body part correlates with the size of the animal. Such studies frequently reveal that the body part does not scale proportionately with body size. That is, body parts tend to scale allometrically, rather than isometrically (Schmidt-Nielsen, 1984). For example, the limb bones of an elephant are much thicker than those of small mammals, even after we account for their differences in length (Figure 1.13). Indeed, across a wide range of tetrapods, limb bone diameter increases faster than limb bone length as one goes to larger and larger species (Alexander, 1979; Christian and Garland, 1996). The ultimate explanation for this disproportionate increase in limb bone diameter is that, under isometric scaling, the limb bones would be too thin to carry the body's weight, which scales with the cube of its linear dimensions (Galilei, 1939). One can call this a physical constraint, but it is probably the result of intense natural selection. Curiously, the exponent that describes how limb bone diameter scales with bone length does vary slightly between lineages (Biewener, 2005). This variation probably reflects other adaptations and constraints that are at work in select lineages. Although scaling rules are commonly presented as universal "scaling laws," evolution has often tweaked them (Agutter and Wheatly, 2004; Herculano-Houzel et al., 2007).

Other studies attempt to correlate specific features of a species with other variables, including habitat and behavior. In one such study, investigators examined how the shape of three different forelimb bones varies with the animal's mode of locomotion (Fabre et al., 2015). The study was carried out in musteloid carnivores (e.g., weasels, otters, and raccoons) because their modes of locomotion vary widely, from fully aquatic to arboreal (Figure 1.14). Bone shape was quantified rigorously, and the data were analyzed using principal component analysis. The major finding was that bone shape does indeed vary with locomotor mode, with the bones of the aquatic and semi-fossorial (i.e., occasionally burrowing) species being more

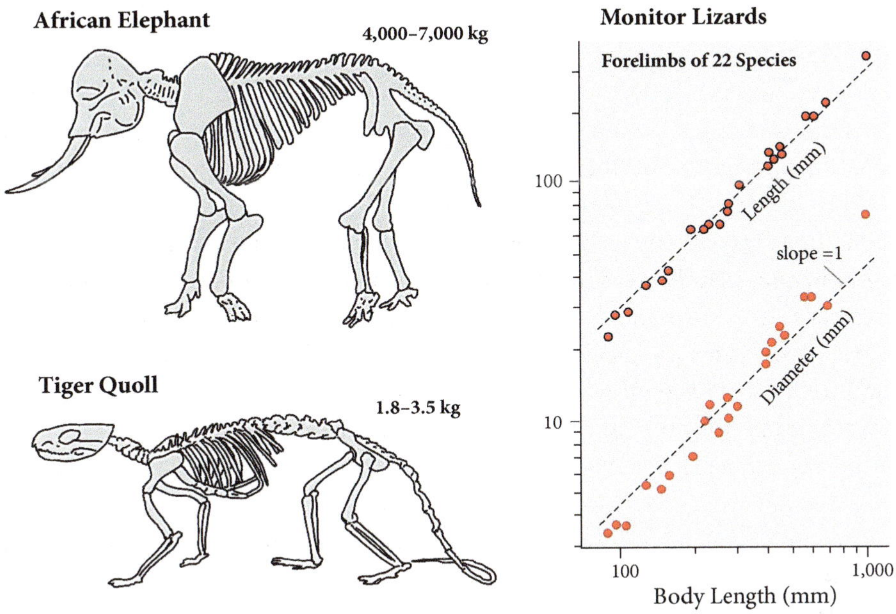

Figure 1.13 Skeletal allometry. The drawings on the left show that the skeleton of an African elephant is proportionately much heavier than that of a small marsupial. The legs, in particular, are proportionately thicker in the larger animal. The graph on the right shows that leg bone diameter also scales allometrically in monitor lizards (i.e., with a slope >1; the slope of the dashed lines equals 1). The most likely explanation for this phenomenon is that leg bone strength increases with the square of bone diameter, whereas body mass increases with the cube of leg bone length.
Skeletons from https://en.wikipedia.org/wiki/Allometry; graphs from Christian and Garland (1996).

robust than those of the arboreal and semi-arboreal species. These findings are consistent with the idea that digging and swimming require strong muscles and bones, whereas climbing trees is easier with a light skeleton. Thus, the correlational data, combined with general biomechanical considerations, strongly suggest that the observed variations in bone shape were created by natural selection.

A general issue with such correlational studies is that the species being compared cannot be treated as statistically independent data points, because they vary in their degree of phylogenetic relatedness, and one would expect closely related species to be quite similar in many traits (Felsenstein, 1985). Consider, for example, the finding that herbivorous mammals tend to have smaller home ranges than carnivorous mammals (Dunstone and Gorman, 1993). This correlation seems sensible, because carnivores tend to eat herbivores and, therefore, need to roam more widely to obtain their food. However, in one large data set showing the correlation between diet and home range size, all the herbivores were ungulates (e.g., horses, cows), whereas all the meat-eaters were from the carnivore lineage. Thus, one can argue that the data set only contains two fully independent data points, one for each major lineage. Indeed, when the data were reanalyzed with statistical

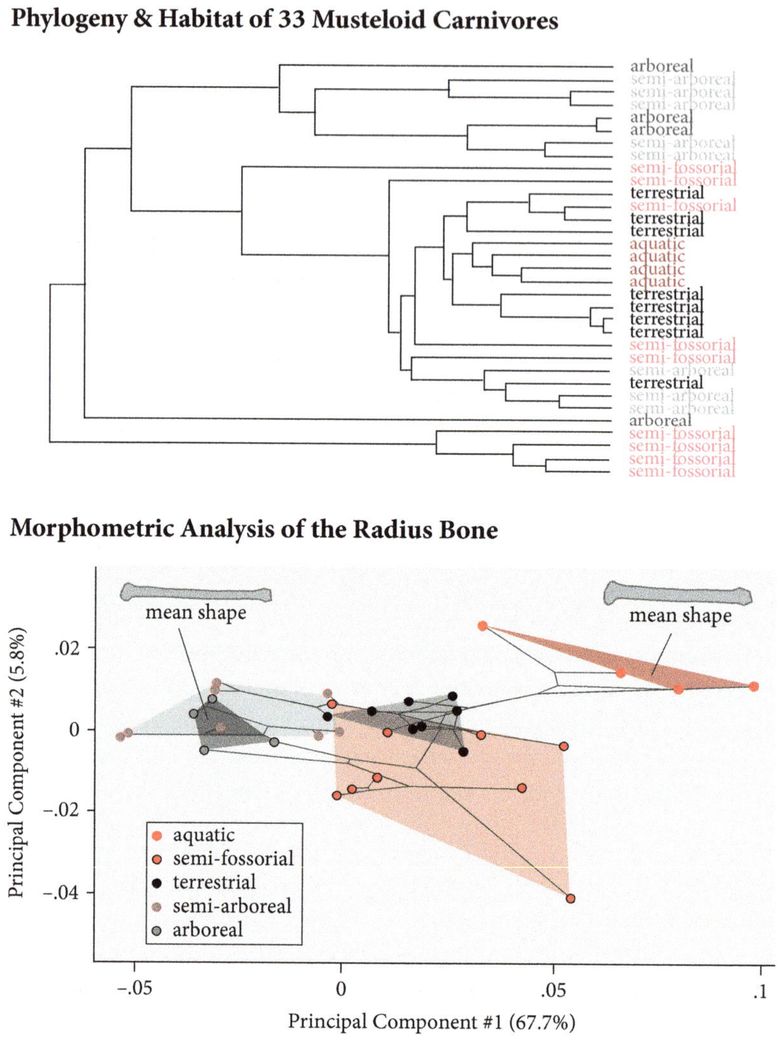

Figure 1.14 Adaptive changes in bone shape. The cladogram at the top shows that musteloid carnivores (e.g., weasels, otters, badgers, red pandas, and raccoons) vary widely in their mode of locomotion and that this variation is largely independent of phylogeny. Comparing the shape of the radius (one of the forelimb bones) across the species reveals a significant correlation with locomotor mode, which is only partially explicable in terms of shared phylogeny (indicated by the thin gray lines). In general, the aquatic species tend to have a thicker radius bone than the arboreal and semi-arboreal species (mean bone shapes shown in the top corners of the graph). Less than 8% of the variation can be attributed to body size. Overall, these data suggest that "forelimb shape is adaptive and evolves, at least partly, in response to the constraints imposed by the different lifestyles" (Fabre et al., 2015, p. 603).

methods that correct for phylogenetic non-independence, the original correlation disappeared (Garland et al., 1993). This does not mean that the original hypothesis is false, just that the original data set could not be used to support it. In contrast, most of the aforementioned correlations between bone shape and locomotor mode in musteloids remained significant even after taking phylogeny into account. The principal reason for this difference is that the pattern of locomotor mode variation in musteloids (Figure 1.14) implies a great deal of independent evolution (i.e., convergence), which is not the case for the ungulate and carnivore data.

The need to take phylogeny into account when conducting comparative correlational analyses has become widely appreciated, and a multitude of different statistical methods are now available (e.g., Rohlfs and Nielsen, 2015; Fuentes-G et al., 2016). However, the approach does have limitations (Martins, 2000). For example, most of the statistical methods consider only evolutionary changes as potential evidence for adaptation, but a character's stability can also provide such evidence, since it may indicate that the character is under continued, stabilizing selection (Hansen, 1997; Butler and King, 2004). Moreover, different lineages might respond to similar selection pressures in different ways, with different solutions. Such divergent responses would not be identified as evidence for adaption in most correlational analyses. In general, the most serious problem with the correlational approach is that it cannot identify instances of adaptation that have occurred only once or a few times, which are precisely the sort of "key innovations" (Hunter, 1998) that occupy us in this book. In such instances, the correlational approach simply lacks statistical power (Iwaniuk, 2004). In short, the comparative correlational approach can provide statistically significant support for hypotheses of adaptation, but it cannot provide strong evidence against them.

The way to overcome these limitations is to collect additional types of evidence. One may, for example, try to determine precisely when and where a particular character evolved and then ask whether this coincides with changes in the environment that might have created novel selective pressures. One may also create biophysical models to show how the character might contribute to an individual's fitness. Even better, one can manipulate the character experimentally to see if this has the predicted effect. Ideally, such manipulations are carried out under natural conditions, so that their effects on survival and reproduction can be measured directly. For example, correlative studies had disagreed about whether the mating success of male peacocks covaries with the number of "eye-spots" on their tail (Petrie et al., 1991; Takahashi et al., 2008), but experimental removal of some eye-spots reduced the mating success of the experimental males during the following mating season (Petrie and Halliday, 1994). Thus, the experimental approach bolstered the adaptive explanation for eye-spots (Loyau et al., 2008). A second, even more convincing example of the experimental approach tested the hypothesis that body coloration in male guppies represents a compromise between selection for being colorful in order to attract females and blending in with the environment to hide from predators. In one experiment, a population of guppies that had been exposed to high levels of predation was introduced into a stream that had few predators (and no guppies). Within two years the introduced guppies had become more colorful, thus supporting the adaptation hypothesis (Endler, 1980).

In summary, it is generally difficult to explain the evolution of specific characters in terms of selective pressures and constraints, but converging lines of evidence can strengthen such efforts substantially. Indeed, evolutionary biologists have now assembled a large number of case studies that go well beyond the speculative "just-so stories" that used to plague the field (Gould and Lewontin, 1979; Lauder, 1996).

1.3. Comparative Approaches in Neurobiology

Comparative studies of the nervous system should, in principle, be just like comparative research on other body parts and organ systems. However, the nervous system is far more complex than other organs insofar as it contains an enormous number of cell types, a vast number of often highly specific connections between those cells, and molecular intricacies that surpass those observed in other parts of the body. Most importantly, the principles of how brains work remain much more mysterious than those for other morphological systems, especially if we consider not just mammalian brains but brains in general. Compared to the principles of biomechanics that biologists can use to understand how bones and muscles work in diverse species (Gans, 1980), the principles of neuronal computation remain largely an unexplored frontier (e.g., Abbott, 2013; Mengistu et al., 2016; Litwin-Kumar et al., 2017). For all of these reasons, comparative research on brains is challenging and relatively rare. Still, the field has made substantial progress since its early days.

1.3.1. Moving Past the Triune Brain Hypothesis

One of the most popular and influential ideas about brain evolution has been the triune brain hypothesis, which states that human brains consist of three major components that were acquired successively during phylogeny (Figure 1.15). Specifically, Paul MacLean began to propose in the 1960s that deep inside the human brain lies an essentially reptilian brain (his R-complex), which controls instinctive behaviors, including feeding, fighting, fleeing, and reproduction (MacLean, 1990). He further argued that all mammals added to this primitive brain an outer layer that comprises the "limbic system," which includes the amygdala and hippocampus and endows mammals with an expanded capacity for prolonged parental care (love) and improved memory. The third component of MacLean's triune brain is the neocortex, which may have originated early in mammalian phylogeny but was expanded greatly in the lineage leading to humans. It provides us with a supposedly unique capacity for rational thought. The most attractive aspect of

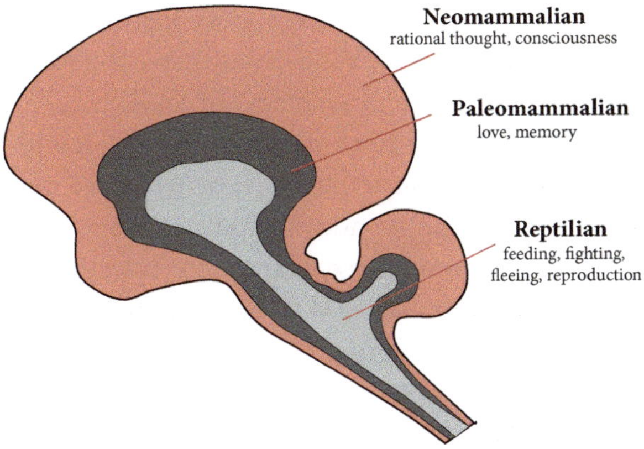

Figure 1.15 The triune brain hypothesis. According to MacLean (1990), human brains consist of three divisions that evolved in succession. The deepest of these divisions is a "reptilian brain" that controls instinctive behaviors. Mammals added a "paleomammalian" shell that consists of the "limbic system" and controls higher emotions (e.g., parental love) and extends memory capacity. All mammals have a neocortex, but it became very large only in select species, notably humans, which it endows with the capacity for rational thought and consciousness. Although MacLean considered these three divisions to have been added sequentially in phylogeny, he stressed that they can interact. Importantly, those interactions often lead to internal conflicts (e.g., guilt or the suppression of "primitive" impulses).
Adapted from MacLean (1990).

MacLean's proposal was his suggestion that the "three brains" can sometimes be in conflict with one another. In this respect, MacLean's proposal is remarkably similar to Plato's ideas about the existence of three separate, sometimes conflicted souls (Smith, 2010). Sigmund Freud's (1923) concepts of id, ego, and super-ego also fit neatly into MacLean's tripartite model of inner conflict. Finally, MacLean's hypothesis probably benefited substantially from the support of Carl Sagan, who wrote a popular book on the subject (Sagan, 1977).

From a neuroanatomical perspective, the triune brain hypothesis never had strong support. Comparative neuroanatomists in the first half of the 20th century had argued that some parts of the forebrain and cerebellum were "new" with mammals (e.g., neocortex), but they already suspected that reptiles and other non-mammals possess homologs of the mammalian hippocampus and amygdala (Edinger, 1908; Kappers et al., 1936). As later chapters will explain, modern research has confirmed that the hippocampus and amygdala are not, as MacLean had claimed, mammalian innovations. Early 20th-century neuroanatomists also thought that reptiles and birds, and even sharks, had at least a small "general pallium" or "dorsal cortex" that was probably homologous to the mammalian neocortex. These hypotheses were extended in the 1960s and 1970s, when comparative neuroanatomists used a variety of new techniques to reveal the detailed connectivity and histochemical organization of non-mammalian brains (Karten, 1968;

Ebbesson and Heimer, 1970; Hall and Ebner, 1970; Pritz, 1974; Northcutt, 1981). Thus, the mammalian neocortex had deeper roots than MacLean had imagined.

Over time, many comparative neuroanatomists came to believe that most features of mammalian brains have rather ancient roots. At least with regard to major brain divisions, such as the hippocampus or neocortex, the currently dominant paradigm holds that brain evolution does not add new components to old ones, as MacLean had argued, but modifies existing ones (Northcutt and Kaas, 1995; Nieuwenhuys et al., 1998; Butler and Hodos, 2005; Striedter, 2005). We will argue in this book that this paradigm tends to mask some important innovations in brain structure and function. However, ours is a minority view, as the idea of evolutionary conservation (rather than innovation) drives the majority of current research in comparative neuroanatomy. The next two sections illustrate this point.

1.3.2. Evo-Devo and the New Neuromorphology

Extensive comparative studies of vertebrate brain development were conducted in the first half of the 20th century by Nils Holmgren, Harry Bergquist, Bengt Källén, and others, who collectively may be referred to as the Swedish School of comparative neuroanatomy (Holmgren, 1922; Bergquist, 1932; Källén, 1951, 1953). One of their main insights was that, across species, embryonic brains are more similar than adult brains. In particular, they focused on similarities at a very early stage when embryonic brains consist mainly of a thin layer of dividing cells that surrounds a relatively large cerebral ventricle filled with cerebrospinal fluid. At this stage, which one might call the "phylotypic period" (Slack et al., 1993; Richardson, 1995), the neural tube forms three more or less distinct bulges, which constitute the embryonic forebrain, midbrain, and hindbrain.

Crucially, the members of the Swedish School recognized that, during the phylotypic period, the neural progenitor cells are divisible into a patchwork of distinct "proliferation zones," which are separated from one another by boundaries that exhibit reduced proliferative activity (Figure 1.16). Specifically, they identified a rostrocaudal series of transverse segments, which are called rhombomeres in the hindbrain, mesomeres in the midbrain, and prosomeres in the forebrain. Orthogonal to these segments are several longitudinal columns, such that the overall pattern resembles a checkerboard that was distorted by the developmental bending of the neural tube (Figure 1.16). Eventually each proliferative zone gives rise to post-mitotic cells, which then migrate away from the intra-cerebral ventricle and differentiate into distinct neuronal cell groups. Although the cells migrating away from different proliferative zones may intermingle, they tend to migrate mainly in the radial direction and, thus, remain largely segregated from one another. As the process of cellular differentiation proceeds, species differences tend to increase, which may then make it difficult to recognize homologies among adult cell groups across the major vertebrate lineages.

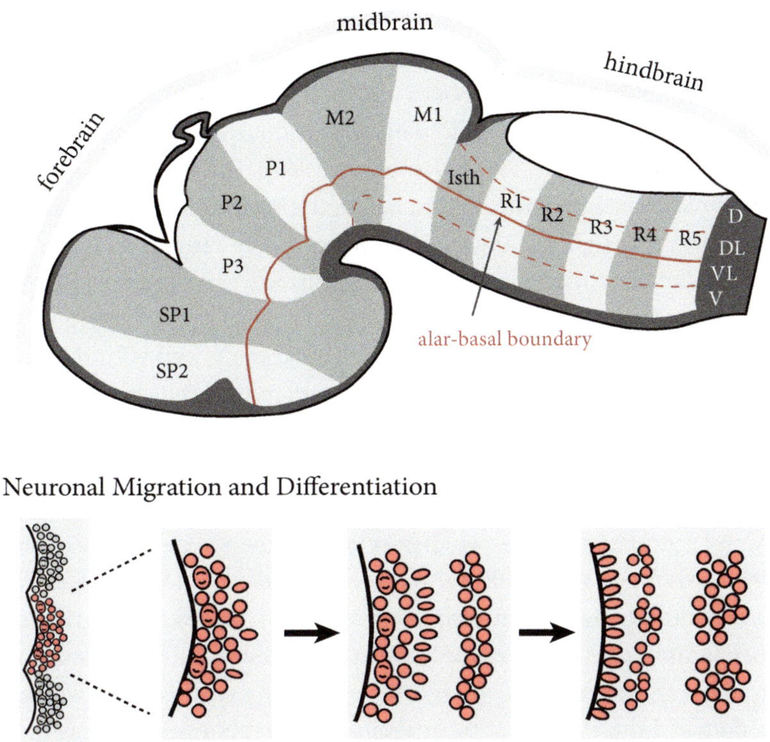

Figure 1.16 The neuromeric model of vertebrate brains. As first proposed by members of the "Swedish School" of comparative neuroembryology, vertebrate brains are divisible into a series of transverse segments (neuromeres), which are called rhombomeres (R1–5) in the hindbrain, mesomeres (M1–2) in the midbrain, prosomeres (P1–3) in the caudal forebrain, and secondary prosomeres (SP1–2) in the rostral forebrain; an isthmic neuromere (Isth) has also been recognized. These neuromeres are further divided into alar and basal regions by a longitudinal boundary that follows the brain's curvature. Additional boundaries divide the hindbrain into dorsal, dorsolateral, ventrolateral, and ventral columns (D, DL, VL, and V). Shown along the bottom is an idealized diagram of how cells in a specific "proliferative zone" of the early embryonic brain migrate away from where they were born and differentiate into adult cell groups.

Adapted from Bergquist and Källén (1954) and Nieuwenhuys and Puelles (2016).

According to the Swedish School, this problem can be solved by tracing the adult cell groups back to their developmental origins. As Källén (1951) wrote: "If it can be proved that two nuclei in different species develop from the same Anlage [precursor area], and in a similar way, they must be looked upon as homologous. The longer the developments of two nuclei are similar, the stricter is the homology between the nuclei" (p. 6). As we review shortly, this idea lives on as "field homology," but it was largely ignored at the time, especially by C. J. Herrick and his American School of comparative neuroanatomy (Herrick, 1948). Herrick and his colleagues tended to ignore comparative embryological findings and focused, instead, on similarities in neural connections. Some American neuroanatomists did remain aware of what the Swedish School had found (e.g., Braford and Northcutt, 1983; Striedter, 1990), but the development of new techniques for tracing neural connections in the 1960s and 1970s created a wealth of new data that encouraged connection-based homology hypotheses (Karten, 1969; Northcutt, 1969). Thus, interest in comparative neuroembryology waned in the second half of the 20th century.

Fortunately, the insights of the Swedish School were rediscovered and extended when molecular biologists started mapping various gene expression patterns in early embryonic brains of diverse vertebrates. First came maps of various *hox* genes, whose gene expression boundaries lined up remarkably well with the boundaries between the rhombomeres identified by the earlier researchers (Wilkinson et al., 1989; Kiecker and Lumsden, 2005). These *hox* genes are not expressed in more rostral brain regions, but the subsequent mapping of other transcription factors corroborated the existence of mesomeres and prosomeres, as well as many of the smaller subdivisions postulated by the Swedish school (Simeone et al., 1992; Rubenstein et al., 1994). Although the resulting "neuromeric model" of vertebrate brain organization has undergone several rounds of revision (Puelles, 2013; Affaticati et al., 2015; Puelles and Rubenstein, 2015; Watson and Puelles, 2016; Puelles et al., 2017), it has provided an extremely useful framework for the synthesis of diverse gene expression data and the experimental analysis of brain patterning. It has thus helped to clarify the hierarchical organization of adult brain regions within a given species by showing how they are developmentally related to one another. Finally, it has proven to be widely applicable to a broad range of vertebrates (Rodríguez-Moldes et al., 2016; González et al., 2017; López et al., 2017; Pombal and Megías, 2017). Thus, the comparative gene expression data have revived the fundamental tenet of the Swedish School, namely that early embryonic brains are divisible into multiple distinct progenitor zones that are broadly conserved across the vertebrates. With this notion at its core, a "New Neuromorphology" (Nieuwenhuys and Puelles, 2016) is ascendant.

1.3.2.1. The Field Homology Concept

A second major tenet of the New Neuromorphology is the concept of field homology, which Hobart Smith (1967) originally defined as "derivation of structures, however similar or dissimilar, from . . . the same ontogenetic source" (p. 102). This form of homology clearly echoes the views of Källén and the other members of the Swedish School (quoted earlier), but it deserves a special name, because field homologies tend to be invoked only when one-to-one homologies between adult structures cannot be found (Figure 1.17). For example, if the adults of one species have a greater number of deep cerebellar nuclei (see Chapter 6) than the adults of another species, strict homologies between those nuclei become difficult to draw. Even if one can homologize some of the nuclei between species, at least one nucleus would remain unaccounted for. Instead of having to argue that the extra nucleus was gained as a new feature in one of the species, or lost in the other, one can use

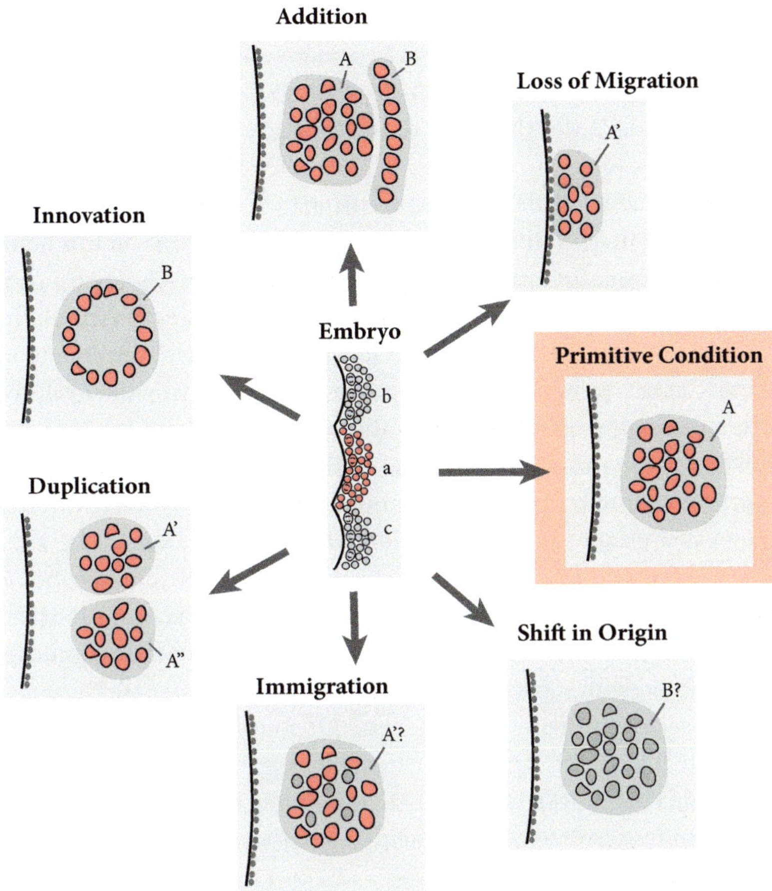

Figure 1.17 Field homology and the evolution of neural ontogenies. Shown in the center is a schematic section through three "proliferative zones" in an early embryonic brain (see also Figure 1.16). The arrows indicate how this embryonic tissue may develop in seven different directions, leading to different numbers and types of adult brain regions. According to the field homology concept, all of the adult derivatives are homologous "as derivatives of a conserved embryonic field," except for the case on the bottom right, in which a structure's embryonic origin shifted during phylogeny. A more traditional (strict) application of the homology concept would acknowledge several innovations (i.e., non-homologous adult forms or "transformational homologies") and the details of its application would vary with the phylogenetic distribution of the observed ontogenies. Whether structure A is homologous to structures A', A", A?, B?, or B is subject to debate.

the field homology concept to sidestep the problem entirely. That is, if all of the adult cell groups under consideration can be traced back to a developmental precursor region that clearly is homologous between the species, then we can say that any of the adult nuclei in one species are field homologs of all the other nuclei in the other species. Thus, the concept of field homology permits homologies when more traditional (strict) homologies cannot be established. This benefit makes the field homology concept very attractive to seekers of homologies, but it has also been controversial (Striedter, 1998; Northcutt, 1999; Cookson, 2001; Puelles and Medina, 2002). This is not the place to review these debates in depth, but a few points are worth making.

First, we are troubled by the fact that invoking field homologies tends to obscure major innovations in adult form, which are our principal focus in this book. In essence, field homology substitutes embryonic homologies for adult non-homologies, relegating the latter to minor significance. Conservation is emphasized, the variation treated as "noise." This one-sided focus on evolutionary conservation (or "diversity denial"; see Murray et al., 2016) probably explains why the field homology concept is not used in systematics. As discussed in Section 1.2.2, phylogenetic systematists are keen to find novel characters that define monophyletic groups (i.e., shared derived characters), but field homologies cannot be used in this way, because they usually represent shared primitive characters (almost by definition). Although systematists may use ontogenetic data to support putative homologies, they are generally not interested in embryonic features that are conserved across all the species of interest. We, too, believe that the study of evolutionary conservation should be balanced with substantive analyses of variation; after all, Darwin referred to evolution as "descent with modification." Of course, and somewhat ironically, we recognize that a superb strategy for such investigations is to identify the developmental origins of adult structures that defy simple homologies. Comparing the relevant ontogenies across species can help us understand how phylogeny has modified ontogenies.

Our second main concern with the field homology concept is that homologous adult characters may derive from non-homologous embryonic precursors, thereby putting field homology and more traditional homology hypotheses in conflict with one another. Indeed, multiple examples of evolutionary shifts in embryonic origin have been reported in the literature. For example, the gut develops from different (though overlapping) sets of embryonic tissues in sharks, lampreys, and amniotes (Jenkinson, 1925; see also Steinmetz et al., 2017), but few would argue that it is therefore not homologous across these lineages. It has been claimed that such evolutionary shifts in embryonic origins only occur before the phylotypic period and are, therefore, not a problem for the field homology concept as practiced by comparative neuroanatomists (Cookson, 2001). However, this argument is weak, because the currently available comparative data on the developmental origins of adult brain regions rarely involve actual fate mapping experiments (but see Puelles et al., 2016). We suspect, for example, that the motor neurons innervating the pectoral fins of fishes are homologous to the neurons innervating tetrapod limbs but develop from different sets of neural tube segments (Ma et al., 2010). If this hypothesis were supported by fate mapping data, it would imply that these motor neurons are "strict homologs" but not "field homologs," unless one wants to claim that homologous precursor regions can arise from different body segments in different species (Smith, 1967). The latter claim is not as outlandish as it may seem at first, but it would raise a host of new questions.

Our third note of caution is that current uses of the field homology concept are based on the assumption that the checkerboard parcellation of vertebrate brains during the phylotypic period (Figure 1.16) is invariant across all vertebrates. Although our understanding of this conserved "Bauplan" of vertebrate brains has evolved over the years, the underlying biological pattern is explicitly assumed to be constant across phylogeny (Puelles and Medina, 2002; Nieuwenhuys and Puelles, 2015). We are not yet convinced of this invariance. For example, the cerebellum is a genuine innovation of jawed vertebrates (see Chapter 3). Furthermore, some vertebrate lineages may have four distinct proliferative zones in their dorsal telencephalon (i.e., four pallial divisions) whereas others have only three (see Chapters 3 and 7). Such a difference would imply a significant evolutionary change in the Bauplan of select vertebrate brains and, thus, run counter to the paradigm of the New Neuromorphology. Part of our argument is that the very idea of a "Bauplan for vertebrate brains" implies that this Bauplan is limited to vertebrates and, therefore, must have evolved from some other, more ancient pattern. Some authors have countered that the Bauplan for vertebrate brains is, in fact, applicable to all bilaterian animals (i.e., all animals with bilateral symmetry), but we argue in Chapter 2 that this hypothesis is not well supported. Others may claim that what we identify as evolutionary novelties were "latent" in the ancestor, waiting only for some additional molecular interactions to reveal what had always been there (e.g., Nagy et al., 2014). We think it is better to invert the perspective and ask, instead, how truly novel characters at one level of biological organization can evolve through novel interactions between lower-level elements (Jacob, 1977; Hall, 2007; Wagner and Lynch, 2010).

In short, we believe that even widely conserved, embryologically grounded patterns of brain organization can change, and almost certainly have changed, over evolutionary time. If this is true, then we cannot simply assume that the Bauplan first identified in mammals and birds necessarily applies to other species. Empirical comparative studies are needed to demonstrate such conservation, as well as any variants.

1.3.3. Molecular and Cellular Homologies

Studies of nervous system evolution have traditionally focused on the phylogeny of entire brain regions, but recent research has shed increasing amounts of light on the phylogeny of neuronal cell types and molecules. Aside from adding substantially to our knowledge base, this work entails some interesting conceptual challenges, especially when it comes to integrating evolutionary changes across multiple levels of biological organization.

1.3.3.1. Gene and Protein Evolution

Research on the evolution of neurotransmitters, receptors, and other molecules integral to neuronal function has revealed a surprising degree of evolutionary conservation across the major vertebrate groups. Some of genes encoding these molecules even predate the origin of neurons (Sakarya et al., 2007; Liebeskind et al., 2011). Despite this deep conservation, most genes involved in neuronal structure and function have changed substantially. For one thing, parts of chromosomes and sometimes entire genomes have duplicated one or more times in diverse lineages, creating complex gene families (Tank et al., 2009; Liebeskind et al., 2015). Frequently, a subset of these duplicated genes has gained novel functions (Zakon et al., 2011; Florio et al., 2015). Some genes appear to have no homologs in other lineages at all and are, therefore, considered "new" (Zhang et al., 2011; Yoshida et al., 2015). Finally, even conserved genes often exhibit substantial variation in their protein-coding region or regulatory sequences (Kim et al., 2015; Mayasich and Clarke, 2016). Some of the molecular changes have occurred repeatedly, in separate lineages (Liu et al., 2014a, b; Lynagh et al., 2015), providing clear evidence for convergent evolution at the molecular level. This book does not deal with molecular evolution at length, but some important gene families, gene regulatory networks, and gene expression patterns are discussed in multiple chapters (e.g., Figures 2.14, 2.20, 2.23, 2.25, 4.15, 6.12, 6.16, and 6.28).

1.3.3.2. Evolution of Cell Types

Biologists have traditionally defined cell types, both in the brain and in other parts of the body, on the basis of their structural and functional features. However, recent advances in molecular techniques have made it possible to examine the gene expression patterns of single cells, and this has sparked a movement to identify cell types in terms of the gene networks that are uniquely expressed in them (e.g., Tosches et al., 2018). This shift has prompted some comparative biologists to argue that homologies between cell types in different species should be based on the conservation of an underlying gene "regulatory signature" (Arendt et al., 2016) or "character identity network" (Wagner, 2007).

A major benefit of this gene-based approach is that it allows cell types to be homologous regardless of how similar they are in terms of structure or function. This flexibility is important, because the idea that characters can be homologous "under every variety of form and function" has been central to the concept of homology since its very inception (Owen, 1843). The gene-based approach also allows for homologous cell types to be located in different, non-homologous parts of the body (of either the same species or different species), as long as homologous gene networks are activated in those cells. In essence, this approach uncouples evolutionary lineage from developmental lineage (Arendt et al., 2016), which is important because some clearly homologous structures have long been known to develop from non-homologous precursor regions (see Section 1.3.2 and Striedter, 1998, 1999). Finally, the gene-based approach makes it possible to reconstruct how cell types might have diversified during the course of evolution, with a single cell type giving rise to multiple "sister cell types" (Arendt, 2008; Duncan and Fritzsch, 2012), much as single genes may duplicate during phylogeny. It is even possible for multiple cell types to "fuse" if their underlying gene networks become causally intertwined (Arendt et al., 2016).

Despite these advantages, the gene-based approach to cellular homologies entails some challenges. For one thing, cell types and gene networks are both hierarchically organized (Peter and Davidson, 2011; Kin, 2015), which means that one must be careful to select the proper set of genes to identify a specific type of cell, rather than some more general cell type. For example, one would not want to use a gene that is expressed in all neurons as evidence that two specific neuronal cell types are homologs of one another. In addition, identifying a set of genes that is co-expressed within a given cell is not the same as showing that these genes form an interacting network of genes. Indeed, identifying and delimiting a gene regulatory network requires extensive experimental evidence, which is at best only available for a subset of the species of interest. Particularly troublesome is the idea that gene networks themselves may change during the course of evolution, potentially losing some components or "co-opting" others. Therefore, establishing homologies between cell-type specific gene networks is not as simple as it may at first appear (Musser and Wagner, 2015; Liang et al., 2018).

Although the issues surrounding cell type homologies are general, they are especially acute for neuronal cell types, because neurons tend to be structurally complex and extremely diverse. Traditionally, the identification of potentially homologous neurons has been based on similarities in cell size, dendritic architecture, and axonal connections (Figure 1.18; Karten, 1969; Major et al., 2000), but similarities in gene expression profiles have recently been used to support a few neuronal homology hypotheses. These efforts have been especially successful in tracing the phylogeny of relatively ancient cell types, such as photoreceptors (Arendt et al., 2004). Molecular data have also been used to support cellular homologies in the telencephalon of various amniotes (e.g., Briscoe et al., 2018; see Chapter 6), but it remains unclear whether the genes in this analysis actually form a coherent network that was retained from a common ancestor (Montiel and Aboitiz, 2018). Thus, it remains possible that some of the observed similarities are due to convergent evolution, rather than homology (Tosches et al., 2018). Ultimately, this possibility can be excluded only by broad taxonomic sampling and the kind of parsimony analysis described in Section 1.2.3.

Another area of concern is that, in our view, comparative neurobiologists tend to underappreciate the possibility that homologous neurons may develop in non-homologous brain regions. When homologous neuron types are found in non-homologous brain regions (e.g., Finger, 1978), neurobiologists tend to assume that those neurons were born in homologous precursor regions and only later migrated into different, non-homologous brain regions. This does happen sometimes (Gilland and Baker, 2005), especially among neurons that tend to migrate

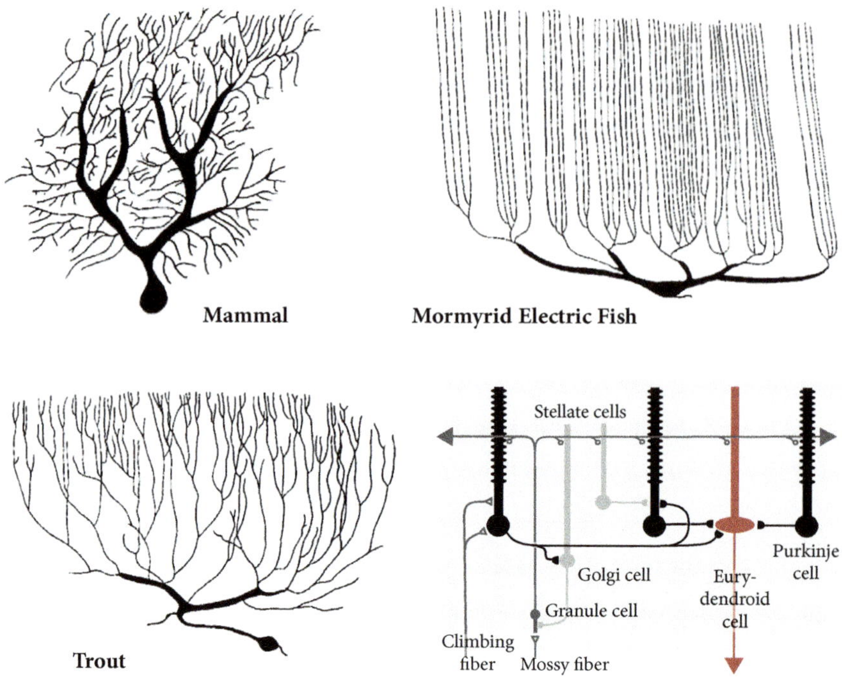

Figure 1.18 Cerebellar cell types in phylogeny. Shown here are Purkinje cells from a mammal and two teleost fishes. Although they differ in dendritic architecture, they share numerous features and are clearly homologous to one another. Shown at the bottom right is the main cerebellar circuitry in a representative teleost. The Purkinje cells project to eurydendroid cells, which are located within the cerebellar cortex but are, nonetheless, thought to be homologous to neurons of the deep cerebellar nuclei in other vertebrates.

Adapted from Meek and Nieuwenhuys (1991) and Pouwels (1978), with permission from Springer Nature and John Wiley & Sons, respectively.

far (Wullimann et al., 2011). However, as noted at the beginning of this section, the gene-based view of cellular homologies explicitly allows homologous cell types to develop in non-homologous tissues, as long as their differentiation involves the activation of homologous gene networks. This phylogenetic uncoupling between neuronal cell types and brain regions becomes a serious problem when one attempts to base hypotheses of brain region homology on the homology of their constituent neurons (e.g., Karten, 1969). Of course, one may counter that it is difficult to imagine how one could homologize two or more brain regions without referring to the attributes of the neurons that they contain.

We address this conundrum by recognizing explicitly that evolution at the molecular, cellular, and regional levels may proceed independently of one another, at least to some extent (Striedter and Northcutt, 1991; Faunes et al., 2015; Tschopp and Tabin, 2016). This lack of one-to-one correspondence across the major levels of biological organization means that one cannot reduce the evolution of brain regions entirely to the phylogeny of specific cell types, or the evolution of cell types to the phylogeny of specific genes and their regulatory elements.

1.3.4. Principles of Variation

Evolutionary conservation of any character, neural or otherwise, is relatively easy to explain as the result of inheritance from a common ancestor, perhaps combined with stabilizing selection. In contrast, variation in a character or the appearance of novel characters is much more difficult to understand. Such variation could be "random noise," but this seems unlikely for most species-typical features. Instead, comparative neurobiologists tend to seek two complementary types of explanations. The first explanatory strategy emphasizes correlations between specific neural characters and the organism's behavior and ecology. Combined with experimental data on the character's functions and some more theoretical considerations, this sort of explanation yields "adaptive scenarios" of how and why the character might have emerged (or stabilized). The second type of explanation focuses on correlations between variation in a neural character and other morphological or physiological aspects of the organism, which one may then explain in terms of various "constraints" and "scaling laws." Both explanatory strategies are important but subject to some serious limitations.

Many studies have explored how brain size varies with body size (van Dongen, 1998; Striedter, 2005; Seid et al., 2011). One general conclusion from this work is that brains tend to scale allometrically, rather than isometrically, with body size. Specifically, brain size tends to decrease in proportion to the rest of the body when body size increases across species. This relationship is most readily apparent in double-logarithmic plots of brain size versus body size, where the data for most taxonomic groups tend to be well-fit by straight lines (Figure 1.19). However, such plots also reveal that the best-fit lines tend to vary across lineages in both their intercept and (to a lesser degree) their slope. Evolutionary changes in the intercepts are usually interpreted as increases or decreases in "encephalization." It is important to realize, however, that increased encephalization need not involve an evolutionary increase in brain size but may, instead, result from a decrease in body size. Little is known about the mechanisms underlying brain-body scaling, but the fact that brains are metabolically expensive to build and use is surely part of the explanation (Isler and van Schaik, 2006; Tsuboi et al., 2015).

Other studies have examined how individual brain components scale against one another and with respect to absolute brain size. These studies have shown that individual brain regions tend also to scale allometrically. For example, in most vertebrate lineages, the proportion of the brain that is occupied by the dorsal telencephalon (e.g., the neocortex in mammals) increases predictably with absolute brain size (Yopak et al., 2010). In general, the regions that enlarge the most are those whose neurons are born (from neuronal precursors) relatively late during

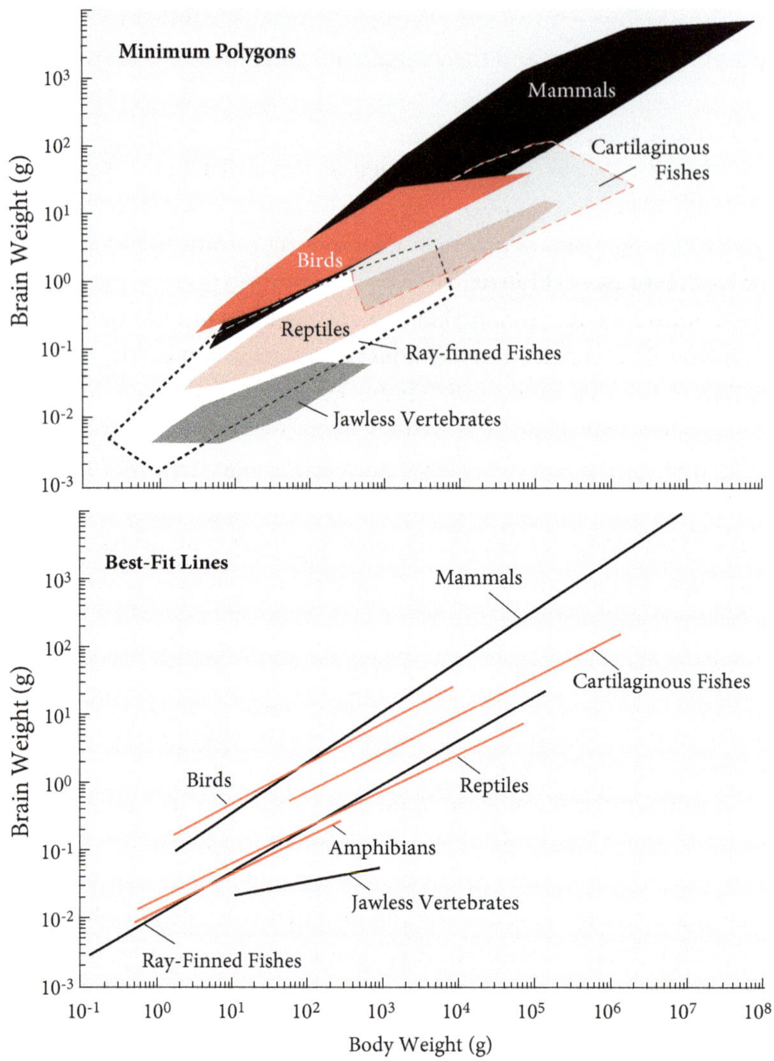

Figure 1.19 Brain-body scaling in vertebrates. Double-logarithmic plots of brain size versus body size reveal clear differences between the major vertebrate lineages when all the data points for a given lineage are enclosed by minimum convex polygons (top). Lines fitted to the data from the various lineages often vary in *y*-intercept or slope (bottom). However, all the slopes tend to be less than one, indicating negative allometry.

Adapted from Jerison (1973), van Dongen (1998), and Salas et al. (2017).

development (Finlay and Darlington, 1995). This observation has led to the suggestion that brains tend to enlarge by stretching their entire developmental schedule without changing the relative timing of developmental events (notably neuronal birth dates). Such a mechanism would cause late-born brain regions to enlarge disproportionately as brain size increases (Finlay and Darlington, 1995; Striedter, 2005). Although this idea is well supported, evolution clearly does sometimes break such rules, which in this case means that it enlarges (or shrinks) specific brain regions far more than one would expect from the ancestral scaling rules (Barton and Harvey, 2000). The anomalous descendants then follow their own, modified scaling rules. Analogous changes in scaling rules are observed when one examines how neuron numbers vary with brain size (Herculano-Houzel et al., 2006; Herculano-Houzel et al., 2007; Olkowicz et al., 2016). By contrast, the scaling rules for non-neuronal cells (mainly glia) are much more conservative (Herculano-Houzel, 2014).

Complementing the research on brain scaling are numerous studies on the functional correlates of variation in neuronal characters. Much of this research correlates variation in the size of the entire brain with aspects of behavior or cognition. A guiding theme is that larger brains allow for more "intelligent" behavior, though measuring intelligence across a range of species remains notoriously difficult. Furthermore, some studies emphasize absolute brain size, while others focus on encephalization (i.e., brain size relative to allometric expectations), or the proportional size of major brain regions (Lefebvre, 2012). Indeed, some researchers have argued that it makes more sense to correlate behavioral capacities with the size of individual brain regions, rather than whole brains, because the functions of a specific brain region must always be more circumscribed than those of an entire brain (Healy and Rowe, 2007). Others have argued that one should examine functional systems rather than individual brain regions (Montgomery et al., 2016), but the highly divergent/convergent nature of neuronal circuits makes defining those systems difficult. In short, "neuroecology" (Bolhuis and Macphail, 2001) is not a trivial enterprise.

That said, all is not lost. As data on the functions of specific behavioral capacities, brain regions, circuits, and genes accrue, correlative studies are providing an ever more detailed picture of how neural features covary with one another and with species behavior and ecology (Iwaniuk, 2016). Among birds, for example, tool use capacity correlates with the relative size of a specific telencephalic area (the nidopallium; Lefebvre et al., 2002). Some studies even address the question of heritability, which is crucial to demonstrating adaptive significance (Airey et al., 2000). Particularly interesting are studies in which neural or behavioral features were modified through artificial selection (Rehkämper et al., 2008). For example, Niclas Kolm and his collaborators bred guppies, which are small teleosts, for either increased relative brain size or decreased relative brain size and then compared the behavior of the two lines (Figure 1.20). They found that the large-brained lineage performed significantly better on some learning tasks, despite some interesting sex differences (Kotrschal et al., 2013, 2015). It is possible that the behavioral differences reflect changes in a specific brain region, rather than overall brain size, but the findings are nonetheless quite intriguing. We suspect that such experiments are just the leading edge of more refined experiments that will go a long way toward clarifying the complex relationships between neuronal and behavioral variation.

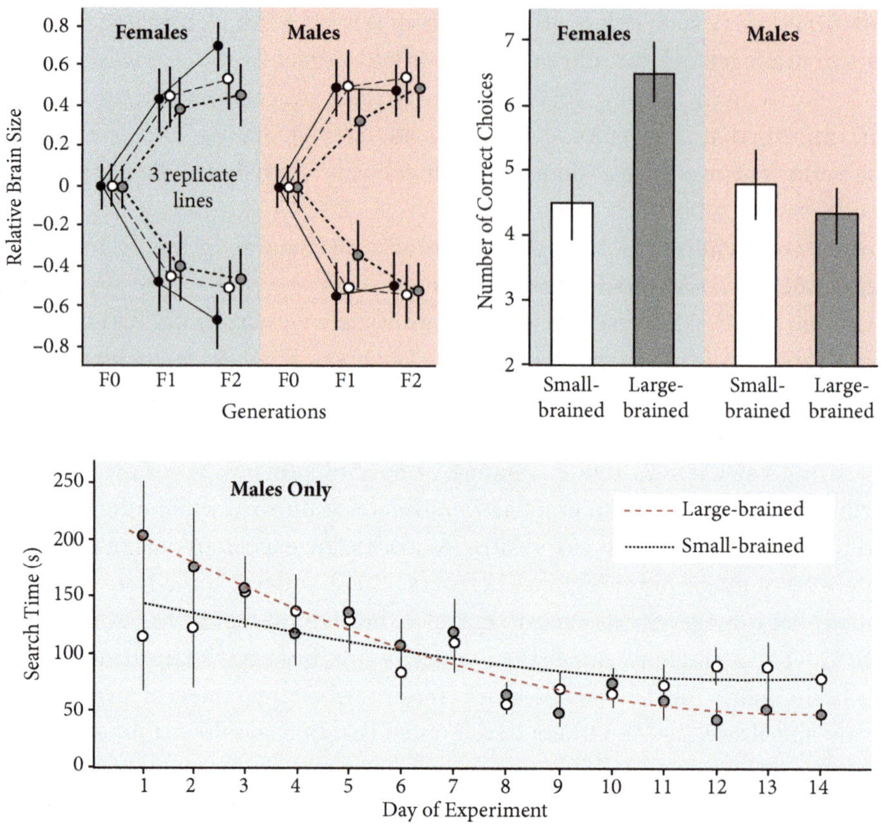

Figure 1.20 Behavioral correlates of artificial selection for brain size. Several populations of the small freshwater teleost *Poecilia reticulata* (i.e., guppies) were bred for either large or small brain size, relative to body size (using optic tectum width and body length as proxies). As shown at the top left, statistically significant differences in relative brain size were apparent in all three replicate lines after just two generations of selection (F0 to F2; the largest observed difference at F2 was 9.3%). The graph at the top right reveals that the females from the large-brained lines performed significantly better in a numerical learning task that required them to associate a specific number of symbols with food. As depicted in the bottom graph, males after three generations of selective breeding showed an analogous difference in performance on a spatial learning task that required them to search for a female in a maze.
Adapted from Kotrschal et al. (2013, 2015).

1.4. The Importance of Natural History

As shown in the preceding overview, comparative neurobiology includes a multitude of approaches, each entailing challenges but also presenting some opportunities for major advances in our understanding of brain function and evolution. Most of the approaches focus either on highly conserved features or on general principles of variation. Thus, they seek to extract regularities that hold across multiple species.

This strategy is time-honored and important. As mentioned previously, most comparative neurobiologists today emphasize the conserved aspects of brain evolution. However, we here pursue a very different aim, namely to understand specific "key innovations" (Hunter, 1998) in vertebrate nervous systems that arose during specific periods of evolutionary history. Philip Ulinski referred to those major changes as "nodal events" (Ulinski, 1989).

Some researchers have argued that the study of unique historical events is inherently futile, that "an explanation that applies to only one case . . . explains nothing" (Cartmill, 1992). However, we agree more closely with Hermann Hesse (1947), who wrote that "to study history one must know in advance that one is attempting something fundamentally impossible, yet necessary and highly important" (p. 169; 1990 edition). After all, what use are all those evolutionary rules and principles if we cannot use them to understand the past (Hempel, 1942), which presumably consists primarily of singular events? For example, it is reasonable to explain the extinction of the dinosaurs as the result of a massive asteroid striking the earth, darkening the skies for several years, killing off most plants, and ultimately causing a global food chain collapse (see Chapter 6). Some additional factors may well have contributed to the dinosaurs' demise, but this does not imply that all such explanations are futile.

When we seek to understand historical events, we must attempt to reconstruct not only what happened when, but also the event's historical context. For our purposes, this means that we must try to reconstruct not only when and how nervous systems changed, but also the condition of other organ systems in the relevant species, their physical environments, and the other species with which they interacted. In essence, we must extend the perspective of integrated organismal biology (Wainwright and Reilly, 1994) to extinct species. Put differently, we are interested in the natural history (Bates, 2014) of the species that underwent substantial changes in their nervous system and ecology at critical junctures of vertebrate phylogeny. How did the key innovations of those species arise, and how did they contribute to the lineage's long-term success? One of us has previously applied this integrative organismal approach to the origin of "a new head" in vertebrates (Gans and Northcutt, 1983; Northcutt and Gans, 1983; Northcutt, 2005), but we here extend it to a wider variety of vertebrate innovations.

A major challenge in synthesizing information from neurobiology, functional morphology, physiology, systematics, paleontology, and paleoecology is that each of these fields is replete with specialized vocabulary. We here try to minimize this jargon and often simplify ideas in order to create a manageable narrative. Our in-text citations are offered as a convenient entry into the relevant literature, rather than as a comprehensive bibliography. In terms of species, we limit our coverage to vertebrates and their immediate relatives, because this is our area of expertise. Most importantly, we do not describe any nervous systems in exhaustive detail, because such coverage would bloat the book. Plus, such descriptions are available elsewhere (Wullimann et al., 1996; Nieuwenhuys et al., 1998; Butler and Hodos, 2005; Kaas,

2007; Puelles et al., 2007; Smeets et al., 2011; Kaas, 2017). Instead, we focus on the major differences among key lineages. Extracting those differences from the existing literature has often been difficult.

Why is it important to understand evolutionary changes in vertebrate nervous systems? It is tempting to reply that "those who cannot remember the past are condemned to repeat it" (Santayana, 1905). Indeed, current concerns about global warming might be informed by what happened at the end of the Permian period, when global temperatures rose rapidly and the earth's largest extinction claimed 96% of all marine species, as well as 70% of the terrestrial vertebrates (Sun et al., 2012). Of course, most details of vertebrate evolution are unlikely to recur, even if convergent evolution is more rampant than most biologists believe (Conway Morris, 2003). Therefore, we do not claim that evolutionary neurobiologists can predict the future course of nervous system evolution with any meaningful degree of certainty (but see Hofman, 2001, for an interesting attempt). Nonetheless, knowing how vertebrates have responded to previous episodes of climate change frames the set of future possibilities. Beyond that, knowledge of past evolutionary changes empowers us to make predictions of a different type: It allows us to make reasonable predictions about the brains of vertebrate species that have not been examined yet, especially if we know both their phylogenetic position and their behavioral ecology.

For example, the relationship between brain size and body size has now been examined in enough species, from a diverse array of families, that future discoveries of species that fall outside their associated polygons in Figure 1.19 seem rather unlikely. Similarly, we can predict that an active and highly social predator will likely have a larger brain, relative to body size, than a slow-moving, solitary herbivore, as long as the two species belong to the same vertebrate order. Looking beyond brain size, an experienced neuroanatomist should be able to examine sections through the brain of an unknown vertebrate species and predict whether the species is a bird, a mammal, a frog and so forth. Experts on the brains of a specific class of vertebrates might even be able to predict the order, if not the family, to which the unknown species belongs (sadly, the number of such experts is small and likely dwindling). Thus, comparative neurobiology is now at a stage similar to that attained when Hertzsprung and Russel developed their diagrams of variation in stellar luminance and temperature (see Figure 1.1). This famous diagram allowed astrophysicists to predict, at least roughly, important attributes of stars in portions of the sky that they had not yet examined. That said, biology is not astrophysics, and species are not like stars or elements in the periodic table; biological evolution seems to delight in breaking rules and provoking surprise. For example, howler monkeys are clearly platyrrhine primates, but the photoreceptors that mediate color vision in this species are much more like those of catarrhine primates than other platyrrhines (Kelber and Jacobs, 2016).

Despite comparative neurobiology's capacity for predictions, our main aim in this book is not to predict anything, but to attain a deeper understanding of the vertebrates living today by reviewing their evolutionary past. We think this is a worthy aim, because the need to "know your animals" (Perrin, 2014) is increasingly acute in today's research climate, where animals are often treated as "materials and supplies" (e.g., in grant applications to the National Institutes of Health; Logan, 2002; Preuss and Robert, 2014). In particular, we hope to raise awareness about important differences between the major animal lineages. It is common nowadays to treat a broad variety of species as "good models" in biomedical research, but discussions about which models are best for a particular purpose are rare or advocate primarily for the species the authors themselves have chosen to study. It is too often the case, we think, that authors exaggerate the similarity of their favorite species to humans and minimize (or ignore) salient differences. These issues should be debated more extensively and more objectively, because species differences might help explain why so many findings obtained in animals fail to translate to humans (van der Worp et al., 2010). For example, it is important to know that the motor cortex and its projections to the spinal cord differ considerably between monkeys and rodents (Lemon, 2008; Rathelot and Strick, 2009). Given these differences, non-human primates should be (and historically have been) more useful than rats or mice when it comes to developing neuroprosthetic devices that can assist people with spinal cord injury (Taylor et al., 2002). In general, we are convinced that a deeper appreciation for (and understanding of) taxonomic differences can help researchers select better "animal models" for the problems that interest them (Bolker, 1995; Preuss, 2000; Krebs, 2005; Manger et al., 2008; Robert, 2008).

Finally, we think that our broad historical analysis of vertebrate nervous systems helps to reveal some general patterns and trends that are less apparent when one adopts a narrower focus. For example, we are struck by how frequently brain size, both relative to body size and in absolute terms, has increased in diverse vertebrate lineages. This pattern is unlikely to have occurred by chance. Moreover, increases in absolute brain size have usually been associated with increases in brain complexity, though the details of that complexity tend to vary across lineages. We do see numerous similarities in brain organization across the vertebrates, but a surprising number of these similarities probably resulted from convergent evolution, rather than shared ancestry. Moreover, some taxonomic differences involve rather fundamental aspects of brain organization, raising important questions about whether the supposed "vertebrate brain Bauplan" is really common to all vertebrates. We return to these overarching questions in Chapter 7.

References

Abbott A. 2013. Neuroscience: solving the brain. *Nature* 499:272–274.

Abzhanov A, Protas M, Grant BR, Grant PR, Tabin CJ. 2004. *Bmp4* and morphological variation of beaks in Darwin's finches. *Science* 305:1462–1465.

Affaticati P, Yamamoto K, Rizzi B, Bureau C, Peyriéras N, Pasqualini C, Demarque M, Vernier P. 2015. Identification of the optic recess region as a morphogenetic entity in the zebrafish forebrain. *Sci Rep* 5:211–213.

Agutter PS, Wheatley DN. 2004. Metabolic scaling: consensus or controversy? *Theoret Biol Med Mod* 1:13.

Airey DC, Castillo-Juarez H, Casella G, Pollak EJ, DeVoogd TJ. 2000. Variation in the volume of zebra finch song control nuclei is heritable: developmental and evolutionary implications. *Proc Roy Soc Lond B* 267:2099–2104.

Alberch P. 1989. The logic of monsters: evidence for internal constraint in development and evolution. *Geobios* 22:21–57.

Alexander RM, Jayes AS, Maloiy GMO, Wathuta EM. 1979. Allometry of the limb bones of mammals from shrews (*Sorex*) to elephant (*Loxodonta*). *J Zool Lond* 189:305–314.

Anquetin J. 2012. Reassessment of the phylogenetic interrelationships of basal turtles (Testudinata). *J Syst Palaeontol* 10:3–45.

Arendt D. 2008. The evolution of cell types in animals: emerging principles from molecular studies. *Nat Rev Genet* 9:868–882.

Arendt D, Musser JM, Baker CVH, Bergman A, Cepko C, Erwin DH, Pavlicev M, Schlosser G, Widder S, Laubichler MD, Wagner GP. 2016. The origin and evolution of cell types. *Nat Ecol Evol* 17:744–757.Arendt D, Tessmar-Raible K, Snyman H, Dorresteijn AW, Wittbrodt J. 2004. Ciliary photoreceptors with a vertebrate-type opsin in an invertebrate brain. *Science* 306:869–871.

Arny T. 1990. The Star Makers: A history of the theories of stellar structure and evolution. *Vistas Astron* 33:211–233.

Avise JC. 2009. Timetrees: beyond cladograms, phenograms, and phylograms. In: Hedges SB, Kumar S (eds.), *The Timetree of Life*. New York: Oxford University Press, pp. 19–25.

Barker FK, Cibois A, Schikler P, Feinstein J, Cracraft J. 2004. Phylogeny and diversification of the largest avian radiation. *Proc Natl Acad Sci USA* 101:11040–11045.

Barrett PH, Gautrey PJ, Herbert S, Kohn D, Smith S. 2009. *Charles Darwin's Notebooks, 1836–1844*. Cambridge, UK: Cambridge University Press.

Barton RA, Harvey PH. 2000. Mosaic evolution of brain structure in mammals. *Nature* 405:1055–1058.

Bates M. 2014. *The Nature of Natural History*. Princeton, NJ: Princeton University Press.

Bateson W. 1892. On numerical variation in teeth, with a discussion of the conception of homology. *Proc Zool Soc London*: 102–115.

Benton MJ. 2009. Dinosaurs. *Curr Biol* 19:R318–R323.

Benton MJ, Donoghue P, Asher RJ. 2009. Calibrating and constraining molecular clocks. In: Hedges SB, Kumar S (eds.), *The Timetree of Life*. New York: Oxford University Press, pp. 35–86.

Bergquist H. 1932. Zur Morphologie des Zwischenhirns bei Niederen Wirbeltieren. *Acta Zoologica* 13:57–303.

Bergquist H, Källén B. 1954. Notes on the early histogenesis and morphogenesis of the central nervous system in vertebrates. *J Comp Neurol* 100:627–659.

Berquist RM, Galinsky VL, Kajiura SM, Frank LR. 2015. The coelacanth rostral organ is a unique low-resolution electro-detector that facilitates the feeding strike. *Sci Rep* 5:229–35.

Bhullar B-AS, Morris ZS, Sefton EM, Tok A, Tokita M, Namkoong B, Camacho J, Burnham DA, Abzhanov A. 2015. A molecular mechanism for the origin of a key evolutionary innovation, the bird beak and palate, revealed by an integrative approach to major transitions in vertebrate history. *Evolution* 69:1665–1677.

Biewener AA. 2005. Biomechanical consequences of scaling. *J Exp Biol* 208:1665–1676.

Bolhuis JJ, Macphail EM. 2001. A critique of the neuroecology of learning and memory. *Trends Cogn Sci* 5:426–433.

Bolker JA. 1995. Model systems in developmental biology. *Bioessays* 17:451–455.

Bolstad GH, Cassara JA, Márquez E, Hansen TF, van der Linde K, Houle D, Pélabon C. 2015. Complex constraints on allometry revealed by artificial selection on the wing of *Drosophila melanogaster*. *Proc Natl Acad Sci USA* 112:13284–13289.

Braford MR Jr, Northcutt RG. 1983. Organization of the diencephalon and pretectum of the ray-finned fishes. In: Davis R, Northcutt RG (eds.), *Fish Neurobiology*, Vol. 2. Ann Arbor: University of Michigan Press, pp. 117–164.
Brigandt I, Love AC. 2010. Evolutionary novelty and the evo-devo synthesis: field notes. *Evol Biol* 37:93–99.
Briscoe SD. 2019. Field homology: still a meaningless concept. *Brain Behav Evol* 93:1–3.
Briscoe SD, Albertin CB, Rowell JJ, Ragsdale CW. 2018. Neocortical association cell types in the forebrain of birds and alligators. *Curr Biol* 28:1–11.
Brower AVZ. 2014. Transformational and taxic homology revisited. *Cladistics* 31:197–201.
Bullard E, Everett JE, Smith AG. 1965. The fit of the continents around the Atlantic. *Phil Trans Roy Soc A* 258:41–51.
Burbidge EM, Burbidge GR, Fowler WA, Hoyle F. 1957. Synthesis of the elements in stars. *Rev Mod Phys* 29: 547–654.
Butler AB, Hodos W. 2005. *Comparative Vertebrate Neuroanatomy*. Hoboken, NJ: John Wiley & Sons.
Butler AB, Saidel WM. 2000. Defining sameness: historical, biological, and generative homology. *Bioessays* 22:846–853.
Butler MA, King AA. 2004. Phylogenetic comparative analysis: a modeling approach for adaptive evolution. *Am Nat* 164:683–695.
Butler RF. 1992. *Paleomagnetism: Magnetic Domains to Geologic Terranes*. New York: Blackwell Scientific.
Campbell CB, Hodos W. 1970. The concept of homology and the evolution of the nervous system. *Brain Behav Evol* 3:353–367.
Cartmill M. 1992. New views on primate origins. *Evol Anthropol* 1:105–111.
Christian A, Garland T Jr. 1996. Scaling of limb proportions in monitor lizards (Squamata: Varanidae). *J Herpetol* 30:219–230.
Collier D. 1993. The comparative method. In: Finifter AW (ed.), *Political Science: The State of the Discipline*. Washington, DC: American Political Science Association, pp. 105–119.
Conway Morris S. 2003. *Life's Solution*. Cambridge, UK: Cambridge University Press.
Cookson K. 2001. Field homology: a meaningful definition. *Eur J Morphol* 39:39–45.
Cracraft J. 2005. Phylogeny and evo-devo: characters, homology, and the historical analysis of the evolution of development. *Zoology* 108:345–356.
Darwin C. 1859. *On the Origin of Species*. London: John Murray.
Darwin C. 1872. *On the Origin of Species*, 6th ed. London: John Murray.
De Beer GR. 1940. *Embryos and Ancestors*. New York: Oxford University Press.
de Queiroz K. 1988. Systematics and the Darwinian revolution. *Philos Sci* 55:238–259.
Demhardt IJ. 2006. Alfred Wegener's hypothesis on continental drift and its discussion in Petermanns Geographische Mitteilungen (1912–1942). *Polarforschung* 75:29–35.
Doyle P, Bennett MR, Baxter AN. 2001. *The Key to Earth History: An Introduction to Stratigraphy*, 2nd ed. New York: John Wiley & Sons.
Drummond AJ, Ho SYW, Phillips MJ, Rambaut A. 2006. Relaxed phylogenetics and dating with confidence. *PloS Biol* 4:e88–12.
Duncan JS, Fritzsch B. 2012. Evolution of sound and balance perception: innovations that aggregate single hair cells into the ear and transform a gravistatic sensor into the organ of Corti. *Anat Rec* 295:1760–1774.
Dunstone N, Gorman ML. 1993. *Mammals as Predators*. Oxford: Oxford University Press.
Ebbesson SO, Heimer L. 1970. Projections of the olfactory tract fibers in the nurse shark (*Ginglymostoma cirratum*). *Brain Res* 17:47–55.
Eddington AS. 1926. *The Internal Constitution of the Stars*. Cambridge, UK: Cambridge University Press.
Edinger L. 1908. The relations of comparative anatomy to comparative psychology. *J Comp Neurol* 18:437–457.
Endler JA. 1980. Natural selection on color patterns in *Poecilia reticulata*. *Evolution* 34:76–91.

Erwin DH, Laflamme M, Tweedt SM, Sperling EA, Pisani D, Peterson KJ. 2011. The Cambrian conundrum: early divergence and later ecological success in the early history of animals. *Science* 334:1091–1097.

Fabre A-C, Cornette R, Goswami A, Peigné S. 2015. Do constraints associated with the locomotor habitat drive the evolution of forelimb shape? A case study in musteloid carnivorans. *J Anat* 226:596–610.

Faunes M, Francisco Botelho J, Ahumada Galleguillos P, Mpodozis J. 2015. On the hodological criterion for homology. *Front Neurosci* 9:223.

Felsenstein J. 1985. Phylogenies and the comparative method. *Am Nat* 125:1–15.

Fernald RD. 2006. Casting a genetic light on the evolution of eyes. *Science* 313:1914–1918.

Finger TE. 1978. Efferent neurons of the teleost cerebellum. *Brain Res* 153:608–614.

Finlay BL, Darlington RB. 1995. Linked regularities in the development and evolution of mammalian brains. *Science* 268:1578–1584.

Florio M, Albert M, Taverna E, Namba T, Brandl H, Lewitus E, Haffner C, Sykes A, Wong FK, Peters J, et al. 2015. Human-specific gene ARHGAP11B promotes basal progenitor amplification and neocortex expansion. *Science* 347:1465–1470.

Freud S. 1923. *Das Ich und das Es*. Leipzig: Internationaler Psychoanalytischer Verlag.

Fuentes-G JA, Housworth EA, Weber A, Martins EP. 2016. Phylogenetic ANCOVA: estimating changes in evolutionary rates as well as relationships between traits. *Am Nat* 188:615–627.

Galilei G. 1939. *Dialogues Concerning Two New Sciences*. New York: Macmillan.

Gans C. 1980. *Biomechanics: An Approach to Vertebrate Biology*. Ann Arbor: University of Michigan Press.

Gans C, Northcutt RG. 1983. Neural crest and the origin of vertebrates: a new head. *Science* 220:268–273.

Gardiner BG. 1979. The salmon, the lungfish and the cow: a reply. *Nature* 277:175–176.

Garland T, Dickerman AW, Janis CM, Jones JA. 1993. Phylogenetic analysis of covariance by computer simulation. *Syst Biol* 42:265–292.

Garstang W. 1922. The theory of recapitulation: a critical re-statement of the biogenetic law. *Zool J Linn Soc* 35:81–101.

Gehring WJ, Ikeo K. 1999. *Pax 6*: mastering eye morphogenesis and eye evolution. *Trends Genet* 15:371–377.

Gilbert SF, Opitz JM, Raff RA. 1996. Resynthesizing evolutionary and developmental biology. *Dev Biol* 173:357–372.

Gilland E, Baker R. 2005. Evolutionary patterns of cranial nerve efferent nuclei in vertebrates. *Brain Behav Evol* 66:234–254.

Gingerich O. 2013. The critical importance of Russel's diagram. In: Way MJ and Hunter D (eds.), *Origins of the Expanding Universe: 1912–1932*. Orem, UT: Astronomical Society of the Pacific, pp. 1–11.

González A, López JM, Morona R, Moreno N. 2017. The organization of the central nervous system of amphibians. In: Kaas JH, Striedter GF (eds.), *Evolution of Nervous Systems*, 2nd ed. Amsterdam: Elsevier, pp. 141–170.

Goswami A. 2012. A dating success story: genomes and fossils converge on placental mammal origins. *EvoDevo* 3:18.

Gould SJ. 1977. *Ontogeny and Phylogeny*. Cambridge, MA: Harvard University Press.

Gould SJ, Lewontin RC. 1979. The spandrels of San Marco and the Panglossian paradigm: a critique of the adaptationist programme. *P Roy Soc Lond B Bio* 205:581–598.

Hall BK. 2000. *Homology*. San Diego: Academic Press.

Hall BK. 2007. Homoplasy and homology: dichotomy or continuum? *J Hum Evol* 52:473–479.

Hall WC, Ebner FF. 1970. Thalamotelencephalic projections in the turtle (*Pseudemys scripta*). *J Comp Neurol* 140:101–122.

Hansen TF. 1997. Stabilizing selection and the comparative analysis of adaptation. *Evolution* 51:1341–1351.

Healy SD, Rowe C. 2007. A critique of comparative studies of brain size. *Proc Roy Soc Lond B* 274:453–464.

Hedges BS, Kumar S. 2004. Precision of molecular time estimates. *Trends Gen* 20:242–247.

Hedges SB, Kumar S. 2009. Discovering the timetree of life. In: Hedges BS, Kumar S (eds.), *The Timetree of Life*. New York: Oxford University Press, pp. 3–18.

Hempel CG. 1942. The function of general laws in history. *J Philosophy* 39:35–48.

Hennig W. 1966. *Phylogenetic Systematics*. Urbana: University of Illinois Press.

Herculano-Houzel S. 2014. Brain scaling in mammalian evolution as a consequence of concerted and mosaic changes in numbers of neurons and average neuronal cell size. *Front Neuroanat* 8:77.

Herculano-Houzel S, Collins CE, Wong P, Kaas JH. 2007. Cellular scaling rules for primate brains. *Proc Natl Acad Sci USA* 104:3562–3567.

Herculano-Houzel S, Mota B, Lent R. 2006. Cellular scaling rules for rodent brains. *Proc Natl Acad Sci USA* 103:12138–12143.

Herrick CJ. 1948. *The Brain of the Tiger Salamander*, Ambystoma Tigrinum. Chicago: University of Chicago Press.

Hess HH. 1962. History of ocean basins. In: *Petrologic Studies: A Volume to Honor A. F. Buddington*. Boulder, CO: Geological Society of America, pp. 599–620.

Hesse H. 1947. *Magister Ludi*. Zürich: Fretz and Wasmuth.

Hodos W, Campbell C. 1969. *Scala naturae*: why there is no theory in comparative psychology. *Psych Review* 76:337–350.

Hofman MA. 2001. Brain evolution in hominids: are we at the end of the road? In: Falk D, Gibson KR (eds), *Evolutionary Anatomy of the Primate Cerebral Cortex*. Cambridge, UK: Cambridge University Press, pp. 113–127.

Holmgren N. 1922. Points of view concerning forebrain morphology in lower vertebrates. *J Comp Neurol* 34:391–459.

Humboldt von A, Bonpland A. 2009. *Essay on the Geography of Plants*. Chicago: University of Chicago Press.

Hunter JP. 1998. Key innovations and the ecology of macroevolution. *Trends Ecol Evolut* 13:31–36.

Isler K, van Schaik CP. 2006. Metabolic costs of brain size evolution. *Biol Lett* 2:557–560.

Iwaniuk AN. 2004. Brood parasitism and brain size in cuckoos: a cautionary tale on the use of modern comparative methods. *Int J Comp Psych* 17:17–33.

Iwaniuk AN. 2016. Functional correlates of brain and brain region sizes in nonmammalian vertebrates. In: Kaas JH, Striedter GF (eds.), *Evolution of Nervous Systems*, Vol. 1, 2nd ed. Amsterdam: Elsevier, pp. 335–348.

Jacob F. 1977. Evolution and tinkering. *Science* 196:1161–1166.

Jarvis ED, Mirarab S, Aberer AJ, Li B, Houde P, Li C, Ho SYW, Faircloth BC, Nabholz B, Howard JT, Suh A, Weber CC, da Fonseca RR, et al. 2014. Whole-genome analyses resolve early branches in the tree of life of modern birds. *Science* 346:1320–1331.

Jenkinson JW. 1925. *Vertebrate Embryology*. Oxford, UK: Clarendon Press.

Jerison H. 1973. *Evolution of The Brain and Intelligence*. San Diego, CA: Academic Press.

Kaas JH (ed.) 2007. *Evolution of Nervous Systems*, 1st ed. San Diego, CA: Academic Press.

Kaas JH (ed.) 2017. *Evolution of Nervous Systems*, 2nd ed. Amsterdam: Elsevier.

Kappers CVA, Huber GC, Crosby EC. 1936. *The Comparative Anatomy of the Nervous System of Vertebrates, Including Man*. New York: Macmillan.

Källén B. 1951. Embryological studies on the nuclei and their homologization in the vertebrate forebrain. *Kungl Fyiogr Sällsk Lund Handl* 62:1–36.

Källén B. 1953. On the nuclear differentiation during ontogenesis in the avian forebrain and some notes on the amniote strioamygdaloid complex. *Acta Anat (Basel)* 17:72–84.

Karten HJ. 1968. The ascending auditory pathway in the pigeon (*Columba livia*). II. Telencephalic projections of the nucleus ovoidalis thalami. *Brain Res* 11:134–153.

Karten HJ. 1969. The organization of the avian telencephalon and some speculations on the phylogeny of the amniote telencephalon. *Ann N Y Acad Sci* 167:164–179.

Kelber A, Jacobs GH. 2016. Evolution of color vision. In: Kremers J, Baraas RC, Marshall NJ (eds.), *Human Color Vision*, Vol. 34, 2nd ed. Basel: Springer International, pp. 317–354.

Kiecker C, Lumsden A. 2005. Compartments and their boundaries in vertebrate brain development. *Nat Rev Neurosci* 6:553–564.

Kim N, Park C, Jeong Y, Song M-R. 2015. Functional diversification of motor neuron-specific *Isl1* enhancers during evolution. *PLoS Genet* 11:e1005560.

Kin K. 2015. Inferring cell type innovations by phylogenetic methods-concepts, methods, and limitations. *J Exp Zool B* 324:653–661.

Kotrschal A, Corral-Lopez A, Amcoff M, Kolm N. 2015. A larger brain confers a benefit in a spatial mate search learning task in male guppies. *Behav Ecol* 26:527–532.

Kotrschal A, Rogell B, Bundsen A, Svensson B, Zajitschek S, Brännström I, Immler S, Maklakov AA, Kolm N. 2013. Artificial selection on relative brain size in the guppy reveals costs and benefits of evolving a larger brain. *Curr Biol* 23:168–171.

Kozmik Z. 2008. The role of Pax genes in eye evolution. *Brain Res Bull* 75:335–339.

Krebs HA. 2005. The August Krogh principle: "for many problems there is an animal on which it can be most conveniently studied." *J Exp Zool* 194:221–226.

Kuratani S. 2009. Modularity, comparative embryology and evo-devo: developmental dissection of evolving body plans. *Dev Biol* 332:61–69.

Lauder GV. 1996. The argument from design. In: Rose MR, Lauder GV (eds.), *Adaptation.* Amsterdam: Elsevier, pp. 55–91.

Lefebvre L. 2012. Primate encephalization. *Prog Brain Res* 195:393–412.

Lefebvre L, Nicolakakis N, Boire D. 2002. Tools and brains in birds. *Behaviour* 139:939–973.

Lemon RN. 2008. Descending pathways in motor control. *Annu Rev Neurosci* 31:195–218.

Liang C, Musser JM, Cloutier A, Prum R, Wagner G. 2018. Pervasive correlated evolution in gene expression shapes cell type transcriptomes. *Genome Biol Evol* 10:538–552.

Liebeskind BJ, Hillis DM, Zakon HH. 2011. Evolution of sodium channels predates the origin of nervous systems in animals. *Proc Natl Acad Sci USA* 108:9154–9159.

Liebeskind BJ, Hillis DM, Zakon HH. 2015. Convergence of ion channel genome content in early animal evolution. *Proc Natl Acad Sci USA* 112:E846–51.

Lijphart A. 1971. Comparative politics and the comparative method. *Am Pol Sci Rev* 65:682–693.

Litwin-Kumar A, Harris KD, Axel R, Sompolinsky H, Abbott LF. 2017. Optimal degrees of synaptic connectivity. *Neuron* 93:1153–1164.

Liu Z, Qi F-Y, Zhou X, Ren H-Q, Shi P. 2014a. Parallel sites implicate functional convergence of the hearing gene prestin among echolocating mammals. *Mol Biol Evol* 31:2415–2424.

Liu Z, Wang W, Zhang T-Z, Li G-H, He K, Huang J-F, Jiang X-L, Murphy RW, Shi P. 2014b. Repeated functional convergent effects of NaV1.7 on acid insensitivity in hibernating mammals. *Proc Roy Soc Lond B* 281:20132950.

Lloyd GT, Davis KE, Pisani D, Tarver JE, Ruta M, Sakamoto M, Hone DWE, Jennings R, Benton MJ. 2008. Dinosaurs and the Cretaceous terrestrial revolution. *Proc Roy Soc Lond B* 275:2483–2490.

Lodders K. 2003. Solar system abundances and condensation temperatures of the elements. *Astrophys J* 591:1220–1247.

Logan CA. 2002. Before there were standards: the role of test animals in the production of empirical generality in physiology. *J Hist Biol* 35:329–363.

Lovejoy AO. 1936. *The Great Chain of Being.* Cambridge, MA: Harvard University Press.

Loyau A, Petrie M, Saint Jalme M, Sorci G. 2008. Do peahens not prefer peacocks with more elaborate trains? *Anim Behav* 76:e5–e9.

López JM, Morona R, Moreno N, González A. 2017. The organization of the central nervous system of lungfishes: an immunohistochemical approach. In: Kaas JH, Striedter GF (eds.), *Evolution of Nervous Systems*, 2nd ed. Amsterdam: Elsevier, pp. 121–139.

Lynagh T, Beech RN, Lalande MJ, Keller K, Cromer BA, Wolstenholme AJ, Laube B. 2015. Molecular basis for convergent evolution of glutamate recognition by pentameric ligand-gated ion channels. *Sci Rep* 5:8558.

Ma L-H, Gilland E, Bass AH, Baker R. 2010. Ancestry of motor innervation to pectoral fin and forelimb. *Nat Commun* 1:49.

Mabee PM. 1989. An empirical rejection of the ontogenetic polarity criterion. *Cladistics* 5:409–416.

MacLean PD. 1990. *The Triune Brain in Evolution*. Berlin/Heidelberg: Springer Science & Business Media.

Major D, Luksch H, Karten H. 2000. Bottlebrush dendritic endings and large dendritic fields: motion-detecting neurons in the mammalian tectum. *J Comp Neurol* 423:243–260.

Manger PR, Cort J, Ebrahim N, Goodman A, Henning J, Karolia M, Rodrigues S-L, Strkalj G. 2008. Is 21st century neuroscience too focussed on the rat/mouse model of brain function and dysfunction? *Front Neuroanat* 2:5.

Margoliash E. 1963. Primary structure and evolution of cytochrome c. *Proc Natl Acad Sci USA* 50:672–679.

Martins E. 2000. Adaptation and the comparative method. *Trends Ecol Evolut* 15:296–299.

Mason RG, Raff AD. 1961. Magnetic survey off the west coast of North America, 32 N. latitude to 42 N. latitude. *Geol Soc Am Bull* 72:1259.

Mayasich SA, Clarke BL. 2016. The emergence of the vasopressin and oxytocin hormone receptor gene family lineage: clues from the characterization of vasotocin receptors in the sea lamprey (*Petromyzon marinus*). *Gen Comp Endocrinol* 226:88–101.

McCune AR, Schimenti JC. 2012. Using genetic networks and homology to understand the evolution of phenotypic traits. *Curr Genomics* 13:74–84.

Meek J, Nieuwenhuys R. 1991. Palisade pattern of mormyrid Purkinje cells: a correlated light and electron microscopic study. *J Comp Neurol* 306:156–192.

Mendeléef D. 1889. The periodic law of the chemical elements. *J Chem Soc* 66:634–656.

Mengistu H, Huizinga J, Mouret J-B, Clune J. 2016. The evolutionary origins of hierarchy. *PLoS Comput Biol* 12:e1004829.

Molnar P. 2015. *Plate Tectonics: A Very Short Introduction*. New York: Oxford University Press.

Montgomery SH, Mundy NI, Barton RA. 2016. Brain evolution and development: adaptation, allometry and constraint. *Proc Royal Soc Lond B* 283: 20160433.

Montiel JF, Aboitiz F. 2018. Homology in amniote brain evolution: the rise of molecular evidence. *Brain Behav Evol* 91:59–64.

Murray E, Wise S, Graham K. 2016. *The Evolution of Memory Systems*. New York: Oxford University Press.

Musser JM, Wagner GP. 2015. Character trees from transcriptome data: Origin and individuation of morphological characters and the so-called "species signal." *J Exp Zool B* 324:588–604.

Nagy LG, Ohm RA, Kovács GM, Floudas D, Riley R, Gácser A, Sipiczki M, Davis JM, Doty SL, de Hoog GS, et al. 2014. Latent homology and convergent regulatory evolution underlies the repeated emergence of yeasts. *Nat Commun* 5:4471.

Near TJ, Sanderson MJ. 2004. Assessing the quality of molecular divergence time estimates by fossil calibrations and fossil-based model selection. *Philos Trans R Soc Lond B* 359:1477–1483.

Nieuwenhuys R, Donkelaar ten HJ, Nicholson C. 1998. *The Central Nervous System of Vertebrates*. Berlin: Springer.

Nieuwenhuys R, Puelles L. 2016. *Towards a New Neuromorphology*. Berlin: Springer.

Northcutt RG. 1969. Discussion of the preceding paper. *Ann NY Acad Sci* 167:180–185.

Northcutt RG. 1981. Evolution of the telencephalon in nonmammals. *Annu Rev Neurosci* 4:301–350.

Northcutt RG. 1984. Evolution of the vertebrate central nervous system: patterns and processes. *Am Zool* 24:701–716.

Northcutt RG. 1992. The phylogeny of octavolateralis ontogenies: a reaffirmation of Garstang's phylogenetic hypothesis. In: Webster DB, Fay RR (eds.), *The Evolutionary Biology of Hearing*. New York: Springer, pp. 21–47.

Northcutt RG. 1997. Evolution of gnathostome lateral line ontogenies. *Brain Behav Evol* 50:25–37.

Northcutt RG. 1999. Field homology: a meaningless concept. *Eur J Morphol* 37:95–99.

Northcutt RG. 2005. The new head hypothesis revisited. *J Exp Zool* 304B:274–297.

Northcutt RG, Gans C. 1983. The genesis of neural crest and epidermal placodes: a reinterpretation of vertebrate origins. *Quart Rev Biol* 58:1–28.

Northcutt RG, Kaas JH. 1995. The emergence and evolution of mammalian neocortex. *Trends Neurosci* 18:373–379.

O'Hara RJ. 1991. Essay-review of Sibley and Ahlquist's 'Phylogeny and Classification of Birds: A Study in Molecular Evolution.' *Auk* 108:990–994.

O'Hara RJ. 1992. Telling the tale: narrative representation and the study of evolutionary history. *Biol Philos* 7:135–160.

Olkowicz S, Kocourek M, Lučan RK, Porteš M, Fitch WT, Herculano-Houzel S, Němec P. 2016. Birds have primate-like numbers of neurons in the forebrain. *Proc Natl Acad Sci USA* 113:7255–7260.

Oreskes N, LeGrand HE. 2001. *Plate Tectonics*. Boulder, CO: Westview Press.

Owen, R. 1843. *Lectures on the Comparative Anatomy and Physiology of the Invertebrate Animals, Delivered at the Royal College of Surgeons*. London: Longman, Brown, Green and Longmans.

Patterson C. 1988. Homology in classical and molecular biology. *Mol Biol Evol* 5:603–625.

Perrin S. 2014. Preclinical research: make mouse studies work. *Nature* 507:423–425.

Peter IS, Davidson EH. 2011. A gene regulatory network controlling the embryonic specification of endoderm. *Nature* 474:635–639.

Petrie M, Halliday T. 1994. Experimental and natural changes in the peacock's (*Pavo cristatus*) train can affect mating success. *Behav Ecol Sociobiol* 35:213–217.

Petrie M, Tim H, Carolyn S. 1991. Peahens prefer peacocks with elaborate trains. *Anim Behav* 41:323–331.

Pollard DA, Iyer VN, Moses AM, Eisen MB. 2006. Widespread discordance of gene trees with species tree in *Drosophila*: evidence for incomplete lineage sorting. *PLoS Genet* 2:e173–14.

Pombal MA, Megías M. 2017. The nervous systems of jawless vertebrates. In: Kaas JH, Striedter GF (eds.), *Evolution of Nervous Systems*, 2nd ed. Amsterdam: Elsevier, pp. 37–57.

Pouwels E. 1978. On the development of the cerebellum of trout, *Salmo Gairdneri*. IV. Development of the pattern of connectivity. *Anat Embryol* 153:55–65.

Preuss TM. 2000. Taking the measure of diversity: comparative alternatives to the model-animal paradigm in cortical neuroscience. *Brain Behav Evol* 55:287–299.

Preuss TM, Robert JS. 2014. Animal models of the human brain: repairing the paradigm. In: Gazzaniga MS, Mangun GR (eds.), *The Cognitive Neurosciences*, 5th ed. Cambridge, MA: MIT Press, pp. 59–66.

Pritz MB. 1974. Ascending connections of a thalamic auditory area in a crocodile, *Caiman crocodilus*. *J Comp Neurol* 153:199–213.

Prum RO, Berv JS, Dornburg A, Field DJ, Townsend JP, Lemmon EM, Lemmon AR. 2015. A comprehensive phylogeny of birds (Aves) using targeted next-generation DNA sequencing. *Nature* 526:569–573.

Puelles L. 2013. Plan of the developing vertebrate nervous system: relating embryology to the adult nervous system (prosomere model, overview of brain organization). In: Rubenstein JLR, Goldman-Rakic P (eds.), *Patterning and Cell Type Specification in the Developing CNS and PNS*. Amsterdam: Elsevier, pp. 187–209.

Puelles L, de la Torre M, Paxinos G, Watson C, Martinez S. 2007. *The Chick Brain in Stereotaxic Coordinates*. Amsterdam: Academic Press.

Puelles L, Medina L. 2002. Field homology as a way to reconcile genetic and developmental variability with adult homology. *Brain Res Bull* 57:243–255.

Puelles L, Medina L, Borello U, Legaz I, Teissier A, Pierani A, Rubenstein JLR. 2016. Radial derivatives of the mouse ventral pallium traced with *Dbx1*-LacZ reporters. *J Chem Neuroanat* 75:2–19.

Puelles L, Rubenstein JLR. 2015. A new scenario of hypothalamic organization: rationale of new hypotheses introduced in the updated prosomeric model. *Front Neuroanat* 9:1229–33.

Puelles L, Sandoval JE, Ayad A, del Corral R, Alonso A, Ferran JL, Martínez-de-la-Torre M. 2017. The pallium in reptiles and birds in the light of the updated tetrapartite pallium model. In: *Evolution of Nervous Systems*, 2nd ed. Amsterdam: Elsevier, pp. 519–555.

Pyron RA, Wiens JJ. 2011. A large-scale phylogeny of Amphibia including over 2800 species, and a revised classification of extant frogs, salamanders, and caecilians. *Mol Phylogenet Evol* 61:543–583.

Rathelot J-A, Strick PL. 2009. Subdivisions of primary motor cortex based on cortico-motoneuronal cells. *Proc Natl Acad Sci USA* 106:918–923.

Reeder TW, Townsend TM, Mulcahy DG, Noonan BP, Wood PL, Sites JW, Wiens JJ. 2015. Integrated analyses resolve conflicts over squamate reptile phylogeny and reveal unexpected placements for fossil taxa. *PLoS ONE* 10:e0118199–22.

Rehkämper G, Frahm HD, Cnotka J. 2008. Mosaic evolution and adaptive brain component alteration under domestication seen on the background of evolutionary theory. *Brain Behav Evol* 71:115–126.

Rice, A. 1996. *Servant of the Bones*. New York: A. Knopf.

Richardson MK. 1995. Heterochrony and the phylotypic period. *Dev Biol* 172:412–421.

Ridley M. 1986. *Evolution and Classification*. London: Longman Scientific and Technical.

Rieppel O, Kearney M. 2002. Similarity. *Biol J Linn Soc* 75:59–82.

Robert JS. 2008. The comparative biology of human nature. *Philosoph Psych* 21:425–436.

Rodríguez-Moldes I, Santos-Durán GN, Pose-Méndez S, Quintana-Urzainqui I, Candal E. 2016. The brains of cartilaginous fishes. In: Kaas JH, Striedter GF (eds.), *Evolution of Nervous Systems*, 2nd ed. Amsterdam: Elsevier, pp. 77–97.

Rohlfs RV, Nielsen R. 2015. Phylogenetic ANOVA: The expression variance and evolution model for quantitative trait evolution. *Syst Biol* 64:695–708.

Roux J, Robinson-Rechavi M. 2008. Developmental constraints on vertebrate genome evolution. *PLoS Genet* 4:e1000311.

Rubenstein JL, Martinez S, Shimamura K, Puelles L. 1994. The embryonic vertebrate forebrain: the prosomeric model. *Science* 266:578–580.

Ruta M, Jeffery JE, Coates MI. 2003. A supertree of early tetrapods. *Proc Roy Soc Lond B* 270:2507–2516.

Sagan C. 1977. *Dragons of Eden*. New York: Random House.

Sakarya O, Armstrong KA, Adamska M, Adamski M, Wang I-F, Tidor B, Degnan BM, Oakley TH, Kosik KS. 2007. A post-synaptic scaffold at the origin of the animal kingdom. *PLoS ONE* 2:e506.

Salas CA, Yopak KE, Lisney TJ, Potter IC, Collin SP. 2017. The central nervous system of jawless vertebrates: encephalization in lampreys and hagfishes. *Brain Behav Evol* 89:33–47.

Santayana G. 1905. *The Life of Reason*, Vol. 1. New York: Charles Scribner's Sons.

Scerri E. 2015. The discovery of the periodic table as a case of simultaneous discovery. *Philos Transact A* 373:20140172–20140172.

Schmidt-Nielsen K. 1984. Scaling. Cambridge, UK: Cambridge University Press.

Seid MA, Castillo A, Wcislo WT. 2011. The allometry of brain miniaturization in ants. *Brain Behav Evol* 77:5–13.

Shubin N, Tabin C, Carroll S. 2009. Deep homology and the origins of evolutionary novelty. *Nature* 457:818–823.

Sibley CG, Ahlquist JE. 1990. *Phylogeny and Classification of Birds*. New York: Berghahn Books.

Simeone A, Acampora D, Gulisano M, Stornaiuolo A, Boncinelli E. 1992. Nested expression domains of four homeobox genes in developing rostral brain. *Nature* 358:687–690.

Slack JM, Holland PW, Graham CF. 1993. The zootype and the phylotypic stage. *Nature* 361:490–492.

Smeets W, Nieuwenhuys R, Roberts BL. 2011. *The Central Nervous System of Cartilaginous Fishes*. Berlin: Springer.

Smith CUMC. 2010. The triune brain in antiquity: Plato, Aristotle, Erasistratus. *J Hist Neurosci* 19:1–14.

Smith HM. 1967. Biological similarities and homologies. *Syst Zool* 16:101–102.

Sokal RR, Sneath PHA. 1963. *Principles of Numerical Taxonomy*. San Francisco, CA: Freeman.

Steinmetz PRH, Aman A, Kraus JEM, Technau U. 2017. Gut-like ectodermal tissue in a sea anemone challenges germ layer homology. *Nat Ecol Evol* 1:1535–1542.

Striedter GF. 1990. The diencephalon of the channel catfish, *Ictalurus punctatus*. I. Nuclear organization. *Brain Behav Evol* 36:329–354.
Striedter GF. 1997. The telencephalon of tetrapods in evolution. *Brain Behav Evol* 49:179–213.
Striedter GF. 1998. Stepping into the same river twice: homologues as recurring attractors in epigenetic landscapes. *Brain Behav Evol* 52:218–231.
Striedter GF. 1999. Homology in the nervous system: of characters, embryology and levels of analysis. *Novartis Found Symp* 222:158–170.
Striedter GF. 2005. *Principles of Brain Evolution*. Sunderland, MA: Sinauer Associates.
Striedter GF, Northcutt RG. 1991. Biological hierarchies and the concept of homology. *Brain Behav Evol* 38:177–189.
Suess HE, Urey HC. 1956. Abundances of the elements. *Rev Mod Phys* 28:53–74.
Sun Y, Joachimski MM, Wignall PB, Yan C, Chen Y, Jiang H, Wang L, Lai X. 2012. Lethally hot temperatures during the Early Triassic greenhouse. *Science* 338:366–370.
Takahashi M, Arita H, Hiraiwa-Hasegawa M, Hasegawa T. 2008. Peahens do not prefer peacocks with more elaborate trains. *Anim Behav* 75:1209–1219.
Tank EM, Dekker RG, Beauchamp K, Wilson KA, Boehmke AE, Langeland JA. 2009. Patterns and consequences of vertebrate *Emx* gene duplications. *Evol Dev* 11:343–353.
Taylor DM, Tillery SIH, Schwartz AB. 2002. Direct cortical control of 3D neuroprosthetic devices. *Science* 296:1829–1832.
Thorne JL, Kishino H. 2002. Divergence time and evolutionary rate estimation with multilocus data. *Syst Biol* 51:689–702.
Tomarev SI, Callaerts P, Kos L, Zinovieva R, Halder G, Gehring W, Piatigorsky J. 1997. Squid *Pax-6* and eye development. *Proc Natl Acad Sci USA* 94:2421–2426.
Tosches MA, Yamawaki TM, Naumann RK, Jacobi AA, Tushev G, Laurent G. 2018. Evolution of pallium, hippocampus, and cortical cell types revealed by single-cell transcriptomics in reptiles. *Science* 360:881–888.
Tschopp P, Tabin CJ. 2016. Deep homology in the age of next-generation sequencing. *Philos Trans R Soc Lond B* 372:20150475.
Tsuboi M, Husby A, Kotrschal A, Hayward A, Buechel SD, Zidar J, Løvlie H, Kolm N. 2015. Comparative support for the expensive tissue hypothesis: big brains are correlated with smaller gut and greater parental investment in Lake Tanganyika cichlids. *Evolution* 69:190–200.
Ulinski PS. 1989. Nodal events in forebrain evolution. *Neth J Zool* 40:215–240.
van der Worp HB, Howells DW, Sena ES, Porritt MJ, Rewell S, O'Collins V, Macleod MR. 2010. Can animal models of disease reliably inform human studies? *PLoS Med* 7:e1000245.
van Dongen PAM. 1998. Brain size in vertebrates. In: Nieuwenhuys R, Donkelaar ten HJ, Nicholson C (ed.), *The Central Nervous System of Vertebrates*. Berlin: Springer, pp. 2099–2134.
Wagner GP. 2000. *The Character Concept in Evolutionary Biology*. San Diego, CA: Academic Press.
Wagner GP. 2007. The developmental genetics of homology. *Nat Rev Genet* 8:473–479.
Wagner GP, Lynch VJ. 2010. Evolutionary novelties. *Curr Biol* 20:R48–R52.
Wainwright PC, Reilly SM. 1994. *Ecological Morphology*. Chicago: University of Chicago Press.
Walker S. 1983. *Animal Thought*. London: Routledge.
Watson C, Puelles L. 2016. Developmental gene expression in the mouse clarifies the organization of the claustrum and related endopiriform nuclei. *J Comp Neurol* 525:1499–1508.
Wiley EO, Lieberman BS. 2011. *Phylogenetics*, 2nd ed. New York: John Wiley & Sons.
Wilkinson DG, Bhatt S, Cook M, Boncinelli E, Krumlauf R. 1989. Segmental expression of *Hox-2* homoeobox-containing genes in the developing mouse hindbrain. *Nature* 341:405–409.
Wullimann MF, Mueller T, Distel M, Babaryka A. 2011. The long adventurous journey of rhombic lip cells in jawed vertebrates: a comparative developmental analysis. *Front Neuroanat* 5:27.
Wullimann MF, Rupp B, Reichert H. 1996. *Neuroanatomy of the Zebrafish Brain*. Basel: Birkhauser.
Yopak KE, Lisney TJ, Darlington RB, Collin SP, Montgomery JC, Finlay BL. 2010. A conserved pattern of brain scaling from sharks to primates. *Proc Natl Acad Sci USA* 107:12946–12951.
Yoshida MA, Ogura A, Ikeo K, Shigeno S, Moritaki T, Winters GC, Kohn AB, Moroz LL. 2015. Molecular evidence for convergence and parallelism in evolution of complex brains of cephalopod molluscs: insights from visual systems. *Integr Comp Biol* 55:1070–1083.

Zakon HH, Jost MC, Lu Y. 2011. Expansion of voltage-dependent Na^+ channel gene family in early tetrapods coincided with the emergence of terrestriality and increased brain complexity. *Mol Biol Evol* 28:1415–1424.

Zhang YE, Landback P, Vibranovski MD, Long M. 2011. Accelerated recruitment of new brain development genes into the human genome. *PloS Biol* 9:e1001179.

Zuckerkandl E, Pauling L. 1962. Molecular disease, evolution and genetic heterogeneity. In: Kasha M, Pullman B (eds.), *Horizons in Biochemistry*. New York: Academic Press, pp. 189–225.

Zuckerkandl E, Pauling L. 1965. Evolutionary divergence and convergence in proteins. In: Bryson V, Vogel HJ (eds.), *Evolving Genes and Proteins*. Amsterdam: Elsevier, pp. 97–166.

2

The Origin of Vertebrates

Invertebrate Chordates and Cyclostomes

The story of vertebrate brain evolution began 500–600 million years ago, with the changes in the nervous system that occurred as the first vertebrates emerged. The invertebrate ancestors of vertebrates clearly had both central and peripheral nervous systems, but the complexity of these systems increased dramatically with the origin of vertebrates. Particularly interesting is that early vertebrates evolved some major brain regions that their ancestors did not possess. To place this story in context, we begin with a discussion of the earliest nervous systems.

2.1. The Origins of Neurons and Nervous Systems

Only animals have neurons and, among the animals, proper neurons (i.e., excitable cells with pre- as well as post-synaptic elements) are only found in Ctenophora (comb jellies), Cnidaria (e.g., sea anemones, jellyfish, corals), and the Bilateria (animals that exhibit bilateral symmetry at least at some point in their development). In contrast, proper neurons are lacking in sponges and placozoans, which are small, flattened, and extremely simple multicellular animals (Meech, 2017). The relationships of these lineages to one another continues to be debated (Pisani et al., 2015), but it seems very likely that neurons evolved at least twice: once in the lineage leading to ctenophores and then again in the last common ancestor of the Cnidaria and Bilateria (Figure 2.1). This hypothesis is supported by the observation that the neurons of ctenophores differ in many fundamental respects from cnidarian and bilaterian neurons. For example, ctenophore neurons do not use serotonin, acetylcholine, dopamine, noradrenaline, octopamine, histamine, or glycine for signaling between neurons; they do, however, use L-glutamate and have evolved an exceptionally large variety of ionotropic glutamate receptors (Moroz et al., 2015).

Although the last common ancestors of all bilaterian animals surely had neurons, it remains uncertain whether those neurons formed diffuse nerve nets distributed across the body or were concentrated into one or more large neuronal clusters that one might call nerve cords, cerebral ganglia, or brains. One way to answer this question is to examine the phylogenetic distribution of diffuse

Brains Through Time. Georg F. Striedter and R. Glenn Northcutt, Oxford University Press (2020). © Oxford University Press.
DOI: 10.1093/oso/9780195125689.001.0001

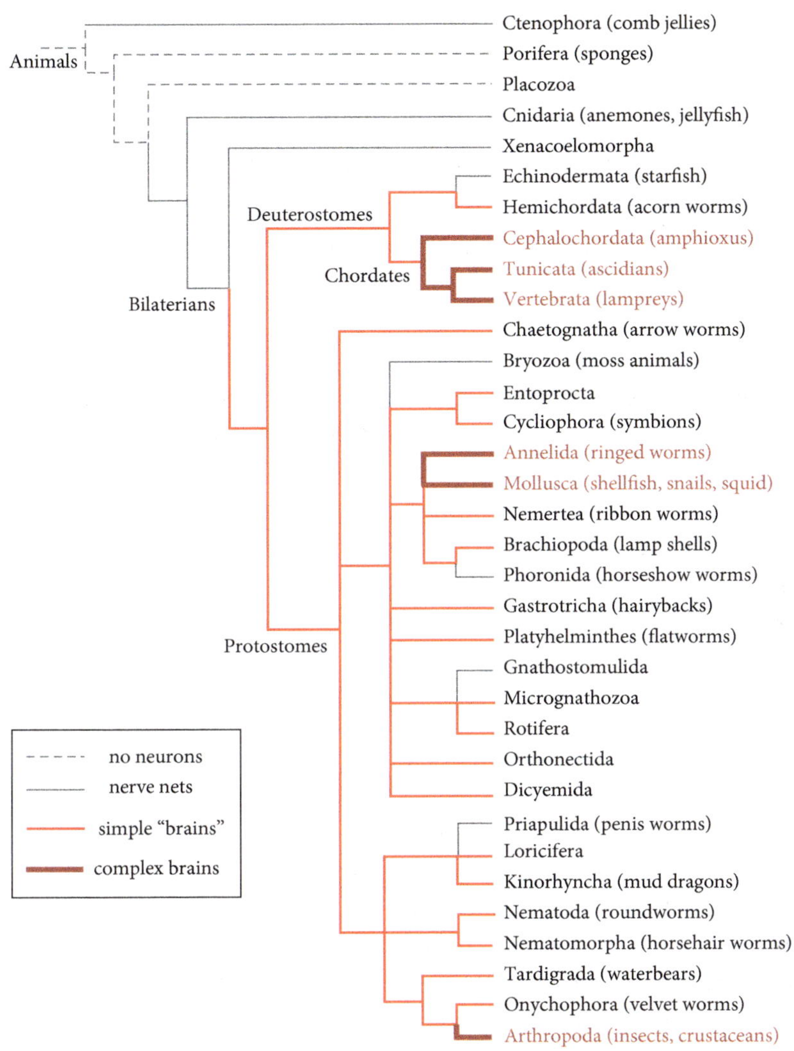

Figure 2.1 Phylogenetic distribution of neurons, nerve nets, and brains. Shown here are the phylogenetic relationships of the major animal lineages, together with the type of nervous system they possess. Note that the representation of time in this diagram is not linear (i.e., it is not a timetree; see Chapter 1). Arthropods, for example, originated at least as early as mollusks.

Adapted from Northcutt (2012), Dunn et al. (2014).

nerve nets versus centralized nervous systems and then use a parsimony analysis (see Chapter 1) to infer at which points in phylogeny centralized nervous systems most likely emerged (Northcutt, 2012). Such an analysis suggests that highly centralized nervous systems with a large rostral brain evolved at least three times: twice in the protostomes, which includes most of the invertebrate taxa, and then again in the lineage leading to chordates (Figure 2.1). Using a similar approach, Moroz (2009) hypothesized that centralized nervous systems evolved at least six times within the animals, namely in ctenophores, box jellies, mollusks, nematodes, arthropods, and chordates. They may even have evolved twice within mollusks (Kocot et al., 2011; Smith et al., 2011). One problem with these analyses is that the criteria for discriminating between decentralized nerve nets, nerve cords, cerebral ganglia, and brains are somewhat arbitrary and likely differ among researchers (Wullimann, 2001). For example, hemichordates, such as the acorn worm illustrated in Figure 2.2, are traditionally described as possessing only a diffuse nerve net, but they contain a small "collar cord" region that resembles the vertebrate spinal cord (Nomaksteinsky et al., 2009); still, few would consider this collar cord a full-fledged brain.

An alternative approach to reconstructing the nervous systems of the earliest bilaterians is to compare the brains of vertebrates and various invertebrates with clearly centralized nerve cords and brains. In particular, comparative neurobiologists have focused on similarities in gene expression patterns between the embryonic brains of insects, annelid worms, and vertebrates. Impressed with the multitude of similarities, some have argued that the last common ancestor of all these species must have already possessed a centralized nervous system exhibiting

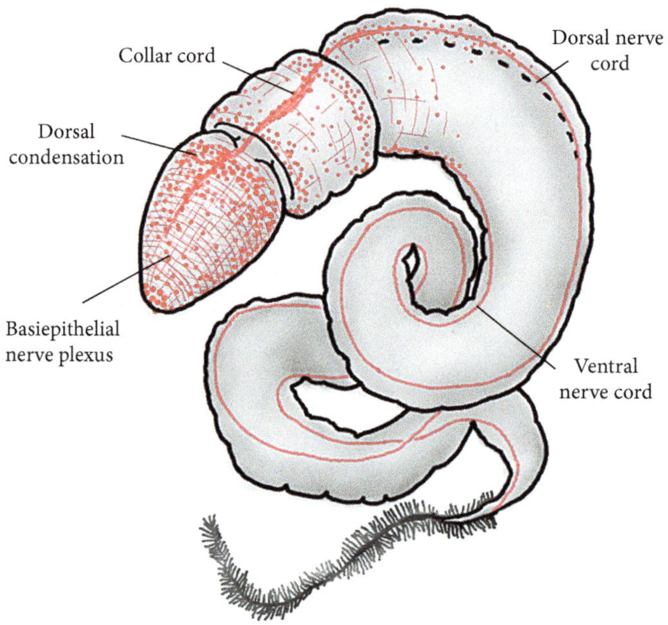

Figure 2.2 The nervous system of an acorn worm. The nervous system (red) of the enteropneust hemichordate *Saccoglossus kowalevskii* includes both a nerve net in the skin (a basiepithelial nerve plexus) and centralized nerve cords that run along the dorsal and ventral midline. The highest concentration of neurons is found in the collar cord.

Adapted from Lowe et al. (2015), with permission from Springer Nature.

those shared features (Denes et al., 2007; Strausfeld and Hirth, 2013a). According to this hypothesis, centralized nervous systems must have been lost repeatedly in diverse lineages (Hirth, 2010). Such rampant decentralization of the nervous system is certainly possible, especially if the ancestors of those lineages adopted a simpler life, but it is not very parsimonious.

Moreover, some of the similarities in gene expression may reflect the conservation of ancient mechanisms responsible for body patterning (e.g., the establishment of rostro-caudal and dorsal-ventral body axes), rather than the generation of central nerve cords and brains per se (Wullimann, 2001; Lowe et al., 2015). Strong evidence for this hypothesis has come from a broader phylogenetic analysis (Martin-Durán et al., 2018), which showed that many of the similarities in neural gene expression patterns between vertebrates, fruit flies, and the annelid worm *Platynereis dumerilii* are not shared with various species that are phylogenetically intermediate between these three taxa (including other annelid worms), nor with the most basal group of bilaterians (Xenacoelomorpha). Although these data are debatable (Arendt, 2018), they provide significant support for the hypothesis that central nerve cords, cerebral ganglia, and brains evolved repeatedly among the Bilateria, and that many of the similarities between them are the result of convergent evolution (Schmidt-Rhaesa et al., 2015; Albertin and Ragsdale, 2018).

2.2. Basal Vertebrates and Their Closest Invertebrate Relatives

In our quest to understand the origin of vertebrates, some lineages are more informative than others. Particularly important are the invertebrate lineages that are most closely related to the vertebrates. According to the currently most widely accepted phylogeny, these invertebrates are the tunicates and the cephalochordates (Figure 2.3). Scientists had thought for many years that cephalochordates were the closest living relatives of vertebrates, but recent molecular analyses have now awarded this distinction to the tunicates (Delsuc et al., 2008; Putnam et al., 2008). Consequently, cephalochordates are now regarded as the closest living relatives of the lineage that includes both tunicates and vertebrates.

The other group of animals that provides us with useful information about the origin of vertebrates is the cyclostomes (lampreys and hagfishes). Although the phylogenetic position of these animals has long been debated, most recent analyses point to lampreys and hagfishes being each other's nearest relatives (Heimberg et al., 2010; Shimeld and Donoghue, 2012). The cyclostomes, in turn, are thought to be the sister group of all other vertebrates (i.e., jawed vertebrates). This general phylogeny (Figure 2.3) is now widely accepted and provides the basis for our comparative analysis. If future analyses end up supporting an alternate phylogeny, then some of our conclusions would have to be modified.

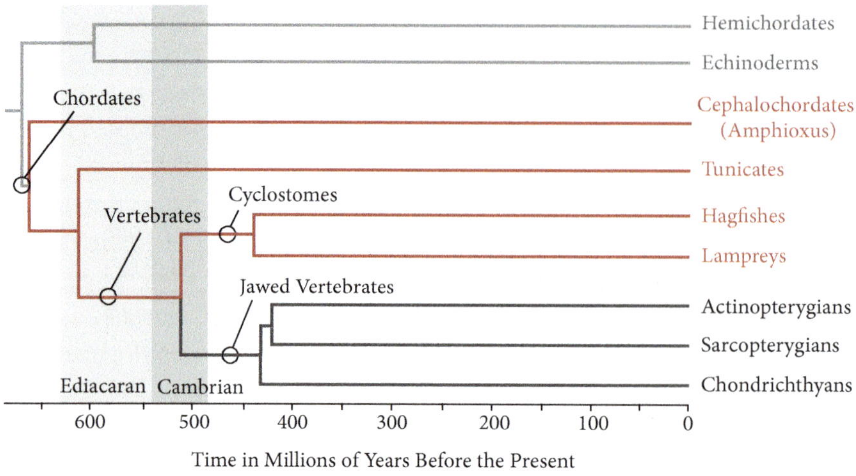

Figure 2.3 Phylogeny of chordates and their closest relatives. Chordates diverged from hemichordates (e.g., acorn worms) and echinoderms (e.g., starfish) more than 650 mya. Vertebrates are thought to have diverged from tunicates during the Ediacaran period, and jawed vertebrates diverged from cyclostomes during the Cambrian.
Adapted from Erwin et al. (2011), Donoghue and Keating (2014).

2.2.1. Extant Basal Chordates

All vertebrates, as well as tunicates and cephalochordates, belong to the phylum Chordata. This lineage is named for the fact that all three groups of animals possess a notochord, which is a stiff cellular rod that lies in the midline, just ventral to the central nervous system, and extends throughout much of the body. The notochord regresses as jawed vertebrates mature, but its remnants form the centers of the cartilaginous vertebral disks (Aszódi et al., 1998). In contrast, the notochord persists in adult lampreys, hagfishes, cephalochordates, and larvacean tunicates as a critical component of their internal skeleton. Because a notochord evolved with the origin of chordates and was then retained in all living chordates, at least at some stage of development, it is a shared derived character for the group. Other features that define chordates are the possession of segmented trunk muscles and a dorsally located central nervous system that forms a hollow neural tube, at least at early stages of development.

2.2.1.1. Cephalochordates

The most basal chordates are the cephalochordates, a group of about 25 living species that are also known as lancelets or, more commonly, amphioxus. These animals reach an adult size of 5–7 cm (Figure 2.4), live in shallow marine environments, and are often buried in the sand so that only their head sticks out. Their bodies are laterally compressed and spindle-shaped (amphioxus is Greek for "both ends pointy"). The notochord of amphioxus (dictionaries notwithstanding, we use "amphioxus"

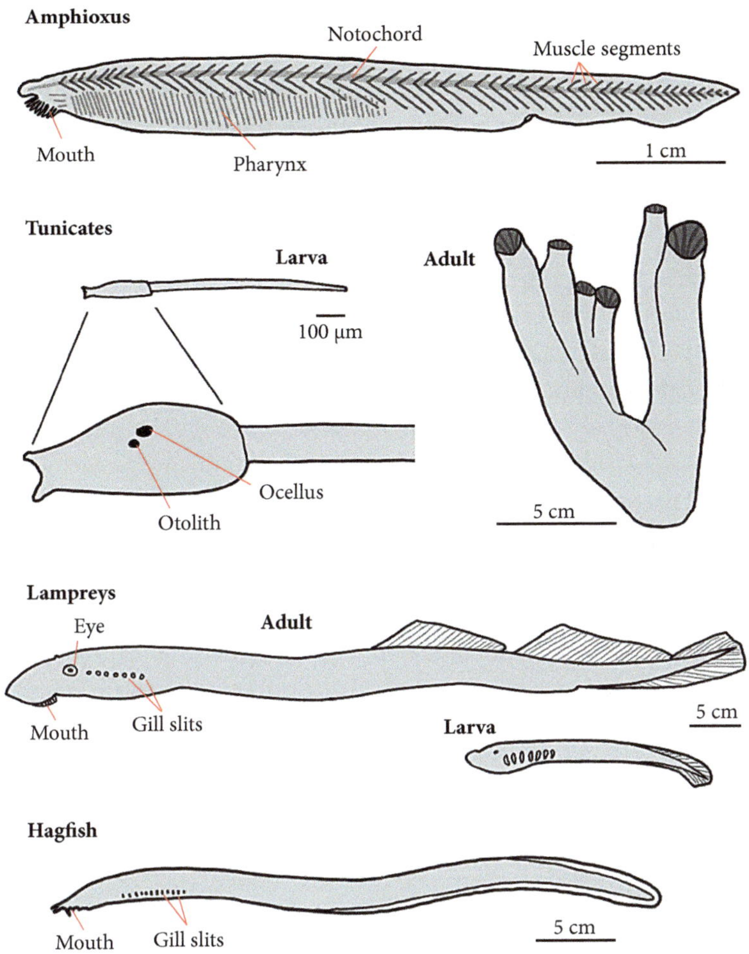

Figure 2.4 Cephalochordates, tunicates, and cyclostomes. Shown here are lateral views of an amphioxus (*Branchiostoma lanceolatum*), an ascidian tunicate (*Ciona intestinalis*; larva and adult), and two cyclostomes, namely a lamprey (*Petromyzon marinus*; adult and larva) and a hagfish (*Myxine glutinosa*). The head of the tunicate larva is also shown at higher magnification.
Adapted from Romer (1962), Mallatt and Chen (2003), Sasakura et al. (2012), Williamson (2012).

for both the singular and the plural form; see Holland and Holland, 2017) is unusual, not only for persisting into adulthood, but also for extending all the way to the tip of the head, which explains the name cephalochordate ("cephalo-" means related to the head). Another surprising feature of the amphioxus notochord is that it contains contractile muscle filaments, which are used to modify the stiffness of the notochord (Suzuki and Satoh, 2000). An increase in stiffness might, for example, make it easier to burrow into dense sand. For these burrowing and swimming movements, amphioxus possesses a series of chevron-shaped muscle segments all along the trunk. The contractile fibers in these muscles are arranged parallel to the body's long axis, so that unilateral contractions of these muscles will cause the trunk to bend sideways (Lacalli, 2012).

The mouth is located on the left side of the head in larval amphioxus but then migrates closer to the midline at metamorphosis. It is roughly circular, lacks jaws, and is surrounded by a ring of "cirri" that look like tentacles and help to keep sand out of the mouth when the animals are burrowing. To feed, amphioxus draws water into the mouth, moves it through the pharynx, and pushes it out through a series of slits in the side of the pharynx. Thus, amphioxus is a filter feeder or, more precisely, a "suspension feeder" that ingests food particles suspended in water. In contrast to most aquatic vertebrates, amphioxus does not use muscles to pump water through the pharynx. Larval amphioxus do have a few muscles around their mouth and pharyngeal slits (Yasui et al., 2013), but water is moved through the pharynx mainly by large ciliated cells inside the pharynx that move their cilia in concert with one another. As food particles enter the pharynx, they tend to get stuck on the mucus that lines the pharynx. This mucus is produced mainly in a ventrally located groove and is then slowly pushed toward the esophagus by ciliary movements. Eventually, the captured food particles enter the gut, where they are digested.

Although the pharyngeal slits of amphioxus are sometimes called "gill slits," this term is misleading because the pharyngeal slits of amphioxus do not exhibit the filamentous extensions and capillary beds that characterize true gills. Indeed, amphioxus performs almost all of its gas exchange across its body skin, not through its "gills." The small body size of amphioxus, with its large surface-to-volume ratio, makes this mode of gas exchange a feasible option that is not readily available to larger animals.

The nervous system of amphioxus has been studied in considerable detail, especially in young larvae that can be cut into extremely thin sections and reconstructed in three dimensions (Lacalli, 1996). In essence, the central nervous system of amphioxus forms a long tube that extends from the tip of the tail to just in front of the mouth (Figure 2.5). It contains roughly 20,000 neurons (Candiani et al., 2012) and is divisible into a brain and a spinal cord, although the boundary between these two divisions is not obvious from the morphology. Perhaps the most striking aspect of the amphioxus brain is that it is no thicker than the spinal cord and tapers rostrally in the adult. Only in the larvae does the rostral portion of the brain exhibit a small enlargement, called the cerebral vesicle. Also intriguing is the fact that amphioxus has only a single eye that is quite small and lies at the brain's rostral tip.

2.2.1.2. Tunicates

The tunicates comprise three taxonomic groups, namely the ascidians (sea squirts), thaliaceans, and larvaceans. With roughly 2,000 species, the ascidians are by far the largest of these groups. Adult ascidians have a sac-like body with two tubular siphons (Figure 2.5). One of these siphons serves to draw water into the animal's pharynx. The water then passes through a set of pharyngeal slits, collects

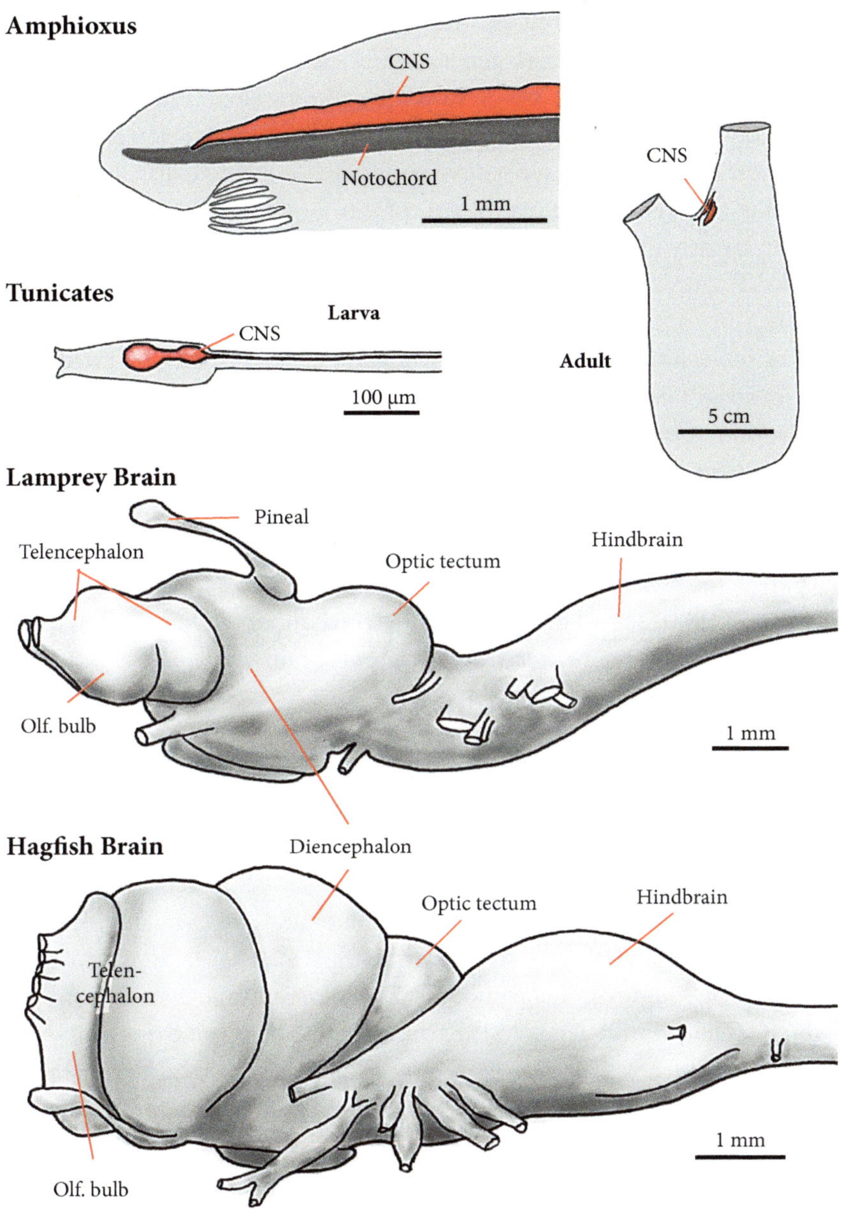

Figure 2.5 The central nervous systems of basal chordates. For amphioxus and an ascidian tunicate (larva and adult) the central nervous system is highlighted in red. Lampreys and hagfish brains were dissected out of the body. Rostral is to the left and dorsal to the top.

Adapted from Nieuwenhuys and Nicholson (1998), Mackie and Burighel (2005), Sasakura et al. (2012).

in a so-called atrium, and eventually flows back out of the animal through the second siphon. As in amphioxus, these water movements are generated by ciliary movements, rather than muscular contractions. Also as in amphioxus, ingested particles get stuck on a large mucous sheet that slowly moves toward the intestine.

Despite these similarities, adult ascidians look very different from amphioxus or vertebrates, obscuring their phylogenetic affinities.

Larval ascidians, however, are quite similar to amphioxus and vertebrates. For example, they have segmented muscles along their tail and an elongate notochord. Moreover, larval ascidians have a central nervous system that is hollow and lies dorsal to the notochord. The rostral end of this larval nervous system consists of two bulges, namely the sensory vesicle and visceral ganglion, separated by a narrow neck region (Figure 2.5). Larval ascidians also have what appears to be a spinal cord, but this region contains very few neuronal cell bodies (larvaceans have a greater number of such spinal cord neurons; Søviknes et al., 2005). Two pigment spots inside the head of larval ascidians are associated with a single eye spot, called the ocellus, and an otolith (aka statocyst) that helps the animals sense their orientation relative to gravity, as well as self-motion. Overall, the central nervous system of larval ascidians contains only 130–150 neurons (Nicol and Meinertzhagen, 1991). Most of these neurons degenerate at metamorphosis. A few cells in the rostral and dorsal part of the central nervous system then proliferate and give rise to the adult ascidian brain (Mackie and Burighel, 2005), which differs radically from that of ascidian larvae and, for that matter, from vertebrate brains.

2.2.1.3. Cyclostomes: Lampreys and Hagfishes

Lampreys and hagfishes both have elongate, eel-like bodies (Figure 2.4), but they are quite different under the hood, especially with regard to their nervous systems. Because of their morphological differences, lampreys are sometimes thought to be more closely related to vertebrates than to hagfishes. However, comparative analyses of diverse genes and microRNAs consistently indicate that lampreys and hagfishes form a monophyletic group (see Heimberg et al., 2010), called "cyclostomes" because of their roughly circular mouths. Debates continue about which data set and which phylogeny get closer to the truth (Near, 2009), but the recent discovery of a hagfish fossil from the late Cretaceous period has provided strong support for the cyclostome monophyly hypothesis even on morphological grounds (Miyashita et al., 2019). These animals all lack jaws, which is why they are sometimes called the agnathans. However, urochordates and cephalochordates, as well as many extinct vertebrates, likewise lack jaws. Therefore, the term "cyclostomes" is more useful when we want to refer selectively to lampreys and hagfishes.

Living lampreys comprise approximately 38 species. They have eel-like bodies with dorsal and caudal fins, but they lack the paired fins typically found in bony or cartilaginous fishes. Even though lampreys are vertebrates, their vertebral column is not made of bony vertebrae, but of cartilaginous "arcualia" that are arranged in a rostrocaudal series just dorsal to the notochord. The circular mouth of lampreys features several concentric rings of teeth that are hardened by enamel and keratin, rather than dentine. Some lamprey species use their teeth to latch onto the skin of larger aquatic animals (even whales) and rasp their way through the skin until they can suck out the animal's fluids or dislodge pieces of flesh; when their prey is close to death, the lampreys move on to find another host. Although lampreys are widely known for this parasitic lifestyle, many lampreys do not feed as adults. They use their teeth mainly to attach themselves to underwater rocks and, thus, to stabilize themselves in fast-flowing rivers.

As you might expect, given that many lampreys do not feed as adults, all lampreys have a larval stage. These larval lampreys, called ammocoetes, tend to be much smaller than adult lampreys, lack the keratinous teeth, and look superficially similar to amphioxus (see Figure 2.4). They typically spend several years buried in river sand so that only their head sticks out. Like amphioxus, they feed by drawing water into their pharynx and extracting suspended food particles that stick to their pharyngeal mucus. However, in contrast to amphioxus and tunicates, ammocoetes actively pump water through their pharynx by rhythmically contracting their pharyngeal muscles. This pumping mechanism greatly increases the rate of water flow through the pharynx. Another important difference between lampreys and invertebrate chordates is that the internal surface of the pharyngeal bars in lampreys is highly folded and contains a capillary bed. Thus, in contrast to amphioxus and tunicates, both ammocoetes and adult lampreys have real gills that are much more efficient than the skin at obtaining oxygen and shedding carbon dioxide. Pumping water across those gills further increases their gas exchange efficiency.

Weighing in at 15–60 mg, adult lamprey brains are significantly larger and more complex than amphioxus brains and, certainly, tunicate brains. No one has counted the number of neurons in lamprey brains, but it is probably at least one order of magnitude larger than in amphioxus. As discussed in more detail later, the forebrain and the midbrain are especially enlarged in lampreys, compared to amphioxus (Figure 2.5). Lampreys do not possess a proper cerebellum (see Section 2.6.1), but they do possess all of the other major brain regions that one can recognize in jawed vertebrates. In that sense, lampreys provide a striking contrast to amphioxus, whose brain looks quite strange from a vertebrate perspective.

Hagfishes are the closest living relatives of lampreys. Like lampreys, they are eel-shaped and relatively large (see Figure 2.4), with the largest species being just over 1 m long. Represented by 50–80 living species, most hagfishes live on the deep ocean floor. Although most people have never seen a live hagfish, these animals exist in large numbers and are harvested for their leathery skin, which can be made into "eel skin" wallets. When threatened, hagfishes secrete copious amounts of slime all over their body surface, which is why they are also known as slime eels. Their natural behavior is difficult to study, but hagfishes appear to feed mainly on polychaete worms and on the sunken carcasses of larger animals. Like lampreys, hagfishes have proper gills and muscles that pump water across them. Another similarity to lampreys is that hagfishes lack jaws. However, hagfishes can open and close their mouth sideways (in the horizontal plane), using numerous keratinized teeth to take a bite out of large food items. In contrast to lampreys, hagfishes do not have a larval stage. They also lack the cartilaginous arcualia that help to stabilize a lamprey's trunk. Hagfishes do, however, have some cartilage surrounding the brain. They also have many of the other characters that help define the vertebrates as a distinct lineage, which means that hagfishes are proper vertebrates even though they lack cartilaginous or bony vertebrae.

Hagfish brains are larger than lamprey brains, at least relative to body size (Salas et al., 2017). Especially enlarged are several parts of the forebrain and parts of the hindbrain related to the trigeminal nerve (see Figure 2.5; an overview of cranial nerve evolution is provided in the Appendix). In contrast to lampreys, hagfishes have very small, degenerate eyes, and their skin is studded with so-called Schreiner's organs, which are probably chemosensory. Although it may be a bit surprising to learn that hagfishes and lampreys are so different from one another in their sense organs and brain anatomy, these two lineages diverged 430–490 million years ago (mya) and have evolved along divergent paths since then (Gess et al., 2006; Hirasawa et al., 2016). Both lineages were successful enough to make it to the present day, largely because of some shared key innovations.

Specifically, lampreys possess an oral disk with teeth made of keratin that can latch onto the skin of other fishes (Figure 2.6). In addition, lampreys have a second set of teeth that they can scrape across the prey's skin using a set of muscular pulleys. This rasping "tongue" allows lampreys to gain access to the prey's blood and, for flesh-eating lampreys, bite off chunks of flesh (Hilliard et al., 1985; Rovainen, 1996). Hagfishes have similar teeth and a similar "tongue apparatus" (Figure 2.6). However, the dental plates of hagfishes open and close mediolaterally, like a book that is opened out flat or snapped shut (Yalden, 1985). These dental plates clearly evolved independently of gnathostome jaws and operate more slowly, but their biting forces are just as strong (Clark and Summers, 2007, 2012). Indeed, hagfishes are not limited to scavenging on deceased, decaying prey; they have been observed to feed on living fish, attacking them in their burrows (Zintzen et al., 2011). Another important innovation of hagfishes is that they discharge copious amounts of slime when they are threatened. This slime clogs up the gills of the attacking predator, encouraging retreat. Secreted slime may also be used to suffocate gill-breathing animals, especially if they have burrowed into the substrate (Zintzen et al., 2011), but this idea remains speculative.

Collectively, these feeding and defensive innovations help to explain why hagfishes and lampreys survived for so many millions of years, when all other jawless vertebrates perished. The surviving cyclostomes did not have jaws as we know them from gnathostomes, but they used their pulley-based tongue apparatus with keratinous teeth to feed (Figure 2.6). Because some extinct jawless vertebrates (notably conodonts and ostracoderms; see Chapter 3, Figure 3.8) apparently possessed a similar pulley-based feeding apparatus (Janvier, 1981; Goudemand et al., 2011), we can infer that such a system was primitive for vertebrates. However, the different lineages equipped this primitive feeding apparatus with non-homologous teeth, made of different materials. In particular, a dental plate with keratinous teeth appears to be a shared derived feature for lampreys and hagfishes (Figure

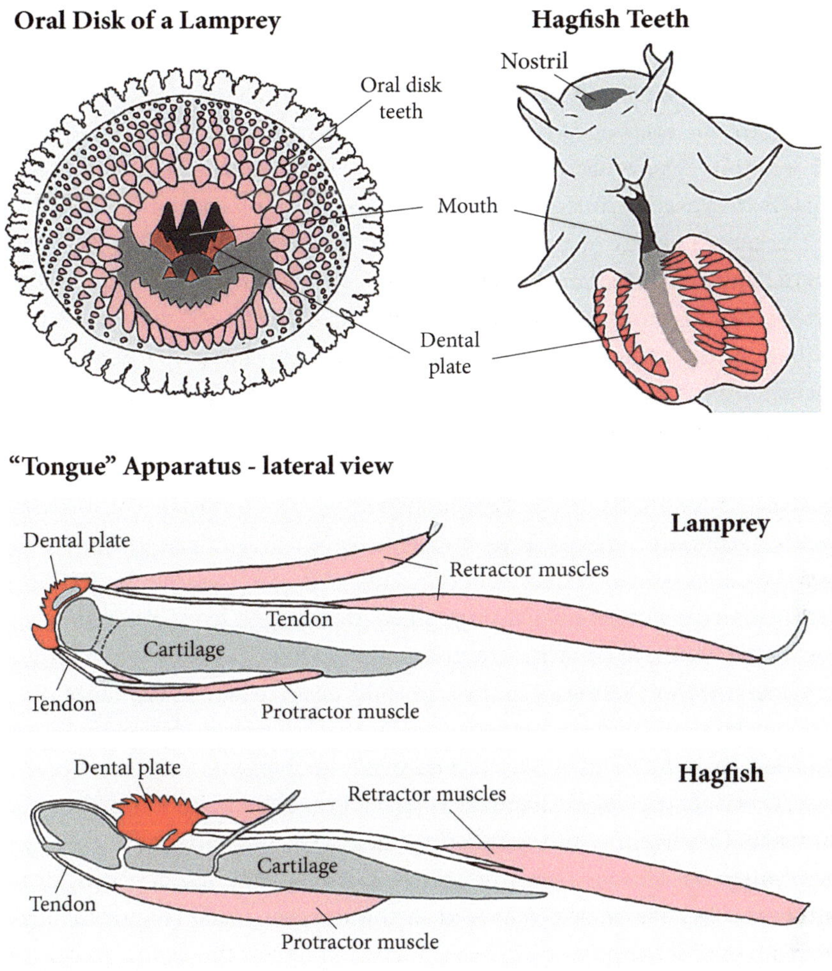

Figure 2.6 The "tongue" apparatus of cyclostomes. Lampreys possess an oral disk that is studded with small keratinous teeth (top left) that can be used to grab onto the skin of other fishes. Lampreys also have a second set of teeth that sit deeper in the mouth, on a dental plate that can be moved up and down by a set of muscular pulleys (middle). This tongue-like apparatus can be used to rasp through the skin of other fishes and, in flesh-feeding species such as the illustrated *Geotria australis*, tear off pieces of meat. Hagfishes have teeth that sit on a pair of dental plates (top right), which open and close mediolaterally and can be used like jaws to bite prey. Although hagfish "jaws" are quite different from a lamprey's rasping "tongue," both involve a similar system of muscles, tendons, and cartilage (bottom).

Adapted from Hilliard et al. (1985), Zintzen et al. (2011), and Yalden (1985); oral disk and tongue apparatus diagrams with permission from John Wiley & Sons and Oxford University Press, respectively.

2.6), supporting the hypothesis that these two lineages form a monophyletic group (Yalden, 1985). We can further conclude that a complex feeding apparatus, capable of processing substantial prey, evolved early in cyclostome phylogeny, before hagfishes and lampreys diverged from one another.

2.2.2. Extinct Basal Chordates

Because early chordates had cartilaginous rather than calcified skeletons, the fossil record contains no early chordate bones. However, paleontologists have discovered several locations where the soft tissues of ancient chordates were well preserved, presumably because the animals perished in anoxic waters and, therefore, did not decay. The most famous of these sites are the Burgess Shale in British Columbia and the Chengjiang Maotianshan Shales in China's Yunan province. The fossils from these sites are thought to be 510 and 535 million years old, respectively, which places them in the early Cambrian period. Even older chordate fossils stem from the Ediacaran period, which lasted from 542 to 630 mya. Fossils from this period were originally discovered in Australia, at a site called Ediacara Hills, but have now been found on all the continents except Antarctica.

Many of the fossils from the Ediacaran and early Cambrian can be assigned to specific taxonomic groups and, therefore, reveal the minimum age of those lineages. When these fossil data are combined with comparative molecular data, they can be used to estimate when those lineages diverged from other lineages (see Chapter 1). Using this approach, scientists estimate that jawed vertebrates diverged from cyclostomes in the mid-Cambrian (~500 mya), whereas tunicates diverged from vertebrates 100 million years earlier, in the Ediacaran (Figure 2.3). Several other lineages originated in between these two branch points, but then became extinct. These extinct species are formally referred to as stem vertebrates, even though they lacked bony vertebrae and some other vertebrate innovations.

One of the most famous stem vertebrates is *Pikaia*, an extinct genus for which numerous fossils have been discovered in the Burgess Shale (Figure 2.7). *Pikaia* specimens resemble amphioxus in body shape but tend to be even smaller (~4 cm long). They apparently had a very thin notochord and zig-zagging trunk muscles (Morris and Caron, 2012). *Pikaia* also had a pair of distinctive tentacles protruding from the head and a small circular mouth, which it may have used to grab small food particles one at a time (Mallatt and Holland, 2013). A series of short appendages extended ventrally away from the body just rostral to the mouth; they look like stubby legs, but their function is unclear. *Pikaia*'s pharynx featured six pairs of slits with small filamentous extensions. If these filaments were used for gas exchange, rather than the filtering of suspended food particles, then it is fair to call them gills (Morris and Caron, 2012). However, there is no evidence that *Pikaia* used pharyngeal muscles to pump water through those slits. As far as one can tell from fossil specimens, *Pikaia* had a dorsally located nerve cord that tapered rostrally, much as it does in amphioxus. Thus, *Pikaia* had at best a very small and simple brain.

A second genus of stem chordates from the Burgess Shale and nearby areas is *Metaspriggina* (Figure 2.7). These animals were spindle-shaped, laterally compressed, and up to 10 cm in length (Morris and Caron, 2014). They had a notochord and segmented muscles, as well as a large pharyngeal cavity that was probably held open by a series of cartilaginous bars. These pharyngeal bars featured small

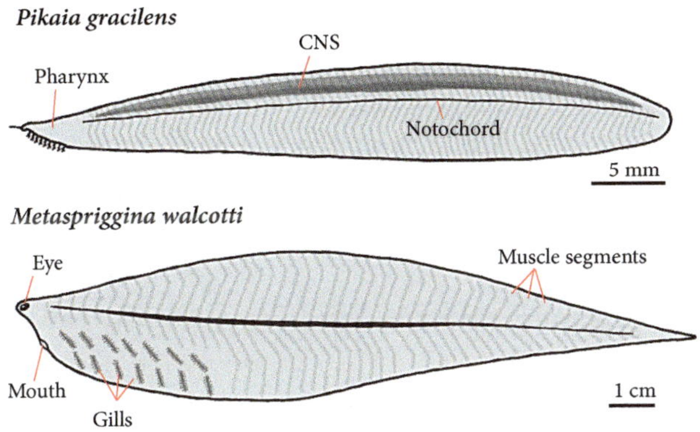

Figure 2.7 Two extinct stem vertebrates. *Pikaia gracilens* and *Metaspriggina walcotti*, which are known only from fossils, are thought to be more closely related to vertebrates than the tunicates are. Therefore, they are considered stem vertebrates.
Adapted from Mallatt and Holland (2013), Morris and Caron (2014).

protuberances that may have functioned in retaining food items, much as "gill rakers" do in modern fishes. They also sported filamentous extensions that probably functioned as gills (Morris and Caron, 2014). Nothing is known about the brain of *Metaspriggina*, but these animals clearly had a pair of large eyes and, between them, a pair of nasal sacs. Based on these features, one can conclude that *Metaspriggina* was probably more closely related to living vertebrates than *Pikaia* had been.

A fascinating group of soft-bodied fossils discovered at the Chengjiang site are the yunnanozoans, which probably include the enigmatic *Haikouella*. According to some authors, these fossils have a number of chordate and even vertebrate features, including a notochord, segmented muscles, respiratory gills, paired eyes, and a fairly large brain (Mallat and Chen, 2003). However, recent analyses have questioned every one of these attributes, leaving the yunnanozoans in phylogenetic limbo (Cong et al., 2014; Janvier, 2015). For example, seven pairs of filamentous "gills" that are attached to stiff rods in the pharyngeal region of *Haikouella* have recently been interpreted as playing a role in filter feeding, rather than respiration. Similar uncertainties surround the interpretation of other early Cambrian fossils that were once considered stem chordates, including *Myllokunmingia* and *Haikouichthys*. Given these uncertainties, we here exclude these specimens from our comparative analysis.

2.3. The Paleoecology of Early Chordates

Because *Pikaia* and *Metaspriggina* are morphologically similar to amphioxus and ammocoetes, it is tempting to imagine them as having lived in a similar ecological setting, exploiting similar opportunities and facing similar challenges. However, the world has changed dramatically since the early Cambrian, in terms of both the physical habitats and the species in those habitats. For example, modern ammocoetes eat mainly diatoms, but these tiny creatures only evolved about 200–250 mya, long after early chordates appeared. Therefore, early chordates must have had a very different diet. To recognize such differences, it helps to reconstruct explicitly the physical and biological environments in which early chordates evolved.

2.3.1. Continental Plates, Sea Levels, and Atmosphere

Unicellular life began in the earth's oceans roughly 4 billion years ago, but multicellular animals did not appear until about 800 mya. During the Ediacaran period, global land mass was probably concentrated in one giant supercontinent (named Rodinia). During the early Cambrian, this supercontinent broke up into four major land masses, called Laurentia, Siberia, Baltica, and Gondwana. As those new continents formed, the total length of the continental coastlines increased dramatically, which in turn increased the amount of coastal marine habitat in which the early chordates lived.

Average global temperatures gradually increased during the Ediacaran. The entire planet at that time was gradually emerging from an extremely cold period, during which the continents and much of the ocean were covered with ice. Indeed, life during this period must have been restricted to relatively small areas near volcanoes and geothermal vents. As this giant snowball melted, sea levels rose dramatically. The rise in sea level, in turn, flooded vast expanses of the continents, which were much flatter in the Ediacaran than they are today. This flooding of the continents further expanded the shallow marine habitat.

The shallow oceans of the Ediacaran were populated mainly by colonial, filamentous cyanobacteria (blue-green algae) that formed vast algal mats. Because these algae engaged in photosynthesis, they consumed carbon dioxide (CO_2) and produced oxygen (O_2). Most of the O_2 would have been confined to the shallow water where it was produced, but chemical analyses of ancient sediments suggests that even the deep oceans became significantly more oxygenated at the close of the Precambrian (Canfield et al., 2007). In any case, much of the O_2 produced by algal photosynthesis would have entered the atmosphere, gradually raising its level of oxygen to 15%–20% by the early Cambrian, which is close to the 21% level of O_2 we find today.

2.3.2. Species Diversity and Food Webs

Because the Ediacaran featured vast shallow oceans that were relatively warm and almost as rich in oxygen as they are today, small aquatic animals could thrive. Most of that life was restricted to the shallow ocean floor, which was covered with the aforementioned algal mats. Sitting on those mats were a variety of relatively immobile organisms, including sponges and frond-shaped colonial creatures related to today's corals and sea anemones (Figure 2.8). A few arthropods and other ancient invertebrates crawled among them and grazed on the algae. The water column contained ancient zooplankton, as well as some floating algae, but most of life was confined to the surface of the ocean floor. Only toward the end of the Ediacaran did diverse animals begin to dig burrows into the algal mats, thereby making life more three-dimensional.

Compared to the Ediacaran, the Cambrian period experienced a dramatic increase in animal diversity, prompting some to call it the "Cambrian explosion." This term is probably an exaggeration, since most major animal phyla had Ediacaran roots (Erwin et al., 2011). However, the diversity of species within those phyla certainly increased during the Cambrian. So did their complexity and, often, size. Particularly interesting are the trilobites, a well-known group of ancient arthropods that became amazingly diverse and numerous during the Cambrian. Most trilobites probably grazed on the algal mats or scavenged for detritus on the ocean floor. Intriguingly, damage to some large trilobite fossils suggests that they had been attacked by ancient predators. One of these large predators was *Anomalocaris*, another arthropod (#10 in Figure 2.8). These animals were up to 1 m in length, could swim, and possessed mouthparts that were specialized for predation.

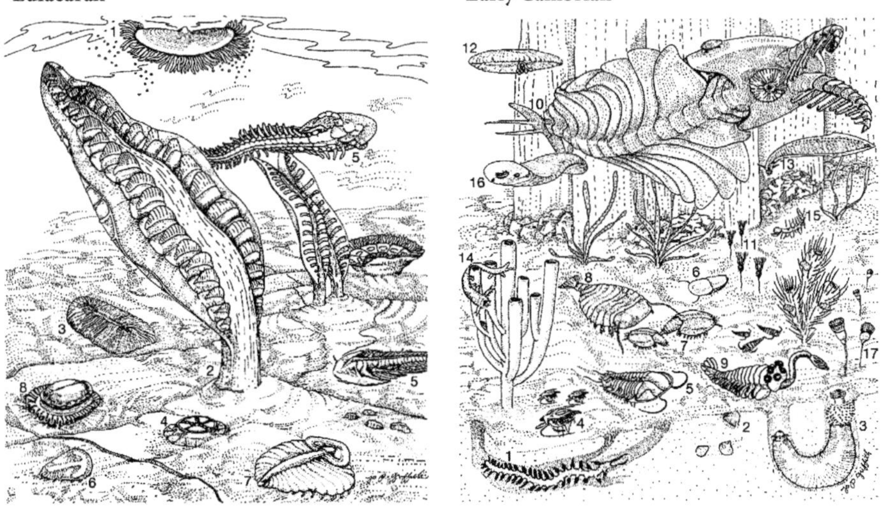

Figure 2.8 Life in the Ediacaran and early Cambrian. The reconstruction of the Ediacaran biota (left) illustrates the maximal degree of morphological complexity at the time; most Ediacaran creatures were probably simpler. The early Cambrian biota (right) is reconstructed from fossils in the Burgess Shale.

Ediacaran specimens: 1 – *Eoporpita*; 2 – *Charniodiscus*; 3 – *Dickinsonia*; 4 – *Arkarua*; 5 – *Spriggina*; 6 – *Praecambridium*; 7 – soft-bodied "trilobite"; 8 – *Kimberella*. Cambrian specimens: 1 – *Burgessochaeta*; 2 – *Lingulella*; 3 – *Ottoia*; 4 – *Marrella*; 5 – *Olenoides*; 6 – *Naraoia*; 7 – *Canadaspis*; 8 – *Sidneyia* – 9 – *Opabinia*; 10 – *Anomalocaris*; 11 – *Gogia*; 12 – *Eldonia*; 13 – *Pikaia*; 14 – *Aysheaia*; 15 – *Hallucigenia*; 16 – *Odontogriphus*; 17 – *Dinomischus*.

From Northcutt (2012), © National Academy of Sciences.

Anomalocaris and other Cambrian predators most likely dined not only on trilobites but also on diverse worms, including segmented worms (annelids), bristle worms (polychaetes), arrow worms (chaetognaths), penis worms (priapulids), and the now extinct lobophobians, which were related to today's velvet worms. Some of these Cambrian worms burrowed into the ground, but most were probably crawling along the ocean floor. Some may have been predators themselves, eating other worms or Cambrian mollusks. The early mollusks were mainly slug-like creatures that moved along the ocean floor on a muscular foot, grazing on the algal mats. Filter-feeding bivalves with hard shells (related to modern mussels and clams) first appeared in the Cambrian, but they truly blossomed only later, as did the remaining major molluskan group, the cephalopods (e.g., squid). Although the Cambrian fauna included many grazers and a few predators, most Cambrian animals were probably suspension feeders, collecting small food particles that floated in the ambient water (Figure 2.9).

How did the earliest chordates fit into this ancient ecosystem? This question is difficult to answer, because no Ediacaran fossils have been conclusively identified as being chordates. Only the comparative molecular data allow us to conclude with reasonable confidence that some early chordates must have predated the Cambrian. These ancient chordates were likely small and inconspicuous. They likely spent much of their time lying on the ocean floor, but they could probably swim by bending their elongate body from side to side. They probably swam mainly to evade predators, but they might also have moved in search of food-rich locations. The first chordates were most likely suspension feeders that moved water through their pharynx using cilia in a manner similar to that observed in modern tunicates and cephalochordates.

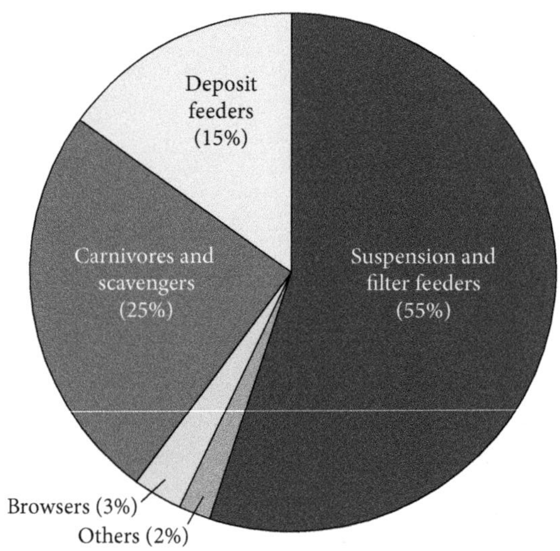

Figure 2.9 Approximate distribution of trophic groups among Cambrian animals.
Adapted from Figure 10.1 in Burzin et al. (2001).

As the Cambrian began, some chordates evolved pharyngeal muscles and cartilage that allowed water to be pumped through the pharynx, greatly increasing the rate at which suspended food could be extracted and consumed. Along with this new mode of feeding came a series of sensory and neurobiological innovations that, collectively, help to define the vertebrates. The sequence in which these various innovations appeared remains unclear, but they almost certainly did not appear in one fell swoop (Northcutt, 2005). Most likely, early vertebrates gradually became more efficient at finding and ingesting food, as well as at detecting and escaping predators. In short, they gradually moved up the ecological food web. As they did so, they faced a variety of challenges and exploited a few key opportunities.

2.3.3. Threats, Constraints, and Opportunities

Within the Ediacaran and early Cambrian ecosystems, the most serious threat to early chordates was probably the ever increasing number of invertebrate predators. One way to escape such threats is to become more mobile and more capable of detecting predators at a distance, using vision or other distance senses. Indeed, early vertebrates had more powerful trunk muscles and larger eyes than their invertebrate ancestors. Additional protection from predators may have been afforded by an increase in body size, as suggested by the fact that today's amphioxus and tunicates are substantially smaller than most extant vertebrates (Figure 2.10). What the first vertebrates did not yet have was protective armor or enameled teeth; these important innovations came later (see Chapter 3).

Although evolutionary increases in body size offer some protection from predators, they generate other significant challenges. Especially problematic is that the body's surface area increases more slowly than its volume. As a result, increases in body size make it more difficult to perform gas exchange exclusively across the skin, which is how invertebrate chordates obtain their oxygen and rid themselves of carbon dioxide. As mentioned previously, the earliest vertebrates solved this problem by evolving gills that were complexly folded (to increase surface area) and vascularized. Using pharyngeal muscles to pump water across those gills further increased the rate of gas exchange in early vertebrates. A related problem is that large and mobile animals require much more energy than small, sessile creatures. Pumping water through the pharynx would have addressed this challenge because, as mentioned earlier, the flow of water would have increased the rate at which suspended food particles get stuck on the pharyngeal mucus. In addition, increased mobility and enhanced distance sensing might have allowed the early vertebrates to forage for food actively. It would be a stretch to call them predators, but they may well have "hunted" small food items in the water column. Thus, some of the innovations that helped early vertebrates escape from predators may also have enabled them to do some hunting of their own and, consequently, boost food intake.

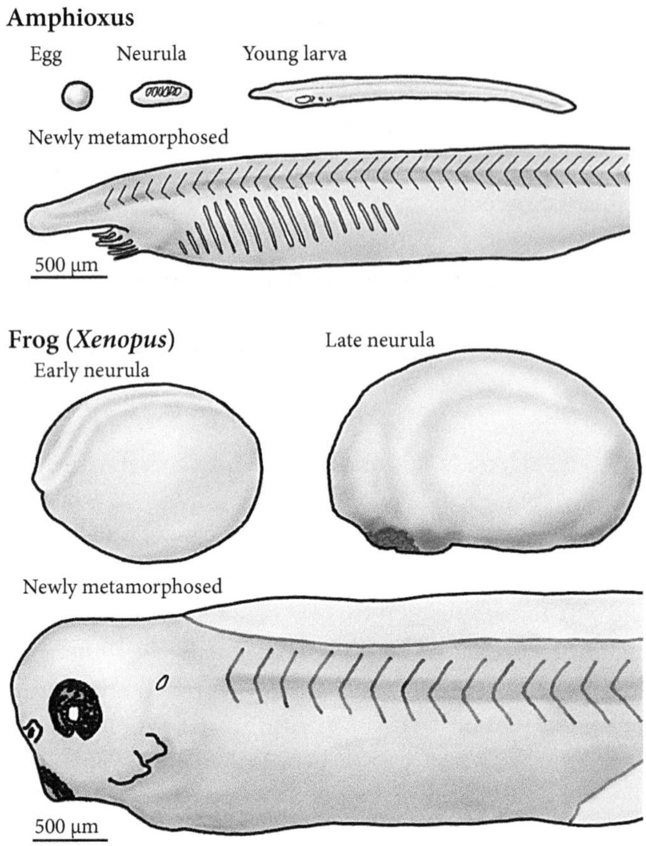

Figure 2.10 Size comparison between amphioxus and a vertebrate. Using the frog *Xenopus* as a representative vertebrate, this figure shows that vertebrates tend to be larger than amphioxus, both as adults and as embryos (same scale for all images). These differences in size presumably require some evolutionary changes in development. Some vertebrates have even larger eggs (e.g., coelacanth eggs can be up to 9 cm in diameter), but the eggs of early vertebrates were probably more similar to those of *Xenopus* in size.
Adapted from Lacalli (2004), with permission from S. Karger AG, Basel.

Increased body size also requires changes in development. For example, the use of diffusible molecular signals to pattern tissues and guide growing axons is limited to very small spatial scales and must be replaced by alternative mechanisms as embryos increase in size. In addition, the evolution of pharyngeal pumping and other vertebrate innovations must have required novel developmental genes, novel interactions among more ancient genes, or both.

Indeed, early vertebrates expanded their genome dramatically. Specifically, it appears that the entire genome duplicated shortly after vertebrates diverged from cephalochordates; it then doubled again, most likely just before jawed vertebrates emerged (Putnam et al., 2008; Smith et al., 2013). The details of this whole genome duplication hypothesis continue to be debated (Jaillon et al., 2009), and it remains possible that the duplications were limited to very large regions of multiple chromosomes (Panopoulou and Poustka, 2005). However, it is clear that numerous gene families expanded considerably during early vertebrate evolution. For example, amphioxus and sea urchins have several different *hox* genes that are arranged in a linear sequence on a single chromosome (Figure 2.11). In contrast, mammals have four sets of *hox* genes, located on four different chromosomes, strongly suggesting that the ancestral set of genes duplicated twice by the time mammals had evolved. In general, vertebrates have multiple homologs of many invertebrate genes. Some of these duplicate genes were lost soon after the initial duplication, but many were retained, presumably because they acquired valuable new functions (Duboule, 2007).

It is certainly reasonable to suppose that the whole genome duplications at the base of vertebrate phylogeny were causally linked to the increase in morphological complexity of early vertebrates. However, this does not mean that morphological complexity increased as soon as the genes had multiplied (Donoghue and Purnell, 2005). Most likely it took some time for the duplicated genes to acquire new molecular functions and to become involved in the construction of novel morphological or physiological features. In essence, the genome duplications opened up a world

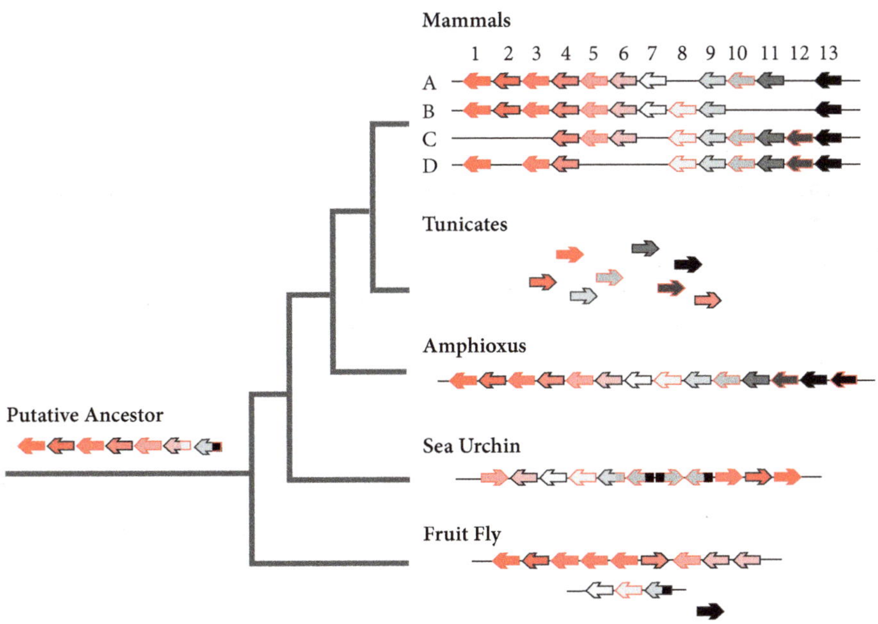

Figure 2.11 Quadruplication of *Hox* genes in vertebrates. Early animals probably had one cluster of multiple *hox* genes located on a single chromosome (each capital letter represents a different chromosome, each arrow represents a single *hox* gene, and the arrow's direction indicates the direction of transcription). This ancestral cluster was disbanded in tunicates, subdivided in flies, and duplicated twice with the origin of vertebrates.

Adapted from Lemons and McGinnis (2006).

of possibilities that evolution then utilized. As the following sections reveal, early vertebrates indeed acquired a wide range of important phenotypic innovations.

2.4. The Major Sense Organs of Early Vertebrates

Our reconstruction of early vertebrate sense organs relies heavily on a comparative analysis of extant chordates, simply because most aspects of those organs do not fossilize. We also emphasize data from amphioxus and lampreys. Hagfishes are less useful for our purposes because many of their features are highly specialized and probably do not reflect the primitive condition even for cyclostomes. Adult ascidians (tunicates) are likewise given little consideration, because their nervous systems are highly degenerate and very different from those of other chordates. Of course, even amphioxus and lampreys exhibit some derived features that are not representative of early vertebrates. As reviewed in Chapter 1, evolutionary biologists have developed several strategies for discriminating such uniquely derived traits from shared derived and primitive ones. We use these strategies whenever possible but, for the sake of brevity, often do so implicitly. Similarly, we do not discuss all aspects of vertebrate sensory and motor systems. Instead, we focus on the most interesting and important aspects. Some systems (e.g., the auditory system) are discussed more thoroughly in later chapters. A summary of the principal sensory and motor innovations of early vertebrates is provided in Table 2.1.

2.4.1. Photoreception

Amphioxus has a single, unpaired eye at the rostral end of the nerve cord (Figure 2.12). Its rostral edge consists of melanin-expressing pigment cells. Just caudal to those cells lies a row of six photoreceptors with long cilia and axons that pass

Table 2.1. Summary of the Major Sensory and Motor Innovations of Early Vertebrates

Vision	Paired lateral eyes with a lens and an expanded retina Rod photoreceptors (in addition to cones) Pineal gland (may have evolved indepently in lampreys and gnathostomes)
Chemical senses	Specialized nasal epithelium Expansion of olfactory receptor gene repertoire Taste buds
Octavolateralis systems	Mechanosensory lateral line system Electroreceptors Paired inner ears with semicircular canals
Motor systems	Pumping water across vascularized gills More efficient swimming Eye movements

ventrally into a fiber tract (Lacalli, 1996). Importantly, these photoreceptors express an opsin gene that is homologous to the opsins expressed in vertebrate photoreceptors (Vopalensky et al., 2012). They also express a homolog of the G protein that is involved in vertebrate photoreceptor signaling, as well as two genes whose homologs in vertebrates (*rx* and *pax6*) play an important role in retinal development. Immediately caudal to the photoreceptors lies a cluster of neurons that do not express opsins and, therefore, are unlikely to be photosensitive; some of these neurons contain serotonin and have long axons that project caudally. Collectively, these data strongly suggest that the frontal eye of amphioxus is homologous to the paired eyes of vertebrates (Lamb, 2013). Further support for this hypothesis comes from the observation that larval tunicates also have an unpaired eye (called the ocellus; see Figure 2.4), which contains a large pigment cell and about 20 ciliary photoreceptors that express a homolog of vertebrate opsins (Kusakabe et al., 2001; Lamb, 2013).

Although the eyes of amphioxus and larval tunicates are almost certainly homologous to vertebrate eyes (as light sensing organs), there are some major differences. For one thing, vertebrate eyes are paired, whereas the frontal eye of amphioxus and the ocellus of tunicates are unpaired structures. They are also much smaller

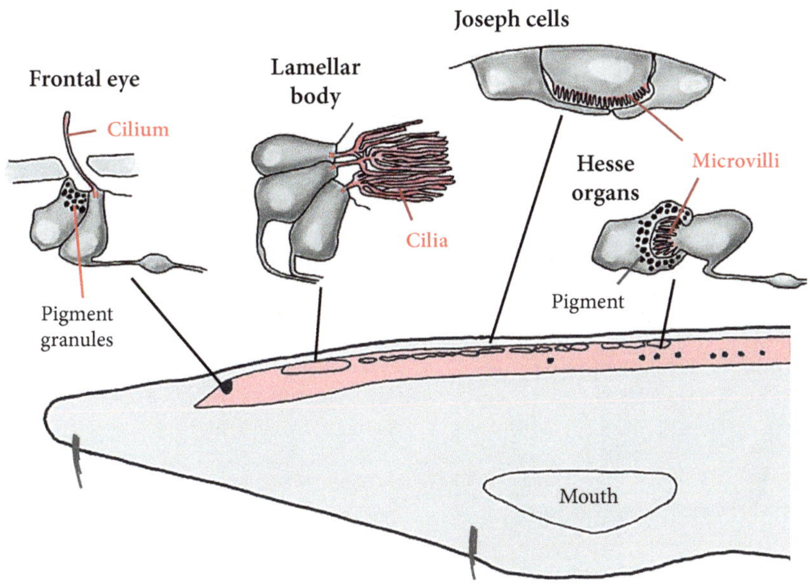

Figure 2.12 Photoreceptor systems in amphioxus. The frontal eye and lamellar body of amphioxus are ciliary photoreceptors, meaning that the light-sensitive parts of the cell are derived from cilia. In contrast, the Joseph cells and Hesse organs are rhabdomeric receptors, in which the light sensing components are derived from microvilli. The lamellar body breaks up in adult animals, and Joseph cells arise late in development. The mouth of the animal is on the left side of the head at the illustrated stage of larval development.

Adapted from Lacalli (2004), with permission from S. Karger AG, Basel.

than vertebrate eyes, contain far fewer cells, and lack clear homologs of bipolar and amacrine cells, two types of neurons that are found in typical vertebrate retinas. Another important difference is that the eye of amphioxus lacks a lens. The ocellus of tunicates does contain three "lens cells" that may help to concentrate incoming light, but tunicates clearly lack the type of image-forming lens that most vertebrate eyes possess. In general, the eyes of amphioxus and tunicates do not allow for the kind of pattern vision that we typically associate with vertebrate eyes. Instead, these animals seem to use their eyes mainly to detect changes in ambient light intensity. However, they may be able to determine the approximate direction of incoming light, which might help them swim toward the light and avoid large, shadowy predators (Nilsson, 2013).

Among cyclostomes, lampreys have large bilateral eyes with typical vertebrate retinas. These retinas contain ciliary photoreceptors (Figure 2.13) and several types of neurons, including horizontal, bipolar, amacrine, and ganglion cells. The photoreceptors in lamprey retinas come in several varieties, each expressing a different opsin gene. Some lampreys possess five different opsin genes, but other

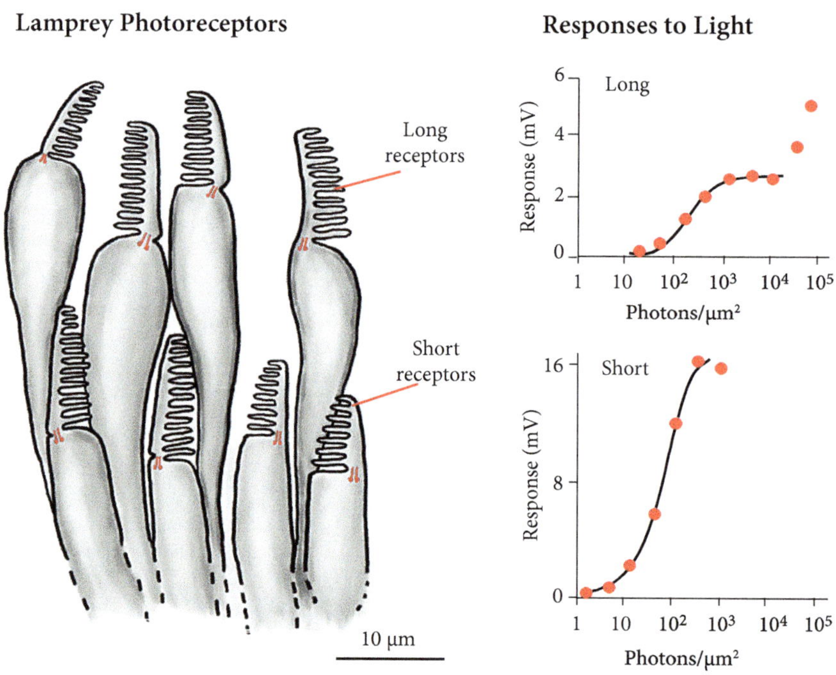

Figure 2.13 Lamprey photoreceptors. The European river lamprey *Lampetra fluviatilis* has long and short types of photoreceptors in its retina. Both types have outer segments that are cone-like insofar as they feature deep invaginations of the cell membrane (rather than the intracellular disks typical of rods). However, the short photoreceptors are much more light-sensitive than the long photoreceptors, making them rod-like. The graphs depict responses to flashes of 520 nm light.
Adapted from Asteriti et al. (2015).

lamprey species seem to have lost three or four of these genes (Collin et al., 2009; Lamb, 2013). At the level of cellular anatomy, all lamprey photoreceptors resemble the cone photoreceptors of other vertebrates insofar as the photopigment-bearing membranes are invaginations of the cell membrane, rather than intracellular disks. However, one type of lamprey photoreceptor is similar to the rods of other vertebrates in expressing rhodopsin (the opsin used by rods) and being extremely light-sensitive. Indeed, this rod-like photoreceptor of lampreys is capable of detecting single photons, which some researchers consider a defining feature of rods (Asteriti et al., 2015; Morshedian and Fain, 2015).

Hagfishes have only rudimentary eyes. Their eyes are paired and clearly homologous to the eyes of other vertebrates, but only 1–1.5 mm in diameter and buried beneath a patch of translucent skin (Lamb, 2013). Hagfish eyes also have no lens, iris, or pigmentation. The hagfish retina contains ciliary photoreceptors, but whether they are cone-like or rod-like remains unclear. Intriguingly, hagfish photoreceptors synapse directly onto neurons with long axons that project into the brain. Thus, hagfish lack the bipolar cells that relay signals from photoreceptors to retinal ganglion cells in other vertebrates. However, the projections from the hagfish retina into the brain are similar to retinal projections in other vertebrates (Wicht and Northcutt, 1990). Given that hagfish eyes are so much smaller and simpler than lamprey eyes, we suspect that hagfish eyes were secondarily simplified, by which we mean that many aspects of their morphology degenerated over evolutionary time.

In general, we can infer that early vertebrates transformed the ancestral median eye into a pair of lateral eyes. In addition, early vertebrates greatly increased the number of photoreceptors in the eye, developed an image-forming lens, and evolved some retinal interneurons, notably amacrine and bipolar cells. Collectively, these innovations made early vertebrates capable of pattern vision, which would have allowed them to identify (or at least distinguish between) small food items and approaching predators. Given their ecology, those traits would have been adaptive. Early vertebrates also evolved rod-like photoreceptors in addition to cones, thereby increasing their ability to see in low-light conditions. Because rods saturate quickly in bright light, it is worth noting that early vertebrates retained their cones and, thus, evolved a "duplex retina" capable of functioning over a broad range of light intensities (Figure 2.13). Two rounds of gene duplications also allowed early vertebrates to diversify their opsin genes, which eventually made color vision possible (e.g., see Chapter 6, Figure 6.12). It remains unclear, however, when exactly color vision arose.

Also unclear is whether early vertebrates had a cluster of ciliated photoreceptors in the dorsal diencephalon that is homologous to the pineal eye of other vertebrates (and the pineal gland of amniotes; see Appendix). Lampreys have a typical anamniote pineal, but hagfishes do not. Amphioxus larvae possess a lamellar body that lies in the dorsal midline and contains ciliated cells (see Figure 2.12), but this structure largely disappears at metamorphosis. More importantly, there is no evidence that the ciliated cells in the lamellar body express opsins or function as photoreceptors. Therefore, the lamellar body's homology to the vertebrate pineal gland remains uncertain. Tunicate larvae do not have a lamellar body, and their only photoreceptors seem to be homologous to the lateral eyes of vertebrates, not the pineal. Given these data, it is equally reasonable to suppose that lampreys and jawed vertebrates evolved a pineal eye independently of one another as it is to think that early vertebrates evolved a pineal eye that hagfishes have lost.

In addition to ciliary photoreceptors, vertebrates also possess microvillar photoreceptors (aka rhabdomeric photoreceptors), in which the photoreceptor-bearing portion of the cell membrane expands by forming microvillar protrusions, rather than modified cilia (Lamb et al., 2013). Such microvillar photoreceptors express a different class of opsin (r-opsins rather than c-opsins), and they are common in invertebrates (Arendt et al., 2004). In vertebrates, microvillar photoreceptors are restricted to a special type of retinal ganglion cell (Provencio et al., 2000). These unusual retinal ganglion cells express the photosensitive pigment melanopsin and are intrinsically sensitive to changes in ambient light levels. They are known to control both pupil diameter and circadian rhythms (Gooley et al., 2003; Lucas, 2013). Amphioxus has no microvillar photoreceptors in its frontal eye, but the skin of amphioxus contains Joseph cells and organs of Hesse (see Figure 2.12), both of which are microvillar in structure and probably light-sensitive (Lacalli, 2004). Such epidermal photoreceptors have not been described in vertebrates, but lampreys are sensitive to illumination of their tail (Ronan and Bodznick, 1991), and the aquatic frog *Xenopus laevis* has melanopsin expressing cells embedded among mechanosensory cells on the body surface (Baker et al., 2015).

Overall, we conclude that the last common ancestor of vertebrates almost certainly had both microvillar and ciliary photoreceptors. Both types were retained as vertebrates evolved, but only the ciliary photoreceptors became more numerous and more diverse in the vertebrate lineage. In contrast, non-chordate invertebrates have few ciliary photoreceptors, opting instead for building complex eyes with microvillar photoreceptors. Especially the arthropods evolved some complex microvillar (rhabdomeric) eyes. These eyes embody a fundamentally different design than vertebrate eyes, but both eye types allow for complex pattern vision. It seems reasonable to speculate that early vertebrates and early arthropods were engaged in an ecological "arms race" in which pattern vision was a critical tool. The two lineages adopted very different strategies to achieve pattern vision, but both designs were highly effective. In this context, it is interesting that at least some arthropods from the early Cambrian had not only well-developed eyes but also relatively large brains resembling those of modern malacostracans (e.g., shrimp; Ma et al., 2012).

2.4.2. Chemical Senses

Most animals possess a wide variety of chemical sensors, most of which can be classified into olfactory and gustatory (taste) sensors. Studying the evolution of these sensors is complicated, because chemosensory cells cannot be identified on purely morphological grounds. Fortunately, comparative genomic analyses have provided substantial information about the evolution of olfactory and taste receptor molecules. In conjunction with the morphological data, these genomic analyses provide some clues about the chemosensory abilities of early vertebrates.

Lampreys have a well-developed sense of smell, which they use for feeding, reproduction, and migration. They have an unpaired nose with sensory neurons that project mainly to a large olfactory bulb. As in other vertebrates, each olfactory sensory neuron in lampreys expresses a specific type of olfactory receptor molecule. All in all, the lamprey genome contains approximately 30 functional olfactory receptor genes that are closely related to the olfactory receptor genes of jawed vertebrates (Figure 2.14; Niimura, 2009). Lampreys also possess two other types of chemosensors that are typical of most jawed vertebrates, namely trace amine-associated receptors (TAARs) and type 1 vomeronasal receptors (V1Rs), which are used to detect pheromones (Hashiguchi and Nishida, 2007; Libants et al., 2009). Whether hagfishes have homologs of these three types of chemoreceptors remains unknown, because their genome has not been sequenced. However, hagfishes do have a large olfactory epithelium with sensory neurons that project directly to a large olfactory bulb.

Amphioxus does not have a clearly identifiable nasal epithelium, but it has numerous cells in its skin that may be chemosensory. Some of these cells have axons; others do not. In any case, the amphioxus genome contains more than 60 homologs of vertebrate olfactory receptor genes, half of which appear to be functional (Churcher and Taylor, 2009). At least one of these olfactory receptor genes is expressed in bipolar neurons that send one process to the skin on the dorsal side of the head and the other to the brain (Satoh, 2005). Thus, amphioxus seems to have an epidermal chemosensory system that is closely related to the olfactory system of vertebrates. Tunicates, in contrast, do not possess any homologs of the vertebrate olfactory receptor genes (Churcher and Taylor, 2009; Niimura, 2009).

Given these data, one might conclude that early chordates possessed a sizable family of olfactory receptor genes that was retained in vertebrates and lost in tunicates. However, all of the amphioxus olfactory receptor genes are more closely related to one another than they are to the olfactory receptor genes of lampreys and other vertebrates (Figure 2.14; Churcher and Taylor, 2009; Niimura, 2009). Therefore, it seems more likely that early chordates had only a single olfactory receptor gene that then multiplied independently in amphioxus and vertebrates. Yet another possibility is that amphioxus and vertebrates evolved their olfactory receptors independently of one another from the large family of rhodopsin-like G protein-coupled receptors, which predates the origin of chordates. In both of these scenarios, early vertebrates would have undergone an explosive expansion of their olfactory receptor repertoire. An interesting twist on this story is that

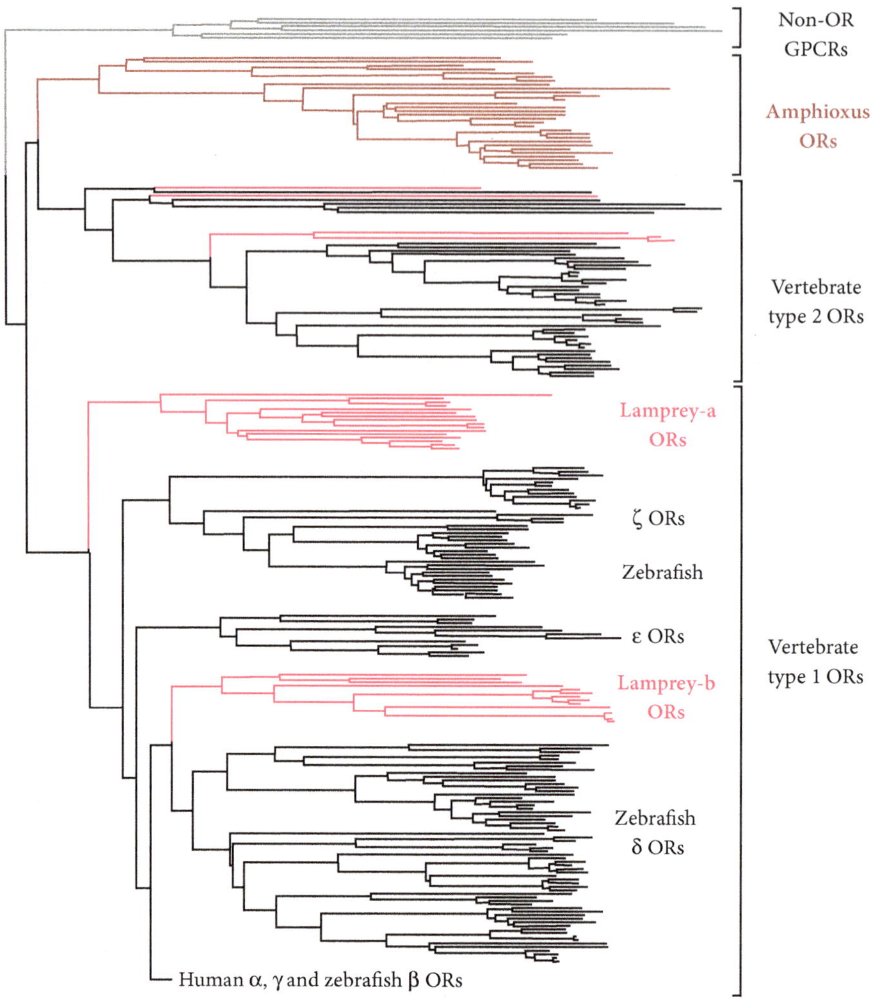

Figure 2.14 Phylogeny of chordate olfactory receptor genes. Shown here is an excerpt from a phylogenetic tree for 615 olfactory receptor-like (OR-like) genes and 6 non-OR G protein-coupled receptor (GPCR) genes from amphioxus (red), lampreys (pink), zebrafish (black), and humans. The illustrated portion of the tree emphasizes the non-human gene subfamilies. Note that the OR genes of amphioxus all cluster with one another, suggesting that they diversified independently of those in vertebrates.
Adapted from Niimura (2009).

vertebrate-like olfactory receptor genes have now been found also in comb jellies and echinoderms (Churcher and Taylor, 2011). These data have been used to argue that vertebrate-like olfactory receptor genes predated the origin of Bilateria and were lost in protostomes. However, the independent evolution hypothesis remains quite plausible, especially since it is widely accepted that arthropods evolved a wide variety of olfactory receptors that are only remotely related to their vertebrate analogs (Wicher, 2012). On balance, we conclude that the olfactory capacities of early vertebrates were significantly greater than those of their immediate ancestors.

Taste buds are found in most vertebrates, including lampreys (Barreiro-Iglesias et al., 2010; Kirino et al., 2013). These taste buds are aggregates of taste cells that release the transmitters ATP and serotonin onto sensory axon terminals. Hagfishes do not possess taste buds, but their skin contains a large number of Schreiner's organs that consist of multiple sensory cells (Braun, 1998; Finger, 2009). These Schreiner's organs may well be gustatory in function, but their innervation pattern does not support a homology with the taste buds of other vertebrates. Amphioxus and tunicates also do not possess any structures that resemble taste buds; nor do they have Schreiner's organs. Given these findings, it seems likely that taste buds are a vertebrate innovation that was lost in hagfishes. Although this hypothesis is well supported, it remains possible that invertebrates have homologs of vertebrate taste cells, but that these cells do not aggregate into distinct taste buds.

2.4.3. Mechanosensory Hair Cells

Most anamniotic vertebrates possess a mechanosensory lateral line system, whose major function is to sense vibrations of the ambient water, including those caused by potential predators or prey. The vibrations are sensed by epithelial cells that extend multiple microvilli and a single cilium (called stereocilia and kinocilium, respectively) into the space around the animal. These so-called hair cells are quite similar to the hair cells in the inner ears of amniotes, but the hair cells in the lateral line system usually aggregate into small clusters, called neuromasts. Moreover, the "hairs" of all the hair cells in a neuromast tend to be covered with a gelatinous dome, called a cupula (Figure 2.15). When water vibrations move the cupula, they bend the hairs, which causes the hair cells to release neurotransmitter molecules onto the peripheral processes of neurons whose central processes course in special lateral line nerves (see Appendix) and transmit the gathered information to the brain. Individual neuromasts tend to be arranged in lines on both sides of the animal, which explains the system's name. Most of the lines are on the head, but some extend along the sides of the trunk.

Lampreys have lines of neuromasts on both the head and the trunk. These neuromasts sit in shallow pits, rather than deep grooves or enclosed canals, as they do in many fishes (Figure 2.15), but they are otherwise quite similar to the neuromasts of other vertebrates (Northcutt, 1989). Hagfishes, in contrast, lack neuromasts. Some hagfish species do have short grooves in their skin that look like an early stage of lateral line development in other vertebrates, and they do contain isolated mechanosensory cells. However, hagfishes lack the multicellular neuromasts typical of other fishes (Braun and Northcutt, 1997; Wullimann and

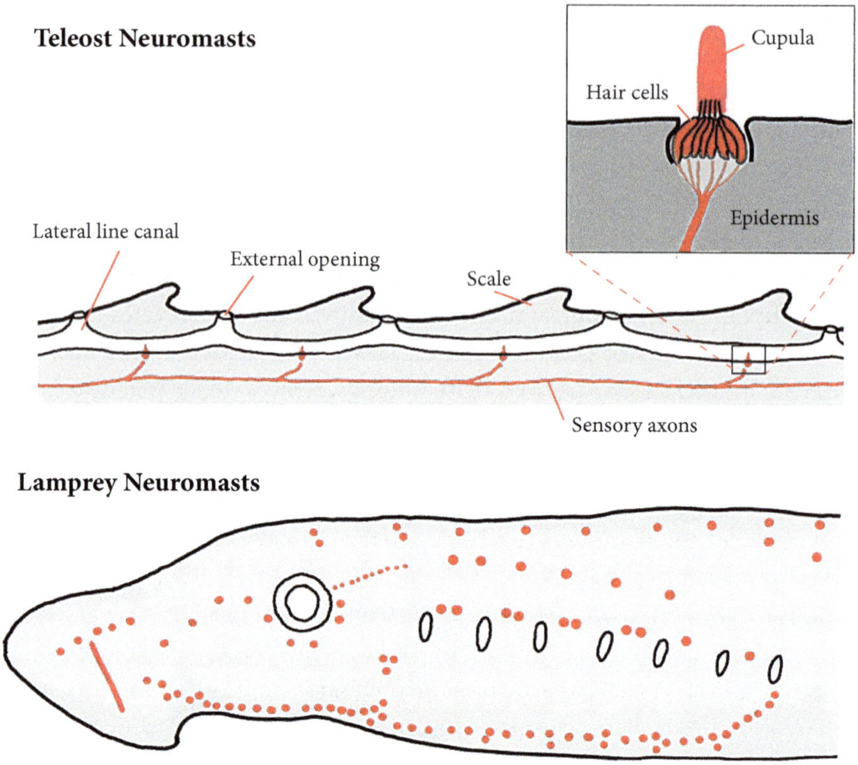

Figure 2.15 The mechanosensory lateral line system. The mechanosensory organs of the lateral line system are clusters of hair cells, called neuromasts, that are covered by a gelatinous cupula. In many fishes the neuromasts are located within a system of canals that open periodically to the outside of the fish (top). Lampreys lack lateral line canals. Instead, their neuromasts are located in shallow pits or lines on the body surface.
Adapted from a drawing by Thomas Haslwanter (top; https://commons.wikimedia.org/wiki/File:LateralLine_Organ.jpg) and from Northcutt (1989).

Grothe, 2013). Tunicates and amphioxus likewise lack neuromasts. Given these data, it seems most likely that the mechanosensory lateral line is a vertebrate innovation that was secondarily simplified in hagfishes.

Although invertebrate chordates lack a mechanosensory lateral line system, they do possess some epithelial cells that may be homologous to vertebrate hair cells. In particular, several species of tunicates have one or more rows of epithelial cells with stereo- and kinocilia at the base of their water intake siphon. These cells are probably used to monitor water flow through the pharynx (Burighel et al., 2011). Importantly, they are secondary sensory cells, meaning that they do not have axons and release neurotransmitter onto the peripheral processes of sensory neurons, just like vertebrate hair cells. In addition, tunicates possess some cells that resemble hair cells but do have axons and are, thus, classified as primary sensory cells. This is interesting because amphioxus also has both primary and secondary hair cell-like cells in its epidermis. Bernd Fritzsch and his collaborators have argued that these two types of sensory cells are homologous to one another and derive from a single type of primary hair cell-like neuron that dates back to the last common ancestor of all bilaterian animals (Duncan and Fritzsch, 2012). Be that as it may, the idea that all Bilateria have hair cells of one type or another does not negate our proposal that only vertebrates have clustered these hair cells into neuromasts and arranged them into a lateral line system. One might say that the character "hair cell" is common to all Bilateria, but that their "character state" of being assembled into neuromasts is unique to the vertebrates. However, we here prefer to consider hair cells and neuromasts as two different characters, located at two different levels of biological organization.

Vertebrates have hair cells not only in the lateral line, but also in the inner ear, where they mediate both hearing and the vestibular sense. There is no experimental evidence that lampreys can hear, and their inner ear lacks a clear homolog of the sensory structures that tetrapods use for hearing. However, both lampreys and hagfishes have a vestibular apparatus that features a set of semicircular canals (Figure 2.16). Whereas jawed vertebrates have three pairs of semicircular canals, arranged in three roughly orthogonal planes, hagfishes have just a single pair (McVean, 1991). Lampreys exhibit a vestibular system of intermediate complexity (Maklad et al., 2014), which they can use to sense head rotations in all three standard planes (i.e., pitch, roll, and yaw). Even hagfishes can sense such rotations, but their sensitivity and precision are low. Comparative developmental data indicate that early stages of semicircular canal development are conserved across all vertebrates, but that hagfishes have simplified later stages of canal development, whereas canal development in lampreys follows a unique trajectory (Higuchi et al., 2019).

In addition to the semicircular canals, hagfishes and lampreys have patches of vestibular hair cells that are probably homologous to the vestibular sacculus and utricle of jawed vertebrates. These hair cells are covered by an otoconial membrane (i.e., a gelatinous mass of calcium carbonate crystals), rather than a solid otolith (ear stone), which is found only in ray-finned fishes. Importantly, these vestibular hair cells allow hagfishes and lampreys to sense linear acceleration and the direction of gravity—two sensory capacities that are quite useful in deep, dark water or at night.

In contrast to the vertebrates, amphioxus does not have paired inner ears. It does, however, have an unpaired cluster of cells with club-shaped cilia that is probably used to sense the direction of gravity and, perhaps, some self-motion. Similarly, larval tunicates lack paired inner ears but have a single otolith (see Figure 2.4) that consists of a large pigment mass, balanced on top of two neurons. This structure, too, seems specialized for sensing gravity and linear acceleration (Dilly, 1962).

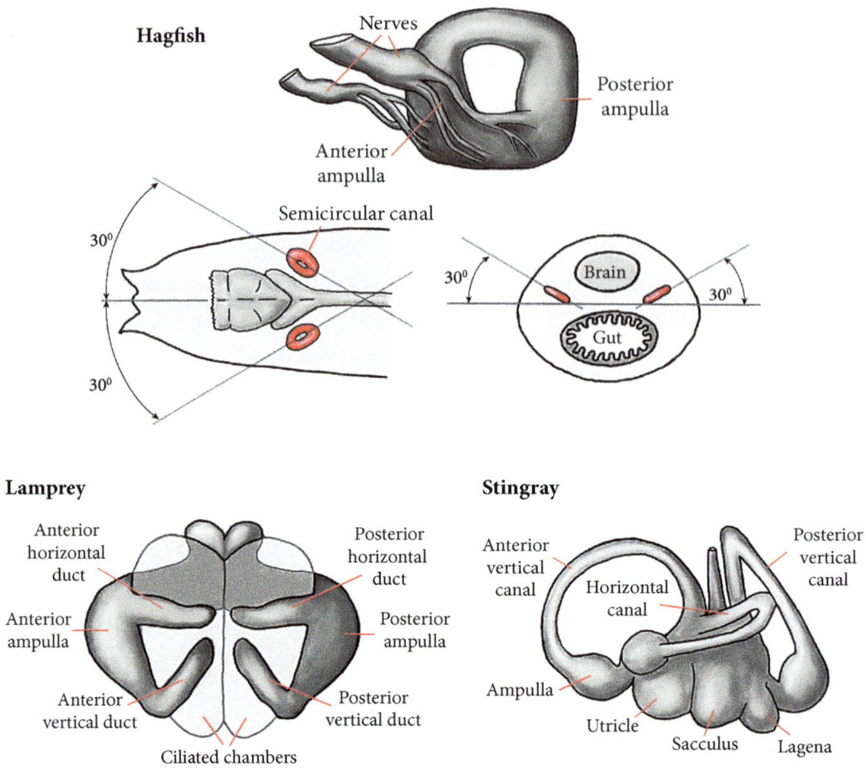

Figure 2.16 The vestibular apparatus. Hagfishes have only a single semicircular canal per side; it features two expansions (ampullae) that house the sensory epithelia. Because of how these canals are arranged in the head, hagfishes can sense head rotations in all of the main orientation planes. Lampreys have a more complex vestibular apparatus, featuring a pair of horizontal ducts that may be functionally analogous (but likely not homologous) to the horizontal canal of jawed vertebrates, here represented by the bluespotted stingray *Neotrygon kuhlii*.

Adapted from Lowenstein and Thornhill (1970), McVean (1991), Evangelista et al. (2010), Maklad et al. (2014); hagfish schematics and lamprey canals with permission from John Wiley & Sons.

Overall, these data indicate that sensors of gravity and linear acceleration predated the origin of vertebrates, but that semicircular canals capable of sensing head rotation evolved only with the origin of vertebrates. These canals then became more complex with the emergence of jawed vertebrates. The use of hair cells to detect sound-induced vibrations (i.e., to hear) probably evolved later, well after the origin of vertebrates.

2.4.4. Electroreception

Many aquatic vertebrates are capable of sensing weak electric fields (Bullock et al., 2006). They use this ability mainly to detect muscle contractions of potential prey that may be hidden or hard to see in murky water or at night (since muscle contractions are driven by electrical currents across the muscle cell membranes). Electroreception may also aid in spatial navigation, because some underwater locations generate electric fields. Moreover, swimming through the earth's magnetic field can generate electric currents that sharks, at least, can sense. The ability to sense such weak electric fields derives from electroreceptors that are typically located near the mechanosensory lateral lines and innervated by lateral line sensory nerves. The most common type of electroreceptor lies deep within the skin, at the bottom of a flask-like, mucus-filled "ampulla" that opens to the external surface (Figure 2.17). The individual electrosensory cells within these ampullary organs resemble hair cells in that they have multiple microvilli and no axon of their own.

Invertebrate chordates reportedly lack electroreceptors, as do hagfishes. Lampreys, however, have electroreceptors on their head and all over their trunk. Curiously, these electroreceptors do not sit at the bottom of ampullae. Instead, they have very elongated cell bodies that span most of the epidermis and extend microvilli directly to the skin surface (Figure 2.17; Baker et al., 2013). Because these so-called end buds of lampreys are innervated by the lateral line nerves (see Appendix) and respond to the same kind of electrical stimuli as ampullary electroreceptors in jawed vertebrates, they are probably homologous to them. If this is true, then we can conclude that ampullary electroreceptors arose with the origin of vertebrates but were lost in the hagfish lineage. Whether electroreceptors were present in stem gnathostomes (e.g., placoderms; see Chapter 3, Section 3.2.1) remains unclear, because these sense organs are not guaranteed to leave an impression in fossilized remains (King et al. 2018).

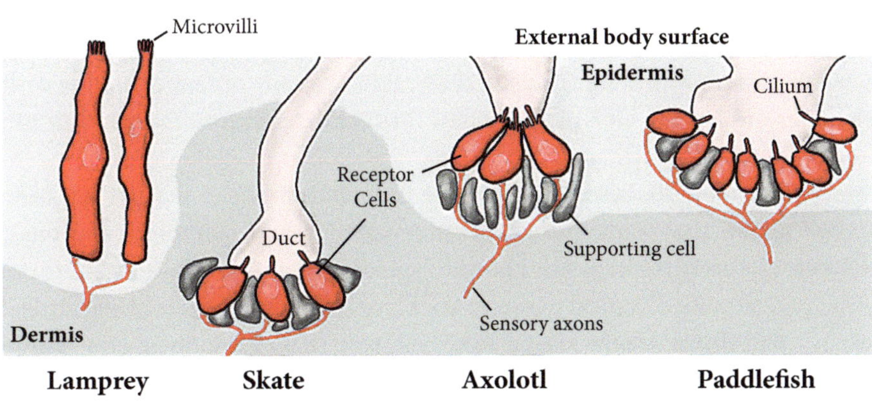

Figure 2.17 Electroreceptors. The electroreceptor cells (red) of lampreys are more elongate than those of other vertebrates, represented here by a skate (a cartilaginous fish), an axolotl (an aquatic amphibian), and a paddlefish (a nonteleost ray-finned fish). The lamprey's electroreceptors also differ from the others in extending microvilli directly to the body surface and lacking a cilium.
Adapted from Baker et al. (2013).

2.5. Movements and Motor Control

Aside from gill movement, locomotion was the most obvious kind of movement in early vertebrates. Lacking paired fins or legs, the early vertebrates would have used mainly their trunk muscles to move from place to place. These muscles are segmentally organized in all chordates, with fibrous septa located between the individual muscle segments. When the trunk muscles on one side of the body contract, the trunk bends toward that side. When the muscles relax, the body straightens again because of the notochord's elastic recoil. This bending of the trunk causes the animal to move sideways, and many fishes still use this kind of C-start locomotion to escape from predators. To achieve more controlled forward locomotion, early vertebrates probably alternated contractions on the two sides of the body, slightly delaying the contractions in successively more caudal body segments. This arrangement creates a traveling wave of alternating sideways bends that pushes water toward the tip of the tail and moves the animal forward. This mode of locomotion is called lateral undulation and is common to many fishes, legless salamanders and lizards, as well as snakes.

Adult amphioxus swim by means of lateral undulation and use it to burrow into sand (Stokes, 1997). Larval ascidians and adult larvaceans also swim by undulating their tail, and the latter clearly use a traveling wave to generate forward propulsion (Kreneisz and Glover, 2015). Combined with the observation that early chordates had a notochord and segmented trunk muscles (see Figure 2.7), these data suggest that lateral undulation with a traveling wave predated the origin of vertebrates. What remains less clear is the shape of the individual muscle segments. They are chevron-shaped in amphioxus, but W-shaped or straight in some of the fossil chordates. These differences are functionally significant, because chevron- and W-shaped muscle segments allow the contractions of individual muscle fibers to sum more effectively than straight muscle segments, thus generating stronger bending forces and faster swimming (Lacalli, 2012). Still, we can conclude that the earliest vertebrates did not invent a radically new form of locomotion, though they might have improved its efficiency.

More substantial variation exists in the innervation of the skeletal muscles. In jawed vertebrates the trunk muscles are innervated by axons of motor neurons that have their cell body in the ventral horn of spinal cord and send axons out through segmentally arranged ventral nerve roots. Lampreys and hagfish spinal cords also contain a ventral horn with skeletal motor neurons that extend their axons through ventral nerve roots. Tunicates, however, have a very different motor system. Larval ascidians do have a spinal cord (Gionti et al., 1998), but it contains only two pairs of inhibitory neurons and numerous ependymal cells (Nishitsuji et al., 2012). Instead, the motor neurons in these animals are all located in the caudal part of the putative brain, which is called the visceral ganglion (Figure 2.18). This lack of spinal motor neurons in larval ascidians is probably a derived feature, because the larvacean spinal cord does contain 10 pairs of cholinergic motor neurons (Søviknes

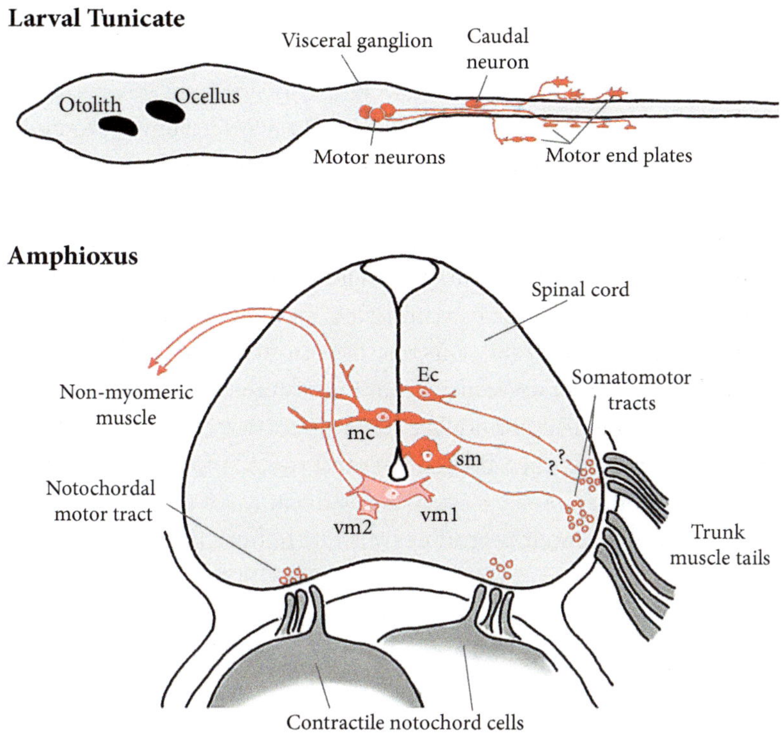

Figure 2.18 Locomotor neurons in larval ascidians and amphioxus. In larval ascidians the motor neurons innervating the trunk muscles have their cell bodies in the visceral ganglion or immediately caudal to it, rather than in the putative spinal cord. Their axons run parallel to the spinal cord for variable distances before terminating as motor end plates on the muscle fibers. Shown at the bottom is a transverse section through an amphioxus spinal cord, highlighting the neurons innervating the trunk muscles (red) and the neurons innervating non-myomeric muscles (pink). The axons of the trunk motor neurons do not exit the spinal cord but, instead, terminate on slender "muscle tails" that extend toward the spinal cord from the trunk muscle fibers.

Abbreviations: Ec – Edinger cell; mc – mid-commissural cell; sm – somatomotor cell; vm1/vm2 – visceromotor cell types 1 and 2.

Adapted from Wicht and Lacalli (2005), Imai and Meinertzhagen (2007), with permission from NRC Research Press (© Canadian Science Publishing or its licensors) and John Wiley & Sons, respectively.

et al., 2007). Regardless of their location, the axons of tunicate motor neurons pass down the spinal cord for varying distances (Imai and Meinertzhagen, 2007), exit the spinal cord, and then release acetylcholine onto the trunk muscle fibers.

Amphioxus has a spinal cord with numerous skeletal motor neurons in its ventral horns, but the axons of these neurons do not exit the spinal cord. Instead, they terminate at the spinal cord's ventrolateral edge, where they synapse onto slender processes of individual muscle fibers, called muscle tails, that contact the spinal cord (Figure 2.18). When the motor axons release transmitter onto the muscle tails, action potentials spread through the entire muscle fiber, triggering a contraction.

Thus, the muscle tails in amphioxus look and function like ventral spinal nerves, but they consist of muscle fibers rather than motor axons. The functional significance of this odd arrangement remains unknown. Similarly unusual are synaptic contacts between the amphioxus spinal cord and the underlying notochord, which may be used to modify the notochord's stiffness (McHenry et al., 1995).

Given this variation in skeletal muscle innervation among chordates, it seems likely that early chordates had a spinal cord containing at least a few motor neurons, which were then lost in ascidian tunicates. Some authors have proposed that the visceral ganglion of larval ascidians is homologous to the vertebrate and amphioxus spinal cord (Dufour et al., 2006), but this hypothesis seems unlikely, because the spinal cord marker gene *hox5* is expressed only caudal to the visceral ganglion in larval ascidians (Gionti et al., 1998). We also suspect that early chordates had motor axons that exit the spinal cord through ventral roots, and that the lack of ventral spinal roots in amphioxus is a uniquely derived feature of these animals. However, the lack of clearly segmented ventral nerve roots in both tunicates and amphioxus raises the possibility that ventral nerve roots arose with the origin of vertebrates.

Vertebrates use skeletal muscles not only to locomote, but also to move their eyes. Jawed fishes have seven pairs of extraocular muscles, and lampreys share six of them. In contrast, hagfishes, amphioxus and tunicates have no eye muscles and, thus, cannot move their eyes independently of the entire head. Based on these data and the degenerate nature of hagfish eyes, we surmise that extraocular muscles evolved in the first vertebrates but were then lost in hagfishes. If this hypothesis is valid, then the evolution of extraocular muscles probably went hand in hand with the evolution of image-forming eyes and a vestibular system that can detect head rotation. This evolutionary coincidence is interesting because it suggests that early vertebrates may have evolved a vestibulo-ocular reflex that uses vestibular inputs to move the eye muscles in such a way that the retinal image is stabilized when the head turns. Because phototransduction is a relatively slow process, this image stabilization would have reduced vision blur and, thus, helped early vertebrates detect small stimuli, such as potential prey or large but distant predators, even when those early vertebrates were swimming and, therefore, moving their head from side to side.

As noted earlier, another major innovation of early vertebrates is the use of muscles to pump water through the pharynx and across the gills. Lampreys and jawed vertebrates use a set of pharyngeal (branchial) muscles to compress the pharynx. When these muscles relax, the pharynx expands again because the pharyngeal bars of vertebrates are made of calcified cartilage and, therefore, elastic. The motor neurons controlling the vertebrate pharyngeal muscles are located in the hindbrain and send their axons to the muscles via cranial, rather than spinal, nerves (see Appendix). Their activity is typically rhythmic and controlled by one or more central pattern generators that are also located in the hindbrain (Kinkead, 2009). Hagfishes, in contrast, pump water through their pharynx using a very different mechanism. They employ antagonistic muscles and a set of cartilaginous bars to alternately curl and uncurl a specialized membrane (called a velum) in their anterior pharynx (Malte and Lomholt, 1998). This velar pumping mechanism is clearly a uniquely derived feature for hagfishes, as amphioxus and tunicates use motile cilia, rather than muscular pumps, to move water through their pharynx.

2.6. Early Vertebrate Brains

Early vertebrates underwent major evolutionary changes not only in their sense organs and motor machinery, but also in their brains. Some of these changes were linked to modifications of the sensory and motor systems. For example, the neural circuits underlying the vestibulo-ocular reflex surely evolved after vertebrates evolved extraocular eye muscles (or, possibly, at the same time). However, the brain also underwent more fundamental changes, adding at least two major divisions that almost certainly did not exist prior to the origin of vertebrates. Before describing these changes, it is useful to recall briefly the major divisions of a typical vertebrate brain (Figure 2.19).

The three main divisions of a vertebrate brain are the forebrain, midbrain, and hindbrain (aka prosencephalon, mesencephalon, and rhombencephalon). Each of these main divisions initially appears as a distinct enlargement of the rostral neural tube in early embryonic development. At those early stages of development, the brain's neural tissue is relatively thin and filled with fluid. Therefore, the

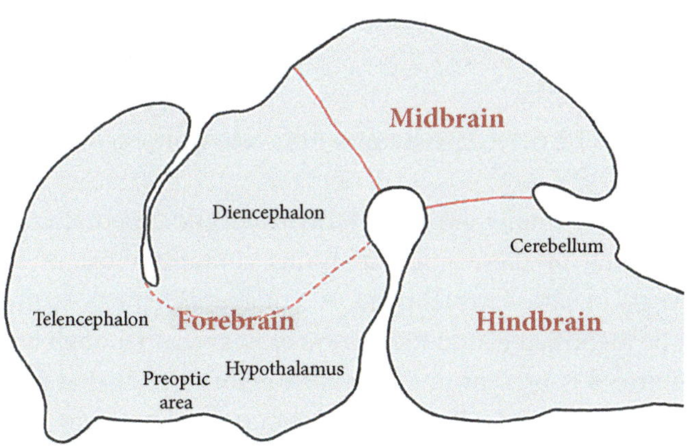

Figure 2.19 Main divisions of a vertebrate brain. Shown here is a sagittal section through the brain of a mammalian embryo. The boundaries between the three principal divisions—forebrain, midbrain, hindbrain—are shown as solid red lines. Within the telencephalon, the dashed red line marks the rostral boundary of the diencephalon. Anterior to it lies the secondary prosencephalon, whose main components are the telencephalon, preoptic area, and hypothalamus. The cerebellum develops from the most rostral and dorsal part of the hindbrain.
Adapted from Rodríguez-Moldes et al. (2017).

three cerebral enlargements result mainly from expansion of the fluid-filled cerebral ventricles and the concomitant ballooning of the cerebral tissue at three distinct locations, which is why those cerebral enlargements are often called "cerebral vesicles."

As development proceeds, the three major divisions of the brain subdivide further. The hindbrain develops six or more distinct swellings, called rhombomeres. The most rostral and dorsal (alar) part of the hindbrain develops into the cerebellum, as well as several cerebellum-like structures, such as the dorsal cochlear nucleus in mammals (see Chapter 6). The midbrain gives rise to the optic tectum and torus semicircularis (the superior and inferior colliculus in mammals), and several more ventral (basal) tegmental regions. The forebrain changes most drastically (Rubenstein et al., 1994; Puelles et al., 2013). Its caudal portion develops into the pretectum, thalamus (aka dorsal thalamus), and prethalamus (aka ventral thalamus). Collectively, these areas are called the diencephalon. Rostral to the diencephalon lies the secondary prosencephalon. It gives rise to the hypothalamus, preoptic area, and retina, which forms as a laterally directed evagination ventral to the preoptic area. The secondary prosencephalon also forms the telencephalon, which is divisible into a subpallium (including the septum and the striatum) and a pallium (*pallium* means "mantle" in Latin). The latter gives rise to the olfactory bulb, hippocampus, parts of the amygdala, and neocortex.

As this entire book will demonstrate, the size and appearance of the various telencephalic divisions vary considerably across the vertebrates. In this chapter, we focus only on the largest, most fundamental divisions of the brain, some of which are clearly vertebrate innovations. Specifically, we argue that invertebrate chordates almost certainly had no telencephalon and no cerebellum, and probably lacked a midbrain as well.

2.6.1. The Brains of Cyclostomes

Lamprey and hagfish brains are clearly divisible into forebrain, midbrain, and hindbrain, and each of these main divisions contains most of the secondary subdivisions seen in jawed vertebrates. A notable exception is the cerebellum, which cannot be identified in adult lampreys or hagfishes (though one of its developmental precursors is present in cyclostome embryos; Sugahara et al., 2016; for more on this, see Chapter 7). Lampreys do have a rudimentary cerebellum with cerebellum-typical granule cells, but lampreys do not possess Purkinje cells or deep cerebellar nuclei (Wicht, 1996; Lannoo and Hawkes, 1997). In fact, the granule cells of lampreys project specifically to the lateral line nuclei of the hindbrain where they form a "cerebellar crest" (Weigle and Northcutt, 1998), which means that they are homologous only to the cerebellum-like structures of other fishes, not to the cerebellum proper (see Chapter 3). In this context, it is interesting that the dorsal part of the most rostral rhombomere in lampreys does not express *pax6*, which is required for cerebellar development in other vertebrates (Murakami et al., 2001).

The forebrain of lampreys is relatively small, but it is clearly divisible into a diencephalon and a secondary prosencephalon, which in turn contains both a hypothalamus and a telencephalon (Figure 2.20). Moreover, the telencephalon of lampreys contains both a pallium and an adjacent subpallium; both have been identified using the expression pattern of genes that are known to pattern the telencephalon in amniotes (e.g., birds and mammals; see Chapters 5 and 6) and have clear homologs in lampreys (e.g., *Pax6* and *Dlx*; Figure 2.20). Sugahara et al. (2016) recently used the expression of *Nkx2.1* to identify an embryonic pallidum (i.e., a medial ganglionic eminence) within the lamprey subpallium, supporting previous reports of a

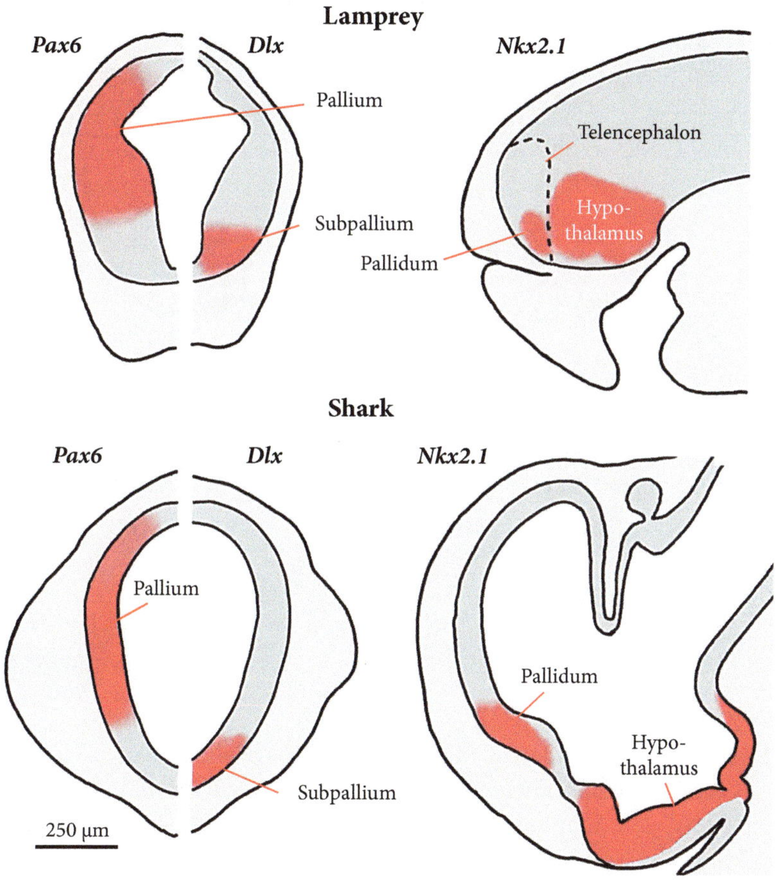

Figure 2.20 Genoarchitecture of the embryonic telencephalon. Shown on the left are transverse sections through the telencephalon of an Arctic lamprey (*Lethenteron japonicum*) and a catshark (*Scyliorhinus torazame*) at mid-pharyngula stages of embryonic development, highlighting the expression of *Pax6* and *Dlx* in the pallium and the subpallium, respectively. Shown on the right are sagittal sections (rostral is to the left) showing the expression of *Nkx2.1* in the hypothalamus and embryonic pallidum.

Adapted from Sugahara et al. (2013, 2016).

putative pallidum in adult lampreys (Stephenson-Jones et al., 2011). In contrast, interpretations of the lamprey pallium remain quite contentious.

In most vertebrates, it is customary to divide the pallium into medial, dorsal, lateral, and ventral divisions. The medial and dorsal pallial divisions correspond to the mammalian hippocampus and neocortex, respectively. The homologies of the lateral and ventral pallial divisions are more controversial, but include the olfactory cortex and pallial amygdala (we will come back to this in Chapter 3). In lampreys, Northcutt and his collaborators identified the evaginated portion of the

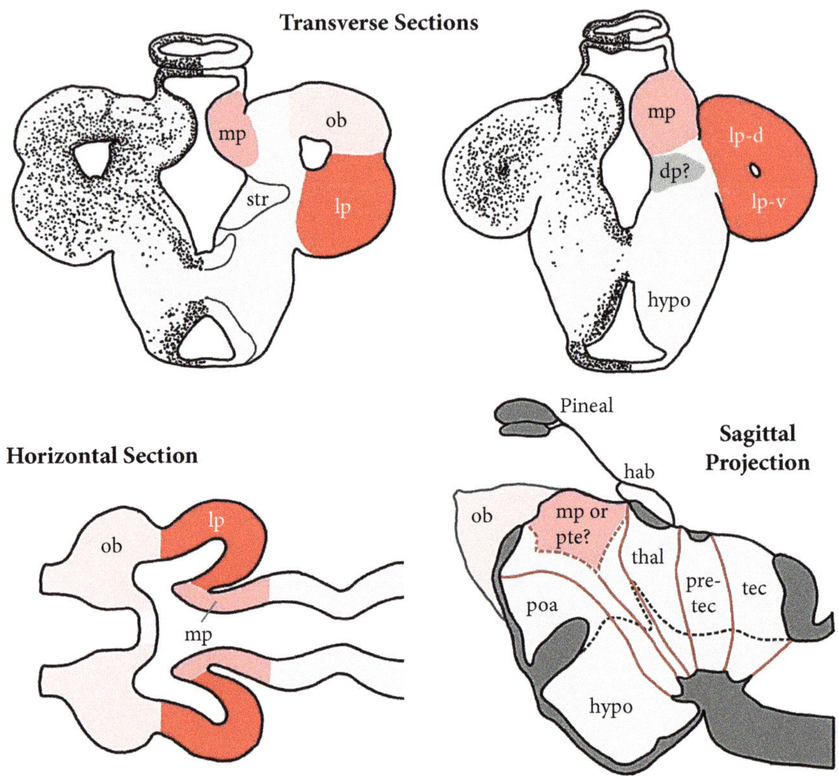

Figure 2.21 The telencephalon of adult lampreys. The telencephalon of lampreys is relatively small, and the homologies of its subdivisions remain controversial. According to our view, lampreys have a large lateral pallium (lp) that is divisible into dorsal and ventral divisions (lp-d and lp-v), as well as an unevaginated medial pallium (mp). Other authors have interpreted this medial pallium as an enlarged prethalamic eminence (pte), which would make it part of the diencephalon, rather than the telencephalon. The issue hinges on how one draws the boundaries between forebrain segments (i.e., neuromeres). As indicated by the dashed red lines in the bottom right panel, critical portions of those boundaries have been drawn differently in different studies. Just as controversial is whether lampreys possess a dorsal pallium (dp).
Other abbreviations: hab – habenula; hypo – hypothalamus; ob – olfactory bulb; poa – preoptic area; pretec – pretectum; str – striatum; tec – optic tectum; thal – thalamus.
Adapted from Northcutt and Wicht (1997), Wicht (1996), Pombal and Puelles (1999), Pombal et al. (2009).

pallium as the lateral pallium (Figure 2.21), based mainly on this region's relative position and massive input from the olfactory bulb (Northcutt and Puzdrowski, 1988; Northcutt and Wicht, 1997); the ventral subdivision of this lateral pallium is probably the ventral pallium. A likely candidate for the lamprey dorsal pallium is a small cell group in the unevaginated portion of the hemispheres, located medial to the lateral pallium and sometimes referred to as the subhippocampal lobe (labeled "dp?" in Figure 2.21). This cell group receives strong input from the thalamus but not from the olfactory bulbs, just as one might expect from a dorsal pallial derivative (Polenova and Vesselkin, 1993; Nieuwenhuys and Nicholson, 1998).

These interpretations have been challenged by the discovery that some neurons in the evaginated portion of the lamprey pallium have long descending projections to the midbrain, medulla, and spinal cord (Ocaña et al., 2015). Moreover, a different set of neurons in the evaginated portion of the lamprey pallium apparently receives thalamic input and responds to visual and somatosensory stimulation (Suryanarayana et al., 2017). Collectively, these data have been used to argue that the evaginated portion of the lamprey pallium (what we call the lateral pallium) is homologous to the mammalian dorsal pallium (i.e., neocortex). We suspect that this hypothesis is incorrect, mainly because the evaginated pallium in lampreys receives olfactory input, which is not characteristic of mammalian neocortex, and because pallial projections to the medulla and spinal cord have not been reported in other anamniotes. In addition, thalamic inputs to the dorsal pallium (or presumed dorsal pallium) are minimal or nonexistent in most anamniotes (see Chapters 3 and 4). Therefore, we conclude that at least some of the reported similarities between the lamprey evaginated palllium and mammalian neocortex are the result of convergent evolution, rather than homology. As we discuss more fully in Chapter 7, it is quite possible that lampreys and other anamniotes lack a proper dorsal pallium entirely.

The identity of the medial pallium in lampreys is likewise subject to significant debate. Northcutt and others identified the dorsal, unevaginated portion of the lamprey pallium as the medial pallium (aka the primordial hippocampus of earlier authors; see Northcutt and Puzdrowski, 1988). Consistent with this interpretation, the medial pallium in lampreys receives a wide variety of inputs, including thalamic ones, and projects to a wide variety of other areas, including the hypothalamus (Northcutt and Wicht, 1997). However, Pombal et al. (2009) considered this region to be the prethalamic eminence (aka thalamic eminence), which means that it would be a dramatically enlarged part of the anterior diencephalon, rather than the medial pallium. This revision was supported by the expression pattern of *lhx15* in lamprey embryos (Osório et al., 2006), but it ignored all of the connectional data, including the lack of thalamic inputs to the prethalamic eminence of amphibians (Krug et al., 1993). Thus, we are left with two very different interpretations, one based mainly on connectional data and one based on embryonic gene expression patterns. This dichotomy is a recurring theme throughout this book (especially in Chapters 5 and 6), but in the present case, the relative paucity of gene expression data causes us to give more credence to the connectional information.

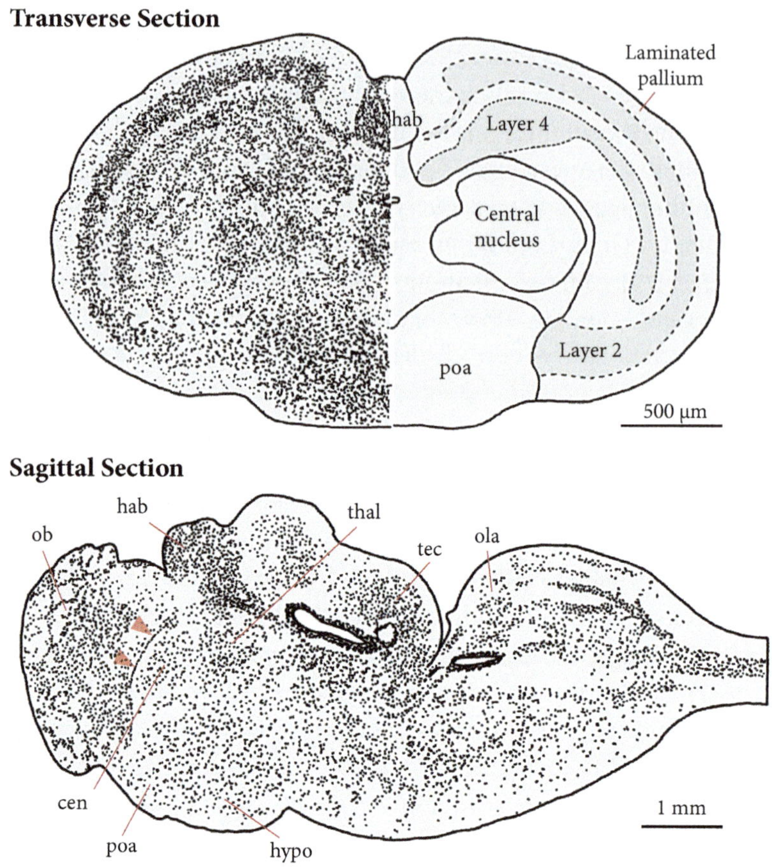

Figure 2.22 The brain of hagfishes. Shown here are transverse and sagittal sections through the telencephalon of an adult hagfish. Particularly interesting is that a large part of the pallium in these animals is clearly laminated (top). It is difficult, however, to homologize specific pallial divisions between hagfishes and other vertebrates, in part because the forebrain ventricles of hagfishes are obliterated during development (light red arrowheads indicate a small remnant).

Abbreviations: cen – central nucleus; hab – habenula; hypo – hypothalamus; ob – olfactory bulb; ola – octavolateralis area; poa – preoptic area; tec – optic tectum; thal – thalamus.

Adapted from Wicht and Northcutt (1992) and Pombal and Mégias (2017).

Challenging as it may be to understand the forebrain of lampreys, the forebrain of hagfishes is even more enigmatic (Figure 2.22). An *Nkx2.1*-positive region in the subpallium of embryonic hagfishes has been identified as a putative pallidum (Sugahara et al., 2016), but the organization of the hagfish pallium remains opaque. Part of the problem is that the forebrain is so massively developed in hagfishes that, in their adult forms, the telencephalic ventricles have become almost completely obliterated (Figure 2.22; Wicht and Northcutt, 1992). The absence of these ventricles makes it impossible to use standard topological criteria to identify the pallium's medial, dorsal, and lateral aspects. In addition, the olfactory bulbs in hagfishes project to all parts of the pallium (Wicht and Northcutt, 1993), which means that these projections cannot be used to distinguish the various pallial divisions from one another. The available immunohistochemical data are likewise of little help (Wicht and Northcutt, 1994). At this point, we suspect that the highly laminated lateral aspect of the hagfish pallium (Figure 2.22) represents a very large lateral pallium, but whether hagfishes possess medial and dorsal pallia remains quite uncertain. Be that as it may, the fact that lampreys and other basal vertebrates do not have a laminated pallium means that pallial lamination in hagfishes must have evolved independently of that observed in amniotes (see Chapter 5).

2.6.2. Invertebrate Chordate Brains

When examined with gross anatomical and histological techniques, the brains of amphioxus and tunicates look very different from vertebrate brains. The rostral end of the neural tube in amphioxus larvae has a slightly expanded ventricle, which is why this region is often called the cerebral vesicle, but in adult amphioxus the cerebral vesicle is no longer apparent, and the neural tube tapers anteriorly (see Figure 2.5). Moreover, the central nervous system of amphioxus contains far fewer neurons than that of any vertebrate, and most of these neurons have their cell bodies located close to the ventricle. Given this anatomy, it is impossible to determine where the brain ends and the spinal cord begins. Nor is it possible to fathom whether amphioxus brains contain the same main divisions as vertebrate brains. Similarly, the brains of tunicates remain inscrutable when examined with standard histological techniques. Major progress in understanding amphioxus and tunicate brains came only when researchers compared embryonic gene expression patterns between those species and vertebrates.

2.6.2.1. Amphioxus Brains

The first of these comparative molecular studies examined the expression of an amphioxus gene that is homologous to the vertebrate gene *hoxB3* (Holland et al., 1992). In vertebrates this gene is expressed in the spinal cord and the posterior hindbrain (Figure 2.23). The exciting finding was that the homolog of this gene in amphioxus embryos has a rostral expression boundary that lies far caudal to the cerebral vesicle. These data suggested that the hindbrain in amphioxus must include this gene expression boundary and that, therefore, the brain as a whole includes far more than the cerebral vesicle. Later studies supported this hypothesis by showing that the rostral expression boundary of *hox2* lies rostral to that of *hox3* in amphioxus, just as it does in vertebrates (Schubert et al., 2006; Takio et al., 2007; Parker et al., 2015). The expression patterns for *hox4* and *hox5* in amphioxus are also consistent with the vertebrate patterns. The only serious discrepancy is that *hox1* has the most rostral expression boundary of all the hox genes in amphioxus, whereas the rostral expression boundary of vertebrate *hox1* lies caudal to that of *hox2*.

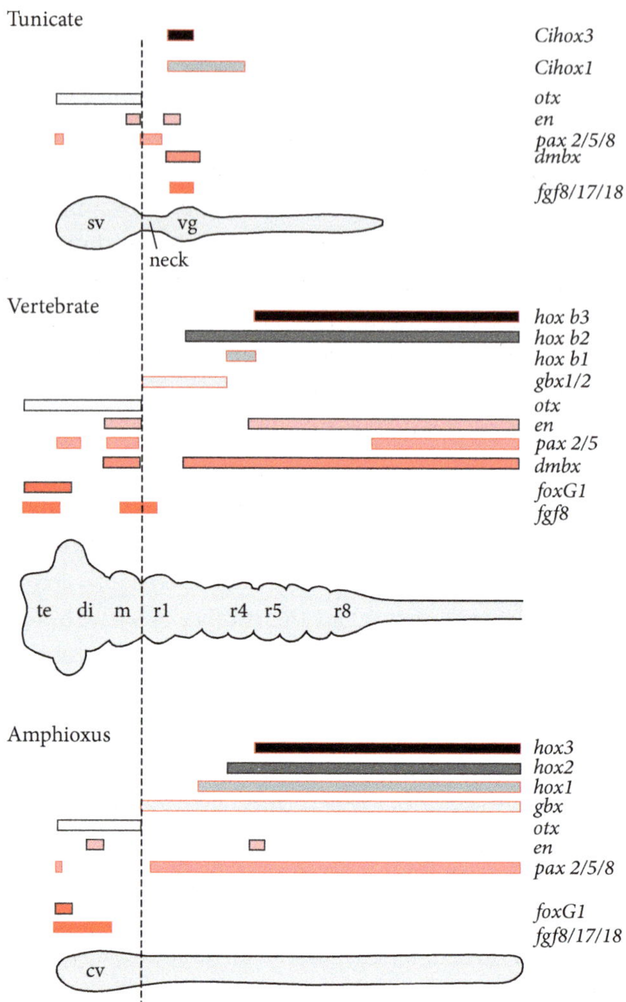

Figure 2.23 Gene expression in the CNS of embryonic chordates. The dashed line indicates the caudal limit of *otx* expression, which coincides with the midbrain-hindbrain boundary in embryonic vertebrate brains (depicted in a schematized dorsal view). The comparative data provide no clear evidence for a midbrain or telencephalon in larval amphioxus and tunicates.

Abbreviations: di – diencephalon; cv – cerebral vesicle; m – midbrain; r1-8 – hindbrain rhombomeres; sv – sensory vesicle; te – telencephalon; vg – visceral ganglion.

See main text and Holland et al. (2013) for sources.

Given these results, it became important to ask how far rostrally the hindbrain of amphioxus extends. One way to answer this question is to examine the expression of *gbx2*, which is expressed up to the midbrain-hindbrain boundary in vertebrate embryos and antagonized by *otx*, a gene that is expressed in the embryonic midbrain and forebrain. Indeed, the amphioxus homologs of these two genes also have complementary expression patterns, with *otx* being expressed in the embryonic cerebral vesicle and *gbx* being expressed caudal to that (Figure 2.23). According to these findings, the rostral hindbrain boundary in amphioxus embryos coincides with the caudal limit of the cerebral vesicle.

However, *engrailed* (*en*) and *fgf8*, two genes that are expressed at or near the midbrain-hindbrain boundary in vertebrates, are not expressed where the *gbx* and *otx* expression domains meet in amphioxus (Castro et al., 2006; Meulemans and Bronner-Fraser, 2007a; Bertrand et al., 2011). Moreover, several markers of the vertebrate midbrain, notably *dmbx* and *pax2/5/8* (i.e., a single amphioxus gene that is homologous to vertebrate *pax2, pax5*, and *pax8*) are not expressed immediately rostral to the *gbx* expression boundary in amphioxus (see Figure 2.23; Kozmik et al., 1999; Takahashi and Holland, 2004). In fact, *dmbx* is not expressed in the central nervous system of amphioxus at all. Moreover, *pax6* is expressed in the forebrain but not the midbrain of lampreys, yet it extends all way to the anterior hindbrain boundary in amphioxus (Suzuki et al., 2015). Additional molecular data suggest that the entire diencephalon, pretectum, and midbrain of vertebrates corresponds to a single, small, and undifferentiated region in the brain of larval amphioxus (Albuixech-Crespo et al., 2017). In light of these data, we conclude that amphioxus lacks a distinct midbrain. Since the dorsal midbrain receives strong retinal projections in vertebrates, this conclusion implies that the central target of the frontal eye in amphioxus is homologous to some other, more rostral brain region (Suzuki et al., 2015).

An interesting related observation is that amphioxus lacks a cerebellum as well as a midbrain. At least, no evidence for a cerebellum in amphioxus has ever been adduced. Given that the midbrain-hindbrain boundary in vertebrates is known to be required for both midbrain and cerebellar development (Martínez et al., 1999; Wurst and Bally-Cuif, 2001), it seems reasonable to speculate that amphioxus lacks a midbrain-hindbrain boundary entirely—or has one that cannot induce a midbrain or cerebellum. This hypothesis is consistent with our broader conclusion that amphioxus brains comprise two large divisions, namely a hindbrain that lies caudal to the cerebral vesicle and a more rostral region that is probably homologous to the vertebrate forebrain (alternatively, it may be some sort of forebrain/midbrain amalgam; see Albuixech-Crespo et al., 2017).

Does the forebrain of amphioxus contain a telencephalon? Again, the gene expression data reveal at least a tentative answer. The gene *foxG1* is expressed at high levels in the embryonic telencephalon of vertebrates, and its homolog is expressed at the anterior pole of the cerebral vesicle in amphioxus, just rostral and ventral to the frontal eye (Figure 2.23; Toresson et al., 1998). This finding suggests that amphioxus might have a telencephalon. However, *foxG1* is also expressed in the mammalian hypothalamus, which is part of the secondary prosencephalon but not included in the telencephalon. Indeed, the observation that amphioxus has a homolog of the vertebrate retina already implies that amphioxus has a secondary prosencephalon; the crucial question is whether, within that region, it possesses a telencephalon. At this point, the answer seems to be negative, mainly because of negative gene expression data. For example, *emx* is selectively expressed in the vertebrate pallium, but it is not expressed in the central nervous system of amphioxus larvae (Tank et al., 2009). According to Albuixech-Crespo et al. (2017), the most rostral region of the larval amphioxus brain is an undifferentiated region homologous to both the hypothalamus and the prethalamus of vertebrates.

Some very recent findings (deposited on an unreviewed pre-print server shortly before this book was accepted: Benito-Gutiérrez et al., 2018) suggest that adult amphioxus may possess a forebrain region resembling the telencephalon of vertebrates insofar as it expresses *foxG1* as well as *emx* and *lhx* homologs. Additional data suggest that this telencephalon-like area in amphioxus develops much later, relative to other brain regions, than the vertebrate telencephalon. If confirmed, these observations would suggest that a telencephalon may have emerged with the origin of chordates and was then lost in tunicates (see later discussion). At this point, however, we consider it more likely that some of the reported similarities are spurious or arose independently in adult amphioxus and embryonic vertebrates.

2.6.2.2. Tunicate Brains

What about tunicates? What do genes reveal about their brain? *Hox1* and *hox3* are both expressed in the visceral ganglion of embryonic ascidians, implying that this region is homologous to the rostral hindbrain of vertebrates (Figure 2.23). In contrast to vertebrates and amphioxus, however, both of these genes in tunicates have the same rostral expression limit (Ikuta et al., 2004). Moreover, *hox2* is not expressed in the nervous system of *Ciona intestinalis*, the most commonly studied ascidian. In general, the *hox* gene family underwent significant changes in the tunicate lineage, losing some members, rearranging their order on the chromosomes (see Figure 2.11), and modifying their spatiotemporal expression patterns (Spagnuolo et al., 2003; Seo et al., 2004). Therefore, we cannot use *hox* gene expression patterns to identify individual subdivisions of the hindbrain (i.e., specific rhombomeres) in tunicates. Still, the visceral ganglion is an excellent candidate for being the hindbrain of larval tunicates.

Just rostral to the visceral ganglion lies the so-called neck region of the larval tunicate brain. In an influential study, Wada et al. (1998) showed that this region in embryonic tunicates expresses *pax2/5/8*, whose vertebrate homologs are widely considered to be good markers for the midbrain-hindbrain boundary, at least during early stages of development. Moreover, the tunicate *en* and *fgf8/17/18* genes are expressed right next to the *pax2/5/8* expression domain, and the caudal expression boundary of the tunicate *otx* gene lines up with the rostral limit of the neck region (Imai et al., 2002). These findings support the hypothesis that the neck region of the tunicate brain is homologous to the midbrain-hindbrain boundary in vertebrates.

However, there are some caveats to this hypothesis. For one thing, *otx* in vertebrates collaborates with *gbx* to establish the location of the midbrain-hindbrain boundary (Millett et al., 1999), but *gbx* was lost in tunicates. Furthermore, in the larvacean tunicate *Oikopleura dioica, pax2/5/8* is expressed only at the rostral end of the sensory vesicle, far from the putative midbrain-hindbrain boundary (Cañestro et al., 2005). Finally, the vertebrate midbrain marker *dmbx* is expressed caudal to the neck region in tunicates (Takahashi, 2005), but rostral to the midbrain-hindbrain boundary in vertebrates. Based on these findings, we suspect that tunicates, like amphioxus, either have no midbrain-hindbrain boundary or have one that lacks the ability to induce midbrain and cerebellar development.

The sensory vesicle of larval tunicates is probably homologous to the forebrain of vertebrates. As noted earlier, it expresses *otx* and contains the light-sensing ocellus, which is probably homologous to the vertebrate retina. In addition, the ventral part of the sensory vesicle expresses *otp, meis, nkx2.1,* and several other genes that are expressed in the hypothalamus of vertebrates (Moret et al., 2005). Although it is remarkable how many hypothalamic markers are expressed in the sensory vesicle, the spatial relationship of their expression domains is somewhat different in tunicates compared to vertebrates, which might explain why vertebrates have two eyes, rather than just one (Moret et al., 2005). What the available gene expression data fail to establish is whether tunicates have a telencephalon. Their sensory vesicle does not seem to express an *fgf8* homolog (Imai et al., 2002), and the expression of *foxG1* has not yet been examined in tunicates. It seems unlikely that tunicates possess anything resembling the late-developing telencephalon-like region reported in amphioxus (Benito-Gutiérrez et al., 2018), but further scrutiny, especially of larvacean brains, seems warranted.

To summarize, the data from amphioxus and tunicates indicate that ancestral invertebrate chordates had a brain that included both a hindbrain and a forebrain. Their forebrain almost certainly included both a diencephalon and a secondary prosencephalon. However, the evidence that they possessed a telencephalon is limited to adult amphioxus and, at this point, should be considered cautiously. Moreover, there is no evidence that invertebrate chordates ever had a cerebellum, and the evidence for them having a midbrain is weak at best. Therefore, it appears that the hindbrain's rostral boundary was modified with the origin of vertebrates so that it became capable of inducing a midbrain rostral to the boundary and cerebellum-like structures caudal to it. Later, with the evolution of jaws, the midbrain-hindbrain boundary was further modified to induce a proper cerebellum in the rostral hindbrain.

2.7. Developmental Mechanisms for Evolving a "New Head"

As the previous sections have shown, the origin of vertebrates entailed a large number of evolutionary innovations, including profound changes in peripheral sensory and motor systems, as well as changes in the brain. Many of these changes are functionally interrelated. For example, the evolution of a midbrain in early vertebrates was probably linked to the evolution of pattern vision, since the midbrain's optic tectum is the major visual sensorimotor area in most anamniotic vertebrates. Similarly, the evolution of a well-developed telencephalon in vertebrates may well be linked to the evolution of an expanded olfactory system in early vertebrates, since the olfactory bulb and its telencephalic projection targets are the principal olfactory centers in vertebrates. Collectively, these changes would have increased the ability of early vertebrates to pursue small prey and to escape from predators. To support this more active life, early vertebrates needed an increased metabolism, which was facilitated by pumping water through the pharynx and the evolution of highly vascularized gills with a large surface area. Complementary changes also occurred in the circulatory system, which evolved a large and chambered heart with valves, a vascular endothelium, and a novel form of hemoglobin (Monahan-Earley et al., 2013; Schwarze et al., 2014).

Just as interesting as the functional linkages between all these evolutionary changes are the developmental relationships. Especially fascinating is that many of the novel vertebrate structures are derived from two embryonic tissues, namely placodes and neural crest, that also have no homologs—or only rudimentary homologs—in amphioxus and tunicates. In the following sections, we review the evolution of these two embryonic tissues and how they contributed to the emergence of novel adult features in vertebrates, especially within the head region. Together with the embryonic mesoderm that gives rise to the pharyngeal muscles (branchiomeres), they "represent a set of embryonic tissues that must have arisen at or close to the transition from protochordates to vertebrates and that produce the structures that prove diagnostic for vertebrates" (Northcutt and Gans, 1983, p. 10). We also discuss what evolutionary changes in development might account for the emergence of the vertebrate telencephalon in early vertebrates.

2.7.1. Placodes

Placodes are thickened patches of non-neural ectoderm that first appear in vertebrate embryos as an arc of tissue adjacent to the anterior end of the central nervous system when it is still flat, rather than tube-shaped (i.e., when it is a neural plate). As development proceeds, this arc of tissue breaks up into multiple placodes, some of which migrate away from their original position (Figure 2.24). Each placode then gives rise to one or more important structures: the lens placode gives rise to the lens of the eye; the olfactory placode gives rise to olfactory sensory neurons and gonadotropin releasing hormone neurosecretory cells, which form part of the terminal nerve (see Appendix; Wray, 2010); the otic placode develops into hair cells of the inner ear and the neurons that innervate them; the lateral line placodes give rise to neuromasts and electroreceptors in the skin, as well as to the neurons that innervate those sensory organs; the epibranchial and trigeminal/profundal placodes give rise to neurons that form the mechanosensory and chemosensory components of several cranial nerves (numbers V, VII, IX, and X; see Appendix);

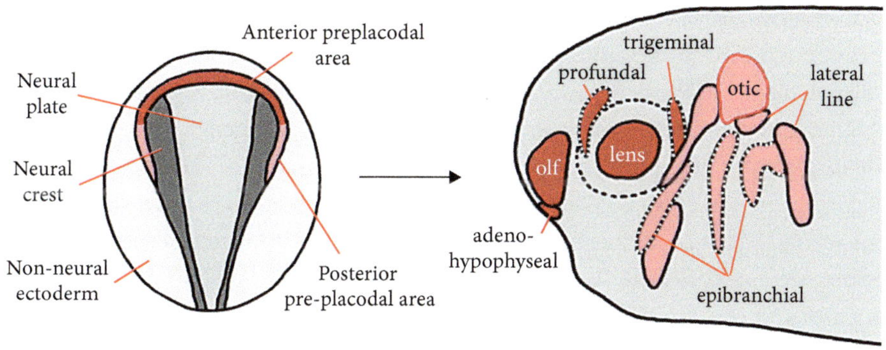

Figure 2.24 Development of vertebrate placodes. Shown at the left is a dorsal view of a very young frog embryo. A horseshoe-shaped band of pre-placodal tissue abuts the anterior end of the neural plate, and its posterior wings lie lateral to the neural crest. As indicated by the color coding, the pre-placodal tissue can be divided into anterior and posterior components. Shown on the right is a lateral view of an older frog embryo (stage 27 of *Xenopus laevis*) in which numerous distinct placodes can be observed as epidermal thickenings. The four lateral line placodes are shaded pink with a solid black outline.

Additional abbreviation: olf – olfactory placode.

Adapted from Schlosser and Northcutt (2000), Park and Saint-Jeannet (2010).

the adenohypophyseal placode, finally, develops into the anterior pituitary gland (Patthey et al., 2014; Schlosser et al., 2014; Sánchez-Arrones et al., 2015).

Since most of the placode-derived adult structures lack definitive homologs in amphioxus and tunicates, it is interesting to ask whether the placodes themselves have homologs in the invertebrate chordates. This question is difficult to answer with purely morphological techniques. However, some members of the *eya* and *six* gene families turn out to be expressed rather selectively in virtually all vertebrate placodes at early stages of embryonic development. Armed with this knowledge, researchers set out to identify homologs of these genes in tunicates and to examine their expression patterns. They found that the *six* and *eya* homologs in tunicates are expressed in an arc-shaped region adjacent to the anterior end of the neural plate, much as they are in vertebrates (Bassham and Postlethwait, 2005). Moreover, markers for anterior and posterior placodes in vertebrates (*pitx* and *pax2/5/8*) are expressed in anterior and posterior parts of the putative tunicate placodes, respectively (Mazet and Shimeld, 2005). Given these similarities, the *six*- and *eya*-positive territory in tunicates is almost certainly homologous to the placodes of vertebrates. However, the placodes in tunicates give rise to a much smaller number and variety of cells, notably a few primary sensory cells in the siphons that probably detect water movements. Thus, placodes did not originate with vertebrates, but early vertebrates greatly increased the number and variety of cell types that derive from them.

In contrast to tunicates, amphioxus does not possess placodes. It has no epidermal thickenings that are morphologically identifiable as placodes, and its *six* and *eva* homologs are not expressed adjacent to the neural plate. Amphioxus does exhibit isolated cells in the epidermis that express *soxB1*, which is expressed in vertebrate placodes, but these cells are not aggregated into a specific region in amphioxus (Meulemans and Bronner-Fraser, 2007b). Moreover, *sox* genes are expressed not only in placodes but also in the neural plate of vertebrates, making *soxB1* a general marker for neural stem cells rather than a good placode marker. In short, amphioxus seems to lack even the rudimentary placodes that tunicates possess. Given this finding, we can conclude that placodes first arose in the last common ancestor of tunicates and vertebrates (collectively referred to as *Olfactores*) and were then elaborated in vertebrates.

2.7.2. Neural Crest

The neural crest comprises a fascinating set of embryonic progenitor cells (Figure 2.25). In the trunk region of early embryos the neural crest occupies the lateral

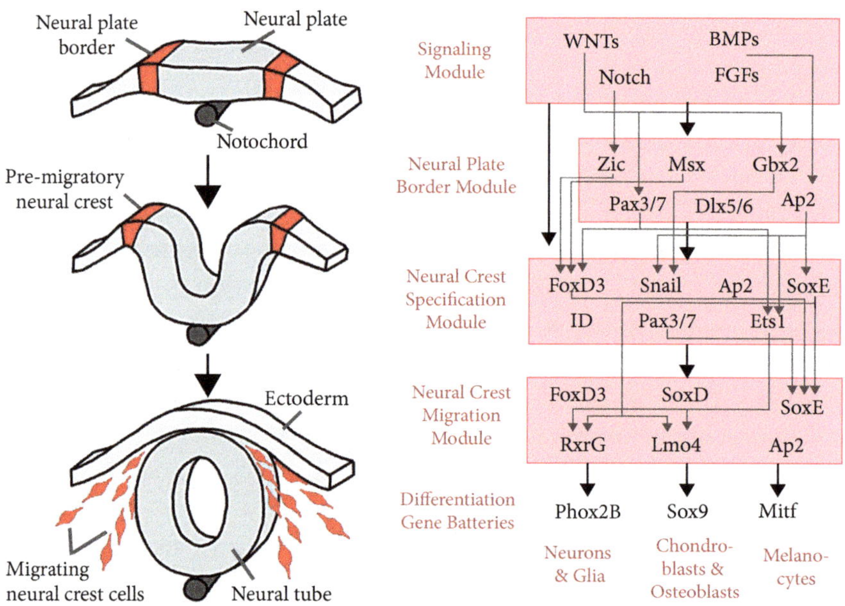

Figure 2.25 Neural crest development. Shown at the left is a series of transverse sections through the neural plate (top), neural fold (middle), and neural tube (bottom) stages of vertebrate neural development. The neural crest is initially located at the lateral border of the neural plate. Once the neural tube has pinched off from the overlying ectoderm, neural crest cells start migrating laterally and ventrally. Shown on the right is a model of the gene regulatory network that controls neural crest development in vertebrates; it is composed of several modules that regulate successive developmental steps.
Adapted from Green et al. (2015), with permission from Springer Nature.

border of the neural plate, just medial to the developing epidermis (i.e., non-neural ectoderm). In the head region, the so-called cephalic neural crest lies sandwiched between the developing brain and the placodes we discussed in the previous section. However, there are no neural crest cells at the rostral edge of the neural plate (Figure 2.24), which means that the placodes in this region directly abut the prospective forebrain. Despite these complexities, we can think of the neural crest as forming initially at the lateral edge of the neural plate. Then, as the neural plate bends upward to form the neural groove, the neural crest cells are carried toward the midline, riding atop the lateral edges of the groove (hence the "crest" in their name). Eventually the neural tissue meets in the midline, converting the neural groove into the neural tube. At that point the neural crest cells separate from their adjacent tissues (Figure 2.25) and begin to migrate to a wide variety of locations, typically dividing several times along the way.

Perhaps the most fascinating aspect of the neural crest is that it gives rise or contributes to a wide variety of tissues. These include epidermal pigment cells, the cranial skeleton, facial skin and connective tissue, the meninges surrounding the forebrain, the Schwann cells that produce myelin and ensheath most peripheral nerves, the sensory ganglia of spinal nerves, the non-placodal parts of cranial nerve ganglia (see Appendix), the autonomic ganglia and enteric nervous system, the adrenal medulla, the heart, and the gills. As evident from this long list, many of the tissues that receive a major developmental contribution from the neural crest are vertebrate innovations (Gans and Northcutt, 1983; Northcutt and Gans, 1983). Therefore, it is important to ask whether the neural crest itself is a vertebrate innovation.

Tunicates do seem to have some sort of neural crest, because a few cells laterally adjacent to the neural plate of tunicates express several genes that are involved in specifying the neural crest of lampreys and other vertebrates, including *snail, id, foxD*, and *etx* (Sauka-Spengler et al., 2007). Some of these putative neural crest cells give rise to pigment cells in the gravity-sensing otolith and the light-sensing ocellus (Abitua et al., 2012). This is important because pigment cells are one of the cell types derived from neural crest in vertebrates. However, the pigment cell precursors in tunicates differ from vertebrate neural crest cells in that they do not migrate far; nor do they multiply along the way. The tunicate cells exhibit this migratory phenotype only when they are experimentally induced to express *twist*, another vertebrate neural crest gene (Abitua et al., 2012).

These findings suggest that tunicates possess a neural crest and that vertebrates modified these cells in such a way that they migrate further and give rise to a greater variety of cells. However, tunicates also possess a more caudal population of cells that express neural crest marker genes, migrate a short distance through trunk mesoderm, and eventually differentiate into neurons that make contact with sensory and motor structures in the tail (Stolfi et al., 2015). The existence of these cells suggests that some neural crest cells in the last common ancestor of tunicates and vertebrates were already capable of migrating and able to give rise to multiple adult cell types. Still, tunicate neural crest cells do not exhibit the full range of genes expressed in vertebrate neural crest and do not generate as many distinct cell types as they do in vertebrates (Green et al., 2015).

Amphioxus does not appear to have a neural crest at all. Its genome does contain homologs of most of the genes that are involved in specifying the vertebrate neural crest (Yu, 2010), but most of these genes are not expressed in the neural plate border region. The sole exception is that *snail* is expressed in a few cells along the neural plate border. However, these cells do not express any of the other neural crest marker genes examined so far and do not migrate actively to their adult location.

Collectively, these data indicate that the neural crest evolved not all at once but in a series of steps near the origin of vertebrates. Most of the 50 or more genes that are involved in specifying the vertebrate neural crest existed long before the origin of vertebrates, but the interactions between these genes were modified as vertebrates evolved (Figure 2.25). As the neural crest gene regulatory network gradually emerged in evolution, cells at the neural plate border acquired new behavioral features, such as the ability to migrate over long distances and give rise to a greater variety of adult cell types.

Thus, evolutionary changes in the neural crest equipped the early vertebrates with an array of features that, in concert with the emergence of placodes and muscles capable of pumping water through the pharynx, were pivotal to the success of vertebrates. By evolving novel embryonic tissues capable of generating novel adult structures, early vertebrates and their immediate ancestors managed to evolve "a new head" (Gans and Northcutt, 1983). Of course, those ancestors already had a head, and the new embryonic tissues give rise to some new structures caudal to the head (e.g., the autonomic nervous system). Nonetheless, the early vertebrates transformed their head so much that calling it "new" seems fair. Since its inception, the new head hypothesis has stimulated a lot of new research on placode and neural crest evolution. Particularly interesting is the discovery of rudimentary placodes and neural crest in tunicates, but not amphioxus, which supports the recently established consensus that tunicates are the closest living relatives of vertebrates.

2.7.3. Developing a Telencephalon

Although placodes and neural crest are causally linked to the evolution of many different vertebrate features, the novel features of vertebrate brains are not so easily explained, because both placodes and the neural crest give rise to tissues that lie almost entirely outside the brain. Still, the evolution of the vertebrate telencephalon does involve both placodes and the neural crest, at least to some extent. To understand this causal nexus, it helps to know exactly which portion of the embryonic neural plate gives rise to the vertebrate telencephalon.

To that end, fate maps of the neural plate have been established in several vertebrate species, with largely concordant results (Couly and Le Douarin, 1987; Eagleson and Harris, 1990; Cobos et al., 2001; Garcia-Lopez et al., 2009). As shown in Figure 2.26, the vertebrate telencephalon at the neural plate stage curves anterolaterally around the tissue that gives rise to the retina and hypothalamus. Its topologically rostral portion becomes the subpallium (mainly striatum and pallidum), whereas its caudolateral portion develops into the pallium (e.g., the cerebral cortex of mammals). Eventually, as the neural plate rolls up into a tube, the telencephalon assumes a position dorsal to the retina and hypothalamus (Figure 2.26). Within the telencephalon, the pallium assumes the most dorsal position.

For our purposes, the most important aspect of these fate mapping results is that the telencephalon's precursor cells are located at the rostrolateral edge of the neural plate, precisely where one would expect to find the neural crest, if it extended around the neural plate's anterior pole. The cells in this region do not migrate long distances and, therefore, do not fulfill a defining criterion for being neural crest. Still, this observation raises the interesting possibility that at least some of the telencephalon's

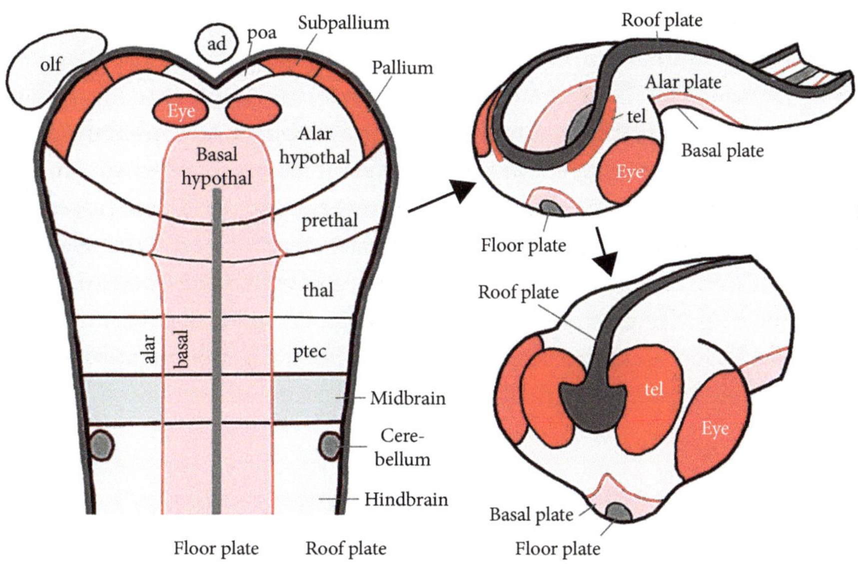

Figure 2.26 Fate maps of the vertebrate forebrain. Shown on the left is a diagram of the neural plate (dorsal view; anterior to the top), indicating which precursor regions will develop into which adult structures. Also shown are the olfactory and adenohypophyseal placodes (olf and ad, respectively). Shown at the right are later stages, shortly before and shortly after anterior neural tube closure. The color coding matches that of the neural plate fate map.

Additional abbreviations: poa – preoptic area; ptec – pretectum; prethal – prethalamus; thal – thalamus. Adapted from Puelles (2001, 2013), with permission from Elsevier.

precursor cells are modified neural crest cells that lack the migratory drive. Some support for this hypothesis derives from the observation that cells from the rostral edge of the neural plate will migrate if they are transplanted into more posterior regions where the neural crest normally originates (Ezin et al., 2014). Thus, one might explain the vertebrate telencephalon as yet another novel derivative of the neural crest. Of course, this hypothesis would have to be modified if amphioxus is confirmed to have a homolog of the vertebrate telencephalon but lack the neural crest. In any case, telencephalon development and evolution are almost certainly a complicated affair.

Particularly interesting is the finding that ablations of the more posterior neural crest prevent the telencephalon's normal development (Creuzet et al., 2006). It appears that some of these posterior neural crest cells migrate rostrally and then induce non-neural ectoderm just rostral to the neural plate—in the anterior neural ridge—to secrete FGF8, a protein that is necessary for normal forebrain development and sufficient to induce telencephalon-specific genes (e.g., *foxG1*) in parts of the brain that do not normally develop into a telencephalon (Shimamura and Rubenstein, 1997; Houart et al., 1998; Cajal et al., 2014). Another interesting observation is that experimental ablation of the olfactory placode greatly impairs telencephalon development, especially in the region of the olfactory bulb. This dependency seems to be mediated by sensory axons growing out of the olfactory epithelium into the presumptive olfactory bulb (Graziadei and Monti-Graziadei, 1992).

Taken together, these findings imply that the development of the vertebrate telencephalon requires the confluence of several different factors that derive from several different tissues, including the cephalic neural crest, the anterior neural ridge, and the olfactory placode. Importantly, many of the tissues and molecules involved in this development seem to have existed before the origin of vertebrates and, thus, have invertebrate homologs (Pani et al., 2012). In general, we conclude that the novel features of vertebrates arose mainly because preexisting elements started to interact in novel ways. This thought deserves some elaboration; hence the next section.

2.8. The Question of Novelty in Evolution

A major challenge in evolutionary neurobiology is to distinguish between what is old and what is new. As we note repeatedly throughout this book—especially in this chapter—evolution does periodically give rise to new morphological structures, to novel physiological functions and behaviors, and even to new genes. However, those novel traits do not appear entirely out of nowhere; they have some evolutionary precursors. Yet one may ask: how can something that has evolutionary precursors be "new"? The key to solving this puzzle is to realize that biological organization is hierarchical and that questions about what is old and what is new can be asked and answered at multiple levels.

2.8.1. Levels of Homology

As we discussed in Chapter 1, homologous characters are those that can be traced back, along a continuous history, to a single ancestral character. Whether two or more characters are homologous to one another is usually determined by reconstructing the character's evolutionary history from its phylogenetic distribution across taxa. This procedure is not controversial among comparative biologists, but how do we define a character in the first place? This question is surprisingly difficult to answer (Striedter, 1998; Wagner, 2000, 2007). Part of the problem is that comparative biologists routinely work with a wide range of different characters, ranging from genes and proteins to morphological structures, physiological processes, and even complex behaviors. Each type of character requires a different kind of definition, based on different sets of attributes (e.g., nucleotide sequence, gene expression, relative position, cellular interactions, participation in some physiological process).

Moreover, the different kinds of characters are hierarchically organized. One way to think about this hierarchy is to recognize three distinct levels of biological organization: molecules, morphological structures, and behaviors (Figure 2.27). As we have pointed out (Striedter and Northcutt, 1991), homologies can be identified at each of these levels, but homology at one level need not imply homology at the other levels. Thus, two morphological structures can be homologous to one another, but not all of the genes expressed in those structures (nor all of the cell types) need be homologous, and they may well participate in non-homologous behaviors. The neural crest of tunicates, for example, is homologous to that of vertebrates, even

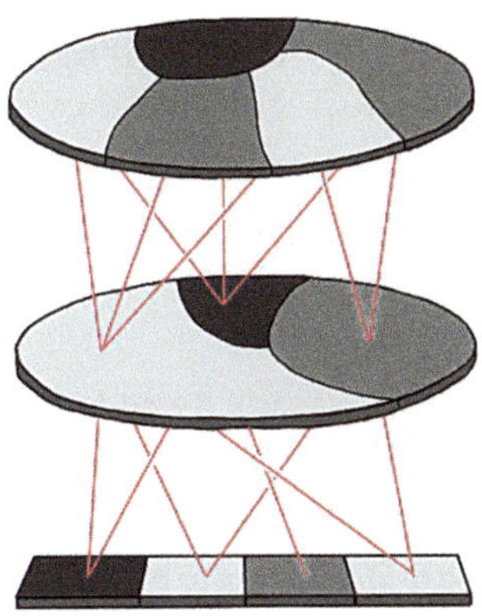

Figure 2.27 Levels of organization and homology. Organisms are hierarchically organized, meaning that they operate at several different, causally interacting levels of biology. Features for comparative analyses can be identified at all of these levels. However, because there are very few one-to-one correspondences between characters at different levels of organization, homologies at one level need not imply homologies at other levels.

Adapted from Striedter and Northcutt (1991).

though it does not express all of the genes that are expressed in vertebrate neural crest. The difference just means that numerous additional genes were recruited into the "neural crest regulatory gene network" as vertebrates evolved (Green et al., 2015). Furthermore, a gene that was important for developing a morphological character in an ancestral species may become unimportant or even lost entirely without imperiling the structural homologies. For example, the gene *gbx* is a good hindbrain marker in vertebrates and amphioxus, but it was lost in tunicates (Wada et al., 2003); few would use this loss to argue that larval tunicates lack a hindbrain.

Gene expression patterns do help us to identify and homologize morphological characters, but the genes that are involved in patterning and specifying a particular region of embryonic tissue may come and go in evolution (Striedter, 1998). The downside of this dynamism is that using the expression of a single conserved gene to homologize morphological structures is ill-advised. For example, *krox20* is expressed in hindbrain rhombomeres 3 and 5 in vertebrates, but its homolog in amphioxus is expressed in the anterior forebrain (Knight et al., 2000). Apparently this gene has changed its expression pattern at least once during chordate evolution. Therefore, arguments of structural homology should be based on the expression patterns of many genes (ideally a unique set of interacting genes and proteins; see Chapter 1, Section 1.3.3), and on additional criteria, such as relative position (i.e., topology), neuronal connectivity, and physiology. For example, the argument that tunicates have a neural crest homolog is based on the expression of several genes, on the location of the candidate cells next to the neural plate, and on their migratory phenotype (Stolfi et al., 2015). In short, seekers of morphological homologies must be eclectic and flexible in their search for conserved attributes, and they must be careful not to confuse the homology of genes with the homology of the morphological structures in which those genes are expressed (see Chapter 1).

Hierarchies also exist within the molecular, morphological, and behavioral levels of biology. Focusing on the morphological level, we know that the brain consists of several major divisions that can themselves be subdivided, and that each brain area typically contains multiple cell types. Again, homology at one of these levels need not imply homology at the other levels. For example, the fact that amphioxus lacks hindbrain neurons innervating eye muscles (which it does not possess) does not imply that the hindbrain of amphioxus is not homologous to that of vertebrates (as a hindbrain). Conversely, the homology of individual cells need not imply that the areas in which those cells are located are necessarily homologous (see Section 1.3.3). For example, the hypothesis that hair cells are homologous (as hair cells) across amphioxus and vertebrates need not imply that amphioxus has an inner ear that is homologous to that of vertebrates. Hair cells are found in diverse multicellular organs, and determining homologies among those organs requires comparisons at the organ level, though it clearly helps to know what kinds of cells constitute those organs and whether those cells are homologous to one another.

Researchers sometimes try to resolve these problems by comparing the developmental origins of morphological structures, arguing that structures that derive from homologous precursors must be homologous (see Chapter 1). Clearly, data on embryonic origins have helped identify many putative homologies, both in general morphology and more specifically in comparative neuroanatomy. One must be careful, however, not to conflate the homology of embryonic precursor regions with the homology of their adult derivatives. For example, the finding that pigment cells in tunicates and the enteric nervous system of vertebrates are both derived from neural crest does not make them homologous to one another (e.g., as pigment cells). Similarly, the observation that the vertebrate midbrain develops from the neural plate, which is common to all chordates, need not imply that all chordates possess a midbrain. Saying that the neural plate of invertebrate chordates is a "field homolog" of the vertebrate midbrain would provide no useful phylogenetic information beyond stating that the neural plate is homologous across chordates (Northcutt, 1999).

In short, the determination of homologies requires an examination of many different features, which must then be analyzed in a phylogenetic and hierarchical framework to determine whether the putative homologs can be traced back (along a continuous history) to a common ancestor. Because this search for homologs can be quite challenging, it is easy to forget that evolution does not deal only in homologies; it also creates real novelties—characters that have no homologs in other species.

2.8.2. Identifying Novelties

The main reason why homologies can be so difficult to ascertain is that characters may change in any of their attributes without losing their identity—that is, without losing their homology to other characters (Striedter, 1998). But if this is the case, how can we ever decide that a specific character is truly new (Wagner and Lynch, 2010; Peterson and Müller, 2016), rather than merely modified?

To answer this question, consider gene evolution. As mentioned earlier, genes often duplicate in evolution, typically by unequal crossing-over or the duplication of entire chromosomes (Ohno, 1970; Zhang, 2003). Immediately after such a duplication event, the duplicated genes are both homologous to the single ancestral gene. As comparative genome biologists would say, the duplicated genes are orthologous to the ancestral gene; in contrast, the duplicated genes are said to be paralogous to one another (Fitch, 1970). So far so good, but what happens when one of the two duplicated genes changes its nucleotide sequence and cellular functions dramatically, presumably with little ill effect because the unchanged gene suffices to perform the old functions. Once that happens, many biologists will say that the unchanged gene remains as the only ortholog or, as Walter Fitch proposed, the "isortholog" (Fitch, 2000). Considerable debate rages about the correct usage of these terms (Jensen, 2001), but for our purposes it is most interesting to note that the highly modified copy of the duplicated gene tends to be neglected in these discussions. In a way, the highly modified gene is more interesting than the conserved one, because it unlocks new functionalities and evolutionary potential. Indeed, one might well call it a "new gene."

Morphological structures can also duplicate in evolution (Allman and Kaas, 1974; Kaas, 1995), although it is more common for them to change substantially without duplicating. Still, a similar problem remains: how much change do we allow before we call a structure "new"? In one sense, evolution never creates anything that is entirely new. Because new characters always arise by the modification or duplication of ancestral developmental pathways, they can always be traced back to ancestral developmental precursors and genes. As Braun and Northcutt put it in 1997: "Morphological structures, and the ontogenies which produce them, do not simply arise from the dust of the earth" (Braun and Northcutt, 1997, p. 263). Or, from a more molecular perspective: "Novelties come from previously unseen association of old material. To create is to recombine" (Jacob, 1977, p. 1163). Just as an author can write a truly novel book by using preexisting letters and words, so evolution can produce truly novel characters by using more ancient elements in novel combinations. As this chapter showed, the origin of vertebrates indeed involves a wide variety of substantial innovations; recognizing them as such seems important to us.

To illustrate the point, consider the telencephalon. As we discussed, *foxG1* is expressed in the vertebrate telencephalon and in the forebrain of amphioxus. However, vertebrates express this gene not only in the telencephalon, but also in the hypothalamus and the preoptic area (i.e., in other components of the secondary prosencephalon; see Figure 2.19). Therefore, the similarities in the expression patterns of this gene can at best be used to argue that amphioxus has a secondary prosencephalon, which includes the telencephalon as well as the hypothalamus and preoptic region. Given the lack of additional evidence for a telencephalon in amphioxus and tunicates, we argue that the telencephalon is a vertebrate innovation. This conclusion is at odds with gene expression studies claiming that insects and annelid worms have homologs of the vertebrate telencephalon (Tomer et al., 2010; Strausfeld and Hirth, 2013b). However, these reports are based on a rather selective view of the evidence and overemphasize similarities at the expense of differences (Farries, 2013). Thus, for now, we remain convinced that the telencephalon emerged with the origin of vertebrates. Recognizing this novelty raises important questions about how that novelty arose—questions that would not be asked if the novelty remains unrecognized.

The quest for understanding the mechanistic origins of vertebrate novelties remains in its infancy. However, the two rounds of genome duplications near the origin of vertebrates were almost certainly a very important factor. Particularly interesting is the finding that many of the duplicated genes are expressed in vertebrate but not invertebrate brains, suggesting that they were recruited into brain development and neural function in the vertebrate lineage. Even more interesting is that "more than 50% and 30% [of the] duplicate genes are expressed in the telencephalon and mid-hindbrain boundary, respectively" (Chen et al., 2011, p. 577). These findings suggest that many duplicated genes changed their expression patterns during phylogeny (compared to their orthologs) and were directly involved in creating novel brain traits. Of course, morphological or physiological innovations may also occur without gene duplication, as even unduplicated genes may alter their functions by changing their protein coding or regulatory sequences over evolutionary time. In this context, it is interesting that the number of regulatory micro-RNA families greatly increased in the chordate lineage after it diverged from hemichordates (Erwin et al., 2011). In short, major changes in the genomes of early chordates and early vertebrates probably facilitated the emergence of most vertebrate innovations.

References

Abitua PB, Wagner E, Navarrete IA, Levine M. 2012. Identification of a rudimentary neural crest in a non-vertebrate chordate. *Nature* 492:104–107.

Albertin CB, Ragsdale CW. 2018. More than one way to a central nervous system. *Nature* 553:34–36.

Albuixech-Crespo B, López-Blanch L, Burguera D, Maeso I, Sánchez-Arrones L, Moreno-Bravo JA, Somorjai I, Pascual-Anaya J, Puelles E, Bovolenta P, Garcia-Fernàndez J, Puelles L, Irimia M, Ferran JL. 2017. Molecular regionalization of the developing amphioxus neural tube challenges major partitions of the vertebrate brain. *Plos Biol* 15:e2001573.

Allman JM, Kaas JH. 1974. A crescent-shaped cortical visual area surrounding the middle temporal area (MT) in the owl monkey (*Aotus trivirgatus*). *Brain Res* 81:199–213.

Arendt D. 2018. Animal evolution: convergent nerve cords? *Curr Biol* 28:R225–R227.

Arendt D, Tessmar-Raible K, Snyman H, Dorresteijn AW, Wittbrodt J. 2004. Ciliary photoreceptors with a vertebrate-type opsin in an invertebrate brain. *Science* 306:869–871.

Asteriti S, Grillner S, Cangiano L. 2015. A Cambrian origin for vertebrate rods. *Elife* 4: e07166.

Aszódi A, Chan D, Hunziker E, Bateman JF, Fässler R. 1998. Collagen II is essential for the removal of the notochord and the formation of intervertebral discs. *J Cell Biol* 143:1399–1412.

Baker CVH, Modrell MS, Gillis JA. 2013. The evolution and development of vertebrate lateral line electroreceptors. *J Exp Biol* 216:2515–2522.

Baker GE, de Grip WJ, Turton M, Wagner H-J, Foster RG, Douglas RH. 2015. Light sensitivity in a vertebrate mechanoreceptor? *J Exp Biol* 218:2826–2829.

Barreiro-Iglesias A, Anadón R, Rodicio MC. 2010. The gustatory system of lampreys. *Brain Behav Evol* 75:241–250.

Bassham S, Postlethwait JH. 2005. The evolutionary history of placodes: a molecular genetic investigation of the larvacean urochordate *Oikopleura dioica*. *Development* 132:4259–4272.

Benito-Gutiérrez E, Stemmer M, Rohr SD, Schuhmacher LN, Tang J, Marconi A, Jékely G, Arendt D. 2018. Patterning of a telencephalon-like region in the adult brain of amphioxus. *bioRxiv* 307629, doi: 10.1101/307629.

Bertrand S, Camasses A, Somorjai I, Belgacem MR, Chabrol O, Escande M-L, Pontarotti P, Escriva H. 2011. Amphioxus FGF signaling predicts the acquisition of vertebrate morphological traits. *Proc Natl Acad Sci USA* 108:9160–9165.

Braun CB. 1998. Schreiner organs: a new craniate chemosensory modality in hagfishes. *J Comp Neurol* 392:135–163.

Braun CB, Northcutt RG. 1997. The lateral line system of hagfishes (Craniata: Myxinoidea). *Acta Zoologica* 78:247–268.

Bullock TH, Hopkins CD, Popper AN, Fay RR (eds.). 2006. *Electroreception*. New York: John Wiley & Sons.

Burighel P, Caicci F, Manni L. 2011. Hair cells in non-vertebrate models: lower chordates and molluscs. *Hear Res* 273:14–24.

Burzin MB, Debrenne F, Zhuravlev AY. 2001. Evolution of shallow-water level-bottom communities. In: Zhuravlev A, Riding A (eds.), *The Ecology of the Cambrian Radiation*. New York: Columbia University Press, pp. 217–237.

Cajal M, Creuzet SE, Papanayotou C, Saberan-Djoneidi D, Chuva de Sousa Lopes SM, Zwijsen A, Collignon J, Camus A. 2014. A conserved role for non-neural ectoderm cells in early neural development. *Development* 141:4127–4138.

Candiani S, Moronti L, Ramoino P, Schubert M, Pestarino M. 2012. A neurochemical map of the developing amphioxus nervous system. *BMC Neurosci* 13:59.

Cañestro C, Bassham S, Postlethwait J. 2005. Development of the central nervous system in the larvacean *Oikopleura dioica* and the evolution of the chordate brain. *Dev Biol* 285:298–315.

Canfield DE, Poulton SW, Narbonne GM. 2007. Late-Neoproterozoic deep-ocean oxygenation and the rise of animal life. *Science* 315:92–95.

Castro LFC, Rasmussen SLK, Holland PWH, Holland ND, Holland LZ. 2006. A Gbx homeobox gene in amphioxus: Insights into ancestry of the ANTP class and evolution of the midbrain/hindbrain boundary. *Dev Biol* 295:40–51.

Chen Y, Ding Y, Zhang Z, Wang W, Chen J-Y, Ueno N, Mao B. 2011. Evolution of vertebrate central nervous system is accompanied by novel expression changes of duplicate genes. *J Genet Genomics* 38:577–584.

Churcher AM, Taylor JS. 2009. Amphioxus (*Branchiostoma floridae*) has orthologs of vertebrate odorant receptors. *BMC Evol Biol* 9:242.

Churcher AM, Taylor JS. 2011. The antiquity of chordate odorant receptors is revealed by the discovery of orthologs in the cnidarian *Nematostella vectensis*. *Genome Biol Evol* 3:36–43.

Clark AJ, Summers AP. 2007. Morphology and kinematics of feeding in hagfish: possible functional advantages of jaws. *J Exp Biol* 210:3897–3909.

Clark AJ, Summers AP. 2012. Ontogenetic scaling of the morphology and biomechanics of the feeding apparatus in the Pacific hagfish *Eptatretus stoutii*. *J Fish Biol* 80:86–99.

Cobos I, Shimamura K, Rubenstein JL, Martínez S, Puelles L. 2001. Fate map of the avian anterior forebrain at the four-somite stage, based on the analysis of quail-chick chimeras. *Dev Biol* 239:46–67.

Collin SP, Davies WL, Hart NS, Hunt DM. 2009. The evolution of early vertebrate photoreceptors. *Philos Trans R Soc Lond B* 364:2925–2940.

Cong P-Y, Hou X-G, Aldridge RJ, Purnell MA, Li Y-Z. 2014. New data on the palaeobiology of the enigmatic yunnanozoans from the Chengjiang Biota, Lower Cambrian, China. *Pap Palaeontol* 58:45–70.

Couly GF, Le Douarin NM. 1987. Mapping of the early neural primordium in quail-chick chimeras. II. The prosencephalic neural plate and neural folds: implications for the genesis of cephalic human congenital abnormalities. *Dev Biol* 120:198–214.

Creuzet SE, Martinez S, Le Douarin NM. 2006. The cephalic neural crest exerts a critical effect on forebrain and midbrain development. *Proc Natl Acad Sci USA* 103:14033–14038.

Delsuc F, Tsagkogeorga G, Lartillot N, Philippe H. 2008. Additional molecular support for the new chordate phylogeny. *genesis* 46:592–604.

Denes AS, Jékely G, Steinmetz PRH, Raible F, Snyman H, Prud'homme B, Ferrier DEK, Balavoine G, Arendt D. 2007. Molecular architecture of annelid nerve cord supports common origin of nervous system centralization in bilateria. *Cell* 129:277–288.

Dilly PN. 1962. Studies on the receptors in the cerebral vesicle of the ascidian tadpole. I. The otolith. *Quart J Micr Sci* 103:393–398.

Donoghue PCJ, Keating JN. 2014. Early vertebrate evolution. *Palaeontology* 57:879–893.

Donoghue PCJ, Purnell MA. 2005. Genome duplication, extinction and vertebrate evolution. *Trends Ecol Evolut* 20:312–319.

Duboule D. 2007. The rise and fall of Hox gene clusters. *Development* 134:2549–2560.

Dufour HD, Chettouh Z, Deyts C, de Rosa R, Goridis C, Joly J-S, Brunet J-F. 2006. Precraniate origin of cranial motoneurons. *Proc Natl Acad Sci USA* 103:8727–8732.
Duncan JS, Fritzsch B. 2012. Evolution of sound and balance perception: Innovations that aggregate single hair cells into the ear and transform a gravistatic sensor into the organ of Corti. *Anat Rec* 295:1760–1774.
Dunn CW, Giribet G, Edgecombe GD, Hejnol A. 2014. Animal phylogeny and its evolutionary implications. *Annu Rev Ecol Evol Syst* 45:371–395.
Eagleson GW, Harris WA. 1990. Mapping of the presumptive brain regions in the neural plate of *Xenopus laevis*. *J Neurobiol* 21:427–440.
Erwin DH, Laflamme M, Tweedt SM, Sperling EA, Pisani D, Peterson KJ. 2011. The Cambrian conundrum: early divergence and later ecological success in the early history of animals. *Science* 334:1091–1097.
Evangelista C, Mills M, Siebeck UE, Collin SP. 2010. A comparison of the external morphology of the membranous inner ear in elasmobranchs. *J Morphol* 271:483–495.
Ezin M, Barembaum M, Bronner ME. 2014. Stage-dependent plasticity of the anterior neural folds to form neural crest. *Differentiation* 88:42–50.
Farries MA. 2013. How "basal"are the basal ganglia. *Brain Behav Evol* 82:211–214.
Finger TE. 2009. The evolution of taste systems. In: Kaas JH (ed.), Evolutionary *Neuroscience*. San Diego, CA: Academic Press, pp. 459–478.
Fitch WM. 1970. Distinguishing homologous from analogous proteins. *Syst Zool* 19:99–113.
Fitch WM. 2000. Homology: a personal view on some of the problems. *Trends Genet* 16:227–231.
Gans C, Northcutt RG. 1983. Neural crest and the origin of vertebrates: a new head. *Science* 220:268–273.
Garcia-Lopez R, Pombero A, Martinez S. 2009. Fate map of the chick embryo neural tube. *Dev Growth Differ* 51:145–165.
Gess RW, Coates MI, Rubidge BS. 2006. A lamprey from the Devonian period of South Africa. *Nature* 443:981–984.
Gionti M, Ristoratore F, Di Gregorio A, Aniello F, Branno M, Di Lauro R. 1998. Cihox5, a new *Ciona intestinalis* Hox-related gene, is involved in regionalization of the spinal cord. *Dev Genes Evol* 207:515–523.
Gooley JJ, Lu J, Fischer D, Saper CB. 2003. A broad role for melanopsin in nonvisual photoreception. *J Neurosci* 23:7093–7106.
Goudemand N, Orchard MJ, Urdy S, Bucher H, Tafforeau P. 2011. Synchrotron-aided reconstruction of the conodont feeding apparatus and implications for the mouth of the first vertebrates. *Proc Natl Acad Sci USA* 108:8720–8724.
Graziadei PP, Monti-Graziadei AG. 1992. The influence of the olfactory placode on the development of the telencephalon in *Xenopus laevis*. *Neuroscience* 46:617–629.
Green SA, Simoes-Costa M, Bronner ME. 2015. Evolution of vertebrates as viewed from the crest. *Nature* 520:474–482.
Hashiguchi Y, Nishida M. 2007. Evolution of trace amine associated receptor (TAAR) gene family in vertebrates: lineage-specific expansions and degradations of a second class of vertebrate chemosensory receptors expressed in the olfactory epithelium. *Mol Biol Evol* 24:2099–2107.
Heimberg AM, Cowper-Sal-lari R, Sémon M, Donoghue PCJ, Peterson KJ. 2010. microRNAs reveal the interrelationships of hagfish, lampreys, and gnathostomes and the nature of the ancestral vertebrate. *Proc Natl Acad Sci USA* 107:19379–19383.
Hilliard RW, Potter IC, Macey DJ. 1985. The dentition and feeding mechanism in adults of the southern hemisphere lamprey *Geotria australis* Gray. *Acta Zool Stockh* 66:159–170.
Hirasawa T, Oisi Y, Kuratani S. 2016. *Palaeospondylus* as a primitive hagfish. *Zool Lett* 2:20.
Hirth F. 2010. On the origin and evolution of the tripartite brain. *Brain Behav Evol* 76:3–10.
Holland LZ, Carvalho JOE, Escriva H, Laudet V, Schubert M, Shimeld SM, Yu J-K. 2013. Evolution of bilaterian central nervous systems: a single origin? *EvoDevo* 4:27.
Holland ND, Holland LZ. 2017. The ups and downs of amphioxus biology: a history. *Int J Dev Biol* 61:575–583.

Holland PW, Holland LZ, Williams NA, Holland ND. 1992. An amphioxus homeobox gene: sequence conservation, spatial expression during development and insights into vertebrate evolution. *Development* 116:653–661.

Houart C, Westerfield M, Wilson SW. 1998. A small population of anterior cells patterns the forebrain during zebrafish gastrulation. *Nature* 391:788–792.

Ikuta T, Yoshida N, Satoh N, Saiga H. 2004. *Ciona intestinalis* Hox gene cluster: its dispersed structure and residual colinear expression in development. *Proc Natl Acad Sci USA* 101:15118–15123.

Imai JH, Meinertzhagen IA. 2007. Neurons of the ascidian larval nervous system in *Ciona intestinalis*: I. Central nervous system. *J Comp Neurol* 501:316–334.

Imai KS, Satoh N, Satou Y. 2002. Region specific gene expressions in the central nervous system of the ascidian embryo. *Mech Dev* 119 (Suppl):S275–S277.

Jacob F. 1977. Evolution and tinkering. *Science* 196:1161–1166.

Jaillon O, Aury J-M, Wincker P. 2009. "Changing by doubling," the impact of Whole Genome Duplications in the evolution of eukaryotes. *C R Biol* 332:241–253.

Janvier P. 1981. The phylogeny of the Craniata, with particular reference to the significance of fossil "agnathans." *J Vert Paleontol* 1:121–159.

Janvier P. 2015. Facts and fancies about early fossil chordates and vertebrates. *Nature* 520:483–489.

Jensen RA. 2001. Orthologs and paralogs—we need to get it right. *Genome Biol* 2B:1002.1–1002.3.

Kaas JH. 1995. The evolution of isocortex. *Brain Behav Evol* 46:187–196.

King B, Hu Y, Long JA. 2018. Electroreception in early vertebrates: survey, evidence and new information. *Palaeontol* 61:325–358.

Kinkead R. 2009. Phylogenetic trends in respiratory rhythmogenesis: Insights from ectothermic vertebrates. *Respir Physiol Neurobiol* 168:39–48.

Kirino M, Parnes J, Hansen A, Kiyohara S, Finger TE. 2013. Evolutionary origins of taste buds: phylogenetic analysis of purinergic neurotransmission in epithelial chemosensors. *Open Biol* 3:130015.

Knight RD, Panopoulou GD, Holland PW, Shimeld SM. 2000. An amphioxus Krox gene: insights into vertebrate hindbrain evolution. *Dev Genes Evol* 210:518–521.

Kocot KM, Cannon JT, Todt C, Citarella MR, Kohn AB, Meyer A, Santos SR, Schander C, Moroz LL, Lieb B, Halanych KM. 2011. Phylogenomics reveals deep molluscan relationships. *Nature* 477:452–456.

Kozmik Z, Holland ND, Kalousova A, Paces J, Schubert M, Holland LZ. 1999. Characterization of an amphioxus paired box gene, *AmphiPax2/5/8*: developmental expression patterns in optic support cells, nephridium, thyroid-like structures and pharyngeal gill slits, but not in the midbrain-hindbrain boundary region. *Development* 126:1295–1304.

Kreneisz O, Glover JC. 2015. Developmental characterization of tail movements in the appendicularian urochordate *Oikopleura dioica*. *Brain Behav Evol* 86:191–209.

Krug L, Wicht H, Northcutt RG. 1993. Afferent and efferent connections of the thalamic eminence in the axolotl, *Ambystoma mexicanum*. *Neurosci Lett* 149:145–148.

Kusakabe T, Kusakabe R, Kawakami I, Satou Y, Satoh N, Tsuda M. 2001. *Ci-opsin1*, a vertebrate-type opsin gene, expressed in the larval ocellus of the ascidian *Ciona intestinalis*. *FEBS Lett* 506:69–72.

Lacalli TC. 1996. Frontal eye circuitry, rostral sensory pathways and brain organization in amphioxus larvae: evidence from 3D reconstructions. *Phil Trans R Soc Lond B* 351:243–263.

Lacalli TC. 2004. Sensory systems in amphioxus: a window on the ancestral chordate condition. *Brain Behav Evol* 64:148–162.

Lacalli TC. 2012. The Middle Cambrian fossil *Pikaia* and the evolution of chordate swimming. *EvoDevo* 3:12.

Lamb TD. 2013. Evolution of phototransduction, vertebrate photoreceptors and retina. *Prog Retin Eye Res* 36:52–119.

Lannoo MJ, Hawkes R. 1997. A search for primitive Purkinje cells: zebrin II expression in sea lampreys (*Petromyzon marinus*). Neurosci Lett 237:53–55.

Lemons D, McGinnis W. 2006. Genomic evolution of Hox gene clusters. *Science* 313:1918–1922.

Libants S, Carr K, Wu H, Teeter JH, Chung-Davidson Y-W, Zhang Z, Wilkerson C, Li W. 2009. The sea lamprey Petromyzon marinus genome reveals the early origin of several chemosensory receptor families in the vertebrate lineage. *BMC Evol Biol* 9:180.

Lowe CJ, Clarke DN, Medeiros DM, Rokhsar DS, Gerhart J. 2015. The deuterostome context of chordate origins. *Nature* 520:456–465.

Lowenstein O, Thornhill RA. 1970. The labyrinth of *Myxine*: anatomy, ultrastructure and electrophysiology. *Proc Roy Soc Lond B* 176:21–42.

Lucas RJ. 2013. Mammalian inner retinal photoreception. *Curr Biol* 23:R125–133.

Ma X, Hou X, Edgecombe GD, Strausfeld NJ. 2012. Complex brain and optic lobes in an early Cambrian arthropod. *Nature* 490:258–261.

Mackie GO, Burighel P. 2005. The nervous system in adult tunicates: current research directions. *Can J Zool* 83:151–183.

Maklad A, Reed C, Johnson NS, Fritzsch B. 2014. Anatomy of the lamprey ear: morphological evidence for occurrence of horizontal semicircular ducts in the labyrinth of *Petromyzon marinus*. *J Anat* 224:432–446.

Mallatt J, Chen J-Y. 2003. Fossil sister group of craniates: predicted and found. *J Morphol* 258:1–31.

Mallatt J, Holland N. 2013. *Pikaia gracilens* Walcott: stem chordate, or already specialized in the Cambrian? *J Exp Zool* 320:247–271.

Malte H, Lomholt JP. 1998. Ventilation and gas exchange. In: Jørgensen JM, Lomholt JP, Weber RE, Malte H (eds.), *The Biology of Hagfishes*. New York: Springer, pp. 223–234.

Martín-Durán JM, Pang K, Børve A, Lê HS, Furu A, Cannon JT, Jondelius U, Hejnol A. 2018. Convergent evolution of bilaterian nerve cords. *Nature* 553:45–50.

Martínez S, Crossley PH, Cobos I, Rubenstein JL, Martin GR. 1999. FGF8 induces formation of an ectopic isthmic organizer and isthmocerebellar development via a repressive effect on *Otx2* expression. *Development* 126:1189–1200.

Mazet F, Shimeld SM. 2005. Molecular evidence from ascidians for the evolutionary origin of vertebrate cranial sensory placodes. *J Exp Zool B* 304B:340–346.

McHenry MJ, Pell CA, Long JH Jr. 1995. Mechanical control of swimming speed: stiffness and axial wave form in undulating fish models. *J Exp Biol* 198:2293–2305.

McVean A. 1991. The semicircular canals of the hagfish *Myxine glutinosa*. *J Zool Lond* 224:213–222.

Meech RW. 2017. The evolution of neurons. In: Shepherd SV (ed.), *The Wiley Handbook of Evolutionary Neuroscience*. Chichester, UK: Wiley-Blackwell, pp. 88–124.

Meulemans D, Bronner-Fraser M. 2007a. Insights from amphioxus into the evolution of vertebrate cartilage. *PLoS ONE* 2:e787.

Meulemans D, Bronner-Fraser M. 2007b. The amphioxus *SoxB* family: implications for the evolution of vertebrate placodes. *Int J Biol Sci* 3:356–364.

Millett S, Campbell K, Epstein D, Losos K, Harris E, Joyner A. 1999. A role for *Gbx2* in repression of *Otx2* and positioning the mid/hindbrain organizer. *Nature* 401:161–164.

Miyashita T, Coates MI, Farrar R, Larson P, Manning PL, Wogelius RA, Edwards NP, Anné J, Bergmann U, Palmer AR, Currie PJ. 2019. Hagfish from the Cretaceous Tethys Sea and a reconciliation of the morphological–molecular conflict in early vertebrate phylogeny. *Proc Natl Acad Sci USA* 116:2146–2151.

Monahan-Earley R, Dvorak AM, Aird WC. 2013. Evolutionary origins of the blood vascular system and endothelium. *J Thromb Haemost* 11 (Suppl 1):46–66.

Moret F, Christiaen L, Deyts C, Blin M, Vernier P, Joly J-S. 2005. Regulatory gene expressions in the ascidian ventral sensory vesicle: evolutionary relationships with the vertebrate hypothalamus. *Dev Biol* 277:567–579.

Moroz LL. 2009. On the independent origins of complex brains and neurons. *Brain Behav Evol* 74:177–190.

Moroz LL, Kocot KM, Citarella MR, Dosung S, Norekian TP, Povolotskaya IS, Grigorenko AP, Dailey C, Berezikov E, Buckley KM, et al. 2015. The ctenophore genome and the evolutionary origins of neural systems. *Nature* 510:109–114.

Morris SC, Caron J-B. 2012. *Pikaia gracilens* Walcott, a stem-group chordate from the Middle Cambrian of British Columbia. *Biol Rev* 87:480–512.

Morris SC, Caron J-B. 2014. A primitive fish from the Cambrian of North America. *Nature* 512:419–422.

Morshedian A, Fain GL. 2015. Single-photon sensitivity of lamprey rods with cone-like outer segments. *Curr Biol* 25:484–487.

Murakami Y, Ogasawara M, Sugahara F, Hirano S, Satoh N, Kuratani S. 2001. Identification and expression of the lamprey *Pax6* gene: evolutionary origin of the segmented brain of vertebrates. *Development* 128:3521–3531.

Near TJ. 2009. Conflict and resolution between phylogenies inferred from molecular and phenotypic data sets for hagfish, lampreys, and gnathostomes. *J Exp Zool B* 312B:749–761.

Nicol D, Meinertzhagen IA. 1991. Cell counts and maps in the larval central nervous system of the ascidian *Ciona intestinalis* (L.). *J Comp Neurol* 309:415–429.

Nieuwenhuys R, Donkelaar ten HJ, Nicholson C. 1998. *The Central Nervous System of Vertebrates*. Berlin: Springer.

Nieuwenhuys R, Nicholson C. 1998. Lampreys, Petromyzontoidea. In: Nieuwenhuys R, Donkelaar ten HJ, Nicholson C (eds.), *The Central Nervous System of Vertebrates*. Berlin: Springer, pp. 397–495.

Niimura Y. 2009. On the origin and evolution of vertebrate olfactory receptor genes: comparative genome analysis among 23 chordate species. *Genome Biol Evol* 1:34–44.

Nilsson D-E. 2013. Eye evolution and its functional basis. *Vis Neurosci* 30:5–20.

Nishitsuji K, Horie T, Ichinose A, Sasakura Y, Yasuo H, Kusakabe TG. 2012. Cell lineage and *cis*-regulation for a unique GABAergic/glycinergic neuron type in the larval nerve cord of the ascidian *Ciona intestinalis*. *Dev Growth Differ* 54:177–186.

Nomaksteinsky M, Rottinger E, Dufour HD, Chettouh Z, Lowe CJ, Martindale MQ, Brunet J-F. 2009. Centralization of the deuterostome nervous system predates chordates. *Curr Biol* 19:1264–1269.

Northcutt RG. 1989. The phylogenetic distribution and innervation of craniate mechanoreceptive lateral lines. In: Coombs S, Görner P, Münz H (eds.), *The Mechanosensory Lateral Line: Neurobiology and Evolution*. New York: Springer, pp. 17–78.

Northcutt RG. 1999. Field homology: a meaningless concept. *Eur J Morphol* 37:95–99.

Northcutt RG. 2005. The new head hypothesis revisited. *J Exp Zool* 304B:274–297.

Northcutt RG. 2012. Evolution of centralized nervous systems: two schools of evolutionary thought. *Proc Natl Acad Sci USA* 109:10626–10633.

Northcutt RG, Gans C. 1983. The genesis of neural crest and epidermal placodes: a reinterpretation of vertebrate origins. *Quart Rev Biol* 58:1–28.

Northcutt RG, Puzdrowski RL. 1988. Projections of the olfactory bulb and nervus terminalis in the silver lamprey. *Brain Behav Evol* 32:96–107.

Northcutt RG, Wicht H. 1997. Afferent and efferent connections of the lateral and medial pallia of the silver lamprey. *Brain Behav Evol* 49:1–19.

Ocaña FM, Suryanarayana SM, Saitoh K, Kardamakis AA, Capantini L, Robertson B, Grillner S. 2015. The lamprey pallium provides a blueprint of the mammalian motor projections from cortex. *Curr Biol* 25:413–423.

Ohno S. 1970. *Evolution by Gene Duplication*. Berlin: Springer.

Osório J, Megías M, Pombal MA, Rétaux S. 2006. Dynamic expression of the LIM-homeodomain gene *Lhx15* through larval brain development of the sea lamprey (*Petromyzon marinus*). *Gene Exp Patt* 6:873–878.

Pani AM, Mullarkey EE, Aronowicz J, Assimacopoulos S, Grove EA, Lowe CJ. 2012. Ancient deuterostome origins of vertebrate brain signalling centres. *Nature* 483:289–294.

Panopoulou G, Poustka AJ. 2005. Timing and mechanism of ancient vertebrate genome duplications—the adventure of a hypothesis. *Trends Genet* 21:559–567.

Park B-Y, Saint-Jeannet J-P. 2010. *Induction and Segregation of the Vertebrate Cranial Placodes*. San Rafael, CA: Morgan & Claypool.

Parker HJ, Bronner ME, Krumlauf R. 2015. A *Hox* regulatory network of hindbrain segmentation is conserved to the base of vertebrates. *Nature* 514:490–493.

Patthey C, Schlosser G, Shimeld SM. 2014. The evolutionary history of vertebrate cranial placodes. I. Cell type evolution. *Dev Biol* 389:82–97.

Peterson T, Müller GB. 2016. Phenotypic novelty in EvoDevo: the distinction between continuous and discontinuous variation and its importance in evolutionary theory. *Evol Biol* 43:314–335.

Pisani D, Pett W, Dohrmann M, Feuda R, Rota-Stabelli O, Philippe H, Lartillot N, Wörheide G. 2015. Genomic data do not support comb jellies as the sister group to all other animals. *Proc Natl Acad Sci USA* 112:15402–15407.

Polenova OA, Vesselkin NP. 1993. Olfactory and nonolfactory projections in the river lamprey (*Lampetra fluviatilis*) telencephalon. *J Hirnforsch* 34:261–279.

Pombal MA, Megías M. 2017. The nervous systems of jawless vertebrates. In: Kaas JH, Striedter GF (eds.), *Evolution of Nervous Systems*, 2nd ed. Amsterdam: Elsevier, pp. 37–57.

Pombal MA, Megías M, Bardet SM, Puelles L. 2009. New and old thoughts on the segmental organization of the forebrain in lampreys. *Brain Behav Evol* 74:7–19.

Pombal MA, Puelles L. 1999. Prosomeric map of the lamprey forebrain based on calretinin immunocytochemistry, Nissl stain, and ancillary markers. *J Comp Neurol* 414:391–422.

Provencio I, Rodriguez IR, Jiang G, Hayes WP, Moreira EF, Rollag MD. 2000. A novel human opsin in the inner retina. *J Neurosci* 20:600–605.

Puelles L. 2001. Brain segmentation and forebrain development in amniotes. *Brain Res Bull* 55:695–710.

Puelles L. 2013. Plan of the developing vertebrate nervous system: relating embryology to the adult nervous system (prosomere model, overview of brain organization). In: Rubenstein JLR, Goldman-Rakic P (eds.), *Patterning and Cell Type Specification in the Developing CNS and PNS*. Amsterdam: Elsevier, pp. 187–209.

Puelles L, Harrison M, Paxinos G, Watson C. 2013. A developmental ontology for the mammalian brain based on the prosomeric model. *Trends Neurosci* 36:570–578.

Putnam NH, Butts T, Ferrier DEK, Furlong RF, Hellsten U, Kawashima T, Robinson-Rechavi M, Shoguchi E, Terry A, Yu J-K, et al. 2008. The amphioxus genome and the evolution of the chordate karyotype. *Nature* 453:1064–1071.

Rodríguez-Moldes I, Santos-Durán GN, Pose-Méndez S, Quintana-Urzainqui I, Candal E. 2017. The brains of cartilaginous fishes. In: Kaas JH (ed.), *Evolution of Nervous Systems*, 2nd ed., vol. 1. Amsterdam: Elsevier, pp. 77–97.

Romer AS. 1962. *The Vertebrate Body*. Philadelphia: Saunders.

Ronan M, Bodznick D. 1991. Behavioral and neurophysiological demonstration of a lateralis skin photosensitivity in larval sea lampreys. *J Exp Biol* 161:97–117.

Rovainen CM. 1996. Feeding and breathing in lampreys. *Brain Behav Evol* 48:297–305.

Rubenstein JL, Martinez S, Shimamura K, Puelles L. 1994. The embryonic vertebrate forebrain: the prosomeric model. *Science* 266:578–580.

Salas CA, Yopak KE, Lisney TJ, Potter IC, Collin SP. 2017. The central nervous system of jawless vertebrates: encephalization in lampreys and hagfishes. *Brain Behav Evol* 89:33–47.

Sánchez-Arrones L, Ferran JL, Hidalgo-Sánchez M, Puelles L. 2015. Origin and early development of the chicken adenohypophysis. *Front Neuroanat* 9:7.

Sasakura Y, Mita K, Ogura Y, Horie T. 2012. Ascidians as excellent chordate models for studying the development of the nervous system during embryogenesis and metamorphosis. *Dev Growth Differ* 54:420–437.

Satoh G. 2005. Characterization of novel GPCR gene coding locus in amphioxus genome: gene structure, expression, and phylogenetic analysis with implications for its involvement in chemoreception. *genesis* 41:47–57.

Sauka-Spengler T, Meulemans D, Jones M, Bronner-Fraser M. 2007. Ancient evolutionary origin of the neural crest gene regulatory network. *Dev Cell* 13:405–420.

Schlosser G, Northcutt RG. 2000. Development of neurogenic placodes in *Xenopus laevis*. *J Comp Neurol* 418:121–146.

Schlosser G, Patthey C, Shimeld SM. 2014. The evolutionary history of vertebrate cranial placodes. II. Evolution of ectodermal patterning. *Dev Biol* 389:98–119.

Schmidt-Rhaesa A, Harzsch S, Purschke GN. 2015. *Structure and Evolution of Invertebrate Nervous Systems*. New York: Oxford University Press.

Schubert M, Holland ND, Laudet V, Holland LZ. 2006. A retinoic acid-*Hox* hierarchy controls both anterior/posterior patterning and neuronal specification in the developing central nervous system of the cephalochordate amphioxus. *Dev Biol* 296:190–202.

Schwarze K, Campbell KL, Hankeln T, Storz JF, Hoffmann FG, Burmester T. 2014. The globin gene repertoire of lampreys: convergent evolution of hemoglobin and myoglobin in jawed and jawless vertebrates. *Mol Biol Evol* 31:2708–2721.

Seo H-C, Edvardsen RB, Maeland AD, Bjordal M, Jensen MF, Hansen A, Flaat M, Weissenbach J, Lehrach H, Wincker P, Reinhardt R, Chourrout D. 2004. *Hox* cluster disintegration with persistent anteroposterior order of expression in *Oikopleura dioica*. *Nature* 431:67–71.

Shimamura K, Rubenstein J. 1997. Inductive interactions direct early regionalization of the mouse forebrain. *Development* 124:2709–2718.

Shimeld SM, Donoghue PCJ. 2012. Evolutionary crossroads in developmental biology: cyclostomes (lamprey and hagfish). *Development* 139:2091–2099.

Smith JJ, Kuraku S, Holt C, Sauka-Spengler T, Jiang N, Campbell MS, Yandell MD, Manousaki T, Meyer A, Bloom OE, et al. 2013. Sequencing of the sea lamprey (*Petromyzon marinus*) genome provides insights into vertebrate evolution. *Nat Genet* 45:415–21.

Smith SA, Wilson NG, Goetz FE, Feehery C, Andrade SCS, Rouse GW, Giribet G, Dunn CW. 2011. Resolving the evolutionary relationships of molluscs with phylogenomic tools. *Nature* 480:364–367.

Søviknes AM, Chourrout D, Glover JC. 2005. Development of putative GABAergic neurons in the appendicularian urochordate *Oikopleura dioica*. *J Comp Neurol* 490:12–28.

Søviknes AM, Chourrout D, Glover JC. 2007. Development of the caudal nerve cord, motoneurons, and muscle innervation in the appendicularian urochordate *Oikopleura dioica*. *J Comp Neurol* 503:224–243.

Spagnuolo A, Ristoratore F, Di Gregorio A, Aniello F, Branno M, Di Lauro R. 2003. Unusual number and genomic organization of *Hox* genes in the tunicate *Ciona intestinalis*. *Gene* 309:71–79.

Stephenson-Jones M, Samuelsson E, Ericsson J, Robertson B, Grillner S. 2011. Evolutionary conservation of the basal ganglia as a common vertebrate mechanism for action selection. *Curr Biol* 21:1–11.

Stokes M. 1997. Larval locomotion of the lancelet. *J Exp Biol* 200:1661–1680.

Stolfi A, Ryan K, Meinertzhagen IA, Christiaen L. 2015. Migratory neuronal progenitors arise from the neural plate borders in tunicates. *Nature* 527:371–374.

Strausfeld NJ, Hirth F. 2013a. Homology versus convergence in resolving transphyletic correspondences of brain organization. *Brain Behav Evol* 82:215–219.

Strausfeld NJ, Hirth F. 2013b. Deep homology of arthropod central complex and vertebrate basal ganglia. *Science* 340:157–161.

Striedter GF. 1998. Stepping into the same river twice: homologues as recurring attractors in epigenetic landscapes. *Brain Behav Evol* 52:218–231.

Striedter GF, Northcutt RG. 1991. Biological hierarchies and the concept of homology. *Brain Behav Evol* 38:177–189.

Sugahara F, Murakami Y, Adachi N, Kuratani S. 2013. Evolution of the regionalization and patterning of the vertebrate telencephalon: what can we learn from cyclostomes? *Curr Opin Genet Dev* 23:475–483.

Sugahara F, Pascual-Anaya J, Oisi Y, Kuraku S, Aota S-I, Adachi N, Takagi W, Hirai T, Sato N, Murakami Y, Kuratani S. 2016. Evidence from cyclostomes for complex regionalization of the ancestral vertebrate brain. *Nature* 531:97–100.

Suryanarayana SM, Robertson B, Wallén P, Grillner S. 2017. The lamprey pallium provides a blueprint of the mammalian layered cortex. *Curr Biol* 27:3264–3277.

Suzuki DG, Murakami Y, Escriva H, Wada H. 2015. A comparative examination of neural circuit and brain patterning between the lamprey and amphioxus reveals the evolutionary origin of the vertebrate visual center. *J Comp Neurol* 523:251–261.

Suzuki MM, Satoh N. 2000. Genes expressed in the amphioxus notochord revealed by EST analysis. *Dev Biol* 224:168–177.

Takahashi T. 2005. The evolutionary origins of vertebrate midbrain and MHB: Insights from mouse, amphioxus and ascidian *Dmbx* homeobox genes. *Brain Res Bull* 66:510–517.

Takahashi T, Holland PWH. 2004. Amphioxus and ascidian Dmbx homeobox genes give clues to the vertebrate origins of midbrain development. *Development* 131:3285–3294.

Takio Y, Kuraku S, Murakami Y, Pasqualetti M, Rijli FM, Narita Y, Kuratani S, Kusakabe R. 2007. *Hox* gene expression patterns in *Lethenteron japonicum* embryos – Insights into the evolution of the vertebrate Hox code. *Dev Biol* 308:606–620.

Tank EM, Dekker RG, Beauchamp K, Wilson KA, Boehmke AE, Langeland JA. 2009. Patterns and consequences of vertebrate *Emx* gene duplications. *Evol Dev* 11:343–353.

Tomer R, Denes AS, Tessmar-Raible K, Arendt D. 2010. Profiling by image registration reveals common origin of annelid mushroom bodies and vertebrate pallium. *Cell* 142:800–809.

Toresson H, Martinez-Barbera JP, Bardsley A, Caubit X, Krauss S. 1998. Conservation of *BF-1* expression in amphioxus and zebrafish suggests evolutionary ancestry of anterior cell types that contribute to the vertebrate telencephalon. *Dev Genes Evol* 208:431–439.

Vopalensky P, Pergner J, Liegertova M, Benito-Gutiérrez È, Arendt D, Kozmik Z. 2012. Molecular analysis of the amphioxus frontal eye unravels the evolutionary origin of the retina and pigment cells of the vertebrate eye. *Proc Natl Acad Sci USA* 109:15383–15388.

Wada H, Saiga H, Satoh N, Holland PW. 1998. Tripartite organization of the ancestral chordate brain and the antiquity of placodes: insights from ascidian *Pax-2/5/8, Hox* and *Otx* genes. *Development* 125:1113–1122.

Wada S, Tokuoka M, Shoguchi E, Kobayashi K, Di Gregorio A, Spagnuolo A, Branno M, Kohara Y, Rokhsar D, Levine M, Saiga H, Satoh N, Satou Y. 2003. A genomewide survey of developmentally relevant genes in *Ciona intestinalis*. II. Genes for homeobox transcription factors. *Dev Genes Evol* 213:222–234.

Wagner GP. 2000. *The Character Concept in Evolutionary Biology*. San Diego, CA: Academic Press.

Wagner GP. 2007. The developmental genetics of homology. *Nat Rev Genet* 8:473–479.

Wagner GP, Lynch VJ. 2010. Evolutionary novelties. *Curr Biol* 20:R48–R52.

Weigle C, Northcutt RG. 1998. On the phylogenetic origin of the cerebellum: tracing studies in the silver lamprey *Ichthyomyzon unicuspis*. *Eur J Neurosci* 10 (Suppl 10): a68.01.

Wicher D. 2012. Functional and evolutionary aspects of chemoreceptors. *Front Cell Neurosci* 6:48.

Wicht H. 1996. The brains of lampreys and hagfishes: characteristics, characters, and comparisons. *Brain Behav Evol* 48:248–261.

Wicht H, Lacalli TC. 2005. The nervous system of amphioxus: structure, development, and evolutionary significance. *Can J Zool* 83:122–150.

Wicht H, Northcutt RG. 1990. Retinofugal and retinopetal projections in the Pacific hagfish, *Eptatretus stouti* (Myxinoidea). *Brain Behav Evol* 36:315–328.

Wicht H, Northcutt RG. 1992. The forebrain of the Pacific hagfish: a cladistic reconstruction of the ancestral craniate forebrain. *Brain Behav Evol* 40:25–64.

Wicht H, Northcutt RG. 1993. Secondary olfactory projections and pallial topography in the Pacific hagfish, *Eptatretus stouti*. *J Comp Neurol* 337:529–542.

Wicht H, Northcutt RG. 1994. An immunohistochemical study of the telencephalon and the diencephalon in a Myxinoid jawless fish, the Pacific hagfish, *Eptatretus stouti*. *Brain Behav Evol* 43:140–161.

Williamson DI. 2012. The origins of chordate larvae. *Cell Dev Biol* 1:1–10.

Wray S. 2010. From nose to brain: development of gonadotrophin-releasing hormone-1 neurones. *J Neuroendocrinol* 22:743–753.

Wullimann, MF. 2001. Brain phenotypes and early regulatory genes: the Bauplan of the metazoan central nervous system. In: Roth G, Mario F W (eds.), *Brain Evolution and Cognition*. New York: John Wiley & Sons, pp. 11–40.

Wullimann MF, Grothe B. 2013. The central nervous organization of the lateral line system. In: Coombs S, Bleckmann H, Fay RR, Popper AN (eds.), *The Lateral Line System*, Vol. 48, *Springer Handbook of Auditory Research*. New York: Springer, pp. 195–251.

Wurst W, Bally-Cuif L. 2001. Neural plate patterning: upstream and downstream of the isthmic organizer. *Nat Rev Neurosci* 2:99–108.

Yalden DW. 1985. Feeding mechanisms as evidence for cyclostome monophyly. *Zool J Linn Soc* 84:291–300.

Yasui K, Kaji T, Morov AR, Yonemura S. 2013. Development of oral and branchial muscles in lancelet larvae of *Branchiostoma japonicum*. *J Morphol* 275:465–477.

Yu J-KS. 2010. The evolutionary origin of the vertebrate neural crest and its developmental gene regulatory network: insights from amphioxus. *Zoology* 113:1–9.

Zhang J. 2003. Evolution by gene duplication: an update. *Trends Ecol Evolut* 18:292–298.

Zintzen V, Roberts CD, Anderson MJ, Stewart AL, Struthers CD, Harvey ES. 2011. Hagfish predatory behaviour and slime defence mechanism. *Sci Rep* 1:131.

3

The Origin of Jaws and Paired Fins

The Age of Fishes

More than 99% of all living vertebrate species have jaws. That is, they are jawed vertebrates (gnathostomes). This chapter focuses specifically on the gnathostome lineages that have always lived in water, the jawed fishes. This includes all fishes except the cyclostomes, which we discussed in Chapter 2. The gnathostomes that made their way onto land, re-entered water, or rose into the air will be our focus in later chapters.

The two major groups of jawed fishes are the cartilaginous fishes and the bony fishes (Figure 3.1). The latter are divided into two large lineages, namely the ray-finned fishes (actinopterygians) and the lobe-finned fishes (sarcopterygians). Because the sarcopterygians include all of the tetrapods, it makes some sense to call this group the lobe-finned vertebrates, rather than lobe-finned fishes, and to refer to the bony fishes as bony vertebrates. However, as stated earlier, we focus in this chapter on fishes, leaving the tetrapods to subsequent chapters.

Aside from jaws, the gnathostomes evolved several other features that their ancestors lacked. Most importantly, they evolved two sets of paired fins. The pectoral fins of jawed fishes transformed into forelimbs as tetrapods evolved, whereas the pelvic fins of early jawed fishes are homologous to tetrapod hindlimbs (Zhu et al., 2012). These paired fins complement the caudal and median (dorsal and/or ventral) fins that cyclostomes and the invertebrate chordates already used to help them swim. Gnathostomes also evolved an adaptive immune system and a few other useful characters, but they are less obviously linked to nervous system evolution and shall not concern us here.

3.1. Extant Jawed Fishes and Their Brains

Most of what we know about nervous system evolution in jawed fishes is based on data from their three main surviving lineages: the cartilaginous fishes, ray-finned fishes, and lobe-finned vertebrates. We here discuss these extant lineages in turn. Then we briefly review what we can learn from examining the fossils of various extinct gnathostomes.

Brains Through Time. Georg F. Striedter and R. Glenn Northcutt, Oxford University Press (2020). © Oxford University Press.
DOI: 10.1093/oso/9780195125689.001.0001

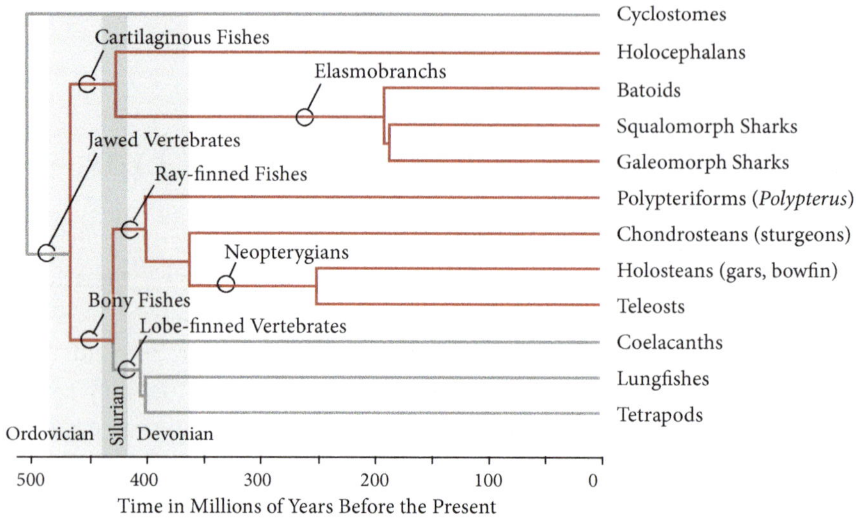

Figure 3.1 Phylogenetic relationships of the major vertebrate groups. The jawed vertebrates comprise the cartilaginous and ray-finned fishes, as well as lobe-finned vertebrates. The red lines indicate the lineages that are discussed extensively in this chapter.

Based on Kriwet et al. (2009), Venkatesh et al. (2014), Friedman (2015).

3.1.1. Cartilaginous Fishes

As their name implies, the cartilaginous fishes have a cartilaginous skeleton. Some of this cartilage is calcified and relatively hard, and some cartilaginous fishes have acellular bone in a few places. Still, as a general rule, the skeleton of cartilaginous fishes is significantly more elastic and less dense than that of bony fishes. The skin of cartilaginous fishes is covered with tooth-like (placoid) scales that contain dentine and are covered by a hard enamel-like substance. Almost all cartilaginous fishes also have multiple rows of teeth, which are replaced continuously. Another typical feature of cartilaginous fishes is that the males have modified part of their pelvic fins to form claspers that are used to inject sperm into the females and, thus, fertilize their eggs internally. Those eggs are relatively large. In more than half of all cartilaginous fishes the fertilized eggs develop inside the body, such that the mothers eventually give birth to live young. The vast majority of cartilaginous fishes live in marine environments, but a few species, notably the freshwater stingrays and bullsharks, can live in brackish water and rivers.

Of the approximately 1,150 cartilaginous fish species, roughly 50 are holocephalans (Chimaeriformes), a lineage that diverged from the other cartilaginous fishes roughly 420 mya (Figure 3.1; Inoue et al., 2010). Members of this lineage include rat-fishes, rabbit-fishes, and elephant-fishes. They typically live at depths below 200 m and feed mainly on bottom-dwelling invertebrates, which they crush with a few large, plate-like teeth. They are odd-looking fish, with a large head and large eyes but a surprisingly small mouth. Their elongate, largely scaleless body sports a long, thin tail and large pectoral fins. The brains of holocephalans also look rather odd (Figure 3.2), mainly because their small telencephalon is separated from more caudal brain regions by a long stalk, consisting mainly of axons. This unusual feature emerges late during development and is correlated with the enlargement of the eyes, which leaves little room for a forebrain between them.

The remaining cartilaginous fishes all belong to the elasmobranchs (Figure 3.1). Roughly 280 mya this lineage split into two large groups, namely the selachians (sharks) and the batoids (skates and rays), which comprise roughly 500 and 600 species, respectively. In general, sharks have streamlined bodies and tend to swim in the open ocean, whereas batoids feed on the ocean floor and are dorsoventrally flattened, with enlarged pectoral fins that fuse in front of the head. Despite this general distinction, some species don't fit the stereotype. Thus, angel sharks (genus *Squatina*) have flattened bodies and enlarged pectoral fins, much as rays do. Conversely, sawfishes (Pristiformes) are rays that resemble saw sharks (Pristiophoriformes).

The sharks themselves are divisible into squalomorph and galeomorph sharks. The former group contains about 130 species, including a large variety of dogfishes (Squaliformes). The galeomorph sharks include roughly 270 different species. They include carcharhiniform sharks (e.g., hammerheads and catsharks) but also several smaller groups, such as lamniform sharks (e.g., great white and megamouth sharks) and carpet sharks (e.g., nurse sharks). As a general rule, galeomorph sharks are larger than squalomorph sharks and fiercer predators. They can protrude their jaws and prey on animals significantly larger than themselves (Wilga et al., 2001; Wilga, 2005). Galeomorph sharks also tend to have larger brains than squalomorph sharks, relative to body size. For example, the brain of a 150 kg hammerhead shark weighs roughly 100 g, more than five times as much as an average squalomorph shark of similar body weight (Figure 3.3).

Batoids, too, include two major lineages, namely ~280 species of skates (Rajiformes) and ~200 species of stingrays (Myliobatiformes). These two taxonomic groups have independently evolved very similar body shapes, featuring a whip-like tail and a disk-shaped body. They propel themselves mainly by flapping the lateral edges of that body disk. Two additional groups of batoids have a thicker tail and swim by means of lateral undulation (Aschliman et al., 2012). These other batoids include ~70 species of electric rays (Torpediniformes), which have electric organs that can deliver up to 220V of electricity, and ~60 species of shovelnose guitarfishes (Rhynchobatiformes). The largest brains among batoids are found in the Myliobatiformes, especially the devil and manta rays (Figure 3.3) (Lisney et al., 2008). These animals have wingspans up to 6.7 m and feed mainly on plankton and small shrimp, which they funnel into their large mouths. Even after accounting for their large body size, these animals have enormous brains. For example, the brain of a 165 kg manta ray weighs roughly 122 g (Ari, 2011). Remarkably, the brain of

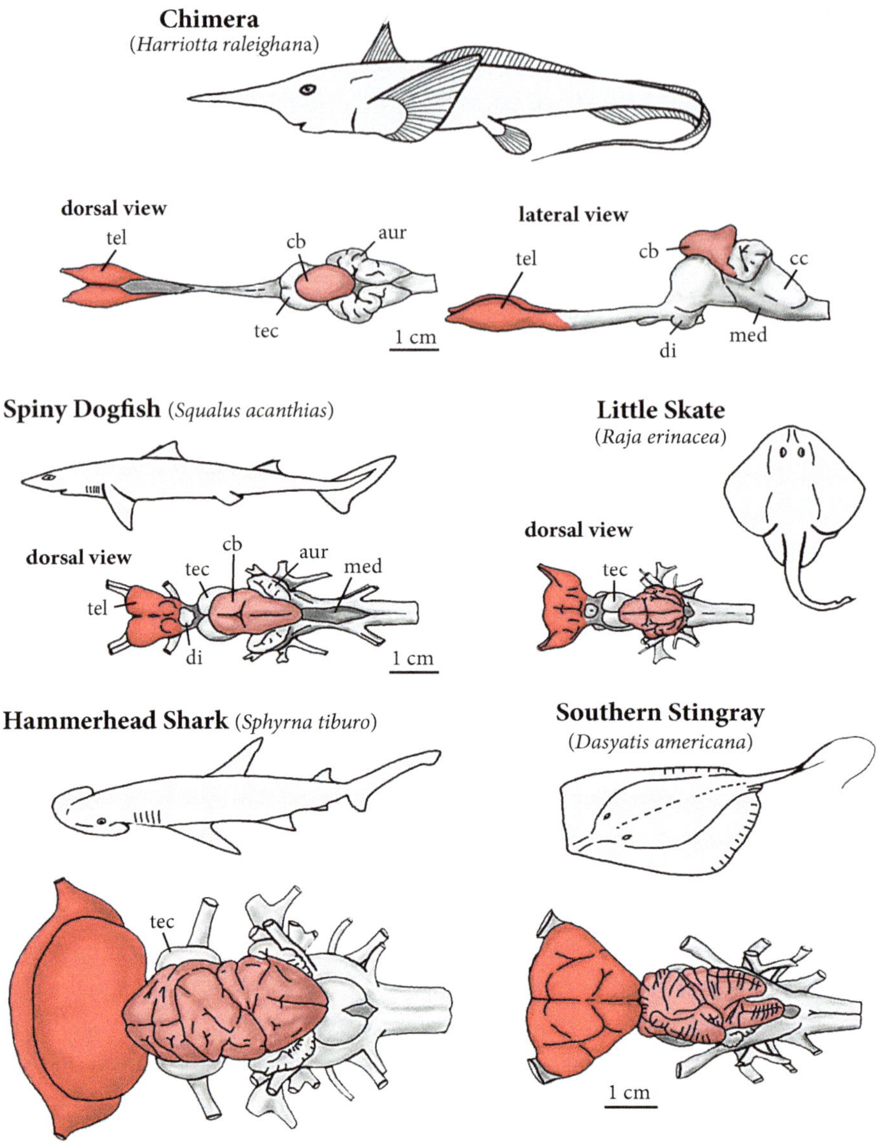

Figure 3.2 Bodies and brains of cartilaginous fishes. The illustrated species demonstrate that brain size and complexity varies considerably within the cartilaginous fishes, and that the telencephalon (red) and cerebellum (pink) vary most. All brains are drawn to roughly the same scale (scale bars provided where available in the originals); the bodies vary in scale.

Abbreviations: aur – auricle; cb – cerebellum; cc – cerebellar crista; di – diencephalon; med – medulla; tec – tectum; tel – telencephalon.

The chimera (holocephalan) brain is adapted from Yopak and Montgomery (2008); the shark and ray brains are modified from Hofmann and Northcutt (2012).

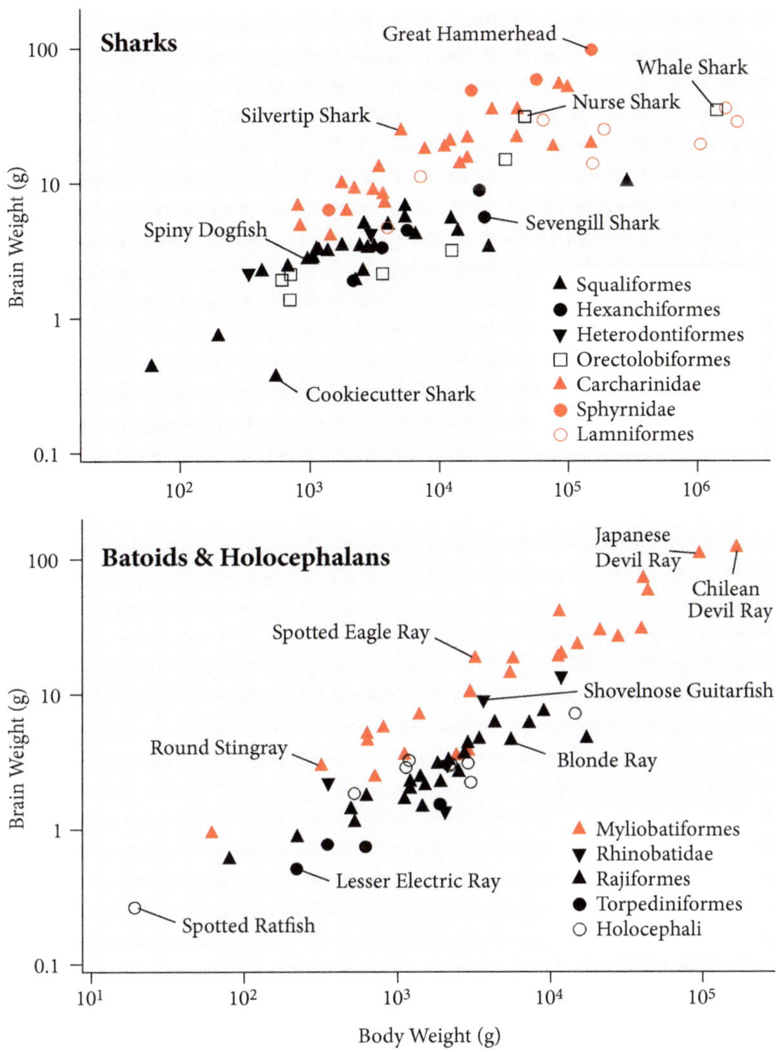

Figure 3.3 Brain-body allometry in cartilaginous fishes. Carcharinidae (requiem sharks), Sphyrnidae (hammerhead sharks), and Lamniformes (mackerel sharks) tend to have larger brains than other types of sharks, relative to body size. Relative brain size also varies within the remaining cartilaginous fishes, with Myliobatiformes having larger brains than other batoids or holocephalans.

Data from Crile and Quiring (1940), Bauchot et al. (1976), Northcutt (1977), Fossi (1984), Myagkov (1991), Bauchot et al. (1995), Yopak et al. (2007), Lisney et al. (2008); additional data kindly provided by Kara Yopak (pers. comm.).

manta rays is surrounded by a complex network of thin arteries and veins, called a *rete mirabile*, that likely warms the brain during deep dives (Thorrold et al., 2014).

There is no doubt that brain size, relative to body size, increased several times within the cartilaginous fishes. Most obvious is that very large brains evolved independently in galeomorph sharks and the myliobatiform rays (Figure 3.3; see Chapter 7 for more details). The most enlarged brain regions in both groups are the telencephalon and the cerebellum. Indeed, the telencephalon in these species may occupy more than 65% of the entire brain (versus ~25% in, e.g., squalomorph sharks) and grow so large that the telencephalic ventricles become highly compressed. The cerebellum of these large-brained sharks and rays also exhibits the kind of complex folding that is otherwise observed only in mammals and birds.

What do these sharks and rays do with their enormous brains? To answer that question, one may note that, after accounting for phylogenetic history, brain size in sharks and batoids covaries with several ecological factors (Yopak et al., 2007; Lisney et al., 2008). In particular, relative brain size correlates positively with living on a reef or the open ocean, rather than on the ocean floor. Most of the large-brained cartilaginous fishes probably use their brains to help them learn when and where to find large amounts of quality food, which is often patchily distributed across both time and space in the open ocean. Great white sharks, for example, consistently visit a network of seasonal feeding grounds, which are often separated by long distances (Jorgensen et al., 2010). Large brain size in cartilaginous fishes also tends to covary with social complexity, though this has not been examined statistically. In particular, we note that the large-brained hammerhead and carcharinid sharks occasionally aggregate into large schools where they exhibit dominance hierarchies and other complex social behaviors; manta rays, too, perform complex social displays. Finally, it is interesting that many of the large-brained sharks evolved a placenta-like structure to supply their embryos with extra nutrition; presumably this extra energy facilitates the building of large brains.

3.1.2. Ray-Finned Fishes

With more than 25,000 species, the ray-finned fishes (actinopterygians) are by far the most successful group of aquatic vertebrates. Their fins are supported by bony fin rays that are stiffer than the fin rays of cartilaginous fishes. Furthermore, the fins of ray-finned fishes are collapsible, which allows the animals to change the forces they exert on the surrounding water. Even more interesting is that their fin rays consist of two halves that can slide past one another, thereby changing the fin's curvature (Lauder, 2015). The muscles controlling these fin movements insert on the bases of the individual fin rays, which allows the fins themselves to be quite thin. They also contain a variety of sensory nerve endings (Williams IV et al., 2013). Collectively, these evolutionary changes in fin structure and control make the ray-finned fishes much more agile than their cartilaginous relatives. For example, they can use their pectoral fins to break forward momentum, swim backwards, or, in some species, move up or down in the water column.

The most basal lineage of ray-finned fishes are the Polypteriformes, which include the reedfish and 11 species of the genus *Polypterus* (Figure 3.4). These animals have elongate bodies and resemble primitive lobe-finned fishes in several respects, including the possession of simple jaws. Nonetheless, most phylogenetic analyses

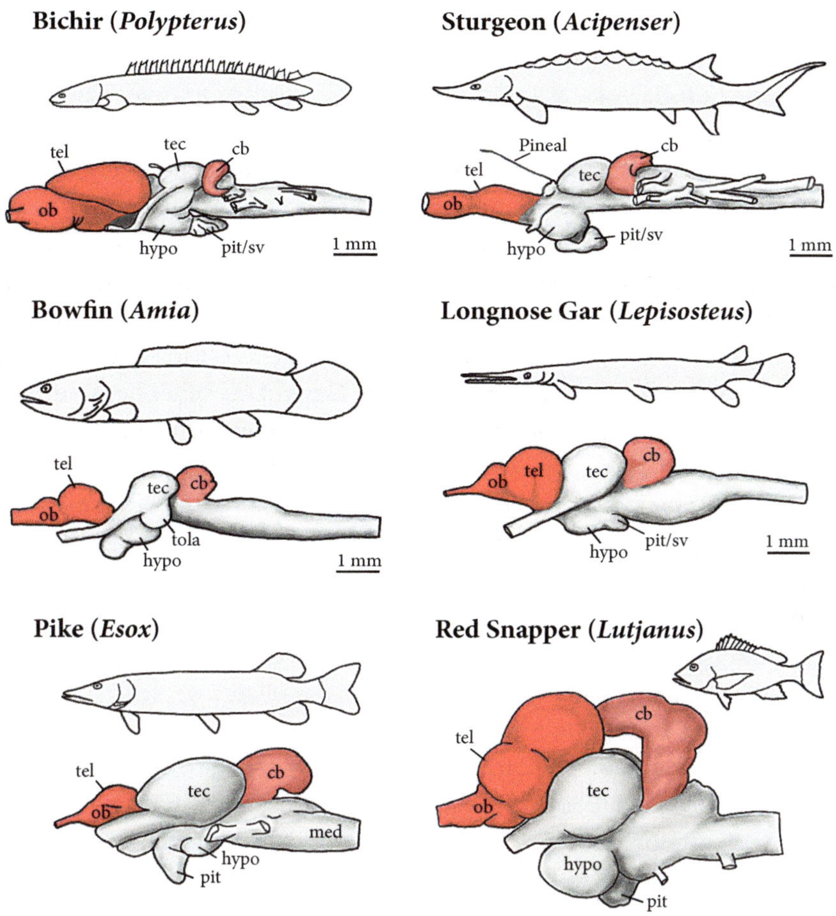

Figure 3.4 Brains and bodies of ray-finned fishes. Bichirs (*Polypterus*) and sturgeons are non-teleost ray-finned fishes (actinopterygians). The bowfin and gars are holostean ray-finned fishes. The pike belongs to a primitive lineage within the teleosts, whereas the red snapper belongs to an advanced teleost lineage. The telencephalon (tel) and cerebellum (cb) are colored dark and light red, respectively, and the olfactory bulb (ob) is the most anterior part of the telencephalon.

Additional abbreviations: hypo – hypothalamus; med – medulla; pit – pituitary; pit/sv – pituitary & saccus vasculosus; tec – tectum; tola – lateral toral nucleus.

The bichir and sturgeon brains are adapted from Nieuwenhuys et al. (1998); the pike and red snapper brains are based on photographs by Michael Hofmann (pers. comm.).

have shown the Polypteriformes to be the sister group of the remaining ray-finned fishes. The next most basal lineages of ray-finned fishes are the Acipenseriformes, which includes ~30 species of sturgeons and paddlefish, and the Holostei, which includes the bowfin (*Amia*) and seven species of gars. All of the other ray-finned fishes are teleosts, an incredibly diverse group that can be found in virtually all aquatic habitats. The two largest lineages of teleosts are the Perciformes and the Ostariophysi. The former group comprises more than 10,000 species and includes cichlids, perches, snappers, barracuda, and swordfish. Most perciform teleosts live in the ocean, but approximately 2,000 of their species occupy freshwater habitats. Sporting roughly 9,000 species, the ostariophysan teleosts are almost as diverse. They include catfish, characins, and electric eels, as well as minnows and goldfish; the vast majority of them live in rivers and lakes.

What features have made the ray-finned fishes, and more specifically the teleosts, such an amazingly successful group? One useful innovation of the ray-finned fishes was the evolution of a gas-filled swim bladder, which increases a fish's buoyancy so that it can float in the water column without constantly battling gravity. Such swim bladders are found in all ray-finned fishes except the Polypteriformes (Longo et al., 2013). They pinch off from the anterior esophagus during development and are thought to be homologous to the primitive lungs of the other bony vertebrates, which, in contrast to true swim bladders, remain connected to the esophagus so that they can be filled with air when the animal extends its head above the water surface (see Chapter 4). Another important characteristic of the ray-finned fishes is that they tend to produce large numbers of eggs that, in contrast to the eggs of cartilaginous fishes, are fertilized externally. Once hatched, the ray-finned fish larvae quickly begin to hunt for food. This strategy of producing immense numbers of young that must fend for themselves is very different from that pursued by cartilaginous fishes, which invest heavily in producing a few offspring.

Morphologically the most important innovations of teleosts involve their jaws, which are far more complex and mobile than those of their ancestors. Thus, teleosts can rapidly protrude their jaws and generate enough suction to draw in small food items. Several groups of teleosts also have a set of "pharyngeal jaws" inside their mouth that help to crush and process food. In moray eels these pharyngeal jaws are so mobile that they are capable of grasping prey and pulling it deep into the pharynx (Mehta and Wainwright, 2008). In addition, teleosts evolved a more complex tail fin, whose dorsal and ventral halves can be moved independently of one another, and lighter, less rigid scales. Whereas *Polypterus*, sturgeons, and gars all have thick, bony, and interlocking (ganoid) scales, *Amia* and all teleosts have much thinner and more flexible, overlapping scales. Collectively, these morphological changes have made the teleosts far more agile than their ray-finned fish ancestors. As a group, they excel at foraging in spatially complex habitats, such as coral reefs, lake shores, and river banks.

Going beyond morphology, it is worth noting that early teleosts almost certainly underwent a third round of whole genome duplication, in addition to the two rounds that occurred at the base of vertebrate evolution (see Chapter 2). Although many of the duplicated genes were subsequently lost (Inoue et al., 2015), the plethora of extra genes must have vastly increased the range of possible phenotypes.

Because teleosts and the other ray-finned fishes do not grow as large as the large cartilaginous fishes, their brains tend to be smaller in absolute size. Thus, a tuna weighing 4.5 kg has a brain that weighs less than 5 g. However, when one compares cartilaginous and ray-finned fishes at the same body size, brain weights for both groups occupy a similar range (see Figure 1.19 in Chapter 1). On average, the cartilaginous fishes do tend to have slightly larger brains, relative to body size (Striedter, 2005), but much of this difference is due to the enormous brains of myliobatiform rays and galeomorph sharks.

Compared to basal lobe-finned fishes (i.e., lungfishes and coelacanths) and basal ray-finned fishes (e.g., *Polypterus*), most teleosts have larger brains, relative to body size (Figure 3.5). Given these data, it seems likely that relative brain size increased, on average, in teleosts. A more detailed phylogenetic analysis (see Chapter 7) indicates that relative brain size actually increased repeatedly with the teleosts, notably in species that hunt in the open ocean or live on coral reefs. An interesting exception to this rule is the family of mormyrid electric fishes, which have enormous brains but dig for prey in mud, using their electric sense (we will come back to these intriguing animals shortly). Relative brain size did decrease in a few teleost lineages, especially in those that are eel-shaped or live in the deep sea (van Dongen, 1998; Iglesias et al., 2015), but these lineages have relatively few living representatives, compared to all the teleosts with enlarged brains.

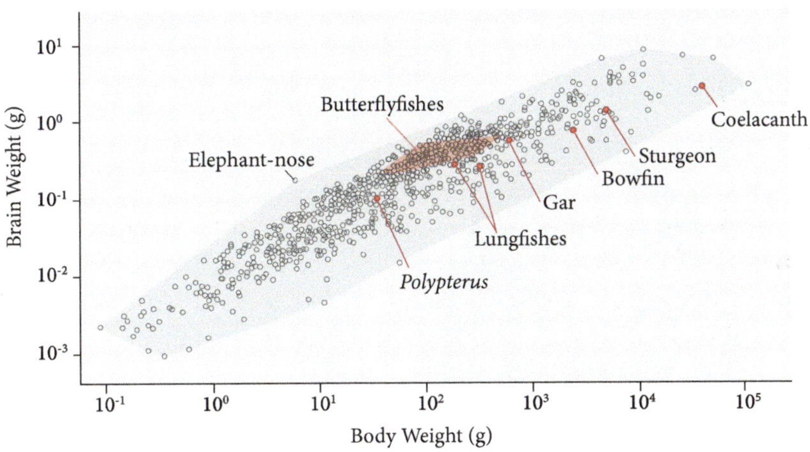

Figure 3.5 Brain-body scaling in basal bony fishes and teleosts. Relative brain sizes for basal bony fishes (red circles) fall in the middle of the range for teleosts, as indicated by the gray minimum convex polygon. However, data for the individual teleosts in this large data set (open gray circles) tend to be located in the top half of the polygon, indicating that most teleosts have brains that are larger, relative to body size, than the brains of their closest living relatives (i.e., basal ray-finned fishes and basal lobe-finned fishes). This observation, in turn, suggests that relative brain increased at least once during the evolution of teleosts (as we discuss in Chapter 7, it probably increased repeatedly within the teleosts). The light red polygon depicts the range of brain-body data for 35 species of butterflyfishes, which live on coral reefs and have relatively large brains. A notable outlier in this data set are the mormyrid electric fishes, represented here by the elephant-nose fish (*Gnathonemus petersii*). As discussed later in this chapter (Figure 3.18), these fishes have an enormous cerebellar valvula.
Adapted from van Dongen (1998) and Bauchot et al. (1989). Elephant-nose data from Nilsson (1996).

An interesting aspect of the brain size increases in teleosts is that they did not involve primarily the telencephalon, as they did in cartilaginous fishes. In teleosts, the telencephalon rarely occupies more than 30% of the entire brain (Bauchot et al., 1989), whereas it reaches more than 50% in the large-brained elasmobranchs (e.g., hammerhead sharks; Northcutt, 1977). Instead, brain size variation among teleosts often results from the dramatic hypertrophy of specific brain regions, with different structures having enlarged in different teleost lineages. For example, the elephant-nose fish has an unusually large brain (see Figure 3.5) because it selectively enlarged the most anterior portion of the teleost cerebellum, the cerebellar valvula. Goldfish have a much smaller cerebellum, but they have elaborated the vagal taste area in the medulla to such an extent that it forms a huge "vagal lobe" (Figure 3.6). An even more dramatic,

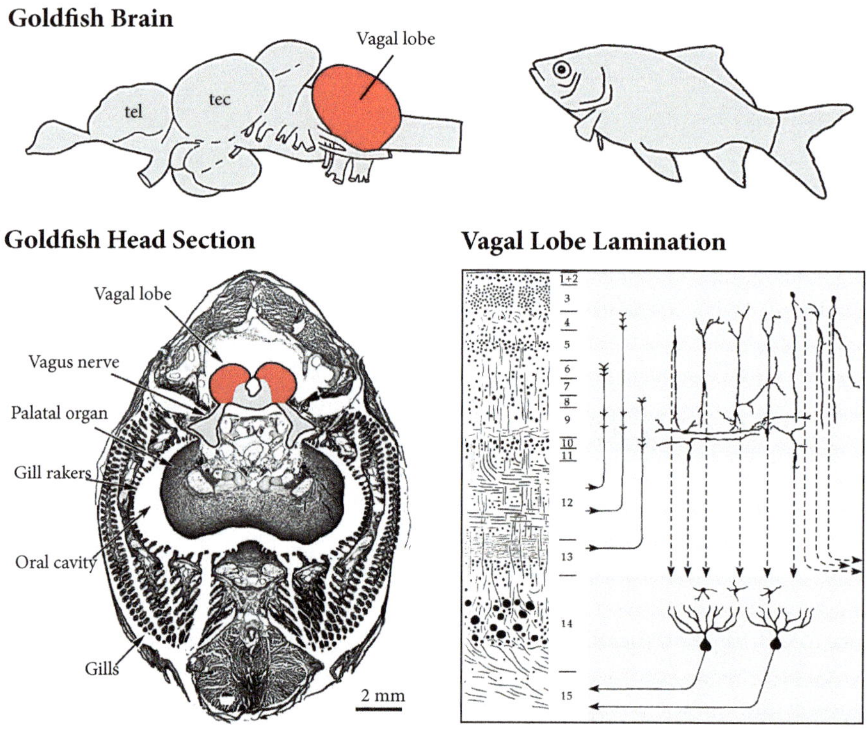

Figure 3.6 The vagal lobe of goldfish. Goldfish and many other cyprinid teleosts have a muscular palatal organ in the roof of their mouth that is used to detect and retain food particles. It is equipped with many taste buds and muscles that are innervated by the vagal nerve. Concomitantly, the vagal sensory and motor regions in the goldfish medulla are enormously hypertrophied, forming a large "vagal lobe." The layers and major cell types of this vagal lobe are shown on the bottom right.

Abbreviations: tec – tectum; tel – telencephalon.

Adapted from Northcutt (1983), Morita et al. (1983, with permission from John Wiley & Sons), Morita and Finger (1985), Nieuwenhuys et al. (1998, with permission from Springer Nature), and Farrell et al. (2002, with permission from John Wiley & Sons).

spiral-shaped enlargement of the vagal taste area has independently evolved in an osteoglossomorph teleost called *Heterotis niloticus* (Braford, 1986). Many other teleosts have enlarged their optic tectum or, in the weakly electric fish, the torus semicircularis. Some teleosts do have a large and complex telencephalon (e.g., red snappers; see Figure 3.4), but species with such large telencephala are rare among the teleosts. An interesting observation is that the telencephalon of *Polypterus* is unusually elongate and occupies 38% of the entire brain, far more than in most teleosts (Northcutt et al., 1978). We suspect that these features are derived for *Polypterus*, possibly linked to the large size of their olfactory bulb and the small size of their cerebellum (see Figure 3.4).

Given that teleosts comprise more than 25,000 species (recent estimates put the count at roughly 30,000 species), such variation in brain region proportions is perhaps not surprising. Still, it seems that individual brain areas vary in size more dramatically among the teleosts than among mammals and birds, which are similarly speciose. If true, this observation might be relevant to the ongoing debate about the extent to which brains evolve mosaically, with individual brain regions changing in size independently of one another (e.g., Finlay and Darlington, 1995; Noreikiene et al., 2015). Of course, variations in brain region size across the teleosts also have functional correlates. Those correlates (or, more precisely, covariates) are fairly obvious when it comes to the enlargement of sensory or motor areas (e.g., the specialized gustatory areas in goldfish and *Heterotis*), but they are more uncertain for the telencephalon or cerebellum. For example, mormyrid teleosts probably use their enlarged cerebellar valvula to facilitate electrolocation, using distortions in the surrounding electric field to sense the presence of objects, but the valvula also receives other forms of sensory input and may well perform additional functions (Finger et al., 1981).

3.1.3. Lobe-Finned Fishes

The lobe-finned fishes are represented nowadays by just two lineages, namely lungfishes and coelacanths (Figure 3.7). Today's lungfishes comprise six species, all of which have elongate bodies and slender, fleshy pectoral and pelvic fins. As the group's name implies, lungfishes possess a highly vascularized outpocketing of the esophagus that they can fill with gulped air and, thus, use as a lung. This ability to obtain oxygen from air is especially useful when a lungfish's habitat dries out. During such times some lungfishes cocoon themselves in mud, so that they don't dry out, and lower their metabolism. In this state of "estivation" they can survive until the rains return.

The coelacanths are very different. They comprise just two living species, *Latimeria chalumnae* and *L. menadoensis*, both of which are endangered, relatively large (up to 2 m in length), and found in deep ocean waters. Underwater films have

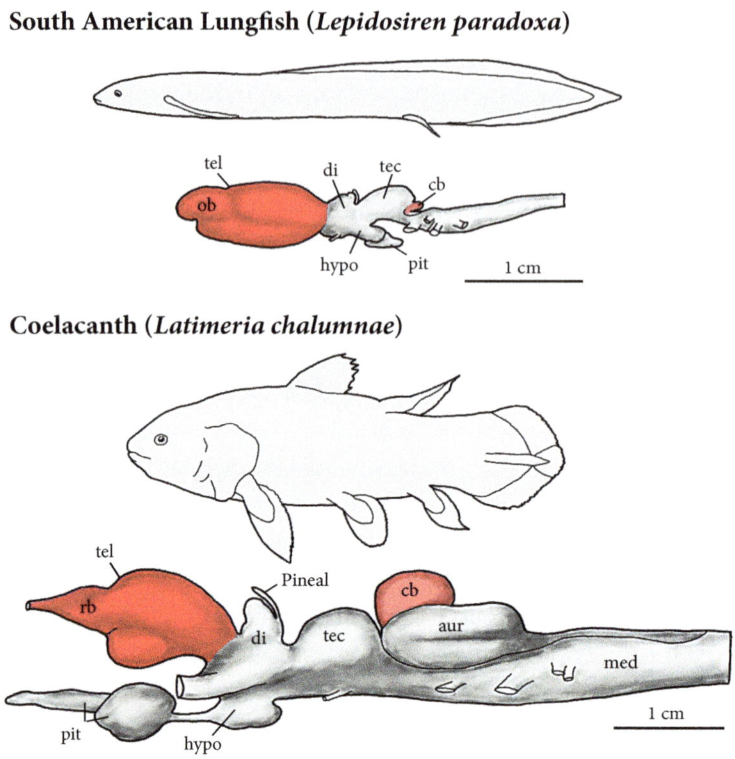

Figure 3.7 Brains and bodies of select lobe-finned fishes. Lungfishes are the extant lobe-finned fishes (sarcopterygians) most closely related to tetrapods. The coelacanth belongs to an even more ancient branch of lobe-finned fishes. The olfactory bulbs of coelacanths lie far rostral to the brain and are not pictured here.

Abbreviations: aur – auricle; cb – cerebellum; di – diencephalon; hypo – hypothalamus; med – medulla; ob – olfactory bulb; pit – pituitary; rb – rostral bodies; tec – tectum; tel – telencephalon.

Both brain drawings are adapted from Nieuwenhuys et al. (1998).

shown that *Latimeria chalumnae* tends to swim slowly and moves its pectoral and pelvic fins in a manner that resembles the quadrupedal gait of vertebrates (Fricke et al., 1987). These animals are thought to rest in underwater caves during the day, feeding on smaller fish and cephalopods during the night.

The brains of coelacanths and lungfishes are small, relative to the large size of these animals, and exhibit a number of unique features (Figure 3.7). For example, coelacanths have a large pituitary gland that extends rostrally, rather than caudally as in most vertebrates, and their telencephalon contains a "rostral body" that has no obvious homolog in other vertebrates. Because coelacanths are extremely rare and nearly impossible to catch alive, little is known about the functional anatomy of their nervous systems (Northcutt and Bemis, 1993; Nieuwenhuys et al., 1998). In this chapter we deal with coelacanth brains only to the extent that they help us reconstruct some features of early gnathostome brains (for more on coelacanths, see Chapter 4).

3.2. The Paleoecology of Early Gnathostomes

Nestled among the extant gnathostomes are a large number of species that have long gone extinct (Figure 3.8). Their fossil record tells us, for example, that lungfishes and holocephalans were once far more diverse than they are now. Also of interest is that ancient cartilaginous fishes (e.g., *Cladoselache*) are more similar to sharks than to modern holocephalans and probably diverged from the other cartilaginous fishes before the holocephalans did. Because these ancient sharks resemble ancient ray-finned fishes, we can surmise that the last common ancestor of all extant jawed fishes probably shared many features with both groups (although it is surely too simple to think of this ancestor simply as an "average" of early bony and cartilaginous fishes; see Davis et al., 2012). This hypothetical ancestor of all crown gnathostomes was probably ~1 m long, likely had an elongate body with long jaws, and presumably lived in a marine environment. The size and shape of its brain are difficult to fathom, but some good endocasts (i.e., casts of the skull cavity in which the brain is housed) from basal cartilaginous and bony fishes indicate that these animals probably had relatively small brains with a surprisingly small telencephalon (Figure 3.9).

3.2.1. Stem Gnathostomes

The extinct lineages that branched off the main vertebrate line after the origin of cyclostomes but before the origin of crown gnathostomes are called stem gnathostomes (Figure 3.8; see also Figure 1.9 in Chapter 1). Some of them had not yet evolved jaws, but they did have paired fins. The most basal of these lineages is the conodonts. These jawless vertebrates first appeared in the fossil record during the late Cambrian and persisted until roughly 200 mya. For a long time they were known only from their fossilized tooth-like (conodont) elements, but several relatively complete fossils of conodont animals have now been described (Briggs et al., 1983; Donoghue et al., 2000). According to these fossils, conodont animals had eel-like bodies and large lateral eyes at the front of the head. They possessed more than a dozen conodont elements within their pharynx and probably used them to grab and shear food. Indeed, they may have had a pulley-based feeding apparatus very similar to that of cyclostomes (see Figure 2.6 in Chapter 2; Goudemand et al., 2011). Relatively late in the evolution of conodont animals, the intra-pharyngeal conodont elements developed a hard enamel-like coating, further increasing their resemblance to the teeth of jawed vertebrates. However, this resemblance is almost certainly the result of evolutionary convergence because the earliest conodonts lacked the enameloid coat (Murdock et al., 2013).

Phylogenetically intermediate between conodonts and crown gnathostomes were a variety of jawless stem gnathostomes that are collectively referred to as

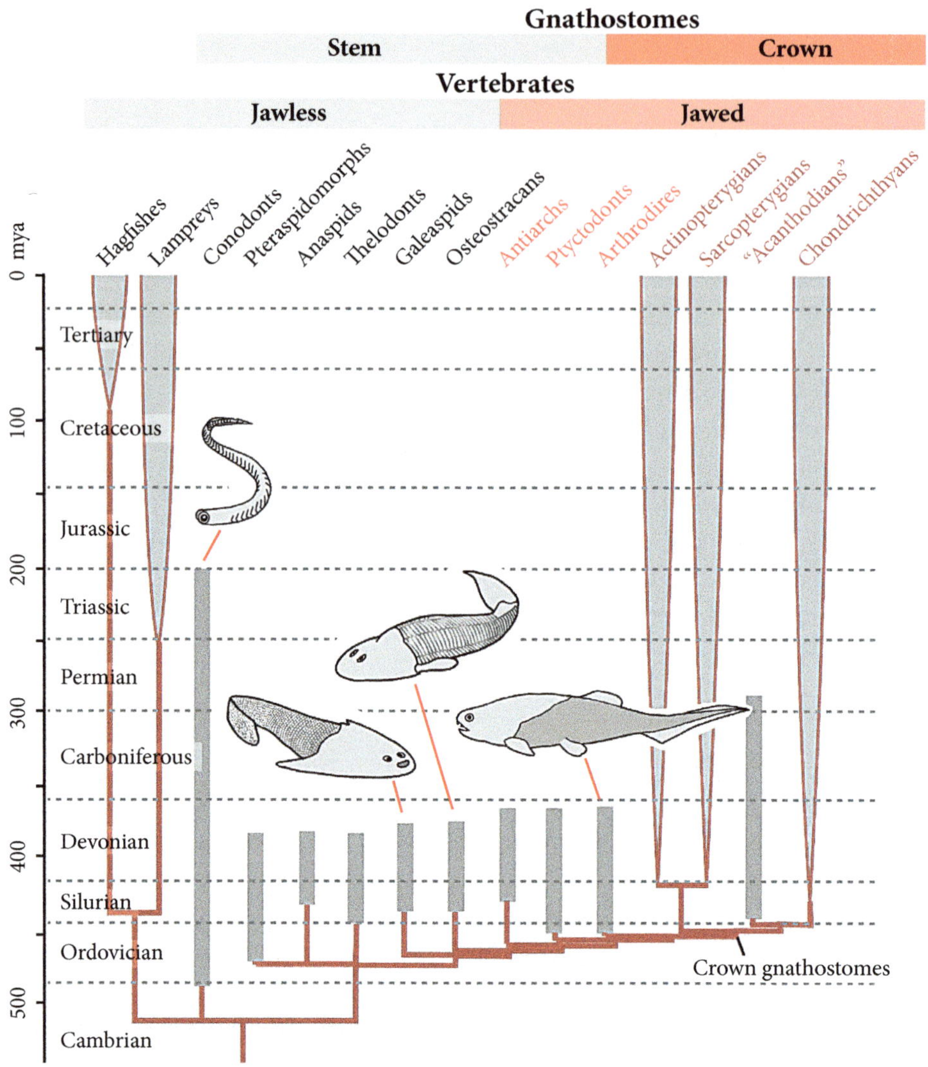

Figure 3.8 Phylogeny of extant and extinct vertebrates. Crown gnathostomes are defined as all extant gnathostomes plus any extinct gnathostomes derived from their last common ancestor. The stem gnathostomes include all other vertebrates, except crown jawless vertebrates (hagfishes and lampreys). Note that some stem gnathostomes lacked jaws, whereas others had them; the former are generally called "ostracoderms," the latter "placoderms." The time of major diversification within the living lineages is indicated by the widening of the branches. The illustrated animals are an advanced conodont, a galeaspid, an osteostracan, and an arthrodire placoderm (left to right).

Phylogeny adapted from Donoghue and Keating (2014). Animal drawings based on Donoghue et al. (2000) and a drawing by Nobu Tamura (Wikimedia).

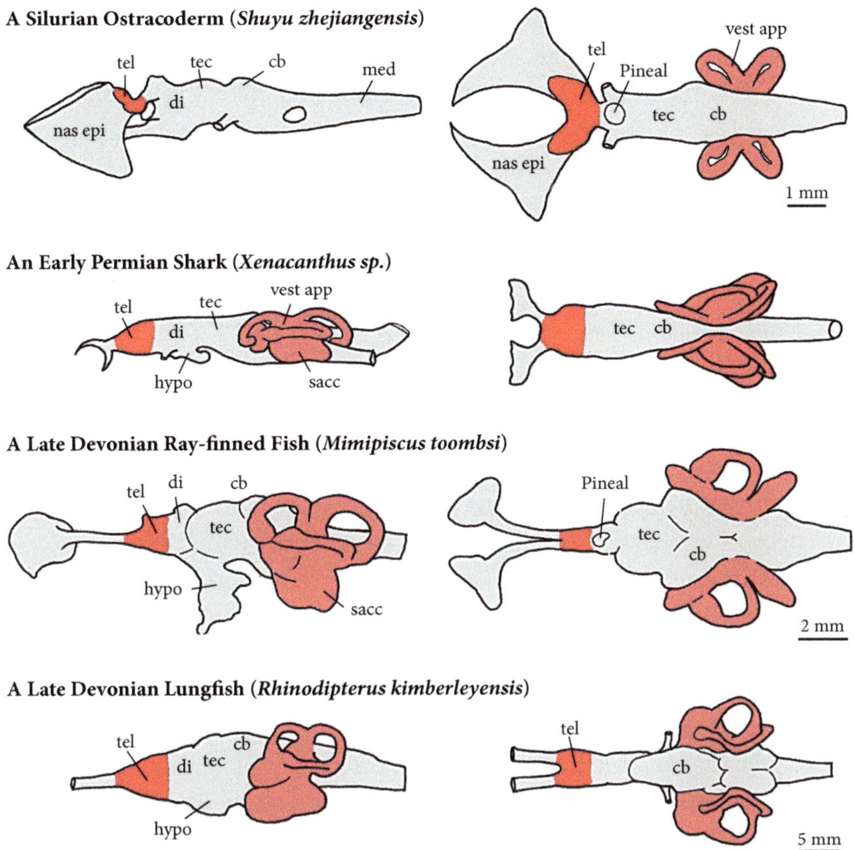

Figure 3.9 Endocasts of early gnathostomes. Shown here are lateral (left) and dorsal (right) views of endocasts from a Silurian ostracoderm (a galeaspid; Gai et al., 2011), an early Permian shark (Schaeffer, 1981), a Devonian ray-finned fish (Giles and Friedman, 2014), and a late Devonian lungfish (Clement and Ahlberg, 2014). The presumed extent of the telencephalon (including the olfactory bulb) is shown in red. The vestibular apparatus is shaded pink (it was removed from the lateral view of the galeaspid endocast). Scale bars are shown when they were provided in the original.
Abbreviations: cb – cerebellum; di – diencephalon; hypo – hypothalamus; med – medulla; nas epi – nasal epithelium; sacc – sacculus; tec – tectum; tel – telencephalon; vest app – vestibular apparatus.

ostracoderms (Figure 3.8). They included a diverse set of lineages (e.g., galeaspids), but most of them, in contrast to conodont animals, had paired pectoral fins, and almost all of them were heavily armored. This armor consisted of large bony plates that formed within the skin around the head and chest. The fact that this dermal armor contained bone implies that bone did not evolve with the origin of bony fishes (as biologists once thought) but very early in the evolution of jawed vertebrates (Wagner and Aspenberg, 2011; Keating et al., 2015). In any case, ostracoderms probably used their heavy armor to protect themselves from predatory arthropods, such as the giant sea scorpions (eurypterids) that roamed the oceans at the time. The ostracoderms themselves probably fed mainly on benthic detritus and invertebrates.

Most of the remaining stem gnathostomes are collectively referred to as placoderms. They resemble ostracoderms insofar as most of them have paired fins and heavy dermal armor encasing the front of the animal. However, placoderms also have jaws. Exactly when vertebrate jaws evolved is somewhat difficult to determine because the diverse placoderms constitute separate branches off the vertebrate tree and may well have evolved jaws more than once. Indeed, it appears that placoderms experimented with a variety of different jaw designs, and later gnathostomes continued to experiment. For example, some lineages fused the upper jaw to the braincase, which generally makes for a powerful bite; others opted to make their jaws more mobile. As mentioned earlier, the teleost fishes evolved especially mobile jaws that frequently contain 20 or more distinct elements that move during feeding. These jaws are good for biting and grinding food, but they also excel at sucking food into the pharynx, which makes it easier to catch small prey.

When gnathostomes close their jaws, water and food are trapped inside the pharynx. The water then exits the pharynx through the gills, while the food is (hopefully) retained. In jawless vertebrates, which cannot close their mouth, this valve-like function is performed by a thin membrane (or velum) at the anterior end of the pharynx. However, closing the mouth with jaws is a more efficient mechanism for maximizing water flow across the gills. This increased efficiency in turn increases the gills' ability to absorb oxygen and release carbon dioxide. Indeed, it has been proposed that jaws first evolved in the service of respiration and were only later used to capture prey (Mallatt, 1996).

Thus, early gnathostomes increased their ability to swim and steer effectively, and they obtained the metabolic energy for all that swimming by improving gas exchange and boosting food intake. Most of the early gnathostomes were efficient hunters, not just of large prey but also small and agile prey. As a result, early gnathostomes were able to exploit the spatially complex habitats provided by growing marine reefs, freshwater wetlands, and streams, whose banks became more stable as land plants with roots evolved in the Devonian. The heavy armor of placoderms and ostracoderms suggests that these stem gnathostomes were themselves hunted, mainly by large arthropods but also, perhaps increasingly, by larger placoderms and other jawed fishes. Indeed, fossilized gut contents indicate that *Cladoselache*, a primitive shark, ate smaller jawed fishes, some shrimp-like creatures, and a few conodonts (Brett and Walker, 2002). Given this fish-eat-fish world, it is not surprising that some Devonian fishes reached enormous body sizes. The placoderm *Dunkleosteus*, for example, was up to 6 m long and weighed up to one ton (it is pictured on the book's cover). These considerations suggest that ostracoderms might have gone extinct because they were outcompeted by jawed fishes, but direct support for this hypothesis is weak (Sansom et al., 2014).

3.2.2. The End-Devonian Mass Extinction

In fact, both ostracoderms and placoderms, as well as many other lineages, went extinct around the same time. Most of these stem gnathostomes had originated shortly after the end of the Ordovician period, after the earth emerged from a global ice age that had eliminated numerous invertebrate lineages and seriously reduced the diversity of conodonts. Stem gnathostomes then thrived throughout the Silurian and most of the Devonian periods, from ~440 to 360 mya, perhaps assisted by a steady rise in oxygen levels during the Silurian (Figure 3.10; Qu et al., 2010). For most of the Devonian, placoderms and ostracoderms coexisted with the early cartilaginous and bony fishes. It was such a good time to be a fish that the Devonian is often called "the Age of Fishes." However, the situation changed dramatically toward the end of the Devonian, which represents one of the "big five" periods of massive, taxonomically broad, and global extinctions. This period most likely featured a series of global disruptions that spanned 10–25 million years and, collectively, killed off roughly 20% of all families and 80% of all species. Among them were all the ostracoderms and placoderms, as well as various "acanthodians" (a paraphyletic group also known as "spiny sharks"; see Figure 3.8) (Brazeau and Friedman, 2015).

The cause of the great extinction near the end of the Devonian remains debatable (Halim and Wignall, 1997; Brannen, 2017). Part of the problem is that there was a series of different catastrophic events that played out over millions of years. One of these cataclysms was probably the near-simultaneous eruption of multiple volcanoes that spewed enormous amounts of CO_2 into the atmosphere, causing extensive global warming through the greenhouse effect. This event was likely followed (and possibly preceded) by a dramatic global cooling brought about by the dramatic proliferation of land plants during the Devonian, including the rise of large spore-producing tree-like plants (especially of the genus *Archaeopteris*).

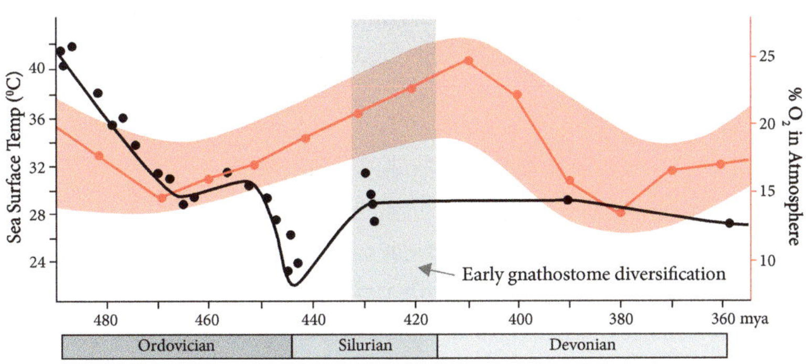

Figure 3.10 Temperatures and oxygen levels from the Ordovician to the Devonian. As this graph shows, early jawed fishes (gnathostomes) diversified when global temperatures were recovering from a very cold period and oxygen levels were relatively high and rising.

Based on data from Berner (2006), Trotter et al. (2008), and Scotese (2008) http://www.scotese.com; adapted from Qu et al. (2010).

Collectively, these plants are thought to have pulled as much as 90% of the existing CO_2 out of the atmosphere (Brannen, 2017), which in turn would have cooled the planet and lowered sea levels dramatically. In addition, the roots of the land plants would have liberated large amounts of phosphorus and other nutrients as they converted rocks to soil. All of this "fertilizer" eventually washed into the rivers and oceans, where it stimulated the formation of enormous algal blooms. These algae would have sucked all the oxygen out of the water, killing most fishes and other aquatic animals.

Although the precise causes of the end-Devonian extinction remain controversial, it seems safe to say that it involved relative rapid and dramatic changes in the habitats of virtually all animals. Under those conditions, early sharks and bony fishes might have persisted because they could migrate to new habitats and hunt over long distances. Early lungfishes seem to have evolved an alternate survival strategy, namely the ability to obtain oxygen from air, which can hold far more oxygen than water. Moreover, African lungfishes evolved the ability to fast for many months or even years. As noted earlier, they can even cocoon themselves in slime and mud when water is scarce, waiting for the rains and floods to return. One could say much more about the early jawed vertebrates and their environment, but our principal interest here is not the fossils or their ecology per se. Instead, our aim is to use this information to better understand the changes in the nervous system that accompanied the evolution of the early jawed fishes.

3.3. The Sense Organs of Early Gnathostomes

The sense organs of early jawed fishes were largely similar to those of their jawless ancestors, but they became somewhat more elaborate. These enhancements involved vision, chemoreception, and the vestibular sense. Whether the sense of hearing evolved with gnathostomes or earlier remains unclear, but it was certainly elaborated in substantial ways within select gnathostome lineages.

3.3.1. Photoreception

Early gnathostomes retained the paired eyes of their immediate ancestors, as well as a pineal gland that probably had both neurosecretory and light-sensing functions. All indications are that the lateral eyes of early gnathostomes were at least as well developed as those of lampreys, showing all the major retinal cell types typical of extant gnathostomes. Molecular data suggest that they contained at least three types of cone opsins and two different rhodopsins, which probably arose by duplication of a single rhodopsin gene in early gnathostomes (Collin et al., 2009; Lagman et al., 2013). Thus, early gnathostomes retained a duplex retina, capable of vision in both dim and bright light. The presence of multiple cone opsins suggests that early gnathostomes might have been capable of color vision, but color vision requires neural circuitry to compare activity across different cone types. Evidence for this kind of circuitry has been obtained from basal ray-finned fishes (Collin, 2007), but evidence for color vision among basal jawed fishes is currently limited to behavioral data on stingrays (Van-Eyk et al., 2011). In short, we suspect that early gnathostomes were capable of color vision, but the evidence is scarce. As reviewed in Chapter 2 (Section 2.4.1), evidence for color vision is also scarce in jawless vertebrates, notably lampreys.

The retina in early jawed vertebrates probably featured a central area of higher cell density, but this increase was not as steep as in the foveae of many amniotes. The area of increased cell density may have been elongated horizontally, suggesting that visual information along the distant ocean floor was of special importance to these animals. Since the retina of lampreys shares these features, they may have been retained from a jawless ancestor. However, lungfishes and paddlefish do not show these specializations, suggesting that they may have evolved independently in lampreys and gnathostomes. Later, within the various gnathostome lineages, the retina evolved a number of additional specializations. For example, damselfishes have specialized photoreceptors that can detect ultraviolet light, some teleosts have a pitted fovea (Collin and Collin, 1988), and sharks reduced their cone opsin diversity to one. Numerous variations in photoreceptor and retinal ganglion cell density have also been described and linked to life in diverse ecological niches (e.g., Collin and Partridge, 1996). At the origin of gnathostomes, however, the retina probably underwent only relatively minor modifications.

3.3.2. Chemical Senses

The nasal epithelium of lampreys and hagfishes sits inside an olfactory sac that connects to the body surface via a single nostril. In hagfishes, but not lampreys, this olfactory sac is connected to the pharynx by a duct through which water can be drawn across the olfactory epithelium (Holmes et al., 2011). Because odorants diffuse very slowly in water (~1 mm in 10 min), this increased rate of flow across the sensory surface is surely adaptive. However, the nasopharyngeal duct is probably a uniquely derived feature of hagfishes, because the nasal sacs in jawed fishes are not connected to the pharynx. Instead, water flows into their nasal sacs through a rostral nostril and out through a more caudal one, pushed along by beating cilia inside the sacs. The flow of water through these nasal sacs is accelerated further when a fish moves forward relative to the surrounding water (Cox, 2008).

In contrast to lampreys and hagfishes, gnathostomes have paired, bilateral nasal sacs. The nasal epithelium inside those sacs contains sensory cells that express one of several different olfaction-related receptor genes. Gnathostomes inherited two of these classes from their jawless ancestors, namely olfactory receptors (ORs) and one kind of vomeronasal receptors (V1Rs). A second class of vomeronasal receptor

(V2Rs) is found in gnathostomes but not in lampreys, implying that it is probably a gnathostome innovation (Grus and Zhang, 2009). Trace amine-associated receptors (TAARs) may also be a gnathostome innovation, although lampreys have functionally similar receptors that are closely related to serotonin receptors (Hashiguchi and Nishida, 2007; Hussain et al., 2009). The ORs are a large family of genes in tetrapods, but there were probably fewer than nine ORs in early gnathostomes; two of these ancestral ORs later gave rise to the expanded OR family in tetrapods (Niimura, 2012). In that context it is interesting that cartilaginous fishes have only one functional OR and, instead, rely mainly on V2Rs for olfaction (Ferrando and Gallus, 2013). Moreover, the vomeronasal receptors in cartilaginous fishes are not segregated into a separate epithelium, as they are in lungfishes and tetrapods. It is not yet clear whether this feature is primitive for gnathostomes or a specialization of the cartilaginous fishes.

What did early gnathostomes smell with their expanded set of olfaction-related receptors? This question is difficult to answer, but early jawed fishes were probably quite sensitive to odorants emitted by prey. We may note, for example, that the basal ray-finned fish *Polypterus* gives a positive food-seeking response when a filtered solution of 10^{-14} g of beef heart per liter of water is dripped into their aquarium (Pfeiffer, 1969). These fishes probably respond mainly to amino acids and nucleotides that are released from dead or dying animal tissue. Even living animals release chemicals such as bile salts into the surrounding water, and early vertebrates could probably smell them. Indeed, many lampreys spend several years in the open ocean before returning to their natal streams to spawn, finding those streams by smelling the bile salts that are released from other members of their own population—an odor on which the young lampreys apparently imprint (Buchinger et al., 2015). Similarly, many reef fishes spend their larval phase in the open ocean but then return to reefs (often the reef on which they hatched), using chemical cues as guides (Barth et al., 2015). Sharks, too, are good at odor-based navigation. Indeed, some are able to detect slight differences in odorant concentrations between their left and right nasal epithelia (Gardiner and Atema, 2010). Last but not least, early vertebrates probably used olfactory cues in courtship and to warn other members of their species of threats (Johnson et al., 2009; Wagner et al., 2011; Stensmyr and Maderspacher, 2012).

In addition to the principal olfactory systems, early jawed fishes probably possessed a set of sensory cells that innervate the olfactory epithelium and project directly to the hypothalamus, bypassing the olfactory bulb (von Bartheld, 2004; Vilensky, 2014). Such a "terminal nerve" is found in lungfishes, *Polypterus*, and lampreys, but its organization is so variable across lineages that it is difficult to say much more about its evolution than that it must have originated early in vertebrate phylogeny (see Appendix). Early fishes must also have possessed taste buds. As noted in the previous chapter, lampreys have taste buds inside their pharyngeal cavity. So do all the jawed fishes, suggesting that this trait is primitive for vertebrates. However, several ray-finned fishes and some sarcopterygians, especially the lungfishes, have taste buds not just in their mouth but also on the outside of their body (e.g., on the barbels of catfish and sturgeons). These external taste buds are always innervated by cranial nerve VII (the facial nerve), which also innervates some of the internal taste buds. Nonetheless, they clearly evolved independently in multiple lineages (Northcutt, 2004).

3.3.3. Vestibular Sensing

As reviewed in the previous chapter, the vestibular apparatus of hagfishes contains only a single semicircular canal per side, whereas lampreys possess a more complex set of vestibular canals and ducts (see Figure 2.16 in Chapter 2; Maklad et al., 2014). Fossil endocasts reveal that placoderms, like all jawed vertebrates, had three orthogonal pairs of semicircular canals, whereas the jawless ostracoderms have only two vertical canals (see Figure 3.9). Therefore, we hypothesize that gnathostomes and lampreys independently elaborated their semicircular canal system to include horizontal components. Moreover, we conclude that having three sets of semicircular canals oriented orthogonally to one another is a gnathostome innovation. It presumably increased the precision with which early gnathostomes could sense head rotations (specifically angular acceleration) in all possible rotation planes.

In addition to the semicircular canals, the inner ear of early gnathostomes contained two fluid-filled chambers, called the utricle and sacculus. Inside these fluid-filled chambers would have been one or more patches of hair cells, called maculae. Lampreys have just one such macula, called the macula communis, but jawed vertebrates have separated this ancestral macula into at least three distinct maculae. The stereo- and kinocilia of the hair cells in all vertebrate maculae protrude into a gelatinous paste containing small crystals of calcium carbonate, called otoconia (ear dust). In ray-finned fishes, the otoconia-filled paste is replaced with one or more solid ear stones or otoliths, likewise made of calcium carbonates. Because the otoliths and otoconia are denser than the rest of the body, inertia causes them to lag behind when the head accelerates; deceleration has the opposite effect. This differential lag bends the hair cell stereocilia, which then causes the hair cells to change their rate of transmitter release. Because hair cells are sensitive to deflections only in one direction and are oriented in different orientations across the various maculae, the animals can sense acceleration (or deceleration) of the body in all directions (Kasumyan, 2004). This system for sensing linear acceleration is probably derived from the statocyst organ of early chordates, which amphioxus and tunicates retained (see Chapter 2), but it has much greater directional sensitivity. Whether the system has greater directional sensitivity in jawed fishes than in cyclostomes remains unknown.

Overall, the vestibular apparatus of jawed fishes seems to have maintained its ancestral functions but increased its precision. Most likely, this improvement in vestibular sensing aided the animals in maintaining their body posture, especially during rapid swimming and turns. To appreciate the importance of this stabilization, consider that dead fish with swim bladders tend to float upside-down, which means that active movements, especially of the pectoral fins, are required to keep the belly down. Indeed, lesions of the vestibular apparatus dramatically impair swimming ability in fishes (Paul and Roberts, 1979; Colgate and Lynch, 2004).

3.3.4. Hearing and Localizing Sounds

Fishes can use their vestibular sacculus and, to a lesser degree, their utricle not only for sensing body acceleration, but also to hear underwater sounds. When a solid object vibrates underwater or rapidly changes velocity, it does two things. One, it generates a traveling pressure wave that is identical to what we call sound waves in air. Because water is much less compressible than air, this pressure wave travels 4.5 times faster and much further in water than in air (Rogers and Cox, 1988); most fishes cannot sense these long-distance pressures waves (Popper and Fay, 2011).

Two, underwater vibrations displace the water around the vibrating object. The resulting currents don't travel nearly as far as the pressure wave, but they can be felt a few meters away (depending on stimulus strength). Those same currents also move any fish in their way, because the density of fishes is nearly identical to that of water. However, the denser otoliths inside a fish lag behind, just as if the fish moved of its own accord. Therefore, the otolithic end organs can be used to hear external sounds, at least over short distances and at low frequencies (up to 700–1,000 Hz). Indeed, sturgeons, paddlefish, and some cartilaginous fishes can hear such low-frequency sounds over short distances, as long as their vestibular apparatus is functional (Lovell et al., 2005; Casper and Mann, 2009). Because the hair cells in the otolithic end organs are arranged in several different orientations, those fishes may be able to determine a sound source's direction; teleosts certainly can (see Walton et al., 2017). Unfortunately, we do not yet know whether lampreys or hagfishes have this near-field hearing ability. Therefore, it remains unclear whether the ability to hear low-frequency underwater sounds evolved with the origin of gnathostomes or earlier.

Although early gnathostomes almost certainly lacked the ability to hear underwater sound pressure waves, this ability did evolve in several lineages of teleosts (Frisch, 1938). These fishes took advantage of the fact that underwater sound pressure waves cause gas-filled swim bladders, which are a specialty of bony fishes, to vibrate in sympathy. Teleosts couple these sound-induced swim bladder vibrations to the inner ear, either though a chain of small bones (the Weberian ossicles of ostariophysine teleosts) or by placing an anterior extension of the swim bladder in close proximity to the inner ear otoliths (Figure 3.11; Braun and Grande, 2008; Schulz-Mirbach et al., 2012). Either way, those mechanisms activate the hair cells of the sacculus and, to a lesser extent, the utricle. Thereby, they greatly extend the

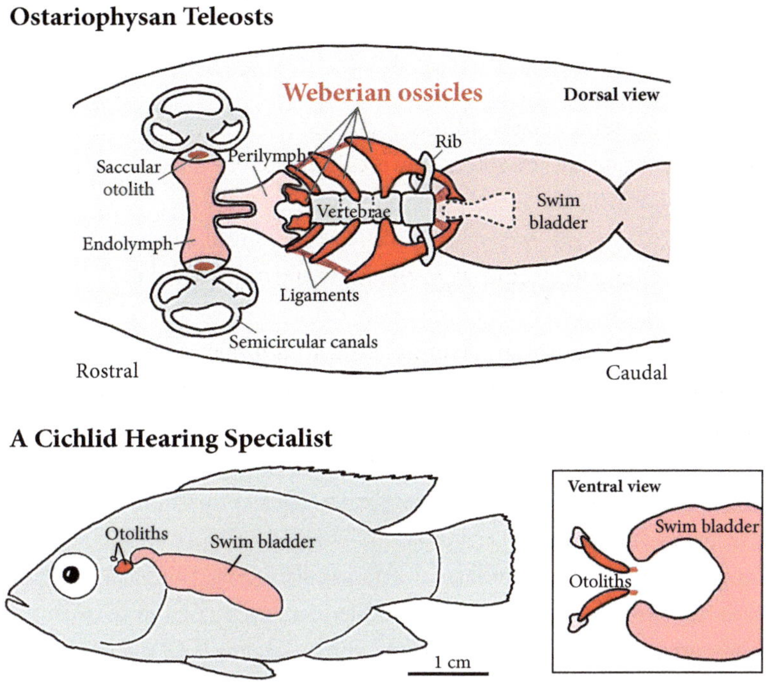

Figure 3.11 Hearing through a swim bladder. In some teleosts, such as the illustrated cichlid *Paratilapia polleni*, the swim bladder has a rostral extension that abuts the otoliths of the inner ear. When sound pressure waves cause the swim bladder to vibrate, those vibrations are transferred to the ear. Ostariophysan teleosts (e.g., goldfish and catfishes) couple their swim bladders to the inner ear through a series of small bones, the Weberian ossicles.
Based on Schulz-Mirbach et al. (2012).

distance over which the fish can hear and the range of frequencies they can detect. In fact, some shads and menhadens (clupeid teleosts) can hear ultrasonic sounds up to 180 kHz (Mann et al., 2001). Whether fishes can use this far-field hearing mechanism to localize external sound sources remains the subject of vigorous debate.

Because both near- and far-field hearing use the same hair cells that also respond to the fish's own movements, the information they encode is potentially confounded. However, swimming fish tend not to move back and forth at frequencies higher than a few Hz. Therefore, sound-related activity can generally be discriminated from movement-related activity within the nervous system by selectively processing different stimulus frequencies.

3.3.5. The Lateral Line Systems

The lateral line system of early gnathostomes included both mechanosensory and electrosensory components, both of which they had inherited from their jawless ancestors (see Chapter 2). However, both divisions of the lateral line were modified in gnathostomes, leading to some substantial changes in sensory abilities.

3.3.5.1. Mechanosensory Lateral Line

As noted in the previous chapter, the mechanosensory lateral line in lampreys contains free neuromasts, which are aggregates of hair cells that extend their stereocilia directly into the water surrounding the animal. Jawed fishes have similar neuromasts, but their stereocilia are covered by a gelatinous cupula (see Figure 2.15 in Chapter 2) that probably makes the hair cells more sensitive to water currents, especially at lower frequencies (Coombs and Montgomery, 1999; Kasumyan, 2003). In addition, most jawed fishes have canal neuromasts, which are located inside canals that run parallel to the skin surface and open to the outside only at regularly spaced locations. These lateral line canals are located mainly on the head, but one or more of them extend along the trunk, which accounts for the word "lateral" in the name of this sensory system.

By comparing the pattern of lateral line canals across the various gnathostome lineages, including placoderms, it is possible to reconstruct, at least in rough outline, the ancestral pattern of lateral line canals (Figure 3.12). Reconstructing the ancestral pattern of free neuromasts is more difficult because these neuromasts are more variable across species and do not fossilize. However, it is fairly clear that the earliest gnathostomes had both free and canal neuromasts (Northcutt, 1989). They

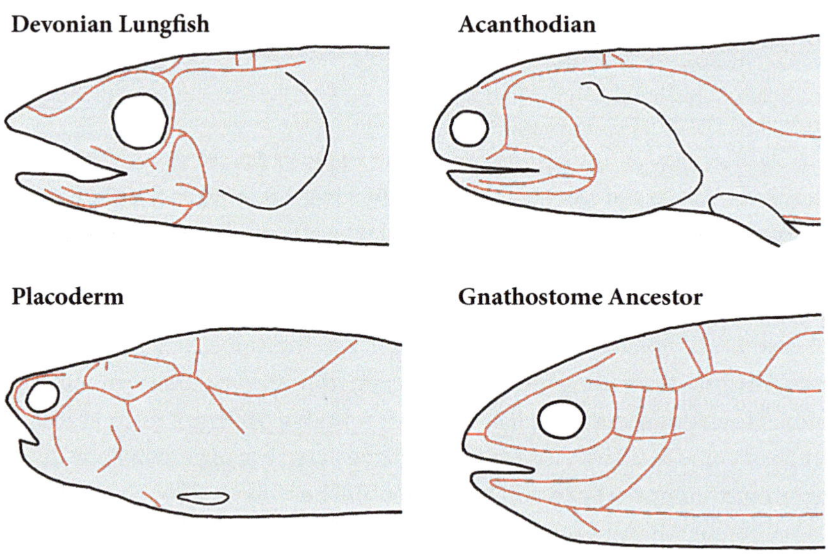

Figure 3.12 The lateral lines of extinct and ancestral gnathostomes. Shown in red are the lateral lines of a Devonian lungfish (genus *Dipterus*), an acanthodian (genus *Euthacanthus*), and an arthrodire placoderm (genus *Coccosteus*). Shown at the bottom right is the most likely distribution of the lateral lines in the last common ancestor of gnathostomes.

Adapted from Northcutt (1989).

were probably innervated by six pairs of lateral line nerves (Song and Northcutt, 1991), which are lacking in amniotes and, therefore, often ignored in standard neuroscience textbook descriptions of "the 12 cranial nerves" (see Appendix).

Although the hair cells in neuromasts are similar to those in the inner ear, they generally respond to different stimuli. Specifically, they tend to respond to "the amplitude, direction and extent of water displacements, applied as a spatial and temporal pattern to the surface of the animal" (Dijkgraaf, 1963, p. 59). The free neuromasts tend to respond preferentially to water currents running parallel to the fish's skin. In contrast, the canal neuromasts respond mainly to pressure differentials between the canal pores, which result from localized water movements perpendicular to the skin. Moreover, free neuromasts tend to encode the velocity of current flows, whereas canal neuromasts encode changes in current velocities (i.e., accelerations). In aggregate, the mechanosensory lateral line system can detect currents produced by small moving objects at a distance of 1–2 m, as well as movements of the fish relative to the surrounding water.

Surprisingly, lesions of the lateral line system seem to have relatively little effect on swimming ability (Dijkgraaf, 1963). Therefore, it appears that the mechanosensory lateral line system is specialized mainly for the short-range detection of moving objects (e.g., potential prey or predators). It may also be used to sense the buildup of water pressure that results when a fish approaches a solid object. Indeed, blinded fish frequently bump into aquarium walls when their lateral line system is rendered non-functional. This explains, at least in part, why blind cave fish have a very well developed mechanosensory lateral line system: they use it to avoid collisions with cave walls (Abdel-Latif et al., 1990; Yoshizawa et al., 2014).

3.3.5.2. Electrosensory Lateral Line

The electrosensory lateral line system also changed as gnathostomes evolved. While lampreys have electroreceptors that sit flush with the skin surface (see Figure 2.17 in Chapter 2), the electroreceptors of jawed fishes are located at the bottom of long mucus-filled channels, also known as ampullae (Baker et al., 2013). Because of this difference in location, and because the ampullae walls are better insulators than the rest of the fish's skin, the ampullary electroreceptors of gnathostomes are more sensitive than the electroreceptors of lampreys. This enhanced sensitivity probably increased the ability of early gnathostomes to sense the weak electric fields generated by potential prey at night, in turbid waters, or when the prey is otherwise invisible (Kalmijn, 1971). Since moving a conductor through a magnetic field induces electric currents, a swimming fish may even be able to use its electroreceptors to sense the earth's magnetic field (Peters et al., 2007). Some marine elasmobranchs use this ability to help them navigate, but the electroreceptors of freshwater fishes are probably not sensitive enough (Peters et al., 2007).

Electroreceptors were lost in vertebrates as they moved onto land (see Chapter 4). This is not surprising, as air is a poor conductor of electricity. Much more surprising is that electroreceptors were also lost in the last common ancestor of teleosts, *Amia*, and gars (i.e., neopterygians). Similar receptors later re-evolved in two teleost lineages, namely in the mormyrids (e.g., elephant-nose fishes) and in a lineage that includes both gymnotoids (e.g., glass knife-fishes) and catfishes. In mormyrids and gymnotoids, this re-evolved electric sense was accompanied by an electric organ, capable of generating weak electric fields (Bullock et al., 2005). These weakly electric fishes can sense distortions in the self-generated electric field that are caused by objects that differ in conductivity from the surrounding water; thus, they can use variations in their electric field to localize and (to some extent) identify nearby objects. The electroreceptive teleosts are also capable of detecting weak electric fields produced by external sources, such as potential prey hidden in mud.

Given these observations, it is difficult to fathom why neopterygians lost their electrosensory abilities. Indeed, no satisfactory hypothesis has been proposed. Some authors have argued that the environment of early neopterygians might have been too turbulent and, therefore, too electrically noisy for electroreceptors to provide useful information (Carrier et al., 1992). Another possibility is that early teleosts could use their improved sense of hearing to detect hidden prey and, thus, had less need for the electric sense. A third possibility is that the evolutionary changes in the scales of early neopterygians somehow interfered with the development of electroreceptors. Currently, none of these hypotheses has strong support.

3.4. Gnathostome Movements and Motor Control

Early gnathostomes retained from their jawless vertebrate ancestors the eel-like form of swimming called lateral undulation (see Chapter 2). However, they probably increased the amplitude of the traveling wave near the tail of the animal, so that most of the propulsion was generated by the tail fin. The other fins would have been used mainly for steering, body stabilization, and smooth deceleration (i.e., braking). As noted earlier, the neural circuits underlying steering and stabilization remain largely unknown. The lateral undulation itself, however, was almost certainly generated by a spinal pattern generator that is similar to the central pattern generator for swimming in lampreys, which has been studied in great detail (Grillner et al., 1995). Descending inputs to this spinal pattern generator in lampreys derive almost exclusively (~90%) from large neurons in the reticular formation of the basal midbrain and hindbrain (Dubuc et al., 2008). By contrast, the spinal cord of jawed vertebrates receives substantial inputs also from the vestibular nuclei, optic tectum, forebrain, and cerebellum-associated nuclei. Thus, the variety of descending projections to the spinal cord probably expanded with the origin of gnathostomes.

One pair of reticulospinal neurons in lampreys and most jawed fishes is much larger than the others. This giant neuron, called the Mauthner cell (Figure 3.13), is involved in several escape behaviors (Eaton et al., 2001). Best studied is its role in the C-start response of teleosts. This behavior begins with a forceful contraction

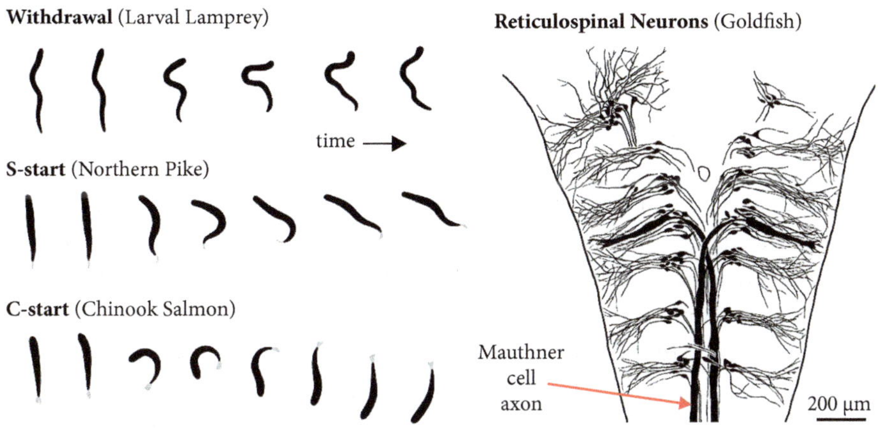

Figure 3.13 Startle responses and reticulospinal neurons. Shown on the left are the responses of a larval lamprey and two different teleosts to a startling stimulus in front of the animal. Although the responses differ in speed, form and effectiveness, they all are thought to involve the activity of reticulospinal neurons, especially the Mauthner cell. Shown on the right is a dorsal view of the reticulospinal neurons in an adult goldfish, retrogradely labeled by tracer injections into the spinal cord. The Mauthner cells are by far the largest reticulospinal neurons and have the thickest axons.
From Liu and Hale (2014) and Lee et al. (1993), with permission from Elsevier and John Wiley & Sons, respectively.

of the trunk muscles on one side of the animal, causing a C-shaped body bend, and is followed by swimming movements that quickly move the animal away from the threatening stimulus (Hale et al., 2002). From stimulus to muscle contraction, the C-start response takes only 12 ms. This speed results mainly from the Mauthner axon's large diameter (50–90 μm in goldfish; Funch et al., 1981). It is also facilitated by the fact that gnathostomes, in contrast to amphioxus and cyclostomes, ensheath their axons in myelin, which further increases axonal conduction velocity (Bullock et al., 1984). Cyclostomes do have glia, but they don't make the proteins associated with myelin sheaths, including myelin basic protein and myelin associated protein (Waehneldt et al., 1986). These proteins appear to be novel with gnathostomes.

Although lampreys have Mauthner cells, they do not exhibit C-starts. Instead, activation of a Mauthner cell in lampreys triggers head withdrawal, which begins with a rapid bending of the animal's front end away from the stimulus (Figure 3.13; Liu and Hale, 2014). Because this head withdrawal response is also seen in eels and in the polypteriform reedfish, we hypothesize that it replaces the C-start in very elongate animals. Since the earliest gnathostomes were probably not nearly as elongated as lampreys, eels, or reedfishes, they probably escaped from sudden threats by means of a C-start. If this suggestion is true, then the head withdrawal response is a later innovation that accompanied body elongation in select lineages. More elongate species also tend to exhibit more bilateral activation of the trunk musculature, and they lack inhibitory connections that, in other species, prevent the simultaneous activation of Mauthner cells on both sides of the body (Bierman et al., 2009). We can conclude, therefore, that the phylogenetic loss of an ancestral set of inhibitory connections allowed elongate animals to evolve a more bilateral mechanism of trunk muscle control.

Much has been written about the origin of vertebrate jaws (e.g., Miyashita, 2016). Almost all authors agree that the main jaws of vertebrates are derived from the first (mandibular) gill arch of jawless vertebrates and from neural crest cells that migrate into it (Mallatt, 2008). Problems arise, however, when one tries to homologize specific elements of the vertebrate jaw to elements in jawless vertebrates. One hypothesis, based mainly on comparative morphology, is that the dorsal part of the first gill arch extended anteriorly as jaws evolved and that most jaw components have homologs in jawless vertebrates (Figure 3.14; Mallatt, 1996). The main alternative hypothesis is that the rostral boundary of *fgf8* expression is shifted caudally in gnathostomes and that this causes a caudal shift in *bmp* expression, which ultimately leads to novel tissue interactions that produce a novel upper jaw element (Kuratani, 2004). The comparative developmental data also suggest that duplication of the *dlx* gene in gnathostomes allowed for more dorsoventral differences in gill arch development. One problem with all of these hypotheses is that early jawless vertebrates probably possessed some kind of complex feeding apparatus (see our discussion of cyclostomes and conodonts in Chapter 2 and Section 3.2.1, respectively) and that it is unclear how this primitive feeding apparatus was transformed as gnathostome jaws evolved.

Because the debate surrounding these hypotheses is not yet resolved, it is difficult to say much about the evolution of the muscles that open and close the jaws. Some of them may be new; others may be derived from muscles in the mouth and pharynx of jawless fishes. In any case, the muscles that control the principal jaws are consistently innervated by the trigeminal motor nucleus, whereas the more caudal pharyngeal jaws are innervated by the motor nuclei of the facial, glossopharyngeal,

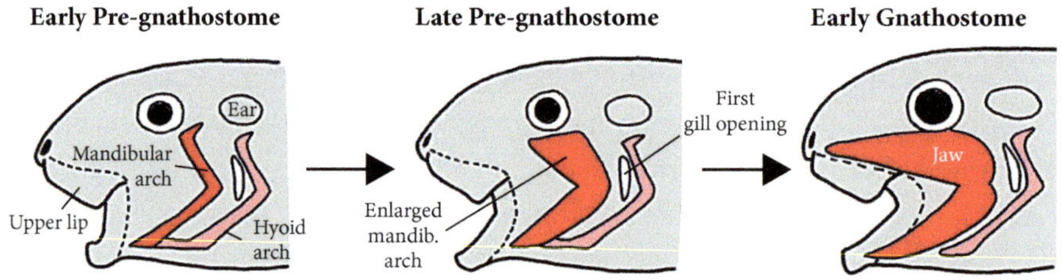

Figure 3.14 Early jaw evolution. It is widely accepted that vertebrate jaws evolved from the most rostral gill arches of jawless vertebrates. During this evolution the upper and lower elements of the first gill arch probably tilted forward and, eventually, extended to the front of the head.

Adapted from Mallatt (1996), with permission from Oxford University Press.

and vagal nerves. Unfortunately, very little is known about the neural mechanisms that control the jaw muscles in gnathostomes.

Given that cyclostomes lack jaws, it is interesting to ask whether these animals also lack the neurons of the mesencephalic trigeminal nucleus (mesV), which relay mechanosensory information from the teeth and their surrounding tissues directly to trigeminal motor neurons (Figure 3.15). In most vertebrates, these mesV neurons have their cell bodies in the optic tectum but otherwise resemble primary sensory neurons in the trigeminal sensory ganglion. In sharks they trigger jaw closing when the teeth are stimulated (Roberts and Witkovsky, 1975), and in tetrapods the jaw-closing reflex can be triggered by the activation of axons innervating muscle spindles in the jaw. Cyclostomes were long thought to lack mesV neurons, but Anadón et al. (1989) identified a group of neurons in the lamprey medulla that resemble the mesV neurons of other vertebrates, even though they are not located in the mesencephalon (see also Northcutt, 1979). Based on these data, it seems reasonable to conclude that mesV neurons evolved with the origin of vertebrates, but then acquired additional jaw-related functions as vertebrates acquired jaws.

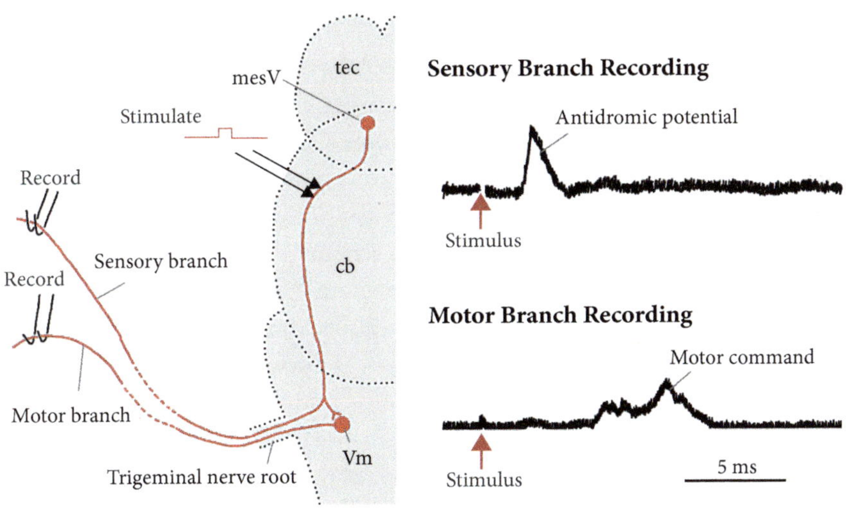

Figure 3.15 The mesencephalic trigeminal nucleus. The neurons of the mesencephalic trigeminal nucleus (mesV) conduct sensory information from the mouth to the trigeminal motor nucleus (Vm), which in turn innervates muscles that close the jaws. Curiously, mesV neurons have their cell bodies in the deep optic tectum, far from where their axon splits. Electrical stimulation of mesV axons near the cell body causes action potentials to travel antidromically (opposite to the normal direction of action potential propagation) into the sensory axon branch. After a brief synaptic delay, the electrical stimulation also evokes activity in the motor branch of the trigeminal nerve.

Adapted from Roberts and Witkovsky (1975).

3.5. The Brains of Early Jawed Fishes

Brain size relative to body size probably increased only slightly with the origin of jaws. However, once jaws were in place, relative brain size increased repeatedly in several different lineages. In the following sections, we review how the internal organization of the brain changed, both with the origin of jaws and in some later lineages. We focus mainly on the brain regions that are vertebrate innovations, namely the cerebellum, midbrain, and telencephalon (see Chapter 2). These regions also account for most of the increases in relative brain size. However, we also cover some aspects of the more conservative brain areas, notably the medulla and hypothalamus.

3.5.1. Medulla

The medulla accounts for roughly half the brain in adult lampreys (Platel and Vesselkin, 1989) and probably occupied at least one-third of the brain in early jawed vertebrates (see Figure 3.9). It is shaped like a rhombus that widens rostrally. Its roof is a vascularized membrane that contributes to the generation of cerebrospinal fluid. Caudally the medulla is continuous with the spinal cord. Indeed, it is useful to think of the spinal cord's ventral and dorsal horns as extending rostrally into the medulla, forming its ventromedial and dorsolateral divisions, respectively. Just as the spinal cord's dorsal horn is generally sensory in function, so is the medulla's dorsolateral region, which is called its alar plate. Conversely, the medulla's ventromedial region, or basal plate, contains numerous motor neurons, just as the ventral horn does. This division of the medulla into alar sensory and basal motor regions, or columns (Figure 3.16), is a good general schema (Nieuwenhuys, 1974; Heijdra and Nieuwenhuys, 1994). However, there are some exceptions to the general pattern. In particular, the basal region of the medulla contains several cell groups that are not directly involved in motor control, such as the serotonergic raphe nuclei and the interpeduncular nucleus. In addition, some motor neuron precursors migrate into the alar plate during embryonic development (Ju et al., 2004).

The structural and functional organization of the medulla is very similar across all jawed fishes, indeed across all vertebrates. However, some medullary nuclei have shifted their adult location during evolution due to altered patterns of migration or, less frequently, shifts in developmental patterning (Gilland and Baker, 2005). Moreover, some specific subdivisions of the medulla have become greatly enlarged (hypertrophied) in select lineages. Torpedo rays, for example, have modified their pectoral fin muscles into a powerful electric organ and greatly expanded the number of motor neurons innervating this structure. Their "electric lobes," which are probably homologous to the motor nuclei of the seventh, ninth, and tenth cranial nerves of other fishes, occupy roughly 60% of the torpedo ray brain (Roberts and Ryan, 2009). Another good example is the previously mentioned "vagal lobe"

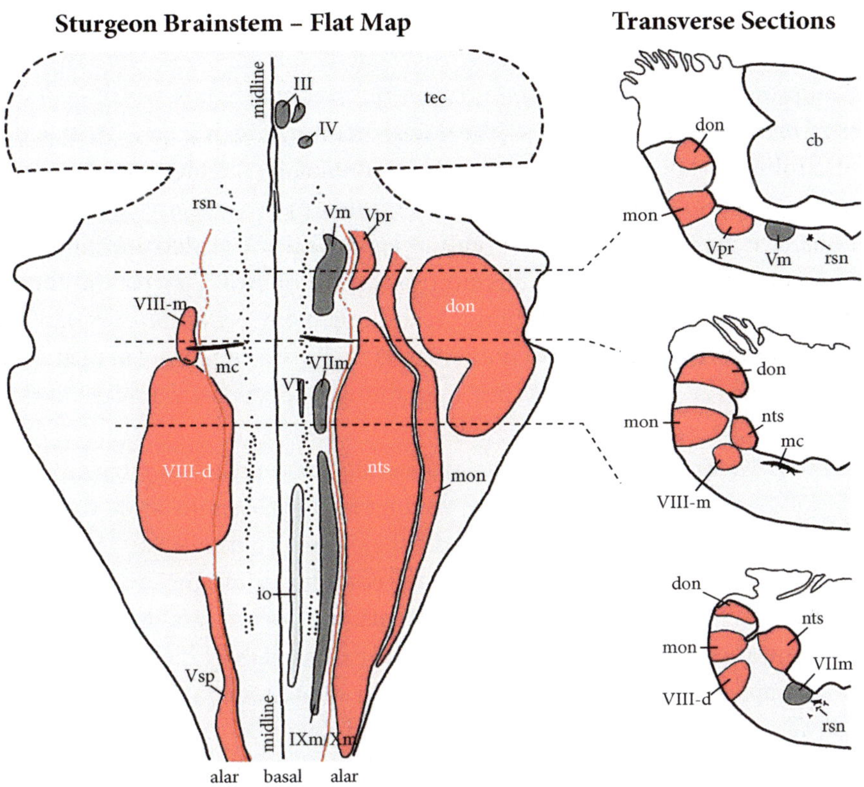

Figure 3.16 The medulla of a basal ray-finned fish. The principal cell groups in a sturgeon's medulla are shown on the left in a "flat map," in which the medulla has been conceptually unrolled. Some cell groups are shown only on the right, others only on the left. Sensory cell groups are shown in red, motor nuclei in dark gray. Reticulospinal neurons are shown as black dots bilaterally. Shown on the right are transverse sections through the medulla at three rostrocaudal levels (dashed lines).

Abbreviations: III – oculomotor nuclei; IV – trochlear nucleus; Vm – trigeminal motor nucleus; Vpr – principal (sensory) trigeminal nucleus; Vsp – spinal trigeminal nucleus; VI – abducens nucleus; VIIm – motor nucleus of the facial nerve; VIII-d – descending octaval nucleus; VIII-m – magnocellular octaval nucleus; IXm/Xm – motor nuclei of the glossopharyngeal and vagus nerves; cb – cerebellum; don – dorsal octavolateralis nucleus; io – inferior olive; mc – Mauthner cell; mon – medial octavolateralis nucleus; nts – nucleus of the tractus solitarius; rsn – reticulospinal neurons; tec – tectum.

Adapted from Nieuwenhuys et al. (1998).

of goldfish and other cyprinid teleosts (see Figure 3.6). It receives information from taste buds on a muscular organ in the pharyngeal roof of cyprinids and sends motor commands back to that same muscle. When the taste buds detect a tasty morsel, the muscle contracts at the same location, forming a bump and pressing the food against the gill rakers. The fish can thus retain the food while it spits out less interesting material.

Significant variation also occurs within the "octavolateralis region" of the medulla's alar plate, which receives input from the eighth (octavo-) and lateral line (lateralis) nerves (see Appendix). In cartilaginous and basal bony fishes, including sturgeons (Figure 3.16), this region is divided into three longitudinal columns. The most dorsal column receives electrosensory lateral line input, whereas the medial (aka intermediate) octavolateralis nucleus receives mechanosensory lateral line input. *Amia*, gars, and teleosts lack the dorsal column, which is consistent with the fact that this lineage lost electroreception. Remarkably, the few teleosts that have re-evolved electroreceptors also re-evolved a dorsal electrosensory column in their medulla, which is called the electrosensory lateral line lobe (Metzner and Juranek, 1997; Meek et al., 1999). Indeed, this electrosensory column of teleosts is quite large and divisible into multiple subregions. According to our definition of homology, it is a novel brain region that is not homologous to the dorsal octavolateralis nucleus of cartilaginous and basal bony fishes, even though it occupies a similar position in the brain.

The most ventral column of the octavolateralis region receives projections from the vestibular apparatus. This ventral column is divisible into three distinct cell groups in lampreys, four in basal jawed fishes, and five in teleosts (McCormick, 1982). Although this increase in structural complexity roughly parallels the increase in complexity of the vestibular apparatus, it is not the case that each division of the ventral column receives input from just one patch of hair cells within the vestibular apparatus. Instead, the outputs of the semicircular canals, sacculus, and utricle all overlap with one another (Straka and Baker, 2013). Nonetheless, some functional segregation probably exists at the level of single cells, as the outputs of the ventral column tend to be functionally distinct. In particular, auditory information is conveyed to part of the midbrain roof (to be discussed later), whereas information about angular and linear acceleration of the animal's head is conveyed primarily to brainstem areas that control eye and body movements (Puzdrowski and Leonard, 1994).

3.5.2. Cerebellum and Cerebellum-Like Structures

As noted in the previous chapter, adult hagfishes and lampreys do not have a proper cerebellum, but they do have cerebellum-like areas, which were retained in all the jawed fishes. These cerebellum-like areas consist of numerous small granule cells that are derived from the embryonic rhombic lip (the hindbrain's rostrodorsal edge) and generally quite similar to the granule cells in a proper cerebellum. Their cell bodies form dense aggregates, called auricles in cartilaginous fishes and eminentiae granularis in ray-finned fishes. Their axons form a molecular layer that covers the dorsal and medial octavolateralis columns we discussed in the previous section. In this molecular layer (often called the cerebellar crest) the granule cell axons contact the long and spiny apical dendrites of large neurons in the octavolateralis area. Although these neurons superficially resemble the Purkinje cells of a proper cerebellum, they are excitatory rather than inhibitory (GABAergic) and clearly not homologous to them (Figure 3.17). Indeed, the absence of proper Purkinje cells

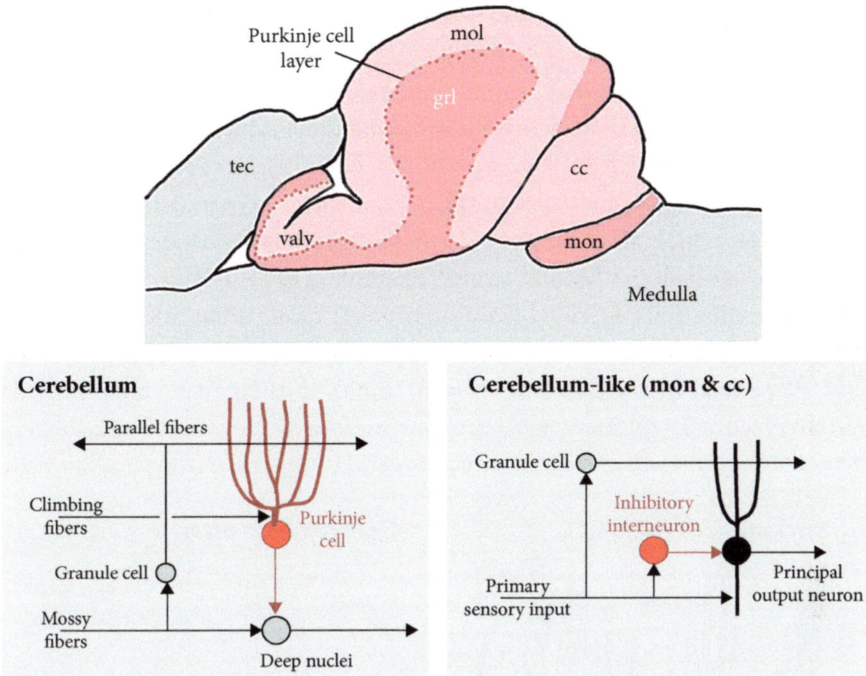

Figure 3.17 Cerebellum versus cerebellum-like structures. Shown at the top is a parasagittal section through the cerebellum of an adult zebrafish. The molecular and granular layers of the cerebellum (mol and grl) are shaded in light and dark pink, with the cell bodies of Purkinje cells depicted as red dots. The cerebellar valvula (valv) protrudes into the ventricle beneath the optic tectum (tec). The cerebellar crest (cc) closely resembles the cerebellum's molecular layer and harbors the dendrites of large neurons in the medial octavolateralis nucleus (mon); together, they form a cerebellum-like structure. The bottom panels indicate the principal circuits of the cerebellum and the cerebellum-like structures. The latter lack inhibitory Purkinje cells and, instead, contain excitatory output cells.
Adapted from Hibi and Shimizu (2012) and Montgomery et al. (2012).

is a major reason why these areas are called cerebellum-like rather than cerebellar. Whereas hagfishes and lampreys have only cerebellum-like areas, jawed vertebrates have both a proper cerebellum and cerebellum-like areas (e.g., the dorsal cochlear nucleus in mammals).

The function of the cerebellum-like areas has been examined in both cartilaginous fishes and some electroreceptive teleosts (Devor, 2000; Bell et al., 2008; Montgomery et al., 2012). The common finding is that the cerebellum-like structures subtract expected sensory input generated by the animal's own movement from the actual input so that information about unexpected, externally generated stimuli can be selectively conveyed to more rostral brain regions. Skates, for example, use the system to subtract away electrical signals that are caused by their own gill movements, making the animals more sensitive to weak signals generated by potential prey. Similarly, electroreceptive teleosts use their cerebellum-like structures to filter out the electric field generated by their own electric organ so that they can detect weak electric signals generated by external sources. The neural mechanisms underlying this adaptive filtering are beyond the scope of our review (see Shadmehr and Wise, 2005; Montgomery et al., 2012), but they include synaptic plasticity at the synapses between the granule cell axons (parallel fibers) and the dendrites of the principal cells in the electrosensory lateral line lobe (Figure 3.18).

The proper cerebellum of gnathostomes contains granule cells and parallel fibers, as well as Purkinje cells. Derived from embryonic tissue adjacent to the rhombic lip (Leto et al., 2016), the Purkinje cells receive synaptic inputs from a very large number of parallel fibers, coursing at right angles to their relatively planar dendritic tree (Figure 3.18). They project either to part of the ventral octavolateralis

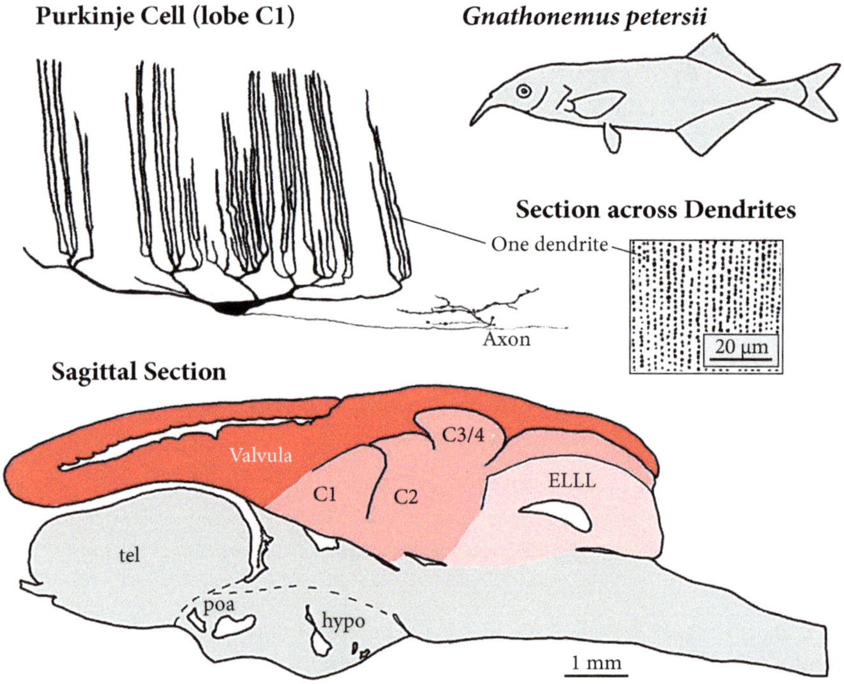

Figure 3.18 The mormyrid cerebellum. The mormyrid electric fishes, exemplified here by *Gnathonemus petersii,* have a magnificent cerebellum. Their valvula is so large that it cannot be contained inside the tectal ventricle. They also have a large electrosensory lateral line lobe (ELLL), which is a cerebellum-like structure. The Purkinje cells of the mormyrid cerebellum are unusual in having highly planar dendritic trees with very straight palisade-like dendrites (shown here in a parasagittal section). In a horizontal section across the dendrites (insert), the individual dendrites appear as rows of dots.

Additional abbreviations: C1–C4 – principal lobes of the cerebellar corpus; hypo – hypothalamus; poa – preoptic area; tel – telencephalon.

Adapted from Meek and Nieuwenhuys (1991, with permission from John Wiley & Sons), and Bell and Szabo (1986).

column (the vestibular nuclei) or to one or more deep cerebellar nuclei. Among gnathostomes, only the teleosts lack deep cerebellar nuclei. They do, however, possess large "eurydendroid" cells that are probably homologous to the deep cerebellar nuclei of other vertebrates (Murakami and Morita, 1987). Intriguingly, the progenitors of the eurydendroid cells in teleosts express a different set of transcription factors than the cells that give rise to the mammalian deep cerebellar nuclei (Kani et al., 2010). Therefore, this may be a case where homologous cells in different species derive from different, possibly non-homologous, embryonic precursors (see Chapter 1, Section 1.3.3)

Another characteristic feature of the gnathostome cerebellum is that it receives so-called climbing fiber inputs from the inferior olive. In jawed fishes these climbing fibers do not "climb" all over the dendritic tree of the Purkinje cells, as they do in mammals, but instead synapse mainly onto the proximal dendrites, close to the cell bodies (Alvarez-Otero et al., 1993; Xue et al., 2008). Still, each climbing fiber synapse is likely much more powerful than a single parallel fiber synapse. Importantly, when a climbing fiber causes a Purkinje cell to fire an action potential, any simultaneously active parallel fiber inputs are weakened. This synaptic weakening, called long-term synaptic depression (LTD), has been observed in both tetrapods and teleosts (Han et al., 2007) and is therefore likely to be primitive for gnathostomes. It is very similar to the synaptic weakening observed in the cerebellum-like structures and probably homologous to it (as a cellular/molecular process). However, the cellular context in which this LTD operates is very different in the two types of structures, as the cerebellum-like structures lack both Purkinje cells and climbing fibers.

This raises an interesting question: did the cerebellum of gnathostomes evolve from the cerebellum-like structures of their jawless ancestors? Clearly, one did not transform into the other, since early gnathostomes possessed both a proper cerebellum and some cerebellum-like structures. However, it is possible that the ancestral cerebellum-like structure duplicated in evolution, just as many genes have, and that one of the duplicates subsequently changed more than the other, thus transforming into the cerebellum (Bell et al., 2008; Montgomery et al., 2012). Since both the cerebellum and the cerebellum-like structures derive from precursor tissue in or near the rhombic lip, one could say that both structures are homologous as derivatives of a homologous embryonic precursor region (i.e., a field homology). Although this proposal has appeal, it does not really explain the origin of cerebellar Purkinje cells or deep cerebellar nuclei. Nor does it deal explicitly with the fact that the rhombic lip also gives rise to the inferior olive (Wullimann et al., 2011), which is quite different from the cerebellum itself. Therefore, it seems safer simply to conclude that the cerebellum and the inferior olive are gnathostome innovations that evolved by modifying the development of portions of an ancestral rhombic lip and the immediately adjacent hindbrain (Sugahara et al., 2016). We return to this topic in Chapter 7.

What benefit did early gnathostomes derive from evolving a proper cerebellum? This question is difficult to answer because very few experimental studies have been performed on cerebellar function in fishes (aside from mormyrid electric fishes; e.g., Alviña and Sawtell, 2014). However, it is likely that the cerebellum of early gnathostomes was similar to the vestibulo- and spinocerebellum of tetrapods insofar as it probably used experience-dependent synaptic plasticity to modulate various reflexes that stabilize the body and the eyes, relative to the external world. It was probably also involved in modulating the kind of orienting movements that early vertebrates would have used to hunt small, agile prey. To perform this kind of adaptive motor control, the cerebellum of early gnathostomes probably received sensory information from a variety of sensory structures (especially from the vestibular apparatus) as well as copies of motor commands; it then sent modulatory outputs to midbrain and medullar motor regions. Moreover, the ancient cerebellum probably adjusted these outputs by modifying the strength of the parallel fiber synapses, using the climbing fiber inputs as error signals (generally indicating when sensory expectations were not met) to determine when additional plasticity was required. This adaptive plasticity would have been useful, for example, when muscle fatigue or changes in water temperature altered how muscles respond to motor commands (Montgomery, 1988), or when some body parts were modified by growth or injury.

Although adaptive motor control was almost certainly the primitive function of the vertebrate cerebellum, the cerebellum probably acquired additional functions as it increased in size and complexity. Much has been written about the purely sensory and cognitive functions of the cerebellum in mammals (Strick et al., 2009), but next to nothing is known about such non-motor functions in non-mammalian vertebrates. Nonetheless, the enormous increases in cerebellum size within cartilaginous fishes (see Figure 3.2) are unlikely to reflect merely enhanced motor control (Yopak et al., 2007; Lisney et al., 2008). Similarly, some teleosts have a very large cerebellum without exhibiting unusually complex motor abilities. For example, in mormyrid electric fishes the anterior division of the cerebellum, called the cerebellar valvula, expanded to such a degree that it covers the entire rest of the brain (Figure 3.18). The function of this valvula remains unclear, but it is unlikely to be purely motor, because mormyrids are not much more agile than other teleosts. Instead, the mormyrid valvula is probably involved in processing electrosensory information, as well as other kinds of sensory input (Finger et al., 1981).

3.5.3. Midbrain Roof and Tegmentum

The most intensively studied part of the midbrain in jawed fishes is the optic tectum, a balloon-shaped structure that forms most of the midbrain's roof (Figure 3.19) and is homologous to the mammalian superior colliculus. In all fishes, the optic tectum receives prominent projections from the contralateral retina, but in many species the rostral tectum, which represents the space directly in front of the animal, also receives some weak input from the ipsilateral retina. Because the optic tectum is by far the largest target of the retina in all fishes, it is reasonable to refer to

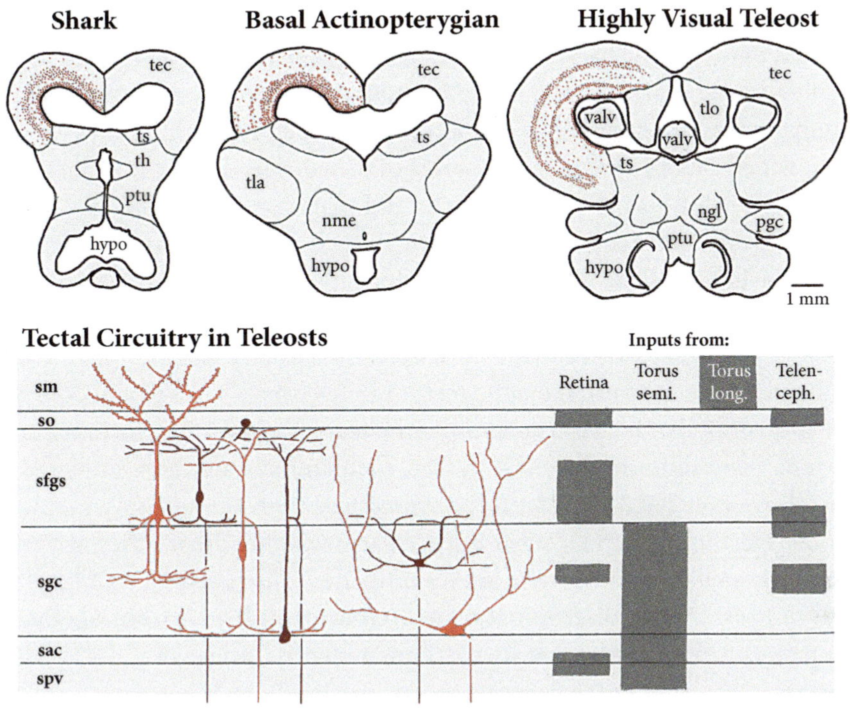

Figure 3.19 The optic tectum and its circuitry. Shown along the top are transverse sections through the brains of a shark (spiny dogfish), a basal ray-finned fish (reedfish, which is closely related to *Polypterus*), and a highly visual teleost (squirrelfish) at the level of the optic tectum. The red dots represent cell bodies and reveal the variable degree of tectal lamination. The bottom diagram illustrates the tectum's inputs (termination zones are shaded gray) and major cell types in a representative teleost. The main tectal layers are called stratum marginale (sm), stratum opticum (so), stratum fibrosum et griseum superficiale (sfgs), stratum griseum centrale (sgc), stratum album centrale (sac), and stratum periventriculare (spv). Note that the spv contains densely packed cells, which are not shown in this diagram.

Additional abbreviations: hypo – hypothalamus; ngl – nucleus glomerulosus; nme – nucleus medianus; pgc – preglomerular complex; ptu – nucleus of the posterior tuberculum; tec – tectum; th – thalamus; tla – torus lateralis; tlo – torus longitudinalis; ts – torus semicircularis; valv – valvula.

Adapted from Schroeder et al. (1980), Vanegas and Ito (1983, with permission from Elsevier), Nieuwenhuys et al. (1998).

it as the *optic* tectum. However, the optic tectum also receives input from other sensory modalities. Specifically, its middle and deeper layers receive mechanosensory lateral line, electrosensory, and auditory information. Although the comparative data are meager, all of these sensory inputs appear to be topographically organized such that the visual projections are "in register" with the non-visual ones. Thus, individual tectal neurons receive multimodal sensory information from a specific location in space (Bodznick, 1990). Tectal outputs target the midbrain reticular formation as well as a variety of other brain regions, most of which are known to be involved in motor control. Broadly speaking, the most widely shared function of the optic tectum is probably to orient the animal's head and eyes toward external stimuli that stand out from the background (i.e., are salient). These orienting movements then allow the animal to scrutinize the stimulus and, as appropriate, attack or flee.

Comparative data suggest that the optic tectum of early gnathostomes was probably no wider than the medulla, that most of its neurons did not migrate far from their site of birth near the tectal ventricle, and that it probably contained no more than 4–5 layers (Northcutt, 1977; Vanegas, 1984). From these humble beginnings, the tectum increased in volume and complexity in diverse cartilaginous fishes (Yopak and Lisney, 2012) and, more dramatically, within the teleosts (Figure 3.19). In some highly visual teleosts, the optic tectum is said to have 15 distinct layers (Northcutt, 1983). Because the optic tectum in teleosts is so large and complex, it is not surprising that bilateral ablation of the tectum causes teleosts to bump into aquarium walls and other objects. In fact, such animals are more impaired than fishes whose eyes have been removed, consistent with the tectum's multimodal function (Northmore, 2011). What is less clear is whether other fishes, with a proportionately smaller optic tectum, are less impaired. Nurse sharks with large tectal lesions can pass a visual discrimination test (Graeber et al., 1973), but whether they are impaired at orienting toward small stimuli remains unclear.

Reciprocally connected to the optic tectum in most bony fishes is nucleus isthmi, which is probably homologous to the mammalian parabigeminal nucleus. This structure is thought to mediate a winner-take-all competition between different tectal neurons so that the animal ends up orienting only to the most salient sensory stimulus (Northmore, 2011). Most ray-finned fishes also possess another midbrain structure that is reciprocally connected to the tectum, namely the torus longitudinalis. This cerebellum-like structure lies adjacent to the dorsal midline of the optic tectum, and its axons course along the tectum's most superficial layer (Figure 3.19). Because the torus longitudinalis is present in all ray-finned fishes except *Polypterus*, it probably evolved very early during ray-finned fish phylogeny, 300–250 mya. It probably helps to stabilize visual perception as ray-finned fishes move their eyes (Northmore, 2017).

Another important component of the alar midbrain in all gnathostomes is the torus semicircularis, which in mammals is called inferior colliculus. Because the torus semicircularis is much smaller than the optic tectum in most fishes, it was likely small in the first gnathostomes. It is best known for conveying auditory information to the optic tectum and some other brain regions, but parts of the torus semicircularis receive inputs from the lateral line and somatosensory systems (Figure 3.20). Indeed, these non-auditory inputs would have been predominant in early gnathostomes, whose sense of hearing was still poorly developed. Within the torus semicircularis, the various sensory modalities were probably segregated, but comparative data on this point are scarce (Yamamoto et al., 2010). What is clear is that this kind of segregation exists in teleosts and that the electrosensory portion of the torus semicircularis has become enormously enlarged in the weakly electric fishes. In gymnotids, for example, the electrosensory division of the torus

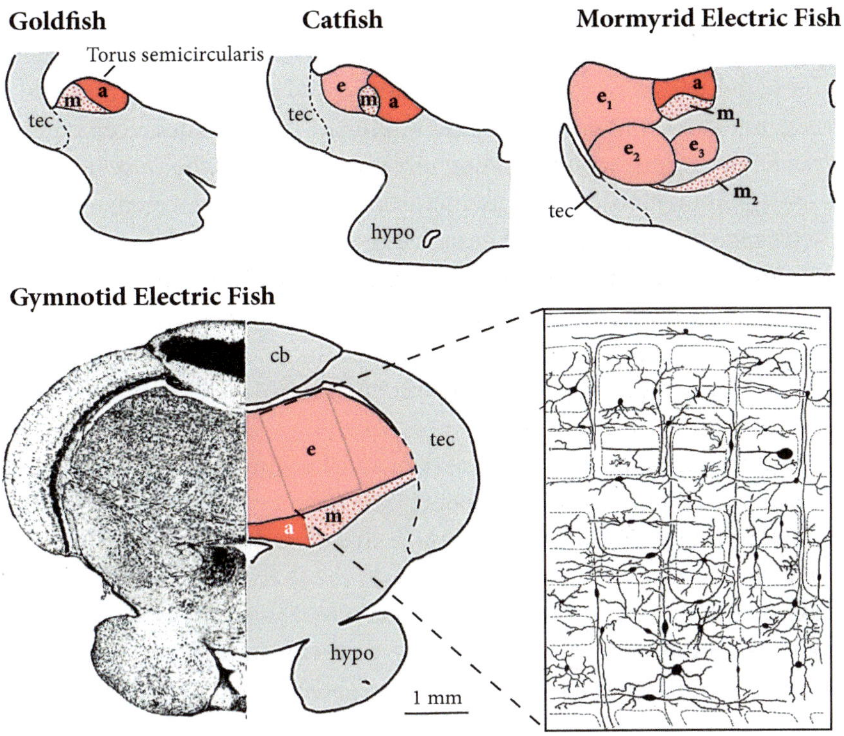

Figure 3.20 The torus semicircularis in diverse teleosts. Shown along the top are schematic transverse sections through the torus semicircularis on one side of the brain; the optic tectum is largely cut away. The auditory (a), mechanosensory lateral line (m), and electrosensory (e) divisions of the torus semicircularis are shaded differently. In mormyrids, which possess an electric organ and electroreceptors, the torus semicircularis contains several large electrosensory subnuclei (e1-3). Shown along the bottom is the torus semicircularis of gymnotid teleosts, which evolved electric organs and electroreception independently of mormyrids. Their torus semicircularis is enormous and exhibits extensive lamination, as illustrated in the composite drawing of Golgi-stained neurons (bottom right).

Additional abbreviations: cb – cerebellum, hypo – hypothalamus, tec – tectum.

Adapted from Carr and Maler (1986, with permission from John Wiley & Sons), McCormick and Braford (1988), Nieuwenhuys et al. (1998, with permission from Springer Nature).

semicircularis fills the entire tectal ventricle and exhibits numerous laminae (Figure 3.20). The electrosensory torus is almost as large in mormyrid electric fishes, but in these fishes the torus semicircularis is divided into several subnuclei, rather than laminae (Wullimann and Grothe, 2013).

Ventral to the optic tectum and torus semicircularis lies the midbrain's basal plate and floor, or tegmentum. This brain region comprises several functionally distinct components, but it is not well understood, especially in fishes. The tegmentum serves a variety of motor functions, including the control of eye movements. In many vertebrates it also contains dopaminergic neurons that project to the telencephalon. Such neurons are found in cartilaginous fishes, gars, lungfishes, and amniotes (González and Northcutt, 2009). In teleosts, however, dopaminergic neurons with projections to the telencephalon are found mainly in the basal diencephalon, not in the midbrain (Rink and Wullimann, 2001). These data suggest that dopaminergic neurons with ascending projections can develop in several different embryonic brain regions and, more importantly, that their developmental origins vary across the major vertebrate lineages. The functional correlates of these evolutionary changes in development remain to be unearthed.

3.5.4. Diencephalon

The diencephalon comprises three rostrocaudal segments, and each of these is divisible into alar and basal components (see Chapter 2, Figure 2.26). The principal alar divisions are called pretectum, thalamus (or dorsal thalamus), and prethalamus (or ventral thalamus). Dorsal to the thalamus lies the epithalamus, whose principal constituents are the habenula and pineal gland. The basal divisions of the diencephalon form mainly the posterior tuberculum. The hypothalamus was traditionally included in the diencephalon, but it develops in a more rostral region of the embryonic brain (the secondary prosencephalon). It is only in later stages of development, after the brain's original longitudinal axis has undergone a series of dorsoventral bends, that the hypothalamus is found at the same rostrocaudal levels as the prethalamus and thalamus in transverse sections (Figure 3.21). That is why we discuss the hypothalamus later, in Section 3.5.5. Within the diencephalon, we focus on the pretectum and posterior tuberculum, because those regions are most variable.

The pretectum was probably a relatively small and simple brain region in the ancestral jawed vertebrates, consisting of just two or three distinct cell groups that received retinal and/or tectal inputs and projected mainly to motor regions in the medulla and to the hypothalamus. In some teleosts, however, the pretectum expanded substantially and is divisible into at least eight distinct cell groups. Most impressive is the "nucleus glomerulosus" of acanthomorph (e.g., perciform) teleosts. This large and roughly spherical nucleus contains numerous glomeruli that house specialized arrangements of dendrites and synapses and, in some species, exhibits multiple shell-like laminae (Figure 3.22; Ito and Kishida, 1977). This nucleus glomerulosus receives visual input via another pretectal nucleus, projects mainly to the inferior lobes of the hypothalamus, and has been suggested to play a role in visual object perception (Wullimann and Meyer, 1990). It appears to be homologous to a much smaller "posterior pretectal nucleus" in other teleosts. A good general hypothesis is that the elaboration of the pretectum in advanced teleosts was part of a suite of characters that adapted these fishes to living on reefs in shallow water during the Cretaceous period (Wullimann, 1997). Both the acanthomorph fishes and the corals they lived on survived a massive extinction event at the end

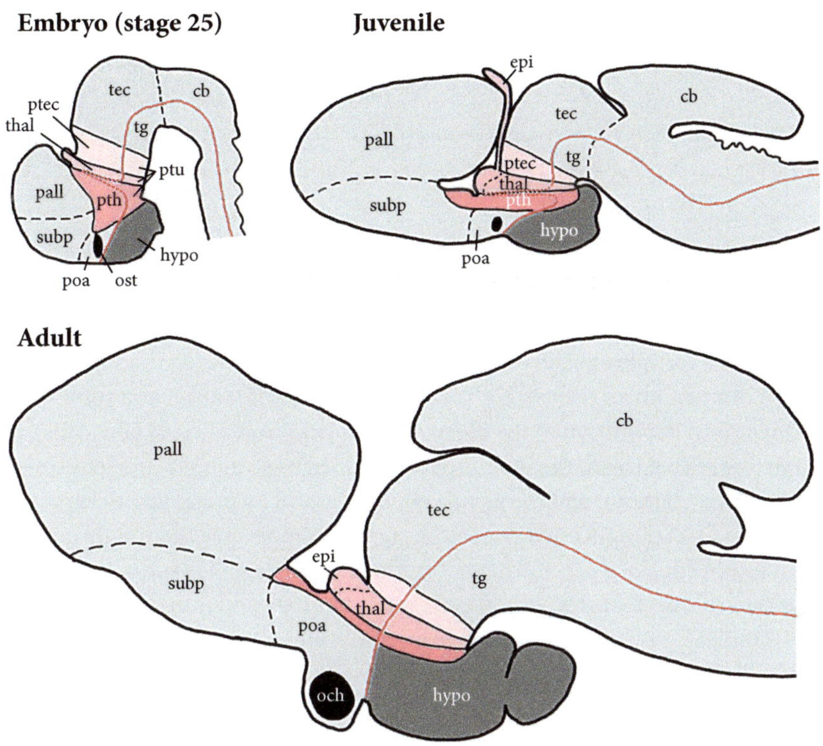

Figure 3.21 Diencephalic development. Shown here are sagittal views of shark brains at three stages of development, highlighting the three embryonic segments that comprise the diencephalon. The red line indicates the brain's longitudinal axis and separates basal regions from alar ones. According to this schema, the tegmentum (tg) and the tectum (tec) are the basal and alar components of the midbrain, respectively. The alar divisions of the diencephalon are the pretectum (ptec), the thalamus (thal), the epithalamus (epi), and the prethalamus (pth). The basal diencephalon consists mainly of the posterior tuberculum (ptu). The hypothalamus (hypo) is traditionally considered to be the ventral part of the diencephalon, but it actually develops rostral to the diencephalon.

Additional abbreviations: cb – cerebellum; och – optic chiasm; ost – optic stalk; subp – subpallium; pall – pallium; poa – preoptic area.

Adapted from Rodríguez-Moldes (2009) and Rodríguez-Moldes et al. (2017).

of the Cretaceous period and then flourished in the Cenozoic era (see Figure 1.3 in Chapter 1).

The posterior tuberculum is generally considered to be the topologically ventral part of the vertebrate diencephalon. In lampreys the posterior tuberculum receives inputs from the olfactory bulb and projects to a "locomotor region" in the midbrain tegmentum (Derjean 2010; Buchinger et al., 2015). Olfactory inputs to the posterior tuberculum are also seen in teleosts and lungfishes (Northcutt and Rink, 2012) and, therefore, were likely primitive for gnathostomes. Although this pathway surely had important functions, such as using olfactory information to navigate, the posterior

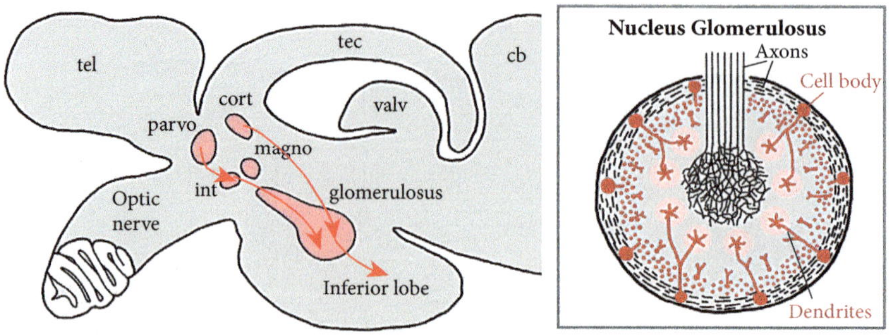

Figure 3.22 The complex pretectum of teleosts. Shown on the left is a sagittal diagram of the brain of a perciform teleost, highlighting five pretectal nuclei and some of their connections. In addition to the illustrated connections, nucleus corticalis (cort) receives strong retinal input, the magnocellular superficial pretectal nucleus (magno) receives dense tectal input, and the parvocellular superficial pretectal nucleus receives both retinal and tectal inputs. Shown on the right is a transverse section through nucleus glomerulosus of a perciform teleost, the dragonet *Callionymus*. In this genus nucleus glomerulosus features several concentric laminae and numerous synaptic glomeruli (indicated by the star-shaped dendritic endings).
Adapted from Wullimann (1997), Ito and Kishida (1977).

tuberculum was likely small in early gnathostomes, with most neurons remaining close to the ventricle. Such a small and relatively indistinct posterior tuberculum is seen not only in lampreys but also in cartilaginous and lobe-finned fishes.

Within the ray-finned fishes, the posterior tuberculum expanded significantly and exhibits enormous variation. In *Polypterus* it consists of a midline nucleus, called nucleus medianus, and a large migrated nucleus called torus lateralis (Figure 3.23). Both of these cell groups project to the telencephalic pallium, and the torus lateralis receives a variety of ascending sensory inputs, especially from the gustatory system (Holmes, 2001; Holmes and Northcutt, 2003; Braford, 2009). Insofar as these two nuclei are diencephalic and project to the pallium, they resemble the thalamus of tetrapods. However, *Polypterus* has a small thalamus that lies dorsal to nucleus medianus, which is positioned exactly where one would expect to find the posterior tuberculum, and dorsomedial to the torus lateralis, whose developmental origin is less certain (Northcutt, 2009a). Based on these observations and on comparisons to other ray-finned fishes, we agree with Braford (2009) that the torus lateralis of *Polypterus* is an enlarged, migrated portion of the posterior tuberculum (alternatively, it could be a midbrain derivative). A substantial torus lateralis is also seen in sturgeons, *Amia*, and gars. A torus lateralis has been identified in cartilaginous fishes (Smeets and Boord, 1985), but its homology to the torus lateralis of ray-finned fishes remains uncertain.

In addition to the torus lateralis and a midline posterior tubercular region, *Amia*, gars, and teleosts (i.e., neopterygians) possess three to six cell groups that have migrated away from the ventricle but lie medial to the torus lateralis. They

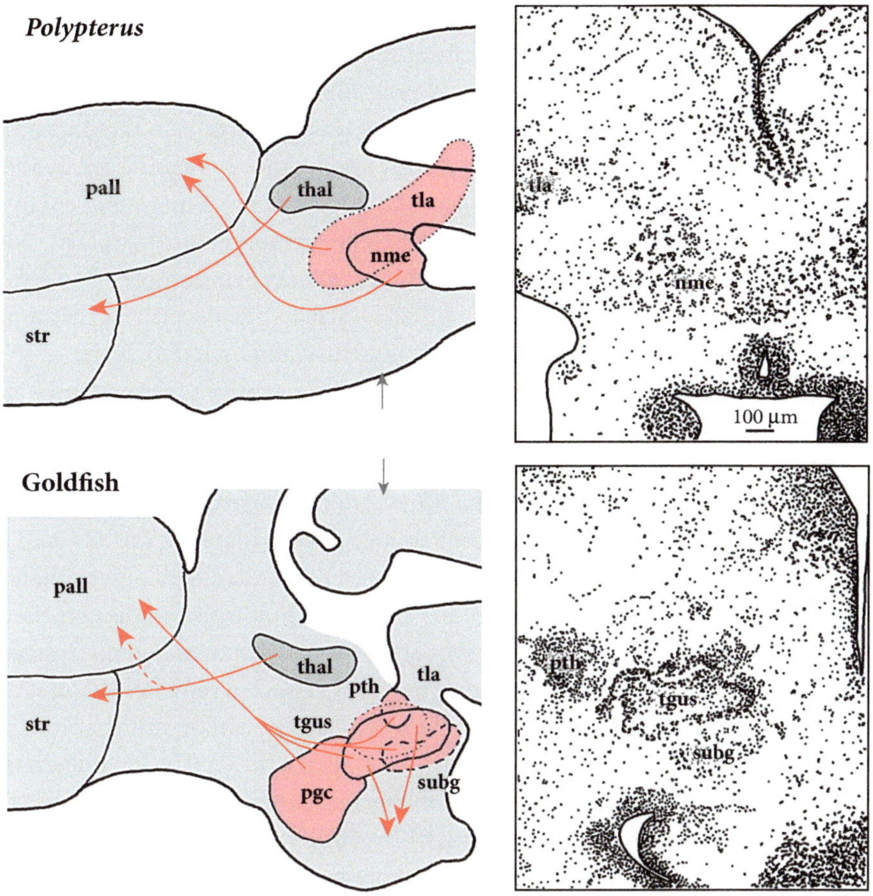

Figure 3.23 The posterior tuberculum in ray-finned fishes. Shown on the left are schematic sagittal sections through the forebrain of a basal ray-finned fish (*Polypterus*) and the goldfish, a teleost. In both species the ascending projections of the thalamus (thal) target mainly the striatum (str). In *Polypterus* ascending projections to the pallium (pall) originate mainly from nucleus medianus (nme) and the torus lateralis (tla). In goldfish they originate from multiple cell groups in the posterior tuberculum, including the preglomerular complex (pgc). Shown on the right are drawings of Nissl-stained transverse sections at the levels indicated by the small gray arrows.

Additional abbreviations: pth – posterior thalamic nucleus; subg – subglomerular nucleus; tgus – tertiary gustatory nucleus.

Adapted from Northcutt (2009a).

are collectively referred to as the preglomerular complex (Figure 3.23). Most of the cell groups in this complex receive various types of sensory input and project to the telencephalic pallium (Striedter, 1991). Given these traits, it is not surprising that the preglomerular complex has been compared to the thalamus of tetrapods (Yamamoto and Ito, 2008). However, all neopterygian fishes have a small thalamus that is clearly distinct from the preglomerular complex (Braford and Northcutt, 1983; Striedter, 1990). Gene expression data have been used to suggest that part of the preglomerular complex is a migrated derivative of the embryonic thalamus (Ishikawa et al., 2007). However, those findings are far from definitive (Northcutt, 2008) and clearly not compatible with a preliminary report indicating that the preglomerular complex of zebrafish develops, at least in part, from mesencephalic precursor cells (Bloch et al., 2017). Therefore, pending more evidence, we continue to interpret the preglomerular complex as being (primarily) a migrated, enlarged, and highly differentiated derivative of the posterior tuberculum (Braford, 2009; Vernier and Wullimann, 2009). Whether the preglomerular complex has a strict homolog outside of the neopterygians remains unclear, but we suspect that it may be homologous to part of midline posterior tubercular region in *Polypterus*.

The thalamus of early gnathostomes was probably relatively simple and small, just as it is in ray-finned fishes and basal lobe-finned fishes (i.e., basal bony fishes). To shore up this hypothesis, it would be good to have more information on diencephalic organization in the cartilaginous fishes. Unfortunately, however, there have been no detailed studies on the diencephalon of any cartilaginous fishes, and the authors of the few available reports have offered quite divergent interpretations (see Northcutt, 1990). Moreover, there have been no published studies on the diencephalon of the most basal clade of cartilaginous fishes, the holocephalans. However, unpublished observations on retinal projections in ratfishes (Figure 3.24) indicate that the thalamus of holocephalans is relatively small and that its anterior portion receives strong retinal inputs. In contrast, the posterior thalamus of ratfishes is devoid of retinal inputs and probably receives projections from the midbrain roof. These data are consistent with the hypothesis that the thalamus of early gnathostomes was similar to that of basal bony fishes living today (e.g., Northcutt and Butler, 1976; Braford and Northcutt, 1983).

3.5.5. Hypothalamus

The hypothalamus is relatively small in cyclostomes, basal ray-finned fishes, and tetrapods. However, it features a large pair of ventrolaterally directed lobes in cartilaginous fishes and teleosts (Figure 3.25). These inferior lobes surround lateral extensions of the hypothalamic ventricle and are likely a primitive feature for gnathostomes. However, they expanded independently in cartilaginous fishes and ray-finned fishes (including sturgeons and teleosts), which feature a large number of migrated neurons in their inferior lobes; some of these neurons originate in the midbrain, at least in teleosts (Bloch et al., 2019). Lesion and electrical stimulation studies in both cartilaginous fishes and teleosts indicate that the inferior lobes play a major role in feeding behavior (Roberts and Savage, 1978; Demski, 2012). Consistent with these behavioral data, the inferior lobes are the principal diencephalic target of ascending gustatory projections, at least in teleosts (Smeets and Boord, 1985; Ahrens and Wullimann, 2002), and receive olfactory inputs from the telencephalon (Figure 3.25). The presence of many CSF-contacting cells (Evan

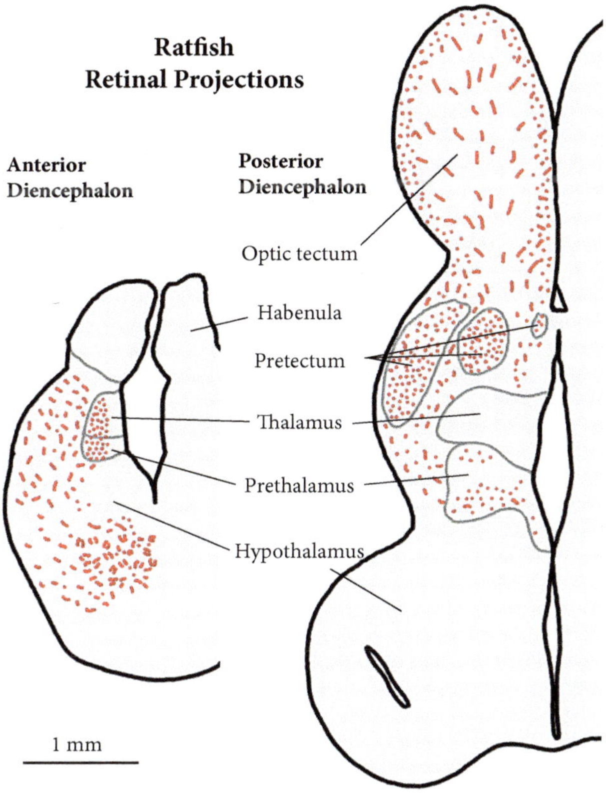

Figure 3.24 The diencephalon of the holocephalan ratfish (*Hydrolagus collei*). Little is known about the diencephalon of holocephalans, but it is probably similar to that of primitive ray-finned fishes. This hypothesis is supported by the illustrated data on retinal projections in ratfishes (unpublished observations by R. G. Northcutt and W. J. A. J. Smeets). As in many ray-finned fishes, the retina in ratfishes projects densely to the anterior thalamus and prethalamus (as well as pretectum, optic tectum, and a few other brain regions that are not shown; see Smeets, 1998, for more data). The posterior thalamus is relatively free of retinal projections and, instead, is thought to receive major projections from the midbrain roof (i.e., optic tectum and torus semicircularis), as it does in ray-finned fishes.

et al., 1976) and neuropeptide receptors in the periventricular portion of the inferior lobes (Demski, 2012) suggests that this region also integrates diverse other signals, some of which may be conveyed via the cerebrospinal fluid. Curiously, mammals seem to be the only major vertebrate lineage that lost the hypothalamic CSF-contacting neurons (Yamamoto et al., 2017).

Protruding between the caudal ends of the inferior lobes is the neural portion of the pituitary gland, the neurohypophysis. This region is quite conservative across the vertebrates, receiving inputs from the hypothalamus and secreting hormone releasing factors into blood vessels that carry those factors to hormone-secreting cells in the anterior pituitary, or adenohypophysis (Meurling, 2011). Although

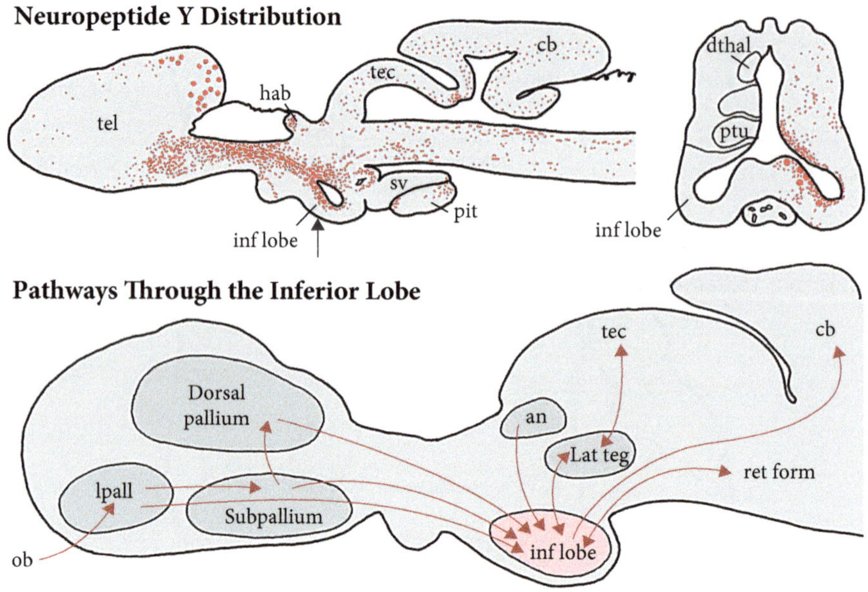

Figure 3.25 The inferior lobe of the hypothalamus. Shown at the top are a sagittal (left) section through the brain of a dogfish (*Scyliorhinus torazame*) and a transverse section (right) taken at the level indicated by the black arrow. Depicted on those sections is the distribution of neuronal cell bodies (red circles) and fibers (red lines and dots) that are immunopositive for neuropeptide Y, which plays a major role in the control of feeding. The bottom diagram summarizes the major connections of the inferior lobe in sharks. Visual, electrosensory, and mechanosensory information are conveyed to the inferior lobe (inf lobe) through the anterior nucleus of the mesencephalic-diencephalic boundary (an) and the lateral tegmental nucleus (lat teg); these pathways probably also convey gustatory information. Olfactory and polysensory information comes from the telencephalon. Major outputs of the inferior lobe target the reticular formation (ret form) and the cerebellum (cb).

Additional abbreviations: hab – habenula; inf lobe – inferior lobe; lpall – lateral pallium; ob – olfactory bulb; pit – pituitary; sv – saccus vasculosus; tec – tectum; tel – telencephalon.

Adapted from Chiba and Honma (1992), Demski (2012).

such a "hypophyseal portal system" is sometimes said to be lacking in lampreys and teleosts, recent reports indicate that even these species exhibit a few blood vessels that could carry releasing factors from hypothalamic axon terminals to the adenohypophysis (Golan et al., 2015). Therefore, some sort of hypophyseal portal system is likely primitive for vertebrates. An interesting twist is that teleosts have, in addition, evolved direct projections from dopaminergic neurons in the preoptic area to the adenohypophysis (Sower, 2015; Fontaine et al., 2015).

One of the most enigmatic components of the hypothalamus in most fishes is the saccus vasculosus, a thin and highly vascularized sac that forms an extension of the hypothalamic ventricle just caudal to the pituitary gland. It is found in both cartilaginous and ray-finned fishes, but not in cyclostomes, suggesting that it is a gnathostome innovation (Sueiro et al., 2007). The saccus vasculosus clearly has some sort of neurosecretory function and likely plays a role in sensing intraventricular pressure and osmolarity. Most intriguingly, specialized glial cells in the saccus vasculosus are photosensitive and can, presumably, be activated by light that passes through the overlying brain (Nakane et al., 2013). Moreover, lesions of this brain region impair an animal's ability to measure seasonal changes in day length. Lobe-finned fishes also have a photoperiodic sense, but they lack a saccus vasculosus. Instead, they use a very different set of photosensitive structures to measure day length, notably the retina and, in some species, the pineal gland (see Appendix).

3.5.6. Telencephalon

The telencephalon is often regarded as the crowning achievement of vertebrate brain evolution, but it began its evolution as a relatively small and simple thing. Only later did it increase in size and complexity, and it did so independently in several different gnathostome lineages. The telencephalon enlarged most spectacularly in birds and mammals (notably primates and cetaceans), which we discuss in Chapters 5 and 6. Here, we focus on the major evolutionary changes in the telencephalon of fishes, especially on its enlargement in cartilaginous fishes and teleosts.

The hypothesis that early gnathostomes had a relatively small telencephalon is well supported by data on the braincases and endocasts of early ray-finned and cartilaginous fishes (see Figure 3.9). In these fossil specimens the midbrain is considerably wider than the telencephalon and most of the telencephalon consists of the olfactory bulbs. Even in early lungfishes, the telencephalon caudal to the olfactory bulbs is very narrow (Clement and Ahlberg, 2014). Indeed, the telencephalon of early fossil fishes is not much larger than the telencephalon of lampreys (see Figure 2.5). Therefore, major increases in telencephalon size must have occurred long after the origin of gnathostomes. As noted in Chapter 2, hagfishes do have a large and complex telencephalon, but their telencephalon is structurally quite different from that of other vertebrates and probably expanded independently of what happened in gnathostomes.

3.5.6.1. Evagination versus Eversion

In most vertebrate lineages the telencephalon develops as a bilateral evagination of the most rostral alar part of the forebrain (Figure 3.21 in Chapter 3). In transverse sections, the evaginated telencephalon looks like two rings of tissue around the left and right telencephalic ventricles (Figure 3.26). This condition was likely primitive for gnathostomes and was retained in most lineages. A striking exception are the ray-finned fishes, in which the telencephalon does not evaginate but, instead, everts (Nieuwenhuys, 1963, 2009). Key to this process of telencephalic eversion is that the midline roof of the embryonic telencephalon becomes very thin in ray-finned fishes

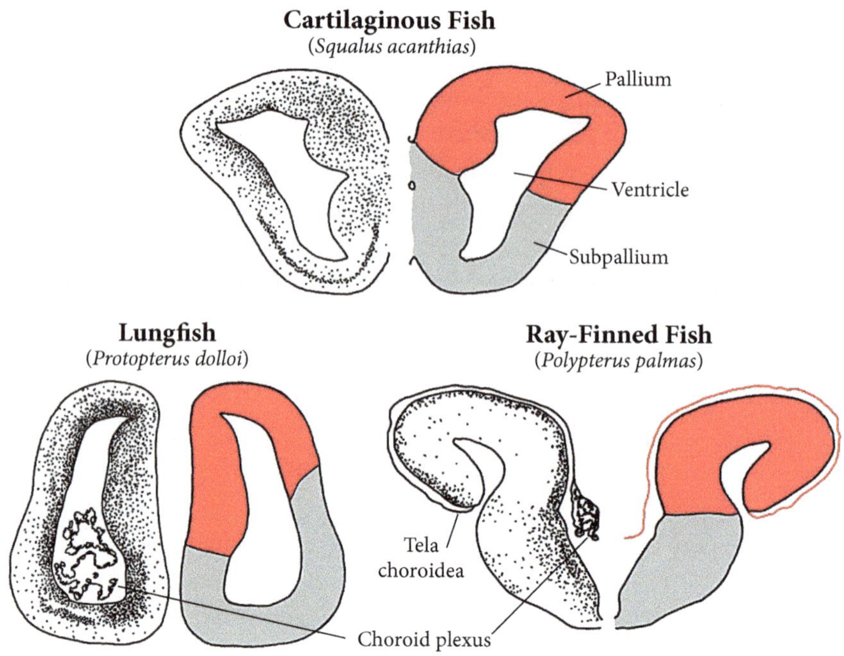

Figure 3.26 Telencephalon of fishes. In most vertebrates the telencephalon evaginates in such a way that transverse sections through the telencephalon form two adjacent ring-shaped pieces of tissue. The dorsal and ventral parts of each telencephalic hemisphere are called the pallium (red) and subpallium (gray), respectively. In ray-finned fishes, here represented by *Polypterus*, the telencephalon does not evaginate but, instead, everts. A tell-tale feature of eversion is that the pallium is covered by a thin membrane, called the tela choroidea.
Adapted from Northcutt (1995, 2009b).

and forms a membranous "tela choroidea" (Figure 3.26). The dorsomedial edges of the left and right telencephalon then move laterally, stretching the tela choroidea so that it ends up covering the entire telencephalon.

Of course, the process of telencephalic eversion is more complex than it appears from cartoonish figures or brief descriptions. For example, the eversion is accompanied by a laterally directed evagination of the area right between the telencephalon and the caudally adjacent diencephalon (Striedter and Northcutt, 2006). Also, the eversion is obscured in teleosts by a dorsally directed expansion of the pallial areas (Folgueira et al., 2012). However, the most basal ray-finned fishes, especially *Polypterus* (Figure 3.26), exhibit telencephalic eversion so clearly that the basic phenomenon can hardly be doubted (Yamamoto et al., 2007; Braford, 2009; Nieuwenhuys, 2011).

Why does the telencephalon of ray-finned fishes evert, rather than evaginate? One possible reason is that the embryos of ray-finned fishes are significantly smaller than those of their ancestors. Because of allometric scaling rules, this decrease in embryonic body size was associated with an increased brain-body ratio. That, in turn, caused the brain and major sense organs to occupy a far greater fraction of the head in embryonic ray-finned fishes than they do in adult fishes or embryos of other, larger embryos. Based on these observations, we have suggested that the telencephalon in embryonic ray-finned fishes may not have sufficient room within the head for a full-fledged, rostrally directed evagination (Striedter and Northcutt, 2006). Specifically, we suggest that telencephalic eversion begins as a response to space constraints within the head at early stages of embryonic development; once the eversion has begun, evagination is no longer a viable option. This hypothesis remains speculative, but it is testable, at least in principle.

3.5.6.2. Pallial Homologies

A structure's position relative to other structures (i.e., topology) is widely recognized as a valuable clue to potential homologies, both in the nervous system and in the rest of the body. Therefore, telencephalic eversion must be taken into account when trying to homologize the telencephalon's main divisions. Indeed, many authors have argued that the dorsomedial edge of an evaginated telencephalon, which develops into the hippocampus, is homologous to the dorsolateral edge of the everted telencephalon (see Figure 3.27). A wide variety of data are consistent with this

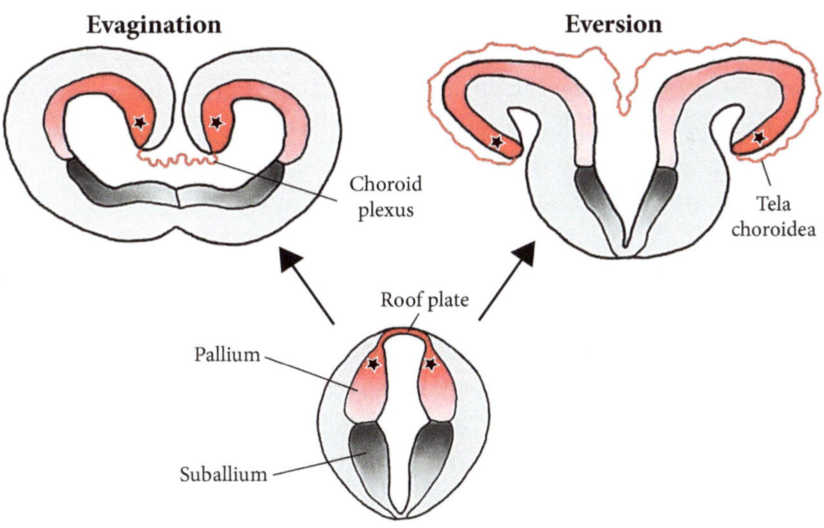

Figure 3.27 Telencephalic evagination versus eversion. Shown at the bottom is a schematic transverse section through the telencephalon of a young vertebrate embryo, depicting the pallium and the subpallium in red and gray, respectively. In most gnathostomes the telencephalon then evaginates (top left), but in ray-finned fishes it everts (top right). The most dorsal part of pallium (the future hippocampus) is indicated by the star; it assumes a dorsomedial position in the evaginated telencephalon, but a dorsolateral position in the everted telencephalon. Telencephalic eversion causes the membranous tela choroidea to expand and cover the entire pallium.

hypothesis, including the finding that the dorsolateral telencephalon of goldfishes is critical for spatial learning and memory, just as the hippocampus is in amniotes (Rodríguez et al., 2002). Similarly, there is little doubt that the ventromedial part of the evaginated telencephalon, which is called the septum in tetrapods, is homologous to the ventralmost portion of the everted telencephalon. However, some homologies between evaginated and everted telencephala are much more controversial; this is true especially for the telencephalon's dorsal division, the pallium.

For many years, neuroanatomists divided the pallium of tetrapods into medial, dorsal, and lateral subdivisions, which correspond roughly to the hippocampus, neocortex, and olfactory cortex of mammals, respectively. Recent studies have further subdivided the old lateral pallium into lateral and ventral pallial components, with the latter corresponding to the pallial amygdala of tetrapods (Puelles et al., 2000). We discuss these divisions more thoroughly in later chapters. For now, suffice it to say that some of the pallial divisions have proven very difficult to homologize across the various fishes. Part of the problem is that the gene expression data that are used to homologize these areas across tetrapods are incomplete or unavailable for fishes. Projections from the olfactory bulbs are often used to identify the ventral and lateral pallial divisions, because they have long been thought to be homologous to the olfactory cortex of amniotes (but see Puelles, 2017; Puelles et al., 2017; and Chapter 6). Similarly, inputs from the thalamus have been used to identify a dorsal pallium in fishes, because such inputs are a characteristic feature of the mammalian neocortex. However, as described in Section 3.6 and later chapters, these connections have changed considerably during vertebrate evolution. Just as troublesome is that the number of morphologically distinct divisions within the pallium varies considerably across the various fish lineages.

To illustrate the latter point, consider the pallium of *Polypterus*. It is relatively thin, and most of its neuronal cell bodies are located close to the ventricle (Figure 3.26). Because the pallium of *Polypterus* (and other polypteriform fishes) is relatively homogeneous in cytoarchitecture, neuroanatomists have long debated how many divisions it contains. The most thorough recent study argues for two major divisions, each of which contains two subdivisions (Holmes and Northcutt, 2003). Based mainly on connectional and topological data, the dorsomedial division has been homologized to the ventral/lateral pallium of tetrapods, whereas the dorsolateral division was interpreted as the medial pallium. An intriguing consequence of this analysis is that the pallium of *Polypterus* does not appear to include a dorsal pallium (Holmes and Northcutt, 2003). This hypothesis is consistent with the observation that the thalamus, which projects to the dorsal pallium in amniotes, has only subpallial projections in *Polypterus* (Holmes, 2001). We will come back to this important point.

Comparing the pallium of *Polypterus* to that of more advanced ray-finned fishes brings further challenges (Figure 3.28). One problem is that the pallium of teleosts is much thicker than that of *Polypterus* and contains large numbers of migrated neurons, making topological comparisons more difficult. The teleost pallium also

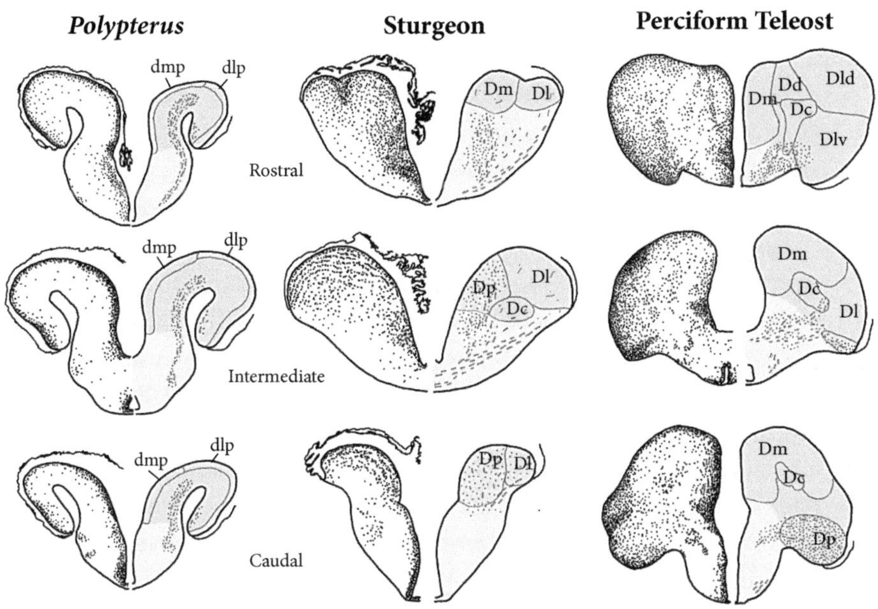

Figure 3.28 Olfactory bulb projections in ray-finned fishes. Shown in red are the projections of the olfactory bulb in three groups of ray-finned fishes. The pallium of *Polypterus* is divided into dorsomedial and dorsolateral divisions (dmp and dlp, respectively), both of which receive at least some olfactory input. Sturgeons have a more elaborate pallium, with olfactory inputs targeting mainly its posterior portion (Dp) and parts of the lateral and central divisions (Dl and Dc). In teleosts the olfactory bulb projections are even more restricted, and the non-olfactory pallial areas are expanded. Note that the neurons in the pallium of *Polypterus* extend their dendrites into the olfactory termination zone and that the pallial divisions in neopterygian fishes are all considered part of "area dorsalis," which is why they are abbreviated D plus one or two letters (Dd – area dorsalis dorsalis; Dm – area dorsalis medialis). For each species, sections are shown for rostral, intermediate, and caudal levels of the telencephalon.

Adapted from Northcutt and Davis (1983); Bartheld and Meyer (1986); Northcutt (2011a).

contains a larger number of distinct subdivisions, many of which are not apparent in *Polypterus* (Braford, 2009). For example, the pallium of teleosts contains a central nucleus of large neurons (called area dorsalis centralis, Dc) that has long descending projections to the midbrain and hypothalamus (Demski, 2013). Such a nucleus is not apparent in *Polypterus* and only barely recognizable in other basal ray-finned fishes (Northcutt, 2011a). In fact, large injections of neuronal tracers into the optic tectum of *Polypterus* fail to retrogradely label any neurons in the pallium (unpublished observations by RGN), supporting the hypothesis that these animals lack a homolog of the teleost Dc (at least, they lack the tectum-projecting portion of Dc). Similarly striking is that basal ray-finned fishes lack the small dorsal division of area dorsalis (abbreviated as Dd; Figure 3.28) that lies dorsally within the pallium of most teleosts and develops in close association with Dc (Mueller et al., 2011, and unpublished observations by RGN).

These observations are important, because both Dd and Dc have been proposed as being homologous to the dorsal pallium of tetrapods (Northcutt, 2011b; Mueller et al., 2011). These hypotheses are unlikely to be valid if Dd and Dc cannot be identified in basal ray-finned fishes. As we discussed in Chapter 1, homology requires that the characters in question are derived from a common ancestral character and were retained with a continuous history. As far as we can tell, neither Dd nor Dc in teleosts shares a continuous evolutionary history with the dorsal pallium of tetrapods. Therefore, the most parsimonious interpretation is that these two structures evolved with the origin of teleosts (for more on this, see Chapter 7). Alternatively, one would have to assume that *Polypterus*, sturgeons, *Amia*, and gars all simplified their pallium and did so independently of one another.

The telencephalon of cartilaginous fishes seems, at first glance, simpler to compare to that of tetrapods. Squalomorph sharks have a relatively small, evaginated telencephalon with large telencephalic ventricles (Figure 3.26), and their pallium is more easily divided into medial, dorsal, lateral, and ventral components. Of these, only the lateral and ventral pallial divisions receive substantial input from the olfactory bulb (Figure 3.29), which means that they are probably homologous to the olfactory cortex and pallial amygdala of amniotes, respectively. The connections of the other pallial divisions have not been studied in squalomorph sharks, but the dorsal pallial division in a galeomorph shark receives ascending sensory information from the diencephalon (Luiten, 1981), suggesting that it may be homologous to the dorsal pallium of amniotes. That leaves the medial pallial division as the likely homolog of the tetrapod medial pallium (i.e., the hippocampus). If this interpretation is correct, then it would be reasonable to conclude that the earliest gnathostomes already had a pallium that was divisible into medial, dorsal, lateral, and ventral divisions, and that this ancient pattern of organization was modified, or at least obscured, in the early ray-finned fishes.

However, interpreting the telencephalon of cartilaginous fishes is more complicated than it at first appears. For one thing, almost all of the connectional data for cartilaginous fishes come from sharks or rays with telencephala that are much larger and more complex than those of squalomorph sharks (Smeets et al., 2011), making them less likely to represent the primitive condition for cartilaginous fishes. For another, the ascending projections to the dorsal pallial division in rays are relatively weak and arise from multiple sources. Some of those sources may be homologous to the thalamic nuclei that provide the major ascending input to the dorsal pallium in amniotes, but most of them are not (Hofmann and Northcutt, 2008, 2012). Moreover, even in amphibians, only a few neurons in the thalamus project to the dorsal pallium, and these probably do not carry the kind of high-resolution sensory information that thalamo-telencephalic pathways carry in amniotes (Roth et al., 2003; see Chapter 4). Because sensory inputs from the thalamus became prominent

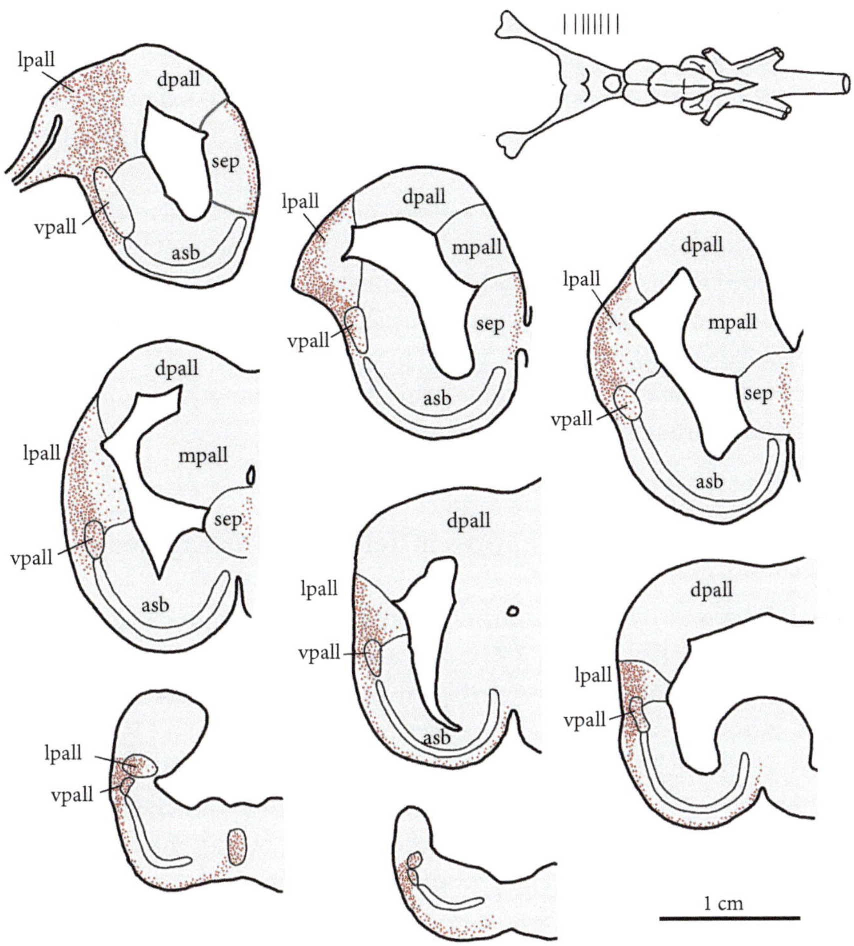

Figure 3.29 Projections of the olfactory bulb in the spiny dogfish. The olfactory bulbs of *Squalus acanthias* project mainly to the lateral pallium (lpall), ventral pallium (vpall), caudal portion of area superficialis basalis (abs), and septum (sep). The dorsal and medial pallial divisions (dpall and mpall) do not receive direct projections from the olfactory bulb. The levels of the indicated sections are shown at the top right, and the most rostral section is shown at the top left.
Based on unpublished observations by RGN.

only with the emergence of amniotes (see Chapters 5 and 6), their presence is not a strong criterion for identifying the dorsal pallium in cartilaginous fishes and other anamniotes.

Unfortunately, alternative criteria for identifying the dorsal pallium are scarce, and gene expression data, which have been used to homologize pallial divisions in amniotes, are just starting to become available for cartilaginous fishes (Quintana-Urzainqui et al., 2012). Another problem is that we know very little about telencephalic organization in the most basal lineage of cartilaginous fishes, the holocephalans (see Figure 3.1; see also Chapter 7). Given this dearth of data, it seems premature to reach firm conclusions on pallial organization in cartilaginous fishes or, for that matter, in early gnathostomes.

Even in the subpallium, some homologies remain quite uncertain. For, example, the largest part of the subpallium in cartilaginous fishes is the area superficialis basalis (Figure 3.29). Parts of this area are probably homologous to the striatum, olfactory tubercle, and pallidum of tetrapods, but there is little data to support more specific hypotheses. In light of these uncertainties, it seems premature to conclude that the circuits of the basal ganglia, including those through the striatum and pallidum, are all conserved across all vertebrates. Recent findings of numerous similarities between this circuitry in lampreys and amniotes (Stephenson-Jones et al., 2011, 2012) are intriguing but, again, there is currently scant evidence that these traits all share a continuous evolutionary history.

3.6. Evolutionary Changes in Telencephalic Connections

One major reason why it is so difficult to homologize the individual subdivisions of the pallium across anamniotes is that the neuronal inputs and outputs of those areas are not as immutable across phylogeny as comparative neuroanatomists often assumed. Although "connectional fingerprints" (Passingham and Wise, 2012) can be a useful tool for identifying putative homologies between some brain regions, especially among closely related species, those connections are themselves subject to evolutionary change. Indeed, the connections of the telencephalon seem to have changed dramatically as the major groups of gnathostomes diverged.

3.6.1. Evolutionary Restriction of Olfactory Inputs

Most fascinating is that the projections of the olfactory bulbs became more restricted during gnathostome evolution. In lampreys and hagfishes the olfactory bulbs project to almost the entire pallium, as well as a few subpallial areas (Northcutt and Puzdrowski, 1988; Wicht and Northcutt, 1993). In lungfishes, too, the olfactory bulbs project to almost the entire telencephalon (Figure 3.30; Northcutt and Rink, 2012). Among the ray-finned fishes, the pattern is more varied. In *Polypterus* the olfactory bulbs project mainly to the dorsomedial pallium, but the dorsolateral pallium also receives at least sparse olfactory inputs (Figure 3.28; Bartheld and Meyer, 1986). Similarly, the olfactory bulbs in sturgeons project to most pallial areas, although they terminate most heavily in a posterior pallial division (Dp in Figure 3.28). In contrast, in all teleosts that have been examined, the olfactory bulbs project almost exclusively to the posterior pallium (Northcutt, 2006), which apparently develops from the pallium's posterolateral pole (Dirian et al., 2014). Given these data, we conclude that the secondary olfactory projections (i.e., the projections of the olfactory bulbs) became more restricted as the teleosts evolved. An analogous

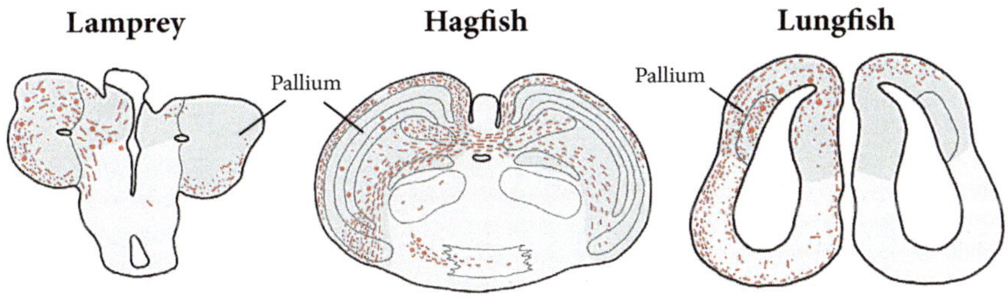

Figure 3.30 Widespread olfactory projections in basal gnathostomes. After injections of neuronal tracers into the olfactory bulb, labeled axons and terminals are seen throughout most of the ipsilateral pallium in lampreys, hagfishes, and lungfishes. These data suggest that the secondary olfactory projections were similarly widespread in early gnathostomes.

Adapted from Northcutt and Puzdrowski (1988), Wicht and Northcutt (1993), Northcutt and Rink (2012).

restriction of the secondary olfactory projections probably occurred with the origin of tetrapods, as the olfactory bulb projections are far more restricted in amphibians and other tetrapods than in the lungfishes (Northcutt, 1981; see Chapter 4). In amniotes they may have become restricted even further, leaving most pallial regions to process other sensory inputs (Puelles, 2017; see Chapter 6).

Less clear is whether the secondary olfactory projections also became more restricted within the cartilaginous fishes. A classic study by Ebbesson and Heimer (1970) showed that the olfactory bulbs in nurse sharks project to just a relatively small portion of the remaining telencephalon, and a similar pattern is evident also in the dogfish (Figure 3.29). These data have been used to argue that restricted olfactory projections are a primitive feature of all vertebrates. However, the more widespread projections in basal ray-finned fishes (Figure 3.28), cyclostomes, and lungfishes (Figure 3.30) suggest that restricted olfactory bulb projections are, instead, a derived feature for cartilaginous fishes, just as they are derived features for teleosts and tetrapods. That is, they evolved independently in those two lineages. This hypothesis is difficult to test, but an examination of secondary olfactory projections in holocephalans would be extremely useful.

In any case, as Hofmann and Northcutt (2008) have pointed out, the telencephalon of cartilaginous fishes is dominated by the sense of smell even though the efferent projections of the olfactory bulb (aka the secondary olfactory projections) are limited, because the tertiary and quaternary olfactory projections within the telencephalon are massive and widespread (Figure 3.31). Combined with the widespread olfactory projections to the pallium in cyclostomes and lungfishes (Figure 3.30), these findings strongly suggest that the telencephalon of early gnathostomes was mainly an olfactory structure. We realize that this hypothesis is a bit of a throwback to earlier proposals made by Ariëns Kappers and others, who argued that the pallium of the earliest vertebrates is merely an olfacto-recipient "paleopallium" (Kappers et al., 1936). These old ideas were challenged in the 1960s and 1970s,

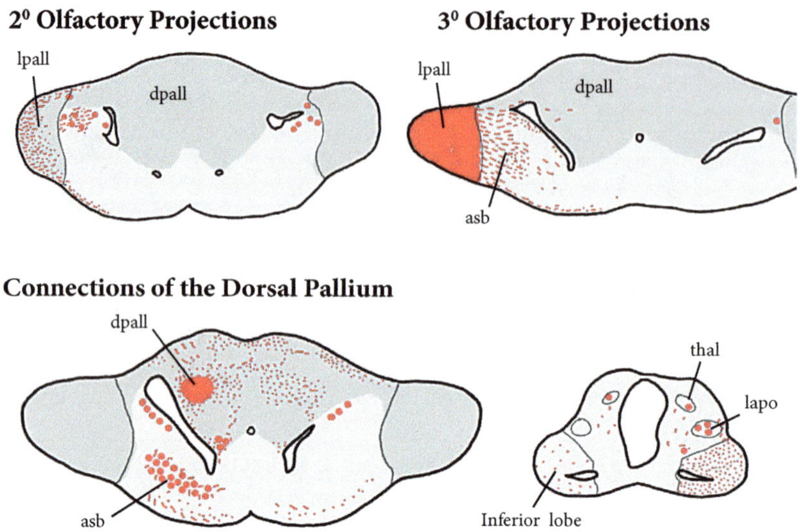

Figure 3.31 Olfactory pathways through the telencephalon in a batoid cartilaginous fish. Shown at the top left are labeled cells and axons after a neuronal tracer injection into the olfactory bulb of the thornback guitarfish (*Platyrhinoidis triseriata*). The olfactory bulb's principal target is the lateral pallium (lpall), whose connections are shown at the top right (injection site is solid red). Note that the lateral pallium projects densely to the area superficialis basalis (asb). Shown along the bottom are results from a tracer injection into the dorsal pallium (dpall), which receives inputs from asb (filled red circles represent labeled cell bodies) and projects bilaterally to the inferior lobe of the hypothalamus (red dots represent labeled axon terminals). The dorsal pallium also receives minor input from the thalamus (thal) and from a lateral posterior thalamic nucleus (lapo). The latter nucleus receives electrosensory input and is probably a migrated component of the posterior tuberculum, rather than the thalamus.
Adapted from Hofmann and Northcutt (2008).

as better tract tracing methods revealed more restricted olfactory projections in sharks, teleosts, and amphibians (Ebbesson, 1972; Northcutt, 1981), but it now appears that wholesale rejection of the old ideas was premature.

3.6.2. Functions of the Telencephalon in Early Gnathostomes

How might early gnathostomes have used their olfaction-dominated telencephalon? As described earlier, olfactory cues decay only slowly in water and, in the presence of currents, can provide useful information over long distances (see Section 3.3.2). However, because turbulence tends to break up odor plumes into discontinuous packets (Figure 3.32), fishes cannot orient toward an odor source just by moving up a smooth odor concentration gradient. Instead, they must integrate over multiple odor packets and swim in the direction that maximizes the

Figure 3.32 Odor plume in a turbulent environment. Shown here is a two-dimensional section through a plume created when a tracer substance is injected into a turbulent stream of water. It mimics the formation of an odor plume in turbulent air. In both cases, the plume doesn't form a continuous concentration gradient but, instead, breaks up into discontinuous packets.
Adapted from Celani et al. (2014, Creative Commons Attribution 3.0 License).

probability of encountering packets of a specific odorant (Vergassola et al., 2007). This kind of analysis requires memory (Murray et al., 2016). In addition, long-term memory for odorants may have been used by early gnathostomes to locate conspecifics or potential prey. Supporting this hypothesis is the finding that lampreys and some teleosts imprint on the odor of their conspecifics and then navigate toward those odorants when they return to spawn in their natal streams (Buchinger et al., 2015). Sharks, too, seem to be using olfaction to navigate through familiar environments over great distances (Nosal et al., 2016). We suspect that early gnathostomes likewise displayed such memory-intensive olfactory behaviors, which required an intact telencephalon.

As olfactory inputs to the telencephalon became more restricted in ray-finned fishes, amniotes, and cartilaginous fishes, non-olfactory inputs to the telencephalon probably expanded or emerged de novo. The latter hypothesis is supported by the observation that the non-olfactory ascending sensory pathways tend to originate from different brain regions in different lineages. In teleosts, non-olfactory inputs reach the telencephalon mainly through the preglomerular complex. Extensive studies have shown that ascending visual, auditory, gustatory, and somatosensory projections in goldfish terminate in largely separate pallial areas (e.g., Northcutt, 2006). Such separate pallial representations for the different sensory modalities are present also in amniotes, but amniotes route most of the ascending sensory information through the thalamus (see Chapters 5–7). Cartilaginous fishes have been studied much less, but ascending visual and electrosensory pathways to their pallium have been described. These ascending pathways were originally thought to pass through the thalamus, but they have also been suggested to involve the pretectum and/or posterior tuberculum and even the midbrain tegmentum (Hofmann and Northcutt, 2012). More work on these pathways in a variety of cartilaginous fishes would clearly be useful but, for now, we conclude that several gnathostome lineages independently evolved, or dramatically expanded, different non-olfactory pathways to the telencephalon. As we discuss in Chapter 7, we further hypothesize that the restriction of olfactory projections to the pallium in these lineages was causally related to the emergence of novel pallial divisions, notably Dd and Dc in teleosts and the dorsal pallium in amniotes.

The principal outputs of the telencephalon in early gnathostomes probably originated in the subpallium, especially the striatum, and targeted mainly the hypothalamus and midbrain tegmentum. Parts of the pallium may have had descending projections, because such pathways have been described in multiple lineages, but the origins of these pathways are variable: in lampreys they originate from the lateral pallium (Ocaña et al., 2015), in cartilaginous fishes from part of the dorsal pallium (Hofmann and Northcutt, 2012), in *Polypterus* from both medial and lateral pallial divisions (Holmes and Northcutt, 2003), and in teleosts primarily from the medial and central pallial areas (Northcutt, 2006; Demski, 2013). This variability, combined with variation in the downstream targets, makes reaching firm conclusions difficult. Some descending pathways from the pallium to extra-telencephalic targets may have been present in the earliest gnathostomes, but they have clearly been modified extensively during subsequent evolution. In comparison, long descending projections from the subpallium have been much more conserved. Indeed, we suspect that in early gnathostomes the striatum, rather than the pallium, was the main origin of long descending projections.

The function of the long descending projections in early gnathostomes is difficult to fathom, but they probably did not contact motor neurons directly or program specific movements. Instead, they probably modulated the activity of various "behavior controllers" in the hypothalamus and tegmentum (Swanson, 2005). The subpallial descending pathways, in particular, may have played a major role in solving the "action selection problem," which is created by the need of organisms to select just one of many possible, potentially competing behaviors at any point in time (Redgrave et al., 1999). According to this view, circuits passing through the striatum and its associated structures (the basal ganglia) engage in a winner-take-all competition where the winning neurons get to "command" the next behavior (see Striedter, 2015). The basal ganglia are thought to function in this kind of action selection in both lampreys and various amniotes, suggesting that this trait is primitive for vertebrates (Stephenson-Jones et al., 2011; 2012). Although the basal ganglia circuits do exhibit some significant variation (to be discussed in later chapters), the core action selection function may well have been conserved. In that context, it is interesting that the selection of appropriate actions requires information about external stimuli and current needs (as signaled by internal stimuli), and is substantially enhanced by memories about which actions had previously brought rewards. Perhaps a major theme in the evolution of the telencephalon has been the emergence of novel pathways that provide the telencephalon with additional kinds of sensory information, which then allows for the formation of new types of memories. Those memories, in turn, likely improved the animals' ability to navigate and, more generally, to select among competing behaviors.

3.7. Functional Synthesis

Two of the most crucial innovations of early gnathostomes were a proper cerebellum and enhanced vestibular system to fine-tune motor control. Together with the evolution of paired fins, these innovations greatly increased the agility of early gnathostomes. In conjunction with the origin of jaws, they made early gnathostomes much more efficient, faster, and more acrobatic predators. As efficient swimmers, early gnathostomes were also capable of moving over longer distances than their jawless ancestors, allowing them to take better advantage of patchily distributed resources. They likely found those resources mainly through their expanded olfactory system, which is a better long-distance sense than vision underwater and at night. Their telencephalon likely received massive olfactory inputs and may have been specialized for comparing current olfactory inputs to remembered ones. Once potential prey or other resources were relatively close, early gnathostomes would have used their improved visual, lateral line, auditory, and gustatory systems to orient toward those objects and discern their identity. Those short-range interactions with stimuli were likely controlled mainly by regions other than the telencephalon, especially the optic tectum.

In general, we suspect that the midbrain (including the optic tectum), hindbrain, and spinal cord in early gnathostomes were "in charge" of reflexive responses to stimuli that are currently close to the animal, and that the telencephalon guided the animal toward (or away from) stimuli that are more distant both in space and time. Although our proposal applies primarily to early gnathostomes, hints of this functional distinction were retained in later gnathostomes. Indeed, massive lesions of the cerebral hemispheres (i.e., the telencephalon) leave most non-mammalian vertebrates capable of performing various machine-like movements but relatively "unconcerned" with future goals (Ferrier, 1876; James, 1890).

Quite early in their history, the gnathostomes diverged into cartilaginous fishes, ray-finned fishes, and lobe-finned fishes. It is tempting to arrange these lineages along a linear scale, because cartilage is often said to be more primitive than bone, and fishes are generally considered the "lowest" type of vertebrate. However, we now know that bone is older than the cartilaginous fishes, and that all three lineages thrived simultaneously in the Devonian and afterward. Even many jawless fishes survived through the Devonian and dwindled in numbers only later. As discussed in Chapter 1, it is generally impossible to arrange species along a single *scala naturae* unless one focuses on just a few, carefully selected attributes. Even basal members of a lineage often have some derived characters. Therefore, reconstructing the evolutionary history of individual features requires a careful, character-by-character, cladistic analysis. When this is done, it becomes apparent that many features of cartilaginous, ray-finned, and lobe-finned fishes have evolved along divergent trajectories.

Most obviously, relative and absolute brain size increased several times within the cartilaginous fishes and, separately, within both the ray-finned fishes and the lobe-finned fishes. Importantly, brain size did not increase uniformly across all brain regions; instead, different brain regions hypertrophied in different lineages. In cartilaginous fishes the increases affected mainly the telencephalon and the cerebellum, and a similar pattern can be discerned in lobe-finned fishes. In the ray-finned fishes, however, increases in brain size were commonly driven by changes in the size of diverse other brain regions. For example, mormyrids enlarged mainly the cerebellar valvula, whereas cyprinids expanded their vagal lobe. In highly visual teleosts, the tectum is usually quite large, whereas coral reef fishes tend to have a large and complex telencephalon. This diversity of regional expansions underscores that brain enlargement did not happen just once, or in a linear sequence, but repeatedly along multiple diverging trajectories. As they increased in size, some brain regions evolved additional subdivisions or multiple laminae; they also changed at least some of their connections. Particularly interesting is that the major lineages of gnathostomes seem to use different, non-homologous pathways to convey non-olfactory sensory information to the telencephalic pallium. Again, the operating principle is evolutionary divergence and convergence, rather than progression through a linear series.

Why did the major radiations of gnathostomes diverge so much? The simplest answer is that their members became adapted to different niches, living in different habitats, eating different kinds of food, and pursuing different reproductive strategies (see Kotrschal, 1998). They may have competed to some extent, but niche specialization was probably a more significant factor than direct competition. In this context, evolving a larger brain was not necessarily "better," but it would have allowed individuals of the more encephalized species to perform behaviors that require more computational power. Those more complex behaviors would have allowed the more encephalized animals to pursue food that is unobtainable for other, less brainy species. The details of these adaptations surely varied from species to species, lineage to lineage, but a good general rule seems to be that animals with larger brains are more efficient at procuring nutritious sustenance when food is scarce. This rule might in fact reflect a hard constraint: since brains are metabolically costly to build and operate (Mink et al., 1981; Liao et al., 2016), larger brains must be accompanied by increased food intake. Social factors, too, may have played a major role in the evolution of larger brains. Enlargements of the telencephalon correlate especially strongly with social complexity, perhaps because complex social behaviors require a good memory, a function for which the telencephalon seems specialized. In each species, the telencephalon is specialized for storing memories related to the species' ecological niche, foraging habits, and reproductive strategy; and it is this function—the storage of specialized forms of memory—that provides the main adaptive value of having a telencephalon.

References

Abdel-Latif H, Hassan ES, Campenhausen von C. 1990. Sensory performance of blind Mexican cave fish after destruction of the canal neuromasts. *Naturwissenschaften* 77:237–239.

Ahrens K, Wullimann MF. 2002. Hypothalamic inferior lobe and lateral torus connections in a percomorph teleost, the red cichlid (*Hemichromis lifalili*). *J Comp Neurol* 449:43–64.

Alvarez-Otero R, Regueira SD, Anadon R. 1993. New structural aspects of the synaptic contacts on Purkinje cells in an elasmobranch cerebellum. *J Anat* 182:13–21.

Alviña K, Sawtell NB. 2014. Sensory processing and corollary discharge effects in posterior caudal lobe Purkinje cells in a weakly electric mormyrid fish. *J Neurophysiol* 112:328–339.

Anadón R, De Miguel E, Gonzalez-Fuentes MJ, Rodicio C. 1989. HRP study of the central components of the trigeminal nerve in the larval sea lamprey: organization and homology of the primary medullary and spinal nucleus of the trigeminus. *J Comp Neurol* 283:602–610.

Ari C. 2011. Encephalization and brain organization of mobulid rays (*Myliobatiformes, Elasmobranchii*) with ecological perspectives. *Open Anat J* 3:1–13.

Aschliman NC, Nishida M, Miya M, Inoue JG, Rosana KM, Naylor GJP. 2012. Body plan convergence in the evolution of skates and rays (Chondrichthyes: Batoidea). *Mol Phylogenet Evol* 63:28–42.

Baker CVH, Modrell MS, Gillis JA. 2013. The evolution and development of vertebrate lateral line electroreceptors. *J Exp Biol* 216:2515–2522.

Barth P, Berenshtein I, Besson M, Roux N, Paramentier E, Banaigs B, Lecchini D. 2015. From the ocean to a reef habitat: how do the larvae of coral reef fishes find their way home? A state of art on the latest advances. *Vie Milieu* 65:91–100.

Bartheld von CS. 2004. The terminal nerve and its relation with extrabulbar "olfactory" projections: lessons from lampreys and lungfishes. *Microsc Res Tech* 65:13–24.

Bartheld von CS, Meyer DL. 1986. Central connections of the olfactory bulb in the bichir, *Polypterus palmas*, reexamined. *Cell Tissue Res* 244:527–535.

Bauchot R, Platel R, Ridet JM. 1976. Brain-body weight relationships in Selachii. *Copeia* 2:305–310.

Bauchot R, Platel R, Ridet JM, Diagne M, Delfini C. 1995. Encéphalisation et adaptations écobiologiques chez les chondrichtyens. *Cybium* 19:153–165.

Bauchot R, Ridet J-M, Bauchot M-L. 1989. The brain organization of butterflyfishes. *Environ Biol Fish* 25:205–219.

Bell CC, Han V, Sawtell NB. 2008. Cerebellum-like structures and their implications for cerebellar function. *Annu Rev Neurosci* 31:1–24.

Bell CC, Szabo T. 1986. Electroreception in mormyrid fish: central anatomy. In: Bullock TH, Heiligenberg W (eds.), *Electroreception*. New York: Springer, pp. 375–421.

Berner RA. 2006. GEOCARBSULF: A combined model for Phanerozoic atmospheric O_2 and CO_2. *Geochimica* 70:5653–5664.

Bierman HS, Zottoli SJ, Hale ME. 2009. Evolution of the Mauthner axon cap. *Brain Behav Evol* 73:174–187.

Bloch S, Manon, T, Yamamoto K. 2017. Mesencephalic origin of the inferior lobe and the pregolomerular nucleus in zebrafish. *Brain Behav Evol* 89:288.

Bloch S, Thomas M, Colin I, Galant S, Machado E, Affaticati P, Jenett A, Yamamoto K. 2019. Mesencephalic origin of the inferior lobe in zebrafish. *BMC Biol* 17:22.

Bodznick D. 1990. Elasmobranch vision: multimodal integration in the brain. *J Exp Zool Suppl* 5:108–116.

Braford MR Jr. 1986. *De gustibus non est disputandem*: a spiral center for taste in the brain of the teleost fish, *Heterotis niloticus*. *Science* 232:489–491.

Braford MR Jr. 2009. Stalking the everted telencephalon: comparisons of forebrain organization in basal ray-finned fishes and teleosts. *Brain Behav Evol* 74:56–76.

Braford MR Jr, Northcutt RG. 1983. Organization of the diencephalon and pretectum of the ray-finned fishes. In: Davis R, Northcutt RG (eds.), *Fish Neurobiology*, Vol. 2. Ann Arbor: University of Michigan Press, pp. 117–164.

Brannen P. 2017. *The Ends of the World*. New York: HarperCollins.

Braun CB, Grande T. 2008. Evolution of peripheral mechanisms for the enhancement of sound reception. In: *Fish Bioacoustics*, Vol. 32, *Springer Handbook of Auditory Research*. New York: Springer, pp. 99–144.

Brazeau MD, Friedman M. 2015. The origin and early phylogenetic history of jawed vertebrates. *Nature* 520:490–497.

Brett CE, Walker SE. 2002. Predators and predation in Paleozoic marine environments. *Paleontol Soc Papers* 4:93–118.

Briggs D, Clarkson E, Aldridge RJ. 1983. The conodont animal. *Lethaia* 16:1–14.

Buchinger TJ, Siefkes MJ, Zielinski BS, Brant CO, Li W. 2015. Chemical cues and pheromones in the sea lamprey (*Petromyzon marinus*). *Front Zool* 12:32.

Bullock TH, Hopkins CD, Popper AN, Fay RR (eds.). 2005. *Electroreception*. New York: Springer.

Bullock TH, Moore JK, Fields RD. 1984. Evolution of myelin sheaths: both lamprey and hagfish lack myelin. *Neurosci Lett* 48:145–148.

Carr C, Maler L. 1986. Electroreception in gymnotiform fish: central anatomy and physiology. In: Bullock TH, Heiligenberg W (eds.), *Electroreception*. New York: John Wiley & Sons, pp. 319–373.

Carrier DR, Heglund NC, Farmer CG. 1992. Speculation on the loss of electroreception in neopterygian fishes. *Am Zool* 32:156A.

Casper BM, Mann DA. 2009. Field hearing measurements of the Atlantic sharpnose shark *Rhizoprionodon terraenovae*. *J Fish Biol* 75:2768–2776.

Celani A, Villermaux E, Vergassola M. 2014. Odor landscapes in turbulent environments. *Phys Rev X* 4:041015.

Chiba A, Honma Y. 1992. Distribution of neuropeptide Y-like immunoreactivity in the brain and hypophysis of the cloudy dogfish, *Scyliorhinus torazame*. *Cell Tissue Res* 268:453–461.

Clement AM, Ahlberg PE. 2014. The first virtual cranial endocast of a lungfish (Sarcopterygii: Dipnoi). *PLoS ONE* 9:e113898.

Colgate JE, Lynch KM. 2004. Mechanics and control of swimming: a review. *IEEE J Ocean Eng* 29:660–673.

Collin SP. 2007. Nervous and sensory systems. In: McKenzie DJ, Farrell AP, Brauner CJ (eds.), *Primitive Fishes*. Amsterdam: Elsevier, pp. 121–179.

Collin SP, Collin HB. 1988. Topographic analysis of the retinal ganglion cell layer and optic nerve in the sandlance *Limnichthyes fasciatus* (Creeiidae, Perciformes). *J Comp Neurol* 278:226–241.

Collin SP, Davies WL, Hart NS, Hunt DM. 2009. The evolution of early vertebrate photoreceptors. *Phil Trans Roy Soc B* 364:2925–2940.

Collin SP, Partridge JC. 1996. Retinal specializations in the eyes of deep-sea teleosts. *J Fish Biol* 49(Suppl A):157–174.

Coombs S, Montgomery JC. 1999. The enigmatic lateral line system. In: Coomgs S, Montogmery JC (eds.), *Comparative Hearing: Fish and Amphibians*, Vol. 11, *Springer Handbook of Auditory Research*. New York: Springer, pp. 319–362.

Cox JPL. 2008. Hydrodynamic aspects of fish olfaction. *J R Soc Interface* 5:575–593.

Crile G, Quiring DP. 1940. A record of the body weight and certain organ and gland weights of 3690 animals. *Ohio J Sci* 40:219–259.

Davis SP, Finarelli JA, Coates MI. 2012. Acanthodes and shark-like conditions in the last common ancestor of modern gnathostomes. *Nature* 486:247–250.

Demski LS. 2012. The neural control of feeding in elasmobranchs: a review and working model. *Environ Biol Fish* 95:169–183.

Demski LS. 2013. The pallium and mind/behavior relationships in teleost fishes. *Brain Behav Evol* 82:31–44.
Derjean D, Moussaddy A, Atallah E, St-Pierre M, Auclair F, Chang S, Ren X, Zielinski B, Dubuc R. 2010. A novel neural substrate for the transformation of olfactory inputs into motor output. *Plos Biol* 8:e1000567.
Devor A. 2000. Is the cerebellum like cerebellar-like structures? *Brain Res Rev* 34:149–156.
Dijkgraaf S. 1963. The functioning and significance of the lateral-line organs. *Biol Rev* 38:51–105.
Dirian L, Galant S, Coolen M, Chen W, Bedu S, Houart C, Bally-Cuif L, Foucher I. 2014. Spatial regionalization and heterochrony in the formation of adult pallial neural stem cells. *Dev Cell* 30:123–136.
Donoghue PCJ, Forey PL, Aldridge RJ. 2000. Conodont affinity and chordate phylogeny. *Biol Rev* 75:191–251.
Donoghue PCJ, Keating JN. 2014. Early vertebrate evolution. *Palaeontology* 57:879–893.
Dubuc R, Brocard F, Antri M, Fénelon K, Gariépy J-F, Smetana R, Ménard A, Le Ray D, Viana Di Prisco G, Pearlstein É, et al. 2008. Initiation of locomotion in lampreys. *Brain Res Rev* 57:172–182.
Eaton RC, Lee RK, Foreman MB. 2001. The Mauthner cell and other identified neurons of the brainstem escape network of fish. *Prog Neurobiol* 63:467–485.
Ebbesson SO. 1972. New insights into the organization of the shark brain. *Comp Biochem Physiol A* 42:121–129.
Ebbesson SO, Heimer L. 1970. Projections of the olfactory tract fibers in the nurse shark (*Ginglymostoma cirratum*). *Brain Res* 17:47–55.
Evan AP, Saland LC, Demski LS. 1976. The structure of the hypothalamic inferior lobes of the blacktip reef shark: scanning and transmission electron microscopic observations. *J Morphol* 150:59–78.
Farrell WJ, Böttger B, Ahmadi F, Finger TE. 2002. Distribution of cholecystokinin, calcitonin gene-related peptide, neuropeptide Y, and galanin in the primary gustatory nuclei of the goldfish. *J Comp Neurol* 450:103–114.
Ferrando S, Gallus L. 2013. Is the olfactory system of cartilaginous fishes a vomeronasal system? *Front Neuroanat* 7:37.
Ferrier D. 1876. *The Functions of the Brain*. London: Smith, Elder.
Finger TE, Bell CC, Russell CJ. 1981. Electrosensory pathways to the valvula cerebelli in mormyrid fish. *Exp Brain Res* 42:23–33.
Finlay BL, Darlington RB. 1995. Linked regularities in the development and evolution of mammalian brains. *Science* 268:1578–1584.
Folgueira MN, Bayley P, Navratilova P, Becker TS, Wilson SW, Clarke JD. 2012. Morphogenesis underlying the development of the everted teleost telencephalon. *Neural Dev* 7:32.
Fontaine R, Affaticati P, Bureau C, Colin I, Demarque M, Dufour S, Vernier P, Yamamoto K, Pasqualini C. 2015. Dopaminergic neurons controlling anterior pituitary functions: anatomy and ontogenesis in zebrafish. *Endocrinol* 156:2934–2948.
Fossi A. 1984. Quelques aspects de la biologie des raies du genre Mobula du Golfe de Californie, vol. Memoire de fin d'etudes. MSc Thesis. Cherbourg, France: Institut national des Techniques de la Mer.
Fricke H, Reinicke O, Hofer H, Nachtigall W. 1987. Locomotion of the coelacanth *Latimeria chalumnae* in its natural environment. *Nature* 329:331–333.
Friedman M. 2015. The early evolution of ray-finned fishes. *Palaeontology* 58:213–228.
Frisch KV. 1938. The sense of hearing in fish. *Nature* 141:8–11.
Funch PG, Kinsman SL, Faber DS, Koenig E, Zottoli SJ. 1981. Mauthner axon diameter and impulse conduction velocity decrease with growth of goldfish. *Neurosci Lett* 27:159–164.
Gai Z, Donoghue PCJ, Zhu M, Janvier P, Stampanoni M. 2011. Fossil jawless fish from China foreshadows early jawed vertebrate anatomy. *Nature* 476:324–327.
Gardiner JM, Atema J. 2010. The function of bilateral odor arrival time differences in olfactory orientation of sharks. *Curr Biol* 20:1187–1191.

Giles S, Friedman M. 2014. Virtual reconstruction of endocast anatomy in early ray-finned fishes (Osteichthyes, Actinopterygii). *J Paleontol* 88:636–651.

Gilland E, Baker R. 2005. Evolutionary patterns of cranial nerve efferent nuclei in vertebrates. *Brain Behav Evol* 66:234–254.

Golan M, Zelinger E, Zohar Y, Levavi-Sivan B. 2015. Architecture of GnRH-gonadotrope-vasculature reveals a dual mode of gonadotropin regulation in fish. *Endocrinol* 156:4163–4173.

González A, Northcutt RG. 2009. An immunohistochemical approach to lungfish telencephalic organization. *Brain Behav Evol* 74:43–55.

Goudemand N, Orchard MJ, Urdy S, Bucher H, Tafforeau P. 2011. Synchrotron-aided reconstruction of the conodont feeding apparatus and implications for the mouth of the first vertebrates. *Proc Natl Acad Sci USA* 108:8720–8724.

Graeber RC, Ebbesson SO, Jane JA. 1973. Visual discrimination in sharks without optic tectum. *Science* 180:413–415.

Grillner S, Deliagina T, Manira El A, Hill RH, Orlovsky GN, Wallén P, Ekeberg Ö, Lansner A. 1995. Neural networks that co-ordinate locomotion and body orientation in lamprey. *Trends Neurosci* 18:270–279.

Grus WE, Zhang J. 2009. Origin of the genetic components of the vomeronasal system in the common ancestor of all extant vertebrates. *Mol Biol Evol* 26:407–419.

Hale ME, Long JH, McHenry MJ, Westneat MW. 2002. Evolution of behavior and neural control of the fast-start escape response. *Evolution* 56:993–1007.

Hallam A, Wignall PB. 1997. *Mass Extinctions and Their Aftermath*. Oxford: Oxford University Press.

Han VZ, Zhang Y, Bell CC, Hansel C. 2007. Synaptic plasticity and calcium signaling in Purkinje cells of the central cerebellar lobes of mormyrid fish. *J Neurosci* 27:13499–13512.

Hashiguchi Y, Nishida M. 2007. Evolution of trace amine associated receptor (TAAR) gene family in vertebrates: lineage-specific expansions and degradations of a second class of vertebrate chemosensory receptors expressed in the olfactory epithelium. *Mol Biol Evol* 24:2099–2107.

Heijdra YF, Nieuwenhuys R. 1994. Topological analysis of the brainstem of the bowfin, *Amia calva*. *J Comp Neurol* 339:12–26.

Hibi M, Shimizu T. 2012. Development of the cerebellum and cerebellar neural circuits. *Dev Neurobiol* 72:282–301.

Hofmann MH, Northcutt RG. 2008. Organization of major telencephalic pathways in an elasmobranch, the thornback ray *Platyrhinoidis triseriata*. *Brain Behav Evol* 72:307–325.

Hofmann MH, Northcutt RG. 2012. Forebrain organization in elasmobranchs. *Brain Behav Evol* 80:142–151.

Holmes PH. 2001. The afferent and efferent connections of the telencephalon in the Senegal bichir. PhD Thesis. University of California, San Diego.

Holmes WM, Cotton R, Xuan VB, Rygg AD, Craven BA, Abel RL, Slack R, Cox JPL. 2011. Three-dimensional structure of the nasal passageway of a hagfish and its implications for olfaction. *Anat Rec* 294:1045–1056.

Holmes PH, Northcutt RG. 2003. Connections of the pallial telencephalon in the Senegal bichir, *Polypterus*. *Brain Behav Evol* 61:113–147.

Hussain A, Saraiva LR, Korsching SI. 2009. Positive Darwinian selection and the birth of an olfactory receptor clade in teleosts. *Proc Natl Acad Sci USA* 106:4313–4318.

Iglesias TL, Dornburg A, Brandley MC, Alfaro ME, Warren DL. 2015. Life in the unthinking depths: energetic constraints on encephalization in marine fishes. *J Evol Biol* 28:1080–1090.

Inoue J, Sato Y, Sinclair R, Tsukamoto K, Nishida M. 2015. Rapid genome reshaping by multiple-gene loss after whole-genome duplication in teleost fish suggested by mathematical modeling. *Proc Natl Acad Sci USA* 112:14918–14923.

Inoue JG, Miya M, Lam K, Tay BH, Danks JA, Bell J, Walker TI, Venkatesh B. 2010. Evolutionary origin and phylogeny of the modern holocephalans (Chondrichthyes: Chimaeriformes): a mitogenomic perspective. *Mol Biol Evol* 27:2576–2586.

Ishikawa Y, Yamamoto N, Yoshimoto M, Yasuda T, Maruyama K, Kage T, Takeda H, Ito H. 2007. Developmental origin of diencephalic sensory relay nuclei in teleosts. *Brain Behav Evol* 69:87–95.

Ito H, Kishida R. 1977. Synaptic organization of the nucleus rotundus in some teleosts. *J Morphol* 151:397–417.

James W. 1890. *The Principles of Psychology*. New York: Henry Holt.

Johnson NS, Yun S-S, Thompson HT, Brant CO, Li W. 2009. A synthesized pheromone induces upstream movement in female sea lamprey and summons them into traps. *Proc Natl Acad Sci USA* 106:1021–1026.

Jorgensen SJ, Reeb CA, Chapple TK, Anderson S, Perle C, Van Sommeran SR, Fritz-Cope C, Brown AC, Klimley AP, Block BA. 2010. Philopatry and migration of Pacific white sharks. *Proc Roy Soc Lond B* 277:679–688.

Ju MJ, Aroca P, Luo J, Puelles L, Redies C. 2004. Molecular profiling indicates avian branchiomotor nuclei invade the hindbrain alar plate. *Neuroscience* 128:785–796.

Kalmijn AJ. 1971. The electric sense of sharks and rays. *J Exp Biol* 55:371–383.

Kani S, Bae Y-K, Shimizu T, Tanabe K, Satou C, Parsons MJ, Scott E, Higashijima S-I, Hibi M. 2010. Proneural gene-linked neurogenesis in zebrafish cerebellum. *Dev Biol* 343:1–17.

Kappers CUA, Huber GC, Crosby EC. 1936. *The Comparative Anatomy of the Nervous System of Vertebrates, Including Man*. New York: Hafner.

Kasumyan AO. 2003. The lateral line in fish: structure, function, and role in behavior. *J Ichthyol* 43:S175–S213.

Kasumyan AO. 2004. The vestibular system and sense of equilibrium in fish. *J Ichthyol* 44: S224–S268.

Keating JN, Marquart CL, Donoghue PCJ. 2015. Histology of the heterostracan dermal skeleton: insight into the origin of the vertebrate mineralised skeleton. *J Morphol* 276:657–680.

Kotrschal K, van Staaden MJ, Huber R. 1998. Fish brains: evolution and environmental relationships. *Rev Fish Biol Fisheries* 8:373–408.

Kriwet J, Kiessling W, Klug S. 2009. Diversification trajectories and evolutionary life-history traits in early sharks and batoids. *Proc Biol Sci* 276:945–951.

Kuratani S. 2004. Evolution of the vertebrate jaw: comparative embryology and molecular developmental biology reveal the factors behind evolutionary novelty. *J Anat* 205:335–347.

Lagman D, Daza DO, Widmark J, Abalo XSM, m GRS, Larhammar D. 2013. The vertebrate ancestral repertoire of visual opsins, transducin alpha subunits and oxytocin/vasopressin receptors was established by duplication of their shared genomic region in the two rounds of early vertebrate genome duplications. *BMC Evol Biol* 13:238.

Lauder GV. 2015. Flexible fins and fin rays as key transformations in ray-finned fishes. In: Dial KP, Shubin N Brainerd EL (eds.), *Great Transformations in Vertebrate Evolution*. Chicago: University of Chicago Press, pp. 31–45.

Lee RK, Eaton RC, Zottoli SJ. 1993. Segmental arrangement of reticulospinal neurons in the goldfish hindbrain. *J Comp Neurol* 329:539–556.

Leto K, Arancillo M, Becker E, Buffo A, Chiang C. 2016. Consensus paper: cerebellar development. *Cerebellum* 15:789–828.

Liao WB, Lou SL, Zeng Y, Kotrschal A. 2016. Large brains, small guts: the expensive tissue hypothesis supported within anurans. *Am Nat* 188:693–700.

Lisney TJ, Yopak KE, Montgomery JC, Collin SP. 2008. Variation in brain organization and cerebellar foliation in chondrichthyans: batoids. *Brain Behav Evol* 72:262–282.

Liu Y-C, Hale ME. 2014. Alternative forms of axial startle behaviors in fishes. *Zoology* 117:36–47.

Longo S, Riccio M, McCune AR. 2013. Homology of lungs and gas bladders: insights from arterial vasculature. *J Morphol* 274:687–703.

Lovell JM, Findlay MM, Moate RM, Nedwell JR, Pegg MA. 2005. The inner ear morphology and hearing abilities of the Paddlefish (*Polyodon spathula*) and the Lake Sturgeon (*Acipenser fulvescens*). *Comp Biochem Physiol A* 142:286–296.

Luiten P. 1981. Two visual pathways to the telencephalon in the nurse shark (*Ginglymostoma cirratum*). II. Ascending thalamo-telencephalic connections. *J Comp Neurol* 196:531–538.

Maklad A, Reed C, Johnson NS, Fritzsch B. 2014. Anatomy of the lamprey ear: morphological evidence for occurrence of horizontal semicircular ducts in the labyrinth of *Petromyzon marinus*. *J Anat* 224:432–446.

Mallatt J. 1996. Ventilation and the origin of jawed vertebrates: a new mouth. *Zool J Linn Soc* 117:329–404.

Mallatt J. 2008. The origin of the vertebrate jaw: neoclassical ideas versus newer, development-based ideas. *Zool Sci* 25:990–998.

Mann DA, Higgs DM, Tavolga WN, Souza MJ, Popper AN. 2001. Ultrasound detection by clupeiform fishes. *J Acoust Soc Am* 109: 3048–3054.

McCormick CA. 1982. The organization of the octavolateralis area in actinopterygian fishes: a new interpretation. *J Morphol* 171:159–181.

McCormick CA, Braford MR Jr. 1988. Central connections of the octavolateralis system: evolutionary considerations. In: Atema J, Fay RR, Popper AN, Tavolga WN (eds.), *Sensory Biology of Aquatic Animals*. New York: Springer, pp. 733–756.

Meek J, Grant K, Bell C. 1999. Structural organization of the mormyrid electrosensory lateral line lobe. *J Exp Biol* 202:1291–1300.

Meek J, Nieuwenhuys R. 1991. Palisade pattern of mormyrid Purkinje cells: a correlated light and electron microscopic study. *J Comp Neurol* 306:156–192.

Mehta RS, Wainwright PC. 2008. Functional morphology of the pharyngeal jaw apparatus in moray eels. *J Morphol* 269:604–619.

Metzner W, Juranek J. 1997. A sensory brain map for each behavior? *Proc Natl Acad Sci USA* 94:14798–14803.

Meurling P. 2011. The vascularization of the pituitary in *Chimaera monstrosa* (Holocephali). *Sarsia* 30:83–106.

Mink JW, Blumenschine RJ, Adams DB. 1981. Ratio of central nervous system to body metabolism in vertebrates: its constancy and functional basis. *Am J Physiol* 241:R203–12.

Miyashita T. 2016. Fishing for jaws in early vertebrate evolution: a new hypothesis of mandibular confinement. *Biol Rev* 91:611–657.

Montgomery JC. 1988. Temperature compensation in the vestibulo-ocular reflex: a novel hypothesis of cerebellar function. *J Theor Biol* 132:163–170.

Montgomery JC, Bodznick D, Yopak KE. 2012. The cerebellum and cerebellum-like structures of cartilaginous fishes. *Brain Behav Evol* 80:152–165.

Morita Y, Finger TE. 1985. Topographic and laminar organization of the vagal gustatory system in the goldfish, *Carassius auratus*. *J Comp Neurol* 238:187–201.

Morita Y, Murakami T, Ito H. 1983. Cytoarchitecture and topographic projections of the gustatory centers in a teleost, *Carassius carassius*. *J Comp Neurol* 218:378–394.

Mueller T, Dong Z, Berberoglu MA, Guo S. 2011. The dorsal pallium in zebrafish, *Danio rerio* (Cyprinidae, Teleostei). *Brain Res* 1381:95–105.

Murakami T, Morita Y. 1987. Morphology and distribution of the projection neurons in the cerebellum in a teleost, *Sebastiscus marmoratus*. *J Comp Neurol* 256:607–623.

Murdock DJE, Dong X-P, Repetski JE, Marone F, Stampanoni M, Donoghue PCJ. 2013. The origin of conodonts and of vertebrate mineralized skeletons. *Nature* 502:546–549.

Murray E, Wise S, Graham K. 2016. *The Evolution of Memory Systems*. Oxford: Oxford University Press.

Myagkov NA. 1991. The brain sizes of living Elasmobranchii as their organization level indicator. I. General analysis. *J Hirnforsch* 32:553–561.

Nakane Y, Ikegami K, Iigo M, Ono H, Takeda K, Takahashi D, Uesaka M, Kimijima M, Hashimoto R, Arai N, et al. 2013. The saccus vasculosus of fish is a sensor of seasonal changes in day length. *Nat Commun* 4:2108.

Nieuwenhuys R. 1963. The comparative anatomy of the actinopterygian forebrain. *J Hirnforsch* 7:171–192.

Nieuwenhuys R. 1974. Topological analysis of the brain stem: a general introduction. *J Comp Neurol* 156:255–276.

Nieuwenhuys R. 2009. The forebrain of actinopterygians revisited. *Brain Behav Evol* 73:229–252.

Nieuwenhuys R. 2011. The development and general morphology of the telencephalon of actinopterygian fishes: synopsis, documentation and commentary. *Brain Struct Funct* 215:141–157.

Nieuwenhuys R, Donkelaar ten HJ, Nicholson C. 1998. *The Central Nervous System of Vertebrates*. Berlin: Springer.

Niimura Y. 2012. Olfactory receptor multigene family in vertebrates: from the viewpoint of evolutionary genomics. *Curr Genomics* 13:103–114.

Nilsson G. 1996. Brain and body oxygen requirements of *Gnathonemus petersii*, a fish with an exceptionally large brain. *J Exp Biol* 199:603–607.

Noreikiene K, Herczeg G, Gonda A, Balázs G, Husby A, Merilä J. 2015. Quantitative genetic analysis of brain size variation in sticklebacks: support for the mosaic model of brain evolution. *Proc Roy Soc Lond B* 282:20151008.

Northcutt RG. 1977. Elasmobranch central nervous system organization and its possible evolutionary significance. *Am Zool* 17:411–429.

Northcutt RG. 1979. Experimental determination of the primary trigeminal projections in lampreys. *Brain Res* 163:323–327.

Northcutt RG. 1981. Evolution of the telencephalon in nonmammals. *Annu Rev Neurosci* 4:301–350.

Northcutt RG. 1983. Evolution of the optic tectum in ray-finned fishes. In: Davis RE, Northcutt RG (eds.), *Fish Neurobiology*, Vol. 2. Ann Arbor: University of Michigan Press, pp. 1–42.

Northcutt RG. 1989. The phylogenetic distribution and innervation of craniate mechanoreceptive lateral lines. In: Coombs S, Görner P, Münz H (eds.), In: *The Mechanosensory Lateral Line: Neurobiology and Evolution*. New York: Springer, pp. 17–78.

Northcutt RG. 1990. Visual pathways in elasmobranchs: organization and phylogenetic implications. *J Exp Zool Suppl* 5:97–107.

Northcutt RG. 1995. The forebrain of gnathostomes: in search of a morphotype. *Brain Behav Evol* 46:275–318.

Northcutt RG. 2004. Taste buds: development and evolution. *Brain Behav Evol* 64:198–206.

Northcutt RG. 2006. Connections of the lateral and medial divisions of the goldfish telencephalic pallium. *J Comp Neurol* 494:903–943.

Northcutt RG. 2008. Forebrain evolution in bony fishes. *Brain Res Bull* 75:191–205.

Northcutt RG. 2009a. Phylogeny of nucleus medianus of the posterior tubercle in ray-finned fishes. *Integr Zool* 4:134–152.

Northcutt RG. 2009b. Telencephalic organization in the spotted African Lungfish, *Protopterus dolloi*: a new cytological model. *Brain Behav Evol* 73:59–80.

Northcutt RG. 2011a. Olfactory projections in the white sturgeon, *Acipenser transmontanus*: an experimental study. *J Comp Neurol* 519:1999–2022.

Northcutt RG. 2011b. Do teleost fishes possess a homolog of mammalian isocortex? *Brain Behav Evol* 78:136–138.

Northcutt RG, Bemis WE. 1993. Cranial nerves of the coelacanth, *Latimeria chalumnae* [Osteichthyes: Sarcopterygii: Actinistia], and comparisons with other craniata. *Brain Behav Evol* 42 (Suppl 1):1–76.

Northcutt RG, Butler AB. 1976. Retinofugal pathways in the longnose gar *Lepisosteus osseus* (Linnaeus). *J Comp Neurol* 166:1–16.

Northcutt RG, Davis RE. 1983. Telencephalic organization in ray-finned fishes. In: Davis RE, Northcutt RG (eds.), *Fish Neurobiology*, Vol. 2. Ann Arbor: University of Michigan Press, pp. 203–236.

Northcutt RG, Neary TJ, Senn DG. 1978. Observations on the brain of the coelacanth *Latimeria chalumnae*: External anatomy and quantitative analysis. *J Morphol* 155:181–192.

Northcutt RG, Puzdrowski RL. 1988. Projections of the olfactory bulb and nervus terminalis in the silver lamprey. *Brain Behav Evol* 32:96–107.

Northcutt RG, Rink E. 2012. Olfactory projections in the lepidosirenid lungfishes. *Brain Behav Evol* 79:4–25.

Northmore D. 2011. Optic tectum. In: Farrell AP (ed.), *Encyclopedia of Fish Physiology: From Genome to Environment*. Amsterdam: Elsevier, pp. 131–142.

Northmore DPM. 2017. Holding visual attention for 400 million years: A model of tectum and torus longitudinalis in teleost fishes. *Vision Res* 131:44–56.

Nosal AP, Chao Y, Farrara JD, Chai F, Hastings PA. 2016. Olfaction contributes to pelagic navigation in a coastal shark. *PLoS ONE* 11:e0143758.

Ocaña FM, Suryanarayana SM, Saitoh K, Kardamakis AA, Capantini L, Robertson B, Grillner S. 2015. The lamprey pallium provides a blueprint of the mammalian motor projections from cortex. *Curr Biol* 25:413–423.

Passingham RE, Wise SP. 2012. *The Neurobiology of the Prefrontal Cortex*. Oxford: Oxford University Press.

Paul DH, Roberts BL. 1979. The significance of cerebellar function for a reflex movement of the dogfish. *J Comp Physiol* 134:69–74.

Peters RC, Eeuwes LBM, Bretschneider F. 2007. On the electrodetection threshold of aquatic vertebrates with ampullary or mucous gland electroreceptor organs. *Biol Rev* 82:361–373.

Pfeiffer W. 1969. Der Geruchssinn der Polypteridae (Pisces, Brachiopterygii). *J Comp Physiol A* 63:151–164.

Platel R, Vesselkin NP. 1989. [Comparative studies of the anatomy of the brain in three species of Agnatha: *Petromyzon marinus, Lampetra fluviatilis* and *Lampetra planeri*]. *J Hirnforsch* 30:23–32.

Popper AN, Fay RR. 2011. Rethinking sound detection by fishes. *Hear Res* 273:25–36.

Puelles L. 2017. Comments on the updated tetrapartite pallium model in the mouse and chick, featuring a homologous claustro-insular complex. *Brain Behav Evol* 90:171–189.

Puelles L, Kuwana E, Puelles E, Bulfone A, Shimamura K, Keleher J, Smiga S, Rubenstein JL. 2000. Pallial and subpallial derivatives in the embryonic chick and mouse telencephalon, traced by the expression of the genes Dlx-2, Emx-1, Nkx-2.1, Pax-6, and Tbr-1. *J Comp Neurol* 424:409–438.

Puelles L, Sandoval JE, Ayad A, del Corral R, Alonso A, Ferran JL, Martinez-de-la-Torre M. 2017. The pallium in reptiles and birds in the light of the updated tetrapartite pallium model. In: Kaas J (ed.), *Evolution of Nervous Systems*, 2nd ed. Amsterdam: Elsevier, pp. 519–555.

Puzdrowski RL, Leonard RB. 1994. Vestibulo-oculomotor connections in an elasmobranch fish, the Atlantic stingray, *Dasyatis sabina*. *J Comp Neurol* 339:587–597.

Qu Q-M, Zhu M, Zhao W-J. 2010. Silurian atmospheric O_2 changes and the early radiation of gnathostomes. *Palaeoworld* 19:146–159.

Quintana-Urzainqui I, Sueiro C, Carrera I, Ferreiro-Galve S, Santos-Durán G, Pose-Méndez S, Mazan S, Candal E, Rodríguez-Moldes I. 2012. Contributions of developmental studies in the dogfish *Scyliorhinus canicula* to the brain anatomy of elasmobranchs: insights on the basal ganglia. *Brain Behav Evol* 80:127–141.

Redgrave P, Prescott TJ, Gurney K. 1999. The basal ganglia: a vertebrate solution to the selection problem? *Neuroscience* 89:1009–1023.

Rink E, Wullimann MF. 2001. The teleostean (zebrafish) dopaminergic system ascending to the subpallium (striatum) is located in the basal diencephalon (posterior tuberculum). *Brain Res* 889:316–330.

Roberts BL, Ryan KP. 2009. Cytological features of the giant neurons controlling electric discharge in the ray, Torpedo. *J Mar Biol Ass* 55:123–131.

Roberts BL, Witkovsky P. 1975. A functional analysis of the mesencephalic nucleus of the fifth nerve in the selachian brain. *Proc Roy Soc Lond B* 190:473–495.

Roberts MG, Savage GE. 1978. Effects of hypothalamic lesions on the food intake of the goldfish (*Carassius auratus*). *Brain Behav Evol* 15:150–164.

Rodríguez F, López JC, Vargas JP, Gómez Y, Broglio C, Salas C. 2002. Conservation of spatial memory function in the pallial forebrain of reptiles and ray-finned fishes. *J Neurosci* 22:2894–2903.

Rodríguez-Moldes I. 2009. A developmental approach to forebrain organization in elasmobranchs: new perspectives on the regionalization of the telencephalon. *Brain Behav Evol* 74:20–29.

Rodríguez-Moldes I, Santos-Durán GN, Pose-Méndez S, Quintana-Urzainqui I, Candal E. 2017. The brains of cartilaginous fishes. In: Kaas J (ed.), *Evolution of Nervous Systems*, 2nd ed. Elsevier, pp. 77–97.

Rogers PH, Cox M. 1988. Underwater sound as a biological stimulus. In: Atema J, Fay RR, Popper AN, Tavolga WN (eds.), *Sensory Biology of Aquatic Animals*. New York: Springer, pp. 131–149.

Roth G, Grunwald W, Dicke U. 2003. Morphology, axonal projection pattern, and responses to optic nerve stimulation of thalamic neurons in the fire-bellied toad *Bombina orientalis*. *J Comp Neurol* 461:91–110.

Sansom RS, Randle E, Donoghue PCJ. 2014. Discriminating signal from noise in the fossil record of early vertebrates reveals cryptic evolutionary history. *Proc Roy Soc Lond B* 282:20142245.

Schaeffer B. 1981. The xenacanth shark neurocranium, with comments on elasmobranch monophyly. *Bull Am Mus Nat Hist* 169:1–66.

Schroeder DM, Vanegas H, Ebbesson SO. 1980. Cytoarchitecture of the optic tectum of the squirrelfish, *Holocentrus*. *J Comp Neurol* 191:337–351.

Schulz-Mirbach T, Metscher B, Ladich F. 2012. Relationship between swim bladder morphology and hearing abilities: a case study on Asian and African cichlids. *PLoS ONE* 7:e42292.

Shadmehr R, Wise SP. 2005. *The Computational Neurobiology of Reaching and Pointing*. Cambridge, MA: MIT Press.

Smeets WJ, Boord RL. 1985. Connections of the lobus inferior hypothalami of the clearnose skate *Raja eglanteria* (Chondrichthyes). *J Comp Neurol* 234:380–392.

Smeets W, Nieuwenhuys R, Roberts BL. 2011. *The Central Nervous System of Cartilaginous Fishes*. Berlin; Heidelberg: Springer.

Song JK, Northcutt RG. 1991. Morphology, distribution and innervation of the lateral-line receptors of the Florida gar, *Lepisosteus platyrhincus*. *Brain Behav Evol* 37:10–37.

Sower SA. 2015. Breaking dogma on the hypothalamic-pituitary anatomical relations in vertebrates. *Endocrinol* 156:3882–3884.

Stensmyr MC, Maderspacher F. 2012. Pheromones: fish fear factor. *Curr Biol* 22:R183–6.

Stephenson-Jones M, Ericsson J, Robertson B, Grillner S. 2012. Evolution of the basal ganglia: dual-output pathways conserved throughout vertebrate phylogeny. *J Comp Neurol* 520:2957–2973.

Stephenson-Jones M, Samuelsson E, Ericsson J, Robertson B, Grillner S. 2011. Evolutionary conservation of the basal ganglia as a common vertebrate mechanism for action selection. *Curr Biol* 21:1–11.

Straka H, Baker R. 2013. Vestibular blueprint in early vertebrates. *Front Neural Circuits* 7:182.

Strick PL, Dum RP, Fiez JA. 2009. Cerebellum and nonmotor function. *Annu Rev Neurosci* 32:413–434.

Striedter GF. 1990. The diencephalon of the channel catfish, *Ictalurus punctatus*. II. Retinal, tectal, cerebellar and telencephalic connections. *Brain Behav Evol* 36:355–377.

Striedter GF. 1991. Auditory, electrosensory, and mechanosensory lateral line pathways through the forebrain in channel catfishes. *J Comp Neurol* 312:311–331.

Striedter GF. 2005. *Principles of Brain Evolution*. Sunderland, MA: Sinauer Associates.

Striedter GF. 2015. *Neurobiology: A Functional Approach*. Oxford: Oxford University Press.

Striedter GF, Northcutt RG. 2006. Head size constrains forebrain development and evolution in ray-finned fishes. *Evol Dev* 8:215–222.

Sueiro C, Carrera IAN, Ferreiro S, Molist P, Adrio F, Anadon R, Rodriguez-Moldes I. 2007. New insights on saccus vasculosus evolution: a developmental and immunohistochemical study in elasmobranchs. *Brain Behav Evol* 70:187–204.

Sugahara F, Pascual-Anaya J, Oisi Y, Kuraku S, Aota S-I, Adachi N, Takagi W, Hirai T, Sato N, Murakami Y, Kuratani S. 2016. Evidence from cyclostomes for complex regionalization of the ancestral vertebrate brain. *Nature* 531:97–100.

Swanson LW. 2005. Anatomy of the soul as reflected in the cerebral hemispheres: neural circuits underlying voluntary control of basic motivated behaviors. *J Comp Neurol* 493:122–131.

Thorrold SR, Afonso P, Fontes J, Braun CD, Santos RS, Skomal GB, Berumen ML. 2014. Extreme diving behaviour in devil rays links surface waters and the deep ocean. *Nat Commun* 5:4274.

Trotter JA, Williams IS, Barnes CR, Lécuyer C, Nicoll RS. 2008. Did cooling oceans trigger Ordovician biodiversification? Evidence from conodont thermometry. *Science* 321:550–554.

van Dongen PAM. 1998. Brain size in vertebrates. In: Nieuwenhuys R, Donkelaar ten HJ, Nicholson C (eds.), *The Central Nervous System of Vertebrates*. Berlin: Springer. pp 2099–2134.

Van-Eyk SM, Siebeck UE, Champ CM, Marshall J, Hart NS. 2011. Behavioural evidence for colour vision in an elasmobranch. *J Exp Biol* 214:4186–4192.

Vanegas H. 1984. *Comparative Neurology of the Optic Tectum*. New York; London: Plenum Press.

Vanegas H, Ito H. 1983. Morphological aspects of the teleostean visual system: a review. *Brain Res* 287:117–137.

Venkatesh B, Lee AP, Ravi V, Maurya AK, Lian MM, Swann JB, Ohta Y, Flajnik MF, Sutoh Y, Kasahara M, et al. 2014. Elephant shark genome provides unique insights into gnathostome evolution. *Nature* 505:174–179.

Vergassola M, Villermaux E, Shraiman BI. 2007. "Infotaxis" as a strategy for searching without gradients. *Nature* 445:406–409.

Vernier P, Wullimann MF. 2009. Evolution of the posterior tuberculum and preglomerular nuclear complex. In: Binder MD, Hirokawa N, Windhorst U (eds.), *Encyclopedia of Neuroscience*. Berlin; Heidelberg: Springer, pp. 1404–1413.

Vilensky JA. 2014. The neglected cranial nerve: nervus terminalis (cranial nerve N). *Clin Anat* 27:46–53.

Waehneldt TV, Matthieu JM, Jeserich G. 1986. Appearance of myelin proteins during vertebrate evolution. *Neurochem Int* 9:463–474.

Wagner CM, Stroud EM, Meckley TD. 2011. A deathly odor suggests a new sustainable tool for controlling a costly invasive species. *Can J Fish Aquat Sci* 68:1157–1160.

Wagner DO, Aspenberg P. 2011. Where did bone come from? *Acta Orthop* 82:393–398.

Walton PL, Christensen-Dalsgaard J, Carr CE. 2017. Evolution of sound source localization circuits in the nonmammalian vertebrate brainstem. *Brain Behav Evol* 90:131–153.

Wicht H, Northcutt RG. 1993. Secondary olfactory projections and pallial topography in the Pacific hagfish, *Eptatretus stouti*. *J Comp Neurol* 337:529–542.

Wilga CD. 2005. Morphology and evolution of the jaw suspension in lamniform sharks. *J Morphol* 265:102–119.

Wilga CD, Hueter RE, Wainwright PC. 2001. Evolution of upper jaw protrusion mechanisms in elasmobranchs. *Am Zool* 41:1248–1257.

Williams R IV, Neubarth N, Hale ME. 2013. The function of fin rays as proprioceptive sensors in fish. *Nat Commun* 4:1729.

Wullimann MF. 1997. Major patterns of visual brain organization in teleosts and their relation to prehistoric events and the paleontological record. *Paleobiology* 23:101–114.

Wullimann MF, Grothe B. 2013. The central nervous organization of the lateral line system. In: Coombs S, Bleckmann H, Fay RR, Popper AN (eds.), *The Lateral Line System*, Vol. 48, Springer Handbook of Auditory Research. New York: Springer, pp. 195–251.

Wullimann MF, Meyer DL. 1990. Phylogeny of putative cholinergic visual pathways through the pretectum to the hypothalamus in teleost fish. *Brain Behav Evol* 36:14–29.

Wullimann MF, Mueller T, Distel M, Babaryka A. 2011. The long adventurous journey of rhombic lip cells in jawed vertebrates: a comparative developmental analysis. *Front Neuroanat* 5:27.

Xue H-G, Yang C-Y, Yamamoto N. 2008. Afferent sources to the inferior olive and distribution of the olivocerebellar climbing fibers in cyprinids. *J Comp Neurol* 507:1409–1427.

Yamamoto N, Ishikawa Y, Yoshimoto M, Xue H-G, Bahaxar N, Sawai N, Yang C-Y, Ozawa H, Ito H. 2007. A new interpretation on the homology of the teleostean telencephalon based on hodology and a new eversion model. *Brain Behav Evol* 69:96–104.

Yamamoto K, Bloch S, Vernier P. 2017. New perspective on the regionalization of the anterior forebrain in Osteichthyes. *Dev Growth Differ* 59:175–187.

Yamamoto N, Ito H. 2008. Visual, lateral line, and auditory ascending pathways to the dorsal telencephalic area through the rostrolateral region of the lateral preglomerular nucleus in cyprinids. *J Comp Neurol* 508:615–647.

Yamamoto N, Kato T, Okada Y, Somiya H. 2010. Somatosensory nucleus in the torus semicircularis of cyprinid teleosts. *J Comp Neurol* 518:2475–2502.

Yopak KE, Lisney TJ. 2012. Allometric scaling of the optic tectum in cartilaginous fishes. *Brain Behav Evol* 80:108–126.

Yopak KE, Lisney TJ, Collin SP, Montgomery JC. 2007. Variation in brain organization and cerebellar foliation in chondrichthyans: sharks and holocephalans. *Brain Behav Evol* 69:280–300.

Yopak KE, Montgomery JC. 2008. Brain organization and specialization in deep-sea chondrichthyans. *Brain Behav Evol* 71:287–304.

Yoshizawa M, Jeffery WR, van Netten SM, McHenry MJ. 2014. The sensitivity of lateral line receptors and their role in the behavior of Mexican blind cavefish (*Astyanax mexicanus*). *J Exp Biol* 217:886–895.

Zhu M, Yu X, Choo B, Wang J, Jia L. 2012. An antiarch placoderm shows that pelvic girdles arose at the root of jawed vertebrates. *Biol Lett* 8:453–456.

4

The Invasion of Land

Lobe-Finned Fishes and Amphibians

For us today it is quite natural to think that vertebrates can thrive out of the water, on the ground and in the air, but the colonization of land was a huge step for vertebrates, as it had been for land plants and invertebrates before them. Vascular plants first appeared in the Silurian period (443–416 mya), and by the late Devonian (~380 mya) forests of large tree-like plants were globally distributed across the land (Gensel and Edwards, 2001). Closely following those plants were arthropods and other invertebrates, starting with millipedes in the Silurian (Jeram et al., 1990). Vertebrates were extremely successful during the Devonian, but most of them remained fully aquatic, breathing with gills and swimming with fins (see Chapter 3). That said, a few lobe-finned fishes became increasingly terrestrial during the late Devonian, evolving complex lungs and limbs that they could use on land. By the end of the Carboniferous period (360–300 mya), some fully terrestrial vertebrates emerged.

4.1. The Lobe-Finned Vertebrates

In the previous chapter we briefly introduced the lobe-finned vertebrates (sarcopterygians), which include the coelacanths and lungfishes (see Figure 3.1). However, the sarcopterygians also include all of the tetrapods, including amphibians and amniotes (see Chapter 5), as well as numerous extinct taxa. We here discuss them all in turn.

4.1.1. Coelacanths

Coelacanths were quite diverse in the Devonian, but all of them were thought to be extinct until 1938, when Marjorie Courtenay-Latimer, working at a natural history museum in South Africa, found an unusual 1.5 m long fish among the animals some local fishermen had caught. She suspected that this specimen was related to extinct coelacanths, a suspicion later confirmed by J. L. B. Smith. In honor of Courtenay-Latimer the new species was named *Latimeria chalumnae*. It took 15 years before a second individual of this rare species was found, preserved, and examined

Brains Through Time. Georg F. Striedter and R. Glenn Northcutt, Oxford University Press (2020). © Oxford University Press.
DOI: 10.1093/oso/9780195125689.001.0001

more thoroughly. In 1997, a second species of living coelacanth, called *Latimeria menadoensis*, was discovered in Indonesia. More recently, molecular analyses have shown that the proteins of coelacanths have, on average, evolved very slowly and that these animals comprise the most basal branch of living sarcopterygians (Figure 4.1).

Given the rarity of living coelacanths, it is not surprising that we know very little about their behavior. We do know, however, that they tend to live in relatively deep water (100–250 m deep) off steep coastal slopes. They tend to hide in caves during the day and eat mainly smaller fish and cephalopods. Underwater video has shown that they usually move very slowly, using their paired fins for steering and stabilization, but they can swim fast by moving their large caudal fin from side to side (Fricke et al., 1987). Remarkably, they tend to alternate pectoral and pelvic fin movements on the left and right sides of the body in a pattern very similar to that employed by tetrapods during four-legged walking.

An examination of pregnant female coelacanths revealed that they give birth to live young, with the eggs hatching inside the mother's body. This trait (ovoviviparity)

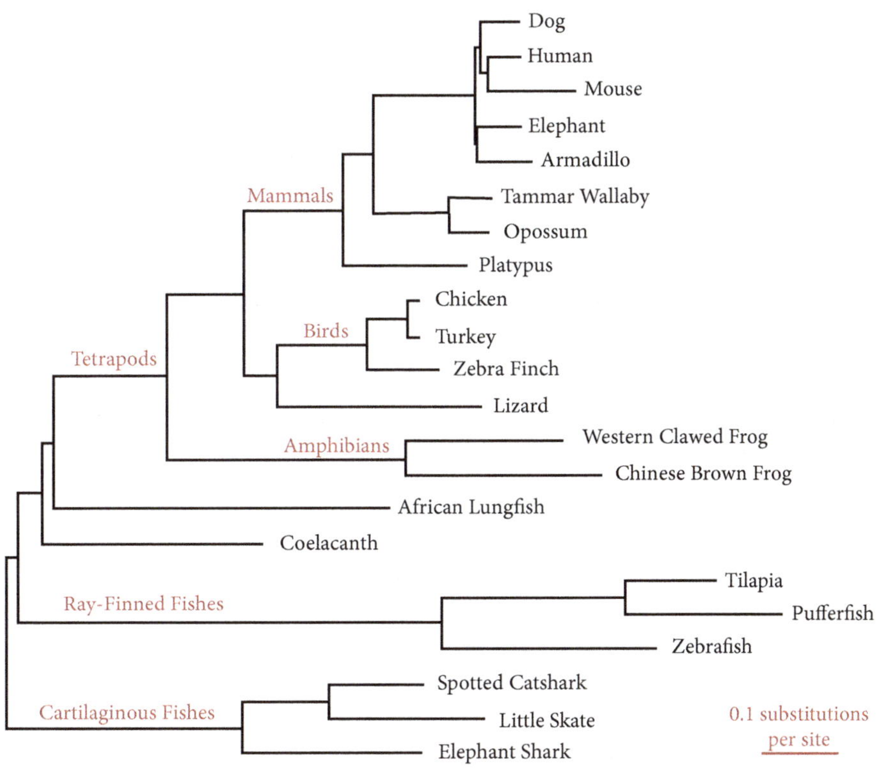

Figure 4.1 Phylogeny of jawed vertebrate proteins. A comparative analysis of 251 protein-coding genes from 22 species supports the hypothesis that lungfishes are the closest living relatives of tetrapods. The rate of amino acid substitution was relatively low in coelacanths, as indicated by their short branch length.
Adapted from Amemiya et al. (2013)

is unusual among fishes, although it has independently evolved in a few teleosts and cartilaginous fishes. Another unusual feature of coelacanths is that their body contains a large fat-filled structure that increases the animal's buoyancy. This fatty organ was long thought to be a modified swim bladder or lung, but juvenile coelacanths posses a separate swim bladder homolog that gradually degenerates as the animals mature (Cupello et al., 2015). Therefore, the identity of the coelacanth's fatty organ remains a mystery.

The brains of coelacanths are relatively small (Northcutt et al., 1978; Nieuwenhuys et al., 1998; Northcutt and González, 2011). The brain of one 30 kg coelacanth reportedly weighed just 1.1–1.5 g, which is roughly what one would expect for an amphibian of similar body size but smaller than the brains of same-sized sharks and teleosts (Northcutt et al., 1978). Curiously, the brains of coelacanths are surrounded by a thick layer of fat and occupy less than 2% of the endocranial cavity, which is far less than what one sees in other extant vertebrates.

Living coelacanths possess a well-developed cerebellum with large auricles and a relatively small optic tectum (Figure 4.2; Nieuwenhuys et al., 1998). Their diencephalon contains a large thalamus, a small saccus vasculosus, and a pituitary gland that protrudes rostrally, rather than ventrally or caudally (which is what it does in all other vertebrates). The coelacanth telencephalon is relatively large and unusual in that its medial portion is very thick and bulges into the ventricle (Northcutt and González, 2011), suggesting that part of it may be "everted," rather than evaginated (see Chapter 3). The coelacanth telencephalon also includes an unusual "rostral body" that probably receives input from the olfactory bulb. Sadly, there are currently no histochemical or experimental studies on coelacanth brains.

The cranial nerves and main sensory organs of coelacanths have been studied in some detail (Fritzsch, 1987; Northcutt and Bemis, 1993). Thus, we know that coelacanth eyes are relatively large and that their retinas are adapted for deep water with little light (Yokoyama et al., 1999). Coelacanths also have a well-developed mechanosensory lateral line system. They lack typical electroreceptors but have an unusual "rostral organ" that is probably electrosensory. The olfactory system of coelacanths is well developed anatomically and features a diverse array of olfactory receptor molecules (Picone et al., 2013). Finally, it is interesting that coelacanths have greatly expanded the T2R family of taste receptors, which is responsible for sensing "bitter" substances in other vertebrates (Syed and Korsching, 2014).

4.1.2. Lungfishes

Lungfishes are probably the closest living relatives of tetrapods (Figure 4.1). Some authors have suggested that lungfishes and coelacanths are sister groups (Forey, 1988; Yokobori et al., 1994), but a comparative analysis of 251 genes suggest that lungfishes are more closely related to tetrapods than coelacanths are (Amemiya et al., 2013; Irisarri and Meyer, 2016).

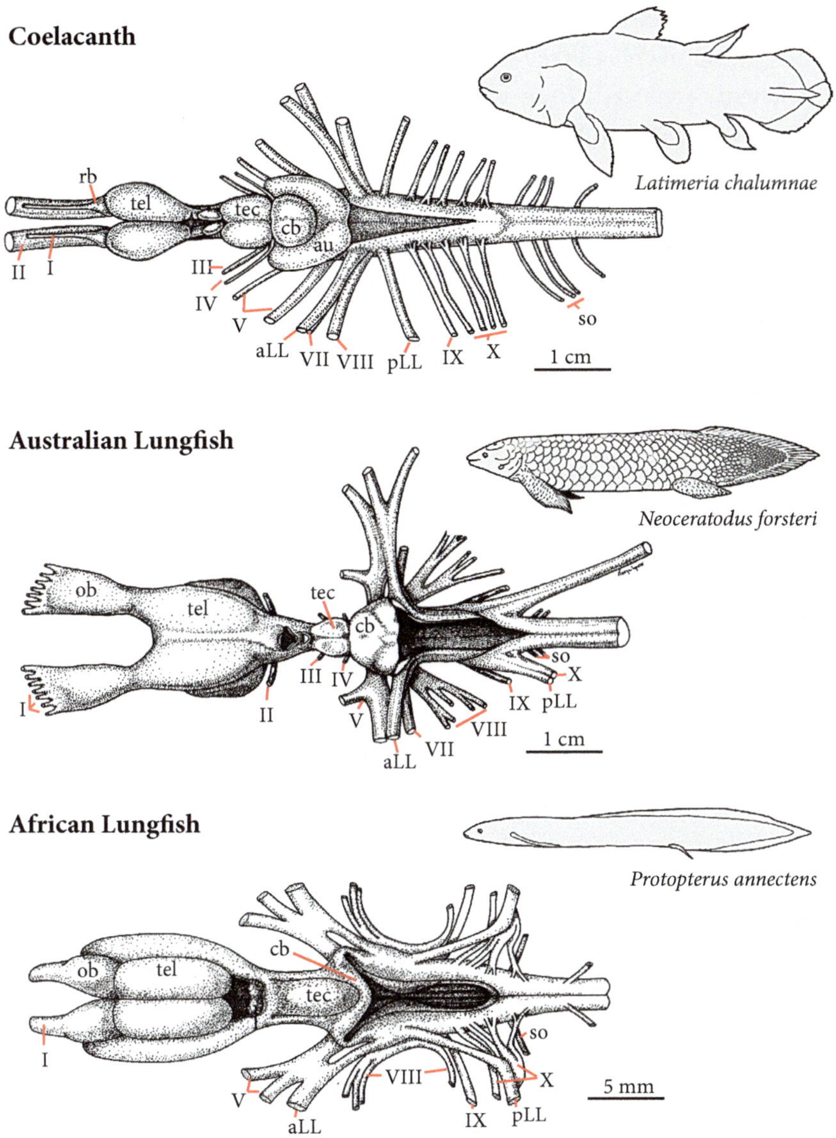

Figure 4.2 Brains and bodies of basal lobe-finned fishes. All three brains are shown from a dorsal perspective. Most of the cranial nerves are numbered with Roman numerals (see the Appendix for more details on the cranial nerves).

Abbreviations: aLL – anterior lateral line nerve; au – auricle; cb – cerebellum; ob – olfactory bulb; pLL – posterior lateral line nerve; rb – rostral body; so – spinooccipital nerves; tect – tectum; tel – telencephalon.

Brain drawings adapted from Northcutt (1986), with permission from John Wiley & Sons.

The living lungfishes comprise only six species in three genera (Jørgensen and Joss, 2011). The genus of Australian lungfishes contains just a single species, *Neoceratodus forsteri* (Figure 4.2). The adults of this species are up to 1.5 m in length and protected by heavy scales. Their elongate paired fins are used mainly for steering. *Neoceratodus* is carnivorous, active mainly at night, and typically lives in deep river water. By comparison, the South American and African lungfishes (collectively referred to as lepidosirenid lungfishes) are much thinner and less heavily armored. Their paired fins are highly elongate and thin, but do allow for something akin to "walking on stilts" under water (Horner and Jayne, 2014); they are also covered with external taste buds, which may help the animals find food. Members of the African genus *Protopterus* are known for the ability to estivate, which is to say that they can encase themselves in a mucus-based cocoon when the lakes or ponds in which they live dry out. A somewhat unusual feature of the South American lungfish, *Lepidosiren paradoxa*, is that it *must* gulp air at the water surface to survive. This constraint probably reflects the generally low oxygen content of the slow-moving rivers and swamps in which these fishes live. Indeed, many ray-finned fishes also gulp air when water oxygen levels are low (Magid, 1966).

Lungfishes have enormous genomes. Whereas diploid cells of most vertebrates contain fewer than 10 picograms of DNA per cell, those of *Neoceratodus* contain roughly 160 pg of DNA, and those of *Lepidosiren* and *Protopterus* harbor 241 and 284 pg, respectively (Thomson, 1972; see also http://www.genomesize.com). These huge genome sizes are probably the result of transposable elements multiplying within the species genome (Elliott and Gregory, 2015). Indeed, roughly 40% of the *Neoceratodus* genome consists of recognizable transposable elements (Metcalfe et al., 2012). Even more interesting is that, across species, genome size tends to correlate positively with cell size (Figure 4.3). Given this correlation, Thomson (1972) used bone cell size in extinct lungfishes to estimate when in lungfish phylogeny genome size expanded. He concluded that cell and genome size increased long after the Devonian heyday of lungfish diversity.

The brains of lungfishes are smaller than those of coelacanths in absolute size, but similar once you account for differences in body size (Striedter, 2005). Of the various lungfishes, the brain of *Neoceratodus* is most similar to that of *Latimeria*. By comparison, the African and South American lungfishes have a much smaller cerebellum, optic tectum, and hypothalamus. In general, these lepidosirenid lungfishes have simpler brains with fewer migrated cells and less-differentiated cell groups. In contrast to *Latimeria,* all lungfishes have a fully evaginated, thin-walled telencephalon. Lungfishes also lack a saccus vasculosus, which means that this hypothalamic derivative (see Chapter 3) was likely lost in the last common ancestor of lungfishes and tetrapods. Although the telencephalon of extant lungfishes is large, relative to the remaining brain (Figure 4.2), recent analyses of lungfish endocasts indicate that early lungfishes had a much smaller telencephalon (Clement and Ahlberg, 2014; Challands, 2015). Specifically, the first lungfishes probably had a telencephalon that was relatively shallow and narrow (see Figure 3.9 in Chapter 3).

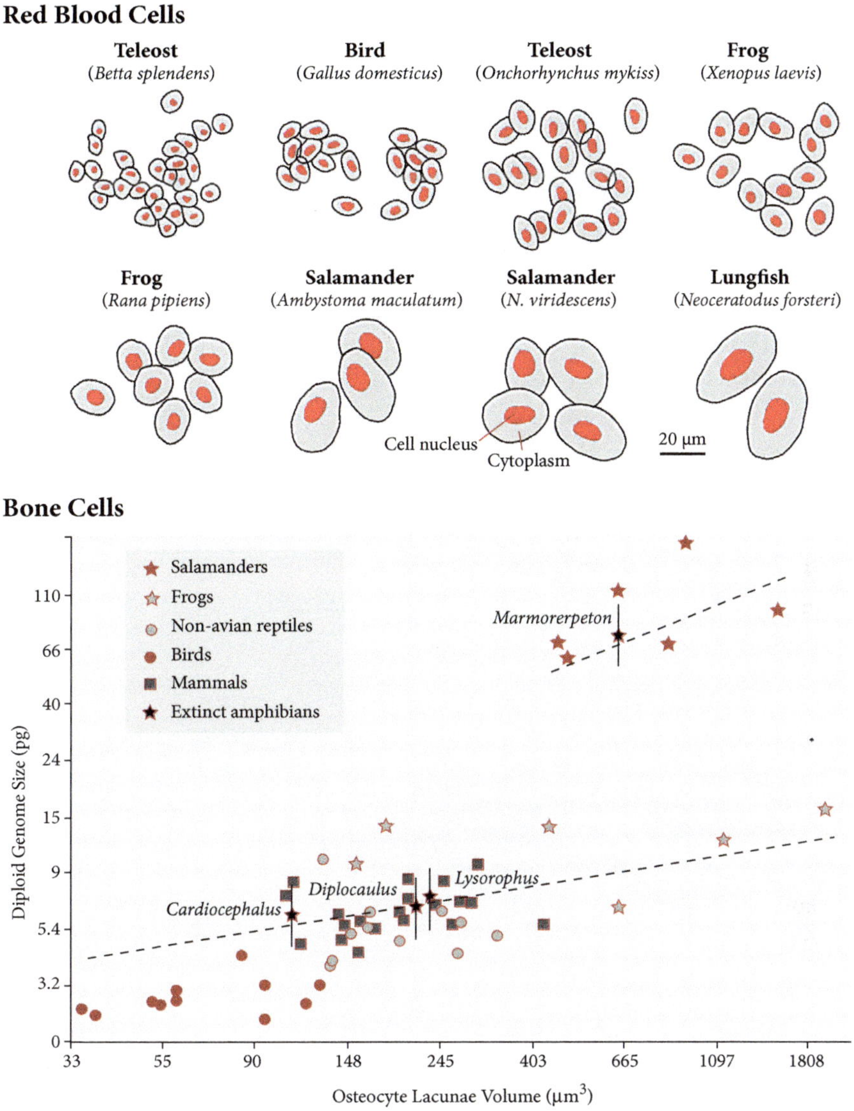

Figure 4.3 Correlated variation in cell and genome size. Shown at the top are red blood cells from eight vertebrate species (cell nuclei in red). The graph shows that diploid genome size in vertebrates correlates positively with bone cell (osteocyte) size, as determined from holes (lacunae) in the bone. Of the examined species, salamanders have the largest genomes (lungfishes have even larger genomes but are not included in the graph). Since *Marmorerpeton,* one of the oldest known salamanders already had very large cells, we can infer that genome size increased early in salamander phylogeny. Estimates of genome size for fossil specimens are shown with 95% confidence intervals.

Adapted from Gregory (2001) and Laurin et al. (2016).

4.1.3. Extant Amphibians

The three major groups of living amphibians, collectively referred to as Lissamphibia (Figure 4.4), are the anurans (frogs and toads), the urodeles (newts and salamanders), and the caecilians. Collectively, they comprise roughly 7,800 species. In contrast to amniotes, all lissamphibians have smooth, non-scaly skin (*lissos* means smooth in ancient Greek). As long as this skin is moist, adult lissamphibians perform significant amounts of gas exchange across their skin, even though they

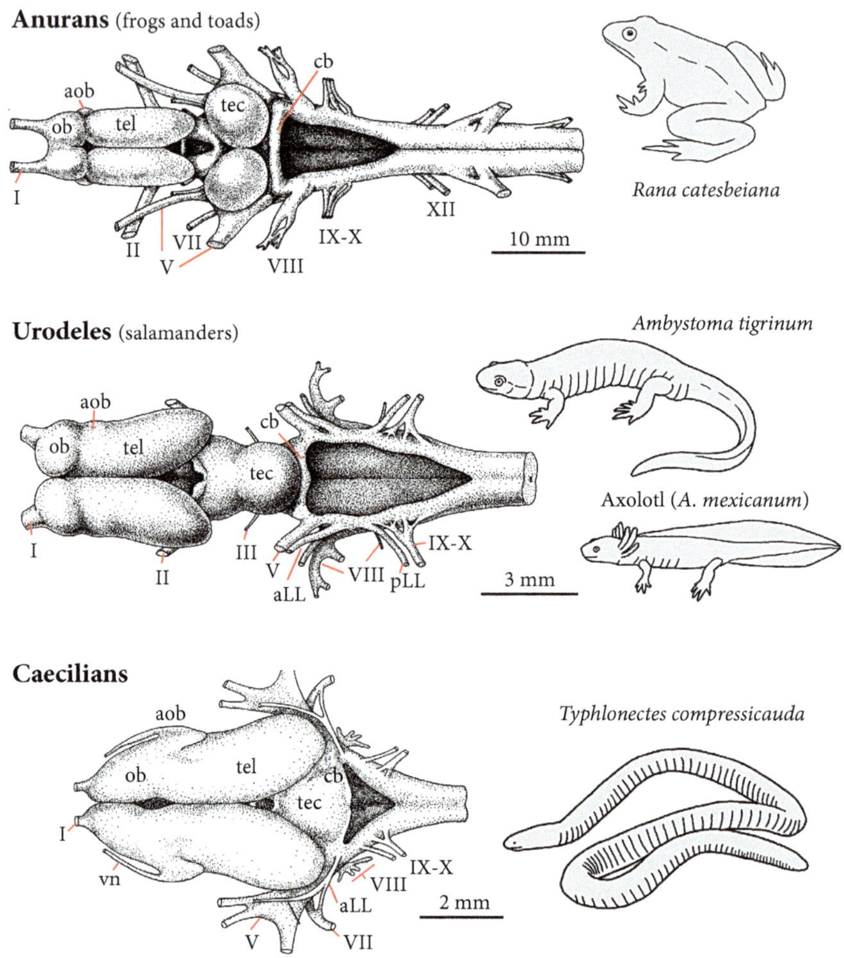

Figure 4.4 Brains and bodies of extant amphibians. All three brains are shown from a dorsal perspective. Most of the cranial nerves are numbered with Roman numerals (see Appendix for more details on the cranial nerves). Note that the axolotl is a neotenic form of the tiger salamander, *Ambystoma tigrinum*, which retains its larval features into adulthood.

Abbreviations: aLL – anterior lateral line nerve; aob – accessory olfactory bulb; cb – cerebellum; ob – olfactory bulb; pLL – posterior lateral line nerve; tect – tectum; tel – telencephalon.

Brain drawings adapted from Northcutt and Kicliter (1980), with permission from Springer Nature.

also possess lungs. To prevent the skin from drying out, and thus impeding gas exchange, all lissamphibians must live in relatively moist environments, such as damp leaf litter or, for most caecilians, wet soil. Furthermore, amphibians must generally lay their eggs in water to prevent desiccation. The exception to this rule are some caecilians (the typhlonectids) that hatch their eggs inside the mother's body, which of course is rather moist, and a few anuran species in which the males let the eggs develop inside their vocal sac (e.g., Darwin's frog *Rhinoderma darwinii*). Despite these constraints, the adults of most species in all three lissamphibian lineages spend much of their life on land.

Anurans are by far the most successful group of lissamphibians, comprising roughly 88% of all their species (Figure 4.5). A recent phylogenomic analysis (Feng et al., 2017) indicates that the vast majority of anurans arose shortly after a major extinction event roughly 66 million years ago (see Chapter 6, Section 6.4.4). Their enormous success since then was probably driven by several key innovations, including the emergence of hopping, which is an efficient and effective way to escape from predators, and the use of vocalizations to attract mates (Gerhardt, 1994), which likely boosted speciation rates. Another, less obvious innovation of anurans is the evolution of tadpole larvae, which in contrast to other amphibian

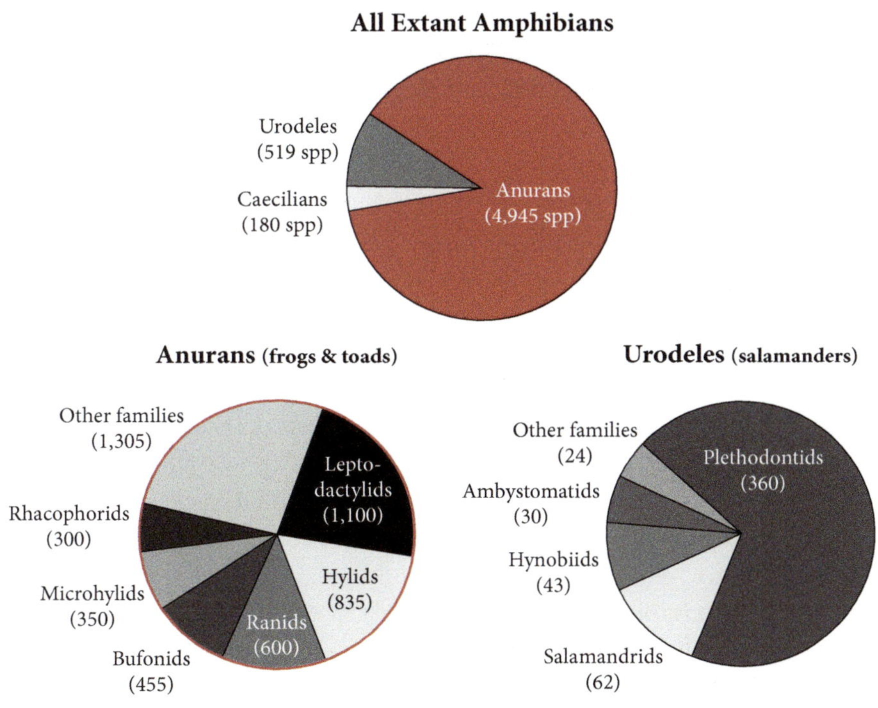

Figure 4.5 Species counts for extant amphibians. Among the lissamphibians, anurans are more speciose, and among urodeles, the plethodontid salamanders are most successful.

Data from Duellman and Trueb (1986).

and fish larvae are capable of eating algae and plants, thereby opening up a whole new range of dietary resources (McDiarmid and Altig, 1999; Pryor, 2014). Finally, many anurans have dispensed with their free-living tadpole stage entirely, instead exhibiting "direct development" (e.g., Hanken et al., 2001; Schlosser, 2008). This remarkable re-programming of embryonic development allows the species with direct development to lay their eggs in small puddles or even entirely out of the water (as long as the environment is moist; Gomez-Mestre, 2012).

With roughly 500 species, urodeles contain just one-tenth as many species as the anurans, and their numbers are dwindling rapidly. Most salamanders are less than 15 cm long, but some giant salamanders (family Cryptobranchidae) can grow to almost 2 m in length. Many salamanders are aquatic or semiaquatic, but the largest family of salamanders, the Plethodontidae, consists mainly of fully terrestrial species. These intriguing salamanders tend to be lungless and very small, performing all of their gas exchange across the skin and the moist membranes in their mouth. Being relatively defenseless, they tend to hide from predators in crevices. Although the genomes of urodeles are not as large as those of lungfishes, they are significantly larger than those of all other vertebrates (Figure 4.3). Given the aforementioned correlation between genome and cell size, it is not surprising that urodeles also have unusually large cells. An analysis of fossil urodeles suggests that these increases occurred very early in urodele phylogeny (Laurin et al., 2016).

With about 200 species, caecilians are the smallest lineage of lissamphibians and by far the least familiar. Most extant caecilians live in the Southern hemisphere or just North of the equator. All of them lack limbs and have extremely elongate bodies, making them look like large worms or small snakes (Figure 4.4). They live mainly in moist soil, burrowing into the ground, and have very small eyes that are covered with skin for protection. Their ears are also reduced in size and functionality (Maddin and Anderson, 2012). Roughly a quarter of caecilian species lay eggs, which the mothers then guard. The remaining 75% of caecilians give birth to live young.

Studies of amphibian brains have focused mainly on a few frogs and toads, as well as two salamanders, namely the tiger salamander, *Ambystoma tigrinum*, and the axolotl, *Ambystoma mexicanum* (Figure 4.4). Adults of the tiger salamander are terrestrial, whereas the axolotl is completely aquatic. However, the two species are close relatives. In fact, the axolotl is thought to be a neotenic form of the tiger salamander, meaning that it retains a large number of ancestral larval characters but is, nonetheless, able to reproduce. Relative to body size, urodeles tend to have smaller brains than anurans, but this may stem at least in part from salamanders having more elongate bodies, which tend to correlate with decreased relative brain size (van Dongen, 1998; Striedter, 2005).

In general, the brains of anurans are remarkably similar to those of squalomorph sharks (see Chapter 3). Like those small-brained sharks, anurans have an elongate, evaginated telencephalon with relatively thin walls and a well developed optic tectum. Their cerebellum, however, is minute by comparison. The majority of neurons in anuran brains have their cell bodies close to the ventricle, where the neurons are born, but anuran brains also contain a substantial number of neurons that are located away from the ventricle. In contrast, salamander brains contain very few such neurons and are, in general, much less differentiated than anuran brains; they also have a significantly smaller optic tectum. From the outside, caecilian brains look like rostrocaudally compressed salamander brains, but few aspects of their internal anatomy have been studied in detail (e.g., González and Smeets, 1994; López et al., 2007).

4.1.4. Extinct Tetrapods

All lissamphibians are tetrapods, as are all amniotes. Some lineages in both taxa have lost their legs and locomote by slithering on land (e.g., caecilians and snakes), some have become bipedal (e.g., humans and birds), and some have returned to an aquatic habitat (e.g., dolphins and whales, ichthyosaurs, and plesiosaurs). Taxonomically, however, they all belong to the large lineage we call the tetrapods. Since most of these tetrapods are at least partly terrestrial, we may ask: When and how did early tetrapods become capable of moving across land? When did their pectoral and pelvic fins transform into forelegs and hind legs? And when did tetrapods begin to get their oxygen by breathing air? To answer these questions one must examine the fossil record. Fortunately, numerous relevant fossils have been discovered and described. These specimens are sometimes called stem tetrapods or tetrapodomorph fishes. Alternatively, we refer to them simply as early tetrapods. Among the features that they share with later tetrapods are paired nostrils that open into the mouth and paired fins with long bones that are homologous to the long bones of tetrapod limbs (e.g., humerus, radius, and ulna).

The most basal stem tetrapods (e.g., *Eusthenopteron*; Figure 4.6) were fish-like insofar as they had well-developed fins with fin rays and were fully aquatic. The next most basal tetrapods, exemplified by *Acanthostega* and *Ichthyostega* (Figure 4.6), had more robust legs that might have allowed them to move on land. However, they clearly did not "walk" on all four legs or lift their entire body off the ground. Instead, they probably used their front legs to drag themselves across the land, much as modern sea lions and walruses do (Clack, 2012; Pierce et al., 2013). Moreover, basal stem tetrapods probably spent most of their time in shallow water and used their hind legs as paddles. Two other distinctions of these early tetrapods are that their pectoral girdle became separated from the skull and that their pelvic girdle became connected to the vertebral column. These evolutionary changes made it easier for an animal to move its head independently of the rest of the body (signaling the emergence of a "neck") and to lift the body off the ground, respectively. Overall, one key conclusion of the work on stem tetrapods is that tetrapod legs and their skeletal support evolved over a long period of time, long before tetrapods became fully terrestrial.

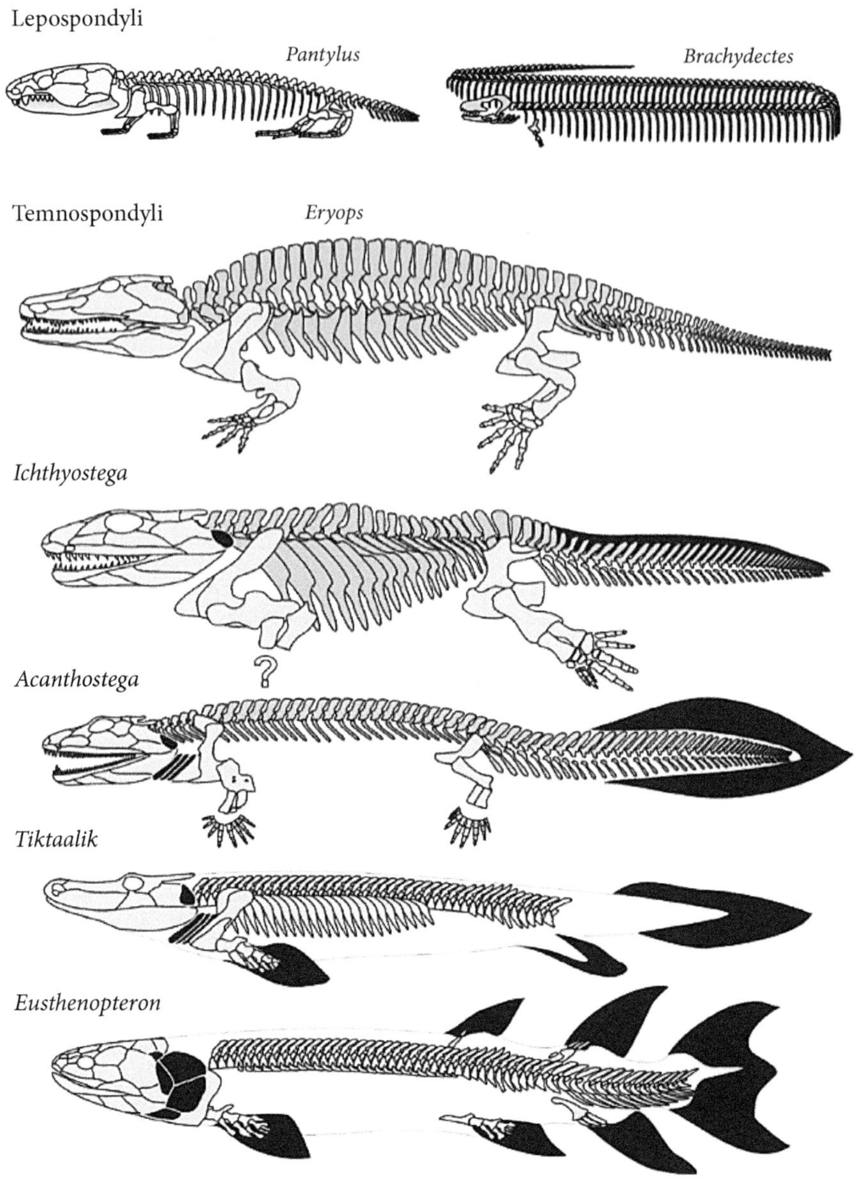

Figure 4.6 The skeletons of stem tetrapods. The most terrestrial species are shown toward the top, the most clearly aquatic ones toward the bottom. Fish-like features, notably fins and opercular bones, are shown in black. *Eusthenopteron* and *Tiktaalik* are technically stem tetrapods, but they are often called "tetrapodomorph fishes" in recognition of the fact that they were fully aquatic.
Adapted from Schoch (2014).

Many early tetrapods were more closely related to modern tetrapods than the stem tetrapods we have discussed so far. Their three main lineages are called temnospondyls, reptiliomorphs, and lepospondyls (though the latter is probably a paraphyletic assemblage; see Figure 4.7 and Chapter 1). The reptiliomorphs

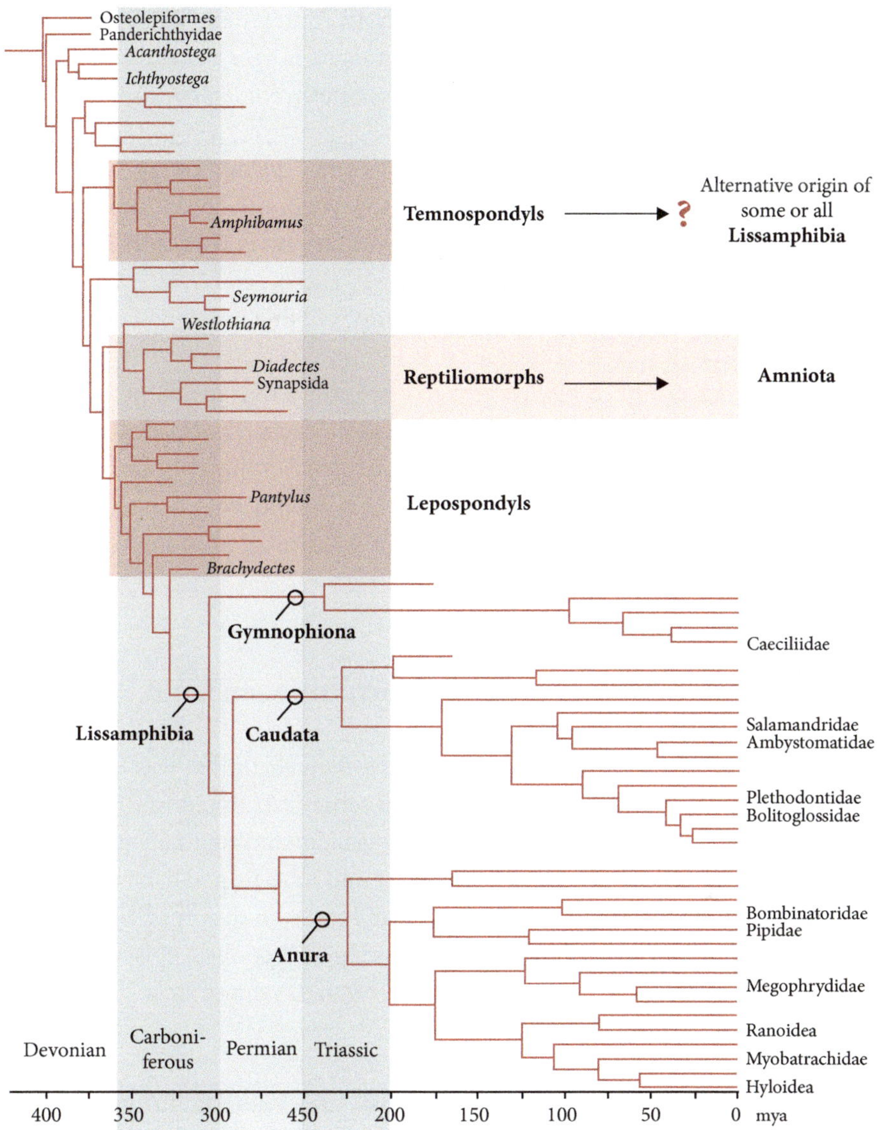

Figure 4.7 Dated phylogeny of early and modern amphibians. The illustrated phylogeny is based on a Bayesian analysis of both molecular and morphological data from extant and extinct species. Confidence intervals for the divergence times have been omitted but are given in the original (Pyron, 2011). According to this phylogeny, lissamphibians emerged from a group of lepospondyls. One should note, however, that other authors have argued that lissamphibians are, instead, derived from temnospondyls.

probably gave rise to amniotes, which we discuss at length in Chapter 5. The temnospondyls had strong limbs with five digits and a vertebral column that was probably rigid enough to keep the body off the ground. They also tended to be significantly larger than most extant amphibians. The temnospondyl *Eryops* (Figure

4.6), for example, grew to at least 3 m in length. In contrast, the lepospondyls were much smaller. *Pantylus*, for instance, was probably no longer than 25 cm. Although most lepospondyls could walk on land, some species were legless and, most likely, aquatic (e.g., *Brachydectes*; Figure 4.6).

A major unresolved question in vertebrate paleontology is whether lissamphibians evolved from temnospondyls, lepospondyls, or both. Based in part on similarities in ear morphology and dentition, several authors have argued that temnospondyls gave rise to lissamphibians (Ruta et al., 2003; Maddin and Anderson, 2012; Schoch, 2014). Others have proposed that lissamphibians are more closely related to lepospondyls (Figure 4.7; Pyron, 2011; Marjanović and Laurin, 2013). Yet another group of researchers has proposed that urodeles and anurans originated from temnospondyls, whereas caecilians emerged from lepospondyls (Lee and Anderson, 2006; Carroll, 2009). This debate is unlikely to be resolved any time soon. What is fairly clear is that lissamphibians did not evolve until at least the Late Carboniferous, around 315 mya (San Mauro, 2010), long after the first tetrapods had made their way onto dry land.

4.2. Challenges and Opportunities on Land

The early evolution of tetrapods spanned a period on earth that was beset by a series of extinction events near the end of the Devonian. As noted in Chapter 3, this prolonged mass extinction ravaged numerous taxa, including many plants, corals, all placoderms, and the vast majority of jawless and lobe-finned fishes. The precise causes of these mass extinctions remain unclear. However, they may well have been caused by an alternating series of global warming and cooling episodes, coupled with periodic algal blooms that depleted the water of oxygen (Clack, 2007).

4.2.1. Air Breathing, Water Loss, and Gas Exchange

Since air typically carries about 20 times more oxygen than water does (not counting the oxygen in water molecules), the anoxic conditions of the water in the late Devonian would have favored the survival of animals that could extract oxygen from air, in addition to water. Indeed, it is now clear that air breathing preceded the origin of tetrapod limbs. As mentioned in Chapter 3, today's lungfishes can obtain oxygen from air, and ancient lungfishes could almost certainly do so as well. In fact, even early ray-finned fishes had paired lungs, which were retained in *Polypterus* but lost (and probably converted into swim bladders) in the last common ancestor of sturgeons and teleosts (Longo et al., 2013; Lambertz and Perry, 2015). Most of these early air breathers gulped air at the water surface and then pushed this air into their primitive lungs. In addition, they could probably breathe through their spiracle, which is an opening behind the eye that leads directly into the pharynx

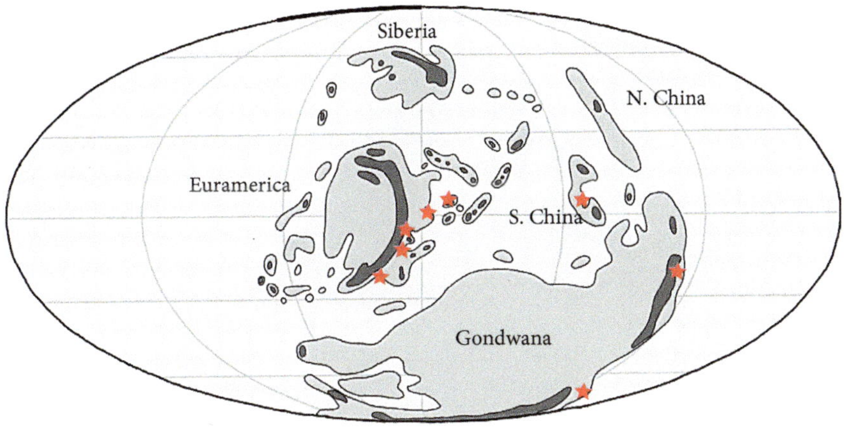

Figure 4.8 Land masses of the Devonian. During the Devonian period (419–359 mya) the earth's land mass was mainly concentrated in a Southern supercontinent called Gondwana, an equatorial Euramerica, and a Northern Siberia. As indicated by the red stars, early tetrapod fossils have been discovered at locations that correspond mainly to coastal Euramerica, though some tetrapod fossils were deposited at the eastern edge of Gondwana, including one that is surprisingly far south, in what is today South Africa (Gess and Ahlberg, 2018). The illustrated map is a Mollweide projection, which represents the surface area of continents more accurately than their shapes.
Adapted from Behrensmeyer et al. (1992) and Clack (2006). Courtesy of Preston Holmes.

(Graham et al., 2014). In addition to the spiracle, some early tetrapods may have breathed air through a single external nostril that became connected to the oral cavity during the early stages of tetrapod phylogeny (Zhu and Ahlberg, 2004). In any case, these early air breathers lived mainly in shallow marine habitats along the edges of the Euramerican continent (Figure 4.8). Over time, early tetrapods increasingly invaded freshwater streams and lakes, perhaps because increased plant growth on land (Figure 4.9) made those habitats increasingly stable, shady, and full of potential prey.

As early tetrapods moved onto land for more than a few minutes at a time, they faced a serious problem, namely dehydration. With the evaporation of water from the animal's surface, skin and gills dry out and become less conducive to gas exchange. Even for the early air breathers, that loss of gas exchange capacity would have been life-threatening. Early amniotes solved this problem by making the lungs more efficient and the skin less permeable, but these changes took millions of years. Early tetrapods instead remained close to water and reduced their body size, which may have helped them find shade and moist environments (e.g., under leaves). They would have also benefited from the increase in average cell size (see Figure 4.3), because larger cells have proportionately less surface area across which ionic gradients must be maintained and, therefore, require less metabolic energy than an equivalent volume of small cells. Indeed, many amphibians (and lungfishes) can survive in very low oxygen environments that are lethal for other vertebrates.

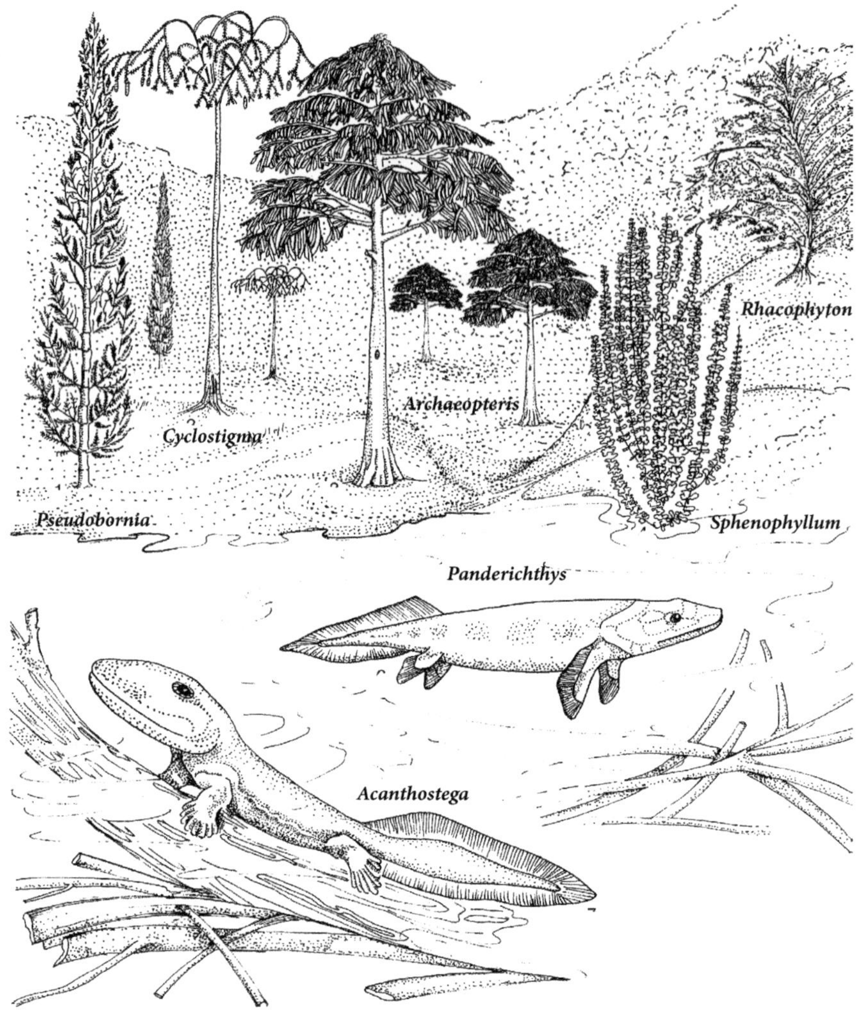

Figure 4.9 Stem tetrapods in their likely environment. *Panderichthys* probably never left the water but *Acanthostega* may have been able to drag itself onto and over dry land. The land would have been occupied by diverse plants, including: *Cyclostigma*, a tall tree-like lycopsid; *Pseudobornia*, a kind of horsetail; *Archaeopteris*, a fern-like vascular plant up to 10 m tall; *Rhacophyton*, an early fern; and *Sphenophyllum*, a small shrub or creeping vine.
Drawing by Jo Griffith.

Desiccation also threatens an animal's ability to reproduce, as embryos also require gas exchange and quickly die if they dry out. To solve this problem, amniotes surround their egg with a membrane that simultaneously allows some gas exchange and limits water loss; this is the amniotic membrane for which amniotes are named (see Chapter 5). Amphibians instead lay their eggs in water or moist soil and, in most species, retain a fully aquatic larval stage.

4.2.2. Moving on Land and Sensing in Air

Once the aquatic amphibian larvae transform into terrestrial adults, they must be able to locomote on land. This is more challenging than swimming in water, because animals (consisting mainly of water) tend to float in water but fall to the ground in air. Normally aquatic animals can drag themselves across land, as many early tetrapods did, but this will cause the abdomen and tail to drag along the ground (Standen et al., 2016). Walking on legs is less problematic but requires legs that are sturdy enough to support the body's weight. Moreover, walking tetrapods need a vertebral column that does not sag and special mechanisms to balance the body when some of the limbs are off the ground. To put the challenge succinctly, animals walking on legs can easily fall down. Moving across land also requires roughly 10 times as much metabolic energy, at any given body size, as swimming in water (Tucker, 1975).

Another major obstacle to the invasion of land by early tetrapods was that many sense organs had evolved for use in water and did not work as well in air. Just as your own vision becomes blurry underwater, so does the vision of aquatic animals in air. Because the refractive index of water is larger than that of air, light rays bend when they hit any air-water interface at an angle. Because of this phenomenon, the curved cornea at the front of the eye bends light much more in air than in water, thereby altering the plane of focus for the image that is projected on the retina (Figure 4.10). Another consequence of the difference in refractive index is that the cornea must be smoother in air than in water; otherwise the retinal image will be blurry. Further complicating the issue is that eyes can easily dry out in air, causing corneal wrinkles. Tetrapods have solved the latter problem by evolving lipid-containing tears and eyelids that can spread those tears across the surface of the eye. Once those challenges were met, terrestrial animals could see much further than their aquatic ancestors, because air does not scatter light as much as water does.

Because bodies are denser than air, airborne sound waves tend to bounce off the surface of terrestrial animals, rather than propagating through them (as they do in most fishes; see Section 3.3.4). Many terrestrial animals have solved this problem by evolving a thin ear drum (tympanic membrane) that vibrates in response to airborne sounds, as well as one or more middle ear bones that transmit those vibrations to hair cell sensors in the inner ear. Although these innovations make it possible to hear high-frequency airborne sounds, most sounds don't propagate as far (nor as fast) in air as in water, which is one of the reasons why the underwater songs of whales (Payne and McVay, 1971) can be heard over many miles. Whereas hearing and vision must be tweaked to work in air, the mechanosensory lateral line system becomes completely inoperable in air, because superficial neuromasts quickly dry out on land and canal neuromasts cannot be activated effectively by airborne vibrations (which, as mentioned previously, tend to bounce off the body surface). Electroreceptors also fail to work in air, because air is a poor conductor of electricity.

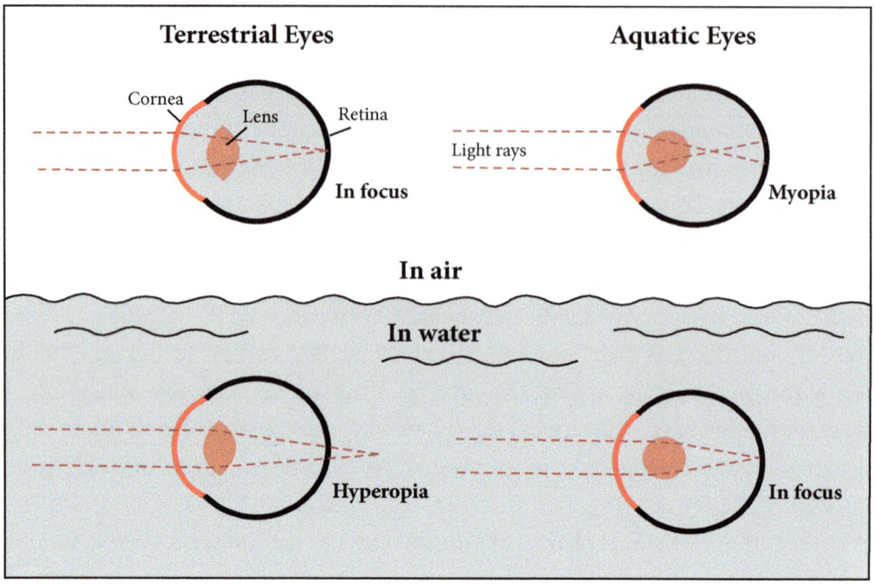

Figure 4.10 Optics of eyes in water versus air. Terrestrial animals tend to have a flattened lens and a steep cornea; the latter is responsible for most of the light refraction. When such air-adapted eyes are used underwater, the optical image tends to be focused behind the retina (causing hyperopia). In contrast, aquatic animals tend to have spherical lenses and flatter corneas, with the former accomplishing most of the refraction. When used in air, such eyes tend to focus the image in front of the retina (causing myopia).
Adapted from Sivak (1988).

Given all these challenges, what were the benefits of becoming terrestrial? Since even aquatic animals can obtain oxygen from air, why was it adaptive for early tetrapods to invade land? One answer to this question is that the water was probably teeming with carnivorous fishes, whereas the land was generally free of predators. Predatory arthropods did exist on land (e.g., scorpions and spiders), but most of them were probably too small to threaten early tetrapods. Reptiles, birds, and mammals that might have preyed on early tetrapods did not evolve until later. Moreover, early tetrapods on land could likely dine on a variety of terrestrial invertebrates, and any tetrapod capable of moving both in water and on land would have been able to prey effectively on smaller fishes and invertebrates trapped in tide pools or small ponds. Because it is difficult to optimize adult bodies and behavior for water and air at the same time, most early tetrapods probably retained an aquatic larval form that underwent substantial changes (e.g., losing gills and growing legs) before climbing onto land as an adult. In the following sections we review how this evolutionary shift away from a purely aquatic life was reflected in early tetrapod sensory systems, motor functions, and brains.

4.3. Sense Organs for Use on Land

As in previous chapters, we discuss evolutionary changes in the sense organs separately from changes in motor control (see Section 4.4). Changes in the central sensory systems are deferred to Section 4.5, where we attempt to reconstruct early tetrapod brains.

4.3.1. Terrestrial Vision

Even before they became fully terrestrial, early tetrapods roughly tripled the size of their eyes, relative to body size, suggesting that these animals spent a lot of time with their eyes just above the water surface, stealthily breathing air and peering onto land for potential prey at the water's edge (MacIver et al., 2017). In concert with this increase in eye size, the cornea of early tetrapods became more highly curved and the lens became less spherical (Figure 4.10). These evolutionary changes optimized the eye for vision in air but would have produced blurred vision in water. Indeed, building an eye that can be used in both environments is difficult and typically requires moving the lens within the eye (i.e., accommodation). However, the first fully terrestrial tetrapods probably adopted a simpler solution: their cornea became more highly curved as the aquatic larvae transformed into terrestrial adults, which is what still happens in terrestrial frogs (Sivak, 1988).

The retina appears to have changed in merely minor ways as early tetrapods invaded land. The closest relatives of tetrapods, the lungfishes, have small eyes with large photoreceptors (and hence low spatial resolution), but in most respects the lungfish retina is far more similar to that of tetrapods than other fishes (Bailes et al., 2006; Hart et al., 2008). Specifically, it contains rod photoreceptors and three morphologically distinct types of cones (Figure 4.11). At the molecular level, lungfishes have one type of opsin for rods and four separate opsins for cones (one of which is lost as the animals mature). Modern anurans and urodeles added to this set a "green rod" that is tuned to blue/UV light (~440 nm) and lost the cone pigment that is tuned to medium wavelengths (Bowmaker, 2008). These observations suggest that the ancestors of early tetrapods already had good color vision.

A related observation is that many lungfish cones contain red oil droplets that filter incoming light and are, therefore, thought to improve color discrimination. Since most lizards and birds also have such colored oil droplets in some of their cone photoreceptors, whereas mammals and extant amphibians do not (Figure 4.11), it seems likely that lungfishes and sauropsids evolved this feature independently of one another. Alternatively, it may have evolved in the last common ancestor of amniotes and lungfishes and was then lost independently in the lineages leading to lissamphibians and mammals. Either way, it was not causally linked to the invasion of land. More likely, it is related to the elaboration of color vision. Unfortunately, we

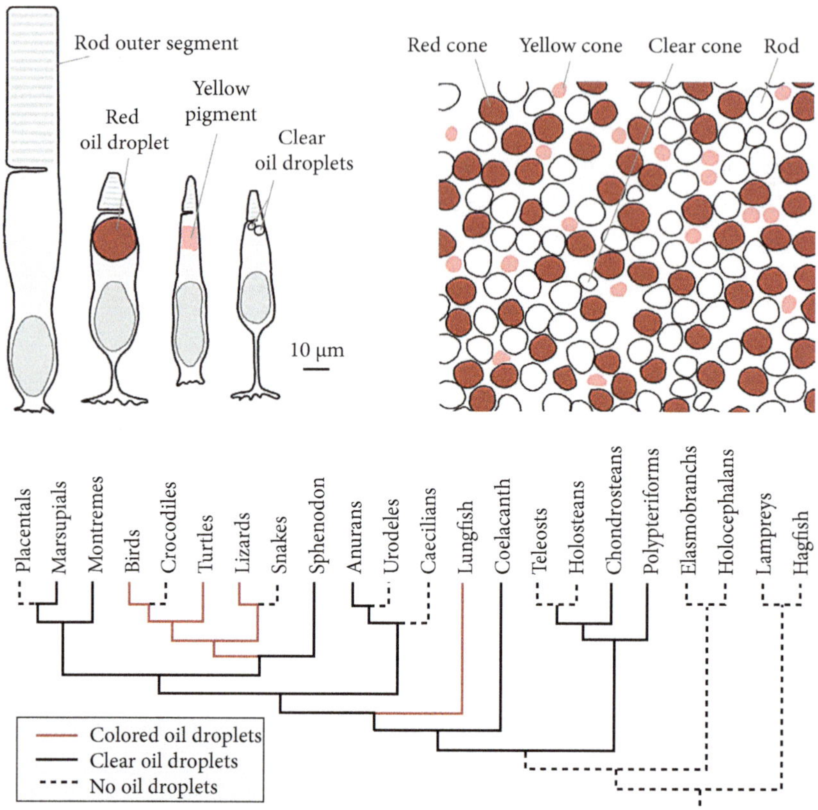

Figure 4.11 Lungfish retina and oil droplets. Shown at the top are the four types of photoreceptors found in Australian lungfishes. The most numerous type of cone contains a red oil droplet; other cones contain either a yellow pigment or colorless oil droplets. The colored inclusions are thought to improve color vision. The distribution of these four types of photoreceptors across the retina of a lungfish is shown at the top right (based on a horizontal section through a flattened retina). The cladogram at the bottom depicts the phylogenetic distribution of colored and colorless oil droplets in vertebrates. It suggests that colored oil droplets evolved independently in lungfishes and sauropsids (lizards and birds), and that colorless oil droplets were lost in placental mammals and a few other lineages.
Adapted from Bailes et al. (2006), Bowmaker (2008).

are not aware of any studies explicitly testing for color vision in lungfishes (Marshall et al., 2016).

All major classes of retinal neurons likewise predate the origin of tetrapods, as they are found in lungfishes and many other fishes. However, the diversity of those retinal neurons may have increased as early tetrapods evolved. For example, only four distinct types of retinal ganglion cells have been described in lungfishes, whereas urodeles reportedly possess 5–7 types, and frogs may have 11 distinct types or more (Figure 4.12; Cajal et al., 1995; Segev, 2005; Pushchin and Karetin, 2009). Since the retinas of lungfishes, urodeles, and frogs have been studied with different

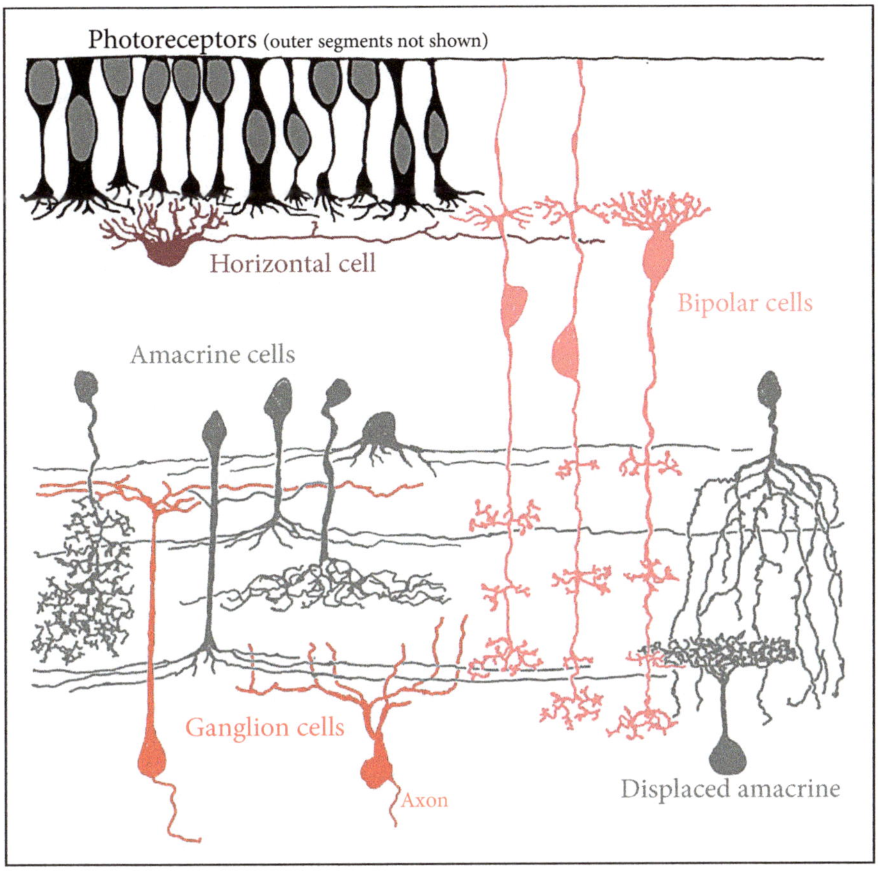

Figure 4.12 A frog's retina. Shown here are Golgi-stained representative neurons in the retina of the frog *Rana esculenta*. The photoreceptors (black) are aligned across the top, and the retinal ganglion cells, whose axons project out of the retina, lie at the bottom of the diagram. The retinal interneurons include horizontal cells, several types of bipolar neurons, and numerous amacrine cells. Frogs probably have at least 11 types of retinal ganglion cells, even though just two are illustrated here.
Adapted from Cajal et al. (1995).

techniques and different levels of intensity, comparing cell type numbers in this way is problematic. Still, the available data suggest that retinal complexity increased soon after tetrapods invaded land and then increased again in the lineage leading to frogs. As we discuss in Chapter 5, retinal complexity increased even more dramatically in amniotes (Sanes and Masland, 2015).

4.3.2. Hearing in Air

As noted in Section 4.2.2, hearing in air is more difficult than hearing in water because airborne sounds tend to bounce off the animal's surface. Terrestrial animals can hear vibrations that travel through the ground (substrate-borne vibrations), but this ability is limited to low-frequency sounds. To solve this problem, terrestrial tetrapods evolved a tympanic membrane that is connected to a middle ear bone, specifically the stapes, which then transmits those vibrations to sensors in the inner ear.

Biologists used to think that such tympanic ears evolved very early in tetrapod phylogeny, because the skulls of many early tetrapods exhibit an "otic notch" (Figure 4.13) that was thought to indicate the presence of a tympanum (see Lombard and Bolt, 1979). However, the stapes of most early tetrapods was short and stout and, therefore, not conducive to high-frequency sound transmission (Clack, 2002, 2016). Only a few temnospondyl amphibians had the kind of thin, elongate stapes that would have enabled high frequency hearing in air (Figure 4.13). Among extant amphibians, only anurans have a tympanum and good terrestrial hearing. Their tympanic membrane is similar to that of amniotes, but anurans and amniotes emerged from separate branches of the tetrapod phylogenetic tree (see Figure 4.7). Moreover, the available data indicate that the last common ancestor of anurans and amniotes did not have a tympanic ear (Lombard and Bolt, 1979; Manley et al., 2004; Tucker, 2016). Therefore, tympanic ears and good terrestrial hearing almost

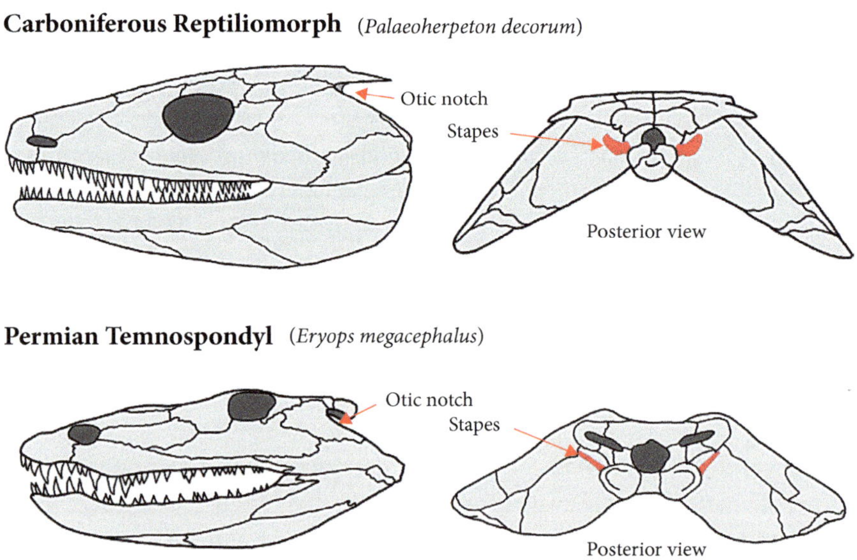

Figure 4.13 The middle ears of early tetrapods. Many early tetrapods had an otic notch, once thought to indicate the presence of a tympanum (ear drum), and a stapes that was probably homologous to the stapes in the middle ear of amniotes. However, the stapes in most early tetrapods was short and stout and, therefore, poorly suited for high-frequency hearing in air. In contrast, some later temnospondyls evolved a long and slender stapes that, in conjunction with a tympanum, might well have allowed these animals to hear high-frequency airborne sounds.
Adapted from Clack and Allin (2004).

certainly evolved independently in anurans and one or more lineages of amniotes (see Chapter 5).

Given this independent evolution, it is not surprising that the inner ear of frogs differs in several important ways from that of mammals and birds, especially in its sensory epithelium. Rather than having a single sensory epithelium devoted to sound, called the cochlea in mammals and birds, amphibians have two such organs, namely the basilar papilla and the amphibian papilla. The basilar papilla of amphibians is homologous to the mammalian cochlea, and even fishes have a homolog of it (Fritzsch et al., 2013). In amphibians, the basilar papilla is a relatively simple structure, whose hair cells respond preferentially to relatively high sound frequencies, ranging from 1–4 kHz and extending into the ultrasonic range for some highly specialized anurans (Feng et al., 1975; Arch et al., 2012). Importantly, for a given species, all of the hair cells in the basilar papilla respond preferentially to the same frequency (Schoffelen et al., 2008). In contrast, the hair cells of the amphibian papilla, which is probably homologous to the macula neglecta of fishes (Lewis and Narins, 1999), are tuned to lower frequencies, down to 100 Hz, with different hair cells responding to different frequencies. Importantly, sound frequencies are represented topographically along the length of the amphibian papilla in many anurans. In this respect, the amphibian papilla and the mammalian cochlea are very similar to one another, even though they evolved independently. That is, they are an excellent example of convergent evolution.

The mechanisms underlying the frequency tuning of auditory hair cells also exhibit both similarities and differences between amphibians and amniotes. In mammals and birds, auditory hair cells sit on a thin membrane, called the basilar membrane, that vibrates in response to sound. The stereocilia of these hair cells extend into an overlying tectorial membrane that slides across the hair cells when the basilar membrane vibrates. This sliding motion causes the stereocilia to bend back and forth, rhythmically modulating their rate of transmitter release (see Striedter, 2015). Because of systematic differences in the mechanical properties of the basilar membrane, different sound frequencies cause vibrations at different points along the membrane and, therefore, activate different sets of hair cells. In contrast, the hair cells in the inner ear of amphibians sit on a bony substrate that does not vibrate (Figure 4.14). Instead, sounds cause movements in the fluid of the inner ear, which move a tall tectorial membrane that then bends the hair cell stereocilia. Thus, the tectorial membrane in amphibians moves against stationary hair cells, rather than vice versa as in mammals and birds. This tectorial membrane is the dominant regulator of frequency tuning in the basilar papilla of amphibians. The amphibian papilla of anurans also contains a tectorial membrane, but its mechanical properties vary along its length. This variation is thought to control the location-dependent frequency tuning of the hair cells in the caudal part of the amphibian papilla. In contrast, hair cells in the rostral part of the amphibian papilla exhibit a different kind of frequency tuning, mediated by systematic variation in the cells' electrical properties (Smotherman and Narins, 1999; Schoffelen et al., 2008). A similar form

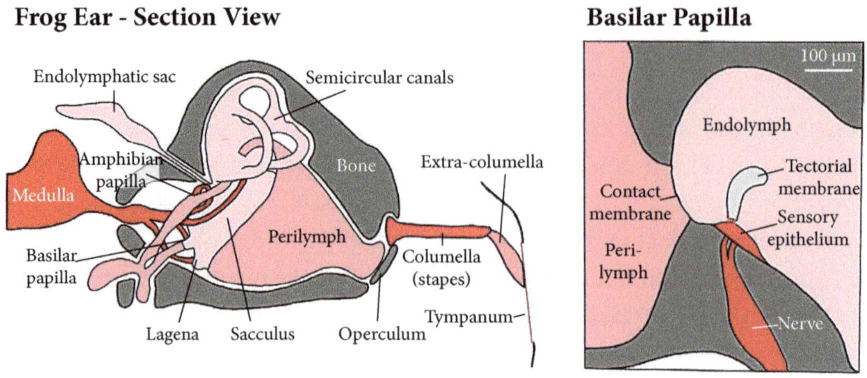

Figure 4.14 A frog's inner ear. Shown on the left is a schematic transverse section through the middle and inner ears of a frog (medial is to the left, dorsal to the top). Vibrations of the tympanum cause the columella (i.e., the stapes) to move, which creates pressure waves inside the perilymph. The diagram on the right focuses on the basilar papilla and its surrounding structures (from a different angle than the left diagram). It shows that the perilymph is separated from the endolymph by a thin membrane, allowing vibrations to pass from the former into the latter. Pressure waves in the endolymph ultimately cause vibrations of the tectorial membrane that overlies the hair cells of the basilar papilla. Those vibrations deflect the hair cell stereocilia, which leads to changes in the rate of transmitter release onto the sensory axons.
Adapted from Frishkopf and Goldstein (1963) and Capranica (1976).

of "electrical tuning" has also been observed in amniotes (Fettiplace and Fuchs, 1999). Several other processes make additional contributions to hair cell tuning in mammals, some of which are also seen in other tetrapods (Hudspeth, 2014).

What kind of hearing did vertebrates possess before their descendants evolved tympanic ears and complex auditory papillae? A tentative answer to this question is provided by lungfishes. They have neither a tympanum nor air-filled cavities next to the inner ear (see Figure 3.11), but they do have "rudimentary aerial hearing" (Christensen et al., 2015). Still, lungfishes can hear airborne sounds only when they are very loud (>85 dB), of low frequency (~80 Hz), and relatively long in duration. This is rudimentary indeed! Of course, they are much better at hearing sounds under water.

4.3.3. Losing the Lateral Line

Early tetrapods probably had both mechanosensory and electrosensory lateral lines. Evidence for this hypothesis comes from stem tetrapod skulls, which exhibit some tubular canals or bony grooves that probably contained some neuromasts. Small pits located near those grooves suggest that these animals also had electroreceptors (Klembara, 1994).

The situation is more complicated in lissamphibians. Most aquatic lissamphibians have a mechanosensory lateral line system, at least on the head (Northcutt, 1989). However, all lissamphibian neuromasts are located within the skin, rather than inside canals. Since canal neuromasts usually begin development in the epidermis (specifically in placodes; see Chapter 2) and only later sink into canals, the superficial location of neuromasts in lissamphibians probably represents the retention of a juvenile character (Northcutt et al., 1994), a process that is called paedomorphosis (see Chapter 1, Section 1.2.4). Indeed, it has been suggested that the origin of lissamphibians involved a substantial amount of paedomorphosis (Long and Gordon, 2004; Kimmel et al., 2009), possibly linked to increased genome size (Gregory, 2002).

Be that as it may, in amphibians with terrestrial adults those superficial neuromasts disappear as the young animals get ready to emerge from the water. Intriguingly, in some terrestrial urodeles they seem not to degenerate but to sink deeper into the skin, only to re-emerge when the animals return to water for reproduction (Fritzsch and Wahnschaffe, 1983). This does not happen in aquatic frogs, such as the commonly studied African clawed frog, *Xenopus laevis*, which retains functional neuromasts throughout life.

Aquatic urodeles and caecilians generally possess ampullary electroreceptors (Hetherington and Wake, 1979; Münz et al., 1984). Some terrestrial urodeles and caecilians retain their electroreceptors with the transition to land, possibly using them to sense potential prey in a very humid leaf-litter habitat, and some fully aquatic caecilians (the typhlonectids) have ampullary electroreceptors but no mechanosensory lateral line, suggesting that their ancestors lived in a very humid or muddy habitat, where electroreceptors were useful but neuromasts were not (Fritzsch and Wake, 1986; Fritzsch and Neary, 1998). In contrast, anuran tadpoles never develop electroreceptors in the first place.

Just as modern amphibians lost their mechanosensory and electrosensory lateral line systems with the invasion of land, so did the amniotes. Monotremes, such as the platypus, do have electroreceptors (Scheich et al., 1986; Pettigrew, 1999), but these receptors evolved independently of those in aquatic anamniotes. Consistent with this interpretation, the electroreceptors of monotremes are innervated by the trigeminal nerve, rather than the lateral line nerves (see Appendix).

4.3.4. Smelling on Land and in the Air

All vertebrates except toothed whales (Kishida et al., 2015) have an olfactory system with sensory neurons that express olfactory receptor molecules and project to the olfactory bulb. In most fishes, the olfactory epithelium is connected to the external environment through two pairs of nostrils. In contrast, lungfishes and tetrapods have only one pair of external nostrils, because the posterior nostril moved into the oral cavity during very early stages of tetrapod evolution, forming an internal nostril (aka choana; Zhu and Ahlberg, 2004; Janvier, 2004). As a result of this anatomical modification, early tetrapods could use negative pressure inside the oral cavity to pull water or air across the olfactory epithelium, thereby increasing the rate of odorant delivery. The African clawed frog (*Xenopus laevis*) and its close relatives actually direct inhaled air toward a specialized part of the olfactory epithelium, separate from the region they use to smell underwater, but this arrangement is derived for these species (Hansen et al., 1998). As stem tetrapods became better at breathing air (see Section 4.2.1), possession of an internal nostril allowed them to smell the inhaled air effectively.

In addition to the main olfactory system, most living tetrapods (except for birds, crocodilians, and catarrhine primates) have a vomeronasal system. The precise location of the vomeronasal epithelium varies considerably across species, but it consistently develops as an evagination of the main nasal epithelium (Eisthen and Polese, 2007; Halpern, 2007). The sensory cells of the vomeronasal epithelium express several types of specialized vomeronasal receptors. Like the olfactory receptors, the vomeronasal receptors belong to the large superfamily of G protein-coupled receptors; however, they are merely distant cousins. The sensory neurons of the vomeronasal epithelium project to the accessory olfactory bulb, which is similar to the main olfactory bulb but structurally simpler and located more caudally and laterally.

It had long been thought that the vomeronasal system is unique to tetrapods and, therefore, linked to the invasion of land. However, we now know that lungfishes also have a vomeronasal epithelium and an accessory olfactory bulb (González et al., 2010). Moreover, vomeronasal receptors are found in both cartilaginous fishes and lampreys (Chang et al., 2013), even though these animals lack a distinct vomeronasal epithelium. Therefore, we can surmise that vomeronasal receptors are primitive for vertebrates (Grus and Zhang, 2009) and that the segregation of these receptors into a distinct vomeronasal epithelium probably occurred with the onset of air-breathing, rather than the invasion of land.

Given this hypothesis, one might guess that the vomeronasal epithelium is specialized for the detection of airborne odorants, but this is not the case. Instead, the vomeronasal system is generally specialized for sensing non-volatile odorants, which are transferred to the vomeronasal epithelium either through direct contact (in frogs and many mammals) or by the tongue (in snakes). In contrast, the olfactory epithelium may detect either airborne or waterborne odorants, depending on the species and the conditions. As noted previously, the olfactory epithelium of the aquatic frog *Xenopus* has two distinct compartments, one for air and one for water, with a muscular valve directing the incoming flow (Freitag et al., 1995). However, even airborne odorants must traverse the mucus layer that covers the olfactory epithelium, either by dissolving into the mucus or binding to specialized transport molecules.

Collectively, the available data indicate that the olfactory receptors became more diverse with the evolution of air breathing, which enabled the animals to smell a broad new class of odorants, namely volatile odorants that are relatively insoluble in water. Some comparative molecular data are consistent with this hypothesis (Figure 4.15; Niimura, 2009; Nikaido et al., 2013). Specifically, the α and γ subfamilies of olfactory receptor genes are selectively expanded in modern tetrapods and coelacanths, which probably breathe air as juveniles (Cupello et al., 2015). Unfortunately, it remains unclear whether the receptors in these expanded olfactory receptor families respond selectively to airborne odorants. Nor do we know whether the olfactory receptors that tetrapods lost were selectively responsive to water-soluble

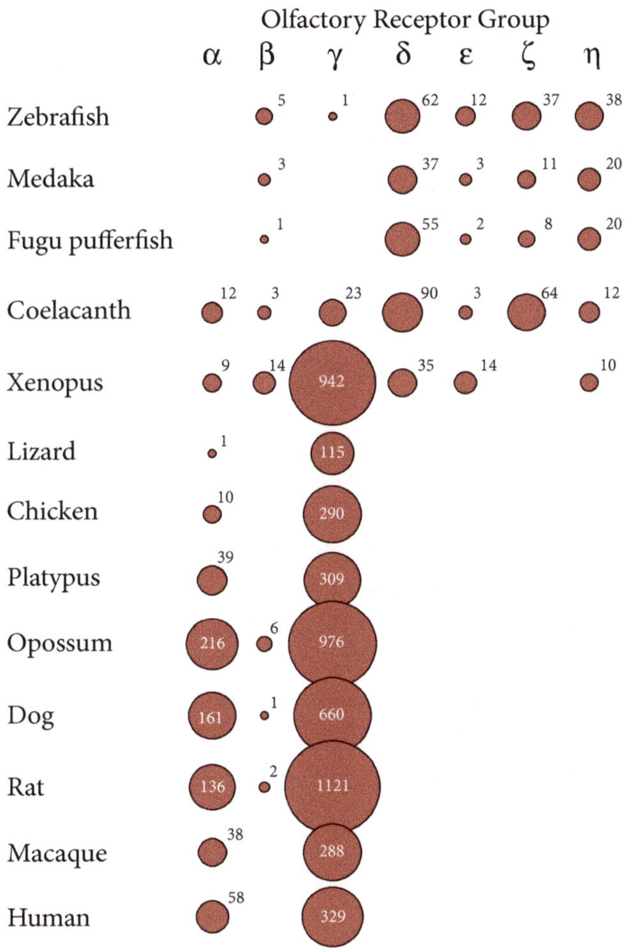

Figure 4.15 Evolution of olfactory receptor repertoires. A comparative analysis of olfactory receptor (OR) genes found them divisible into seven major groups, named α, β, γ, δ, ε, ζ, and η. Groups α and γ expanded dramatically in tetrapods (circle size is proportional to OR group size), whereas OR genes from groups δ, ε, ζ, and η were lost. Intriguingly, coelacanths have a moderate number of OR genes in all seven groups. Note that this diagram excludes pseudogenes but includes truncated OR genes (which may have been sequenced incompletely).

Adapted from Niimura (2012) with additional data from Nikaido et al. (2013).

odorants. Similarly, it remains unknown whether olfactory receptors that respond preferentially to airborne odorants evolved from receptors that in ancestral fishes were tuned to water-soluble odorants. An interesting alternative is that they might have evolved from ancestral receptors tuned to volatile but hydrophobic molecules (e.g., some terpenoids). Such odorants cannot be smelled at a distance underwater, but fishes might still benefit from smelling them upon contact (Mollo et al., 2014).

4.4. Movement on Land

As one might expect, the adoption of a terrestrial habitat had a profound effect on tetrapod motor behavior, especially locomotion. Whereas fishes swim mainly by lateral undulation (see Chapter 3), fully terrestrial tetrapods tend to walk, hop, or fly. What about the early tetrapods, the ones that successfully invaded land? Most likely, they swam by means of lateral undulation when in water (either during their larval stage or as adults) but were capable of walking on four legs as terrestrial adults. As explained earlier, this transition required strong legs and a stiff vertebral column. It also required the evolution of new muscles that can rotate the limbs forcefully, make compensatory adjustments along the trunk, and provide resistance against gravity. Assisting in these tasks was a relatively new type of sensor, called the muscle spindle afferent, which encodes changes in muscle length. These sensory structures are found in all amniotes and mediate a variety of reflexes that, among other things, counteract gravity. Lissamphibians also have muscles spindles, but they are less specialized than those of amniotes (Ottoson, 1976). In contrast, fishes generally lack muscle spindles, although some very simple muscle spindles have been reported in a jaw muscle of teleosts (Maeda et al., 1983).

When a salamander walks on land, the bending movements of its trunk seem similar to those of a swimming fish or, for that matter, a swimming salamander (Figure 4.16). However, the body of a swimming salamander forms an S-shaped wave that travels from the head toward the tip of the tail, whereas the body of a walking salamander forms a standing wave, with alternating sideways bends. This distinction has important implications for the neural circuits that control those movements (Chevallier et al., 2008; Bicanski et al., 2013). As research on lampreys and amphibians has shown, swimming by lateral undulation involves a central pattern generator (CPG) that is distributed along the length of the spinal cord. Walking on land involves additional CPGs that control the limb movements. It has been proposed that, in salamanders, both kinds of CPG can be activated by descending inputs from a locomotor region in the midbrain. When this descending input is weak, the limb CPGs force the swimming CPG to generate a standing wave. However, when the descending input exceeds a certain threshold, the limbs are folded back and the swimming CPG creates a traveling wave that propagates toward the tail (Figure 4.16). This model of salamander locomotion remains somewhat hypothetical, but it has been implemented in a salamander-like robot (Ijspeert et al., 2007). Analogous

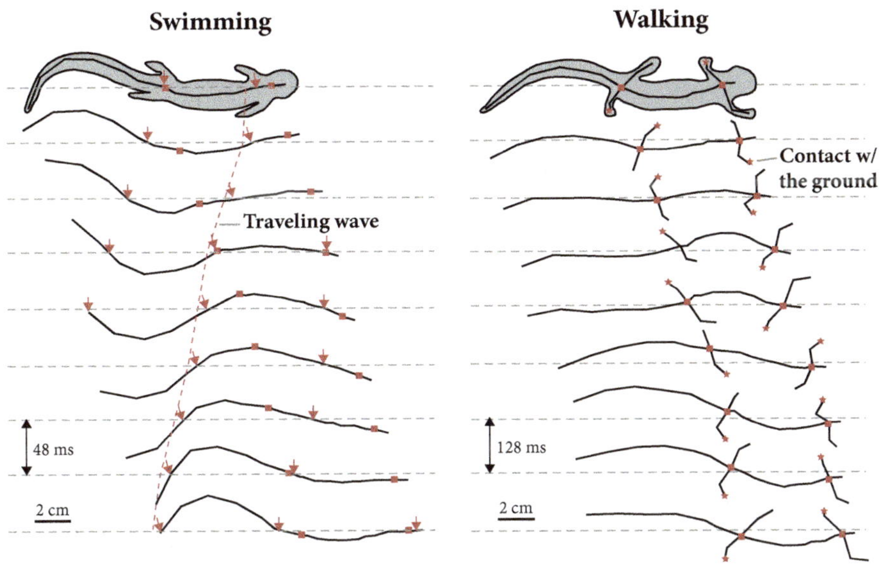

Figure 4.16 Swimming versus walking. Adult salamanders were videotaped as they either swam in water or walked on land. The former involves a wave of body flexion that travels toward the tail, as indicated by the dashed red line; red arrows indicate points of minimum lateral displacement from the mean direction of forward travel. In contrast, walking involves a series of alternating body bends that form a standing wave. Mathematical and robot modeling studies have shown that these two forms of locomotion can be produced by varying the amount of excitatory input (drive) to separate central pattern generators for spine and limb movements. The red stars represent estimates of when each foot contacts the ground.
Adapted from Ijspeert et al. (2007).

mechanisms may well have been involved in the evolutionary shift from obligatory swimming in fishes to walking-plus-swimming in early tetrapods (Standen et al., 2016). The ability of modern anurans to hop with their hind legs must have involved major changes to the limb CPGs, such that the left and right hind legs can be extended simultaneously, rather than alternately. This transition from alternate to synchronous hind leg movements has been studied in anuran tadpoles (Combes et al., 2004), but the neurobiological details remain largely unknown.

Anatomically, the transition to land was accompanied by the evolution of additional motor neurons that innervate the limb muscles. In amniotes, these limb motor neurons are located in the lateral portion of the spinal cord's ventral horn at brachial (aka cervical) and lumbar levels. At these locations the spinal cord is noticeably enlarged (Figure 4.17). Similar but smaller spinal enlargements are evident in the lungfish *Neoceratodus*, but other fishes do not exhibit them; the condition in coelacanths remains unclear (Antony and Millot, 1965). Given these data, we conclude that well developed brachial and lumbar enlargements are a derived feature for lobe-finned fishes and, thus, predated the origin of tetrapods. The number of

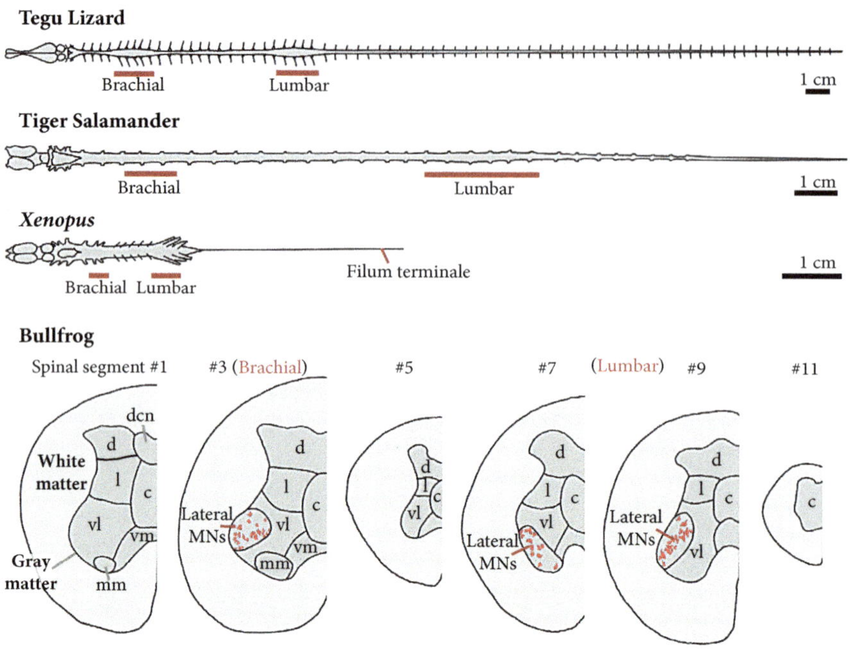

Figure 4.17 Spinal cord organization and variation. Shown at the top are the brains and spinal cords of three tetrapods in dorsal view. The spinal cord of the lizard exhibits clear brachial (aka cervical) and lumbar enlargements at the level of the forelimbs and hindlimbs, respectively. These enlargements are less obvious in amphibians, especially in the brachial region. However, as shown in the series of transverse sections, the size and shape of the gray matter does vary along the length of a frog's spinal cord (segments are numbered from rostral to caudal). In particular, the large motor neurons that innervate the limbs (the lateral MNs; red dots) are found only in the brachial and lumbar regions.

Abbreviations: c – central field; d – dorsal field; dcn – dorsal column nucleus; l – lateral field; mm – medial motor field; vl – ventrolateral field; vm – ventromedial field.

Adapted from Nieuwenhuys et al. (1998) and Ebbesson (1976).

limb motor neurons then increased further as tetrapods evolved more robust limbs and walked on land.

Recent developmental data have shown that activation of the gene *hox-9* at thoracic levels of the spinal cord suppresses the genes *hox-6* and *hox-10*, which are needed to specify the limb motor neurons in the brachial and lumbar regions, respectively (Figure 4.18). Thus these genes, together with *foxP1*, help to specify both the phenotype and the position of the limb motor neurons (Jung et al., 2014). Given these observations, it seems reasonable to speculate that evolutionary changes in the expression of these genes helped to reorganize the spinal cord as tetrapods emerged. As part of this transformation, the genes *hox-6* and *hox-9* probably shifted their expression caudally, relative to the condition in ray-finned and cartilaginous fishes (Figure 4.19). Given that the pectoral fins of fishes are homologous to the

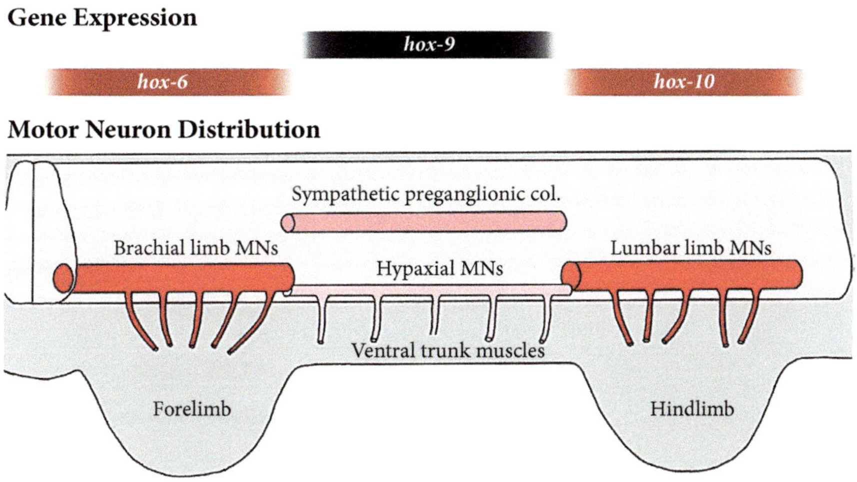

Figure 4.18 Hox genes and motor neurons. This lateral view of an embryonic tetrapod spinal cord shows that the brachial and lumbar motor neurons innervating the limbs are present only in embryonic domains that express *hox-6* or *hox-10*. The intervening, thoracic region expresses *hox-9*, which suppresses *hox-6* and *hox-10*, as well as *foxP1*; the latter gene is required for the limb motor neuron development. Instead of limb motor neurons, this intervening region of the spinal cord contains motor neurons that innervate ventral trunk muscles (i.e., hypaxial motor neurons) and the preganglionic neurons of the sympathetic nervous system.
Adapted from Murakami and Tanaka (2011).

forelimbs of tetrapods, this caudal shift in gene expression could explain why the motor neurons that innervate the pectoral fins in fishes are located much more anteriorly than the forelimb motor neurons of tetrapods (Ma et al., 2010).

The invasion of land required changes not only in locomotion, but also in feeding. The biggest problem with feeding on land is that opening the mouth in air creates much less suction than opening the mouth underwater, and therefore fails to suck prey in (furthermore, the prey are no longer neutrally buoyant in air). To overcome this dilemma, early tetrapods probably lunged at their prey, using their legs and their emergent neck. To some extent, they also used their motile, muscular tongue, which is another shared derived feature of living tetrapods (Iwasaki, 2002). Some adult lissamphibians still hunt by lunging at their prey, but many lissamphibians have evolved an elongated tongue that they can rapidly extend and use to strike at distant prey (Combes et al., 2004). This remarkable behavior evolved independently in several lineages of frogs and salamanders, which differ in the detailed mechanics of tongue protrusion (Deban et al., 2007). Tongue-based hunting certainly contributed to the enormous success of frogs and plethodontid salamanders (see Figure 4.5). It evolved again in some lizards, notably chamaeleons (Meyers and Nishikawa, 2000).

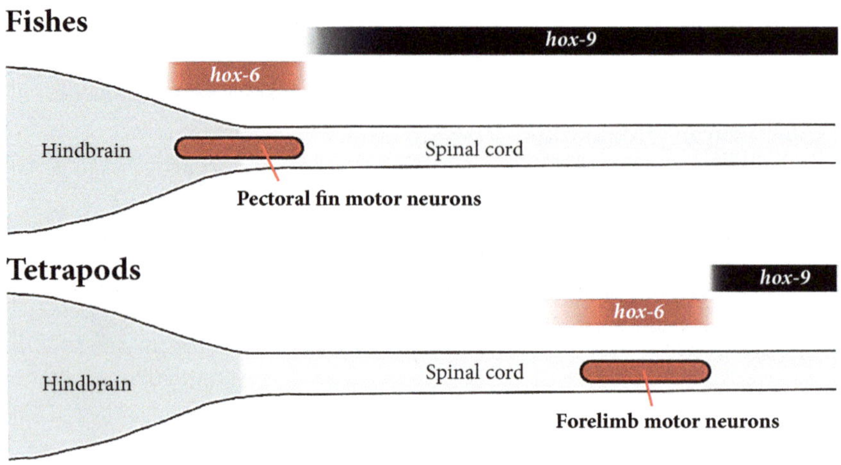

Figure 4.19 Evolutionary transposition of homologous motor neurons. The pectoral fins of cartilaginous and ray-finned fishes are innervated by motor neurons in the medulla and most anterior spinal cord (top). The forelimbs of tetrapods are thought to be homologous to the pectoral fins of fishes, and so are the motor neurons innervating them. However, the forelimb motor neurons in tetrapods are shifted caudally into the brachial region of the spinal cord. This caudal transposition of homologous neurons is correlated with a caudal shift in the gene expression of the *hox* genes that are involved in limb motor neuron development (see Figure 4.18).
Adapted from Ma et al. (2010).

4.5. The Brains of Early Tetrapods

As we try to reconstruct the brains of early tetrapods, which living species might give us the best clues? Among the extant tetrapods, lissamphibians are the best (though hardly ideal) candidates, because they are the only tetrapod anamniotes and, therefore, lack the innovations that characterize amniotes (rather than all tetrapods). But which amphibians might be most representative of early tetrapods? This question is important to our task, because the brains of anuran amphibians are substantially more complex than those of urodeles. For example, the midbrain's optic tectum is relatively large and contains multiple laminae in anurans, yet small and simple in all urodeles (Figure 4.20). Caecilian brains exhibit an intermediate level of complexity, featuring a simple optic tectum but a large telencephalon with numerous migrated cells (in at least some species; Schmidt and Wake, 1997). Given this variation among amphibians, we might look for guidance to their closest anamniote relatives. However, those species also exhibit substantial variation. Specifically, the lepidosirenid lungfishes have extremely simple brains that look in many ways like those of urodeles, whereas the Australian lungfish (*Neoceratodus*) and coelacanths have brains that are at least as complex as those of anurans. For example, the cerebellum and cerebellum-like structures of *Neoceratodus* and coelacanths are much better developed than those of frogs and toads.

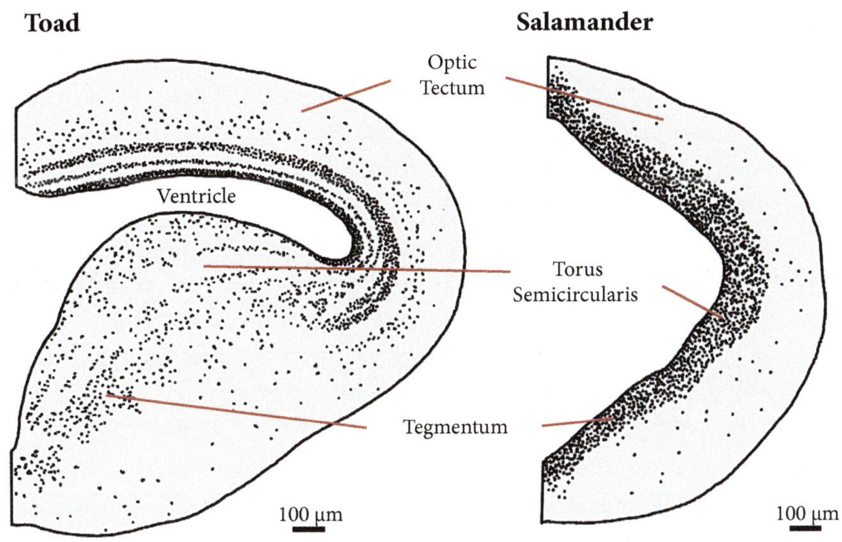

Figure 4.20 Variation in the midbrain roof of amphibians. The optic tectum and torus semicircularis are relatively large and laminated in anurans (e.g., the common toad *Bufo bufo*), but small and simple in urodeles (e.g., *Salamandra salamandra*), where the vast majority of cell bodies remain clustered near the ventricle. The simplicity of the urodele tectum appears to be an example of secondary simplification, more specifically of paedomorphosis.
Adapted from Schmidt and Wake (1997).

Confronted with all this variation, C. Judson Herrick, a leading figure in comparative neuroanatomy during the first half of the 20th century, chose to focus on the species with the simplest brains. In particular he focused on the brains of urodeles, which were easier to obtain and work with than lepidosirenid brains. In his landmark book, *The Brain of the Tiger Salamander*, Herrick (1948) described the brains of salamanders in impressive detail and used his findings to make inferences about the structure of early tetrapod brains. Herrick considered the possibility that salamander brains might have become simpler over the course of evolutionary time, potentially making them even simpler than the brains of early tetrapods, but he was not too troubled by this risk. In that case, he argued, salamander brains would be a good model for even more ancient brains, specifically the brains of early gnathostomes. This approach seems reasonable as long as one supposes that evolutionary simplification always retraces its antecedent steps, reverting to the condition of some old ancestor. However, this need not be the case (Gould, 1977). Moreover, Herrick's hypothesis is inconsistent with the observation that the brains of cyclostomes (i.e., jawless vertebrates) are more complex than those of urodeles. To resolve this puzzle, we need to delve more deeply into the potential mechanisms underlying evolutionary simplification.

4.5.1. Paedomorphosis and the Brain

As noted briefly in Section 4.3.3, paedomorphosis occurs when the adults of a descendent species retain juvenile or embryonic features of their ancestors (Gould, 1977). In essence, paedomorphosis (aka neoteny) occurs when an ancestral developmental process is cut short so that development in the descendent species does not proceed as far as in the ancestors. Of course, for this to work, the gonads of the descendants must fully mature, even if other body parts do not. This is the case in the axolotl, for example, because this urodele is capable of reproduction but also retains many features of aquatic urodele larvae, such as external gills (see Figure 4.4). Other urodeles exhibit many even more highly paedomorphic features (Wake, 1966), as do the lepidosirenid lungfishes (Bemis, 1984).

Given this context, it is reasonable to hypothesize that the relatively simple brains of urodeles and lepidosirenid lungfishes are also the result of paedomorphosis (Roth et al., 1993, 1997). Thus, the clustering of neuronal cell bodies near the ventricles in these species (Figure 4.20) can be seen as a failure of those neurons to undergo their vertebrate-typical migration away from where they were born (i.e., became post-mitotic). Similarly, the dearth of cytoarchitecturally defined cell groups in urodeles and lepidosirenids (Northcutt, 1986; Wicht and Himstedt, 1988) can be viewed as a failure of cytoarchitectural differentiation, which is a relatively late developmental process in other vertebrates. A third good example is the vestigial nature of the cerebellum in these species, since the cerebellum is one of the last brain regions to grow and mature in other vertebrates. In short, the simplicity of urodele and lepidosirenid brains can be viewed as a feature of ancestral embryonic brains that both lineages retained into adulthood, independently of one another. Unfortunately, this tells us little about the condition of the ancestral *adult* brains.

Intriguing support for this hypothesis comes from a positive correlation between paedomorphosis and genome size. As reviewed in Section 4.1, urodeles and lepidosirenid lungfishes have exceptionally large genomes (see Figure 4.3). Increased genome size is thought to be causally linked to the size of a cell's nucleus (Gregory, 2001), which in turn is bound to increase overall cell size. Such large cells might well have difficulty migrating through tissue, which might help to explain the lack of migrated neurons in the urodeles and lepidosirenids (Roth et al., 1993, 1994; Gregory, 2002). More importantly, increased DNA content would slow down progression through the cell cycle, thereby reducing the rate of embryonic cell division and, other things being equal, reducing the number of cells in the adult. This would explain why urodeles appear to have far fewer neurons in their brains than other vertebrates (Roth et al., 1993). In general, urodeles and lepidosirenid lungfishes develop very slowly and, it appears, many of the features in their brain never develop as far as in their ancestors.

In light of these observations and hypotheses, we conclude that salamander and lepidosirenid brains are significantly simpler than those of early tetrapods. This leaves anuran brains as the best "model" for early tetrapod brains. Of course, anuran brains surely exhibit various features that are unique to their own lineage, especially in the auditory and locomotor systems. An even more important caveat is that anuran brains are simpler than the brains of coelacanths, Australian lungfishes, basal ray-finned fishes, and squalomorph sharks in at least some respects. Most notably, their cerebellum and cerebellum-like structures are unusually small for gnathostomes. These observations suggest that anurans may themselves be somewhat paedomorphic, compared to their ancestors (Chipman et al., 2001). Consistent with this hypothesis, anurans also have larger genomes than most other vertebrates, excepting urodeles and lungfishes (Figure 4.3).

These considerations raise the intriguing possibility that early tetrapods themselves were at least somewhat paedomorphic. Why might that have happened? It is possible that retroviruses or other self-replicating elements invaded the genomes of early tetrapods, caused their genomes and cell sizes to balloon, and that this then forced a slowdown in development. In this case the paedomorphosis might have been a non-adaptive consequence of selfish "junk DNA" proliferation (Ohno, 1972). However, it is interesting to note that highly paedomorphic salamanders and lungfishes are much better than most other vertebrates at surviving in low oxygen environments. One reason for this anoxia resistance is that large cells, with their low surface-to-volume ratios, require less metabolic energy than the equivalent volume of smaller cells in order to maintain the ion gradients across their cell membranes. Thus, at any given body and brain size, increasing cell size reduces oxygen requirements substantially (Szarski, 1983). Given that early tetrapods evolved in relatively anoxic environments, it seems reasonable to speculate that paedomorphosis in early tetrapods might have been adaptive from a metabolic perspective. The decrease in neuron number might have made early tetrapods behaviorally less complex (Herculano-Houzel, 2011), but even brains with a relatively low number of large neurons can exhibit surprising degrees of structural complexity (Roth et al., 1999), which might in turn support complex behaviors.

4.5.2. Medulla

The medulla of anurans is similar in most respects to that of jawed fishes, as exemplified by the medulla of sturgeons depicted in the previous chapter (see Figure 3.16). It contains the principal motor areas and most of the sensory nuclei associated with the cranial nerves, as well as serotonergic raphe nuclei and a noradrenergic locus coeruleus. The most significant changes from the ancestral condition have occurred in the octavolateralis region of the medulla.

As reviewed in Chapter 3, basal jawed fishes have dorsal and medial (aka intermediate) octavolateralis columns that receive electrosensory and mechanosensory lateral line inputs, respectively. These cell groups disappeared as tetrapods became fully terrestrial and lost their electroreceptors and neuromasts. Some amphibians retain one or both of the lateral line systems during their aquatic larval phase, but those receptors and their associated medullary targets degenerate with the transition to land. Only *Xenopus* and a few other aquatic frogs retain mechanosensory lateral line receptors and their octavolateralis area into adulthood (Figure 4.21). None of this is very surprising, given that the mechanosensory and electrosensory lateral line systems become essentially useless in air.

The more interesting question is how the central projections of the inner ear changed as anurans evolved tympanic ears capable of hearing high-frequency airborne sounds. In basal jawed fishes the axons of the eighth cranial nerve, which innervate the auditory and vestibular sensors of the inner ear, project to the ventral octavolateralis column (or zone). This ventral octavolateralis column contains

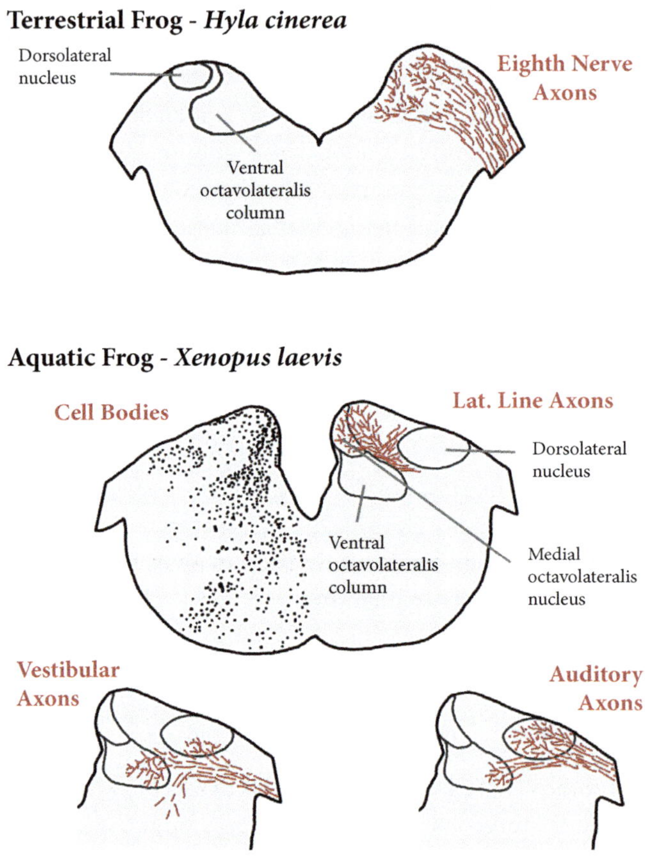

Figure 4.21 The octavolateralis region of the anuran medulla. In adult terrestrial frogs the eighth cranial nerve transmits auditory and vestibular information to the dorsolateral nucleus and ventral column of the octavolateralis region. The condition is different in aquatic frogs, which retain their mechanosensory lateral line system into adulthood. In these animals the lateral line axons project to the medial octavolateralis nucleus, which lies dorsal and medial to the vestibular and auditory projection targets. The dorsolateral nucleus, which is probably unique to anurans, is the primary target of the axons that innervate the auditory hair cells in the inner ear.
Adapted from Will et al. (1985) and Will and Fritzsch (1988).

at least four distinct cell groups that receive predominantly auditory inputs (from the sacculus; see Chapter 3), predominantly vestibular inputs, or a mixture of both (McCormick and Braford, 1988). Amphibians retain this ventral octavolateralis column, including its auditory components. However, anurans evolved an additional cell group that receives only auditory inputs. On account of its dorsolateral position, this cell group is called the dorsolateral nucleus (Figure 4.21).

Is the dorsolateral nucleus of anurans homologous to the dorsal octavolateralis nucleus of fishes? That is, did the ancestral electrosensory lateral line nucleus of the medulla become transformed into an auditory nucleus as the electrosensory system was lost and novel auditory receptors appeared? Developmental data had once favored this hypothesis (Larsell, 1934; Herrick, 1948), but later studies showed that the dorsolateral nucleus in frogs develops separately from the lateral line nuclei (Jacoby and Rubinson, 1983). It is still possible that the electrosensory lateral line nucleus of ancestral jawed fishes was phylogenetically (rather than developmentally) transformed into the acoustic dorsolateral nucleus of anurans (Fritzsch et al., 1984; Fritzsch, 1988), but we now believe that the dorsolateral nucleus is an innovation of anurans that is dedicated preferentially to sensing the new kinds of sounds that early anurans could hear on account of their "new" tympanic ears. Of course, the dorsolateral nucleus did not evolve "out of nothing" (see Section 2.8 in Chapter 2); it simply seems to be the case that this nucleus has no specific homolog in the nervous systems of adult fishes; nor is it homologous to the cochlear nuclei of amniotes (see Chapters 5 and 6).

4.5.3. Cerebellum

The cerebellum of anurans is significantly smaller than that of coelacanths (see Figure 3.7 in Chapter 3) but larger than the cerebellum of urodeles and caecilians. Large variations in cerebellum size are also seen in lungfishes, because the cerebellum of the Australian lungfish, *Neoceratodus*, is similar in size to that of coelacanths (Northcutt, 2011), whereas the cerebellum of the lepidosirenid lungfishes (see Figure 3.7) is about as small as that of urodeles. As noted earlier, we suspect that these variations in cerebellum size resulted in large measure from extensive paedomorphosis in urodeles and lepidosirenids, and that the cerebellum of early tetrapods was roughly as large, relative to the remaining brain, as that of today's anurans. However, if early tetrapods were slightly paedomorphic, compared to their fully aquatic ancestors (see Section 4.5.1), then their cerebellum, being a very late-developing structure, may well have been somewhat reduced in relative size.

Despite some variations in the cerebellum's size, its internal structure did not change substantially with the invasion of land. Anurans have the two principal types of cerebellar neurons, Purkinje and granule cells, and the connections between these neuron types are basically similar to those in other vertebrates. The cerebellum's extrinsic connections are likewise highly conserved. Thus, the cerebellum in anurans receives inputs mainly from the vestibular nuclei and spinal cord, as well as the inferior olive in the ventral medulla. Its outputs are funneled, at least in part, through a deep cerebellar nucleus and target mainly the vestibular system and basal midbrain regions that, in turn, project to motor neurons in the medulla and spinal cord.

The most significant change that seems to have accompanied the transition to life on land was a dramatic reduction of the cerebellum-like structures, which are very large in coelacanths and most other fishes, but poorly developed in terrestrial amphibians. As discussed in Chapter 3, the cerebellum-like systems in fishes are associated primarily with the lateral line, vestibular, and fledgling auditory systems. With the loss of electroreceptors and lateral line neuromasts in fully terrestrial vertebrates (see Section 4.3.3), those cerebellum-like regions were reduced to their vestibular and auditory components. That said, early amphibians probably had aquatic larvae, which means that they probably lost their lateral line systems only at metamorphosis.

Little is known about the functions of the cerebellum and cerebellum-like regions in amphibians or basal jawed fishes, but a mutant variant of the axolotl (an aquatic urodele) has abnormal cerebellum-like regions and exhibits abnormal swimming behavior (Ide et al., 1977; Elbert et al., 1983). Furthermore, anatomical data strongly suggest that the cerebellum in anurans is involved in the control of tongue protrusion (Anderson, 2001), and lesion studies indicate a major role in diverse forms of motor coordination (ten Donkelaar, 1998). The latter hypothesis is supported by the observation that arboreal frogs tend to have a slightly larger cerebellum than frogs living in habitats where locomotion is presumably a simpler affair (Taylor et al., 1995). Of course, the earliest tetrapods were exceedingly unlikely to have been arboreal, and only a few anurans have taken to the trees. Therefore, the slight expansion of the cerebellum in arboreal frogs probably occurred relatively late in anuran phylogeny.

4.5.4. Midbrain

Early tetrapods probably had a midbrain tegmentum similar to that of jawed fishes, but the tegmentum has received relatively little attention in comparative studies. Most likely, it continued to play a major role in various aspects of motor control, as it does in all examined vertebrates. It may also have contained some of the dopaminergic neurons that project to the striatum, although the location of these neurons is surprisingly variable across phylogeny (Yamamoto and Vernier, 2011; Wullimann, 2014).

Much more information is available on the midbrain roof, especially the optic tectum. Since anurans and most jawed fishes have a well-developed optic tectum, we suspect that this structure was also fairly large in early tetrapods. As in all living gnathostomes, the optic tectum of early tetrapods surely received inputs from the retina. In addition, it probably received somatosensory and auditory inputs. Inputs from the lateral line were probably lost as early tetrapods became fully terrestrial. As mentioned previously (Figure 4.20), the tectum's internal organization is highly variable across amphibians and, therefore, challenging to reconstruct for early tetrapods. For example, 95% of its neurons remain in a periventricular position in plethodontid salamanders, whereas the corresponding percentage in ranid frogs is 70% (Roth et al., 1993). However, despite the lack of cell migration in the optic tectum of urodeles, the dendrites of their tectal neurons arborize in very specific layers (Figure 4.22), and most types of tectal neurons seen in frogs also exist in urodeles (Roth et al., 1999). Furthermore, tectal neurons in frogs and urodeles respond very similarly to a wide range of stimuli (Dicke and Roth, 2007). Given these similarities, we can infer that early tetrapods probably possessed a relatively large optic tectum with multiple layers and a large diversity of neuronal cell types.

Functionally, the optic tectum of early tetrapods was probably involved in orienting behaviors, just as it is in jawed fishes. Large tectal lesions in anurans cause profound deficits in spatial orientation behavior (Ewert, 1970; Ingle, 1970). Moreover, the visual and other sensory inputs to the optic tectum appear to be

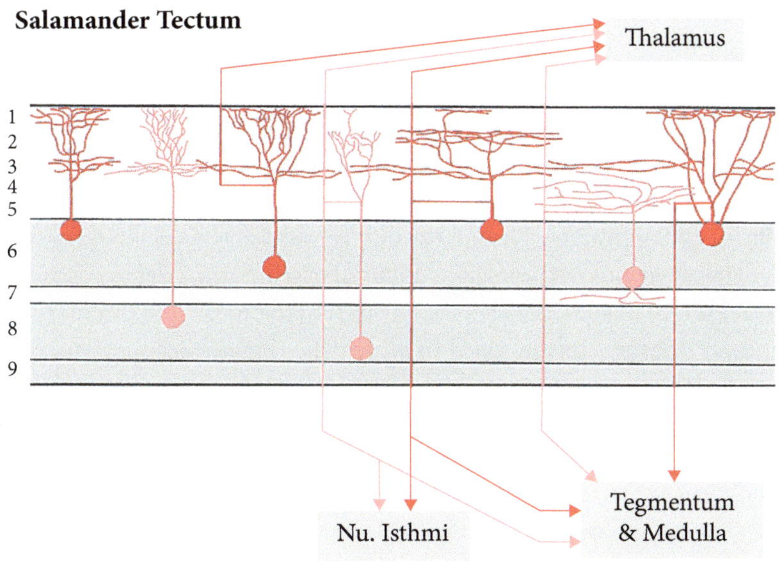

Figure 4.22 Surprising complexity in the optic tectum of salamanders. The optic tectum of urodeles appears very simple insofar as the vast majority of its neuronal cell bodies (shaded gray) don't migrate far from their site of birth adjacent to the ventricle. However, intracellular labeling of tectal cells in salamanders (mainly *Plethodon jordani*) reveals that the dendrites of the tectal neurons (red or pink) arborize in very specific layers within the tectal neuropil. Based on these dendritic patterns, as well as physiological data, we can conclude that salamanders possess a large variety of different tectal neuron types.

Adapted from Roth et al. (1999), with permission from John Wiley & Sons.

topographically organized in all vertebrates, and electrical stimulation of specific tectal neurons causes toads to orient toward (and often snap at) the locations that correspond to the stimulated neurons' spatial receptive fields. Anatomical data suggest that tectal activity elicits these orienting movements by means of descending projections to the tegmentum, medulla, and rostral spinal cord. Also critical are reciprocal connections between the optic tectum and nucleus isthmi, which is located just caudal to the midbrain (Figure 4.22; Gruberg et al., 1991). The specific function of nucleus isthmi in frogs remains mysterious, but its homolog in ray-finned fishes is thought to mediate a winner-take-all competition among tectal neurons (see Chapter 3), and the avian and mammalian homologs of nucleus isthmi are thought to be involved in spatial attention (Gruberg et al., 2006).

The second major component of the midbrain roof is the torus semicircularis, which originally develops immediately caudal to the optic tectum but ends up protruding into the tectal ventricle in species with a large tectum (see Figure 4.23, bottom). The torus semicircularis is found in all jawed fishes and is homologous to the mammalian inferior colliculus. It is best known as the midbrain target of ascending auditory pathways, even in species that lack tympanic ears, but it also receives ascending vestibular, somatosensory, and lateral line inputs, which are then passed on to the tectum and several other brain regions. As early tetrapods moved onto land, the torus semicircularis would have lost the lateral line inputs, but the other sensory inputs were almost certainly retained (Wilczynski, 1981). The torus semicircularis of early tetrapods was probably larger than that of extant urodeles but smaller than that of anurans (see Figure 4.20), which probably expanded it as they evolved tympanic ears and became highly vocal during their reproductive season (Emerson and Boyd, 1999).

In anurans and many jawed fishes, the different sensory modalities are processed in different subdivisions of the torus semicircularis, but some degree of multimodal convergence is also observed. Whether any part of the torus semicircularis is "new" in the anuran lineage—in the sense in which the dorsolateral nucleus of the anuran medulla is new—remains unclear. However, it is certainly possible that a new nucleus at one level of the nervous system projects to an "old" nucleus at higher levels, with the latter being merely reorganized to handle the additional input (Wilczynski, 1984).

4.5.5. Diencephalon

The diencephalon of tetrapods can be divided into three rostrocaudal divisions, each of which is divisible into alar and basal components, just as we discussed for jawed fishes in Chapter 3 (see Figure 3.21). All of these major divisions, including the pretectum, epithalamus, thalamus, prethalamus, and posterior tuberculum, are homologous between tetrapods and jawed fishes. Even many of the smaller subdivisions within these areas can be homologized across anamniotes, though the

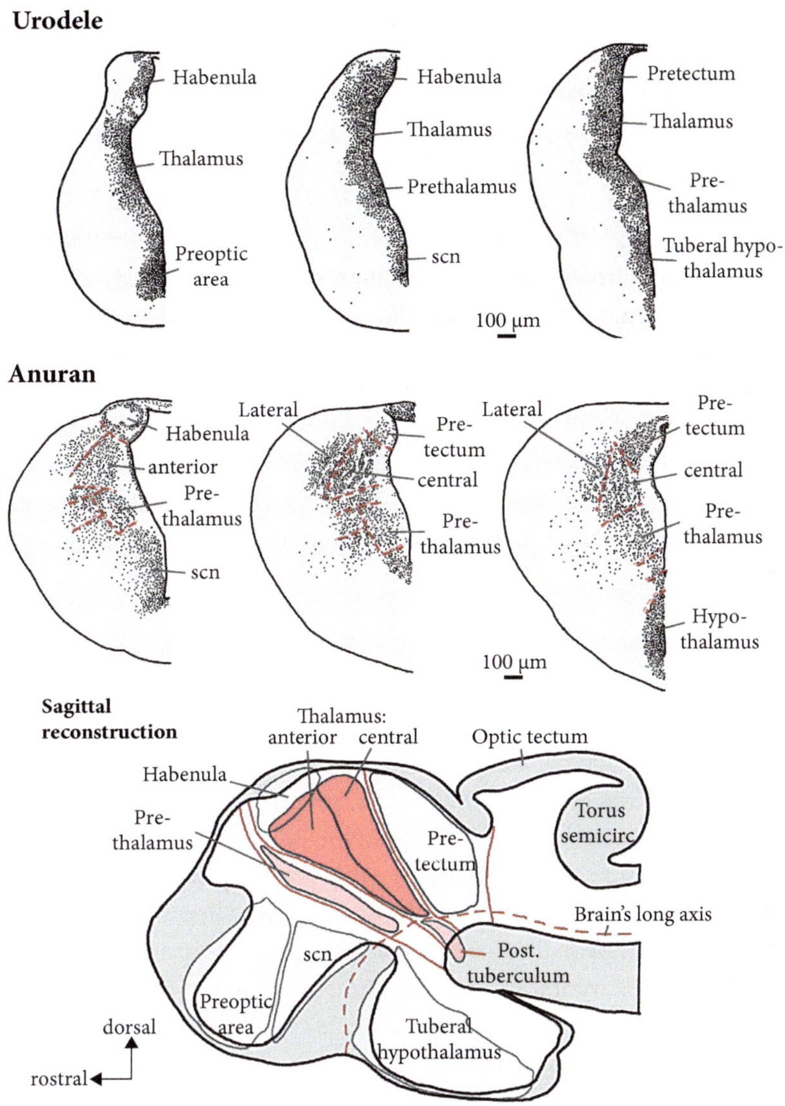

Figure 4.23 The diencephalon of amphibians. Shown at the top are Nissl-stained transverse sections through the diencephalon of a newt (*Triturus alpestris*); only one side of the brain is shown, and the section on the left is most rostral. Shown below those images are three equivalent sections through the diencephalon of a toad (*Bombina orientalis*). Obviously, the anuran diencephalon contains far more migrated cells and more clearly differentiated cell groups (boundaries are indicated by red dashed lines). The image at the bottom is a sagittal reconstruction of the diencephalon in a frog (*Rana perezi*). The boundaries of the three major diencephalic segments are indicated by solid red lines. The suprachiasmatic nucleus (scn) is considered to be the dorsal (alar) part of the hypothalamus. Migrated cell groups are not shown.

Adapted from Wicht and Himstedt (1988), Roth et al. (2003), Neary and Northcutt (1983), and Puelles et al. (1996).

evidence in some cases is tenuous (Northcutt, 1995). Indeed, it seems that the diencephalon did not undergo any major changes as early tetrapods evolved, especially in comparison to the dramatic transformation of the posterior tuberculum in teleosts (see Chapter 3).

Among amphibians, the urodeles have a much simpler diencephalon than anurans. Especially their thalamus and prethalamus exhibit far fewer migrated cells and less histological differentiation than in anurans (Figure 4.23). Within the thalamus of anurans, researchers have identified anterior, central, and lateral thalamic nuclei, and numerous studies have examined their neural connections (see Neary and Northcutt, 1983; Puelles et al., 1996). Because most thalamic neurons have long dendrites that extend far beyond the boundaries of the individual cell groups (Figure 4.24), it is difficult to determine for certain what kinds of inputs these neurons receive. However, intracellular labeling and physiological studies suggest that most of the thalamic nuclei in anurans receive mainly multimodal sensory input from the midbrain roof; direct retinal projections are quite limited and target primarily the prethalamus (Neary and Northcutt, 1983; Roth et al., 2003; Wilczynski and Endepols, 2006). The outputs of the thalamus are directed mainly at the optic tectum, tegmentum, and spinal cord, but the thalamus also projects to the telencephalon. In particular, the anterior thalamic nucleus projects to the septum and the pallium (Figure 4.24), whereas the central nucleus projects mainly to the striatum (Roth et al., 2003; Laberge and Roth, 2007a).

The functions of the thalamus in anurans remain largely unknown, but it seems mainly to modulate neurons in the midbrain. In particular, lesions of the thalamus and adjacent pretectum in toads disinhibit prey-catching behavior in toads, which means that the lesioned toads become less selective in what they attack; they even snap at stimuli that normally elicit escape behavior (Ewert, 1968). It remains unclear whether this disinhibition involves direct projections from the thalamus to the optic tectum, pathways through the prethalamus and pretectum, loops through the telencephalon, or some combination of all of the above (Ewert et al., 1999). GABAergic neurons are found in all these thalamic target areas and may well be involved in the lesion-induced disinhibition. In this context, it is worth noting that the largest concentration of inhibitory neurons in the anuran diencephalon is found in the prethalamus (Brox et al., 2003), which projects to a wide variety of lower brain regions, including the thalamus (Roth et al., 2003).

4.5.6. Hypothalamus

The hypothalamus is a very complicated, hard-to-study brain region that has only just begun to receive extensive attention from comparative neurobiologists (Alvarez-Bolado et al., 2015; Domínguez et al., 2015). Based on gene expression patterns in young embryos, researchers have divided the hypothalamus of selected vertebrates into a large number of subdivisions. As part of that effort, they have

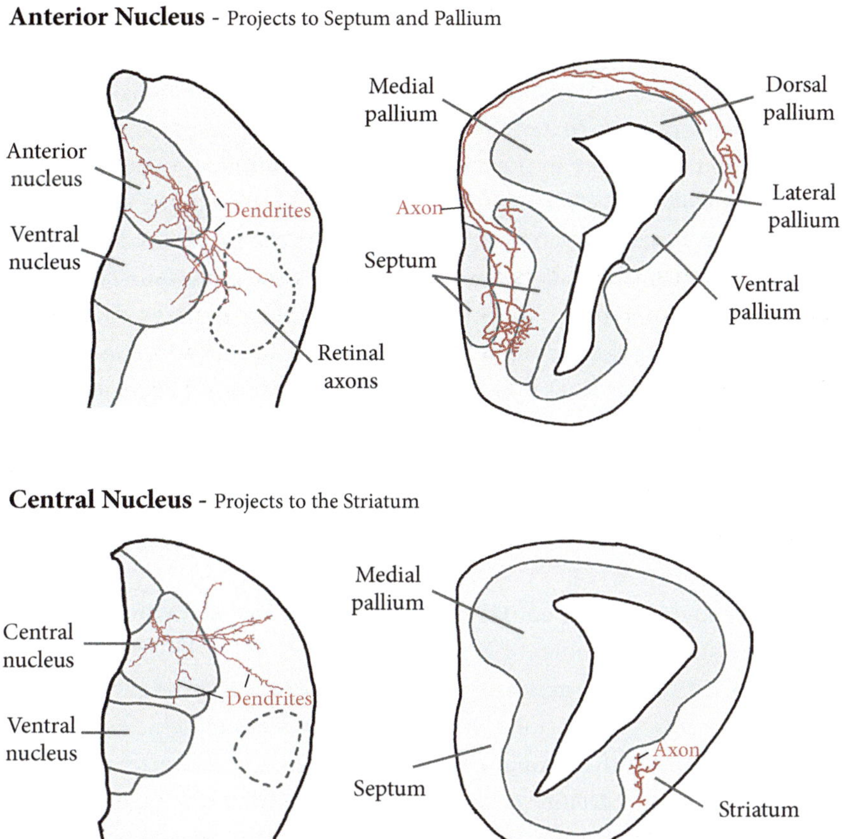

Figure 4.24 Telencephalic projections of the anuran thalamus. Gerhard Roth and his collaborators injected more than 145 individual neurons in the thalamus and prethalamus of toads with a neuronal tracer. Shown at the top is a labeled neuron in the anterior nucleus of the thalamus that has far-reaching dendrites (left) and extends its axon (right) mainly to the septum, but also to the pallium. The illustrated neuron extends its axon to the dorsal pallium, but most anterior nucleus neurons limit their pallial projections to the medial pallium. Shown at the bottom are the dendrites (left) and axon (right) of a labeled neuron in the central nucleus of the thalamus; its ascending projections target mainly the striatum.
Adapted from Roth et al. (2003), with permission from John Wiley & Sons.

revised many traditional ideas about hypothalamic organization, especially with regard to what is ventral, dorsal, rostral, and caudal within the hypothalamus and how its position in the brain relates to that of other brain regions (Puelles and Rubenstein, 2015). Because of how the brain's long axis bends during development, the hypothalamus as a whole does not lie "below the thalamus," as the traditional name implies. Instead, it lies topologically ventral to the telencephalon, but calling it a "hypotelencephalon" (Puelles and Rubenstein, 2015) would perhaps be too radical a change in terminology.

The published studies that compare hypothalamic gene expression patterns across species tend to emphasize that those patterns are highly conserved. This emphasis reinforces the traditional idea that evolution of the hypothalamus has been far more conservative than that of the thalamus or telencephalon. However, some of the examined genes do vary in their hypothalamic expression patterns among the major vertebrate lineages (Domínguez et al., 2015). Moreover, the sizes of the various hypothalamic domains vary significantly across the major taxonomic groups. For example, a neurogenetic domain that surrounds the optic stalk (i.e., the evagination that gives rise to the retina and optic tract; see Figure 3.21 in Chapter 3) is much larger in teleosts than amniotes (Affaticati et al., 2015). Once this size difference is recognized, it becomes apparent that an important set of peptidergic neuroendocrine cells (called the supraoptic and paraventricular nuclei in mammals) arises from this domain in both teleosts and amniotes, even though these neurons are traditionally assigned to the alar hypothalamus in amniotes and to the "preoptic" region in teleosts (Herget et al., 2014; Puelles and Rubenstein, 2015; Moreno et al., 2016; Yamamoto et al. 2017).

Given this pattern of both conservation and variation, we hypothesize that the preoptic/optic-stalk/hypothalamic region of early tetrapods was similar to that of basal ray-finned fishes and anurans. That is, it was probably relatively small and contained just a few subdivisions. It surely featured a suprachiasmatic nucleus, which controls daily and seasonal rhythms in essentially all vertebrates, and the aforementioned neurosecretory cells, but it did not exhibit the large inferior lobes of teleosts or many cartilaginous fishes (Northcutt, 1995). In addition, the saccus vasculosus, which is needed for seasonal changes in the gonads of ray-finned fishes (see Chapter 3; Nakane et al., 2013) was lost in the last common ancestor of lungfishes and tetrapods. The tuberal part of the anterior pituitary in birds and mammals is known to be involved in the seasonal regulation of gonadal function and may, therefore, have assumed some functions of the saccus vasculosus as the latter structure was lost.

4.5.7. The Telencephalon

The telencephalon of the earliest tetrapods was probably elongate and tubular. It may have looked roughly like the endocast of *Eusthenopteron*, an osteolepiform sarcopterygian from the late Devonian (Figure 4.25). Unfortunately, the fossil record for endocasts from early tetrapods is scant and difficult to interpret (Lu et al., 2012). Therefore, a comparative analysis of lissamphibians and basal jawed fishes is more informative. Such an analysis suggests that early tetrapods indeed had slender, elongated telencephalic hemispheres. Moreover, those hemispheres probably formed by developmental evagination, rather than eversion (see Chapter 3, Figure 3.26), just as they do in all extant amphibians and lungfishes. The only major wrinkle in this story is that coelacanths have a large telencephalon in which only the

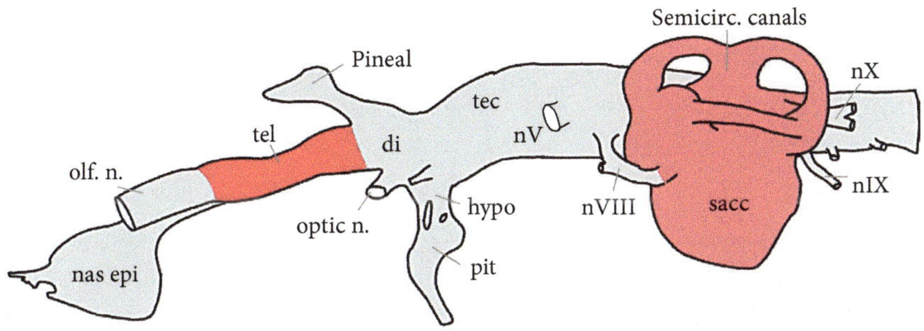

Figure 4.25 Endocast of *Eusthenopteron*. This is a lateral view of a reconstructed endocast of *Eusthenopteron foordi*, a tetrapodomorph fish (and early stem tetrapod) from the late Devonian (see Figure 4.6). The putative telencephalon (including the olfactory bulb) is shaded red, and the vestibular apparatus is shaded pink.

Abbreviations: di – diencephalon; hypo – hypothalamus; nVIII–nX – numbered cranial nerves; nas epi – nasal epithelium; olf. n. olfactory nerve; optic n. – optic nerve; pit – pituitary; sacc – sacculus; semicirc – semicircular; tec – tectum; tel – telencephalon.

Adapted from Stensiö (1963), who based his figure on a reconstruction by Erik Jarvik.

ventral (subpallial) portion is fully evaginated. Their pallium forms a prominent intraventricular bulge that causes the tela choroidea to become stretched across the dorsal pallial surface, much as it does in teleosts (see Figure 4.26; Northcutt, 1986; Nieuwenhuys et al., 1998; Northcutt and González, 2011). Therefore, the pallium of coelacanths can be viewed as being at least partly everted, although it may be better to think of it as simply being inordinately thick. We consider these features to be derived for coelacanths.

4.5.7.1. Subpallial Derivatives

Going beyond gross morphology, we can say that the telencephalon of early tetrapods must have been divided into pallial and subpallial divisions, as it is in all vertebrates. These divisions, long recognized by embryologists, have recently received substantial support from gene expression studies.

For example, the gene *distal-less* (*dlx*) is expressed only in the subpallium of frogs during early development (Brox et al., 2003). In adult frogs, some *dlx*-expressing cells are found also in the pallium, but these cells are thought to have migrated into the pallium from their subpallial place of birth (Moreno et al., 2008). Since the expression pattern for *dlx* is very similar to that of the synthetic enzyme for the inhibitory neurotransmitter GABA (Figure 4.27), we can infer that most subpallial neurons, as well as the subpallial immigrants within the pallium, are inhibitory (Brox et al., 2003). This general pattern of inhibitory neuron distribution and migration has also been demonstrated in birds and mammals (Cobos et al., 2001). In lungfishes and teleosts, most inhibitory neurons are also found in the subpallium, although a spattering of GABAergic neurons exists within the pallium (González and Northcutt, 2009; Mueller and Guo, 2009). Assuming that these pallial inhibitory neurons also originate from the subpallium, we can infer that the migration of

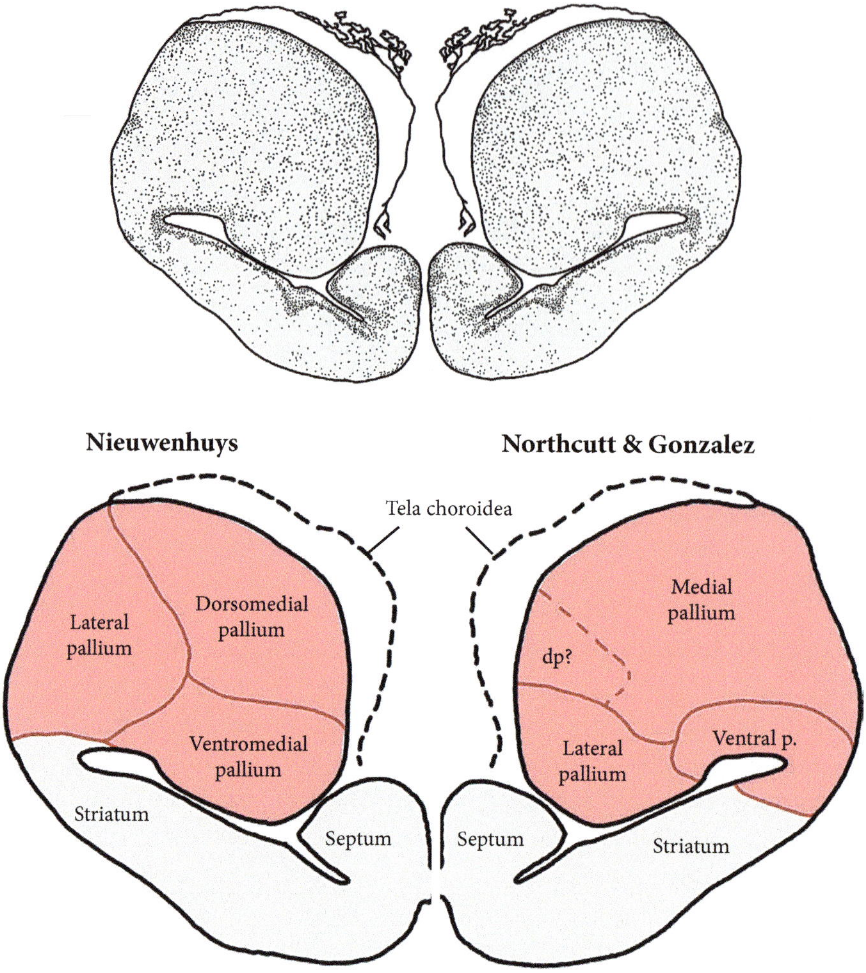

Figure 4.26 The telencephalon of coelacanths. Shown at the top is a transverse section through the telencephalon of an adult coelacanth (*Latimeria chalumnae*), stained to reveal the cell bodies. The extreme thickening of the dorsal telencephalon makes it difficult to determine whether this telencephalon is evaginated, everted, or a bit of both (see Figure 3.27). Shown at the bottom are two different interpretations of the coelcanth's pallium (shaded red). According to Nieuwenhuys (1965), the pallium can be divided into three major divisions that were named for their position, rather than their presumed homology (left). In contrast, Northcutt and Gonzalez (2011) proposed that the pallium of coelacanths extends further ventrally and deserves a new, homology-based nomenclature (right). In particular, they identified a ventral pallium (ventral p.) and a dorsal pallium (dp?). We now think that the latter region could easily be a subdivision of the medial pallium or, less likely, the lateral pallium (which is probably the main target of olfactory bulb projections). Therefore, it remains an open question whether coelacanths have a dorsal pallium that is homologous to the dorsal pallium of amniotes.

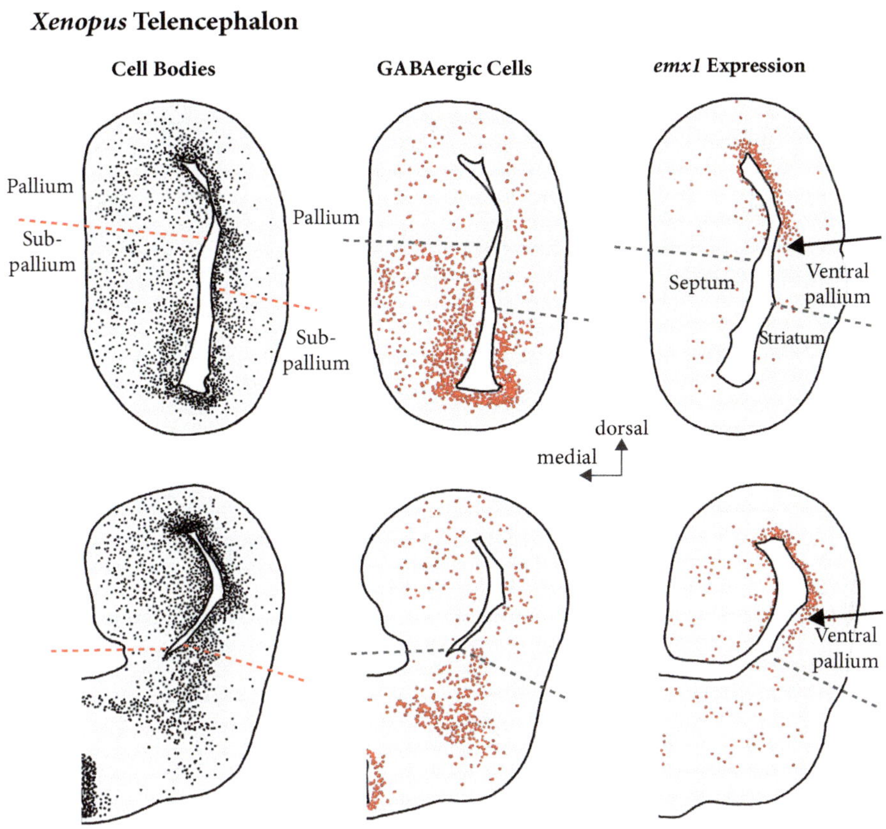

Figure 4.27 Subpallium and ventral pallium of frogs. Shown on the left are transverse Nissl-stained sections through one telencephalic hemisphere of *Xenopus laevis* at intermediate and caudal levels (top and bottom, respectively). The middle column depicts corresponding sections stained with the GABA synthetic enzyme GAD67. The column on the right shows the expression pattern of *emx1*, which densely labels most of the pallium (especially adjacent to the ventricle) except for its ventrolateral edge (ventral to the arrows). The *emx1*-negative pallial region has come to be called the ventral pallium.

Adapted from Brox et al. (2003) and Brox et al. (2004).

inhibitory neurons from the subpallium into the pallium is an ancient feature that evolved before the origin of tetrapods (Martinez-de-la-Torre, 2011). Even if this is not the case, the predominance of GABAergic neurons in the subpallium is almost certainly a feature that early tetrapods retained from their vertebrate ancestors.

The medial portion of the subpallium in tetrapods consists mainly of the septum, which comprises multiple subdivisions. The septum as a whole is well developed in both lungfishes and amphibians (Figure 4.27; Moreno et al., 2018), and its connections are very similar in both taxa (Northcutt, 1995; Endepols et al., 2005; Northcutt and Westhoff, 2011). Specifically, the septum receives input from the olfactory bulb, overlying pallium, amygdala, preoptic area, hypothalamus, thalamus, prethalamus, torus semicircularis, and tegmentum. In turn, the septum projects back to most of these areas, as well as to the optic tectum and the habenula, a highly conserved (but poorly understood) structure in the dorsal diencephalon (see Figure 4.23). Given this plethora of inputs and outputs, the septum's functions are difficult to fathom. Extensive research in mammals indicates that the septum is involved in diverse social and reproductive behaviors, but almost nothing is known about septal functions in anamniotes. All we can say with reasonable certainty is that, neuroanatomically, the septum has changed relatively little across the fish-tetrapod divide (Lanuza and Martinez-Garcia, 2009).

The ventrolateral portion of the subpallium comprises mainly the striatum and the pallidum, as well as the preoptic area (e.g., Domínguez et al., 2015). In mammals the striatum projects to the adjacent pallidum, which then projects to more distant targets. The situation is similar in anurans, but in these animals the striatum itself also has long descending projections (Figure 4.28). Thus, the striatum and the pallidum are not as structurally distinct in amphibians as in mammals. Similarly, the striatopallidal complex of anurans can be divided into dorsal and ventral divisions, just as in mammals, but the connections of these two divisions are not as different from one another in anurans as they are in mammals (Marin et al., 1997a; 1998). Given these data, we speculate that the striatum and pallidum were not well segregated from one another in early tetrapods, but then became more distinct in the lineage leading to amniotes. Additional comparative data, especially on lungfishes, would help to shore up this hypothesis.

The dorsolateral subpallium in tetrapods contains parts of the amygdala. One of these subpallial amygdalar components is the medial amygdala, which in amphibians receives mainly vomeronasal input from the accessory olfactory bulb and projects heavily to the preoptic area and hypothalamus (Moreno and González, 2003). The other principal component of the subpallial amygdala is the central

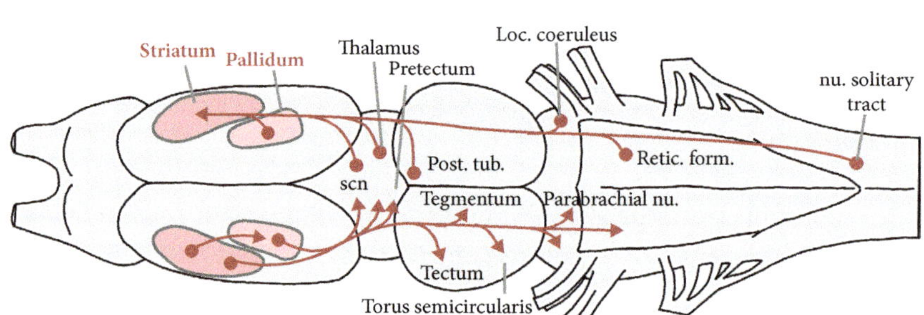

Figure 4.28 Connections of the striatum and pallidum in amphibians. The top half of the illustrated anuran brain (dorsal view, rostral to the left) shows the striatum's principal inputs. The bottom half depicts the main outputs of the striatum and pallidum, which overlap extensively.

Abbreviations: loc. coeruleus – locus coeruleus; nu. solitary tract – nucleus of the solitary tract; post. tub. – posterior tuberculum; retic. form. – reticular formation; scn – suprachiasmatic nucleus.

Based on Marin et al. (1997a).

amygdala, which projects heavily to autonomic regions in the brainstem, including the nucleus of the solitary tract and the parabrachial nucleus (Moreno and Gónzalez, 2006; López et al., 2017).

4.5.7.2. Divisions of the Pallium

The olfactory bulbs that form the rostral end of the anuran telencephalon derive mainly from the embryonic pallium. However, like other pallial regions, they include inhibitory interneurons that probably migrated from the subpallium (Brox et al., 2003). In both amphibians and amniotes the olfactory bulbs are sessile, meaning that they are "sitting" at the front end of the telencephalic hemispheres. In contrast, the olfactory bulbs of coelacanths and the Australian lungfishes are connected to the more caudal telencephalon by long axon tracts (i.e., they are "pedunculated"). Which of these two conditions is primitive for tetrapods is difficult to determine, because the lepidosirenid lungfishes have sessile olfactory bulbs, while both patterns are seen in cartilaginous and ray-finned fishes. In any case, there is no known functional significance to having the olfactory bulbs located very close to their telencephalic targets or far away from them. The remaining pallium is divisible into medial, dorsal, lateral, and ventral divisions (Puelles et al., 2000).

The medial pallium is very large in amphibians and lepidosirenid lungfishes (Figure 4.29). This region seems also to be very large in coelacanths (see Figure 4.26), but this hypothesis must remain tentative as long as we have no developmental or experimental data for these species (Northcutt and Gonzalez, 2011). In any case, the medial pallium contains numerous neurons that migrated radially away from the ventricle (where they were likely born), even in lungfishes and urodeles, which contain few such neurons in the rest of the brain. The medial pallium receives limited inputs from the olfactory bulb but extensive inputs from other pallial areas. In amphibians it also receives major inputs from the septum and some ascending multimodal sensory inputs from the anterior nucleus of the thalamus (see Figure 4.24; Northcutt and Ronan, 1992). The latter projection is not observed in lungfishes or ray-finned fishes (Holmes and Northcutt, 2003; Northcutt and Westhoff, 2011), but similar projections are observed in cartilaginous fishes, lampreys, and hagfishes. Therefore, we suspect that projections from the thalamus to the medial pallium were lost in early bony fishes and then re-evolved in tetrapods. Alternatively, they might have been lost independently in ray-finned fishes and lungfishes, but we discount this hypothesis because the thalamic nuclei projecting to the medial pallium in amphibians are likely not homologous to those in cartilaginous fishes.

The medial pallium has descending projections to the septum, ventral striatum, preoptic area, and hypothalamus (Northcutt and Ronan, 1992). Because most of these connections are similar between lungfishes and amphibians (Northcutt and Westhoff, 2011), they probably did not change substantially with the invasion of land. The connections of the medial pallium in amphibians and lungfishes also bear some similarity to the connections of the hippocampus in amniotes, supporting the hypothesis that these two structures are homologous. This homology hypothesis is

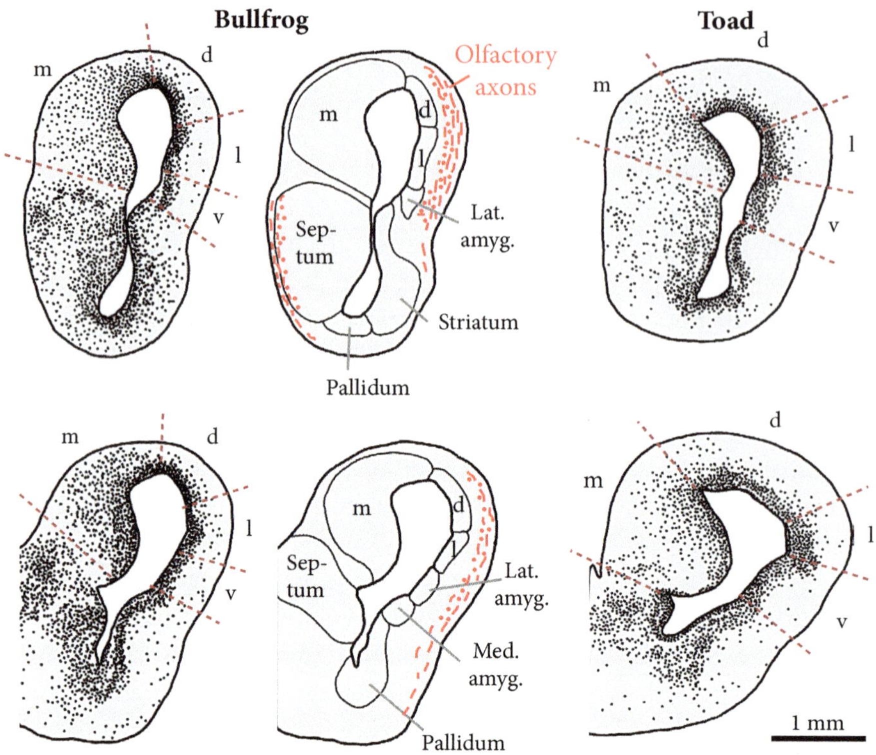

Figure 4.29 Pallial organization and olfactory projections in anurans. The left column presents two transverse sections through one telencephalic hemisphere of an American bullfrog (*Rana catesbeiana*), highlighting the putative division of the pallium into medial, dorsal, lateral, and ventral sectors (m, d, l, and v). The middle column shows the approximate distribution of olfactory bulb projections in this same species (red dashes and dots). The rightmost column depicts comparable sections through the telencephalon of a toad (*Bombina orientalis*). The basic anatomy is very similar, but the researchers working on this species drew the boundary between the medial and dorsal pallial divisions more medially than those working on the bullfrog, suggesting that the boundary is not self-evident.

Adapted from Neary and Northcutt (1983), Marin et al. (1998), Neary (1990), Roth et al. (2007).

further supported by data implicating the anuran medial pallium in hippocampus-typical functions, including spatial and navigational memory (Sotelo et al., 2016), behavioral adjustments to changes in reward magnitude (Papini et al., 1995), and stimulus-specific habituation (Figure 4.30). However, in anamniotes the medial pallium is the pallium's principal sensorimotor area, whereas in amniotes this status is assumed by the dorsal or ventral pallial sectors (see Chapters 5 and 6).

The lateral pallium of anuran amphibians is relatively thin, with most of the cell bodies lying close to the ventricle (see Figure 4.29) and extending their dendrites into the superficial neuropil, where they contact mainly axons coming from the olfactory bulbs. The lateral pallium also receives some inputs from the

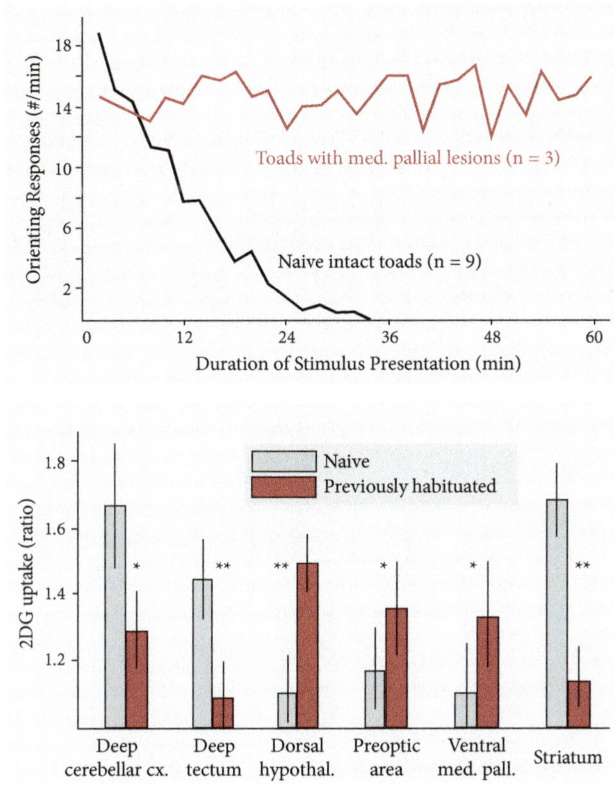

Figure 4.30 The medial pallium's role in stimulus-specific habituation. When continuously presented with a moving worm-like stimulus, intact toads gradually stop responding to the stimulus with turning movements. One day later, three of these toads received bilateral lesions in the medial pallium and, after one day of recovery, they were once again tested with the moving dummy prey. As the graph shows, the lesioned animals did not habituate; in contrast, two sham-operated toads exhibited normal habituation (data not shown). The histogram shows the uptake of 2-deoxyglucose (2DG) during 60 minutes of exposure to the moving dummy prey in toads that had previously habituated to the moving stimulus, and a naive control group that had not received this pre-treatment. Larger values indicate increased 2DG uptake, relative to a reference structure, and are thought to indicate increased neural activity. Out of 20 examined structures, only the 6 illustrated regions showed differences that were statistically significant at $p<0.01$ or $p<0.001$ (double asterisks). These data suggest that habituation to this stimulus requires an intact medial pallium and involves changes in the activity of several brain regions.
Adapted from Finkenstädt and Ewert (1988).

anterior thalamus (see Figure 4.24) and projects to the vomeronasal amygdala (to be discussed shortly) as well as the preoptic area. However, the vast majority of its connections are intrinsic to the lateral pallium or with other pallial areas (Roth et al., 2007). Since these intra-pallial connections presumably distribute olfactory information throughout the pallium, it seems fair to conclude that the amphibian pallium is, like that of basal cartilaginous and bony fishes, dominated by olfactory information (see Chapter 3). Evoked potential recordings in toads are consistent with this hypothesis (Laberge and Roth, 2007a). Although it is widely assumed that the amphibian lateral pallium is homologous to that of amniotes, Puelles et al. (2017) claim that the lateral pallium of amniotes does not receive direct inputs from the olfactory bulbs (which instead project mainly to the ventral pallium). If this is true (see Wullimann, 2017), then we would argue that the olfactory bulb projections must have become even more restricted as amniotes emerged, relative to the early tetrapod condition. In other words, we are proposing that inputs from the olfactory bulb are not an "essential feature" of the lateral pallium in all vertebrate lineages. Rather, they may have been lost in a subset of lineages without destroying the homologies of the pallial divisions being compared (see Chapter 1, Section 1.2.3).

Sandwiched between the medial and lateral pallial division lies the dorsal pallium. This pallial sector has been recognized by all students of the pallium in anurans and lungfishes, but its supposed boundaries have varied across researchers (e.g., Figure 4.29). Indeed, no generally accepted criteria for defining the dorsal pallium in anamniotes have been proposed. In amniotes, the dorsal pallium is often defined as receiving thalamic sensory inputs but lacking projections from the olfactory bulbs. The problem with applying this criterion to amphibians is that their putative dorsal pallium receives thalamic input (Figure 4.24) as well as olfactory bulb projections (Figure 4.29). Moreover, in contrast to the dorsal pallium of amniotes, the putative dorsal pallium of amphibians does not project outside of the telencephalon (Figure 4.31).

Given these problems, one might wonder whether amphibians possess a dorsal pallium at all. Since the early days of comparative neuroanatomy, researchers have generally assumed that amphibians and other anamniotes have a homolog of the amniote dorsal pallium. However, we already concluded in Chapter 3 that a dorsal pallium is difficult to find in basal ray-finned fishes and may have evolved independently in teleosts and (possibly) in cartilaginous fishes. Its presence in coelacanths is likewise debatable (see Figure 4.26). These considerations cause us to suggest that amphibians lack a homolog of the amniote dorsal pallium and that the structure traditionally identified as an amphibian dorsal pallium is actually a "transition area" between the medial and lateral pallial sectors (Kicliter and Ebbesson, 1976; Westhoff and Roth, 2002). Similarly, Bruce and Braford (2009) proposed that the so-called dorsal pallium of amphibians is homologous to the amniote lateral pallium and, thus, not homologous to the mammalian neocortex. Additional data, especially comparative molecular data, will be needed to test these hypotheses. Even if amphibians do possess a small dorsal pallium (Roth et al., 2007), the connections of this division must have changed dramatically as amniotes evolved. In particular, it must have lost its direct inputs from the olfactory bulbs, elaborated its thalamic sensory inputs, and gained some long descending projections. We come back to these issues in later chapters, especially Chapter 7.

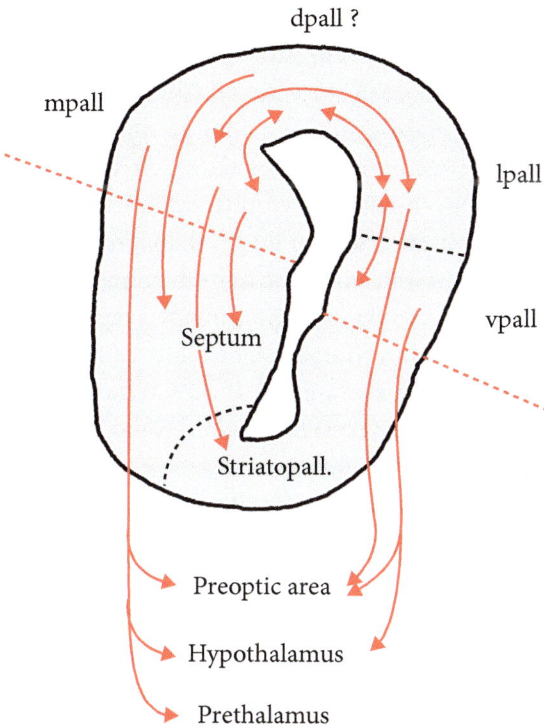

Figure 4.31 Outputs of the amphibian pallium. This schematic section through the right hemisphere of an anuran telencephalon shows the main intrinsic and efferent projections of the pallium. Among other things, it reveals that the projections to targets outside the telencephalon arise from the medial, lateral, and ventral divisions of the pallium (mpall, lpall, and vpall). We suspect that anurans do not have a well-defined homolog of the amniote dorsal pallium, but even if they do, this putative dorsal pallium (dpall?) would lack projections to the striatopallium and to extratelencephalic targets (both of which are prominent in amniotes). Projections from the pallium to the olfactory bulbs are not shown.
Based on data in Roth et al. (2007).

In contrast to the dorsal pallium, the ventral pallium is readily apparent in amphibians. It was first recognized as a distinct pallial division on the basis of gene expression, specifically the lack of *Emx1* expression at the pallium's ventrolateral edge (Smith Fernandez et al., 1998; Puelles et al., 2000). The original studies described the ventral pallium in mammals, birds, and frogs, but a ventral pallium has now been identified also in lungfishes, coelacanths, and, with less certainty, other fishes (González and Northcutt, 2009; Northcutt and González, 2011). An intriguing aspect of the ventral pallium in anurans is that some of its neurons migrate into the subpallium, while some subpallial neurons migrate into the ventral pallial territory (Moreno and González, 2007a, b). This finding suggests that the ventral pallium may be a distinct pallial division in embryos but becomes much less distinct during subsequent development. Be that as it may, one major component of the amphibian ventral pallium is the lateral amygdala. This structure is reciprocally interconnected with the olfactory bulbs, but, in contrast to the lateral pallium, it has descending projections to the hypothalamus (Moreno and González, 2006). Whether the lateral amygdala receives ascending thalamic inputs remains debated in the literature (Laberge and Roth, 2007b; Moreno and González, 2007b; Laberge et al., 2008).

In summary, the telencephalon of early tetrapods was probably quite similar to that of modern anurans, both in its gross anatomy and its internal wiring. Ascending inputs from the thalamus probably targeted mainly the septum and the striatum, and these two structures also provided most of the long projections coming out of the telencephalon. The pallium of early tetrapods probably contained three divisions: medial, lateral, and ventral. If early tetrapods had a dorsal pallium homologous to that of amniotes, then this region would likely have been small and poorly differentiated from the adjacent areas. Moreover, the early tetrapod pallium was probably dominated by olfactory inputs (either directly from the olfactory bulbs or indirectly via intra-pallial connections) and received only minor inputs from the thalamus. Any non-olfactory thalamic input that did reach the pallium of early tetrapods was almost certainly multimodal, rather than segregated into distinct sensory modalities (Wilczynski and Endepols, 2006). Most pallial divisions projected to the preoptic area or hypothalamus, but information from the dorsal pallium could leave the telencephalon only through the septum or through other pallial areas. In all of these respects, the telencephalon of early tetrapods was far more similar to that of basal ray-finned fishes and lungfishes than amniotes. Thus, we can conclude that the telencephalon of early tetrapods changed only in relatively minor ways as these animals invaded land. Furthermore, we can anticipate that major changes in telencephalic organization must have accompanied the evolution of amniotes.

4.6. Functional Organization of Early Tetrapod Brains

When discussing the brain one region at a time, as we have done in the preceding sections, it is difficult to grasp the brain's overall functional organization. Indeed, gaining such an understanding is an ambitious goal for any species (Swanson, 2005). Nonetheless, neuroanatomy without functional hypotheses is sterile (Edinger, 1908) and tedious. Therefore, it is important to note that, according to the comparative neuroanatomical data, the brain's functional organization most likely changed considerably during vertebrate evolution. Much of that change occurred within the amniotes, which we consider in the following chapters, but now is a good time to think about the primitive condition that amniotes took as their departure point. In particular, let us briefly examine the functional relationships between the telencephalon and the rest of the brain in early tetrapods.

One person who considered this topic carefully was William James, who in the second chapter of his *Principles of Psychology* (1890) summarized the then-available literature on the effects of large telencephalic lesions in diverse vertebrates. One of his main conclusions was that lesions of the cerebral hemispheres (i.e., the telencephalon) have far more severe and permanent effects in primates than in frogs and other "lower" species. In particular, James noted that frogs without a telencephalon act more machine-like than intact frogs and exhibit less spontaneous, much more predictable behavior. Based on these observations, James inferred that the central nervous system below the telencephalon in frogs is capable of responding appropriately to a wide variety of stimuli, and that the hemispheres are "superadded organs for breaking up the various reflexes performed by the lower centers, and combining their motor and sensory elements in novel ways" (p. 72). Further, James concluded that only animals with an intact telencephalon can be guided by memories, e.g., to search for food where it had previously been located. As James put it, "the difference between the hemisphereless animal and the whole one may be concisely expressed by saying that the one obeys absent, the other only present, objects" (p. 20).

Although they were penned more than 100 years ago, James's conclusions remain consistent with most of the evidence. In general, telencephalic lesions in anurans inhibit or disinhibit behaviors that are thought to be generated by lower brain regions (Ewert, 1970), and the telencephalon itself seems not to be involved in direct sensorimotor control. A potential challenge to this view is that the striatum in anurans does receive a variety of sensory inputs and has long descending projections (Veenman et al., 1989). However, such connections would also be needed to modulate the lower brain regions in context-appropriate ways. Indeed, a large literature now considers the striatum to have this kind of "action selection" function in mammals (Mink, 1996; Redgrave et al., 1999; Prescott et al., 2006; Striedter, 2015). According to this view, the striatum's output modulates the activity of other brain regions (through inhibition and disinhibition) but does not generate specific motor commands. Moreover, the inputs to the striatum can be modified by experience and, thus, provide learned information about which action is most appropriate in a given behavioral context. We suspect that this kind of modulatory action selection was the main function of the telencephalon in early tetrapods. Many forms of navigation can be subsumed under this general function, as moving toward or away from stimuli involves selection of approach or avoidance behaviors, respectively.

One implication of our proposal is that the septum's long projections to the preoptic area and hypothalamus are also involved in action selection, rather than direct sensorimotor control (Veenman et al., 1989). This prediction is difficult to evaluate, because the septum's functions remain poorly understood, especially in the anamniotes. However, diverse authors have observed that the septal complex resembles the striatopallidal complex in several respects, including the fact that its efferent projections originate from medium spiny GABAergic neurons (Swanson, 2005). Therefore, it is quite possible that the septum is an "extended striatum" and that these two brain regions perform analogous functions. Accordingly, we suggest that the septal complex in early tetrapods was probably involved in selecting context-appropriate behaviors related to the sleep-wake cycle, reproduction, social aggression, and appetite regulation. In contrast, the (traditionally defined) striatopallidal complex was probably much more involved in the selection of context-appropriate skeletal and eye movements.

A second implication of our hypothesis is that the functions of the telencephalon in early tetrapods were intimately tied to olfaction, much as in early gnathostomes (see Chapter 3). As noted in the previous chapter, olfactory memories are very useful for navigating in water. However, olfaction can also guide navigation on land, as evidenced by snakes following odor trails (Lemaster et al., 2001) and birds using odorants to find their way (Wallraff, 2014; Pollonara et al., 2015). We suspect that early tetrapods also depended heavily on odor cues to navigate, though they may have supplemented the olfactory information with cues from other sensory modalities (e.g., visual gradients; Jacobs and Schenk, 2003; Murray et al., 2017). In addition, early tetrapods probably made extensive use of olfactory stimuli to set the behavioral context for subsequent actions. For example, they may have used olfactory stimuli to clarify whether another individual is a potential mate or enemy, whether a place is familiar or strange, and whether food is delicious or putrid. These context-setting decisions would then have guided subsequent action selection. Since these functions would often have been critical for the organism's survival and reproduction, it is not surprising that the neuronal machinery for navigation and action selection came to be located within the telencephalon, where it has ready access to olfactory input. Despite this olfactory dominance, the striatum and medial pallium of early tetrapods probably received some non-olfactory inputs, which the animals most likely used to complement the role of olfaction.

Most importantly, our hypothesis implies that the pallium of early tetrapods was not involved in the kind of fine-grained sensory analysis and direct sensorimotor control that is the hallmark of mammalian neocortex or, for that matter, avian pallium (see Chapters 5 and 6). Moreover, we question the widely held assumption that anamniotic vertebrates possess a homolog of mammalian neocortex; that is, we doubt that they have a homolog of the amniote dorsal pallium. We first questioned this assumption in Chapter 3 by pointing out that early gnathostomes probably lacked a dorsal pallium. We went further in the present chapter by proposing that even early tetrapods may not have had a dorsal pallium. In effect, we are hypothesizing that the dorsal pallium is an innovation of amniotes, even though a superficially similar dorsal pallium evolved independently in ray-finned fishes and, we suspect, in cartilaginous fishes. As we discuss in Chapter 6, mammals transformed this dorsal pallium into a six-layered cortex, which we refer to as neocortex. In contrast, the dorsal pallium of reptiles and birds assumed a much less laminar disposition (see Chapter 5).

Compared to the pallium, the subpallium and its associated basal ganglia circuits are relatively similar across all vertebrates. Most striking is that many features of basal ganglia organization in mammals have now been described in lampreys (see Grillner and Robertson, 2016). This surprising degree of similarity has prompted some to argue that "the organisation of the basal ganglia is almost identical throughout vertebrate phylogeny—from lamprey to primates" (Grillner and Robertson, 2016, p. R1091). However, data from intervening lineages suggest that this statement is exaggerated. For example, the location of dopaminergic neurons varies substantially across vertebrate lineages (Rink and Wulliman, 2001; Yamamoto and Vernier, 2011), and the neurons thought to be homologous to the mammalian ventral tegmental area and substantia nigra in amphibians are intermingled rather than spatially segregated (Marin et al., 1997b). Similarly, amphibians possess neurons that are quite similar to those of the mammalian subthalamic nucleus, but these neurons in amphibians are scattered across multiple cell groups, rather than forming a discrete nucleus (Maier et al., 2010). In addition, only birds and mammals seem to possess a "re-entrant" pathway from the basal ganglia back to the pallium via the thalamus (Wulliman, 2014). Finally, comparative genomic data have shown that the molecules involved in dopamine signaling and synthesis vary significantly across the major vertebrate lineages (Candy and Collet, 2005; Yamamoto et al., 2010, 2015). Therefore, it is probably best to conclude, for now, that the basal ganglia circuits are highly conserved but do exhibit some significant variation.

References

Affaticati P, Yamamoto K, Rizzi B, Bureau C, Peyriéras N, Pasqualini C, Demarque M, Vernier P. 2015. Identification of the optic recess region as a morphogenetic entity in the zebrafish forebrain. *Sci Rep* 5:211–13.

Alvarez-Bolado G, Grinevich V, Puelles L. 2015. Editorial: Development of the hypothalamus. *Front Neuroanat* 9:83.

Amemiya CT, Alföldi J, Lee AP, Fan S, Philippe H, Maccallum I, Braasch I, Manousaki T, Schneider I, Rohner N, et al. 2013. The African coelacanth genome provides insights into tetrapod evolution. *Nature* 496:311–316.

Anderson CW. 2001. Anatomical evidence for brainstem circuits mediating feeding motor programs in the leopard frog, *Rana pipiens*. *Exp Brain Res* 140:12–19.

Antony J, Millot J. 1965. Anatomie de *Latimeria Chalumnae*, Tome 2: *Systeme nerveux & organes des sens*. Centre National de la Recherche Scientifique (France).

Arch VS, Simmons DD, Quiñones PM, Feng AS, Jiang J, Stuart BL, Shen J-X, Blair C, Narins PM. 2012. Inner ear morphological correlates of ultrasonic hearing in frogs. *Hear Res* 283:70–79.

Bailes HJ, Robinson SR, Trezise AEO, Collin SP. 2006. Morphology, characterization, and distribution of retinal photoreceptors in the Australian lungfish *Neoceratodus forsteri* (Krefft, 1870). *J Comp Neurol* 494:381–397.

Behrensmeyer AK, Damuth JD, DiMichele WA, Potts R, Sues HD, Wing SL 1992. *Terrestrial Ecosystem Through Time*. Chicago: University of Chicago Press.

Bemis WE. 1984. Paedomorphosis and the evolution of the Dipnoi. *Paleobiology* 10:293–307.

Bicanski A, Ryczko D, Cabelguen J-M, Ijspeert AJ. 2013. From lamprey to salamander: an exploratory modeling study on the architecture of the spinal locomotor networks in the salamander. *Biol Cybern* 107:565–587.

Bowmaker JK. 2008. Evolution of vertebrate visual pigments. *Vision Res* 48:2022–2041.

Brox A, Puelles L, Ferreiro B, Medina L. 2003. Expression of the genes *GAD67* and *Distal-less-4* in the forebrain of *Xenopus laevis* confirms a common pattern in tetrapods. *J Comp Neurol* 461:370–393.

Brox A, Puelles L, Ferreiro B, Medina L. 2004. Expression of the genes *Emx1, Tbr1*, and *Eomes* (*Tbr2*) in the telencephalon of *Xenopus laevis* confirms the existence of a ventral pallial division in all tetrapods. *J Comp Neurol* 474:562–577.

Bruce LL, Braford MR Jr. 2009. Evolution of the limbic system. In: Squire LR (ed.), *Encyclopedia of Neuroscience*, San Diego: Academic Press, pp. 43–55.

Cajal SRY, Swanson N, Swanson LW. 1995. *Histology of the Nervous System of Man and Vertebrates*. New York: Oxford University Press.

Candy J, Collet C. 2005. Two tyrosine hydroxylase genes in teleosts. *Biochimica et Biophysica Acta* 1727:35–44.

Capranica RR. 1976. Morphology and physiology of the auditory system. In: Llinás R, Precht W (eds), *Frog Neurobiology*. Berlin; Heidelberg: Springer, pp. 551–575.

Carroll RL. 2009. *The Rise of Amphibians: 365 Million Years of Evolution*. Baltimore, MD: Johns Hopkins University Press.

Challands TJ. 2015. The cranial endocast of the Middle Devonian dipnoan *Dipterus valenciennesi* and a fossilized dipnoan otoconial mass. *Pap Palaeontol* 1:289–317.

Chang S, Chung-Davidson Y-W, Libants SV, Nanlohy KG, Kiupel M, Brown CT, Li W. 2013. The sea lamprey has a primordial accessory olfactory system. *BMC Evol Biol* 13:172.

Chevallier S, Jan Ijspeert A, Ryczko D, Nagy F, Cabelguen J-M. 2008. Organisation of the spinal central pattern generators for locomotion in the salamander: biology and modelling. *Brain Res Rev* 57:147–161.

Chipman AD, Khaner O, Haas A, Tchernov E. 2001. The evolution of genome size: what can be learned from anuran development? *J Exp Zool* 291:365–374.

Christensen CB, Christensen-Dalsgaard J, Madsen PT. 2015. Hearing of the African lungfish (*Protopterus annectens*) suggests underwater pressure detection and rudimentary aerial hearing in early tetrapods. *J Exp Biol* 218:381–387.

Clack JA. 2002. Patterns and processes in the early evolution of the tetrapod ear. J *Neurobiol* 53:251–264.

Clack JA. 2006. The emergence of early tetrapods. *Palaeogeo Palaeoclim Palaeoecol* 232:167–189.

Clack JA. 2007. Devonian climate change, breathing, and the origin of the tetrapod stem group. *Integr Comp Biol* 47:510–523.

Clack JA. 2012. *Gaining Ground*. Bloomington: Indiana University Press.

Clack JA. 2016. Vertebrate diversity in a sensory system: the fossil record of otic evolution. In: Clack JA, Fay RR, Popper AN (eds.), *Evolution of the Vertebrate Ear*, Vol. 59, 2nd ed., *Springer Handbook of Auditory Research*. Basel: Springer, pp. 1–16.

Clack JA, Allin E. 2004. The evolution of single- and multiple-ossicle ears in fishes and tetrapods. In: Manley GA, Popper AN, Fay RR (eds), *Evolution of the Vertebrate Auditory System*. New York: Springer, pp. 128–163.

Clement AM, Ahlberg PE. 2014. The first virtual cranial endocast of a lungfish (Sarcopterygii: Dipnoi). *PLoS ONE* 9:e113898.

Cobos I, Puelles L, Martínez S. 2001. The avian telencephalic subpallium originates inhibitory neurons that invade tangentially the pallium (dorsal ventricular ridge and cortical areas). *Dev Biol* 239:30–45.

Combes D, Merrywest SD, Simmers J, Sillar KT. 2004. Developmental segregation of spinal networks driving axial- and hindlimb-based locomotion in metamorphosing *Xenopus laevis*. *J Physiol* 559:17–24.

Cupello C, Brito PM, Herbin M, Meunier FJ, Janvier P, Dutel H, Clément G. 2015. Allometric growth in the extant coelacanth lung during ontogenetic development. *Nat Commun* 6:8222.

Deban SM, O'Reilly JC, Dicke U, van Leeuwen JL. 2007. Extremely high-power tongue projection in plethodontid salamanders. *J Exp Biol* 210:655–667.

Dicke U, Roth G. 2007. Evolution of the amphibian nervous system. In: Kaas JH, Bullock TH (eds.), *Evolution of Nervous Systems*, Vol. 2. San Diego: Academic Press, pp. 61–124.

Domínguez L, González A, Moreno N. 2015. Patterns of hypothalamic regionalization in amphibians and reptiles: common traits revealed by a genoarchitectonic approach. *Front Neuroanat* 9:3.

Duellman WE, Trueb L. 1986. *Biology of Amphibians*. New York: McGraw-Hill.

Ebbesson SOE. 1976. Morphology of the spinal cord. In: Llinas R, Precht W (eds.), *Frog Neurobiology*. Berlin; Heidelberg: Springer, pp. 679–706.

Edinger L. 1908. The relations of comparative anatomy to comparative psychology. *J Comp Neurol* 18:437–457

Eisthen HL, Polese G. 2007. Evolution of vertebrate olfactory subsystems. In: Kaas JH (ed.), *Evolution of Nervous Systems*, Vol. 2. San Diego: Academic Press, pp. 355–406.

Elbert OE, Ide CF, Duda M, Hunt RK. 1983. *Spastic* mutant axolotl: identification of a phenocopy pathway with implications for the control of axolotl swimming by the vestibulocerebellum. *J Comp Neurol* 220:97–105.

Elliott TA, Gregory TR. 2015. What's in a genome? The C-value enigma and the evolution of eukaryotic genome content. *Phil Trans Roy Soc B* 370:20140331.

Emerson SB, Boyd SK. 1999. Mating vocalizations of female frogs: control and evolutionary mechanisms. *Brain Behav Evol* 53:187–197.

Endepols H, Roden K, Walkowiak W. 2005. Hodological characterization of the septum in anuran amphibians: II. Efferent connections. *J Comp Neurol* 483:437–457.

Ewert JP. 1968. Der Einfluß von Zwischenhirndefekten auf die Visuomotorik im Beute- und Fluchtverhalten der Erdkröte (*Bufo bufo* L.). *J Comp Physiol A* 61:41–70.

Ewert JP. 1970. Neural mechanisms of prey-catching and avoidance behavior in the toad (*Bufo bufo* L.). *Brain Behav Evol* 3:36–56.

Ewert JP, Buxbaum-Conradi H, Glagow M, Rottgen A, Schurg-Pfeiffer E, Schwippert WW. 1999. Forebrain and midbrain structures involved in prey-catching behaviour of toads: stimulus-response mediating circuits and their modulating loops. *Eur J Morphol* 37:172–176.

Feng AS, Narins PM, Capranica RR. 1975. Three populations of primary auditory fibers in the bullfrog (*Rana catesbeiana*): their peripheral origins and frequency sensitivities. *J Comp Physiol A* 100:221–229.

Feng Y-J, Blackburn DC, Liang D, Hillis DM, Wake DB, Cannatella DC, Zhang P. 2017. Phylogenomics reveals rapid, simultaneous diversification of three major clades of Gondwanan frogs at the Cretaceous–Paleogene boundary. *Proc Natl Acad Sci USA* 114:E5864–E5870.

Fettiplace R, Fuchs PA. 1999. Mechanisms of hair cell tuning. *Annu Rev Physiol* 61:809–834.

Finkenstädt T, Ewert JP. 1988. Stimulus-specific long-term habituation of visually guided orienting behavior toward prey in toads: a ^{14}C-2DG study. *J Comp Physiol A* 163:1–11.

Forey PL. 1988. Golden jubilee for the coelacanth *Latimeria chalumnae*. *Nature* 336:727–732.

Freitag J, Krieger J, Strotmann J, Breer H. 1995. Two classes of olfactory receptors in *Xenopus laevis*. *Neuron* 15:1383–1392.

Fricke H, Reinicke O, Hofer H, Nachtigall W. 1987. Locomotion of the coelacanth *Latimeria chalumnae* in its natural environment. *Nature* 329:331–333.

Frishkopf LS, Goldstein MH Jr. 1963. Responses to acoustic stimuli from single units in the eighth nerve of the bullfrog. *J Acoust Soc Am* 35:1219–1228.

Fritzsch B. 1987. Inner ear of the coelacanth fish *Latimeria* has tetrapod affinities. *Nature* 327:153–154.

Fritzsch B. 1988. Phylogenetic and ontogenetic origin of the dorsolateral auditory nucleus of anurans. In: Fritzsch B, Ryan MJ, Walter W, Hetherington TE, Walkowiak W (eds.), *The Evolution of the Amphibian Auditory System*. New York: Wiley, pp. 561–585.

Fritzsch B, Neary T. 1998. The octavolateralis system of mechanosensory and electrosensory organs. In: Heatwole Y (ed.), *Amphibian Biology*, Vol 3. Chipping Norton, Australia: Surrey Beatty and Sons, pp. 878–922.

Fritzsch B, Nikundiwe AM, Will U. 1984. Projection patterns of lateral-line afferents in anurans: a comparative HRP study. *J Comp Neurol* 229:451–469.

Fritzsch B, Pan N, Jahan I, Duncan JS, Kopecky BJ, Elliott KL, Kersigo J, Yang T. 2013. Evolution and development of the tetrapod auditory system: an organ of Corti-centric perspective. *Evol Dev* 15:63–79.

Fritzsch B, Wahnschaffe U. 1983. The electroreceptive ampullary organs of urodeles. *Cell Tissue Res* 229:483–503.

Fritzsch B, Wake MH. 1986. The distribution of ampullary organs in Gymnophiona. J *Herpetol* 20:90–93.

Gensel PG, Edwards D. 2001. *Plants Invade the Land*. New York: Columbia University Press.

Gerhardt HC. 1994. The evolution of vocalization in frogs and toads. *Ann Rev Ecol Syst* 25:293–324.

Gess R, Ahlberg PE. 2018. A tetrapod fauna from within the Devonian Antarctic Circle. *Science* 360:1120–1124.

Gomez-Mestre I, Pyron RA, Wiens JJ. 2012. Phylogenetic analyses reveal unexpected patterns in the evolution of reproductive modes in frogs. *Evolution* 66:3687–3700.

González A, López JM, Morona R, Moreno N. 2017. The organization of the central nervous system of amphibians. In: Kaas JH, Striedter GF (eds.), *Evolution of Nervous Systems*, Vol 1, 2nd ed. Amsterdam: Elsevier, pp. 141–170.

González A, Morona R, López JM, Moreno N, Northcutt RG. 2010. Lungfishes, like tetrapods, possess a vomeronasal system. *Front Neuroanat* 4:130.

González A, Northcutt RG. 2009. An immunohistochemical approach to lungfish telencephalic organization. Brain Behav Evol 74:43–55.

González A, Smeets WJ. 1994. Distribution of tyrosine hydroxylase immunoreactivity in the brain of *Typhlonectes compressicauda* (Amphibia, Gymnophiona): further assessment of primitive and derived traits of amphibian catecholamine systems. *J Chem Neuroanat* 8:19–32.

Gould SJ. 1977. *Ontogeny and Phylogeny*. Cambridge, MA: Harvard University Press.

Graham JB, Wegner NC, Miller LA, Jew CJ, Lai NC, Berquist RM, Frank LR, Long JA. 2014. Spiracular air breathing in polypterid fishes and its implications for aerial respiration in stem tetrapods. *Nat Commun* 5:3022.

Gregory TR. 2001. Coincidence, coevolution, or causation? DNA content, cell size, and the C-value enigma. *Biol Rev* 76:65–101.

Gregory TR. 2002. Genome size and developmental complexity. *Genetica* 115:131–146.

Grillner S, Robertson B. 2016. The basal ganglia over 500 million years. *Curr Biol* 26:R1088–R1100.

Gruberg E, Dudkin E, Wang Y, Marín G, Salas C, Sentis E, Letelier J, Mpodozis J, Malpeli J, Cui H, Ma R, Northmore D, Udin S. 2006. Influencing and interpreting visual input: the role of a visual feedback system. *J Neurosci* 26:10368–10371.

Gruberg ER, Wallace MT, Caine HS, Mote MI. 1991. Behavioral and physiological consequences of unilateral ablation of the nucleus isthmi in the leopard frog. *Brain Behav Evol* 37:92–103.

Grus WE, Zhang J. 2009. Origin of the genetic components of the vomeronasal system in the common ancestor of all extant vertebrates. *Mol Biol Evol* 26:407–419.

Halpern M. 2007. The evolution of the vomeronasal system. In: Kaas J, Bullock TH (eds.), *Evolution of Nervous Systems*, Vol. 2. San Diego: Academic Press, pp. 407–415.

Hanken J, Carl TF, Richardson MK, Olsson L, Schlosser G, Osabutey CK, Klymkowsky MW. 2001. Limb development in a "nonmodel" vertebrate, the direct-developing frog *Eleutherodactylus coqui*. *J Exp Zool* 291:375–388.

Hansen A, Reiss JO, Gentry CL, Burd GD. 1998. Ultrastructure of the olfactory organ in the clawed frog, *Xenopus laevis*, during larval development and metamorphosis. J *Comp Neurol* 398:273–288.

Hart NS, Bailes HJ, Vorobyev M, Marshall NJ, Collin SP. 2008. Visual ecology of the Australian lungfish (*Neoceratodus forsteri*). *BMC Ecol* 8:(21)1–14.

Herculano-Houzel S. 2011. Brains matter, bodies maybe not: the case for examining neuron numbers irrespective of body size. *Ann N Y Acad Sci* 1225:191–199.

Herrick CJ. 1948. *The Brain of the Tiger Salamander*, Ambystoma tigrinum. Chicago: University of Chicago Press.

Hetherington TE, Wake MH. 1979. The lateral line system in larval *Ichthyophis* (Amphibia: Gymnophiona). *Zoomorphol* 93:209–225.

Holmes PH, Northcutt RG. 2003. Connections of the pallial telencephalon in the Senegal bichir, *Polypterus*. *Brain Behav Evol* 61:113–147.

Horner AM, Jayne BC. 2014. Lungfish axial muscle function and the vertebrate water to land transition. *PLoS ONE* 9:e96516.

Hudspeth AJ. 2014. Integrating the active process of hair cells with cochlear function. *Nat Rev Neurosci* 15:600–614.

Ide CF, Tompkins R, Miszkowski N. 1977. Neuroanatomy of *spastic*, a behavior mutant of the Mexican axolotl: Purkinje-cell distribution in adult cerebellum. *J Comp Neurol* 176:373–386.

Ijspeert AJ, Crespi A, Ryczko D, Cabelguen J-M. 2007. From swimming to walking with a salamander robot driven by a spinal cord model. *Science* 315:1416–1420.

Ingle D. 1970. Visuomotor functions of the frog optic tectum. *Brain Behav Evol* 3:57–71.

Irisarri I, Meyer A. 2016. The identification of the closest living relative(s) of tetrapods: phylogenomic lessons for resolving short ancient internodes. *Syst Biol* 65:1057–1075.

Iwasaki S-I. 2002. Evolution of the structure and function of the vertebrate tongue. *J Anat* 201:1–13.

Jacobs LF, Schenk F. 2003. Unpacking the cognitive map: the parallel map theory of hippocampal function. *Psych Rev* 110:285–315.

Jacoby J, Rubinson K. 1983. The acoustic and lateral line nuclei are distinct in the premetamorphic frog, *Rana catesbeiana*. *J Comp Neurol* 216:152–161.

James W. 1890. *The Principles of Psychology*. New York: Holt.

Janvier P. 2004. Wandering nostrils. *Nature* 432:23–24.

Jeram AJ, Selden PA, Edwards D. 1990. Land animals in the silurian: arachnids and myriapods from Shropshire, England. *Science* 250:658–661.

Jørgensen JM, Joss J (eds.). 2011. *The Biology of Lungfishes*. Boca Raton, FL: CRC Press.

Jung H, Mazzoni EO, Soshnikova N, Hanley O, Venkatesh B, Duboule D, Dasen JS. 2014. Evolving *hox* activity profiles govern diversity in locomotor systems. *Dev Cell* 29:171–187.

Kicliter E, Ebbesson SOE. 1976. Organization of the "nonolfactory" telencephalon. In: Llinas RR, Precht W (eds.), *Frog Neurobiology*. Berlin; Heidelberg: Springer, pp. 946–972.

Kimmel CB, Sidlauskas B, Clack JA. 2009. Linked morphological changes during palate evolution in early tetrapods. *J Anat* 215:91–109.

Kishida T, Thewissen J, Hayakawa T, Imai H, Agata K. 2015. Aquatic adaptation and the evolution of smell and taste in whales. *Zool Lett* 1:9.

Klembara J. 1994. Electroreceptors in the Lower Permian tetrapod *Discosauriscus austriacus*. *Palaeontol* 37:609–626.

Laberge F, Mühlenbrock-Lenter S, Dicke U, Roth G. 2008. Thalamo-telencephalic pathways in the fire-bellied toad *Bombina orientalis*. *J Comp Neurol* 508:806–823.

Laberge F, Roth G. 2007a. Organization of the sensory input to the telencephalon in the fire-bellied toad, *Bombina orientalis*. *J Comp Neurol* 502:55–74.

Laberge F, Roth G. 2007b. Is there a structure equivalent to the mammalian basolateral amygdaloid complex in amphibians? *J Anat* 211:830–831.

Lambertz M, Perry SF. 2015. Phylogeny, anatomy and physiology of ancient fishes. In: Zaccone G, Dabrowski K, Hedrick MS, Fernandes JMO, Icardo JM (eds.), *Phylogeny, Anatomy and Physiology of Ancient Fishes*. Boca Raton: CRC Press, pp. 201–211.

Lanuza E, Martinez-Garcia F. 2009. Evolution of septal nuclei. In: Binder MD, Hirokawa N, Windhorst U (eds.), *Encyclopedia of Neuroscience*. New York: Springer, pp. 1270–1278.

Larsell O. 1934. The differentiation of the peripheral and central acoustic apparatus in the frog. *J Comp Neurol* 60:473–527.

Laurin M, Canoville A, Struble M, Organ C. 2016. Early genome size increase in urodeles. *C R Palevol* 15:74–82.

Lee MSY, Anderson JS. 2006. Molecular clocks and the origin(s) of modern amphibians. *Mol Phylogenet Evol* 40:635–639.

Lemaster MP, Moore IT, Mason RT. 2001. Conspecific trailing behaviour of red-sided garter snakes, *Thamnophis sirtalis parietalis*, in the natural environment. *Anim Behav* 61:827–833.

Lewis ER, Narins PM. 1999. The acoustic periphery of amphibians: anatomy and physiology. In: Fay RR, Popper AN (eds.), *Evolution of the Vertebrate Auditory System*. New York: Springer, pp. 101–154.

Lombard RE, Bolt JR. 1979. Evolution of the tetrapod ear: an analysis and reinterpretation. *Biol J Linn Soc* 11:19–76.

Long JA, Gordon MS. 2004. The greatest step in vertebrate history: a paleobiological review of the fish-tetrapod transition. *Physiol Biochem Zool* 77:700–719.

Longo S, Riccio M, McCune AR. 2013. Homology of lungs and gas bladders: insights from arterial vasculature. *J Morphol* 274:687–703.

López JM, Moreno N, Morona R, Muñoz M, Domínguez L, González A. 2007. Distribution of somatostatin-like immunoreactivity in the brain of the caecilian *Dermophis mexicanus* (Amphibia: Gymnophiona): comparative aspects in amphibians. *J Comp Neurol* 501:413–430.

López JM, Morona R, Moreno N, González A. 2017. The organization of the central nervous system of lungfishes: an immunohistochemical approach. In: Kaas JH, Striedter GF (eds.), *Evolution of Nervous Systems*, Vol. 1, 2nd ed. Amsterdam: Elsevier, pp. 121–139.

Lu J, Zhu M, Long JA, Zhao W, Senden TJ, Jia L, Qiao T. 2012. The earliest known stem-tetrapod from the Lower Devonian of China. *Nat Commun* 3:1160.

Ma L-H, Gilland E, Bass AH, Baker R. 2010. Ancestry of motor innervation to pectoral fin and forelimb. *Nat Commun* 1:49.

MacIver MA, Schmitz L, Mugan U, Murphey TD, Mobley CD. 2017. Massive increase in visual range preceded the origin of terrestrial vertebrates. *Proc Natl Acad Sci USA* 114:E2375–E2384.

Maddin HC, Anderson JS. 2012. Evolution of the amphibian ear with implications for lissamphibian phylogeny: insight gained from the caecilian inner ear. *Fieldiana Life Earth Sci* 5:59–76.

Maeda N, Miyoshi S, Toh H. 1983. First observation of a muscle spindle in fish. *Nature* 302:61–62.

Magid AM. 1966. Breathing and function of the spiracles in *Polypterus senegalus*. *Anim Behav* 14:530–533.

Maier S, Walkowiak W, Luksch H, Endepols H. 2010. An indirect basal ganglia pathway in anuran amphibians? *J Chem Neuroanat* 40:21–35.

Manley GA, Popper AN, Fay RR. 2004. *Evolution of the Vertebrate Auditory System*. New York: Springer.

Marin O, González A, Smeets W. 1997a. Basal ganglia organization in amphibians: efferent connections of the striatum and the nucleus accumbens. *J Comp Neurol* 380:23–50.

Marin O, Smeets WJ, González A. 1997b. Basal ganglia organization in amphibians: catecholaminergic innervation of the striatum and the nucleus accumbens. *J Comp Neurol* 378:50–69.

Marin O, Smeets WJ, González A. 1998. Basal ganglia organization in amphibians: evidence for a common pattern in tetrapods. *Prog Neurobiol* 55:363–397.

Marjanović D, Laurin M. 2013. The origin(s) of extant amphibians: a review with emphasis on the "lepospondyl hypothesis." *Geodiversitas* 35:207–272.

Marshall J, Collin S, Hart N, Bailes H. 2016. Vision in lungfish. In: Jørgensen JM, Joss J (eds.), *The Biology of Lungfishes*. Boca Raton, FL: CRC Press, pp. 447–476.

Martinez-de-la-Torre M, Pombal MA, Puelles L. 2011. Distal-less-like protein distribution in the larval lamprey forebrain. *Neuroscience* 178:270–284.

McCormick CA, Braford MR Jr. 1988. Central connections of the octavolateralis system: evolutionary considerations. In: Atema J, Fay RR, Popper AN, Tavolga WN (eds.), *Sensory Biology of Aquatic Animals*. New York: Springer, pp. 733–756.

McDiarmid RW, Altig R (eds.). 1999. *Tadpoles: The Biology of Anuran Larvae*. Chicago: University of Chicago Press.

Metcalfe CJ, Filée J, Germon I, Joss J, Casane D. 2012. Evolution of the Australian lungfish (*Neoceratodus forsteri*) genome: a major role for CR1 and L2 LINE elements. *Mol Biol Evol* 29:3529–3539.

Mink JW. 1996. The basal ganglia: focused selection and inhibition of competing motor programs. *Prog Neurobiol* 50:381–425.

Mollo E, Fontana A, Roussis V, Polese G, Amodeo P, Ghiselin MT. 2014. Sensing marine biomolecules: smell, taste, and the evolutionary transition from aquatic to terrestrial life. *Front Chem* 2:92.

Moreno N, González A. 2003. Hodological characterization of the medial amygdala in anuran amphibians. *J Comp Neurol* 466:389–408.

Moreno N, González A. 2006. The common organization of the amygdaloid complex in tetrapods: new concepts based on developmental, hodological and neurochemical data in anuran amphibians. *Prog Neurobiol* 78:61–90.

Moreno N, González A. 2007a. Development of the vomeronasal amygdala in anuran amphibians: hodological, neurochemical, and gene expression characterization. *J Comp Neurol* 503:815–831.

Moreno N, González A. 2007b. Evolution of the amygdaloid complex in vertebrates, with special reference to the anamnio-amniotic transition. *J Anat* 211:151–163.

Moreno N, González A, Rétaux S. 2008. Evidences for tangential migrations in *Xenopus telencephalon*: developmental patterns and cell tracking experiments. *Dev Neurobiol* 68:504–520.

Moreno N, López JM, Morona R, Lozano D, Jiménez S, González A. 2018. Comparative analysis of Nkx2.1 and Islet-1 expression in urodele amphibians and lungfishes highlights the pattern of forebrain organization in early tetrapods. *Front Neuroanat* 12:42.

Moreno N, Morona R, López JM, González A. 2016. The diencephalon and hypothalamus of nonmammalian vertebrates: evolutionary and developmental traits. In: Kaas JH, Striedter GF (eds.), *Evolution of Nervous Systems*, Vol. 1, 2nd ed. Amsterdam: Elsevier, pp. 409–426.

Mueller T, Guo S. 2009. The distribution of *GAD67*-mRNA in the adult zebrafish (teleost) forebrain reveals a prosomeric pattern and suggests previously unidentified homologies to tetrapods. *J Comp Neurol* 516:553–568.

Münz H, Claas B, Fritzsch B. 1984. Electroreceptive and mechanoreceptive units in the lateral line of the axolotl *Ambystoma mexicanum*. *J Comp Physiol A* 154:33–44.

Murakami Y, Tanaka M. 2011. Evolution of motor innervation to vertebrate fins and limbs. *Dev Biol* 355:164–172.

Murray EA, Wise SP, Graham KS. 2017. Representational specializations of the hippocampus in phylogenetic perspective. *Neurosci Lett* 680:4–12.

Meyers JJ, Nishikawa KC. 2000. Comparative study of tongue protrusion in three iguanian lizards, Sceloporus undulatus, Pseudotrapelus sinaitus and Chamaeleo jacksonii. *J Exp Biol* 203:2833–2849.

Nakane Y, Ikegami K, Iigo M, Ono H, Takeda K, Takahashi D, Uesaka M, Kimijima M, Hashimoto R, Arai N, et al. 2013. The saccus vasculosus of fish is a sensor of seasonal changes in day length. *Nat Commun* 4:2108.

Neary TJ. 1990. The pallium of anuran amphibians. In: Jones EG, Peters A (eds.), *Cerebral Cortex*, Part I, Vol. 8A. Boston, MA: Springer, pp. 107–138.

Neary TJ, Northcutt RG. 1983. Nuclear organization of the bullfrog diencephalon. *J Comp Neurol* 213:262–278.

Nieuwenhuys R, Donkelaar ten HJ, Nicholson C. 1998. *The Central Nervous System of Vertebrates*. Berlin: Springer.

Niimura Y. 2009. On the origin and evolution of vertebrate olfactory receptor genes: comparative genome analysis among 23 chordate species. *Genome Biol Evol* 1:34–44.

Niimura Y. 2012. Olfactory receptor multigene family in vertebrates: from the viewpoint of evolutionary genomics. *Curr Genomics* 13:103–114.

Nikaido M, Noguchi H, Nishihara H, Toyoda A, Suzuki Y, Kajitani R, Suzuki H, Okuno M, Aibara M, Ngatunga BP, et al. 2013. Coelacanth genomes reveal signatures for evolutionary transition from water to land. *Genome Res* 23:1740–1748.

Northcutt RG. 1986. Lungfish neural characters and their bearing on sarcopterygian phylogeny. *J Morphol* 190:277–297.

Northcutt RG. 1989. The phylogenetic distribution and innervation of craniate mechanoreceptive lateral lines. In: Coombs S, Görner P, Münz H (eds.), *The Mechanosensory Lateral Line: Neurobiology and Evolution*. New York: Springer, pp 17–78.

Northcutt RG. 1995. The forebrain of gnathostomes: in search of a morphotype. *Brain Behav Evol* 46:275–318.

Northcutt RG. 2011. The central nervous system of lungfishes. In: Jørgensen JM, Joss J (eds.), *The Biology of Lungfishes*. Boca Raton, FL: CRC Press, pp. 393–446.

Northcutt RG, Bemis WE. 1993. Cranial nerves of the coelacanth, *Latimeria chalumnae* [Oste ichthyes: Sarcopterygii: Actinistia], and comparisons with other craniata. *Brain Behav Evol* 42(Suppl 1):1–76.

Northcutt RG, Catania KC, Criley BB. 1994. Development of lateral line organs in the axolotl. *J Comp Neurol* 340:480–514.

Northcutt RG, González A. 2011. A reinterpretation of the cytoarchitectonics of the telencephalon of the Comoran coelacanth. *Front Neuroanat* 5:9.

Northcutt RG, Kicliter EE. 1980. Organization of the amphibian telencephalon. In: Ebbesson SOE (ed.), *Comparative Neurology of the Telencephalon*. New York: Plenum, pp. 203–255.

Northcutt RG, Neary TJ, Senn DG. 1978. Observations on the brain of the coelacanth *Latimeria chalumnae*: External anatomy and quantitative analysis. *J Morphol* 155:181–192.

Northcutt RG, Ronan M. 1992. Afferent and efferent connections of the bullfrog medial pallium. *Brain Behav Evol* 40:1–16.

Northcutt RG, Westhoff G. 2011. Connections of the medial telencephalic wall in the spotted African lungfish. *Brain Behav Evol* 77:14–32.

Ohno S. 1972. So much "junk" DNA in our genome. In: Smith HH (ed.), *Evolution of Genetic Systems*. New York: Gordon and Breach, pp. 366–370.

Ottoson D. 1976. Morphology and physiology of muscle spindles. In: Llinas RR, Precht W (eds.), *Frog Neurobiology*. Berlin; Heidelberg: Springer, pp. 643–675.

Papini MR, Muzio RN, Segura ET. 1995. Instrumental learning in toads (*Bufo arenarum*): reinforcer magnitude and the medial pallium. *Brain Behav Evol* 46:61–71.

Payne RS, McVay S. 1971. Songs of humpback whales. *Science* 173:585–597.

Pettigrew JD. 1999. Electroreception in monotremes. *J Exp Biol* 202:1447–1454.

Picone B, Hesse U, Panji S, Van Heusden P, Jonas M, Christoffels A. 2013. Taste and odorant receptors of the coelacanth: a gene repertoire in transition. *J Exp Zool B* 322:403–414.

Pierce SE, Hutchinson JR, Clack JA. 2013. Historical perspectives on the evolution of tetrapodomorph movement. *Integ Comp Biol* 53:209–223.

Pollonara E, Luschi P, Guilford T, Wikelski M, Bonadonna F, Gagliardo A. 2015. Olfaction and topography, but not magnetic cues, control navigation in a pelagic seabird: displacements with shearwaters in the Mediterranean Sea. *Sci Rep* 5:16486.

Prescott TJ, Montes González FM, Gurney K, Humphries MD, Redgrave P. 2006. A robot model of the basal ganglia: behavior and intrinsic processing. *Neural Netw* 19:31–61.

Pryor GS. 2014. Tadpole nutritional ecology and digestive physiology: implications for captive rearing of larval anurans. *Zoo Biology* 33:502–507.

Puelles L, Javier Milán F, Martinez-de-la-Torre M. 1996. A segmental map of architectonic subdivisions in the diencephalon of the frog *Rana perezi*: acetylcholinesterase-histochemical observations. *Brain Behav Evol* 47:279–310.

Puelles L, Kuwana E, Puelles E, Bulfone A, Shimamura K, Keleher J, Smiga S, Rubenstein JL. 2000. Pallial and subpallial derivatives in the embryonic chick and mouse telencephalon, traced by the expression of the genes *Dlx-2, Emx-1, Nkx-2.1*, *Pax-6*, and *Tbr-1*. *J Comp Neurol* 424:409–438.

Puelles L, Rubenstein JLR. 2015. A new scenario of hypothalamic organization: rationale of new hypotheses introduced in the updated prosomeric model. *Front Neuroanat* 9:1229–33.

Puelles L, Sandoval JE, Ayad A, del Corral R, Alonso A, Ferran JL, Martinez-de-la-Torre M. 2017. The pallium in reptiles and birds in the light of the updated tetrapartite pallium model. In: Kaas JH and Striedter GF (eds.), *Evolution of Nervous Systems*, Vol 1., 2nd ed. Amsterdam: Elsevier, pp. 519–555.

Pushchin II, Karetin YA. 2009. Retinal ganglion cells in the eastern newt *Notophthalmus viridescens*: topography, morphology, and diversity. *J Comp Neurol* 516:533–552.

Pyron RA. 2011. Divergence time estimation using fossils as terminal taxa and the origins of Lissamphibia. *Syst Biol* 60:466–481.

Redgrave P, Prescott TJ, Gurney K. 1999. The basal ganglia: a vertebrate solution to the selection problem? *Neurosci* 89:1009–1023.

Rink E, Wullimann MF. 2001. The teleostean (zebrafish) dopaminergic system ascending to the subpallium (striatum) is located in the basal diencephalon (posterior tuberculum). *Brain Res* 889:316–330.

Roth G, Blanke J, Wake DB. 1994. Cell size predicts morphological complexity in the brains of frogs and salamanders. *Proc Natl Acad Sci USA* 91:4796–4800.
Roth G, Dicke U, Grunwald W. 1999. Morphology, axonal projection pattern, and response types of tectal neurons in plethodontid salamanders. II. Intracellular recording and labeling experiments. *J Comp Neurol* 404:489–504.
Roth G, Grunwald W, Dicke U. 2003. Morphology, axonal projection pattern, and responses to optic nerve stimulation of thalamic neurons in the fire-bellied toad *Bombina orientalis*. *J Comp Neurol* 461:91–110.
Roth G, Laberge F, Mühlenbrock-Lenter S, Grunwald W. 2007. Organization of the pallium in the fire-bellied toad *Bombina orientalis*. I. Morphology and axonal projection pattern of neurons revealed by intracellular biocytin labeling. *J Comp Neurol* 501:443–464.
Roth G, Nishikawa KC, Naujoks-Manteuffel C, Schmidt A, Wake DB. 1993. Paedomorphosis and simplification in the nervous system of salamanders. *Brain Behav Evol* 42:137–170.
Roth G, Nishikawa KC, Wake DB. 1997. Genome size, secondary simplification, and the evolution of the brain in salamanders. *Brain Behav Evol* 50:50–59.
Ruta M, Coates MI, Quicke D. 2003. Early tetrapod relationships revisited. *Biol Rev* 78:251–345.
San Mauro D. 2010. A multilocus timescale for the origin of extant amphibians. *Mol Phylogenet Evol* 56:554–561.
Sanes JR, Masland RH. 2015. The types of retinal ganglion cells: current status and implications for neuronal classification. *Annu Rev Neurosci* 38:221–246.
Scheich H, Langner G, Tidemann C, Coles RB, Guppy A. 1986. Electroreception and electrolocation in platypus. *Nature* 319:401–402.
Schlosser G. 2008. Development of the retinotectal system in the direct-developing frog *Eleutherodactylus coqui* in comparison with other anurans. *Front Zool* 5:9.
Schmidt A, Wake MH. 1997. Cellular migration and morphological complexity in the caecilian brain. *J Morphol* 231:11–28.
Schoch RR. 2014. *Amphibian Evolution: The Life of Early Land Vertebrates*. New York: Wiley.
Schoffelen RLM, Segenhout JM, Van Dijk P. 2008. Mechanics of the exceptional anuran ear. *J Comp Physiol A* 194:417–428.
Segev R. 2005. Functional organization of ganglion cells in the salamander retina. *J Neurophysiol* 95:2277–2292.
Sivak JG. 1988. Optics of amphibious eyes in vertebrates. In: Atema J, Fay RR, Popper AN, Tavolga W (eds.), *Sensory Biology of Aquatic Animals*. New York: Springer, pp. 467–485.
Smith Fernandez A, Pieau C, Repérant J, Boncinelli E, Wassef M. 1998. Expression of the *Emx-1* and *Dlx-1* homeobox genes define three molecularly distinct domains in the telencephalon of mouse, chick, turtle and frog embryos: implications for the evolution of telencephalic subdivisions in amniotes. *Development* 125:2099–2111.
Smotherman MS, Narins PM. 1999. The electrical properties of auditory hair cells in the frog amphibian papilla. *J Neurosci* 19:5275–5292.
Sotelo MI, Daneri MF, Bingman VP, Muzio RN. 2016. Telencephalic neuronal activation associated with spatial memory in the terrestrial toad *Rhinella arenarum*: participation of the medial pallium during navigation by geometry. *Brain Behav Evol* 88:149–160.
Standen EM, Du TY, Laroche P, Larsson HCE. 2016. Locomotor flexibility of *Polypterus senegalus* across various aquatic and terrestrial substrates. *Zoology* 119:447–454.
Stensiö EA. 1963. The Brain and the Cranial Nerves in Fossil, Lower Craniate Vertebrates. Oslo: Skrift Norsk Videnskaps-Akad.
Striedter GF. 2005. *Principles of Brain Evolution*. Sunderland, MA: Sinauer Associates.
Striedter GF. 2015. *Neurobiology: A Functional Approach*. New York: Oxford University Press.
Swanson LW. 2005. Anatomy of the soul as reflected in the cerebral hemispheres: neural circuits underlying voluntary control of basic motivated behaviors. *J Comp Neurol* 493:122–131.
Syed AS, Korsching SI. 2014. Positive Darwinian selection in the singularly large taste receptor gene family of an "ancient" fish, *Latimeria chalumnae*. *BMC Genomics* 15:650.
Szarski H. 1983. Cell size and the concept of wasteful and frugal evolutionary strategies. *J Theor Biol* 105:201–209.
Taylor GM, Nol E, Boire D. 1995. Brain regions and encephalization in anurans: adaptation or stability. *Brain Behav Evol* 45:96–109.

ten Donkelaar HJ. 1998. Anurans. In: Nieuwenhuys R, Donkelaar ten HJ, Nicholson C. (eds.), *The Central Nervous System of Vertebrates*. Berlin: Springer, pp. 1151–1314.

Thomson KS. 1972. An attempt to reconstruct evolutionary changes in the cellular DNA content of lungfish. *J Exp Zool* 180:363–371.

Tucker AS. 2016. Major evolutionary transitions and innovations: the tympanic middle ear. *Philos Trans R Soc Lond B* 372:20150483.

Tucker VA. 1975. The energetic cost of moving about. *Am Sci* 63:413–419.

van Dongen PAM. 1998. Brain size in vertebrates. In: Nieuwenhuys R, ten Donkelaar HJ, Nicholson C (eds.), *The Central Nervous System of Vertebrates*. Berlin: Springer, pp. 2099–2134.

Veenman CL, Crzan D, Kern H, Rickmann M, Wahle P, van Mier P. 1989. The anatomical substrate for telencephalic function. *Adv Anat Embryol Cell Biol* 117:1–110.

Wake DB. 1966. Comparative osteology and evolution of the lungless salamanders, family Plethodontidae. *Mem So Calif Acad Sci* 4:1–111.

Wallraff HG. 2014. Do olfactory stimuli provide positional information for home-oriented avian navigation? *Anim Behav* 90:e1–e6.

Westhoff G, Roth G. 2002. Morphology and projection pattern of medial and dorsal pallial neurons in the frog *Discoglossus pictus* and the salamander *Plethodon jordani*. *J Comp Neurol* 445:97–121.

Wicht H, Himstedt W. 1988. Topologic and connectional analysis of the dorsal thalamus of *Triturus alpestris* (Amphibia, Urodela, Salamandridae). *J Comp Neurol* 267:545–561.

Wilczynski W. 1981. Afferents to the midbrain auditory center in the bullfrog, *Rana catesbeiana*. *J Comp Neurol* 198:421–433.

Wilczynski W. 1984. Central neural systems subserving a homoplasous periphery. *Am Zool* 24:755–763.

Wilczynski W, Endepols H. 2006. Central auditory pathways in anuran amphibians: the anatomical basis of hearing and sound communication. In: Narins PN, Feng AS, Fay RR, Popper AN (eds.), *Hearing and Sound Communication in Amphibians*, Vol. 28, *Springer Handbook of Auditory Research*. New York: Springer, pp. 221–249.

Will U, Fritzsch B. 1988. The eighth nerve of amphibians: peripheral and central distribution. In: Fritzsch B, Hetherington T, Ryan MJ, Walkowiak W, Wilczynski W (eds), *The Evolution of the Amphibian Auditory System*. New York: Wiley, pp. 159–183.

Will U, Luhede G, Görner P. 1985. The area octavo-lateralis in *Xenopus laevis*. *Cell Tissue Res* 239:147–161.

Wullimann MF. 2014. Ancestry of basal ganglia circuits: new evidence in teleosts. *J Comp Neurol* 522:2013–2018.

Wullimann MF. 2017. Should we redefine the classic lateral pallium? *J Comp Neurol* 525:1509–1513.

Yamamoto K, Bloch S, Vernier P. 2017. New perspective on the regionalization of the anterior forebrain in Osteichthyes. *Dev Growth Differ* 59:175–187.

Yamamoto K, Fontaine R, Pasqualini C, Vernier P. 2015. Classification of dopamine receptor genes in vertebrates: nine subtypes in osteichthyes. *Brain Behav Evol* 86:164–175.

Yamamoto K, Ruuskanen JO, Wullimann MF, Vernier P. 2010. Two tyrosine hydroxylase genes in vertebrates: new dopaminergic territories revealed in the zebrafish brain. *Mol Cell Neurosci* 43:394–402.

Yamamoto K, Vernier P. 2011. The evolution of dopamine systems in chordates. *Front Neuroanat* 5:21.

Yokobori S, Hasegawa M, Ueda T, Okada N, Nishikawa K, Watanabe K. 1994. Relationship among coelacanths, lungfishes, and tetrapods: a phylogenetic analysis based on mitochondrial cytochrome oxidase I gene sequences. *J Mol Evol* 38:602–609.

Yokoyama S, Zhang H, Radlwimmer FB, Blow NS. 1999. Adaptive evolution of color vision of the Comoran coelacanth (*Latimeria chalumnae*). *Proc Natl Acad Sci USA* 96:6279–6284.

Zhu M, Ahlberg PE. 2004. The origin of the internal nostril of tetrapods. *Nature* 432:94–97.

5

The Conquest of Land

Amniote Origins and the Age of Reptiles

As we reviewed in Chapter 4, the earliest tetrapods were forced to live in close association with water, just like today's amphibians, because they easily dried out on land, both as adults and at earlier stages of development. This constraint was overcome by early amniotes, which made both their body skin and their eggs more water-resistant. Together with major changes in the respiratory system, these innovations allowed amniotes to live far away from water, even in deserts, and to lay their eggs on land. Thus, they did not merely invade land; they conquered it.

Very early in their evolution, the amniotes divided into two major branches, one leading to modern reptiles and birds, the other leading to mammals. In the present chapter, we will focus on the reptilian branch, which is called the sauropsid lineage; mammals will be our focus in Chapter 6. Birds are included in the present chapter, because they are descended from dinosaurs (Figure 5.1). However, birds and mammals exhibit many similarities in both their physiology and some key aspects of their brains; these convergent features will be discussed in Chapter 6.

5.1. Early Amniotes and Extant Sauropsids

Roughly 270 million years ago (mya), during the Permian period, the sauropsid lineage split into two major branches (Figure 5.1). One of them includes lizards, snakes, and a single living species of the genus *Sphenodon*, known as the tuatara; together, they are referred to as squamates. The other major branch of the sauropsid lineage comprises turtles and archosaurs (Latin for "ruling lizards"). The archosaurs, in turn, are divisible into crocodilians and dinosaurs, which comprise a large number of extinct species but also include birds.

Given this phylogeny, it becomes apparent that terms like "reptiles," "lizards," and "dinosaurs" refer to paraphyletic, rather than monophyletic, groups (see Chapter 1) and should, therefore, be used cautiously. Specifically, the terms "reptile" and "dinosaur" are problematic because birds could be called "winged dinosaurs" or, more generally, "flying reptiles." Similarly, the fact that snakes evolved deep within the lepidosaur lineage (i.e., all lizards and snakes) makes the term "lizard" problematic because, in a sense, snakes could be called "legless lizards." Indeed, several other

Brains Through Time. Georg F. Striedter and R. Glenn Northcutt, Oxford University Press (2020). © Oxford University Press.
DOI: 10.1093/oso/9780195125689.001.0001

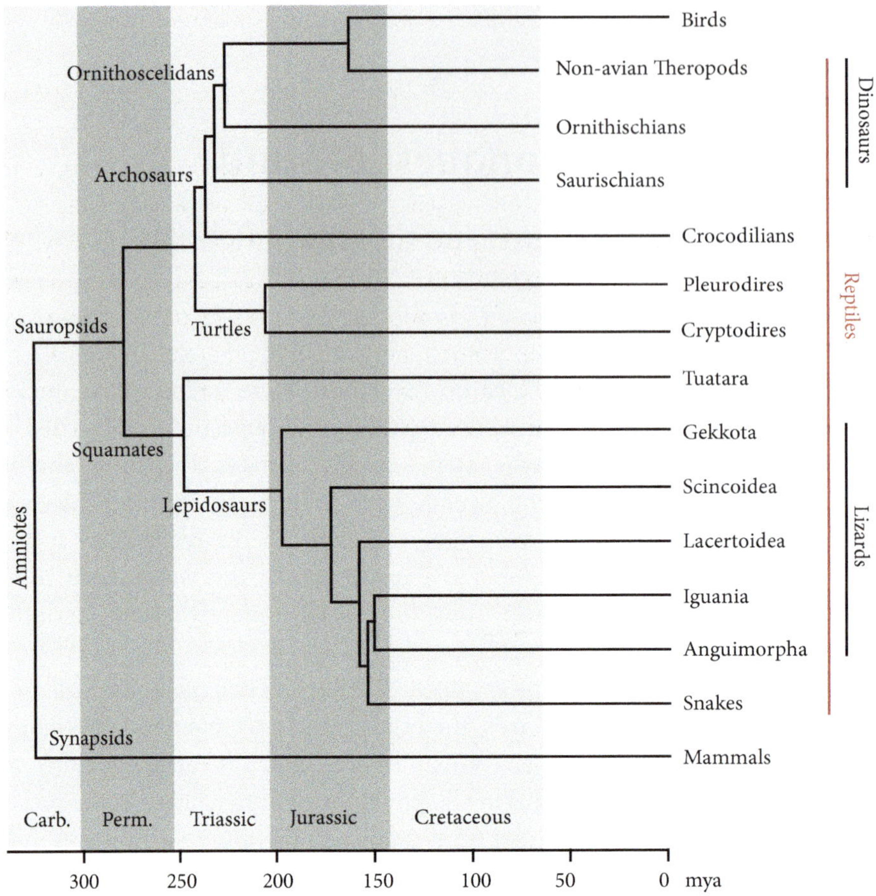

Figure 5.1 Phylogeny of amniotes. As shown in this diagram, the two major groups of amniotes, synapsids and sauropsids, diverged more than 300 million years ago (mya). The synapsids will be discussed in Chapter 6. The sauropsids split into squamates, turtles, and archosaurs, with the latter two being most likely sister groups. Several groups of extinct dinosaurs evolved among the archosaurs, and one of these gave rise to birds. Given this phylogeny, the traditional terms "reptile" and "dinosaur" refer to paraphyletic groups, rather than monophyletic lineages, because they do not include all of the descendants of their last common ancestor. Similarly, "lizards" is a paraphyletic term because it does not include snakes, which evolved deep within the lepidosaur lineage.

Based on Joyce (2007), Shedlock and Edwards (2009), Pyron et al. (2013), Reeder et al. (2015), Benton et al. (2015), Baron et al (2017).

lepidosaur groups have also lost their legs (e.g., anguids and pygopodids), yet are referred to as lizards, rather than snakes. These semantic issues can create imprecision, if not outright confusion, but familiar paraphyletic terms can be convenient and useful, as long as their use is deliberate. In that spirit, we do not always refrain from using the terms "lizard," "reptile," and "dinosaur."

5.1.1. The Tuatara, Lizards, and Snakes

One of the most important species for reconstructing early amniote biology is the tuatara (*Sphenodon punctatus*). This species is the sole survivor of a once successful order (Rhynchocephalia) that diverged from other squamates about 250 mya and is now found only on some small, rocky islands off the coast of New Zealand. With body lengths up to 80 cm, tuataras look like large lizards (Figure 5.2), but they have several features that are not found in lizards and snakes. More importantly, several of these features are characteristic of anamniotes, which is why tuataras are sometimes called "living fossils." It is important to note, however, that the tuatara also has a number of uniquely derived features, especially in tooth structure, skull anatomy, and genome organization (Rest et al., 2003). Therefore, one cannot assume that all features of the tuatara are primitive.

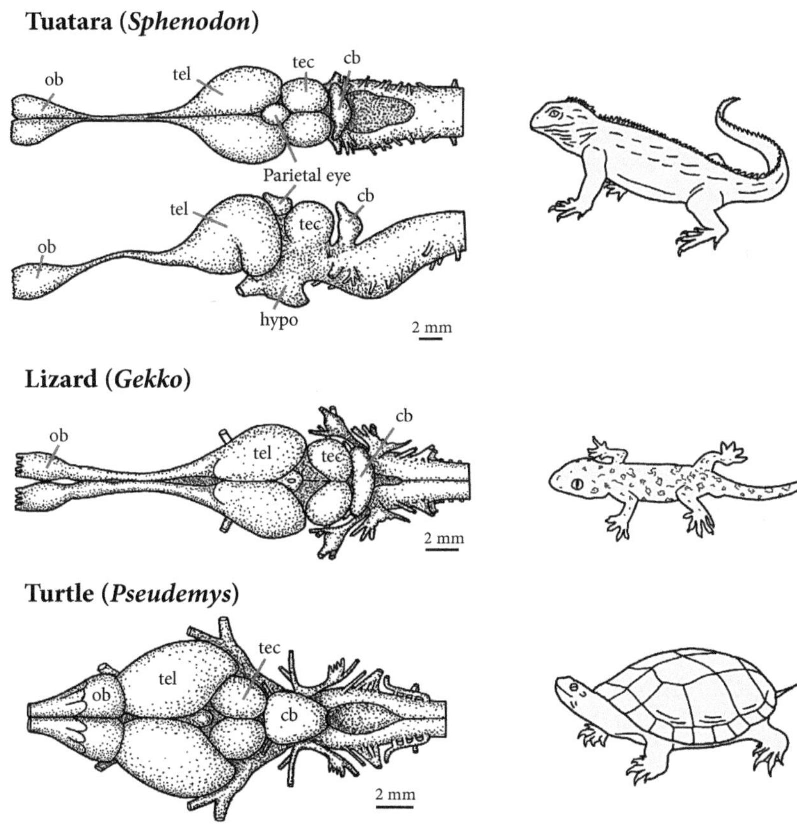

Figure 5.2 Brains and bodies of non-archosaurian reptiles. The tuatara (*Sphenodon punctatus*) is the only surviving member of the most basal lineage of squamate reptiles (see Figure 5.1). Its brain is here shown in dorsal and lateral views. Also shown are dorsal views of the brain of a gekkonid lizard and a cryptodire turtle.
Abbreviations: cb – cerebellum; ob – olfactory bulb; tec – optic tectum; tel – telencephalon.
The tuatara brain is adapted from Platel (1989).

The brain of tuataras features a telencephalon that is relatively wide, compared to that of most amphibians, and it includes olfactory bulbs that are separated from the more caudal telencephalon by long and narrow peduncles (Figure 5.2). The tectum is also well developed in tuataras, and the cerebellum is slightly larger than it is in lungfishes and amphibians. One of the most interesting features of the tuatara brain is its large parietal "third eye," which develops as a dorsal evagination of the diencephalon. A parietal eye is present in most reptiles and many anamniotes (see Appendix), but it is especially well developed in the tuatara, featuring a built-in lens, a cornea, and rod-like photoreceptors (Quay, 1979; Schwab, 2012). It is probably involved in the detection of polarized light and helps animals navigate (Freake, 2001; Foà et al., 2009). In conjunction with the closely associated pineal gland, the parietal eye may also help to regulate circadian rhythms. The parietal eye was lost in mammals and, independently, with the origin of turtles, crocodilians, and birds (Quay, 1979).

The lepidosaurs are the closest living relatives of tuataras. With more than 7,500 species (Figure 5.3), they are almost as speciose as birds and more diverse than mammals. Roughly one-third of all lepidosaurs are snakes; the rest are lizards. At the macroscopic level, the brains of lizards and snakes are similar to one another and to those of tuataras (Figure 5.2). The size of lepidosaur brains, relative to body size, is similar to that of anuran amphibians (Figure 5.4). However, relative brain size has increased, compared to the primitive lepidosaur condition, in several lizard lineages, including the tegus and the varanids (i.e., monitor lizards; Northcutt, 1978). Much of that variation is due to expansion of the forebrain in the large-brained lizards. In contrast, relative brain size was reduced in snakes. However, this change is probably related to snakes having increased their body size by adopting an elongate body shape. Indeed, it seems that eel-shaped animals in general tend to have smaller brains, relative to body size, than their less elongate relatives (Striedter, 2005), but this hypothesis remains to be tested empirically.

5.1.2. Turtles

The sister group of lepidosaurs consists of turtles and archosaurs (crocodilians and birds). This group (aka Archelosauria) has become well established only in the last decade or so, because turtles were long thought to be the most basal group of sauropsids. However, a long line of molecular and morphological studies have now shown that this was incorrect (deBraga and Rieppel, 2008; Gilbert and Corfe, 2013; Field et al., 2014). The revised phylogeny required a reinterpretation of many turtle characters. For example, the lack of openings in the skull's temporal bone (i.e., an "anapsid" skull) was long thought to be a primitive feature of turtles, shared with the earliest, long-extinct amniotes. However, it now appears that early stem turtles had

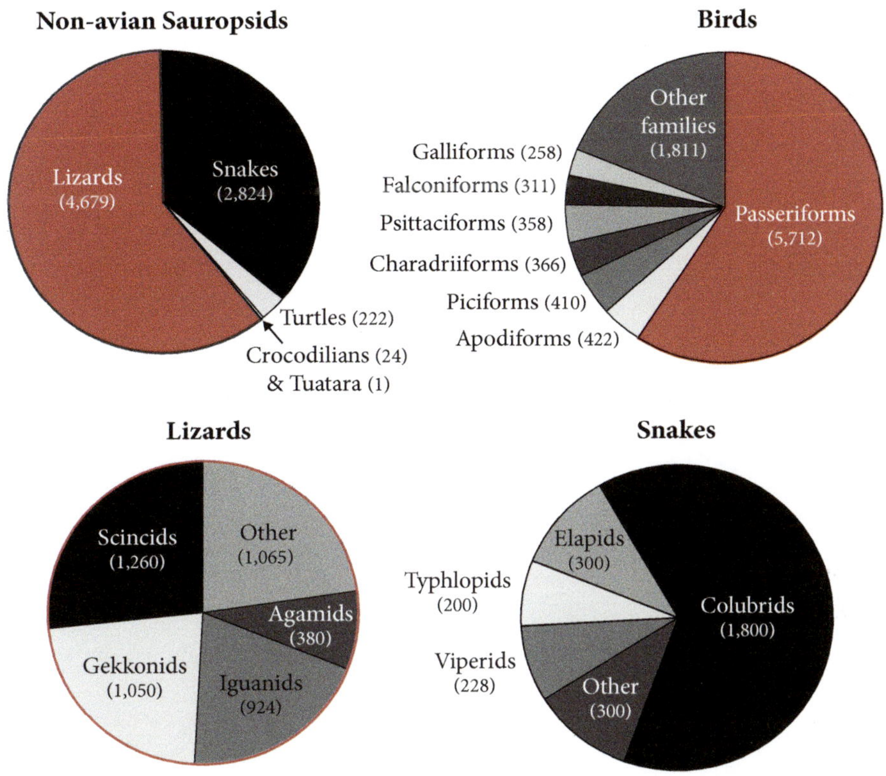

Figure 5.3 Species counts for extant sauropsids. Among non-avian sauropsids, lizards and snakes are each far more diverse than turtles and crocodilians combined. Among birds, the passeriform birds (including songbirds and flycatchers) account for nearly 60% of all species. Note that the colubrid snakes are no longer considered to be a monophyletic group.

Data from Pough et al. (2004) and Gill (1994).

a "diapsid" skull with two temporal bone openings (aka temporal fenestrae) and that the anapsid skull of modern turtles represents an evolutionary reversal (Bever et al., 2015; Schoch and Sues, 2015).

The last common ancestor of all extant turtles was probably aquatic, but the earliest stem turtles were almost certainly terrestrial (Joyce and Gauthier, 2004), which implies that turtles re-entered the water after their ancestors had invaded the land. Some time in the late Triassic or early Jurassic period, the turtle lineage diverged into cryptodire and pleurodire turtles (Joyce et al., 2015). The latter are often called side-neck turtles because they fold their necks sideways into the shell when they feel threatened. In contrast, the cryptodires fold their necks vertically. Pleurodire turtles are interesting from a neurobiological perspective, because their forebrain is significantly larger and more complex than that of cryptodires (Figure 5.5; Riss et al., 1969; Striedter, 2015). Unfortunately,

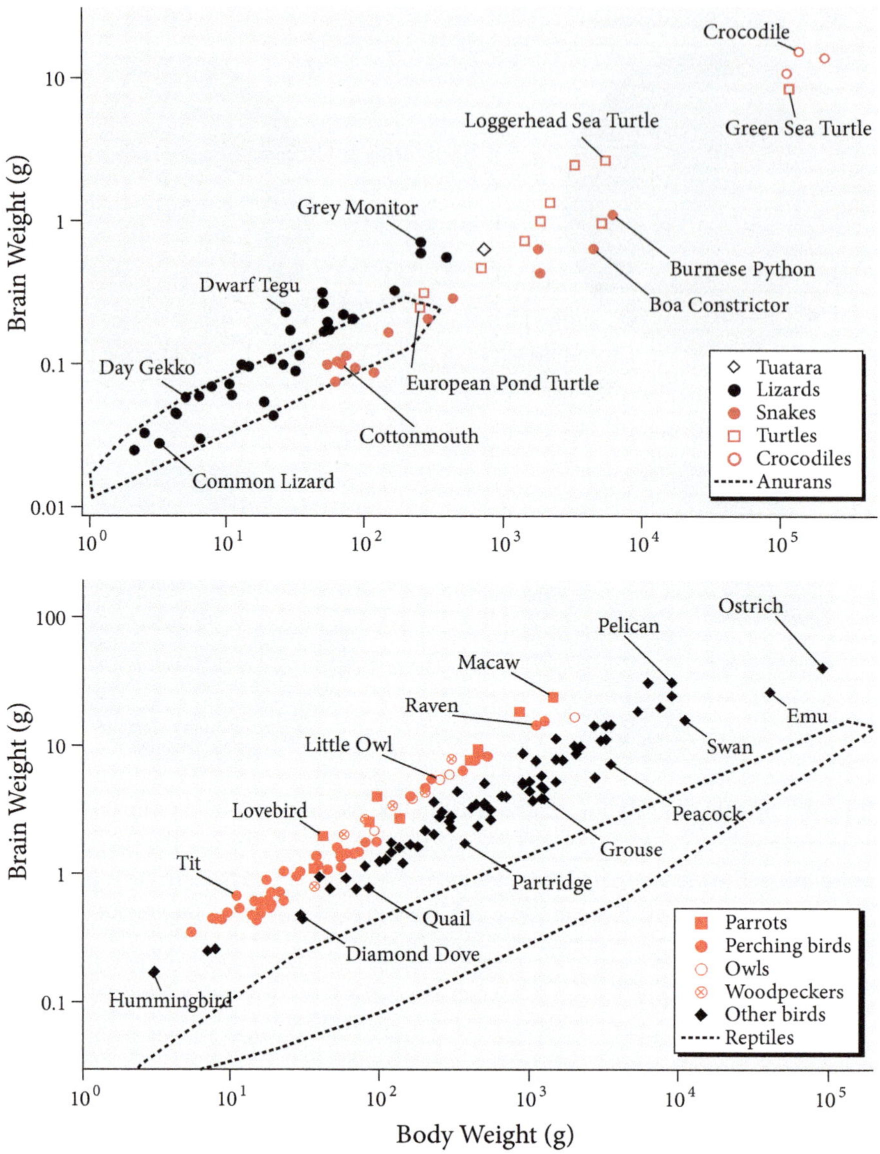

Figure 5.4 Brain-body scaling in reptiles and birds. Relative brain size is quite variable within lizards and larger in large lizards than in snakes of comparable body size; it is intermediate in the tuatara. Birds have much larger brains than non-avian sauropsids of similar body size. Within birds, the parrots and the perching birds (Passeriformes) tend to have the largest brains, relative to body size. The dotted lines represent minimum convex polygons encompassing all the data points for the indicated groups.

Data from Portmann (1947), Platel (1979) and Northcutt (2013).

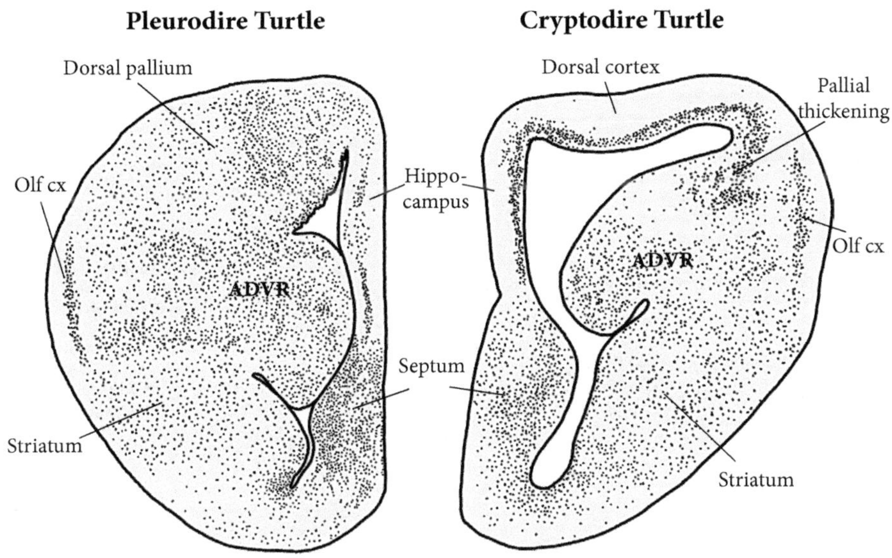

Figure 5.5 The telencephalon of pleurodire versus cryptodire turtles. Almost all studies of turtle brains have utilized cryptodire turtles, but pleurodire turtles have a much larger, more complex telencephalon. Shown on the left is a transverse section through the left telencephalon of a pleurodire turtle (*Podocnemis unifilis*); shown on the right is an equivalent section through the right telencephalon of a cryptodire turtle (*Pseudemys scripta*). Clearly, the pallium is much enlarged in the pleurodire species; this is true especially for the dorsal pallium, which does not exhibit the laminar organization seen in the dorsal cortex of cryptodire turtles and other non-avian sauropsids.

Abbreviations: ADVR – anterior dorsal ventricular ridge; olf cx – olfactory cortex.

The drawing on the right is based on a photograph kindly provided by Cosme Salas and Fernando Rodriguez.

there has been only one experimental study on pleurodire turtle brains (Bass et al., 1973).

The brains of cryptodire turtles are similar to those of the tuatara, but cryptodires have a proportionately larger telencephalon that is wider than the optic tectum and extends further caudally so that its caudal pole lies lateral to the rostral tectum (see Figure 5.2). In addition, the olfactory bulbs of cryptodire turtles are connected to the rest of the telencephalon through very short peduncles (i.e., they are sessile). The cerebellum of cryptodire turtles is similar in size to that of lizards, but it leans backward, toward the medulla, rather than extending vertically or leaning rostrally as in squamates. Although turtles may seem slow and relatively dull to the casual observer, they are capable of complex learning when properly motivated (Wilkinson et al., 2010; Wilkinson and Huber, 2012), can navigate effectively over long distances (Lohmann et al., 2004; Collett and Collett, 2011), and emit a variety of vocalizations (Ferrara et al., 2013).

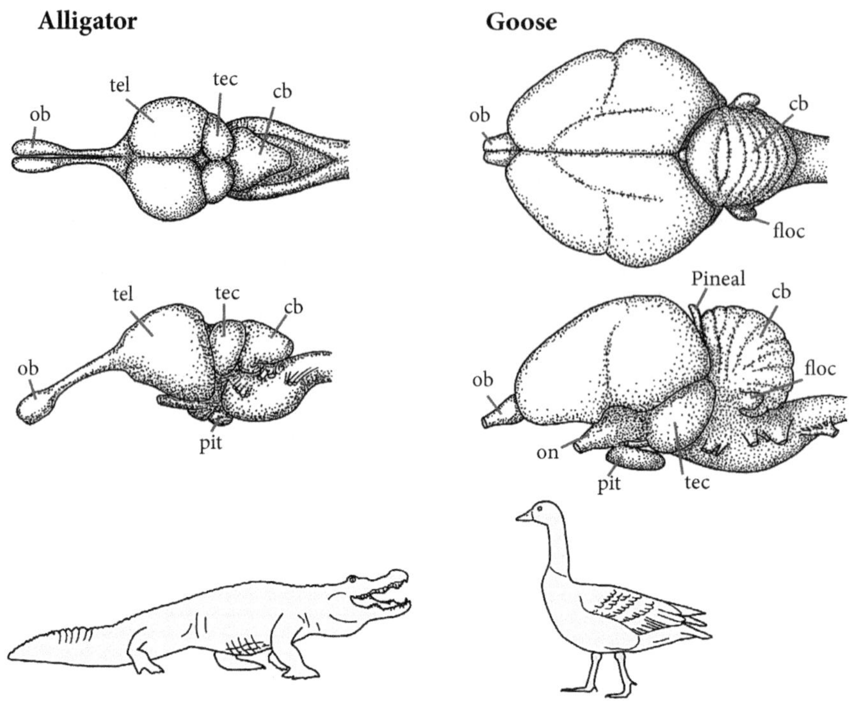

Figure 5.6 Brains and bodies of selected archosaurs. The brain of an alligator, here shown in dorsal and lateral views, is significantly smaller than that of a goose, even though the body of an adult alligator is much heavier. Note that the telencephalon (tel) and cerebellum (cb) are especially enlarged in birds.

Additional abbreviations: floc – cerebellar flocculus; ob – olfactory bulb; on – optic nerve; pit – pituitary; tec – tectum.

Brains after Romer and Parsons (1977).

5.1.3. Crocodilians and Birds

Today's crocodilians include alligators, crocodiles, caiman, and gharials (aka gavials). They tend to be large and carnivorous. Modern crocodilians spend a lot of time in the water but, like turtles, they descended from ancestors that were far more terrestrial. In fact, modern crocodilians are just a small remnant of a much larger group that included many smaller, terrestrial forms. Moreover, those ancient crocodilians were less lizard-like than today's crocodiles and walked mainly on their hind legs (Parrish, 1987), much like birds, *Tyrannosaurus rex*, and diverse other dinosaurs. Although behavioral research on crocodilians is rather limited, they are known to be territorial, stalk prey, and vocalize. Like turtles, they are probably much smarter than most people think. For example, it has been reported that some crocodilians cover themselves with sticks to attract birds, which they then eat (Dinets et al., 2015).

The brains of extant crocodilians are similar to those of turtles and large-brained lizards in terms of relative size, but they are much larger in absolute terms (Figure

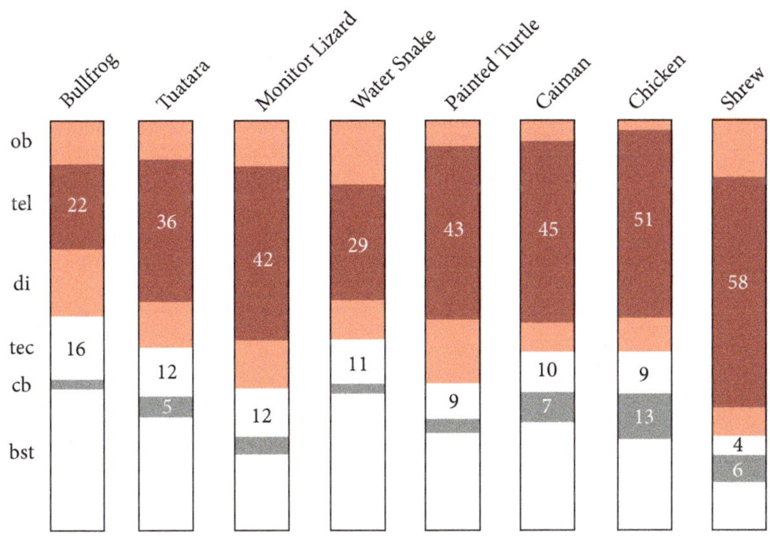

Figure 5.7 Brain region proportions in diverse tetrapods. Shown here are the proportional sizes of the olfactory bulbs (ob), remaining telencephalon (tel), diencephalon (di), tectum (tec), cerebellum (cb), and remaining brainstem (bst) in single specimens of eight different amniotes. The species, from left to right, are *Rana catesbeiana, Sphenodon punctatus, Varanus bengalensis, Nerodia sipedon, Chrysemys picta, Caiman crocodilus, Gallus gallus domesticus,* and *Blarina brevicauda.* The numbers indicate percentages taken up by selected brain regions. For the remaining percentages see Northcutt (2013).

5.4). Indeed, crocodilian brains continue to grow throughout most of adulthood, mainly by adding glial cells and increasing the size of their neurons (although they do exhibit some adult neurogenesis; Ngwenya et al., 2013, 2016, 2018).The general structure of crocodilian brains is similar to that of turtles, but the olfactory bulbs of crocodilians are clearly pedunculated, lying close to the nasal epithelium in the rostral tip of their long snouts (Figure 5.6). Furthermore, the cerebellum is significantly larger in crocodilians than in turtles, lizards, or snakes (Figure 5.7). It exhibits two transverse grooves (fissures), which divide the cerebellum into anterior, middle, and posterior lobes.

With roughly 10,000 extant species (Figure 5.3), birds are the second most successful group of vertebrates, outnumbered only by the teleosts (with their more than 20,000 species). As mentioned earlier, birds diverged from (other) dinosaurs about 165 mya, in the Jurassic period. Within birds, the most basal lineage is called the paleognaths (Figure 5.8). It contains fewer than 60 species of flightless birds, including kiwis, emus, and ostriches, as well as one lineage of flying birds, the tinamous. Phylogenetic data suggest that the earliest paleognaths could fly and that this ability was lost repeatedly in several paleognath lineages near the end of the Cretaceous period (Harshman et al., 2008; Phillips et al., 2010).

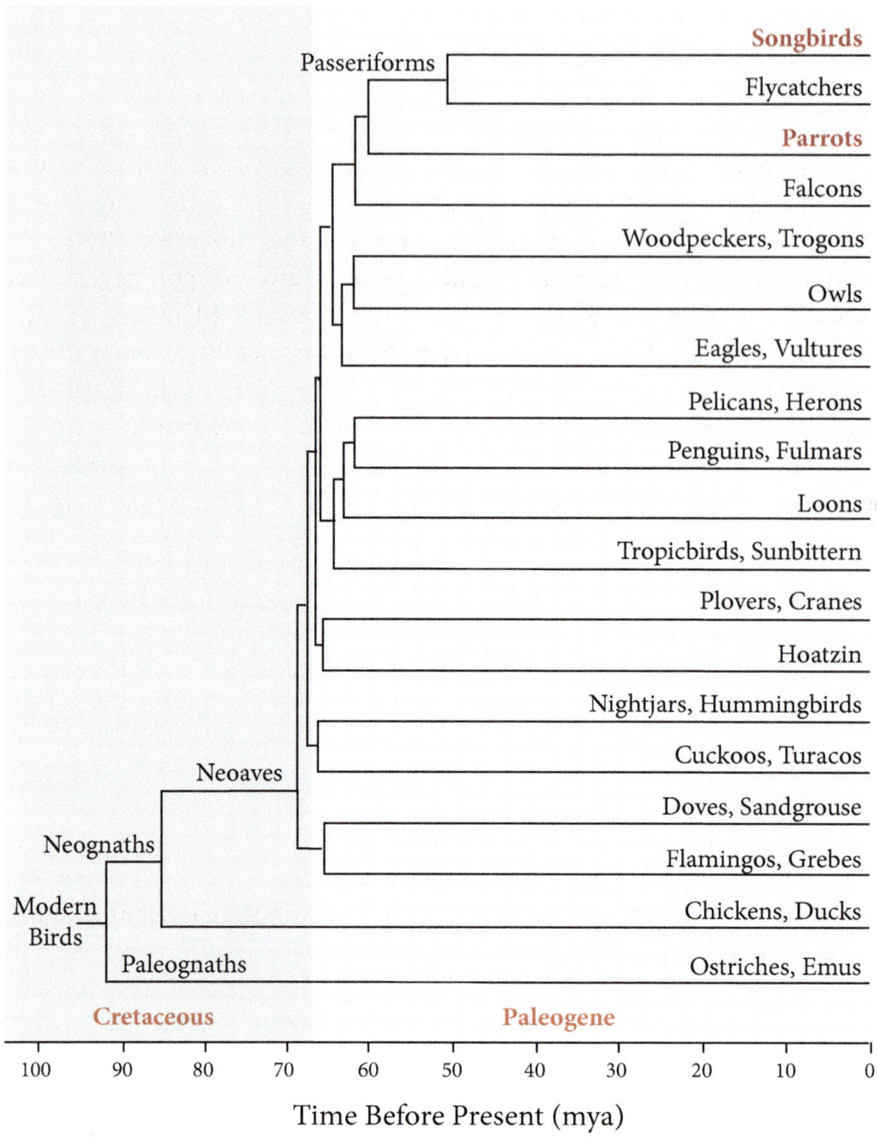

Figure 5.8 Dated phylogeny of birds. This phylogeny of modern birds (Neornithes) was based on a fossil-calibrated Bayesian analysis of two genes from 230 species; it is similar to a phylogeny based on more extensive molecular data from a smaller number of species (Jarvis et al., 2014). According to these analyses, birds diversified explosively shortly after the end-Cretaceous extinction event, but the major lineages of modern birds originated before then. Songbirds and parrots are highlighted in red because they have larger brains, relative to body size, than other birds.

Adapted from Claramunt and Cracraft (2015).

The other, much larger group of living birds is called the neognaths, which diverged from the paleognaths 120–130 mya (Figure 5.8). The most basal neognath lineage includes chickens and ducks, together with their closest relatives. The remaining birds are collectively called Neoaves. Roughly half of all the species in this group are members of the perching birds (passerines), which includes all the songbirds and their smaller sister group, the flycatchers. Their closest relatives appear to be the parrots and parakeets (Hackett et al., 2008; Jarvis et al., 2014). They are among the most intelligent birds, equaled or outperformed only by the corvids (crows and their relatives), which emerged within the songbird lineage roughly 25 mya (Barker et al., 2004; Emery and Clayton, 2004; Striedter, 2013; Emery, 2016).

An average avian brain is roughly 10 times as large, relative to body size, as the brain of any other sauropsid (Figure 5.4). Much of that size difference is due to an expansion of the telencephalon. In chickens, roughly half of the entire brain mass is telencephalon (Figure 5.7), but that percentage rises to 62% in geese and nearly 80% in large parrots and corvids (Iwaniuk et al., 2005). To accommodate such a large telencephalon, the optic tectum is displaced laterally in birds so that it lies below the caudal telencephalon (Figure 5.6). The cerebellum is also enlarged in birds, relative to their reptilian ancestors. It accounts for only about 13% of the brain's mass in chickens but contains roughly half of all its neurons (mainly cerebellar granule cells; Olkowicz et al., 2016). In parallel with this increase in volume and neuron number, the avian cerebellum has become much more folded. It still contains the three main lobes seen in crocodilians, but each of these lobes is thrown into additional folds, creating a total of 11 or more smaller lobules (Figure 5.9; Iwaniuk et al., 2007). Although most brain regions are larger in birds than in their ancestors, the olfactory bulbs tend to be reduced in birds, especially in the parrots and perching birds (Bang and Cobb, 1968; Zelenitsky et al., 2011).

5.1.4. Dinosaurs and Other Extinct Reptiles

So far, we have discussed mainly the sauropsids that still exist today, but this lineage has a rich fossil record. Best known are the dinosaurs, which originated in the Triassic period and dominated life on land during the Jurassic and Cretaceous periods. They are divided into two main lineages, namely the saurischians and the ornithoscelidans; the latter in turn comprise ornithischians and theropods (Baron et al., 2017; see Figure 5.1). One of the most famous theropods is *Tyrannosaurus rex*, but most theropods were not nearly as large. In fact, body size seems to have shrunk fairly consistently along the theropod lineage leading to early birds, which likely weighed less than 1 kg (Lee et al., 2014). In parallel with this shrinkage in body size, theropods decreased their absolute brain size (Figure 5.10). It is difficult to determine the brain's shape in the non-avian theropods, because the brain in most reptiles does not completely fill the endocranial cavity. Indeed, it fills only about half of the endocranium in tuataras and modern turtles. However, the shape

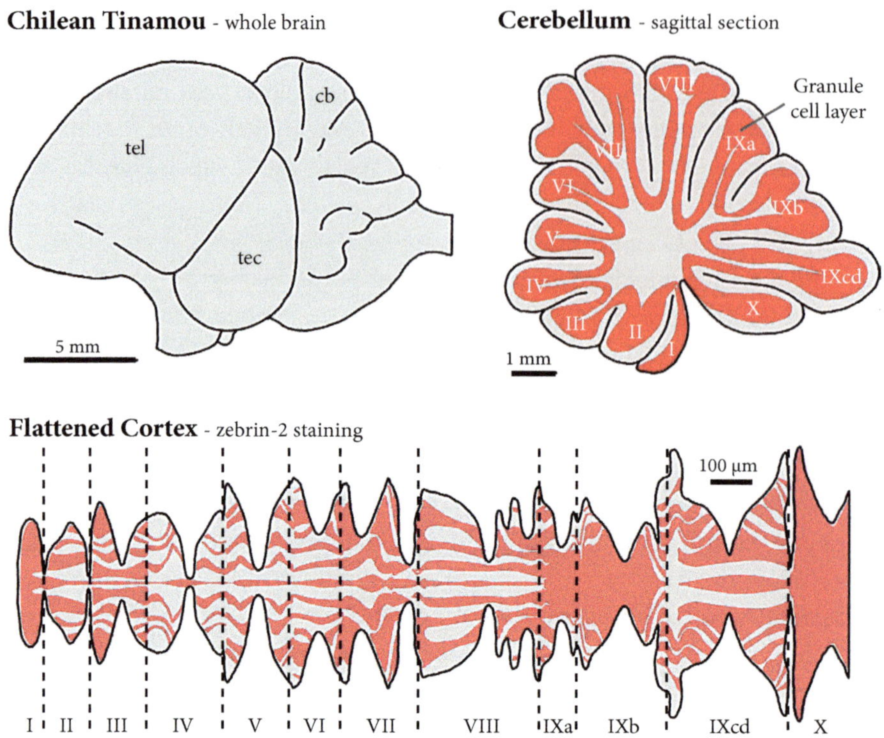

Figure 5.9 Cerebellum of a bird. The brain of the Chilean tinamou (*Nothoprocta perdicaria*) is shown in a lateral view (top left), together with a sagittal section through its cerebellum (top right; anterior to the left). Also illustrated are the cerebellum's granule cell layer (red) and individual lobules (labeled with Roman numerals). Depicted along the bottom is a flattened map of the tinamou's cerebellar cortex. The red color indicates a series of sagittal stripes that stain with antibodies against zebrin-2, which indicates the presence of aldolase C; the functional significance of these stripes remains an issue of debate (e.g., Pakan et al., 2011).
Adapted from Corfield et al. (2015a).

of the cranial endocasts taken from stem birds, such as *Archaeopterix* (Figure 5.10), is similar to that of primitive modern birds (Alonso et al., 2004). These data are consistent with the hypothesis that relative brain size increased in stem birds. Although researchers continue to debate when exactly powered flight evolved within the avian lineage, it probably evolved after stem birds increased their relative brain size (Balanoff et al., 2013, 2016).

Closely related to the dinosaurs were the pterosaurs (Witton, 2013), which originated in the late Triassic, roughly 230 mya, and went extinct together with the non-avian dinosaurs at the end of the Cretaceous period (see Section 5.2.2). Despite their name, these "flying lizards" were more closely related to crocodiles than to lizards. They could, however, fly using a membranous flap of skin that extended from the ankle joint to the tip of a spectacularly elongated fourth finger. Early pterosaurs

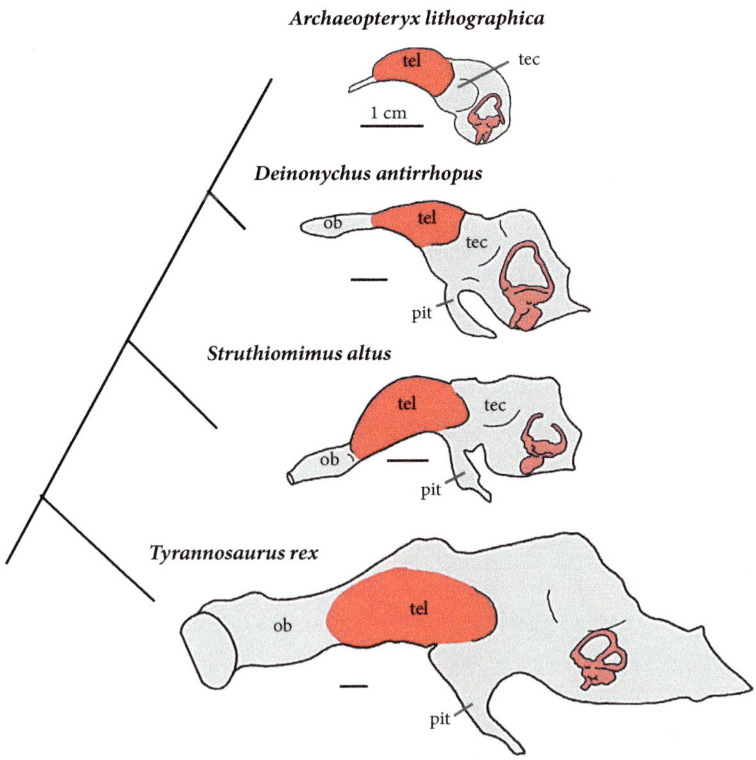

Figure 5.10 Endocasts of theropod dinosaurs, including *Archaeopteryx*. Shown here are lateral views of endocasts from *Archaeopteryx* (a close relative of living birds) and three other theropods. Given the phylogenetic relationships of these species, it is apparent that absolute brain size decreased within the theropods along the lineage leading to birds, paralleling a dramatic decrease in body size (Lee et al., 2014). The smaller the endocast, the more clearly it reflects the size and shape of the underlying brain. The approximate size and shape of the cerebral hemispheres is shown in red; the vestibular apparatus is depicted in pink.
Adapted from Witmer and Ridgely (2009) and Holloway et al. (2013).

probably used these wings to glide through the air, but later pterosaurs were capable of powered flight. Consistent with this hypothesis is the finding that late pterosaurs had very lightweight (pneumatic) bones and air sacs similar to those of birds (see Chapter 6, Section 6.4.3), which likely allowed pterosaurs to breathe very efficiently and, thus, sustain the high levels of metabolic activity needed for active flight. The bodies of many pterosaurs were covered with feather- or hair-like "pycnofibers" (Yang et al., 2019), which probably acted as insulation against the cold. All of these observations suggest that pterosaurs may have been capable of generating their own body heat. If this idea is correct, then endothermy—like flight—would have evolved three times independently among the vertebrates, namely in mammals, birds, and pterosaurs. Little is known about pterosaur brains, but fossil endocasts suggest that they were slightly smaller than those of modern birds but larger than those of modern lizards and turtles, relative to body size (Witmer et al., 2003). In terms of shape, pterosaur endocasts are similar to those of crocodilians and birds. Curiously, pterosaurs seem to have had an enormous cerebellar flocculus, which might have processed information coming from the vestibular apparatus, which is also relatively large in pterosaurs.

Even older than the dinosaurs and pterosaurs were the stem amniotes (aka reptiliomorphs; see Figure 4.7 in Chapter 4). A good example is *Diadectes*, which was a large herbivore with short but sturdy legs (Figure 5.11) that lived during the early Permian and likely spent much of its time on land. Then there are the earliest

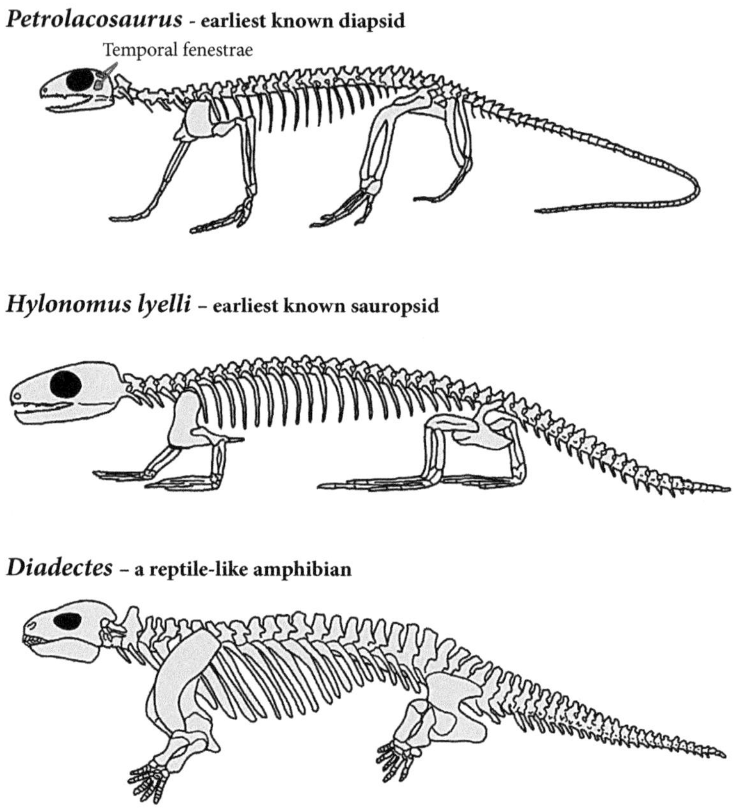

Figure 5.11 Skeletons of extinct early sauropsids and reptile-like amphibians. *Diadectes* was a large, terrestrial amphibian that lived during the early Permian. It is thought to be one of the closest relatives of amniotes. *Hylonomus* was a much smaller animal (~20 cm long) that lived during the late Carboniferous period. It was discovered at the Joggins paleontological site and is considered the earliest known sauropsid. *Petrolacosaurus* was about 40 cm long and lived during the late Carboniferous. It has two separate openings in the posterior portion of its skull (two temporal fenestrae), making it the earliest known diapsid reptile. As illustrated by these species, the skeleton became much less robust in the lineage leading to diapsids. Note especially the thin curved ribs and long, thin legs in the diapsid.
Adapted from Janis and Keller (2001), Carroll and Baird (1972).

sauropsids, most notably *Hylonomus lyelli* (Figure 5.12). This species is known from a fossil specimen discovered in 315 million-year-old coal deposits at the Joggins site in today's Nova Scotia. As illustrated in Figure 5.11, *Hylonomus* had a much lighter skeleton than *Diadectes*, which would have made it easier to lift its body off

Figure 5.12 Life in the late Carboniferous at the Joggins site. At this time, the site was a low coastal swamp, dominated by seed ferns and trees with diamond-shaped leaf scars (*Lepidodendron*, left of figure) or round leaf scars (*Sigillaria*, right of figure). Numerous arthropods, such as land scorpions, millipedes, mayflies, and roaches had already evolved, as had giant dragonflies (*Meganeura*). The fauna also included the earliest known reptile, *Hylonomus* (center of background) and large, fish-eating enbolomere tetrapods, such as *Calligenthlon* (center), and temnospordyl tetrapods, such as *Dendrerpeton* (forefront).
© Tanya Young 2018 tanyayoungart.com

the ground during terrestrial locomotion. *Hylonomus* was also much smaller than *Diadectes* (20 cm versus up to 3 m in length) and probably ate millipedes and other small terrestrial invertebrates.

Another instructive early sauropsid is *Petrolacosaurus* (Figure 5.11), which is the closest known relative of modern reptiles. Its skeleton was lighter than that of *Hylonomus* and its legs were considerably longer. In addition, it had a smaller, lighter skull. Particularly important is that *Petrolacosaurus* had two openings (or fenestrae) in the temporal portion of its skull. The function of these fenestrae continues to be debated, but they probably reflect an increase in the mass and force of major jaw muscles. In any case, because *Petrolacosaurus* has two such temporal fenestrae, it is classified as a "diapsid" reptile, rather than an anapsid like *Hylonomous*. Indeed, it was probably one of the earliest diapsid reptiles. Almost nothing is known about the brains of all these early sauropsids and reptile-like amphibians because, as noted earlier, their brains occupied only a relatively small fraction of the endocranial cavity (Hopson, 1979).

5.2. Ecological Challenges for Early Amniotes

When the first amniotes evolved, the earth's continents were coalescing into a massive supercontinent called Pangea. More specifically, Gondwana in the South merged with Laurasia in the North, creating a vast mountain range between the two (Figure 5.13). This collision of the continental plates greatly reduced the amount of shallow coastal habitat and, instead, expanded the inland continental areas, which probably had a typical continental climate with pronounced wet and dry seasons. These global changes would have greatly reduced the amount of habitat in which amphibians, with their dependence on water, could thrive. Early amniotes overcame the limitations of amphibians by evolving water-resistant eggs and skin.

5.2.1. Water Loss and Gas Exchange

A major early step in the evolution of amniotic eggs was the removal of the thick jelly-like layer that surrounds amphibian eggs and acts as a diffusion barrier to gas exchange (Szarski, 1968). Instead, amniotic eggs became surrounded by a soft and fibrous shell membrane that contains a multitude of tiny pores through which gases diffuse easily. Of course, such a membrane is also permeable to water vapor, but it would have been at least somewhat water-resistant. In addition, amniotic eggs evolved the amniotic membrane, which encloses the fluid surrounding the embryo, and the allantois, a membranous sac that holds the metabolic waste produced by the growing embryo. Another important extra-embryonic membrane of amniotic eggs is the chorion, which lies just beneath the shell membrane. It is highly vascularized and, together with the allantois, is responsible for most of the egg's gas exchange.

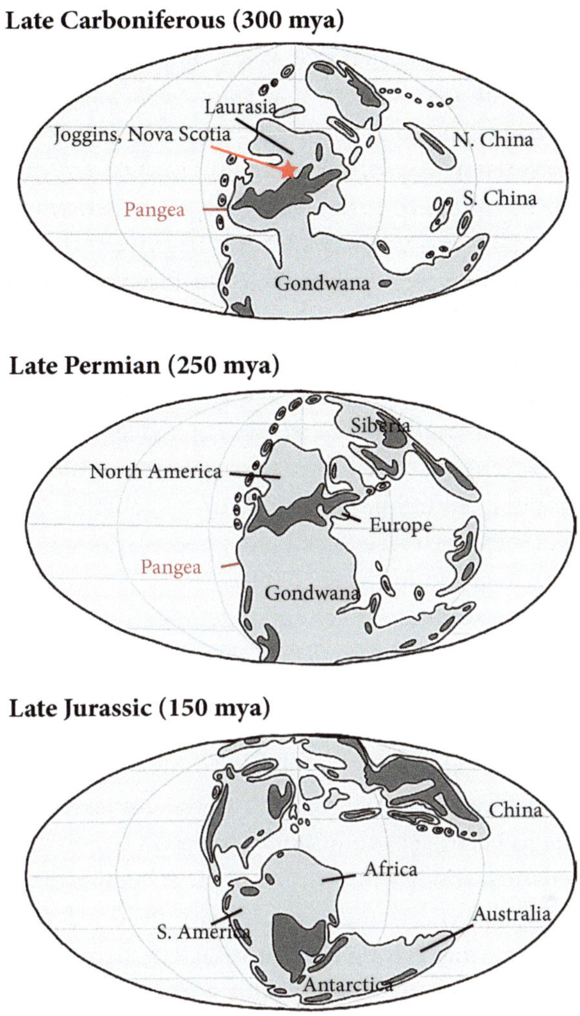

Figure 5.13 Maps of the continents from 300 to 150 million years ago. When continental drift caused Gondwana to collide with Laurasia, roughly 300 mya, the giant supercontinent Pangea was formed. It started to break apart in the Jurassic period, approximately 175 mya. Mountain ranges are depicted in dark gray. The Joggins site, where the first known sauropsids were found, is shown in red. Its original location was probably far inland and flooded seasonally, as a result of runoff from a mountain range. Based on the PALEOMAP Project by C.R. Scotese, http://www.scotese.com.

With all these adaptations, early amniotes no longer had to lay their eggs in water. Their eggs still required a moist environment, but that could be provided by burying the eggs in moist soil (as extant turtles and crocodilians still do) or laying them in shady, humid locations (Packard and Packard, 1980). Such places have the added benefit of hiding the eggs from predators.

Once hatched, the early amniotes required skin that was more water-resistant than that of their ancestors. They solved this problem by evolving scales made of keratin (the main constituent of fingernails, claws, and hair) and covering them with lipid molecules. Such scales reduce water loss across the skin, but they also impede the exchange of gases across that skin. Obtaining oxygen was not the worst problem, because oxygen is relatively plentiful in air (compared to water) and atmospheric oxygen levels were relatively high when early amniotes evolved. However, shedding of carbon dioxide (CO_2) was a very serious challenge because the accumulation of CO_2 makes tissues acidic. To solve this dilemma, early amniotes evolved larger and more complex lungs, giving them more surface area for gas exchange. They also evolved longer, hinged ribs (Figure 5.11) and muscles that could move those ribs alternately inward and outward, pushing air out of the lungs and drawing it in, respectively. Earlier air-breathers had pushed air into their lungs by means of "buccal pumping," which involves alternately raising and lowering the oral cavity's floor. In comparison, rib-based "costal aspiration" is far more efficient (Janis and Keller, 2001). Additional mechanisms to facilitate aspiration breathing (e.g., a muscular diaphragm) later evolved independently in several lineages of amniotes (Gans, 1970; Carrier and Farmer, 2000).

These changes in the respiratory system, together with the evolution of lighter skeletons with longer limbs (Figure 5.11), allowed early amniotes to move efficiently across land. They probably could not move rapidly for long (Carrier, 1987), but over short distances they could have pounced on insects and other invertebrates. Increased terrestrial mobility also allowed the early amniotes to move long distances in search of food, both in the course of any given day and across the seasons, as ecological conditions changed. These early amniotes most likely had few predators, though they may well have been hunted by other, larger amniotes. Daily and seasonal changes would have been a problem for the early amniotes, because they were ectothermic, unable to generate significant amounts of internal body heat. However, they probably addressed this problem behaviorally, much as modern reptiles do, by warming up in the sunshine and cooling off in shady spots or underground. As we discuss in Chapter 6, the origins of endothermy in mammals and, independently, in birds were major evolutionary innovations, but they came long after the first amniotes appeared.

5.2.2. Mass Extinctions and Recovery

Reptiles first emerged in the late Carboniferous and then flourished throughout the Permian period. Their success came to a screeching halt with the Earth's greatest mass extinction at the end of the Permian. This extinction eliminated 96% of all marine species, including acanthodians and placoderms, 63% of all terrestrial tetrapod families, six of nine amphibian families, and all large herbivorous reptiles (Benton, 1995). It was also the only mass extinction that included insects (Labandeira and Sepkoski, 1993). The causes of this decimation remain uncertain (Brannen, 2017), but extreme global warming likely played a major role, with average sea surface temperatures reaching 38°C by the early Triassic (vs. the 25°–30°C we see near the equator today). Average temperatures on land near the equator probably exceeded 38°C (Brannen, 2017). The most likely driver of this extreme global warming was a massive series of volcanic eruptions in Siberia that, over the course of about 1 million years, covered the land with more than a million cubic kilometers of new rock (called the Siberian Traps; Hallam and Wignall, 1997). Crucially, those eruptions would have spewed enormous amounts of CO_2 and ash into the atmosphere. Since CO_2 is a well-established greenhouse gas, it would have caused global warming. It would also have acidified the oceans and killed land plants through acid rain. Other released gases would have destroyed the planet's ozone layer, leading to mutation-inducing levels of radiation on earth. In addition, the atmospheric ash would have darkened the skies across the globe for months at a time, killing off most algae and plants, thereby reducing oxygen levels even further (though perhaps offering a brief respite from global warming). Under these harsh conditions, it is not surprising that less than half of all the tetrapods survived the end of the Permian.

For the next 50 million years, most of the planet stayed lethally hot, but global temperatures stabilized at slightly more tolerable levels in the mid-Triassic (Sun et al., 2012). Life on land during this time was dominated by a few large "stem mammals" (notably the therapsids; see Chapter 6) and various carnivorous reptiles, including the early relatives of modern crocodilians and tuataras. However, another major extinction at the end of the Triassic period, likely driven by another series of volcanic eruptions, eliminated most of the early mammals, as well as many of the large terrestrial reptiles. These extinctions opened up a large number of terrestrial niches that were gradually filled during the Jurassic and Cretaceous periods, especially by dinosaurs (Figure 5.14). Many of the dinosaurs of this post-Triassic "golden age of reptiles" became quite large but, as noted earlier, some of them reduced their body size and, ultimately, gave rise to birds.

For dinosaurs, the golden age came to a close at the end of the Cretaceous period, when all of them, except for birds, became extinct. All of the pterosaurs also died off. A likely cause for this mass extinction was a 10–14 km wide asteroid that hit the earth, shooting impact debris into the atmosphere, creating a global firestorm, and darkening the skies for many months (see Chapter 6; Longrich et al., 2012). Around the same time, geologically speaking, an enormous set of volcanic eruptions in Western India (and possibly elsewhere) probably triggered severe global warming and ocean acidification (Brannen, 2017). Remarkably, some early birds and mammals survived these compounding catastrophes and then diversified quite rapidly; so did the ray-finned fishes in the aquatic habitats (Figure 5.14). Thus we see a general pattern: mass extinctions wipe out vast numbers of species, but the survivors eventually diversify again, largely replacing what was lost (albeit differing in many important respects). In fact, animal diversity increased considerably over the long stretch of evolutionary time, despite repeated massive extinctions (Figure 5.14).

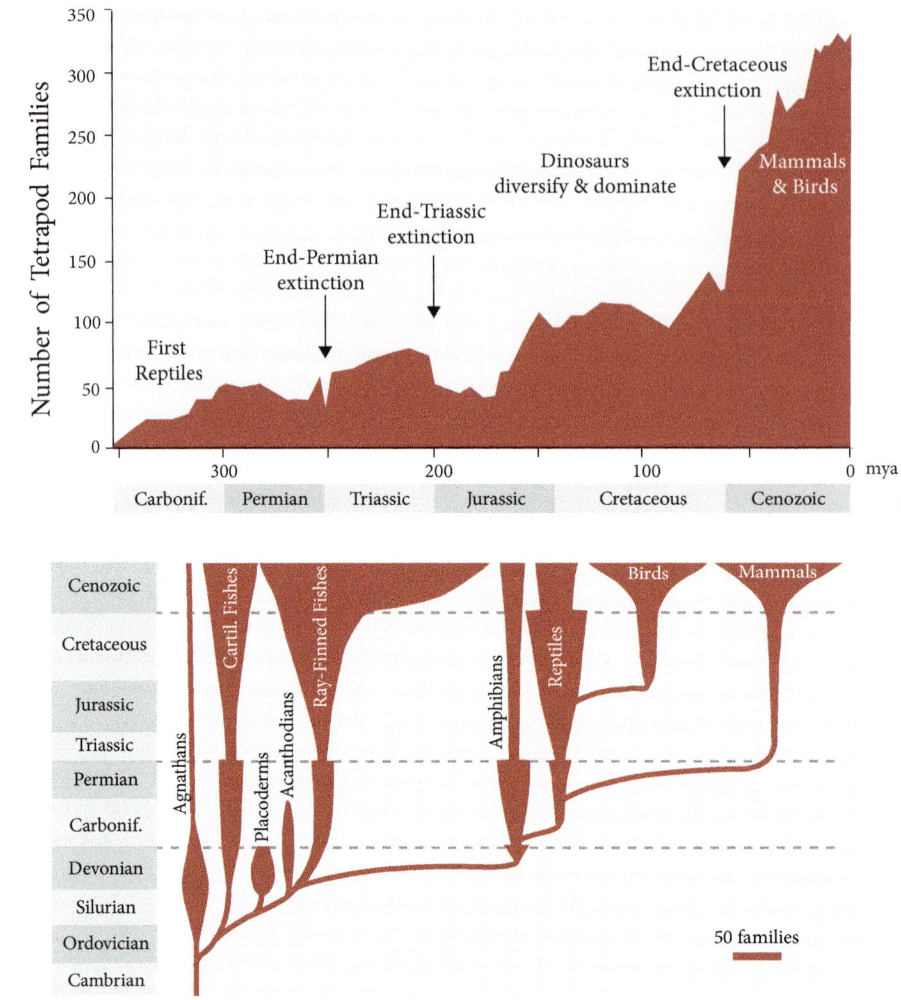

Figure 5.14 Diversification and extinction of vertebrates. The graph at the top shows the history of changes in the number of tetrapod families ($n = 840$) over the last 350 million years. The bottom graph also charts diversity at the family level but spans a longer period and includes non-tetrapods. Both graphs are based on data compiled in (Benton, 1993).

Adapted from Sahney et al. (2010) and Benton (1998).

5.3. Enhanced Sense Organs

Because sauropsids and synapsids diverged from one another relatively soon after the origin of amniotes, it is difficult to reconstruct the features of early amniotes without discussing mammals as well as reptiles and birds. However, to keep the discussion manageable, our focus in this chapter is on the sauropsids, with a discussion of mammals deferred to Chapter 6. In particular, we highlight the main evolutionary changes that made the sauropsids so successful, despite the periods of extinction. With regard to sensory systems, our discussion emphasizes vision, hearing, and the chemical senses. By the time the early amniotes had become fully terrestrial, the mechanosensory and electrosensory lateral line systems that had been so important in the life of primitive fishes and aquatic amphibians had disappeared entirely. Interesting variation also exists in the somatosensory and proprioceptive systems, but comparative data on these systems are so scarce that little can be said about their evolution in amniotes, especially in sauropsids.

5.3.1. Vision, from Infrared to Ultraviolet

Most living sauropsids have excellent vision and large eyes. Therefore, it is likely that early sauropsids already had large eyes. Eye size may have increased during the early stages of amniote and sauropsid phylogeny, but those changes were probably minor. In contrast, eye size must have increased substantially in the lineage leading to birds. The eye of an ostrich is an astounding 5 cm in diameter, the largest of any land vertebrate (Walls, 1942). Even relative to body size, birds have unusually large eyes (Hall and Heesy, 2010). In fact, the eyes are so large in many birds that there is little room within the eye socket for extrinsic eye muscles, which probably became less important in birds as their entire head became increasingly mobile (perched as it is on top of a long, flexible neck). With these large eyes, birds maximize their ability to see in dim light, see fine detail on nearby objects, and see small objects at great distances. As noted in Chapter 4, such fine-grained distance vision is essentially impossible underwater, because light is scattered so much more in water than in air.

In contrast to the other sauropsids, snakes have relatively small eyes, most likely stemming from an evolutionary "bottleneck" (defined as a temporary period of perilously small population size for a species or larger lineage) during which all snakes lived underground. Some more recent snakes abandoned the burrowing life and now rely more heavily on vision, but most of them still retain relatively small eyes (Liu et al., 2012). Consistent with this underground bottleneck hypothesis, all snakes have translucent, immobile eyelids that cover the eyes and protect them against abrasion. Amazingly, several lineages of snakes (rattlesnakes and other pit vipers, most boas, and pythons) independently evolved the ability to detect infrared radiation using modified temperature sensors in their trigeminal somatosensory system (Molenaar, 1992; Goris, 2011). They use their infrared sensors to hunt for small warm-blooded creatures, such as mice, at night and underground. This ability would have been especially useful after the extinction of the non-avian dinosaurs, when grasses evolved and small rodents diversified. Indeed, that is when pit vipers arose and became remarkably diverse (Hsiang et al., 2015). Some snakes may even be able to use this functionally vision-like system to target their strikes at the warmer, most vulnerable parts of the body (Kardong and Mackessy, 1991).

They can also use their infrared sensors to find places where it is nice and warm (Krochmal, 2004).

5.3.1.1. Cornea, Lens, and Papillary Cone

The internal structure of sauropsid eyes is clearly optimized for vision in air, rather than water. Most importantly, the cornea is steeper in sauropsids than in amphibians and is responsible for most of the refraction (light bending) in sauropsid eyes. Since this feature is shared with synapsids, it probably occurred with the origin of amniotes. Remarkably, the curvature of the lens in turtles varies dorsoventrally in such a way that a nearby object on the ground can be in focus at the same time as a potential threat more distant in the sky (Figure 5.15). Similar variation in corneal curvature has also been described in birds and some amphibians (Schaeffel et al., 1994). Therefore, some degree of variation in corneal curvature probably evolved early in sauropsid phylogeny. However, this variation was probably modified repeatedly within the sauropsids, especially in species that keep their eyes at a relatively constant distance from the ground (e.g., turtles).

The lens is less spherical in amniotes than in anamniotes and can change its shape under the influence of ciliary eye muscles (Ott, 2005). Those changes in shape alter the lens curvature and, thus, adjust the focal plane of the incoming light. This mechanism, referred to as accommodation, allows the animals to change their visual focus from nearby objects to distant ones, and vice versa. In contrast, most anamniotes change visual focus by moving the lens forward or backward in the eye. Curiously, the ciliary muscles are striated in most sauropsids but smooth (i.e., relatively slow and not under voluntary control) in mammals and most other vertebrates. Furthermore, the arrangement of the ciliary muscles differs between sauropsids and mammals, such that their contraction makes the lens less spherical in sauropsids but more spherical in mammals. We consider it unwise to call the sauropsid method of accommodation "better" than the others in any general sense, but it is worth noting that the range of accommodation in cormorants, which are fish-hunting birds, is roughly three times the human range (up to 60 diopters; Katzir and Howland, 2003). This extreme ability to change the eye's focus is due to a very large and powerful ciliary muscle that acts on the shape of the lens and allows these birds to have good vision both in water and in air, with the cornea bending light rays only in air.

A striking feature of the eye in most sauropsids is a large, highly vascularized structure that protrudes from the optic nerve head into the fluid-filled space behind the lens. It is called the papillary cone (*conus papillaris*) in lizards and the pecten in birds. Although the pecten is larger and more complex than the papillary cone (Gültiken et al., 2011), the two are probably homologous to one another. That said, turtles and crocodilians lack a papillary cone (at least in adulthood), raising the possibility that these structures evolved independently in birds and lizards. The function of the pecten and the conus are likewise unclear. Because the inner layers of the sauropsid retina are not covered by blood vessels (such vessels do exist in

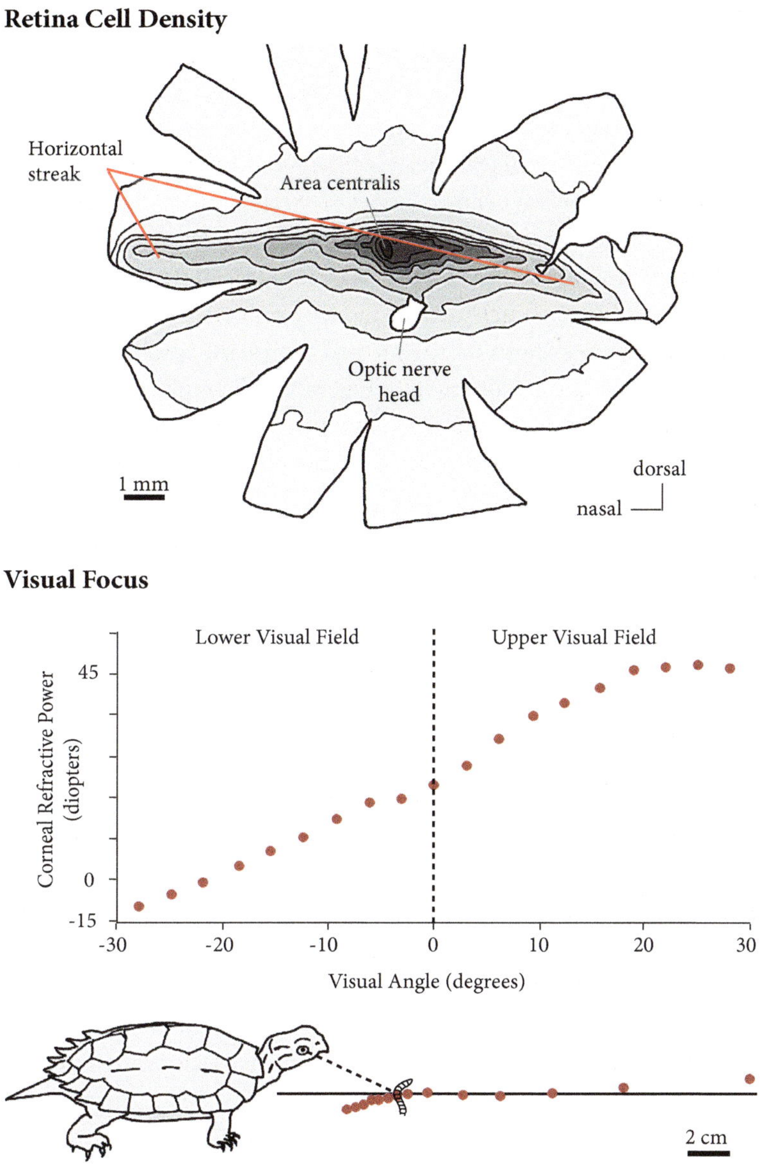

Figure 5.15 Adaptions of a turtle's retina and cornea. A cut and flattened retina of a turtle is shown at the top, with the density of retinal ganglion cells represented by different levels of gray. The diagram reveals both a central area and a horizontal streak, which is indicative of increased spatial resolution along the horizon. The middle graph shows that the refractive power of a turtle's retina increases steadily across the visual field, from bottom to top. As a result, when a turtle looks straight ahead, objects on the ground will tend to appear in focus regardless of their distance, which should aid in prey detection. This phenomenon is illustrated in the bottom diagram, where the red circles represent the retina's focal length at various distances along the ground (black horizontal line) while the turtle is focused on a worm 4 cm away.

Adapted from Peterson (1992) and Henze et al. (2004, with permission from Springer Nature).

mammals and amphibians), it is tempting to speculate that the pecten and conus provide those retinal layers with oxygen *via* diffusion through the intraocular fluid (Yu et al., 2009). However, the inner retinal layers of sauropsids are devoid of mitochondria (Hughes et al., 1972), implying that they have little or no capacity for oxidative phosphorylation. An intriguing alternative is that the conus and pecten may prevent the accumulation of lactic acid in the intraocular fluid, which would be a byproduct of obtaining energy through glycolysis (Brach, 1977). This hypothesis is supported by the observation that lesions of the pecten in birds lower the intraocular pH (Brach, 1976). In any case, the lack of blood vessels and mitochondria in the inner retina of sauropsids is probably a derived feature for this group and maximizes the amount of light reaching the photoreceptors (since blood and mitochondria absorb photons).

5.3.1.2. Sophisticated Retinas

The retinal photoreceptors and neurons are quite similar between sauropsids, amphibians, and lungfishes. They all have rods and cones, including some double-cones that probably arise developmentally by the fusion of two separate cones. Sauropsid cones express one of four different types of opsin genes, and each opsin responds optimally to different wavelengths of light. One of these opsins (SWS1) is tuned to very short wavelengths, extending into the ultraviolet (UV) range. Because homologs of this UV-sensitive opsin are found in many anamniotes, including amphibians, lungfishes, and teleosts, it is probably a primitive feature for amniotes (Hunt and Peichl, 2014; Cronin and Bok, 2016). However, the sensitivity spectrum of SWS1 shifted out of the UV and into the violet frequencies in several tetrapod lineages, including frogs and many mammals. Among birds, ostriches and a few more derived avian linages lack UV sensitivity, but others (e.g., hummingbirds) can see UV quite well (Chen et al., 1984). Based on these observations, it has been suggested that the sensitivity of SWS1 shifted toward violet light in early birds but then shifted back in multiple lineages of modern birds (Davies et al., 2012). This hypothesis is supported by the finding that such shifts can be accomplished by single amino acid substitutions (Yokoyama, 2000).

Aside from the photoreceptors, sauropsids also have the same principal types of retinal neurons that are found in lungfishes and amphibians, and they have multiple subtypes within each of these major cell types. As mentioned in Chapter 4, early amniotes probably had a greater diversity of retinal neurons than their amphibian ancestors. Retinal neuron diversity probably increased further in the sauropsid lineage. In particular, it is likely that the retinas of birds and diurnal lizards are more complex than those of turtles. It is difficult to substantiate such claims, however, because comprehensive comparative analyses of retinal cell types are scarce (Walls, 1942; Cajal, 1972). Indeed, we know much more about the retinal anatomy and physiology of turtles (Ammermuller and Kolb, 1996) than any other sauropsid, presumably because turtle nervous systems can function in low oxygen and are, therefore, ideal for in vitro physiological studies.

Clear species differences do, however, exist in the distribution of cells across the retina. Turtles and many lizards, for example, have a band of increased photoreceptor and retinal ganglion cell density that stretches horizontally across the retina and is, therefore, referred to as a horizontal streak (Figure 5.15). This feature is thought to provide the animal with increased visual acuity along the horizon, where animals that walk along the ground might expect important objects to appear most frequently. In addition, turtles and most other sauropsids tend to have a roughly circular area of increased cell density right in the center of their eye, the so-called *area centralis* (Figure 5.15). Similar areas are also found in amphibians and many anamniotes, suggesting that this feature is primitive for sauropsids. Whether the horizontal streaks are likewise primitive is more difficult to determine, because their existence is highly variable and correlated with an animal's ecology (Collin and Pettigrew, 1988a, b).

One feature of the retina that appears to be derived in sauropsids is the fovea, an area of high photoreceptor density in which the other retinal layers are pushed aside, thereby thinning the retina and reducing light scattering. Amphibians and lungfishes do not possess a fovea, but tuataras, most lizards, and birds do (Schwab, 2012). So do some teleosts and many primates (see Chapter 6), but these species evolved their foveae independently of sauropsids. Birds are interesting because many of the predatory species (e.g., falcons, eagles, swallows, and kingfishers) have two foveae per eye (Fite and Rosenfield-Wessels, 1975; Moroney and Pettigrew, 1987; Tucker, 2000; Tyrrell and Fernández-Juricic, 2017). In these birds, one fovea is usually located in the temporal retina and aimed at the binocular visual field in front of the animal. The second fovea tends to be located in the central retina and is, therefore, aimed more laterally. Intriguingly, this central fovea often has much steeper walls than the temporal fovea, much like the single fovea of tuataras (Figure 5.16). The functional significance of having steep foveal walls remains unclear, but they probably bend incoming rays of light (because their refractive index is different from that of the intraocular fluid). This refraction magnifies the image projected onto the foveal photoreceptors and increases visual resolution in a small part of the central visual field (Locket, 1992). In addition, it introduces small distortions in the projected image, which the animals may use to bring that image into focus rapidly (Harkness and Bennet-Clark, 1978). The fact that most turtles, snakes, and crocodilians lack foveae probably reflects a history of reduced vision, most likely when they became aquatic or, in the case of snakes, went underground.

We conclude that eyes were already well developed by the time the amniotes emerged, but then evolved a few additional features, such as the fovea and papillary cone. They also lost some structural features, such as the pre-retinal vasculature. Overall, though, most of the evolutionary changes in the retina of sauropsids served to improve vision on land, especially in terms of spatial resolution. The major exception are the snakes, which reduced their visual capacity, at least during the early stages of their phylogeny.

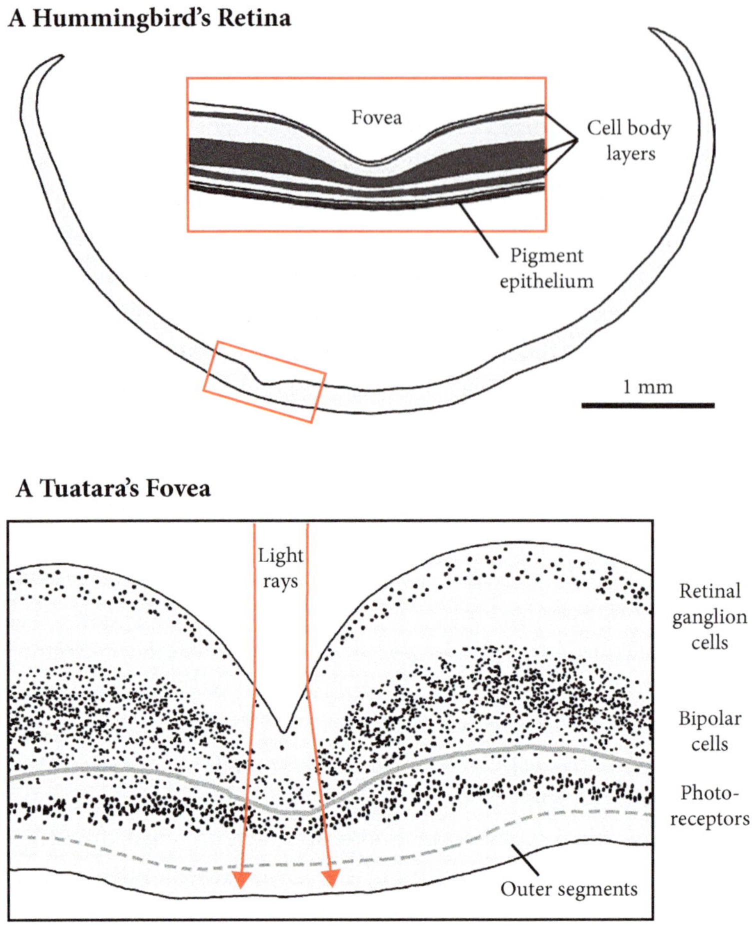

Figure 5.16 Sauropsid retinas with foveae. Many reptiles and birds have an indentation in their retina, called a fovea. Shown here are sections through the central foveae of an Anna's hummingbird (top) and the tuatara (*Sphenodon*). Both of them have relatively steep foveal walls. The function of such steep-walled foveae remains uncertain, but probably involves a slight refraction of light rays (red arrows) at the interface between the retina and the intraocular fluid. This refraction would change the incoming light's plane of focus and potentially increase spatial resolution.
Adapted from Lisney et al. (2015) and Schwab (2012).

5.3.2. Tympanic Ears and High-Frequency Hearing

The sense of hearing underwent important changes after amniotes became fully terrestrial. As we discussed in Chapter 4, tympanic membranes that allowed for high frequency hearing in air did not evolve in early tetrapods, as people used to think. Rather, they evolved independently in anurans and amniotes. Indeed, fossil data on the morphology of the stapes and the skull bones surrounding a putative tympanum strongly suggest that tympanic ears evolved 50–100 million years after the origin of amniotes, and did so independently in synapsids, lepidosaurs, and archosaurs (Clack and Allin, 2004; Clack, 2012). Consistent with this hypothesis, comparative developmental data have shown that the tympanic membrane develops in different locations, using different molecular pathways, in birds and mammals (Kitazawa et al., 2015). This phylogenetic independence explains why many aspects of both the middle and the inner ear are very different between the major amniote lineages. In the present chapter we focus on the evolution of tympanic ears in sauropsids.

The earliest sauropsids did not have a tympanum, but most modern sauropsids do (Manley and Clack, 2004). In turtles and most lizards, the tympanic membrane lies close to the surface of the head and is protected from the outside world by a thin layer of skin. Tuataras do not have a fully developed tympanic membrane, but they possess a similar structure that is covered with skin and probably functions like an eardrum (Wever, 1978). A few groups of lizards and all birds modified this primitive arrangement by moving the membrane deeper into the head and connecting it to the surface of the head via an external ear canal. In contrast, snakes have lost their tympanum, mainly because they evolved loosely connected (rather than tightly jointed) jaw bones that are, presumably, incompatible with the retention of a tympanum. The loss of eardrums in snakes may also be linked to this group's evolutionary history of living underground and, therefore, having little opportunity to hear airborne vibrations. Nonetheless, some extant snakes are capable of hearing airborne vibrations quite well, up to ~500 Hz (Young, 2003).

In contrast to early amniotes (see Figure 4.13 in Chapter 4) and early sauropsids, modern sauropsids have an extremely long and slender stapes, called the columella. In collaboration with a cartilaginous "extra-columella" that attaches to the tympanum's interior surface, the columella can convey vibrations up to ~10 kHz to the membrane-covered oval window, through which those vibrations enter the cochlea. This arrangement is functionally similar to the mammalian condition, but mammals have three middle ear bones instead of one (see Chapter 6). Like mammals, crown sauropsids evolved a round window that can relieve sound-induced pressure in the intracochlear fluid and, thereby, make hearing more sensitive. One major difference between mammals and sauropsids is that the left and right middle ears of most sauropsids, especially lizards, are coupled through the oral cavity (Christensen-Dalsgaard and Manley, 2008). This interaural coupling allows sound waves to pass through the head and drive each tympanum from the inside as well as the outside of the head, an arrangement that makes the ear's response to sounds much more directional. Birds have reduced this interaural coupling (which was probably a feature of all tympanic ears when they initially evolved, regardless of lineage; Christensen-Dalsgaard and Carr, 2008), but mammals have eliminated it almost entirely. Instead, they increased the ear's directionality by evolving external ear flaps.

In amniotes, the sensory epithelium that is primarily responsible for sensing sounds, rather than head movements or gravity, is called the basilar papilla (BP). Because amphibians and coelacanths also have a BP (Fritzsch, 1987), this sense organ was probably present in early amniotes. However, the BP in crown amniotes lies on top of a membrane, called the basilar membrane, rather than a solid foundation. When this basilar membrane vibrates up or down in response to sound-induced pressure waves, the stereocilia of the hair cells that sit on top of this membrane bend back and forth, converting the sound energy into neural signals. This general arrangement probably evolved in conjunction with the evolution of tympanic ears. However, early sauropsids probably had a relatively small BP, less than 1 mm long and containing fewer than 1,000 hair cells. This primitive condition has been retained in cryptodire turtles and tuataras (Manley et al., 2017). The auditory hair cells in early sauropsids responded preferentially to specific sound frequencies, but this frequency tuning was probably not based on the mechanical properties of the basilar membrane, as it is in mammals. Instead, it probably derived from the specific type and number of ion channels that those hair cells express (Fettiplace and Fuchs, 1999). This kind of "electrical tuning" of hair cells is found in many sauropsids and, as we noted in Chapter 4, amphibians. Overall, early sauropsids could likely hear quite well in air, as long as the sound frequencies were between ~100 and 1,000 Hz.

The ability to hear sound frequencies higher than 1 kHz, and to discriminate those sounds from one another, evolved independently in squamates and archosaurs. In lizards this expansion of the hearing range correlates with an increase in the number of cochlear hair cells (to a maximum of about 2,000 cells) and a division of the BP into multiple subregions, some of which are specifically dedicated to sensing high frequencies (Miller, 1992). The auditory hair cells in lizards also exhibit systematic structural variation, such as differences in the length of their stereocilia. This variation in the physical properties of the stereocilia is thought to generate a form of micromechanical tuning that probably replaced the primitive electrical tuning mechanisms. Lizards lengthened their BP, relative to the primitive sauropsid condition, but it still remains relatively short (maximum length ~2 mm).

Archosaurs pursued a somewhat different path. They also evolved micromechanical tuning but, in contrast to lizards, they lengthened their BP considerably. It is roughly 4 mm long in extant crocodiles and more than 10 cm long in some large dinosaurs, as estimated from the length of their fossilized cochlear duct (Figure 5.17; Gleich et al., 2004). This elongation of the BP correlates with body size and hearing range (Walsh et al., 2009), such that larger animals have longer BPs and, in general, a reduced ability to hear high frequencies. This finding seems counterintuitive, as archosaurs tend to have better high-frequency hearing than other sauropsids, including most lizards. The puzzle is at least partially resolved once BP length is adjusted for body size. This procedure reveals that, relative to body size, sauropsids with longer BPs tend to have better high-frequency hearing and, thus, a broader hearing range (Figure 5.17). Birds in particular have lengthened their BP and extended their high-frequency hearing limit. It has been estimated that early birds had a BP that was about 4 mm long and covered with 10,000 hair cells, allowing these animals to hear sounds from 80 to 5,000 Hz (Gleich et al., 2004). Some

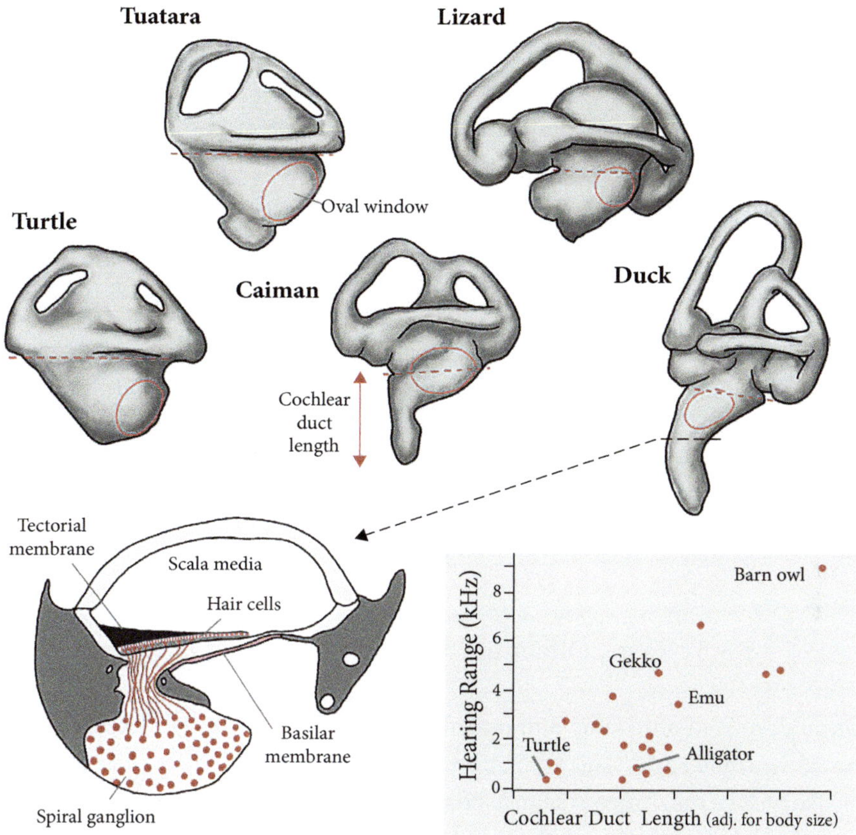

Figure 5.17 Sauropsid inner ears. Shown here in lateral views are the vestibular apparatus and cochlear ducts (aka lagenar recesses; Fritzsch, 1987) of five different sauropsid species (*Sphenodon punctatus, Gambelia wislizenii, Chelydra serpentina, Caiman crocodilus*, and *Aythya fuligula*), as reconstructed from micro-CT scans. The graph shows that the length of the cochlear duct (adjusted for body size) correlates with the ability to hear a broad range of frequencies. In essence, this means that, body weight being equal, species with long cochlear ducts tend to have better high frequency hearing than species with shorter cochleas. The diagram at the bottom left represents a cross section through the cochlea of a chicken, showing the neural elements in red and the basilar membrane in pink.

Adapted from Walsh et al. (2009) and Gleich et al. (2004).

owls, which are specialized for hearing the sounds of prey at night, have even more specialized ears. Their BP is around 9.5–11.5 mm long (Smith et al., 1985), and they can hear sound frequencies up to ~10 kHz.

What benefits did lizards and archosaurs derive from expanding their hearing range to higher frequencies? One possibility is that it may have allowed them to hear the buzzing of flying insects, which diversified substantially during the Cretaceous period, in conjunction with the evolution of flowering plants. Another possibility is that high-frequency hearing evolved together with high-frequency vocalizations, such that animals with high-frequency hearing could take advantage of an essentially "private channel" of the sound spectrum. This function was probably a major factor at least in geckos and crocodilians, which have the most complex vocalizations among non-avian sauropsids. Yet another, complementary hypothesis is that high-frequency hearing improved the ability to localize sounds. Many animals localize sounds by comparing the intensity of the signals detected at the two ears (Walton et al., 2017), but low sound frequencies are attenuated relatively little by the intervening head, especially if that head is small (as in most lizards and birds). This problem can be solved by hearing higher frequencies. Barn owls, for example, can localize the rustling of a mouse running through leaf litter by listening to the high-frequency elements of that sound and comparing their amplitude between the two ears. Owls can also localize sounds by comparing when sound waves arrive at the two ears (Konishi, 2003), but this mechanism works well only at low frequencies. Thus, even for owls, adding information from the high-frequency channel improves localization accuracy.

5.3.3. Taste, Olfaction, and the Vomeronasal Sense

As animals moved onto land, and into air, they were exposed to different odors and tastes. One would expect this transition to be associated with major changes in chemosensory systems. Indeed, sauropsids have experienced significant evolutionary changes in the sensory structures of the taste, olfactory, and vomeronasal systems.

The genes for sensing salty and sour substances are broadly conserved across sauropsids, but a gene needed for tasting sweet substances (T1R2) is missing in most birds. Birds also have a very small number of bitter receptors (T2Rs), relative to crocodilians, lizards, and mammals (Dong et al., 2009; Wang and Zhao, 2015). Penguins have lost these receptors entirely (Zhao et al., 2015), but the T2R family of genes proliferated in perching birds, expanding to 18 family members in some sparrows (Davis et al., 2010). Another interesting observation is that penguins lost the umami taste receptors (T1R1 and T1R3), which are generally used to sense proteins (Zhao et al., 2015). This gene loss may have been linked to the penguin habit of eating fish whole, leaving little opportunity for tasting in the oral cavity. This idea is supported by the recent finding that umami receptors, as well as several other taste receptor types, were lost in whales, which also gulp their food (Feng et al., 2014). Finally, a recent study showed that hummingbirds "re-purposed" their umami receptor into a carbohydrate receptor, thus essentially replacing the sweet receptor that early birds had lost (Baldwin et al., 2014). This novel sweet receptor presumably helps hummingbirds find sweet nectar, their main food source.

The olfactory receptor (OR) genes are more diverse in turtles and crocodilians than in lizards (Khan et al., 2015; Vandewege et al., 2016). Particularly interesting is that different groups of sauropsids selectively expanded or reduced different OR subfamilies. Thus, turtles selectively expanded a group of OR genes that is thought to encode receptors for aqueous odorants (Wang et al., 2013), and most ORs in birds belong to a receptor family that is unique to birds (Steiger and Kuryshev, 2009). The latter finding suggests that birds may be capable of sensing diverse odorants that we and most other vertebrates cannot detect. Little is known about those odorants, but sea birds can smell dimethyl sulfide, which is produced by injured phytoplankton (Nevitt, 2008; Dell'Ariccia et al., 2014). Most likely, seabirds use this odorant to locate and then prey on large schools of fish. Similarly, some birds can smell compounds released by plants when they are infested by herbivorous insects (Amo et al., 2013). This ability allows the birds to find and eat those insects, thereby aiding the plants. Finally, numerous studies have suggested that homing pigeons can use olfactory cues to navigate (Wallraff, 2014). Unfortunately, the molecular identities of those olfactory cues remain unknown.

The available genomic and behavioral data clearly support the hypothesis that birds in general have a well-developed, if highly specialized, sense of smell. This finding runs counter to the old idea that birds must have a poor sense of smell because they tend to have proportionately small olfactory bulbs (see Figures. 5.6 and 5.7). However, the fact that birds greatly enlarged their telencephalon and cerebellum implies that most other brain regions must be proportionately small. Moreover, turkey vultures, kiwis, and seabirds (e.g., petrels and albatrosses) do have large olfactory bulbs (Corfield et al., 2014, 2015b; Grigg et al., 2017). A comparative analysis of non-avian theropods, extinct birds, and extant birds also revealed no major reductions in olfactory bulb size, relative to body size (Zelenitsky 2011). It was mainly in the perching birds and a few smaller avian orders (e.g., penguins) that relative olfactory bulb size decreased significantly (Corfield et al., 2015b).

Aside from the main olfactory system, early amniotes had an accessory olfactory system, more commonly known as the vomeronasal system. This sensory system used to be thought of as being specialized for sensing pheromones, but it was then reframed as being dedicated mainly to sensing non-volatile odorants. This formulation may also be too simple, and it may be better to think of the vomeronasal system as being specialized for sensing relatively large molecules that are not carried far in air (Baxi et al., 2006). As we discussed in Chapter 4, the sensory cells of the vomeronasal system are intermingled with the olfactory sensory neurons in many fishes, but segregated into a separate epithelium, called the vomeronasal organ, in amphibians (Eisthen, 2000). The vomeronasal receptor molecules are G protein-coupled receptors, but they are only distantly related to olfactory receptors or opsins, the other main types of G protein-coupled sensory receptors in vertebrates. Of the two main vomeronasal receptor gene families, the V2 receptors outnumber the V1 receptors in all major groups of vertebrates, except in mammals, which greatly expanded the V1 receptor gene family (Figure 5.18; Shi and Zhang, 2007; Brykczynska et al., 2013). A comparative analysis of V1 receptor genes indicates that they multiplied independently in multiple mammalian lineages and that a surprisingly large fraction of these genes became nonfunctional (Young et al., 2010). The behavioral significance of all this variation is unclear.

Among sauropsids, the vomeronasal system is best developed in squamates. In these animals (but not in tuataras!) the vomeronasal organ is entirely separate from the nasal epithelium and not connected to the outside world through the external nostrils. Instead, the vomeronasal organ of squamates is connected to the oral cavity through a long, slender duct (Figure 5.18). It is through this duct that chemicals in the environment gain access to the vomeronasal epithelium. Snakes, for example, use their long forked tongues to pick up odorants from the ground and (to a lesser extent) the air. They then insert the tongue tips into the vomeronasal ducts, depositing the odorants onto the epithelium (Halpern and Kubie, 1980). The fact that the left and right tongue tips transfer odorants separately onto the left and right vomeronasal epithelia probably provides directional information that helps the snakes locate an odor source. Monitor lizards also have forked tongues and engage in "tongue flicking." However, most lizards tongue flick without having forked tongues. Indeed, their tongues are often too thick to be inserted into the

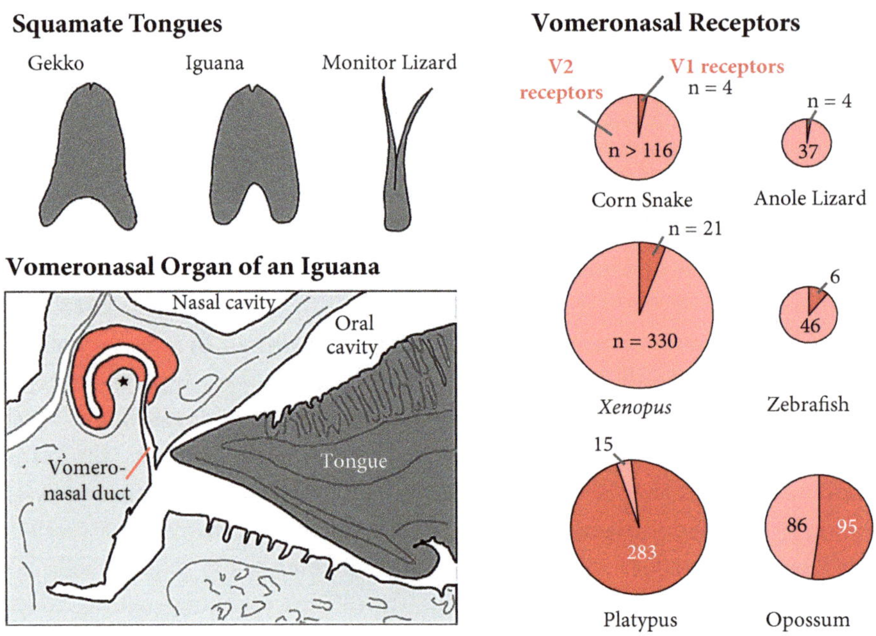

Figure 5.18 The vomeronasal system of reptiles. As illustrated in the parasagittal section at the bottom left, the vomeronasal epithelium in reptiles is separate from the nasal epithelium and is connected to the oral cavity by a narrow duct (rostral is to the left, dorsal to the top). Snakes and a few lizards have split tongues with slender tips that they can stick into this duct to convey odorants to the sensory neurons. However, most lizards have broad tongues that they use like pistons to push fluid into the vomeronasal organ, which lies atop a spongy structure (asterisk) and is therefore elastic. The vomeronasal receptors are mainly of the V2 type in lizards and snakes, but of the V1 type in mammals. Having a proportionately small number of V1 type receptors appears to be the primitive condition.
Adapted from Filoramo and Schwenk (2009) and Brykczynska et al. (2013).

vomeronasal ducts (Figure 5.18). In these species, odorants are brought into the oral cavity via the tongue or by direct contact with the snout (Graves and Halpern, 1989), and the tongue is then used like a piston to push the intraoral fluid into the vomeronasal organ (Filoramo and Schwenk, 2009). This pumping action is aided by a spongy mushroom-like body (asterisk in Figure 5.18) that is compressed when the tongue pushes the intraoral fluid in. The mushroom-like body then recoils when the tongue retracts, thereby squeezing the fluid back out of the vomeronasal organ.

The vomeronasal system was lost in archosaurs. The reason for this loss of an entire sensory system remains unclear, but it may have to do with the bipedal stance of early crocodilians and birds (see Section 5.4), since such a stance might make it difficult for the tongue to make regular contact with the ground. Whether turtles, the sister group of archosaurs, have a vomeronasal system remains debatable. Some authors have claimed that turtles have no vomeronasal receptors at all (Taniguchi and Taniguchi, 2014), but others report that they do have a distinct vomeronasal epithelium. This epithelium seems to be continuous with the main olfactory epithelium in some species, but totally separate in others (Eisthen and Polese, 2007).

5.4. Changes in Motor Patterns and Control

Breathing is one kind of movement that changed dramatically when amniotes emerged. As noted in Section 5.2.1, early amniotes evolved the ability to rotate their ribs both inward and outward, thereby drawing air into their lungs and pushing it back out, respectively. This innovation clearly increased breathing efficiency. However, it is likely that early amniotes, just like today's lizards, could not run and breathe deeply at the same time (Carrier, 1987). Specifically, the alternating sideways bending of the trunk during running compresses the left and right halves of the lung alternately, pushing air back and forth between the two sides, rather than exchanging air with the outside (Figure 5.19). Therefore, early amniotes most likely ran only for short stretches and then paused to "catch their breath." Mammals overcame this mechanical constraint by evolving a muscular diaphragm (see Chapter 6), and turtles independently evolved an analogous arrangement of muscles (Brainerd and Owerkowicz, 2006). Archosaurs, too, evolved a set of muscles that is functionally equivalent to the mammalian diaphragm (Carrier and Farmer, 2000). Aside from breathing, the other forms of movement that changed dramatically as amniotes evolved are locomotion and feeding.

5.4.1. Locomotor Innovations

Terrestrial locomotion was facilitated in early amniotes by the evolution of a lighter skeleton and longer limbs (see Figure 5.11). In addition, lifting the body off the ground became easier when early amniotes evolved tighter connections between

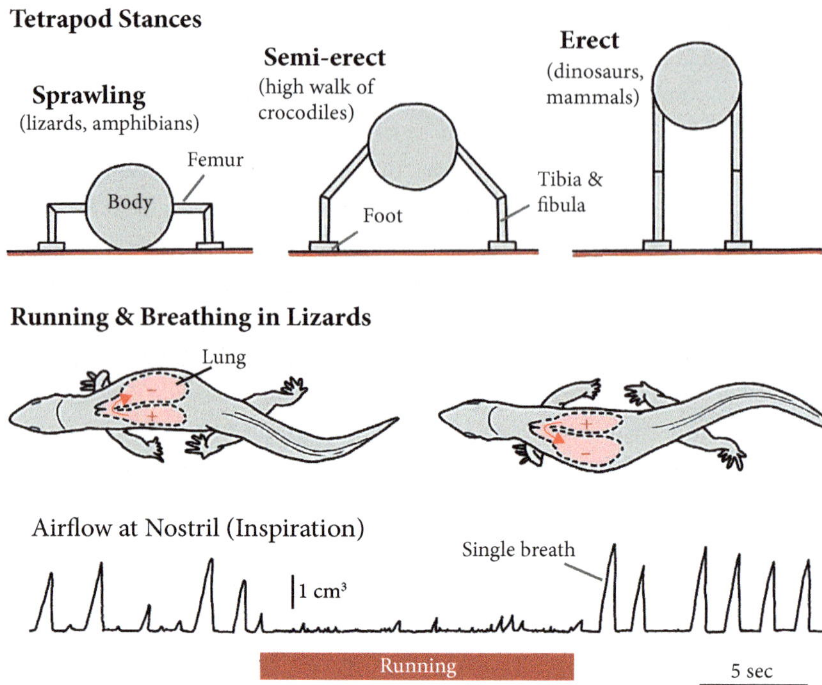

Figure 5.19 Standing, running and breathing in sauropsids. Most lizards and amphibians have a sprawling type of stance, with the legs extended laterally and the body close to the ground. In contrast, birds and mammals have an erect stance in which the legs are held underneath the body; so did most extinct dinosaurs. A conceptually intermediate "semi-erect" stance is exhibited by modern crocodiles when they are "high-walking." The bottom portion of this figure shows that lizards do not, and likely cannot, breathe while they are running. This constraint presumably arises because the sideways flexing of the body during running alternately compresses the left and right sides of the thorax, pumping air from the left lung into the right, and vice versa during the next step, with very little air being exchanged with the outside.
Adapted from Charig (1972) and Carrier (1987, with permission from Cambridge UP).

individual vertebrae. These "intervertebral articulations" stiffened the spine and, thus, made the trunk less likely to sag during terrestrial locomotion.

Along with evolutionary changes in the skeletal system came major changes in the trunk and limb muscles. As we discussed briefly in Chapter 2, the trunk muscles of the earliest vertebrates were divided into a series of rostrocaudal segments, called myomeres, that are separated by thin bands of connective tissue (see Figures 2.4 and 2.7). The individual muscle fibers in those myomeres run parallel to the body's long axis and flex the body laterally when they contract. This primitive myomeric arrangement is retained in all anamniotes with only relatively minor modifications; even in salamanders, most trunk muscles form myomeres. In amniotes, however, the trunk muscles are much more complex and diverse. They include, for instance, muscles that rotate the ribs and several sheets of muscle that crisscross the trunk diagonally. The two principal functions of the various trunk muscles in amniotes are to ventilate the lungs and to stabilize the body, not to bend it laterally as in anamniotes. The trunk of many amniotes does bend leftward and rightward (alternately) as the animal walks, but those lateral bends are generated primarily by movements of the legs, not by contractions of the trunk muscles.

The increased complexity of the trunk and limb muscles in amniotes was accompanied by major changes in their innervation. Most significant is that the motor neurons in the spinal cord of amniotes form distinct "motor pools," with all the neurons in a given pool innervating the same muscle. This segregation of the motor neurons presumably allows them to be controlled independently of one another by descending motor systems. Moreover, the spinal motor neurons are topographically arranged such that, within a given body segment, adjacent motor pools innervate muscles that are derived from adjacent embryonic precursors. This kind of topography has not been observed in any anamniote (including urodeles) and is, therefore, most likely an amniote innovation (Fetcho, 1987, 1992). Once established, the general relationship between spinal motor neuron and the muscles they innervate changed very little among amniotes, even as body form and modes of locomotion were modified extensively (e.g., in snakes; Ryan et al., 1998; Goslow et al., 2000).

Another major change in the evolution of vertebrate motor systems was the shift from a sprawling stance, in which the upper limbs are extended laterally away from the body, to an erect stance, in which the legs extend under the body (Figure 5.19). This shift toward an erect stance, along with a corresponding shift toward a parasagittal gait (in which the limbs rotate primarily in a parasagittal, rather than horizontal, plane) occurred in mammals and, independently, in archosaurs (Padian et al., 2010). It is not difficult to see that birds tend to keep their legs below their body as they walk, but most dinosaurs and even the early crocodilians and pterosaurs did so as well (Bakker, 1971; Alexander, 1985; Gauthier et al., 2011; Witton, 2013). Today's crocodilians reverted to a more sprawling stance, but they too are capable of a semi-erect "high-walk" (Figure 5.19) during which the feet are kept much closer to the trunk than during the more familiar sprawling mode of crocodile locomotion. The major benefit of keeping the legs directly under the body while walking or standing on land is probably that it makes it possible to use the limbs' long bones as weight-supporting struts, thereby minimizing the muscular effort required to battle gravity and reducing the bending forces acting on those bones (Biewener, 1989).

Early archosaurs not only stood erect, most of them (but not the pterosaurs) used only their hind legs to walk or run. Bipedal locomotion is advantageous because it frees the forelimbs for other purposes and elevates the head far off the ground, improving an animal's ability to see and smell distant predators or prey. However, balancing the body on two legs is challenging, even if the tail can be used as a third means of support. The problem is especially severe in long-necked animals (notably birds) in which the vestibular apparatus is far away from the body's center of gravity. Intriguingly, bipedal archosaurs may have solved this problem by evolving a specialized sense organ in an expanded region of the vertebral canal directly above the legs. This intriguing structure resembles a semicircular canal and contains mechanosensory cells (Rosenberg and Necker, 2002; Necker, 2005, 2006). Closely associated with this putative balancing organ is a large collection of glycogen-filled astrocytes, called the glycogen body. It probably provides the hindlimb motor neurons with extra metabolic energy in times of need (e.g., when being chased by predators).

Birds, of course, can fly as well as run. Much has been written about the evolutionary origin of avian flight, especially about whether the first birds flew "from the trees down" or "from the ground up" (see Lewin, 1983). We favor the latter hypothesis because the archosaurian ancestors of birds were much better adapted for running on the ground than climbing in trees, but the issue is far from settled. For example, recent studies suggest that early birds used their "incipient" wings to help them run up steep inclines and control their descent when jumping off a precipice (Heers and Dial, 2012). For present purposes, the most interesting question is what kind of neural innovations may have accompanied the origin of flight. Surely, the evolution of flight must have been associated with major changes in the central pattern generators controlling limb movements. Specifically, birds must have uncoupled the wing controllers from those driving the legs, a separation that may have begun with earlier dinosaurs that walked primarily on their hind legs but could not fly (i.e., theropods). Additional modifications would have ensured that the two wings are moved in concert with one another, rather than alternately. Unfortunately, these issues have not yet been addressed by comparative neurobiologists.

5.4.2. Necks, Jaws, and Catching Prey

Early amniotes separated their skull from the pectoral girdle and, thus, evolved a distinct neck and a more mobile head. The neck then elongated in several archosaur lineages, especially in pterosaurs and birds. Importantly, the combination of a long neck and light-weight skull allows most birds to capture elusive prey with rapid head strikes. Pterosaurs may have had similar abilities, though their heads were relatively large and frequently adorned with large "crests" that probably functioned in social displays (Witton, 2013). An important correlate of birds having long necks is that these animals possess exquisite reflexes for stabilizing the head, relative to the environment, as the animals walk on the ground or fly (Necker, 2007). The neck movements involved in these behaviors must involve neural pathways that descend from the brain to the cervical spinal cord. Although a variety of such pathways has been described (Zeier and Karten, 1971; Karten et al., 1973; Correia et al., 1983), their respective functions remain poorly understood. In particular, it is often difficult to disentangle the neck control pathways from those that control movements of the legs, which some birds also use to catch and "handle" food.

Once food is caught, it must be ingested. As part of this process, most mammals chew their food; in contrast, amphibians just "grab and gulp." Tuataras move their lower jaw forward and backward to shred food (Jones et al., 2012), but most lizards reduce their food simply by biting down on it repeatedly. However, squamates do possess another important feeding-related adaptation, namely an unusually mobile set of jaws (i.e., a kinetic skull; Herrel et al., 2007). Specifically, many lizards and all snakes can elevate their upper jaw, which greatly increases the size of their maximum gape. In addition, many squamates can move different parts of their jaws independently of one another, which helps to move the food into the digestive tract. Snakes represent the extreme end of this trend, as they can eat prey larger than their head, move that prey into their gut without chewing, and then digest it over the course of several days or weeks (Gans, 1961). We strongly suspect that the kinetic skull was a "key innovation" that accounts for the enormous evolutionary success of lepidosaurs, which represent more than 95% of all extant non-avian sauropsids.

5.5. Changes in the Brains of Sauropsids

Relative brain weight probably changed only slightly with the origin of amniotes, and some of this modest increase would likely have resulted from a decrease in the skeleton's weight (see Figure 5.11), rather than an increase in brain size. However, relative brain size did increase substantially at later stages of amniote phylogeny, especially in birds (Figure 5.20) and in mammals. We will return to this topic in Chapters 6 and 7. Here we focus on evolutionary changes in brain structure and, to some extent, function. As in the rest of this chapter, we emphasize changes in the sauropsid lineage.

5.5.1. Hindbrain Auditory Circuits

The hindbrain of amniotes is similar to that of amphibians, but some innovations are evident, especially in the auditory circuits. This is not surprising, given the dramatic evolutionary changes in the middle and inner ear (see Section 5.3). As we discussed in Chapter 4, the neurons innervating the inner ear in early tetrapods probably projected to a ventral octavolateralis zone. Anurans later evolved an additional auditory hindbrain area called the dorsolateral nucleus. Functionally similar cell groups evolved in synapsids and sauropsids, where they are called the cochlear nuclei (Figure 5.21). However, given that tympanic ears evolved independently in anurans and amniotes (see Chapter 4, Section 4.3.2) and the lack of dorsolateral nuclei in non-anuran lissamphibians, we conclude that the cochlear nuclei of amniotes are probably not homologous to the dorsolateral auditory nucleus of frogs. In sauropsids, the cochlear nuclei are generally divisible into a large-celled nucleus magnocellularis and a separate nucleus angularis (Tang et al.,

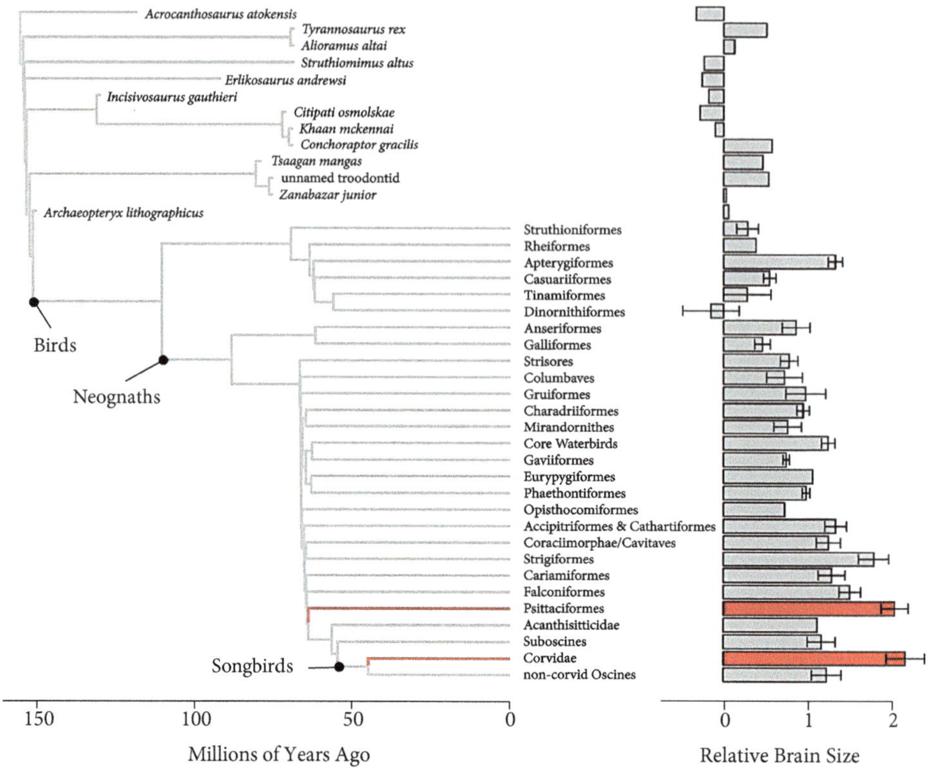

Figure 5.20 Relative brain size in birds and other theropods. A phylogenetic regression of log-transformed brain and body sizes for 308 extant birds, 22 extinct birds, and 12 non-avian theropods reveals several major evolutionary changes in relative brain size. A phylogenetic ANCOVA (Smaers and Rohlf, 2016) indicates that relative brain size increased significantly in parrots and, independently, in corvids (e.g., ravens and crows).

Based on unpublished analyses by Ksepka DT, Balanoff AM, Smith NA, Bever GS, Braun EL, Burleigh G, Clarke JA, Colbert MW, Corfield JR, Degrange FJ, De Pietri VL, Early CM, Field DJ, Gignac PM, Gold MEL, Kimball RT, Lefebvre L, Marugán-Lobón J, Norell MA, Scofield RP, Tambussi CP, Torres CR, van Tuinen M, Walsh SA, Watanabe A, Witmer LM & Smaers JB. Courtesy of Daniel Ksepka, Amy Balanoff, Adam Smith, and Jeroen Smaers.

2012). Nucleus angularis projects directly to the auditory midbrain area, but nucleus magnocellularis projects to an adjacent cell group called nucleus laminaris. The latter nucleus develops in close association with the cochlear nuclei but does not receive direct inputs from the inner ear.

In chickens and other basal birds, nucleus laminaris comprises a single layer of neurons that receive input from both the ipsilateral and the contralateral nucleus magnocellularis (Figure 5.21; Kubke and Carr, 2006). The latter axons vary systematically in length, such that action potentials from the inner ear take longer to reach progressively more lateral neurons in nucleus laminaris. Because of this axonal "delay line," each laminaris neuron receives coincident ipsilateral and contralateral synaptic input at a specific interaural time delay, which corresponds to

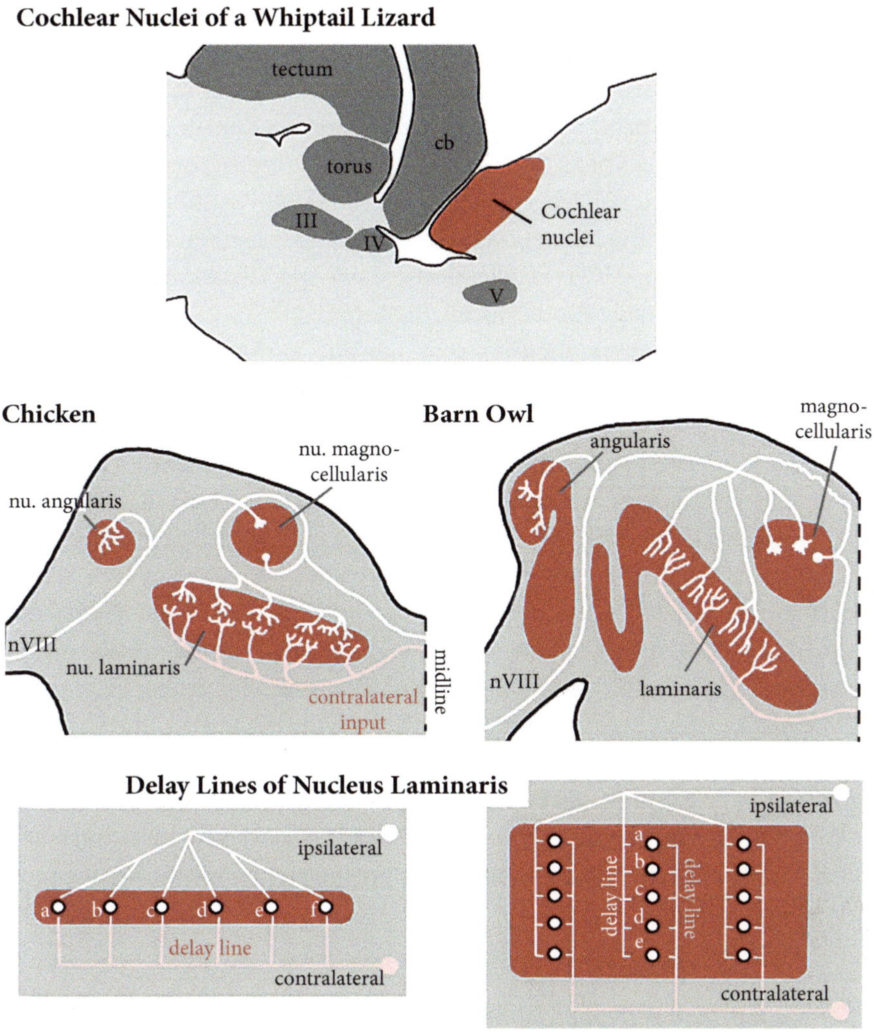

Figure 5.21 The cochlear nuclei of sauropsids. Shown at the top is a sagittal section through the midbrain and hindbrain of a lizard (*Cnemidophorus tigris*); the cochlear nucleus complex is depicted in red. The middle diagrams illustrate the cochlear nuclei (in transverse sections through the left brainstem) and some of their key connections in domestic chickens and barn owls (*Tyto alba*). The bottom diagrams show the same connections in a more schematized form, emphasizing the neuronal "delay lines" that allow neurons in nucleus laminaris to respond selectively to specific interaural time differences and, hence, to sounds coming from specific locations in space. The neurons labeled (a) respond to sounds that arrive earlier at the contralateral ear than the ipsilateral one (i.e., from the right side of the animal's head). As a sound source moves toward the animal's midline, neurons (b) through (f) are activated in sequence. Note that sound source location is mapped along the horizontal axis of nucleus laminaris (left to right) in chickens, but along its vertical axis in owls.

Abbreviations: cb – cerebellum; torus – torus semicircularis; III, IV, V – motor nuclei of cranial nerves III–V. Adapted from Miller (1975), Kubke and Carr (2000).

a specific sound source location (Jeffres, 1948). These time delays and sound source locations are mapped systematically along the rostromedial-caudolateral axis of nucleus laminaris (sound frequency is mapped along the rostrocaudal dimension). Analogous delay lines exist in owls (Carr and Konishi, 1990; Kuba et al., 2005), but the cell bodies of an owl's nucleus laminaris are not confined to a monolayer. Moreover, the axonal delay lines in owls run dorsoventrally through nucleus laminaris, rather than mediolaterally, and involve the ipsilateral as well as the contralateral inputs (Figure 5.21). Collectively, these specializations help to explain why owls are exceptionally good at localizing sounds.

A comparative analysis suggests that nucleus laminaris and its delay-line mechanism for processing interaural time differences are sauropsid innovations. Homologs of the avian nucleus laminaris have been described in crocodilians, turtles, and squamates, although they are quite small in most lizards (Miller, 1980; Szipr et al., 1995; Carr et al., 2009; Yan et al., 2010; Willis et al., 2013). Mammals process interaural time differences in a hindbrain nucleus called the medial superior olive (MSO), but they accomplish this function by means of specialized inhibitory inputs, rather than axonal delay lines (Grothe and Pecka, 2014). Indeed, the functional similarities between nucleus laminaris and the mammalian MSO are almost certainly the result of convergent evolution (see Chapter 6). Even the main cochlear nuclei, which receive inputs directly from the inner ear, differ so much between sauropsids and mammals, in both cellular organization and developmental origin, that they have probably evolved independently in these two lineages (Ryugo and Parks, 2003; Farago et al., 2006). Most likely, their evolution was linked to the emergence of tympanic ears and, as noted previously, improved high-frequency hearing in air.

5.5.2. Cerebellar Expansions and Novelties

Because the cerebellum is small and simple in tuataras, lizards, and turtles, as well as amphibians, we can infer that it must have been a relatively small brain region in the earliest amniotes. However, it clearly increased in both relative and absolute size among the archosaurs, reaching its pinnacle in birds (Iwaniuk and Hurd, 2005), and it expanded independently in the synapsid lineage. Given that the cerebellum is involved in a variety of different behaviors, ranging from the control of eye movements and body posture to cognition (at least in primates), there is probably no single functional explanation for the convergent cerebellar expansion in archosaurs and synapsids. However, it is surely no coincidence that cerebellar enlargement in these lineages was associated with the evolutionary shift toward an erect body posture, a parasagittal gait, and an enormous increase in the complexity of limb and trunk muscles. The expanded cerebellum probably helped these animals maintain their erect posture by fine-tuning many different postural reflexes. This hypothesis is certainly consistent with the profound deficit in postural control observed after cerebellar lesions in mammals, crocodilians, and birds (Nieuwenhuys et al., 1998, Chapters 20–22). The evolution of a longer, more flexible neck may also have provided an impetus for cerebellar enlargement, especially as it relates to the control of head and eye position during complex locomotor behavior.

The cerebella of birds and mammals are remarkably similar, not only in size but also in structure. Most obviously, both are highly folded and divisible into 10 or more lobules. The developmental mechanisms of cerebellar foliation remain poorly understood (Corrales, 2006; Sudarov and Joyner, 2007), but they probably involve a disproportional expansion of the cerebellar surface. Cerebellar foliation may also be linked to evolutionary changes in the external granular layer, a superficial sheet of cells that gives rise to many cerebellar granule cells. Comparative developmental studies have revealed that cells in this external granular layer continue to proliferate (i.e., undergo "transit amplification") in chicks and mice, but not in frogs or sharks (Gona, 1976; Chaplin et al., 2010). These data suggest that proliferation within the external granular layer is an evolutionary innovation of amniotes. However, a proliferative external granular layer has recently been discovered also in teleosts (Biechl et al., 2016), suggesting that it may have evolved independently in teleosts and amniotes. Its presence in birds and mammals may also have resulted from two independent origins, but data from lizards or turtles will be needed to test this hypothesis.

A more subtle, probably convergent similarity between the cerebella of mammals and birds is that in both taxa the "climbing fiber" inputs to the cerebellar cortex "climb" all over the Purkinje cell dendrites, rather than terminating mainly on their cell bodies, as they do in anamniotes and non-avian sauropsids (Yopak et al., 2017). Yet another convergent similarity is that the cerebellum of both mammals and birds is divisible into a series of longitudinal (parasagittal) stripes that vary in their histochemical properties (see Figure 5.9) and connectivity (Marzban and Hawkes, 2011). Comparative histochemical data indicate that these stripes do not exist in crocodilians, turtles, or snakes, implying that they probably evolved independently in birds and mammals. However, a somewhat simpler set of longitudinal stripes has recently been described in an agamid lizard, raising the possibility that those stripes are a more ancient amniote trait that has been lost repeatedly within the sauropsids (Wylie et al., 2017).

One of the most interesting similarities between the cerebella of birds and mammals is that both receive inputs from so-called pontine nuclei, which lie in the rostral brainstem ventral and caudal to the cerebellum. In mammals these pontine nuclei are large and receive massive projections from the neocortex, as well as the optic tectum and spinal cord. Their connections in birds remain poorly studied but seem to include inputs from both the optic tectum (Dubbeldam, 1998) and a pallial telencephalic region called the arcopallium (Zeier and Karten, 1971). Despite these fascinating similarities, the pontine nuclei of birds and mammals have probably evolved independently of one another, as no pontine nuclei have been identified in any non-avian sauropsids.

Thus, although cerebellum size and organization are strikingly similar in birds and mammals, many of these similarities probably evolved independently in the two lineages. By comparison, it is more difficult to identify evolutionary changes in cerebellar organization that occurred earlier, with the origin of amniotes. One good example is provided by cerebellar basket cells, which are a type of inhibitory neuron in the cerebellar cortex. Such cells have been described in mammals, birds, and reptiles, but not in fishes or amphibians (Cajal, 1995; Yopak et al., 2017). This phylogenetic distribution suggests that basket cells were probably added as a cerebellar cell type with the origin of amniotes. Other cerebellar interneurons, notably Golgi cells, may share the same evolutionary path, but their phylogenetic distribution has not yet been examined in detail.

5.5.3. Midbrain Expansion and Complexity

The largest, most impressive component of the midbrain in sauropsids is the optic tectum. In birds and many lizards (notably iguanids, varanids, and chamaeleons) the optic tectum is large and intricately laminated, featuring 14–15 separate layers (Northcutt, 1978). Other lizards, snakes, tuataras, turtles, and crocodilians have a somewhat smaller and simpler tectum (Northcutt, 1984; Reiner, 1994), but even in these species tectal lamination is remarkably complex (Figure 5.22). The sauropsids with the simplest optic tecta are the snakes and various groups of limbless lizards (Senn and Northcutt, 1973). Given these data, as well as data on the tectum of amphibians and lungfishes, it seems likely that the tectum's size and complexity increased slightly with the origin of amniotes, increased dramatically in birds and highly visual lizards, and decreased in snakes and burrowing lizards. The optic tectum was also modified in the lineage leading to mammals, where it is known as the superior colliculus, but we defer this discussion to Chapter 6.

In all sauropsids the superficial layers of the optic tectum receive topographic input from the retina, and most or all of that input is crossed (Bravo and Pettigrew, 1981; Larsson, 2011). The deeper layers receive auditory and somatosensory inputs, which tend to be functionally in register with the overlying visual inputs (Chalupa and Rhoades, 1977). In pit vipers, most boas, and pythons with trigeminal infrared sensors (see Section 5.3.1), the deep layers also receive input from a part of the medullary trigeminal complex that is dedicated to processing information provided by those infrared sensors (Kaldenbach et al., 2016). In addition to these ascending sensory inputs, the tectum of most sauropsids receives a wide variety of descending inputs, including major projections from the pretectum, prethalamus, and telencephalon.

Tectal outputs are likewise varied and widespread (Reiner, 1994). In general, the tectum of all amniotes contains a wide variety of neuron types, which tend to have different dendritic architectures, neurophysiological response profiles, and axonal projections. Some of these cell types, such as tectal ganglion cells with "bottlebrush"

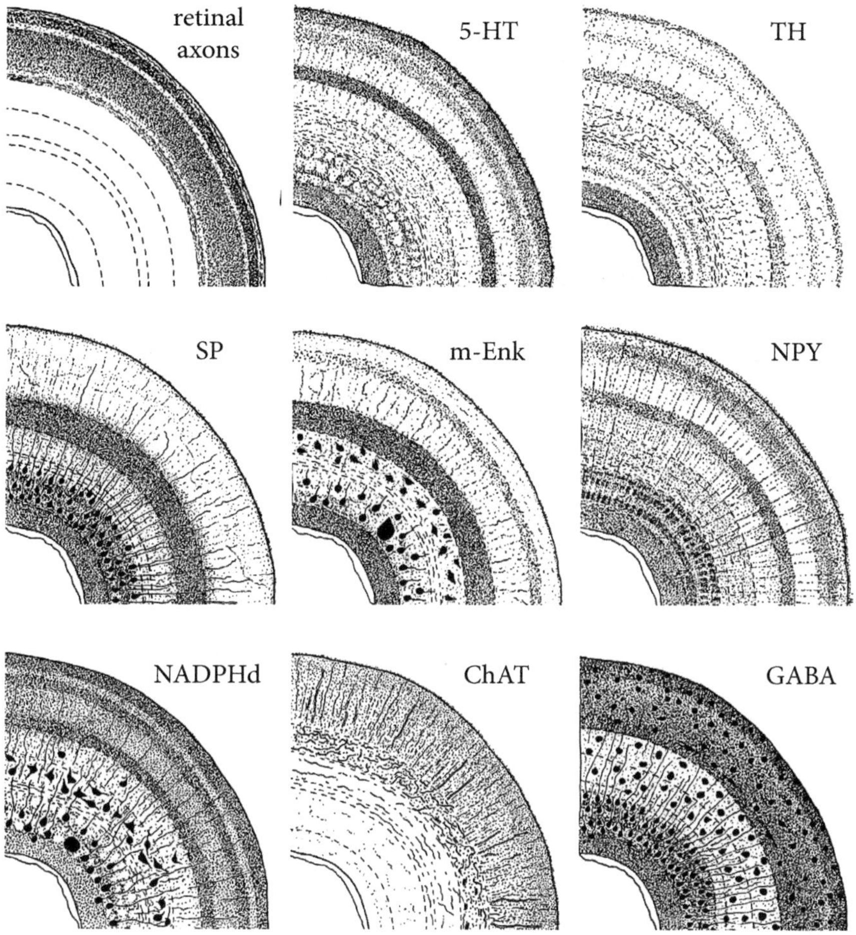

Figure 5.22 Laminar organization of the optic tectum in turtles. The optic tecta of several turtles (mainly *Testudo horsfieldi*) were sectioned transversely and labeled to reveal retinal inputs as well as a variety of neuroactive enzymes, peptides, and neurotransmitters. In aggregate, the staining patterns demonstrate the tectum's intricate laminar organization. The largest black spots in the images of met-enkephalin (m-ENK) and NADPH-diaphorase (NADPHd) staining are the cell bodies of mesencephalic trigeminal neurons.

Other abbreviations: 5-HT – serotonin; ChAT – choline acetyltransferase; GABA – gamma-aminobutyric acid; SP – substance P; TH – tyrosine hydroxylase.

Adapted from Kenigfest and Belekhova (2012), with permission from Springer Nature.

dendritic endings (Figure 5.23), are broadly conserved (Luksch et al., 1998; Major et al., 2000; Báez et al., 2003). Others are harder to homologize across species. The increased complexity of the optic tectum in birds probably involved changes in the development of some homologous cell types (especially changes in the adult position of their cell bodies; Báez et al., 2003). Whether it also involved the evolution of novel cell types remains unclear, as most analyses have focused on cell type conservation, rather than innovation.

The principal function of the optic tectum in sauropsids is to determine which location in the animal's external environment is most salient and then to convey this information to other neurons, which ultimately prompt the animal to direct its attention, eyes, head, or entire body toward the selected location. Given this overarching function, it is not surprising that tectal neurons tend to respond robustly to highly salient stimuli, such as visual stimuli that move toward the animal (looming stimuli). To select the single most salient stimulus from what might be a large variety of stimuli, the tectum projects topographically to a set of "isthmic nuclei" that then project back to the tectum (Figure 5.23). Importantly, the neurons of the magnocellular isthmic nucleus (aka magnocellular preisthmic nucleus; see Hidalgo-Sánchez et al., 2005) are inhibitory and project globally throughout the tectum, except to those tectal neurons that provide them with input (Wang et al., 2003). This circuitry mediates a global winner-take-all competition that lets the most active tectal neurons suppress all the others; the winning neurons then represent the most salient location (Marín et al., 2005; Mysore and Knudsen, 2013; Goddard et al., 2014). A nucleus with similar connections and seemingly similar functions is found in turtles, mammals, and amphibians (Gruberg et al., 2006; Belekhova and Kenigfest, 2014). Therefore, the magnocellular isthmic nucleus probably predates the origin of amniotes.

Another major component of the isthmic nuclear complex in sauropsids is the isthmo-optic nucleus. It receives input from the optic tectum and projects to the retina, forming a topographically organized retino-tecto-isthmo-retinal loop. This system is involved in regulating visual spatial attention, as it tends to boost retinal signal processing at specific locations (Li et al., 1998; Wilson and Lindstrom, 2011). Intriguingly, the isthmo-optic nucleus is largest in birds that peck for food on the ground (e.g., chickens and pigeons) and relatively small in birds that catch their food on the wing. It appears to be lacking entirely in ibises and pelicans (Repérant et al., 1989; Gutiérrez-Ibáñez et al., 2012). The most basal birds (e.g., ostriches, kiwis, and tinamous) also lack a distinct isthmo-optic nucleus, but recent experimental work revealed that tinamous, at least, have an indistinct cluster of neurons that is homologous to the isthmo-optic nucleus of other birds (Krabichler et al., 2017). Outside of birds, small or indistinct homologs of the isthmo-optic nucleus can be identified in most sauropsids, but not snakes. Mammals and most anamniotes also have neurons in the brain that project to the retina, but most of these are located in other brain regions, such as the hypothalamus, and are therefore unlikely to be homologous to the isthmo-optic nucleus of sauropsids (Repérant et al., 2006). A group of retinopetal neurons in the isthmic region has been described in basal ray-finned fishes, but this structure most likely had an independent evolutionary origin. Thus, we conclude that the isthmo-optic nucleus is a sauropsid innovation that increases sensitivity in specific regions of the visual field.

The torus semicircularis lies ventral to the optic tectum in all sauropsids, as it does in amphibians (at least when one examines standard transverse sections through adult brains; see Figure 4.20 in Chapter 4). Only in mammals does its

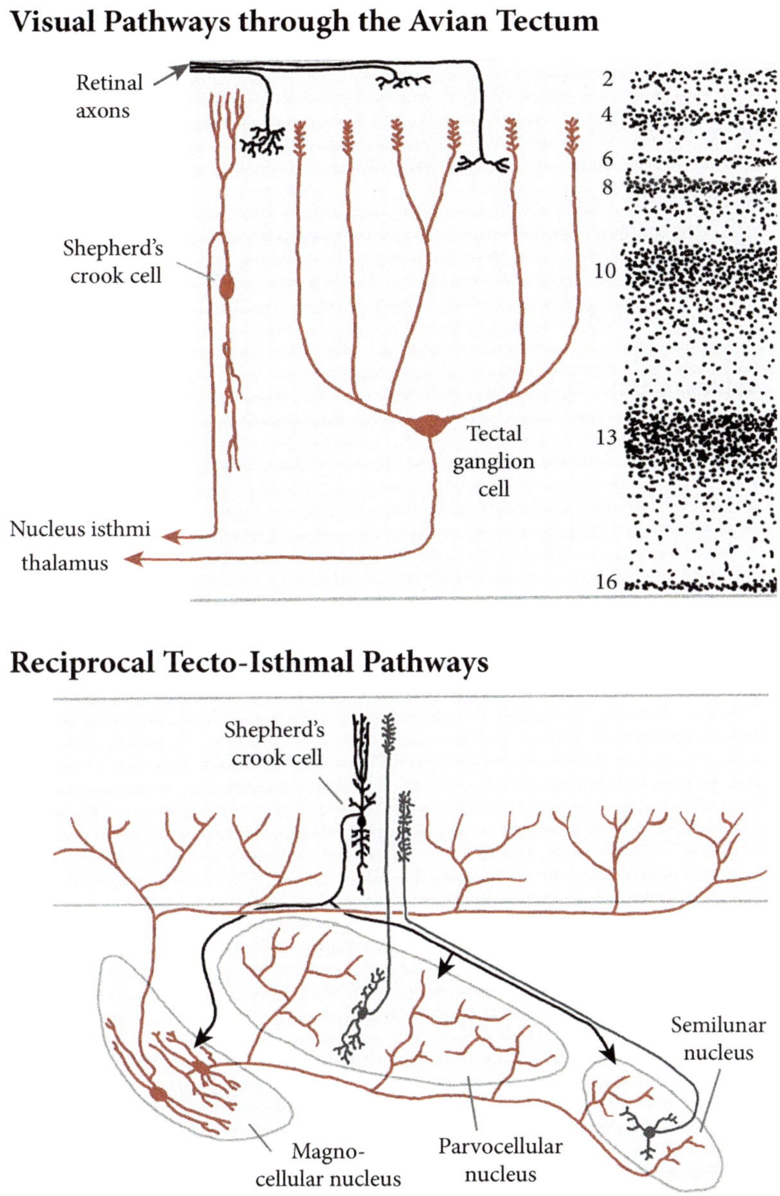

Figure 5.23 Tectal circuitry in birds. The top diagram shows how retinal axons terminate in specific superficial layers of a pigeon's optic tectum. It also depicts two major types of tectal neurons that receive retinal input and project out of the tectum. The 16 tectal laminae are illustrated along the right side of the diagram. The bottom diagram illustrates the reciprocal connections between the tectum and the isthmal complex, which includes magnocellular, parvocellular, and semilunar divisions. The neurons shown in red are GABAergic. When activated, they inhibit large areas of the tectum, except for the cells that activated the GABAergic neurons in the first place. As a result, the network exhibits a winner-take-all competition in which only one tectal location can be active at a time.

Adapted from Agarwala and Ragsdale (2009) and Wang et al. (2003, with permission from John Wiley & Sons).

homolog, the inferior colliculus, lie caudal to the optic tectum in adulthood (see Chapter 6, Figure 6.20). In general, the torus semicircularis receives major inputs from the hindbrain auditory nuclei and projects to both the optic tectum and the thalamus. Its anatomy and physiology have been studied extensively in owls (Knudsen and Konishi, 1978), and that basic organization is thought to be widely conserved across the sauropsids (Puelles et al., 1994). Indeed, almost all comparative studies on the torus semicircularis stress its conserved features, suggesting that it has changed relatively little, even as its principal inputs, the hindbrain auditory nuclei, have undergone profound transformations (see Section 5.5.1 and Section 4.5.2 [in Chapter 4]). Based on these and other, similar observations, Wilczynski (1984) proposed that major evolutionary innovations in the peripheral nervous system may often be accommodated by the reorganization of ancestral brain regions, rather than the evolution of new brain regions. This idea is interesting and plausible, but it is also possible that major innovations are more difficult to identify inside the brain than in the sense organs or musculoskeletal system. Specifically, we are not yet convinced that the torus semicircularis really evolved no novel subdivisions as tympanic ears and good high-frequency hearing evolved in anurans and amniotes.

5.5.4. A Large but Strange Forebrain

The forebrain, including both telencephalon and diencephalon, as well as a few smaller areas, occupies a much larger fraction of the brain in amniotes than in amphibians and lungfishes, their closest relatives. In sauropsids, most of this forebrain hypertrophy is concentrated in two areas, namely the thalamus and the lateral wall of the telencephalon's pallium. Somewhat similar changes have occurred in the synapsid lineage, but in mammals a different component of the pallium was expanded. We will come back to this in Chapter 6. For now, we focus, as we have throughout this chapter, on the evolutionary changes in the earliest amniotes and sauropsids.

When brain regions expand substantially during phylogeny, they often acquire an increased number of subdivisions, and those subdivisions can be difficult to homologize across the species being compared. This problem is especially severe in comparing the thalamus and lateral telencephalon between sauropsids and mammals or, more importantly for our present purposes, between sauropsids and amphibians. In considering this issue, it is natural to ask which criteria are most useful for establishing brain region homologies. Some have argued that similarity in neural connections are ideal signposts of homology (Karten and Shimizu, 1989). Others believe that embryonic origin, as determined through fate mapping and comparative "genoarchitectonics" (Puelles and Ferran, 2012), is a surer guide to potential homologies. We prefer to take an eclectic approach, considering all sorts of evidence. Which types of similarities are most useful is ultimately an empirical question and may vary across types of characters (Striedter, 1999). That said, it seems to us that, for vertebrate forebrains, embryonic origins have varied less than neural connections (Aboitiz, 1993) and are, therefore, more useful indicators of homology (see also Chapter 6). In particular, our analysis suggests that olfactory and thalamic inputs to the pallium have changed substantially as amniotes evolved. As we discuss in Section 5.6, those changes in connectivity have profoundly altered how information flows through the forebrain of amniotes. However, before we consider those larger, systems-level modifications, we introduce the most relevant brain regions separately.

5.5.4.1. Thalamus

The thalamus (aka dorsal thalamus) of most anamniotes is small and simple. In anurans, for example, the thalamus comprises just three principal cell groups: the anterior, central, and lateral nuclei (Figure 4.23 in Chapter 4; Neary and Northcutt, 1983; Puelles et al., 1996). In contrast, the thalamus of lizards harbors at least eight major cell groups (Kenigfest et al., 1997; Davila et al., 2000), and that of birds is home to at least 20 named nuclei (Redies et al., 2000; Puelles et al., 2007). Given these species differences in cytoarchitectural complexity, it is not surprising that homologies between the individual thalamic nuclei are difficult to determine. However, the lateral nucleus of amphibians is the principal thalamic target of the optic tectum, which makes it similar to a structure called "nucleus rotundus" in sauropsids and the lateral posterior-pulvinar complex in mammals. Moreover, nucleus centralis of amphibians is the principal target of the torus semicircularis, making it similar to nucleus medialis of lizards, nucleus ovoidalis in birds, and the medial geniculate nucleus of mammals. Of course, if we accept these similarities as indicators of homology, then only the amphibian anterior nucleus remains a potential homolog for all the other thalamic nuclei of amniotes.

Starting with these observations, Ann Butler (1994) proposed that all of the thalamic nuclei in amniotes that do not receive substantial input from the midbrain roof—what she called the lemnothalamus—are homologous "as a field" to the anterior nucleus of amphibians and other anamniotes. Moreover, she proposed that some of the lemnothalamic nuclei in amniotes became less multimodal and, thus, more functionally specialized, than the amphibian nucleus anterior. A case in point is the emergence in amniotes of a dorsal lateral geniculate nucleus that receives direct retinal inputs and projects to the telencephalon's pallium. A likely homolog of this cell group exists in sauropsids (where it goes by a variety of names), but such a cell group does not seem to exist in amphibians or lungfishes. Therefore, it is arguably "new" with amniotes. By the same reasoning, we suggest that the other components of the lemnothalamus in sauropsids are likewise innovations that arose either in early amniotes or, if they lack mammalian homologs, within the sauropsids. This general scenario of thalamic evolution in amniotes is plausible, but comparative developmental data for the thalamus are too scarce to be conclusive.

5.5.4.2. Dorsal Ventricular Ridge

The telencephalon of amniotes has changed even more dramatically than the thalamus. As reviewed in Chapter 4, the telencephalon in anuran amphibians consists mainly of evaginated hemispheres with a relatively thick medial wall and a thin lateral wall. The dorsal portion of each hemisphere—the pallium—is traditionally divisible into medial, dorsal, lateral, and ventral divisions (Puelles et al., 2000, 2017). Important for present purposes is that the dorsal, lateral, and ventral divisions of the pallium in anurans contain only a few migrated cells and receive direct projections from the main olfactory bulb (see Figure 4.29). The caudal portion of the ventral pallium comprises the "pallial amygdala," which has strong connections with the striatum and hypothalamus. Directly ventral to the pallial amygdala lies the subpallial amygdala, which in anurans consists mainly of a region that receives vomeronasal inputs from the accessory olfactory bulb (Moreno and González, 2007). The other principal components of the anuran subpallium are the striatum, the pallidum, and, more medially, the septum. This general pattern is probably a good representative of the primitive condition against which the telencephalons of amniotes should be compared (see Chapter 4).

Compared to this primitive condition, the telencephalon's lateral wall grows much thicker in all amniotes, albeit to different degrees in different telencephalic subregions. In mammals, the thickening involves primarily the lateral subpallium, which gives rise mainly to the striatum and the pallidum (Figure 5.24; see also Chapter 6). In sauropsids, the most highly thickened region of the lateral telencephalic wall is the ventral pallium (it is unfortunate that pallium and pallidum are such similar names), which ends up forming a large, elongate ridge that protrudes into the ventricle (Figure 5.24). This dorsal ventricular ridge (DVR) is divisible into anterior and posterior divisions, called the anterior DVR (ADVR) and posterior DVR (PDVR), respectively (Figure 5.25; Ulinski, 1983).

The PDVR, which is probably homologous to the pallial amygdala of mammals and amphibians (Medina et al., 2017; Tosches et al., 2018), has strong connections with the hypothalamus and receives direct projections from the main and accessory (vomeronasal) olfactory bulbs (Figure 5.26; Martinez-Marcos, 1999; Martinez-Garcia et al., 2007). The target of the vomeronasal projections is dramatically enlarged in snakes and lizards with well developed vomeronasal organs (Lohman and Smeets, 1993; Lanuza and Halpern, 1998). This enlarged area is roughly spherical in shape (it is called nucleus sphericus; Figure 5.26) and features a prominent cellular layer surrounding a cell-poor core. Such a nucleus is not apparent in turtles or archosaurs, which have lost the vomeronasal system. However, archosaurs and turtles do retain other components of the PDVR, which in birds is generally referred to as the arcopallium (Reiner et al., 2004).

The ADVR is considered by some investigators to be homologous to the endopiriform component of the claustroamygdalar complex in mammals (i.e., a region that includes the claustrum, endopiriform nucleus, and pallial amygdala; see

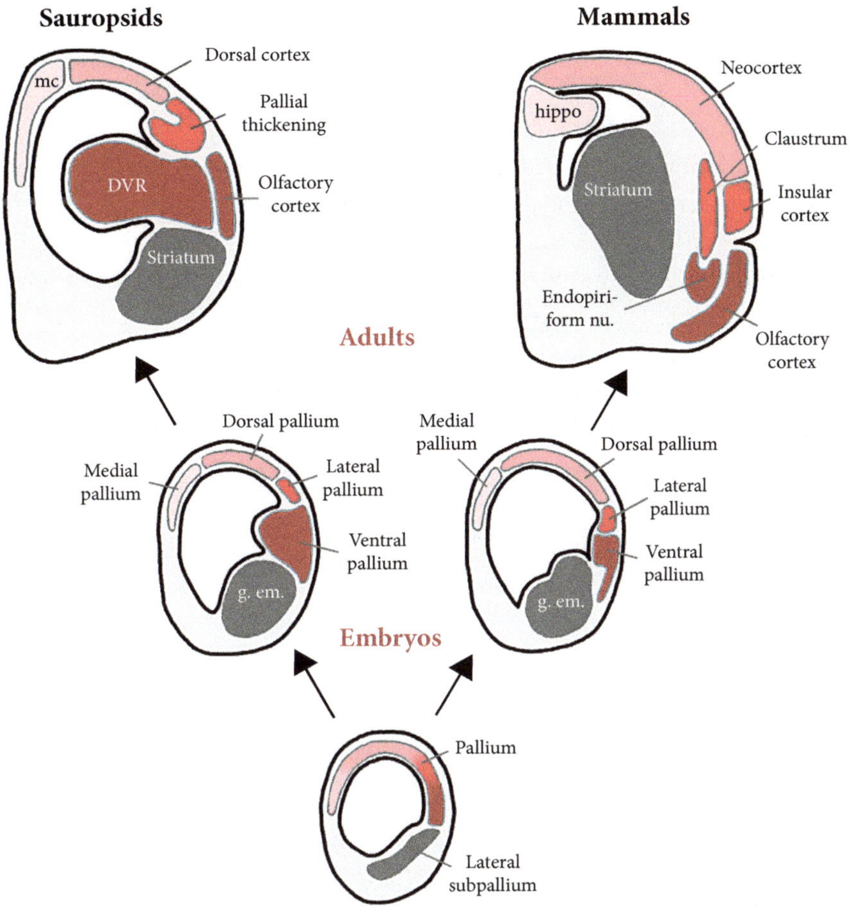

Figure 5.24 Divergent development of the telencephalon in sauropsids versus mammals. The ventral pallium grows much larger in sauropsids (i.e., reptiles and birds) than in mammals, bulging into the ventricle and forming the dorsal ventricular ridge (DVR) of the adult brain; in both lineages, its most superficial portion gives rise to the olfactory cortex. The lateral pallium develops into the pallial thickening of sauropsids (probably homologous to the mesopallium of archosaurs) and into the claustrum and insular cortex of mammals (Puelles et al., 2017). The dorsal pallium remains relatively small in sauropsids, but it expands dramatically in mammals, forming most of the neocortex in adults. The lateral subpallium in both lineages develops into the ganglionic eminences (g. em.), which eventually form the striatum (as well as the pallidum) and contribute inhibitory neurons to the neocortex. The diagrams represent transverse sections through the telencephalon of a cryptodire turtle (left) and a marsupial (right).

Additional abbreviations: mc – medial cortex.

Adapted from Striedter (1997) and Puelles et al. (2017).

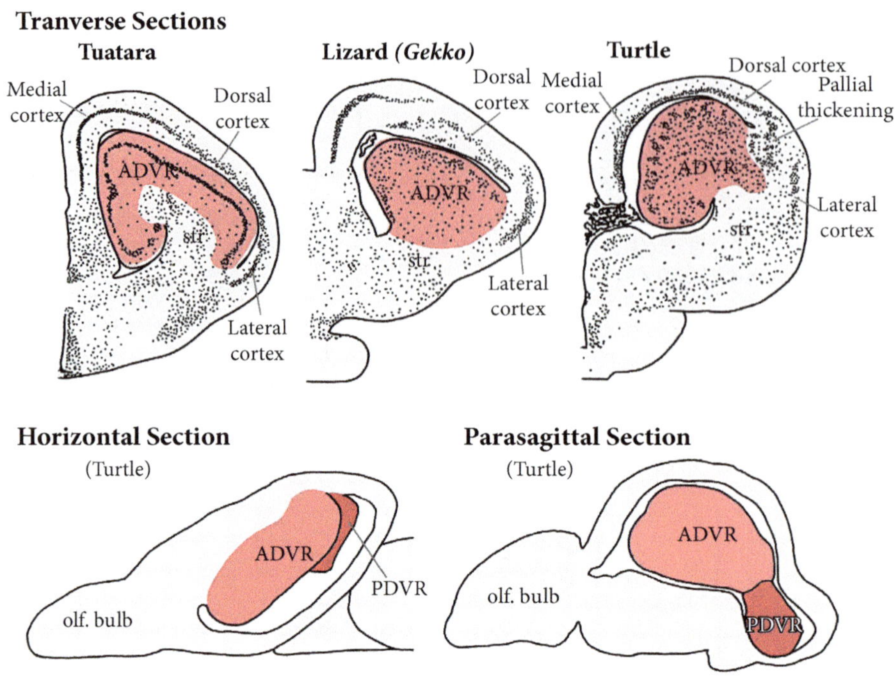

Figure 5.25 The dorsal ventricular ridges of the reptilian telencephalon. As shown along the top, the anterior dorsal ventricular ridge (ADVR) lies dorsal to the striatum (str) and bulges into the telencephalic ventricle, which consequently becomes very narrow (lateral is to the right, dorsal to the top). The ADVR features a prominent cell layer in *Sphenodon*, but this layer is broken up into cell clusters in turtles and is relatively indistinct in most lizards. As shown along the bottom, reptiles also have a posterior DVR (PDVR), which lies posterior and ventral to the ADVR. The medial cortex is thought to be homologous, at least in part, to the mammalian hippocampus. The lateral cortex receive dense inputs from the olfactory bulbs and is thought to be homologous to the mammalian olfactory cortex.
Adapted from Balaban (1978) and Balaban and Ulinski (1981b).

Striedter, 1997). Others think that it is homologous to part of the mammalian neocortex. We will come back to this debate later in this chapter and in Chapter 6. For now, we focus just on sauropsids. In this lineage, the ADVR is typically larger than the PDVR, but its size and cytoarchitecture vary considerably among the extant sauropsids. In the tuatara, the ADVR features a prominent cell-dense lamina that lies close to the ventricular surface and is continuous with a similar cell layer in the lateral pallium (Figure 5.25; Hines, 1923). Other sauropsids do not have such a well-defined cell layer in their ADVR, but cryptodire turtles and many lizards do exhibit numerous cell clusters near the ADVR's ventricular surface (Northcutt, 1978). These periventricular cell clusters do not exist in crocodilians and birds, which have a much larger ADVR that is more uniformly filled with neuronal cell bodies. As we discuss shortly, the avian ADVR does contain a number of subdivisions that are structurally and functionally distinct, but these do not correspond to individual cell clusters in the DVR of lizards or turtles.

The sensory inputs to the ADVR have been examined in several sauropsid taxa. In all of them, the main olfactory bulb projects only to a thin strip of cells at the lateral surface of the ADVR (and PDVR; Figure 5.26). Thus, in contrast to the amphibian

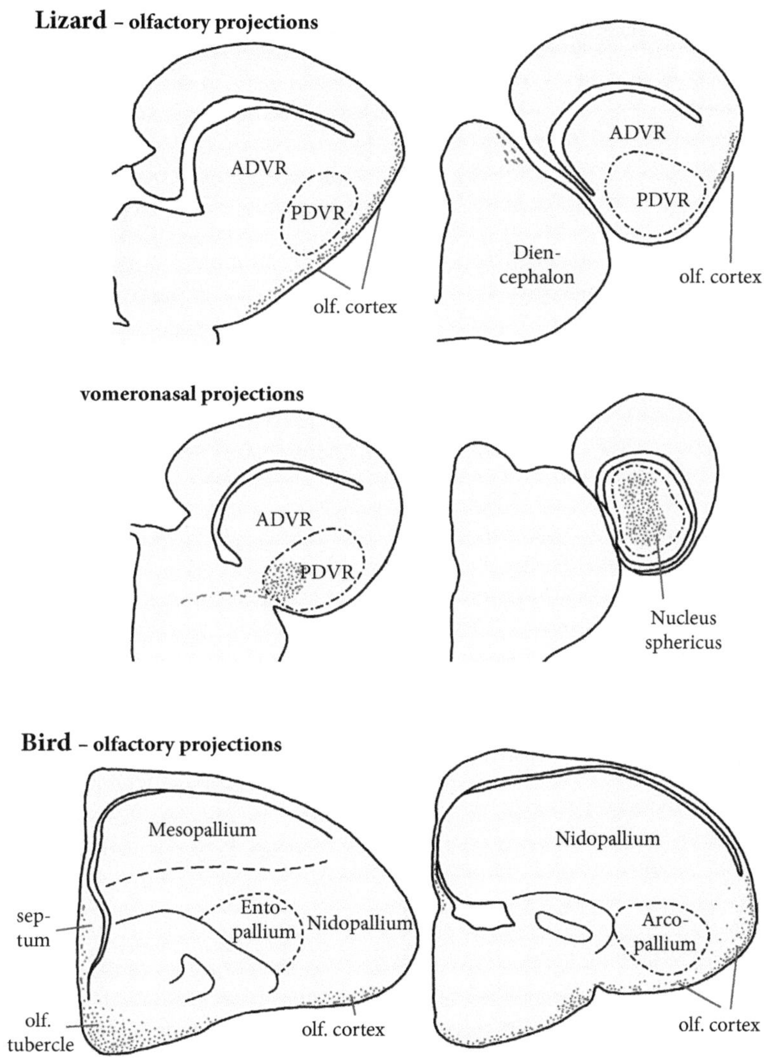

Figure 5.26 Olfactory and vomeronasal projections. Shown in red are the projections of the main and accessory olfactory bulbs, which receive inputs from the olfactory and vomeronasal epithelia, respectively. The lizard data were obtained from geckos, the avian data from pigeons. Note that birds lack the vomeronasal system. Nucleus sphericus lies within the posterior dorsal ventricular ridge (PDVR) of squamates. The avian arcopallium is thought to be homologous, at least in part, to the reptilian PDVR.
Other abbreviations: ADVR – anterior dorsal ventricular ridge; olf – olfactory.
Adapted from Lohman and Smeets (1993) and Reiner and Karten (1985).

condition, most of the neurons in the anterior portion of the sauropsid ventral and lateral pallium do not receive direct olfactory input. Instead, the sauropsid ventral pallium receives substantial inputs from other sensory modalities. In lizards, turtles, and crocodilians, the auditory, somatosensory, and tectorecipient visual nuclei of the thalamus project to medial, intermediate, and lateral regions of the ADVR, respectively (Figure 5.27; Hall and Ebner, 1970; Pritz, 1974, 1975; Balaban and Ulinski, 1981a; Guirado et al., 2000). Collectively, these thalamic sensory inputs fill most of the ADVR in the non-avian sauropsids.

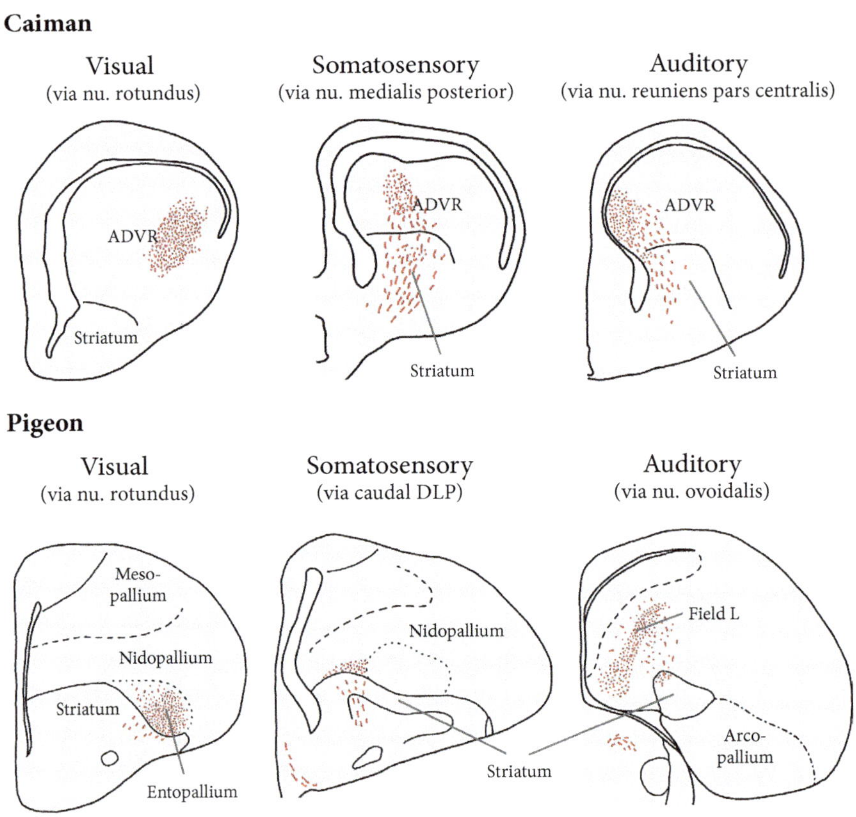

Figure 5.27 Visual, somatosensory, and auditory pathways to the ADVR. The red lines and dots represent axons and terminal arborizations (respectively) that are labeled in the anterior dorsal ventricular ridge (ADVR) of crocodilians (*Caiman crocodilus*) and birds (*Columba livia*) after tracer injections into thalamic nuclei associated with the main ascending sensory systems. Some of the thalamic nuclei have different names in the two taxa, despite being apparently homologous. Also note that the ADVR of birds is divided into multiple areas, including nidopallium, entopallium, and field L. The main point of the figure is that the different sensory modalities are represented in separate portions of the ADVR.

Adapted from Pritz (1974, 1975), Pritz and Stritzel (1994), Karten (1968), Karten and Hodos (1970), Wild (1987).

In birds, however, the thalamic sensory inputs are confined to a series of small-celled areas whose neurons, in turn, project to surrounding portions of the ADVR (Figures 5.26 and 5.27). This internal connectivity has been studied most carefully in the visual part of the avian ADVR, where a cell-dense region called the entopallium receives direct inputs from nucleus rotundus of the thalamus and then projects to an adjacent band of neurons, which project to yet other portions of the ADVR (Karten and Hodos, 1970; Krützfeldt and Wild, 2005). Similar connections have also been described in the auditory portion of the avian ADVR (Karten, 1968; Wang et al., 2010). This pattern of projections from primary targets of the thalamic inputs to secondary areas and, from there, to tertiary areas within the DVR appears to be unique to birds. Even if they are not truly new, we can conclude that the higher-order sensory areas have expanded tremendously in birds, relative to the non-avian sauropsids. Thus, we strongly suspect that the DVR is not only larger in birds than in reptiles, but also contains a greater number of subdivisions and more complex internal circuitry.

In addition to the thalamo-telencephalic sensory pathways, birds possess a somatosensory pathway that courses directly from the medulla to the frontal pole of the ADVR, bypassing the thalamus (Veenman and Gottschaldt, 1986). Having reached its small-celled target area in the anterior ADVR, called nucleus basorostralis (Figure 5.28), somatosensory information is conveyed to adjacent parts of the ADVR by relatively short axonal connections. Roughly parallel to this somatosensory pathway is an auditory pathway that originates from a brainstem auditory nucleus (called the lateral lemniscal nucleus), bypasses the thalamus, and terminates in the anterior ADVR (Arends and Zeigler, 1986; Striedter, 1994; Wild et al., 2001). Given the available data, this auditory pathway seems to have evolved or expanded dramatically within the avian lineage. Whether the somatosensory pathway to the anterior ADVR is likewise unique to birds remains unclear, because hints of such a pathway have been reported in some non-avian sauropsids (Nieuwenhuys et al., 1998, Chapter 20). In any case, it is clearly much better developed in birds than in any reptile.

The circuitry we just described within the avian ADVR is similar to that of the mammalian neocortex insofar as it has a laminar organization. That is, the avian ADVR is divisible into layers of cells (e.g., basorostralis, nidopallium, and mesopallium in Figure 5.28) that are connected to one another by means of short connections, which course largely orthogonal to those layers. However, in contrast to the laminar organization of mammalian neocortex (see Chapter 6), neurons within the individual layers of the avian ADVR tend not to extend their dendrites into the other layers; its interlaminar connections are mostly axonal (Wang et al., 2010; Ahumada-Galleguillos et al., 2015). Moreover, the cells within a given layer of the avian ADVR were born in roughly the same location (i.e., a small band of progenitor cells in the embryonic "ventricular zone"; e.g., Briscoe et al., 2018), whereas the layers in mammalian neocortex are distinguished by when they were born, rather than where (see Chapter 6). Finally, a comparative analysis of transcription

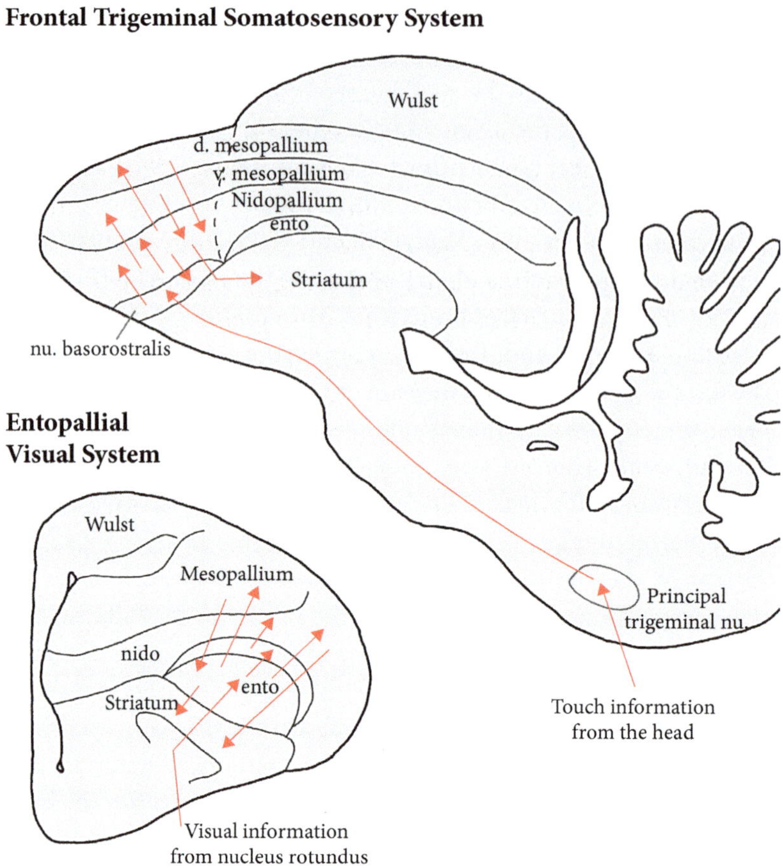

Figure 5.28 Interlaminar circuits in the avian ADVR of birds. The anterior dorsal ventricular ridge (ADVR) of birds contains several distinct zones (entopallium and nucleus basorostralis, nidopallium, mesopallium, and ventral hyperpallium) that are arranged in parallel (sometimes concentric) sheets. Moreover, these cellular sheets tend to be interconnected by relatively short connections (red arrows) that course orthogonal to those sheets, creating a pattern that is superficially similar to that of the mammalian neocortex. This pattern is here illustrated for a somatosensory region in the telencephalon's frontal pole (top, shown in a sagittal section) and the main visual region of the ADVR (bottom, transverse section).

Abbreviations: ento – entopallium; nu – nucleus; v – ventral.

Adapted from Veenman and Gottschaldt (1986) and Krützfeldt and Wild (2003).

factor expression levels revealed that cells in the ADVR of lizards and turtles have more in common with the claustrum and lateral amygdala of mammals than with mammalian neocortex (Tosches et al., 2018). Collectively, these differences indicate that the laminar organization of the avian ADVR evolved independently of that in the mammalian neocortex (Medina, 2007). This point is underscored by the fact that no such laminar organization has been described in the DVR of non-avian sauropsids.

5.5.4.3. Other Pallial Sectors

The region dorsal to the ADVR is highly variable across sauropsids in both structure and connectivity. In lizards and cryptodire turtles, this region consists mainly of the pallial thickening, a cluster of migrated neurons between the dorsolateral edge of the telencephalic ventricle and the brain surface (Figure 5.29; see also Figure 5.5). It is known to receive ascending visual inputs from the dorsal lateral geniculate nucleus (Heller and Ulinski, 1987; Kenigfest et al., 1997). The avian and crocodilian homolog of the pallial thickening appears to be the mesopallium (Bruce and Braford, 2009; Puelles et al., 2015; Briscoe et al., 2018), which lies dorsal to the ADVR and fuses with it so completely that it is usually considered part of the avian and crocodilian DVR (Jarvis et al., 2013; Briscoe et al., 2018). Unlike the pallial thickening, the mesopallium does not receive direct ascending visual inputs. Instead, it has extensive, largely reciprocal connections with the underlying, higher-order sensory divisions of the ADVR (Atoji and Wild, 2012). According to Luis Puelles et al. (2015, 2017) the avian mesopallium is homologous to the pallial

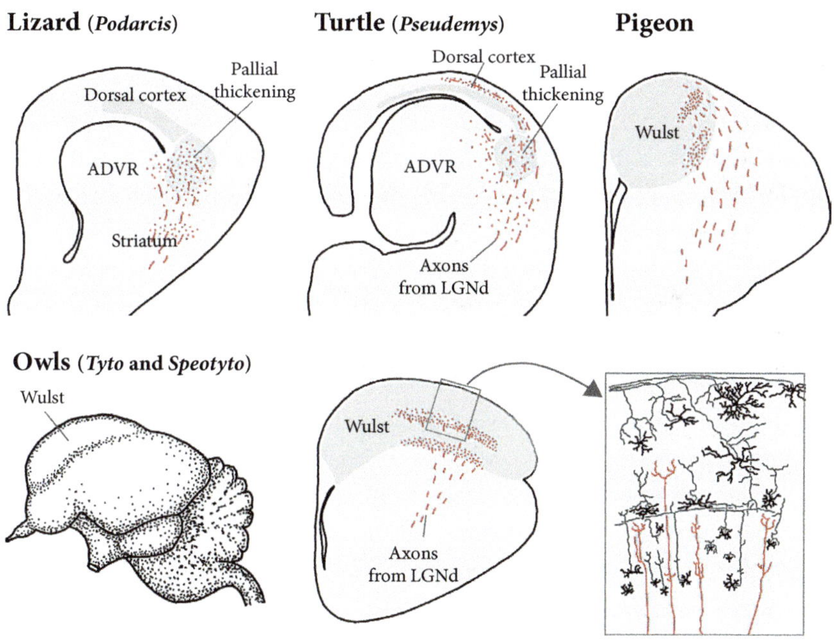

Figure 5.29 Visual projections to the dorsal cortex and Wulst. Shown along the top are transverse sections through the right telencephalon of three different sauropsids. The red lines and dots represent axons and terminals (respectively) that were labeled after tracer injections into a retinorecipient thalamic nucleus thought to be homologous to the mammalian dorsal lateral geniculate nucleus (LGNd). Shown along the bottom are an owl brain (lateral view) and a transverse section through its telencephalon. Axons and terminals labeled after tracer injections into the LGNd are again shown in red. The inset depicts some Golgi-labeled neurons and highlights that the Wulst of owls has a distinctly laminar organization.

Adapted from Kenigfest et al. (1997), Heller and Ulinski (1987), Karten et al. (1973), Pettigrew (1979).

thickening of lizards and turtles, and all of these structures are lateral pallial derivatives and, as such, homologous to the insular cortex and claustrum of mammals (see Figure 5.24). This hypothesis is interesting and may well be valid, but it is a significant departure from previous hypotheses that considered the pallial thickening to be a lateral division of the dorsal pallium. Curiously, many of the genes selectively expressed in the avian mesopallium are also expressed in the mammalian neocortex (in neurons that do not project to extratelencephalic targets) and in the lateral, dorsal, and medial pallial divisions of crocodilians (Briscoe et al., 2018); reconciling these data with earlier findings and theories (Karten and Shimizu, 1989; Puelles et al., 2017) is difficult. In any case, the comparative single-cell gene expression data indicate that the sauropsid mesopallium (aka pallial thickening) is homologous to part of the mammalian neocortex (Tosches et al., 2018).

The telencephalon's most dorsal component (dorsal to the pallial thickening or mesopallium) is also highly variable across the sauropsids. In lizards and cryptodire turtles, this region consists of a thin sheet of tissue that contains a single cell-dense layer sandwiched between two cell-sparse layers; that is, it is a simple trilaminar cortex. In turtles, but not in lizards, this "dorsal cortex" receives visual inputs from the same dorsal lateral geniculate neurons that also project to the pallial thickening (Figure 5.29; Bruce and Butler, 1984; Heller and Ulinski, 1987). The avian dorsal pallium, aka the hyperpallium, does receive visual inputs from the dorsal lateral geniculate nucleus, but it is neither thin nor trilaminar. Instead, the most dorsal part of the avian pallium thickens during development, especially at rostral levels, until it forms a definite ridge on the dorsal surface of the brain. This Wulst, as it is often called, is divided into several dorsal-ventral zones, two of which receive ascending visual inputs (Figure 5.29; Karten et al., 1973), but within each of those zones the neurons are distributed quite homogeneously. Thus, there is no trace of the trilaminar organization seen in non-avian sauropsids. Moreover, the visual receptive fields of neurons in the avian Wulst tend to be much smaller than those in the dorsal cortex of turtles (Pettigrew, 1979; Mulligan and Ulinski, 1990). The rostral portion of the avian Wulst receives somatosensory, rather than visual, information from the thalamus (Wild et al., 2008). Such a somatosensory projection to the rostral dorsal pallium has also been reported in lizards (Desfilis et al., 2002) and may, therefore, have evolved early during sauropsid phylogeny.

The dorsal cortex of turtles and the avian Wulst are generally thought to be homologous to the mammalian neocortex, both because they develop dorsally within the pallium and because they tend to receive unimodal sensory input from the thalamus (e.g., Medina, 2007). Because the neocortex is far more complex that the dorsal cortex of non-avian sauropsids and structurally quite different from the avian Wulst, it is reasonable to say that the neocortex is a mammalian innovation, just as the Wulst is an avian innovation. However, it is also sensible to claim that the neocortex, dorsal cortex, and Wulst are all homologous to one another as the principal derivatives of the embryonic dorsal pallium (e.g., Tosches et al., 2018). An important caveat to this hypothesis is the recent finding that the dorsal cortex of lizards expresses a set of genes that is characteristic of the medial pallium, rather than the dorsal pallium, and that only a relatively small rostral portion of the lizard pallium appears to be homologous to the avian Wulst and mammalian neocortex (Desfilis et al., 2017).

The hypothesis that mammalian neocortex and the avian Wulst are homologous to one another would be strengthened if amphibians, as the outgroup to mammals and sauropsids, were to possess a similar dorsal pallial division. However, as we discussed in the previous chapter, there are real questions about the identity and attributes of the amphibian dorsal pallium. It is poorly defined, receives only relatively minor multimodal inputs from the thalamus, and receives at least some inputs from the olfactory bulbs. We suspect that the line of stem tetrapods that gave rise to amniotes did possess some sort of dorsal pallium, but if they did not, then it is possible that the dorsal pallium in today's sauropsids evolved independently of that in synapsids (e.g., today's mammals). We return to this issue in Chapters 6 and 7.

In light of these issues, it is interesting that the Wulst of some birds, especially the owls, is remarkably similar to the primary visual cortex of mammals, at least in some respects. These similarities include a highly laminar organization and similar responses to visual stimuli (Pettigrew and Konishi, 1976; Pettigrew, 1979). Fascinating as they are, these similarities almost certainly resulted from convergent evolution, because the dorsal cortex of reptiles lacks the complex laminar circuits and responds very differently to visual stimuli (Mazurskaya, 1973). Moreover, the anatomical data suggest that the dorsal cortex of lizards does not (in contrast to the dorsal cortex of turtles) receive visual inputs from the dorsal lateral geniculate nucleus (Figure 5.29), though it does receive input from a different, more multimodal thalamic nucleus. Thus, even if the neocortex of mammals is homologous to the avian Wulst *as a dorsal pallium*, the connections and functions of these brain regions must have changed dramatically, and in some respects convergently, as the synapsid and sauropsid lineages diverged (for more on this, see Chapter 6).

The most medial portion of the pallium is called the medial cortex in reptiles and the hippocampus in mammals and birds. This region is relatively thin in all sauropsids and stretched across the dorsomedial and caudal surface of the DVR in crocodilians and birds (Puelles et al., 2007; Striedter, 2015). We will revisit this topic in Chapter 6, when we discuss the evolution of the mammalian hippocampus. Similarly, we refrain in the present chapter from reviewing the details of the subpallium. This region is relatively small in sauropsids, compared to the overlying DVR, but it is larger and more highly differentiated than in anamniotes (Kuenzel et al., 2011). We review some of its most important features in Chapter 6 as part of a more general review of basal ganglia evolution.

For now, the important point is that sauropsids have two main pallial sectors that receive thalamic sensory inputs and have long descending outputs. One of these is the dorsal pallium, which is significantly larger and more highly differentiated in birds than in other sauropsids, but clearly homologous between them; its mammalian homolog is the neocortex. The second pallial sector with extratelencephalic connections is the ventral pallium, which forms most of the DVR in sauropsids (recall that the dorsal part of the DVR in crocodilians and birds represents the lateral pallium; Puelles et al., 2015; Briscoe et al., 2018) and is homologous to the pallial amygdala and endopiriform nucleus of mammals. Given these homologies, we can look in more detail at some of the evolutionary changes in neuronal connections and functional circuits that accompanied the evolution of the sauropsid forebrain.

5.6. Novel Forebrain Circuits and Functions

Now that we have reviewed the connections of the individual forebrain regions in diverse sauropsids, we can try to paint a picture of how information flowed through the forebrain of early sauropsids and how this changed as sauropsids gave rise to birds. Let us begin with a quick reprise of the condition in amphibians.

In frogs and other close relatives of amniotes, ascending sensory information from the thalamus (aka dorsal thalamus) reaches mainly the striatum, septum, and medial pallium, with only sparse projections to the dorsal aspect of the pallium (see Chapter 4). The remaining pallial areas have strong connections with the medial pallium and hypothalamus (see Figure 4.31 in Chapter 4), but they receive at best minor non-olfactory inputs from lower brain regions. Moreover, these minor non-olfactory inputs probably terminate on dendrites that also receive strong projections from the main olfactory or accessory olfactory bulb (Mühlenbrock-Lenter et al., 2005). Thus, it is very unlikely that the lateral and ventral pallia received unimodal visual, auditory, or somatosensory information in the ancestors of amniotes. Instead, those parts of the pallium were mainly concerned with processing chemosensory information. This situation changed in amniotes, especially in the sauropsid lineage.

As the ventral pallium expanded in early sauropsids to form the DVR, the axons coming from the olfactory bulbs remained in a superficial position (see Figure 5.26), leaving most of the neurons in the DVR without direct chemosensory input. Instead, most neurons of the ADVR started to receive other kinds of sensory input via the thalamus (see Figure 5.27). Importantly, most of those neurons were probably unimodal in function. Many of them probably projected heavily to the PDVR, where this unimodal information would have been integrated with other information streams, before being conveyed to the preoptic area and hypothalamus (Figure 5.30). In addition, both the ADVR and the PDVR of early sauropsids almost certainly had strong projections to the striatum. It is likely that the striatopallidal complex of early sauropsids retained its ancestral role of mediating action selection (see Chapter 4, Section 4.6), but the novel inputs from the ADVR would have provided it with more fine-grained sensory information about specific objects and events in the animal's environment. Presumably, this additional information would have allowed the early sauropsids to make better, more informed decisions about which behaviors to select in any given context.

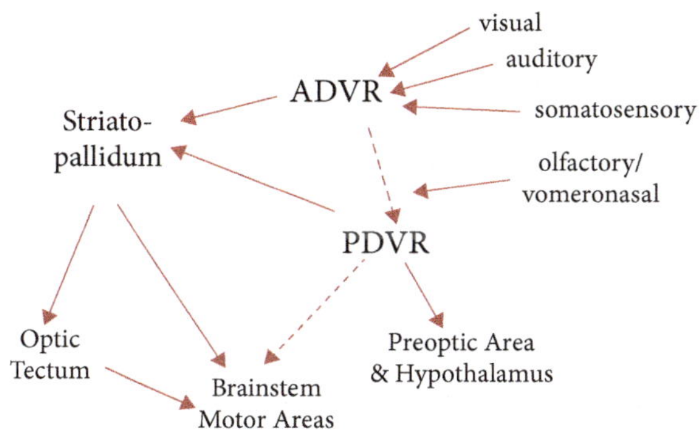

Figure 5.30 Major pathways through the DVR. The red arrows indicate major neuronal projections. The dashed arrow emanating from the PDVR indicates a projection that seems to be present only in birds. The other dashed arrow represents a pathway that is direct in most reptiles but indirect in birds (involving multiple synaptic relays).
Adapted from Ulinski (1983).

Unfortunately, we know very little about the functions of the DVR in non-avian sauropsids. Lesions of the PDVR impair courtship and mating behaviors (Greenberg et al., 1984). Lesions of the underlying striatum also affect courtship, but they impair a broader variety of species-typical behaviors, including aggressive social displays (Greenberg et al., 1979). Intriguingly, some lesions in the PDVR or striatum cause disinhibition of behavior, rather than deficits (Krohmer and Crews, 1987), which is consistent with the idea that these structures are not needed for the execution of movements, but for the regulation and sequencing of them. Functional studies on the reptilian ADVR are largely limited to lesions of its visual component in cryptodire turtles. These lesions tend to impair performance on intensity and pattern discrimination tasks (Reiner and Powers, 1983). In a natural context, these functions would be needed for learning to recognize specific food items, locations in space, and other individuals. Compared to lesions of the ADVR, lesions of the dorsal cortex in cryptodire turtles cause more subtle deficits (Reiner and Powers, 1983; Moran et al., 1998).

With the evolution of birds, the ADVR increased both in size and neuron number (Olkowicz et al., 2016). As it did so, the circuits coursing from the thalamus to the striatum and PDVR became more complex by the interposition of several additional processing steps within the ADVR (see Figure 5.28). In addition, the reciprocal connections between the avian ADVR and the overlying mesopallium provided an additional layer of modulatory influence. Non-avian sauropsids do have some interneurons within their ADVR (Ulinski, 1983), but these neurons have not been studied carefully and probably do not form obligatory links in the forebrain's sensorimotor circuitry. In any case, it seems safe to say that the avian DVR performs far more sophisticated sensory processing than the reptilian DVR, including a greater number of processing stages and a greater number of functionally specialized regions. Birds also extended the descending projections of their amygdalar homolog (the PDVR or arcopallium) to brainstem motor and premotor regions that do not receive such descending inputs in reptiles (Zeier and Karten, 1971; Medina, 2007). These long descending projections have been studied most thoroughly in songbirds, which rely on a projection from the arcopallium to medullary vocal motor neurons to produce learned songs (Wild, 2004).

In addition to changing their DVR, birds modified the circuits passing through their dorsal pallium. As mentioned earlier, axons from the dorsal lateral geniculate nucleus (LGNd) extend into the dorsal cortex in turtles but reach only the pallial thickening in lizards. Assuming the lizard condition is primitive, the visual thalamic inputs must have "invaded" the dorsal cortex in the last common ancestor of archosaurs and turtles. Birds then elaborated this pathway by thickening the dorsal cortex, thereby forming the visual Wulst. They also evolved projections back from the visual Wulst to the LGNd (Figure 5.31). A similar back-projection has been documented in turtles, but not in lizards (Kenigfest et al., 1997). Recent neurophysiological data indicate that the dorsal cortex in cryptodire turtles performs some kind of global, non-topographic analysis of space (Fournier et al., 2018). In contrast, the avian visual Wulst analyzes space in a highly topographic manner, analogous to the computations performed by the mammalian visual cortices (Liu and Pettigrew, 2003). Thus, the structural differences between the dorsal pallium of cryptodire turtles and birds are reflected in some important functional differences.

A more anterior portion of the avian Wulst has reciprocal connections with a somatosensory nucleus in the thalamus, called nucleus DIVA. Again, reptiles possess a homologous somatosensory region in their dorsal pallium (in the rostral part of their dorsal cortex), but this region does not reciprocate the thalamic input. Furthermore, the reptilian dorsal cortex does not have very long descending projections, whereas the avian somatosensory Wulst has projections down to the medulla and cervical spinal cord (Figure 5.31). Another difference in connectivity between birds and non-avian sauropsids is that DIVA in birds receives inputs from the deep cerebellar nuclei and pallidum, whereas its reptilian homolog (called DLV) does not. An intriguing aspect of these changes in the circuits through the dorsal pallium in birds is that they resemble some changes that occurred, albeit independently, with the evolution of mammalian neocortex (see Chapter 6).

It has been suggested that the expansion of the DVR and dorsal pallium in birds was causally related to the evolution of an erect stance and improvement in jaw mechanics (Ulinski, 1989). However, these changes in the skeletomuscular system were more likely linked to changes in the midbrain, cerebellum, and hindbrain. Instead, we suspect that the evolutionary expansion of the pallium in sauropsids was linked mainly to the processing of visual, auditory, and somatosensory information about external stimulus objects. Moreover, the cerebral expansion probably improved not only the perception of objects in the here and now, but also the ability of birds to remember those objects and to be guided by their memory. A good example of this important capacity is filial imprinting, which refers to the ability

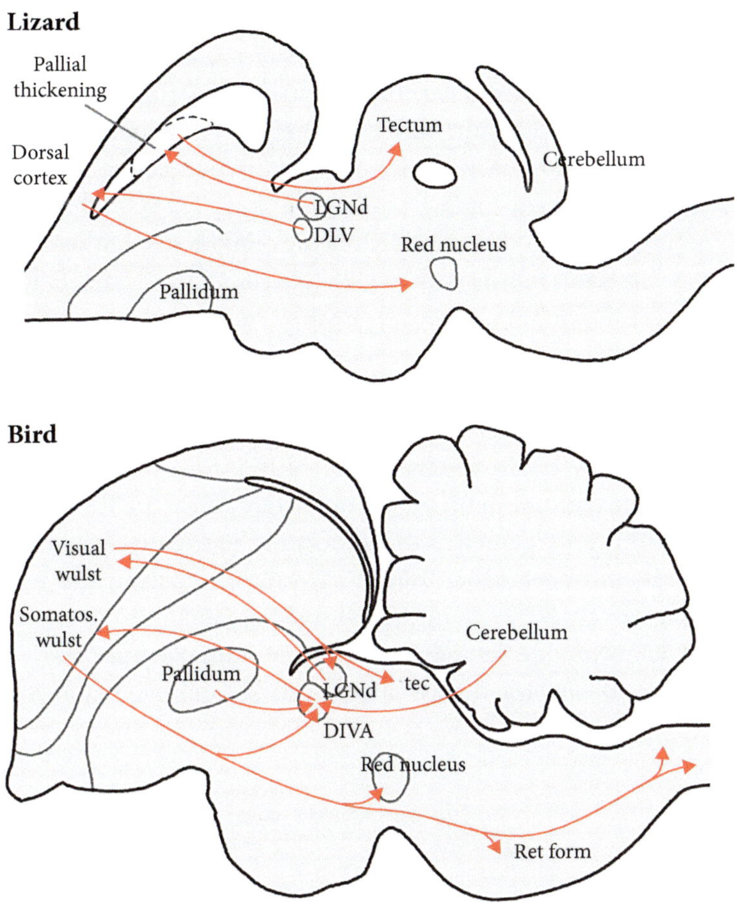

Figure 5.31 Pathways to and from the dorsal pallium. The dorsal pallium in lizards consists of the dorsal cortex. It receives visual and somatosensory information from the thalamus (from LGNd and DLV, respectively). Although birds have homologs of these pallial areas in their pallial Wulst and thalamus, a comparison of the two diagrams reveals some major differences in connectivity.
Adapted from Medina (2007).

of chicks to remember their parents and to follow them. Another example is the ability of parrots and songbirds to imitate heard sounds; those sounds are stored in memory and then used to guide the production of the birds' own vocalizations (Nottebohm, 1972; Pepperberg, 1999; Hile et al., 2000). All of these behaviors are known to involve specific components of the avian DVR and dorsal pallium (Nottebohm et al., 1976; Horn and Johnson, 1989; Striedter, 1994; Nakamori et al., 2013). Non-avian sauropsids and amphibians can also be guided by memories, especially place memories, but we suspect that their memories are less detailed and more dominated by chemosensory information than they are in birds.

One of the most fascinating aspects of avian behavior is the ability of many birds to find novel solutions to unique problems, often using tools to accomplish their goal (see Striedter, 2013). Some non-avian sauropsids, especially those with a large DVR, are also skilled problem-solvers (Manrod et al., 2007), but songbirds and parrots, in particular, are expert innovators. Comparing such cognitive capacities across species is difficult, but researchers have analyzed how frequently various avian species have been reported to obtain food using novel, innovative approaches (Lefebvre et al., 2002). These studies have revealed that the best predictor of feeding innovation rate in birds is the relative size of the mesopallium, which in turn correlates tightly with the relative size of the avian DVR and Wulst (Timmermans et al., 2000). Moreover, diverse functional data indicate that a specific region of the avian ADVR, called the caudolateral nidopallium (Figure 5.32), is important for complex cognition in birds (Güntürkün, 2011). A homolog of this highly integrative region might exist also in lizards (Andreu et al., 1996), but it is clearly much better developed in birds. Thus, the evolutionary enlargement of the pallium in birds had important, measurable effects on avian behavior. Since this enlargement had its roots in the expansion of the ADVR in early sauropsids, we suspect that those early sauropsids were also more intelligent, more cognitively facile, than their immediate anamniotic ancestors. That said, we fully acknowledge that having a large pallium need not imply increased intelligence, since the enlarged areas might be used to support improved sensory

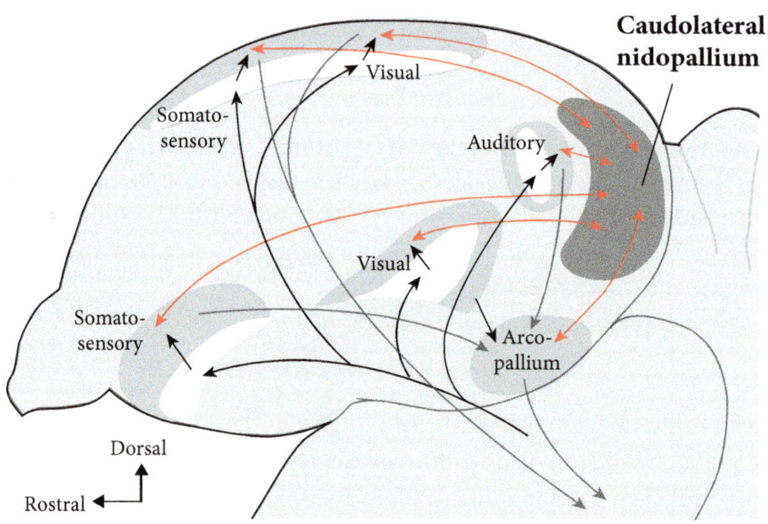

Figure 5.32 DVR circuits for complex cognition in birds. In this schematic sagittal section through an avian telencephalon, the black and gray arrows indicate how information is thought to flow through the various sensory regions of the ADVR and Wulst (projections to the striatum are omitted). The red arrows show that the caudolateral nidopallium (which is part of the avian ADVR) has reciprocal connections with all of the higher order sensory regions. These connections are thought to mediate this region's known role in complex forms of cognition.
Adapted from Striedter (2013).

perception or motor control, rather than behavioral flexibility (e.g., see Murray et al., 2016). Ideally, links between brain or brain region size and intelligence should be established through rigorous empirical analyses and control for as many confounding factors as possible.

References

Aboitiz F. 1993. Further comments on the evolutionary origin of the mammalian brain. *Med Hypotheses* 41:409–418.

Agarwala S, Ragsdale CW. 2009. Midbrain patterning. In: Squire LR (ed.), *Encyclopedia of Neuroscience*. Amsterdam: Elsevier, pp. 879–886.

Ahumada-Galleguillos P, Fernández M, Marín GJ, Letelier JC, Mpodozis J. 2015. Anatomical organization of the visual dorsal ventricular ridge in the chick (*Gallus gallus*): Layers and columns in the avian pallium. *J Comp Neurol* 523:2618–2636.

Alexander RM. 1985. Mechanics of posture and gait of some large dinosaurs. *Zool J Linn Soc* 83:1–25.

Alonso PD, Milner AC, Ketcham RA, Cookson MJ, Rowe TB. 2004. The avian nature of the brain and inner ear of *Archaeopteryx*. *Nature* 430:666–669.

Ammermuller J, Kolb H. 1996. Functional architecture of the turtle retina. *Prog Ret Eye Res* 15:393–433.

Amo L, Jansen JJ, van Dam NM, Dicke M, Visser ME. 2013. Birds exploit herbivore-induced plant volatiles to locate herbivorous prey. *Ecol Letters* 16:1348–1355.

Andreu MJ, Davila JC, Real MA, Guirado S. 1996. Intrinsic connections in the anterior dorsal ventricular ridge of the lizard *Psammodromus algirus*. *J Comp Neurol* 372:49–58.

Arends JJ, Zeigler HP. 1986. Anatomical identification of an auditory pathway from a nucleus of the lateral lemniscal system to the frontal telencephalon (nucleus basalis) of the pigeon. *Brain Res* 398:375–381.

Atoji Y, Wild JM. 2012. Afferent and efferent projections of the mesopallium in the pigeon (*Columba livia*). *J Comp Neurol* 520:717–741.

Báez J, Monzón-Mayor M, Yanes C, del Mar Romero-Alemán M, Francisco Arbelo-Galván J, Puelles L. 2003. Neuronal differentiation patterns in the optic tectum of the lizard *Gallotia galloti*. Brain Res 975:48–65.

Bakker RT. 1971. Dinosaur physiology and the origin of mammals. Evolution 25:636.

Balaban CD. 1978. Structure of anterior dorsal ventricular ridge in a turtle (*Pseudemys scripta elegans*). *J Morphol* 158:291–322.

Balaban CD, Ulinski PS. 1981a. Organization of thalamic afferents to anterior dorsal ventricular ridge in turtles. I. Projections of thalamic nuclei. *J Comp Neurol* 200:95–129.

Balaban CD, Ulinski PS. 1981b. Organization of thalamic afferents to anterior dorsal ventricular ridge in turtles. II. Properties of the rotundo-dorsal map. *J Comp Neurol* 200:131–150.

Balanoff AM, Bever GS, Rowe TB, Norell MA. 2013. Evolutionary origins of the avian brain. *Nature* 501:93–96.

Balanoff AM, Smaers JB, Turner AH. 2016. Brain modularity across the theropod-bird transition: testing the influence of flight on neuroanatomical variation. *J Anat* 229:204–214.

Baldwin MW, Toda Y, Nakagita T, O'Connell MJ, Klasing KC, Misaka T, Edwards SV, Liberles SD. 2014. Evolution of sweet taste perception in hummingbirds by transformation of the ancestral umami receptor. *Science* 345:929–933.

Bang B, Cobb S. 1968. The size of the olfactory bulb in 108 species of birds. *Auk* 85:55–61.

Barker FK, Cibois A, Schikler P, Feinstein J, Cracraft J. 2004. Phylogeny and diversification of the largest avian radiation. *Proc Natl Acad Sci USA* 101:11040–11045.

Baron MG, Norman DB, Barrett PM. 2017. A new hypothesis of dinosaur relationships and early dinosaur evolution. *Nature* 543:501–506.

Bass AH, Pritz MB, Northcutt RG. 1973. Effects of telencephalic and tectal ablations on visual behavior in the side-necked turtle, *Podocnemis unifilis*. *Brain Res* 55:455–460.

Baxi KN, Dorries KM, Eisthen HL. 2006. Is the vomeronasal system really specialized for detecting pheromones? *Trends Neurosci* 29:1–7.

Belekhova MG, Kenigfest NB. 2014. Turtle isthmic complex of visual nuclei: immunohistochemistry of gamma-aminobutyric acid, choline acetyltransferase, calcium-binding proteins and histochemistry of cytochrome oxidase activity. *J Evol Biochem Physiol* 50:435–447.

Benton MJ (ed.). 1993. *The Fossil Record 2*. New York: Springer.

Benton MJ. 1995. Diversification and extinction in the history of life. *Science* 268:52–58.

Benton MJ. 1998. The quality of the fossil record of the vertebrates. In: Donovan SK, Paul CR (eds.), *The Adequacy of the Fossil Record*. New York: Wiley, pp. 269–303.

Benton MJ, Donoghue P, Asher RJ, Friedman M, Near TJ, Vinther J. 2015. Constraints on the timescale of animal evolutionary history. *Paleontol Electronica* 18.1.1FC:1–106.

Bever GS, Lyson TR, Field DJ, Bhullar B-AS. 2015. Evolutionary origin of the turtle skull. *Nature* 525:239–242.

Biechl D, Dorigo A, Köster RW, Grothe B, Wullimann MF. 2016. Eppur Si Muove: evidence for an external granular layer and possibly transit amplification in the teleostean cerebellum. *Front Neuroanat* 10:2081–2022.

Biewener AA. 1989. Scaling body support in mammals: limb posture and muscle mechanics. *Science* 245:45–48.

Brach V. 1976. Structure and function of the ocular conus papillaris of *Anolis equestris* (Sauria: Iguanidae). *Copeia* 1976:552–558.

Brach V. 1977. The functional significance of the avian pecten: a review. *Condor* 79:321–327.

Brainerd EL, Owerkowicz T. 2006. Functional morphology and evolution of aspiration breathing in tetrapods. *Respir Physiol Neurobiol* 154:73–88.

Brannen P. 2017. *The Ends of the World*. New York: HarperCollins.

Bravo H, Pettigrew JD. 1981. The distribution of neurons projecting from the retina and visual cortex to the thalamus and tectum opticum of the barn owl, *Tyto alba*, and the burrowing owl, *Speotyto cunicularia*. *J Comp Neurol* 199:419–441.

Briscoe SD, Albertin CB, Rowell JJ, Ragsdale CW. 2018. Neocortical association cell types in the forebrain of birds and alligators. *Curr Biol* 28:1–11.

Bruce LL, Braford MR Jr. 2009. Evolution of the limbic system. In: Squire LR (ed.), *Encyclopedia of Neuroscience*. San Diego: Academic Press, pp. 43–55.

Bruce LL, Butler AB. 1984. Telencephalic connections in lizards. I. Projections to cortex. *J Comp Neurol* 229:585–601.

Brykczynska U, Tzika AC, Rodriguez I, Milinkovitch MC. 2013. Contrasted evolution of the vomeronasal receptor repertoires in mammals and squamate reptiles. *Genome Biol Evol* 5:389–401.

Butler, AB. 1994. The evolution of the dorsal thalamus of jawed vertebrates, including mammals: cladistic analysis and a new hypothesis. *Brain Res Rev* 19:29–65.

Cajal SRY. 1972. *The Structure of the Retina*. Springfield, IL: C. C. Thomas.

Cajal SRY. 1995. *Histology of the Nervous System of Man and Vertebrates*. Translated by Swanson N, Swanson LW. New York: Oxford University Press.

Carr CE, Konishi M. 1990. A circuit for detection of interaural time differences in the brain stem of the barn owl. *J Neurosci* 10:3227–3246.

Carr CE, Soares D, Smolders J, Simon JZ. 2009. Detection of interaural time differences in the alligator. *J Neurosci* 29:7978–7990.

Carrier DR. 1987. The evolution of locomotor stamina in tetrapods: circumventing a mechanical constraint. *Paleobiology* 13:326–341.

Carrier DR, Farmer CG. 2000. The integration of ventilation and locomotion in archosaurs. *Am Zool* 40:87–100.

Carroll RL, Baird D. 1972. Carboniferous stem-reptiles of the family Romeriidae. *Bull Mus Comp Zool* 143:321–363.

Chalupa LM, Rhoades RW. 1977. Responses of visual, somatosensory, and auditory neurones in the golden hamster's superior colliculus. *J Physiology* 270:595–626.

Chaplin N, Tendeng C, Wingate RJT. 2010. Absence of an external germinal layer in zebrafish and shark reveals a distinct, anamniote ground plan of cerebellum development. *J Neurosci* 30:3048–3057.

Charig AJ. 1972. The evolution of the archosaur pelvis and hindlimb: an explanation in functional terms. Studies in vertebrate evolution. In: Joysey KA, Kemp TS (eds.), *Studies in Vertebrate Evolution*. New York: Winchester, pp. 121–155.

Chen D-M, Collins JS, Goldsmith TH. 1984. The ultraviolet receptor of bird retinas. *Science* 225:337–340.

Christensen-Dalsgaard J, Carr CE. 2008. Evolution of a sensory novelty: tympanic ears and the associated neural processing. *Brain Res Bull* 75:365–370.

Christensen-Dalsgaard J, Manley GA. 2008. Acoustical coupling of lizard eardrums. *J Assoc Res Otolaryngol* 9:407–416.

Clack JA. 2012. *Gaining Ground*. Bloomington: Indiana University Press.

Clack JA, Allin E. 2004. The evolution of single- and multiple-ossicle ears in fishes and tetrapods. In: Manley GA, Fay RR (eds.), *Evolution of the Vertebrate Auditory System*. New York: Springer, pp. 128–163.

Claramunt S, Cracraft J. 2015. A new time tree reveals Earth history's imprint on the evolution of modern birds. *Sci Adv* 1:e1501005.

Collett TS, Collett M. 2011. Animal navigation: following signposts in the sea. *Curr Biol* 21:R843–6.

Collin SP, Pettigrew JD. 1988a. Retinal topography in reef teleosts. I. Some species with well-developed areae but poorly-developed streaks. *Brain Behav Evol* 31:269–282.

Collin SP, Pettigrew JD. 1988b. Retinal topography in reef teleosts. II. Some species with prominent horizontal streaks and high-density areae. *Brain Behav Evol* 31:283–295.

Corfield JR, Eisthen HL, Iwaniuk AN, Parsons S. 2014. Anatomical specializations for enhanced olfactory sensitivity in kiwi, *Apteryx mantelli*. *Brain Behav Evol* 84:214–226.

Corfield JR, Kolominsky J, Marín GJ, Craciun I, Mulvany-Robbins BE, Iwaniuk AN, Wylie DR. 2015a. Zebrin II Expression in the cerebellum of a paleognathous bird, the Chilean tinamou (*Nothoprocta perdicaria*). *Brain Behav Evol* 85:94–106.

Corfield JR, Price K, Iwaniuk AN, Gutiérrez-Ibáñez C, Birkhead T, Wylie DR. 2015b. Diversity in olfactory bulb size in birds reflects allometry, ecology, and phylogeny. *Front Neuroanat* 9:102.

Corrales JD. 2006. The level of sonic hedgehog signaling regulates the complexity of cerebellar foliation. *Development* 133:1811–1821.

Correia MJ, Eden AR, Westlund KN, Coulter JD. 1983. A study of some of the ascending and descending vestibular pathways in the pigeon (*Columba livia*) using anterograde transneuronal autoradiography. *Brain Res* 278:53–61.

Cronin TW, Bok MJ. 2016. Photoreception and vision in the ultraviolet. *J Exp Biol* 219:2790–2801.

Davies WIL, Collin SP, Hunt DM. 2012. Molecular ecology and adaptation of visual photopigments in craniates. *Mol Ecol* 21:3121–3158.

Davila JC, Guirado S, Puelles L. 2000. Expression of calcium-binding proteins in the diencephalon of the lizard *Psammodromus algirus*. *J Comp Neurol* 427:67–92.

Davis JK, Lowman JJ, Thomas PJ, Hallers ten BFH, Koriabine M, Huynh LY, Maney DL, de Jong PJ, Martin CL, NISC Comparative Sequencing Program, Thomas JW. 2010. Evolution of a bitter taste receptor gene cluster in a New World sparrow. *Genome Biol Evol* 2:358–370.

deBraga M, Rieppel O. 2008. Reptile phylogeny and the interrelationships of turtles. *Zool J Linn Soc* 120:281–354.

Dell'Ariccia G, Célérier A, Gabirot M, Palmas P, Massa B, Bonadonna F. 2014. Olfactory foraging in temperate waters: sensitivity to dimethylsulphide of shearwaters in the Atlantic Ocean and Mediterranean Sea. *J Exp Biol* 217:1701–1709.

Desfilis E, Abellán A, Sentandreu V, Medina L. 2017. Expression of regulatory genes in the embryonic brain of a lizard and implications for understanding pallial organization and evolution. *J Comp Neurol* 526:166–202.

Desfilis E, Font E, Belekhova M, Kenigfest N. 2002. Afferent and efferent projections of the dorsal anterior thalamic nuclei in the lizard *Podarcis hispanica* (Sauria, Lacertidae). *Brain Res Bull* 57:447–450.

Dinets V, Brueggen JC, Brueggen JD. 2015. Crocodilians use tools for hunting. *Ethol Ecol Evo* 27:74–78.

Dong D, Jones G, Zhang S. 2009. Dynamic evolution of bitter taste receptor genes in vertebrates. *BMC Evol Biol* 9:12.

Dubbeldam JL. 1998. Birds. In: Nieuwenhuys R, Donkelaar Ten HJ (eds.), *The Central Nervous System of Vertebrates*, Vol. 3. Berlin: Springer, pp. 1525–1636.

Eisthen HL. 2000. Presence of the vomeronasal system in aquatic salamanders. *Philos Trans R Soc Lond B* 355:1209–1213.

Eisthen HL, Polese G. 2007. Evolution of vertebrate olfactory subsystems. In: Kaas JH, Bullock TH (eds.), *Evolution of Nervous Systems*, Vol. 2. San Diego, CA: Academic Press, pp 355–406.

Emery N. 2016. *Bird Brain: An Exploration of Avian Intelligence*. Princeton, NJ: Princeton University Press.

Emery NJ, Clayton NS. 2004. The mentality of crows: convergent evolution of intelligence in corvids and apes. *Science* 306:1903–1907.

Farago AF, Awatramani RB, Dymecki SM. 2006. Assembly of the brainstem cochlear nuclear complex is revealed by intersectional and subtractive genetic fate maps. *Neuron* 50:205–218.

Feng P, Zheng J, Rossiter SJ, Wang D, Zhao H. 2014. Massive losses of taste receptor genes in toothed and baleen whales. *Genome Biol Evol* 6:1254–1265.

Ferrara CR, Vogt RC, Sousa-Lima RS. 2013. Turtle vocalizations as the first evidence of posthatching parental care in chelonians. *J Comp Psychol* 127:24–32.

Fetcho JR. 1987. A review of the organization and evolution of motoneurons innervating the axial musculature of vertebrates. *Brain Res* 434:243–280.

Fetcho JR. 1992. The spinal motor system in early vertebrates and some of its evolutionary changes. *Brain Behav Evol* 40:82–97.

Fettiplace R, Fuchs PA. 1999. Mechanisms of hair cell tuning. *Annu Rev Physiol* 61:809–834.

Field DJ, Gauthier JA, King BL, Pisani D, Lyson TR, Peterson KJ. 2014. Toward consilience in reptile phylogeny: miRNAs support an archosaur, not lepidosaur, affinity for turtles. *Evol Dev* 16:189–196.

Filoramo NI, Schwenk K. 2009. The mechanism of chemical delivery to the vomeronasal organs in squamate reptiles: a comparative morphological approach. *J Exp Zool* 311A:20–34.

Fite KV, Rosenfield-Wessels S. 1975. A comparative study of deep avian foveas. *Brain Behav Evol* 12:97–115.

Foà A, Basaglia F, Beltrami G, Carnacina M, Moretto E, Bertolucci C. 2009. Orientation of lizards in a Morris water-maze: roles of the sun compass and the parietal eye. *J Exp Biol* 212:2918–2924.

Fournier J, Müller CM, Schneider I, Laurent G. 2018. Spatial information in a non-retinotopic visual cortex. *Neuron* 97:164–180.

Freake MJ. 2001. Homing behaviour in the sleepy lizard (*Tiliqua rugosa*): the role of visual cues and the parietal eye. *Beh Ecol Sociobiol* 50:563–569.

Fritzsch B. 1987. Inner ear of the coelacanth fish *Latimeria* has tetrapod affinities. *Nature* 327:153–154.

Gans C. 1961. The feeding mechanism of snakes and its possible evolution. *Am Zool* 1:217–227.

Gans C. 1970. Respiration in early tetrapods: the frog is a red herring. *Evolution* 24:723–734.

Gauthier JA, Nesbitt SJ, Schachner ER. 2011. The bipedal stem crocodilian *Poposaurus gracilis*: inferring function in fossils and innovation in archosaur locomotion. *Bull Peab Mus Nat Hist* 52:107–126.

Gilbert SF, Corfe I. 2013. Turtle origins: picking up speed. *Dev Cell* 25:326–328.

Gill FB. 1994. *Ornithology*, 2nd ed. New York: Freeman.

Gleich O, Fischer FP, Köppl C, Manley GA. 2004. Hearing organ evolution and specialization: archosaurs. In: Manley GA, Fay RR (eds.), *Evolution of the Vertebrate Auditory System*. New York: Springer, pp. 224–255.

Goddard CA, Mysore SP, Bryant AS, Huguenard JR, Knudsen EI. 2014. Spatially reciprocal inhibition of inhibition within a stimulus selection network in the avian midbrain. *PLoS ONE* 9:e85865.

Gona AG. 1976. Autoradiographic studies of cerebellar histogenesis in the bullfrog tadpole during metamorphosis: the external granular layer. *J Comp Neurol* 165:77–87.

Goris RC. 2011. Infrared organs of snakes: an integral part of vision. *J Herpetol* 45:2–14.

Goslow GE, Wilson D, Poore SO. 2000. Neuromuscular correlates to the evolution of flapping flight in birds. *Brain Behav Evol* 55:85–99.

Graves BM, Halpern M. 1989. Chemical access to the vomeronasal organs of the lizard *Chalcides ocellatus*. *J Exp Zool* 249:150–157.

Greenberg N, MacLean PD, Ferguson JL. 1979. Role of the paleostriatum in species-typical display behavior of the lizard (*Anolis carolinensis*). *Brain Res* 172:229–241.

Greenberg N, Scott M, Crews D. 1984. Role of the amygdala in the reproductive and aggressive behavior of the lizard, *Anolis carolinensis*. *Physiol Behav* 32:147–151.

Grigg NP, Krilow JM, Gutiérrez-Ibáñez C, Wylie DR, Graves GR, Iwaniuk AN. 2017. Anatomical evidence for scent guided foraging in the turkey vulture. *Sci Rep* 7:17408

Grothe B, Pecka M. 2014. The natural history of sound localization in mammals – a story of neuronal inhibition. *Front Neural Circuits* 8:116.

Gruberg E, Dudkin E, Wang Y, Marín G, Salas C, Sentis E, Letelier J, Mpodozis J, Malpeli J, Cui H, Ma R, Northmore D, Udin S. 2006. Influencing and interpreting visual input: the role of a visual feedback system. *J Neurosci* 26:10368–10371.

Guirado S, Davila JC, Real MA, Medina L. 2000. Light and electron microscopic evidence for projections from the thalamic nucleus rotundus to targets in the basal ganglia, the dorsal ventricular ridge, and the amygdaloid complex in a lizard. *J Comp Neurol* 424:216–232.

Gültiken ME, Yıldız D, Onuk B, Karayiğit MÖ. 2011. The morphology of the pecten oculi in the common buzzard (*Buteo buteo*). *Vet Ophthalm* 15:72–76.

Güntürkün O. 2011. The convergent evolution of neural substrates for cognition. *Psychol Res* 76:212–219.

Gutiérrez-Ibáñez C, Iwaniuk AN, Lisney TJ, Faunes M, Marín GJ, Wylie DR. 2012. Functional implications of species differences in the size and morphology of the isthmo optic nucleus (ION) in birds. *PLoS ONE* 7:e37816.

Hackett SJ, Kimball RT, Reddy S, Bowie RCK, Braun EL, Braun MJ, Chojnowski JL, Cox WA, Han K-L, Harshman J, et al. 2008. A phylogenomic study of birds reveals their evolutionary history. *Science* 320:1763–1768.

Hall MI, Heesy CP. 2010. Eye size, flight speed and Leuckart's Law in birds. *J Zool Lond* 283:291–297.

Hall WC, Ebner FF. 1970. Thalamotelencephalic projections in the turtle (*Pseudemys scripta*). *J Comp Neurol* 140:101–122.

Hallam A, Wignall PB. 1997. *Mass Extinctions and Their Aftermath*. New York: Oxford University Press.

Halpern M, Kubie JL. 1980. Chemical access to the vomeronasal organs of garter snakes. *Physiol Behav* 24:367–371.

Harkness L, Bennet-Clark HC. 1978. The deep fovea as a focus indicator. *Nature* 272:814–816.

Harshman J, Braun EL, Braun MJ, Huddleston CJ, Bowie RCK, Chojnowski JL, Hackett SJ, Han K-L, Kimball RT, Marks BD, et al. 2008. Phylogenomic evidence for multiple losses of flight in ratite birds. *Proc Natl Acad Sci USA* 105:13462–13467.

Heers AM, Dial KP. 2012. From extant to extinct: locomotor ontogeny and the evolution of avian flight. *Trends Ecol Evolut* 27:296–305.

Heller SB, Ulinski PS. 1987. Morphology of geniculocortical axons in turtles of the genera *Pseudemys* and *Chrysemys*. *Anat Embryol* 175:505–515.

Henze MJ, Schaeffel F, Ott M. 2004. Variations in the off-axis refractive state in the eye of the Vietnamese leaf turtle (*Geoemyda spengleri*). *J Comp Physiol A* 190:131–137.

Herrel A, Schaerlaeken V, Meyers JJ, Metzger KA, Ross CF. 2007. The evolution of cranial design and performance in squamates: consequences of skull-bone reduction on feeding behavior. *Integ Comp Biol* 47:107–117.

Hidalgo-Sánchez M, Martínez-de-la-Torre M, Alvarado-Mallart R-M, Puelles L. 2005. A distinct preisthmic histogenetic domain is defined by overlap of *Otx2* and *Pax2* gene expression in the avian caudal midbrain. *J Comp Neurol* 483:17–29.

Hile AG, Plummer TK, Striedter GF. 2000. Male vocal imitation produces call convergence during pair bonding in budgerigars, *Melopsittacus undulatus*. *Anim Behav* 59:1209–1218.

Hines M. 1923. The development of the telencephalon in *Sphenodon punctatum*. *J Comp Neurol* 35:483–537.

Holloway WL, Claeson KM, O'keefe FR. 2013. A virtual phytosaur endocast and its implications for sensory system evolution in archosaurs. *J Vert Paleontol* 33:848–857.

Hopson JA. 1979. Paleoneurology. In: Gans C (ed.), *Biology of the Reptilia*, Vol. 9 A. San Diego: Academic Press, pp. 39–146.

Horn G, Johnson MH. 1989. Memory systems in the chick: dissociations and neuronal analysis. *Neuropsychologia* 27:1–22.

Hsiang AY, Field DJ, Webster TH, Behlke AD, Davis MB, Racicot RA, Gauthier JA. 2015. The origin of snakes: revealing the ecology, behavior, and evolutionary history of early snakes using genomics, phenomics, and the fossil record. *BMC Evol Biol*:1–22.

Hughes JT, Jerrome D, Krebs HA. 1972. Ultrastructure of the avian retina: an anatomical study of the retina of the domestic pigeon (*Columba liva*) with particular reference to the distribution of mitochondria. *Exp Eye Res* 14:189–195.

Hunt DM, Peichl L. 2014. S cones: evolution, retinal distribution, development, and spectral sensitivity. *Vis Neurosci* 31:115–138.

Iwaniuk AN, Dean KM, Nelson JE. 2005. Interspecific allometry of the brain and brain regions in parrots (Psittaciformes): comparisons with other birds and primates. *Brain Behav Evol* 65:40–59.

Iwaniuk AN, Hurd PL. 2005. The evolution of cerebrotypes in birds. *Brain Behav Evol* 65:215–230.

Iwaniuk AN, Hurd PL, Wylie DRW. 2007. Comparative morphology of the avian cerebellum: II. Size of folia. *Brain Behav Evol* 69:196–219.

Janis CM, Keller JC. 2001. Modes of ventilation in early tetrapods: costal aspiration as a key feature of amniotes. *Acta Palaeontol Pol* 46:137–170.

Jarvis ED, Mirarab S, Aberer AJ, Li B, Houde P, Li C, Ho SYW, Faircloth BC, Nabholz B, Howard JT, et al. 2014. Whole-genome analyses resolve early branches in the tree of life of modern birds. *Science* 346:1320–1331.

Jarvis ED, Yu J, Rivas MV, Horita H, Feenders G, Whitney O, Jarvis SC, Jarvis ER, Kubikova L, Puck AEP, et al. 2013. Global view of the functional molecular organization of the avian cerebrum: mirror images and functional columns. *J Comp Neurol* 521:3614–3665.

Jeffres LA. 1948. A place theory of sound localization. *J Comp Physiol Psychol* 41:35–39.

Jones MEH, O'higgins P, Fagan MJ, Evans SE, Curtis N. 2012. Shearing mechanics and the influence of a flexible symphysis during oral food processing in *Sphenodon* (Lepidosauria: Rhyncho cephalia). *Anat Rec* 295:1075–1091.

Joyce WG. 2007. Phylogenetic relationships of Mesozoic turtles. *Bull Peab Mus Nat Hist* 48:3–102.

Joyce WG, Gauthier JA. 2004. Palaeoecology of triassic stem turtles sheds new light on turtle origins. *Proc Roy Soc Lond B* 271:1–5.

Joyce WG, Parham JF, Lyson TR, Warnock RCM, Donoghue PCJ. 2015. A divergence dating analysis of turtles using fossil calibrations: an example of best practices. *J Paleontol* 87:612–634.

Kaldenbach F, Bleckmann H, Kohl T. 2016. Responses of infrared-sensitive tectal units of the pit viper *Crotalus atrox* to moving objects. *J Comp Physiol A* 202:389–398.

Kani S, Bae Y-K, Shimizu T, Tanabe K, Satou C, Parsons MJ, Scott E, Higashijima S-I, Hibi M. 2010. Proneural gene-linked neurogenesis in zebrafish cerebellum. *Dev Biol* 343:1–17.

Kardong KV, Mackessy SP. 1991. The strike behavior of a congenitally blind rattlesnake. *J Herpetol* 25:208–211.

Karten HJ. 1968. The ascending auditory pathway in the pigeon (*Columba livia*). II. Telencephalic projections of the nucleus ovoidalis thalami. *Brain Res* 11:134–153.

Karten HJ, Hodos W. 1970. Telencephalic projections of the nucleus rotundus in the pigeon (*Columba livia*). *J Comp Neurol* 140:35–51.

Karten HJ, Hodos W, Nauta WJ, Revzin AM. 1973. Neural connections of the "visual wulst" of the avian telencephalon: experimental studies in the piegon (*Columba livia*) and owl (*Speotyto cunicularia*). *J Comp Neurol* 150:253–278.

Karten HJ, Shimizu T. 1989. The origins of neocortex: connections and lamination as distinct events in evolution. *J Cog Neurosci* 1:291–301.

Katzir G, Howland HC. 2003. Corneal power and underwater accommodation in great cormorants (*Phalacrocorax carbo sinensis*). *J Exp Biol* 206:833–841.

Kenigfest NB, Belekhova MG. 2012. Neurochemical characteristics of the turtle optic tectum: comparison with other reptilian species and birds. *J Evol Biochem Physiol* 48:83–100.

Kenigfest N, MartinezMarcos A, Belekhova M, Font C, Lanuza E, Desfilis E, MartinezGarcia F. 1997. A lacertilian dorsal retinorecipient thalamus: a re-investigation in the old-world lizard *Podarcis hispanica*. *Brain Behav Evol* 50:313–334.

Khan I, Yang Z, Maldonado E, Li C, Zhang G, Gilbert MTP, Jarvis ED, O'Brien SJ, Johnson WE, Antunes A. 2015. Olfactory receptor subgenomes linked with broad ecological adaptations in Sauropsida. *Mol Biol Evol* 32:2832–2843.

Kitazawa T, Takechi M, Hirasawa T, Adachi N, me NN-NE, Kume H, Maeda K, Hirai T, Miyagawa-Tomita S, Kurihara Y, Hitomi J, Levi G, Kuratani S, Kurihara H. 1AD. Developmental genetic bases behind the independent origin of the tympanic membrane in mammals and diapsids. *Nat Commun* 6:1–7.

Knudsen EI, Konishi M. 1978. A neural map of auditory space in the owl. *Science* 200:795–797.

Konishi M. 2003. Coding of auditory space. *Annu Rev Neurosci* 26:31–55.

Krabichler Q, Vega-Zuniga T, Carrasco D, Fernández M, Gutiérrez-Ibáñez C, Marín G, Luksch H. 2017. The centrifugal visual system of a palaeognathous bird, the Chilean Tinamou (*Nothoprocta perdicaria*). *J Comp Neurol* 525:2514–2534.

Krochmal AR. 2004. Heat in evolution's kitchen: evolutionary perspectives on the functions and origin of the facial pit of pitvipers (Viperidae: Crotalinae). *J Exp Biol* 207:4231–4238.

Krohmer RW, Crews D. 1987. Facilitation of courtship behavior in the male red-sided garter snake (*Thamnophis sirtalis parietalis*) following lesions of the septum or nucleus sphericus. *Physiol Behav* 40:759–765.

Krützfeldt NOE, Wild JM. 2003. Definition and connections of the entopallium in the zebra finch (*Taeniopygia guttata*). *J Comp Neurol* 468:452–465.

Krützfeldt NOE, Wild JM. 2005. Definition and novel connections of the entopallium in the pigeon (*Columba livia*). *J Comp Neurol* 490:40–56.

Kuba H, Yamada R, Fukui I, Ohmori H. 2005. Tonotopic specialization of auditory coincidence detection in nucleus laminaris of the chick. *J Neurosci* 25:1924–1934.

Kubke MF, Carr CE. 2000. Development of the auditory brainstem of birds: comparison between barn owls and chickens. *Hear Res* 147:1–20.

Kubke MF, Carr CE. 2006. Morphological variation in the nucleus laminaris of birds. *Intl J Comp Psychol* 19:83–97.

Kuenzel WJ, Medina L, Csillag A, Perkel DJ, Reiner A. 2011. The avian subpallium: new insights into structural and functional subdivisions occupying the lateral subpallial wall and their embryological origins. *Brain Res* 1424:67–101.

Labandeira C, Sepkoski J. 1993. Insect diversity in the fossil record. *Science* 261:310–315.

Lanuza E, Halpern M. 1998. Efferents and centrifugal afferents of the main and accessory olfactory bulbs in the snake *Thamnophis sirtalis*. *Brain Behav Evol* 51:1–22.

Larsson M. 2011. Binocular vision and ipsilateral retinal projections in relation to eye and forelimb coordination. *Brain Behav Evol* 77:219–230.

Lee MSY, Cau A, Naish D, Dyke GJ. 2014. Sustained miniaturization and anatomical innovation in the dinosaurian ancestors of birds. *Science* 345:562–566.

Lefebvre L, Nicolakakis N, Boire D. 2002. Tools and brains in birds. *Behaviour* 139:939–973.

Lewin R. 1983. How did vertebrates take to the air? Aerodynamic analysis has produced novel support for the cursorial, as opposed to the arboreal, theory of the origin of birds. *Science* 221:38–39.

Li JL, Xiao Q, Fu YX, Wang SR. 1998. Centrifugal innervation modulates visual activity of tectal cells in pigeons. *Vis Neurosci* 15:411–415.

Lisney TJ, Wylie DR, Kolominsky J, Iwaniuk AN. 2015. Eye morphology and retinal topography in hummingbirds (Trochilidae: Aves). *Brain Behav Evol* 86:176–190.

Liu GB, Pettigrew JD. 2003. Orientation mosaic in barn owl's visual Wulst revealed by optical imaging: comparison with cat and monkey striate and extra-striate areas. *Brain Res* 961:153–158.

Liu Y, Ding L, Lei J, Zhao E, Tang Y. 2012. Eye size variation reflects habitat and daily activity patterns in colubrid snakes. *J Morphol* 273:883–893.

Locket NA. 1992. Problems of deep foveas. *Aust N Z J Ophthalmol* 20:281–295.

Lohmann KJ, Lohmann CMF, Ehrhart LM, Bagley DA, Swing T. 2004. Animal behaviour: geomagnetic map used in sea-turtle navigation. *Nature* 428:909–910.

Lohman A, Smeets W. 1993. Overview of the main and accessory olfactory bulb projections in reptiles. *Brain Behav Evol* 41:147–155.

Longrich NR, Bhullar B-AS, Gauthier JA. 2012. Mass extinction of lizards and snakes at the Cretaceous-Paleogene boundary. *Proc Natl Acad Sci USA* 109:21396–21401.

Luksch H, Cox K, Karten H. 1998. Bottlebrush dendritic endings and large dendritic fields: motion-detecting neurons in the tectofugal pathway. *J Comp Neurol* 396:399–414.

Major D, Luksch H, Karten H. 2000. Bottlebrush dendritic endings and large dendritic fields: motion-detecting neurons in the mammalian tectum. *J Comp Neurol* 423:243–260.

Manley GA, Clack JA. 2004. An outline of the evolution of vertebrate hearing organs. In: Manley GA, Fay RR (eds.), *Evolution of the Vertebrate Auditory System*. New York: Springer, pp. 1–26.

Manley GA, Köppl C, Christensen-Dalsgaard J, Wilson M. 2017. Evolutionary trends in land vertebrate hearing organs. In: Kaas JH, Striedter GF (eds.), *Evolution of Nervous Systems*, Vol. 1, 2nd ed. Amsterdam: Elsevier, pp. 277–290.

Manrod JD, Hartdegen R, Burghardt GM. 2007. Rapid solving of a problem apparatus by juvenile black-throated monitor lizards (*Varanus albigularis albigularis*). *Anim Cogn* 11:267–273.

Marín G, Mpodozis J, Mpdozis J, Sentis E, Ossandón T, Letelier JC. 2005. Oscillatory bursts in the optic tectum of birds represent re-entrant signals from the nucleus isthmi pars parvocellularis. *J Neurosci* 25:7081–7089.

Martinez-Garcia F, Novejarque A, Lanuza E. 2007. Evolution of the amygdala in vertebrates. In: Kaas JH, Bullock TH (eds.), *Evolution of Nervous Systems*, Vol. 2. Amsterdam: Elsevier, pp. 255–334.

Martinez-Marcos A, Lanuza E, Halpern M. 1999. Organization of the ophidian amygdala: chemosensory pathways to the hypothalamus. *J Comp Neurol* 412:51–68.

Marzban H, Hawkes R. 2011. On the architecture of the posterior zone of the cerebellum. *Cerebellum* 10:422–434.

Mazurskaya PZ. 1973. Organization of receptive fields in the forebrain of *Emys orbicularis*. *Neurosci Behav Physiol* 6:311–318.

Medina L. 2007. Do birds and reptiles possess homologues of mammalian visual, somatosensory, and motor cortices? In: Kaas JH, Bullock TH (eds.), *Evolution of Nervous Systems*, Vol. 2. Amsterdam: Elsevier, pp. 163–194.

Medina L, Abellán A, Vicario A, Castro-Robles B, Desfilis E. 2017. The Amygdala. In: Kaas JH, Striedter GF (eds.), *Evolution of Nervous Systems*, Vol. 1, 2nd ed. Amsterdam: Elsevier, pp. 427–478.

Miller MR. 1975. The cochlear nuclei of lizards. *J Comp Neurol* 159:375–406.

Miller MR. 1980. The cochlear nuclei of snakes. *J Comp Neurol* 192:717–736.

Miller MR. 1992. The evolutionary implications of the structural variations in the auditory papilla of lizards. In: Webster DB, Popper AN, Fay RR (eds.), *The Evolutionary Biology of Hearing*. New York: Springer, pp. 463–487.

Molenaar GJ. 1992. Anatomy and physiology of infrared sensitivity of snakes. In: Gans C, Ulinski PS (eds.), *Biology of the Reptilia*, Vol. 17C. Chicago: University of Chicago Press, pp. 367–453.

Moran A, Wojcik L, Cangiane L, Powers AS. 1998. Dorsal cortex lesions impair habituation in turtles (*Chrysemys picta*). *Brain Behav Evol* 51:40–47.

Moreno N, González A. 2007. Development of the vomeronasal amygdala in anuran amphibians: hodological, neurochemical, and gene expression characterization. *J Comp Neurol* 503:815–831.

Moroney MK, Pettigrew JD. 1987. Some observations on the visual optics of kingfishers (Aves, Coraciformes, Alcedinidae). *J Comp Physiol A* 160:137–149.

Mühlenbrock-Lenter S, Roth G, Grunwald W. 2005. Connectivity and cytoarchitecture of telencephalic centers in the fire-bellied toad *Bombina orientalis*. *Brain Res Bull* 66:270–276.

Mulligan KA, Ulinski PS. 1990. Organization of geniculocortical projections in turtles: isoazimuth lamellae in the visual cortex. *J Comp Neurol* 296:531–547.

Murray E, Wise S, Graham K. 2016. *The Evolution of Memory Systems*. New York: Oxford University Press.

Mysore SP, Knudsen EI. 2013. A shared inhibitory circuit for both exogenous and endogenous control of stimulus selection. *Nat Neurosci* 16:473–478.

Nakamori T, Maekawa F, Sato K, Tanaka K, Ohki-Hamazaki H. 2013. Neural basis of imprinting behavior in chicks. *Dev Growth Differ* 55:198–206.

Neary TJ, Northcutt RG. 1983. Nuclear organization of the bullfrog diencephalon. *J Comp Neurol* 213:262–278.

Necker R. 2005. The structure and development of avian lumbosacral specializations of the vertebral canal and the spinal cord with special reference to a possible function as a sense organ of equilibrium. *Anat Embryol* 210:59–74.

Necker R. 2006. Specializations in the lumbosacral vertebral canal and spinal cord of birds: evidence of a function as a sense organ which is involved in the control of walking. *J Comp Physiol A* 192:439–448.

Necker R. 2007. Head-bobbing of walking birds. *J Comp Physiol A* 193:1177–1183.

Nevitt GA. 2008. Sensory ecology on the high seas: the odor world of the procellariiform seabirds. *J Exp Biol* 211:1706–1713.

Ngwenya A, Patzke N, Herculano-Houzel S, Manger PR. 2018. Potential adult neurogenesis in the telencephalon and cerebellar cortex of the nile crocodile revealed with doublecortin immunohistochemistry. *Anat Rec* 301:659–672.

Ngwenya A, Patzke N, Manger PR, Herculano-Houzel S. 2016. Continued growth of the central nervous system without mandatory addition of neurons in the nile crocodile (*Crocodylus niloticus*). *Brain Behav Evol* 87:19–38.

Ngwenya A, Patzke N, Spocter MA, Kruger J-L, Dell L-A, Chawana R, Mazengenya P, Billings BK, Olaleye O, Herculano-Houzel S, Manger PR. 2013. The continuously growing central nervous system of the Nile crocodile (*Crocodylus niloticus*). *Anat Rec* 296:1489–1500.

Nieuwenhuys R, Donkelaar ten HJ, Nicholson C. 1998. *The Central Nervous System of Vertebrates*. Berlin: Springer.

Northcutt RG. 1978. Forebrain and midbrain organization in lizards and its phylogenetic significance. In: Greenberg NL, MacLean PD (eds.), *Behavior and Neurology of Lizards*. Bethesda, MD: US Department of Health, Education and Welfare, pp. 11–64.

Northcutt RG. 1984. Anatomical organization of the optic tectum in reptiles. In: Vanegas H (ed.), *Comparative Neurology of the Optic Tectum*. New York: Plenum Press, pp. 547–600.

Northcutt RG. 2013. Variation in reptilian brains and cognition. *Brain Behav Evol* 82:45–54.

Nottebohm F. 1972. The origins of vocal learning. *Am Naturalist* 106:116–140.

Nottebohm F, Stokes TM, Leonard CM. 1976. Central control of song in the canary, *Serinus canarius*. *J Comp Neurol* 165:457–486.

Olkowicz S, Kocourek M, Lučan RK, Porteš M, Fitch WT, Herculano-Houzel S, Němec P. 2016. Birds have primate-like numbers of neurons in the forebrain. *Proc Natl Acad Sci USA* 113:7255–7260.

Ott M. 2005. Visual accommodation in vertebrates: mechanisms, physiological response and stimuli. *J Comp Physiol A* 192:97–111.

Packard GC, Packard MJ. 1980. Evolution of the cleidoic egg among reptilian antecedents of birds. *Am Zool* 20:351–362.

Paidan K, Li C, Pchelnikova J. 2010. The trackmaker of Apatopus (Late Triassic, North America): implications for the evolution of archosaur stance and gait. *Palaeontology* 53:175–189.

Pakan JMP, Graham DJ, Gutiérrez-Ibáñez C, Wylie DR. 2011. Organization of the cerebellum: correlating zebrin immunochemistry with optic flow zones in the pigeon flocculus. *Vis Neurosci* 28:163–174.

Parrish JM. 1987. The origin of crocodilian locomotion. *Paleobiology* 13:396–414.

Pepperberg IM. 1999. *The Alex Studies*. Cambridge, MA: Harvard University Press.

Peterson EH. 1992. Retinal structure. In: Gans C, Ulinski PS (eds.), *Biology of the Reptilia*, Vol. 17C. Chicago: University of Chicago Press, pp. 1–135.

Pettigrew JD. 1979. Binocular visual processing in the owls telencephalon. *Proc Roy Soc Lond B* 204:435–454.

Pettigrew JD, Konishi M. 1976. Neurons selective for orientation and binocular disparity in visual Wulst of barn owl (*Tyto alba*). *Science* 193:675–678.

Phillips MJ, Gibb GC, Crimp EA, Penny D. 2010. Tinamous and moa flock together: mitochondrial genome sequence analysis reveals independent losses of flight among ratites. *Syst Biol* 59:90–107.

Platel R. 1979. Brain weight-body weight relationships. In: Gans C, Northcutt RG, Ulinski P (eds.), *Biology of the Reptilia*, Vol. 9A. London: Academic Press, pp. 147–171.

Platel R. 1989. [Anatomy of the brain of the New Zealand Gray *Sphenodon punctatus* (Sphenodontidae): a quantitative study of the principle subdivisions of the brain]. *J Hirnforsch* 30:325–337.

Portmann A. 1947. Études sur la cérébralisation chez les oiseaux. III. Cérébralisation et mode ontogenique. *Alauda* 15:161–171.

Pough FH, Andrews RM, Cadle JE, Crump ML, Savitzky AH, Wells KD. 2004. *Herpetology*, 3rd ed. New York: Prentice Hall.

Pritz MB. 1974. Ascending connections of a thalamic auditory area in a crocodile, *Caiman crocodilus*. *J Comp Neurol* 153:199–213.

Pritz MB. 1975. Anatomical identification of a telencephalic visual area in crocodiles: ascending connections of nucleus rotundus in *Caiman crocodilus*. *J Comp Neurol* 164:323–338.

Pritz MB, Stritzel ME. 1994. Anatomical identification of a telencephalic somatosensory area in a reptile, *Caiman crocodilus*. *Brain Behav Evol* 43:107–127.

Puelles L, Ayad A, Alonso A, Sandoval JE, Martinez-de-la-Torre M, Medina L, Ferran JL. 2015. Selective early expression of the orphan nuclear receptor *Nr4a2* identifies the claustrum homolog in the avian mesopallium: impact on sauropsidian/mammalian pallium comparisons. *J Comp Neurol* 524:665–703.

Puelles L, Ferran JL. 2012. Concept of neural genoarchitecture and its genomic fundament. *Front Neuroanat* 6:47.

Puelles L, Javier Milán F, Martinez-de-la-Torre M. 1996. A segmental map of architectonic subdivisions in the diencephalon of the frog *Rana perezi*: acetylcholinesterase-histochemical observations. *Brain Behav Evol* 47:279–310.

Puelles L, Kuwana E, Puelles E, Bulfone A, Shimamura K, Keleher J, Smiga S, Rubenstein JL. 2000. Pallial and subpallial derivatives in the embryonic chick and mouse telencephalon, traced by the expression of the genes *Dlx-2, Emx-1, Nkx-2.1, Pax-6*, and *Tbr-1*. *J Comp Neurol* 424:409–438.

Puelles L, Martinez-de-la-Torre M, Paxinos G, Watson C, Martinez S. 2007. *The Chick Brain in Stereotaxic Coordinates: An Atlas Featuring Neuromeric Subdivisions and Mammalian Homologies*. Amsterdam: Academic Press.

Puelles L, Robles C, Martinez-de-la-Torre M, Martínez S. 1994. New subdivision schema for the avian torus semicircularis: neurochemical maps in the chick. *J Comp Neurol* 340:98–125.

Puelles L, Sandoval JE, Ayad A, del Corral R, Alonso A, Ferran JL, Martinez-de-la-Torre M. 2017. The pallium in reptiles and birds in the light of the updated tetrapartite pallium model. In: Kaas

JH, Striedter GF (eds.), *Evolution of Nervous Systems*, Vol. 1, 2nd ed. Amsterdam: Elsevier, pp. 519–555.

Pyron RA, Burbrink FT, Wiens JJ. 2013. A phylogeny and revised classification of Squamata, including 4161 species of lizards and snakes. *BMC Evol Biol* 13:93.

Quay WB. 1979. The parietal eye-pineal complex. In: Gans C, Northcutt RG, Ulinski PS (eds.), *Biology of the Reptilia*, Vol. 9A. London: Academic Press, pp. 245–406.

Redies C, Ast M, Nakagawa S, Takeichi M, Martinez-de-la-Torre M, Puelles L. 2000. Morphologic fate of diencephalic prosomeres and their subdivisions revealed by mapping cadherin expression. *J Comp Neurol* 421:481–514.

Reeder TW, Townsend TM, Mulcahy DG, Noonan BP, Wood PL, Sites JW, Wiens JJ. 2015. Integrated analyses resolve conflicts over squamate reptile phylogeny and reveal unexpected placements for fossil taxa. *PLoS ONE* 10:e0118199–22.

Reiner A. 1994. Laminar distribution of the cells of origin of ascending and descending tectofugal pathways in turtles: implications for the evolution of tectal lamination. *Brain Behav Evol* 43:254–292.

Reiner A, Karten HJ. 1985. Comparison of olfactory bulb projections in pigeons and turtles. *Brain Behav Evol* 27:11–27.

Reiner A, Perkel DJ, Bruce LL, Butler AB, Csillag A, Kuenzel W, Medina L, Paxinos G, Shimizu T, Striedter G, et al. 2004. Revised nomenclature for avian telencephalon and some related brainstem nuclei. *J Comp Neurol* 473:377–414.

Reiner A, Powers AS. 1983. The effects of lesions of telencephalic visual structures on visual discriminative performance in turtles (*Chrysemis picta picta*). *J Comp Neurol* 218:1–24.

Repérant J, Miceli D, Vesselkin NP, Molotchnikoff S. 1989. The centrifugal visual system of vertebrates: a century-old search reviewed. *Int Rev Cytol* 118:115–171.

Repérant J, Ward R, Miceli D, Rio JP, Medina M, Kenigfest NB, Vesselkin NP. 2006. The centrifugal visual system of vertebrates: a comparative analysis of its functional anatomical organization. *Brain Res Rev* 52:1–57.

Rest JS, Ast JC, Austin CC, Waddell PJ, Tibbetts EA, Hay JM, Mindell DP. 2003. Molecular systematics of primary reptilian lineages and the tuatara mitochondrial genome. *Mol Phylogenet Evol* 29:289–297.

Riss W, Halpern M, Scalia F. 1969. The quest for clues to forebrain evolution: the study of reptiles. *Brain Behav Evol* 2:1–50.

Romer AS, Parsons TS. 1977. *The Vertebrate Body*, 5th ed. Philadelphia: Saunders.

Rosenberg J, Necker R. 2002. Ultrastructural characterization of the accessory lobes of Lachi in the lumbosacral spinal cord of the pigeon with special reference to intrinsic mechanoreceptors. *J Comp Neurol* 447:274–285.

Ryan JM, Cushman J, Jordan B, Samuels A, Frazer H, Baier C. 1998. Topographic position of forelimb motoneuron pools is conserved in vertebrate evolution. *Brain Behav Evol* 51:90–99.

Ryugo DK, Parks TN. 2003. Primary innervation of the avian and mammalian cochlear nucleus. *Brain Res Bull* 60:435–456.

Sahney S, Benton MJ, Ferry PA. 2010. Links between global taxonomic diversity, ecological diversity and the expansion of vertebrates on land. *Biol Lett* 6:544–547.

Schaeffel F, Hagel G, Eikermann J, Collett T. 1994. Lower-field myopia and astigmatism in amphibians and chickens. *J Opt Soc Am A* 11:487–495.

Schoch RR, Sues H-D. 2015. A Middle Triassic stem-turtle and the evolution of the turtle body plan. *Nature* 523:584–587.

Schwab IR. 2012. Evolution's Witness: How Eyes Evolved. New York: Oxford UP.

Senn DG, Northcutt RG. 1973. The forebrain and midbrain of some squamates and their bearing on the origin of snakes. *J Morphol* 140:135–151.

Shedlock AM, Edwards SV. 2009. Amniotes (Amniota). In: Hedges SB, Kumar S (eds.), *The Timetree of Life*. New York: Oxford University Press, pp. 375–379.

Shi P, Zhang J. 2007. Comparative genomic analysis identifies an evolutionary shift of vomeronasal receptor gene repertoires in the vertebrate transition from water to land. *Genome Res* 17:166–174.

Smaers JB, Rohlf FJ. 2016. Testing species' deviation from allometric predictions using the phylogenetic regression. *Evolution* 70:1145–1149.

Smith CA, Konishi M, Schuff N. 1985. Structure of the barn owl's (*Tyto alba*) inner ear. *Hear Res* 17:237–247.

Steiger SS, Kuryshev VY. 2009. A comparison of reptilian and avian olfactory receptor gene repertoires: species-specific expansion of group γ genes in birds. *BMC Genomics* 10:446.

Striedter GF. 1994. The vocal control pathways in budgerigars differ from those in songbirds. *J Comp Neurol* 343:35–56.

Striedter GF. 1997. The telencephalon of tetrapods in evolution. *Brain Behav Evol* 49:179–213.

Striedter GF. 1999. Homology in the nervous system: of characters, embryology and levels of analysis. *Novartis Found Symp* 222:158–70.

Striedter GF. 2005. *Principles of Brain Evolution*. Sunderland, MA: Sinauer Associates.

Striedter GF. 2013. Bird brains and tool use: beyond instrumental conditioning. *Brain Behav Evol* 82:55–67.

Striedter GF. 2015. Evolution of the hippocampus in reptiles and birds. *J Comp Neurol* 524:496–517.

Sudarov A, Joyner AL. 2007. Cerebellum morphogenesis: the foliation pattern is orchestrated by multi-cellular anchoring centers. *Neural Dev* 2:26.

Sun Y, Joachimski MM, Wignall PB, Yan C, Chen Y, Jiang H, Wang L, Lai X. 2012. Lethally hot temperatures during the Early Triassic greenhouse. *Science* 338:366–370.

Szarski H. 1968. The origin of vertebrate foetal membranes. *Evolution* 22:211.

Szpir MR, Wright DD, Ryugo DK. 1995. Neuronal organization of the cochlear nuclei in alligator lizards: a light and electron-microscopic investigation. *J Comp Neurol* 357:217–241.

Tang Y, Christensen-Dalsgaard J, Carr CE. 2012. Organization of the auditory brainstem in a lizard, *Gekko gecko*. I. Auditory nerve, cochlear nuclei, and superior olivary nuclei. *J Comp Neurol* 520:1784–1799.

Taniguchi K, Taniguchi K. 2014. Phylogenic studies on the olfactory system in vertebrates. *J Vet Med Sci* 76:781–788.

Timmermans S, Lefebvre L, Boire D, Basu P. 2000. Relative size of the hyperstriatum ventrale is the best predictor of feeding innovation rate in birds. *Brain Behav Evol* 56:196–203.

Tosches MA, Yamawaki TM, Naumann RK, Jacobi AA, Tushev G, Laurent G. 2018. Evolution of pallium, hippocampus, and cortical cell types revealed by single-cell transcriptomics in reptiles. *Science* 360:881–888.

Tucker VA. 2000. The deep fovea, sideways vision and spiral flight paths in raptors. *J Exp Biol* 203:3745–3754.

Tyrrell LP, Fernández-Juricic E. 2017. The hawk-eyed songbird: retinal morphology, eye shape, and visual fields of an aerial insectivore. *Am Nat* 189:709–717.

Ulinski PS. 1983. *Dorsal Ventricular Ridge*. New York: Wiley-Interscience.

Ulinski PS. 1989. Nodal events in forebrain evolution. *Neth J Zool* 40:215–240.

Vandewege MW, Mangum SF, Gabaldón T, Castoe TA, Ray DA, Hoffmann FG. 2016. Contrasting patterns of evolutionary diversification in the olfactory repertoires of reptile and bird genomes. *Genome Biol Evol* 8:470–480.

Veenman CL, Gottschaldt KM. 1986. The nucleus basalis-neostriatum complex in the goose (*Anser anser* L). *Adv Anato Embryol Cell Biol* 96:1–85.

Wallraff HG. 2014. Pigeon homing from unfamiliar areas: an alternative to olfactory navigation is not in sight. *Commun Integr Biol* 7:e28565.

Walls GL. 1942. *The Vertebrate Eye and Its Adaptive Radiation*. Bloomfield Hills, MI: Cranbrook Institute of Science.

Walsh SA, Barrett PM, Milner AC, Manley G, Witmer LM. 2009. Inner ear anatomy is a proxy for deducing auditory capability and behaviour in reptiles and birds. *Proc Roy Soc Lond B* 276:1355–1360.

Walton PL, Christensen-Dalsgaard J, Carr CE. 2017. Evolution of sound source localization circuits in the nonmammalian vertebrate brainstem. *Brain Behav Evol* 90:131–153.

Wang K, Zhao H. 2015. Birds generally carry a small repertoire of bitter taste receptor genes. *Genome Biol Evol* 7:2705–2715.

Wang Y, Brzozowska-Prechtl A, Karten HJ. 2010. Laminar and columnar auditory cortex in avian brain. *Proc Natl Acad Sci USA* 107:12676–12681.

Wang Y, Major DE, Karten HJ. 2003. Morphology and connections of nucleus isthmi pars magnocellularis in chicks (*Gallus gallus*). *J Comp Neurol* 469:275–297.

Wang Z, Pascual-Anaya J, Zadissa A, Li W, Niimura Y, Huang Z, Li C, White S, Xiong Z, Fang D, et al. 2013. The draft genomes of soft-shell turtle and green sea turtle yield insights into the development and evolution of the turtle-specific body plan. *Nat Genet* 45:701–706.

Wever EG. 1978. *The Reptile Ear*. Princeton, NJ: Princeton University Press.

Wilczynski W. 1984. Central neural systems subserving a homoplasous periphery. *Am Zool* 24:755–763.

Wild JM. 1987. The avian somatosensory system: connections of regions of body representation in the forebrain of the pigeon. *Brain Res* 412:205–223.

Wild JM. 2004. Functional neuroanatomy of the sensorimotor control of singing. *Ann N Y Acad Sci* 1016:438–462.

Wild JM, Kubke MF, Carr CE. 2001. Tonotopic and somatotopic representation in the nucleus basalis of the barn owl, *Tyto alba*. *Brain Behav Evol* 57:39–62.

Wild JM, Kubke MF, Peña JL. 2008. A pathway for predation in the brain of the barn owl (*Tyto alba*): projections of the gracile nucleus to the "claw area" of the rostral wulst via the dorsal thalamus. *J Comp Neurol* 509:156–166.

Wilkinson A, Huber L. 2012. Cold-blooded cognition: reptilian cognitive abilities. In: Shackelford TK, Vonk J (eds.), *The Oxford Handbook of Comparative Evolutionary Psychology*. New York: Oxford University Press, pp. 129–143.

Wilkinson A, Kuenstner K, Mueller J, Huber L. 2010. Social learning in a non-social reptile (*Geochelone carbonaria*). *Biol Lett* 6:614–616.

Willis KL, Christensen-Dalsgaard J, Carr CE. 2013. Auditory brain stem processing in reptiles and amphibians: roles of coupled ears. In: Köppl C, Manley GA, Popper AN, Fay RR (eds.), *Insights from Comparative Hearing Research*. Berlin: Springer, pp. 194–225.

Wilson M, Lindstrom SH. 2011. What the bird's brain tells the bird's eye: the function of descending input to the avian retina. *Vis Neurosci* 28:337–350.

Witmer LM, Chatterjee S, Franzosa J, Rowe T. 2003. Neuroanatomy of flying reptiles and implications for flight, posture and behaviour. *Nature* 425:950–953.

Witmer LM, Ridgely RC. 2009. New insights into the brain, braincase, and ear region of tyrannosaurs (Dinosauria, Theropoda), with implications for sensory organization and behavior. *Anat Rec* 292:1266–1296.

Witton MP. 2013. *Pterosaurs*. Princeton, NJ: Princeton University Press.

Wylie DR, Hoops D, Aspden JW, Iwaniuk AN. 2017. Zebrin II is expressed in sagittal stripes in the cerebellum of dragon lizards (*Ctenophorus* sp.). *Brain Behav Evol* 88:1–10.

Yan K, Tang Y-Z, Carr CE. 2010. Calcium-binding protein immunoreactivity characterizes the auditory system of *Gekko gecko*. *J Comp Neurol* 518:3409–3426.

Yang Z, Jiang B, McNamara ME, Kearns SL, Pittman M, Kaye TG, Orr PJ, Xu X, Benton MJ. 2018. Pterosaur integumentary structures with complex feather-like branching. *Nat Ecol Evol* 3:24–30.

Yokoyama S, Radlwimmer FB, Blow NS. 2000. Ultraviolet pigments in birds evolved from violet pigments by a single amino acid change. *Proc Natl Acad Sci USA* 97:7366–7371.

Yopak KE, Pakan JMP, Wylie D. 2017. The cerebellum of nonmammalian vertebrates. In: Kaas JH, Striedter GF (eds.), *Evolution of Nervous Systems*, Vol. 1, 2nd ed. Amsterdam: Elsevier, pp. 373–385.

Young BA. 2003. Snake bioacoustics: toward a richer understanding of the behavioral ecology of snakes. *Q Rev Biol* 78:303–325.

Young JM, Massa HF, Hsu L, Trask BJ. 2010. Extreme variability among mammalian V1R gene families. *Genome Res* 20:10–18.

Yu CQ, Schwab IR, Dubielzig RR. 2009. Feeding the vertebrate retina from the Cambrian to the Tertiary. *J Zool Lond* 278:259–269.
Zeier H, Karten HJ. 1971. The archistriatum of the pigeon: organization of afferent and efferent connections. Brain Res 31:313–326.
Zelenitsky DK, Therrien F, Ridgely RC, McGee AR, Witmer LM. 2011. Evolution of olfaction in non-avian theropod dinosaurs and birds. *Proc Roy Soc Lond B* 278:3625–3634.
Zhao H, Li J, Zhang J. 2015. Molecular evidence for the loss of three basic tastes in penguins. *Curr Biol* 25:R141–2.

6

The Rise of Endothermy

Mammals, but also Birds

Mammals and birds are the most diverse groups of terrestrial vertebrates on the planet today. According to traditional criteria, contemporary birds comprise at least 10,000 species (Barrowclough et al., 2016), while the number of mammalian species likely exceeds 5,400 (Wilson and Reeder, 2005). Because we have already discussed the evolution of avian nervous systems in the preceding chapter, we here focus on mammals. However, mammals and birds have independently evolved a large number of remarkable similarities, including the extremely useful ability to generate body heat "from within" (i.e., endothermy). We highlight many of these convergent similarities because they extend not only to behavior and physiology, but also to the brain. Indeed, we argue that comparisons between the brains of mammals and birds require that we take convergent evolution into account. As in the previous chapter, we begin with a discussion of the major players in mammalian phylogeny.

6.1. Extant Mammals

Extant mammals are divisible into three major lineages: monotremes, marsupials, and placental mammals. The latter two lineages are sister groups and are collectively referred to as therians (Figure 6.1). Fossils of all three mammalian lineages first appear in the Cretaceous period, but molecular data consistently suggest that the three main lineages of extant mammals had long "ghost lineages" (see Chapter 1) that are not yet documented in the fossil record. Specifically, the monotremes are thought to have separated from the therian lineage approximately 220 mya, in the Triassic period, while placental and marsupial mammals diverged in the Jurassic period. Thus, the main groups of living mammals originated long before the famous "extinction of the dinosaurs" 66 mya, at the end of the Cretaceous period. If this is true, then mammalian evolution had a "long fuse," defined as a long period of relatively little diversification followed by an "explosive" radiation after the dinosaurs had gone.

The living monotremes include four species of spiny anteater, or echidna, and the platypus. These enigmatic mammals share many similarities with other mammals, including hair and mammary glands. However, they differ from therian mammals in a number of respects. For one thing, they have highly specialized toothless snouts

Brains Through Time. Georg F. Striedter and R. Glenn Northcutt, Oxford University Press (2020). © Oxford University Press.
DOI: 10.1093/oso/9780195125689.001.0001

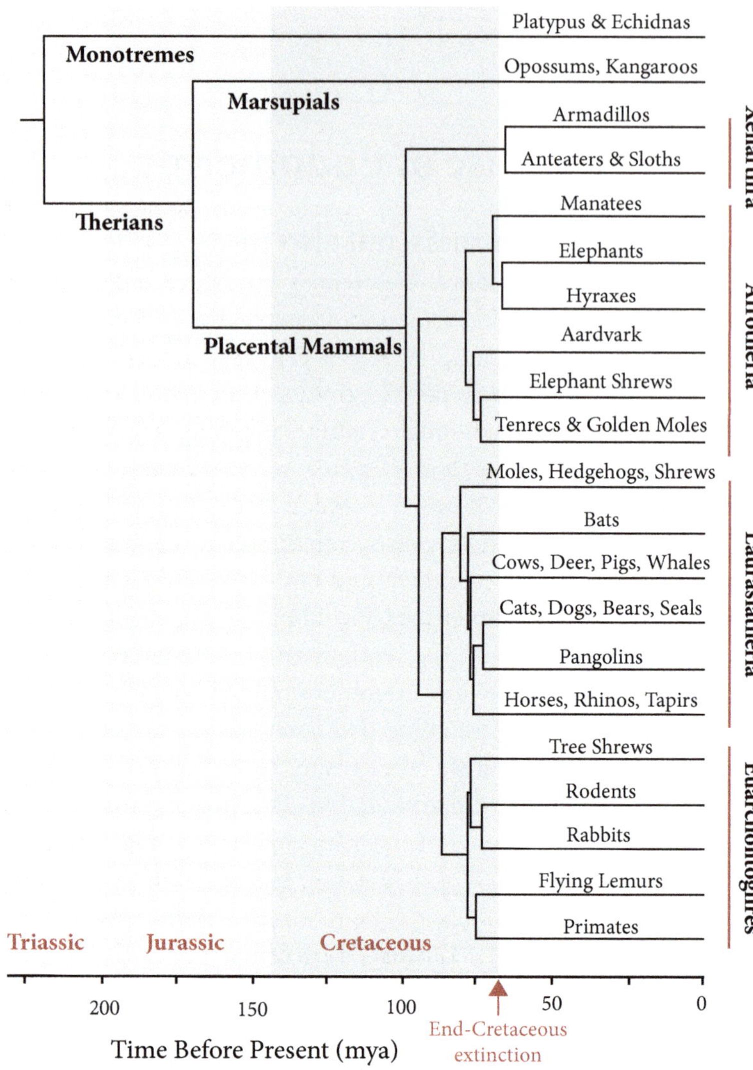

Figure 6.1 Mammalian phylogeny. The illustrated phylogeny is based on an analysis of 26 genes in 286 species and calibrated against fossil data. According to this and many other analyses, placental mammals are divisible into four monophyletic groups, called Xenarthra, Afrotheria, Laurasiatheria, and Euarchontoglires. However, the relationships of these four groups to one another remain controversial. This study also suggests that the major placental lineages all originated long before the mass extinction at the end of the Cretaceous period. This "long fuse" model of placental mammal diversification is supported by most molecular analyses, but morphological data tend to support a later, more explosive radiation.

Adapted from Foley et al. (2016).

that are equipped with highly specialized receptors, many of which can sense weakly electric fields (see Section 6.5). They also retain several primitive features, such as the use of a single opening (the cloaca) for urination, defecation, and reproduction. More dramatically, monotremes lay soft-shelled eggs, just as their amniote ancestors surely did. Echidnas incubate their eggs in an abdominal pouch, keeping them warm and moist. In contrast, the platypus has no pouch. Instead, it lies on its back inside a burrow, incubating the eggs on its abdomen. Either way, when a monotreme's offspring hatch, they are extremely immature and receive extensive maternal care, including milk-like secretions from the mother's mammary glands.

Contemporary marsupials comprise roughly 470 species (Wilson and Reeder, 2005), most of which live in Australia or South America. In contrast to monotremes, marsupials do not lay eggs. Like their placental cousins, marsupials give birth to live offspring. However, their young are very immature and small (the size of a jelly bean) and quickly crawl into the mother's pouch, where they attach tightly to her nipples. In this protected place they grow and develop for many months, until they can maintain a stable body temperature and venture out on their own. Marsupials also differ from placental mammals in their reproductive organs, with males sporting a bifurcated penis and females a matching bilateral vagina. Within marsupials, morphological diversity is substantial, but body size among extant marsupials is limited to about 100 kg.

With almost 5,000 extant species, placental mammals are far more diverse than the marsupials or monotremes (Figure 6.2). They range in body size from a few grams (e.g., small bats) to more than 100 tons (i.e., large whales). They are found on all continents, but indigenous placental mammals are largely absent from New Zealand and rare in Australia, where marsupials predominate. Placental mammals are named for having a well-developed placenta, which allows the embryos to develop much longer inside the mother. Many placental mammals are born in a relatively mature state and, compared to marsupials, require relatively little parental care. The crown group of placental mammals probably originated in the Late Cretaceous (90–100 mya), but some authors dispute such an early origin. As noted earlier with respect to mammals generally, the problem is that no fossils of placental mammals are older than about 66 my, which is when the end-Cretaceous extinction occurred (Wible et al., 2007). Of course, one might counter that the earliest placental mammals were so scarce that fossil representatives have not yet been unearthed.

Within the placental mammals, four main lineages are generally recognized, namely the Xenarthra, Aftrotheria, Laurasiatheria, and Euarchontoglires; the last of these includes both rodents and primates (see Figure 6.1). The phylogenetic relationships between these four main groups remain somewhat controversial, but many analyses support the hypothesis that the Xenarthra form the most basal clade of extant placental mammals. Alternatively, Aftrotheria and Xenarthra might be more closely related to one another than to other mammals. As most readers

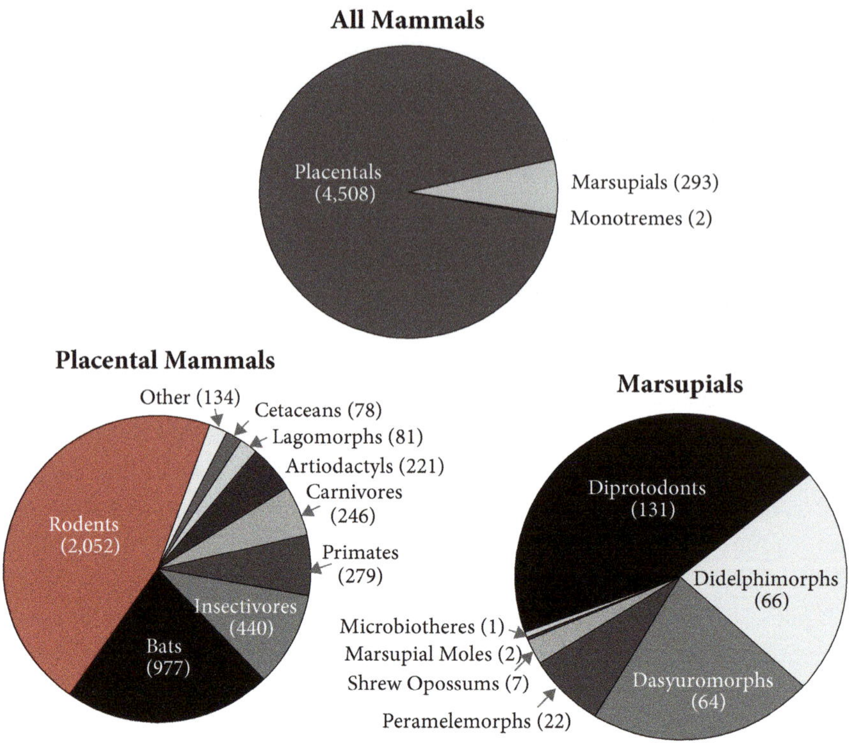

Figure 6.2 Species counts for extant mammals. Placental mammals are far more diverse than marsupials and monotremes combined. Within placental mammals, rodents and bats are the most diverse orders (note that mammalian insectivores are not a monophyletic group). Among marsupials, the diprotodonts (e.g., kangaroos, wombats, koalas) are the most diverse orders, followed by didelphimorphs (opossums) and dasyuromorphs (e.g., quolls and dunnarts).
Data from Nowak (1999).

know, placental mammals are extremely diverse, occupying a bewildering array of ecological niches. Perhaps most remarkably, cetaceans (dolphins and whales) and pinnipeds (seals and sea lions) have reverted to an aquatic niche. Just as impressive is the evolution of powered flight in bats.

6.2. Stem Mammals

Several major lineages of stem therians went extinct near the end of the Cretaceous period (Figure 6.3 top). Particularly interesting were the multituberculates, which resembled rodents in numerous respects but went extinct before the true rodents emerged. In the following discussion, we largely ignore these extinct stem therians. Instead, we focus our attention on the various stem mammals that branched off

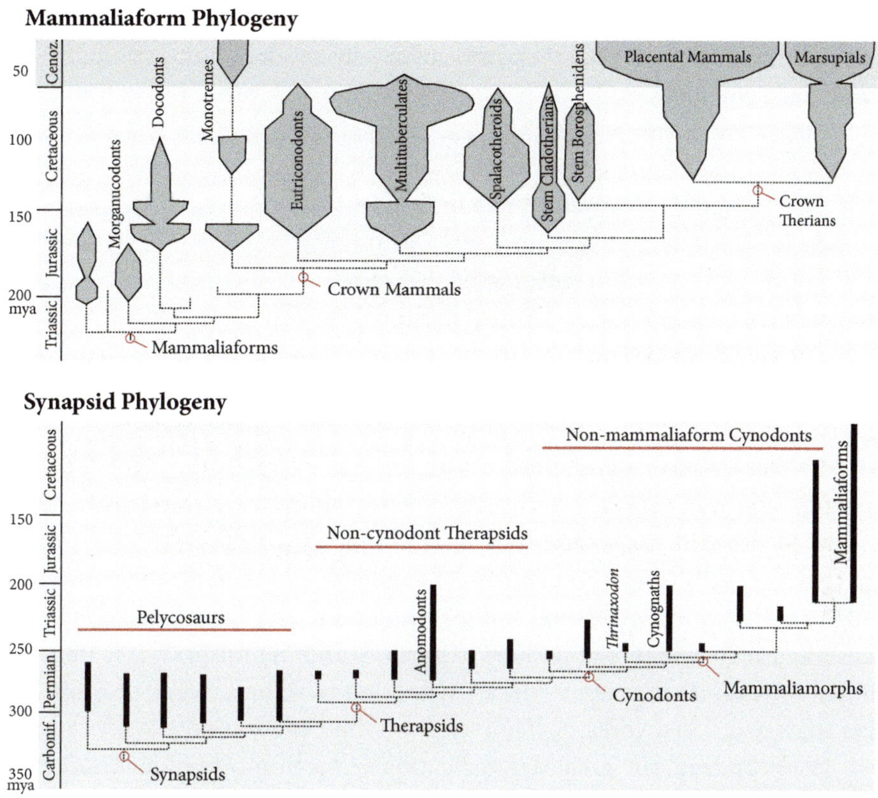

Figure 6.3 The synapsid fossil record. The top diagram depicts the putative phylogeny of both living and extinct crown mammals, as well as their closest mammaliaform relatives. The morphological diversity of each lineage is represented by the width of the gray polygons. The bottom diagram illustrates when the various mammaliaforms and stem mammal lineages appear in the fossil record and how they are most likely related to one another.

Adapted from Luo (2007, with permission from Springer Nature, and from figures by Thomas Holtz at https://www.geol.umd.edu/~tholtz/G331/lectures/331vertsII.html (bottom, with permission).

before the origin of crown mammals but after the divergence between synapsids and sauropsids (which we covered in Chapter 5).

A major challenge in discussing stem mammals is that there are so many of them, distributed over a very long period of time (Figure 6.3 bottom). Moreover, their nomenclature is complex. In essence, stem mammals form a nested set, such that crown mammals are embedded within mammaliaforms, which are part of the mammaliamorph clade, which is one of several cynodont lineages. The cynodonts (not conodonts!), in turn, are just one group of therapsids, which are one of several lineages that collectively comprise the synapsids. In addition to these monophyletic group names (Rowe, 1988), evolutionary biologists have often used paraphyletic terms (see Chapter 1), such as the term "pelycosaurs" for non-therapsid synapsids (Figure 6.3). Because this terminology is rather cumbersome and alien to most readers, we sometimes simply distinguish between early, middle, late, and recent stem mammals, meaning non-therapsids (aka pelycosaurs), non-cynodonts, non-mammalian cynodonts, and non-mammalian mammaliaforms, respectively.

The early stem mammals of the Late Carboniferous and Early Permian (i.e., pelycosaurs) are well represented by members of the genus *Dimetrodon*. Depending on the species, *Dimetrodon* was 0.5 to 5 m long and likely preyed mainly on smaller tetrapods. Its most distinctive feature was a prominent "sail" along the dorsal midline, supported by dorsal extensions of the vertebrae. The functions of this sail remain obscure but might include social displays and temperature regulation. Superficially, *Dimetrodon* resembled sailbacked dinosaurs, but it clearly belongs in the synapsid lineage, as indicated by the presence of just one, rather than two, openings (fenestra) in the posterior portion of its skull (see Chapter 5).

Middle stem mammals of the Middle to Late Permian (i.e., non-cynodont therapsids) are well represented by *Lycaenops* (Figure 6.4). These animals were wolf-like in size and shape, and they were equipped with powerful jaws and diverse types of teeth (incisors, canines, pre-molars, and molars) that would have been good at grasping, crushing, and grinding prey. They likely walked long distances in an increasingly hot and dry climate, searching for prey. It is important to note, however, that many Permian synapsids were herbivores or omnivores, rather than strict carnivores, as deduced from their tooth morphology (Sues and Reisz, 1998). Indeed, the great diversification of tooth morphology that characterizes the synapsid lineage is undoubtedly associated with a substantial diversification in dietary types.

Late stem mammals (non-mammaliaform cynodonts) first appeared in the Late Permian but, in contrast to most other synapsids, they did not go extinct at the end of the Permian. Indeed, they diversified into several lineages during the ensuing Mesozoic Era (i.e., the Triassic, Jurassic, and Cretaceous). A good representative of these late stem mammals is *Thrinaxodon*, which was significantly smaller than *Lycaenops* and lacked abdominal ribs (Figure 6.4). The fenestrae in its skull were also larger, suggesting stronger jaw muscles. In addition, the oral cavity of late stem mammals was separated from their nasal cavity by a bony palate that earlier stem mammals lacked. This hard palate probably facilitated the chewing of food but also made it possible for these animals to breathe and chew at the same time.

The most recent group of stem mammals are the non-mammalian mammaliaforms. A well-studied representative of this assemblage is *Morganucodon* (Figure 6.4), a small, furry creature that was probably arboreal and dined mainly on insects. Even more closely related to mammals was *Hadrocodium*, which is remarkable for weighing a mere 2–3 g. As we discuss more fully in Section 6.3, this species probably represents the terminal extreme of a long-running trend toward decreasing body size in stem mammals. Reducing body size to this extent had a number of important anatomical, physiological, and ecological implications, which we discuss shortly.

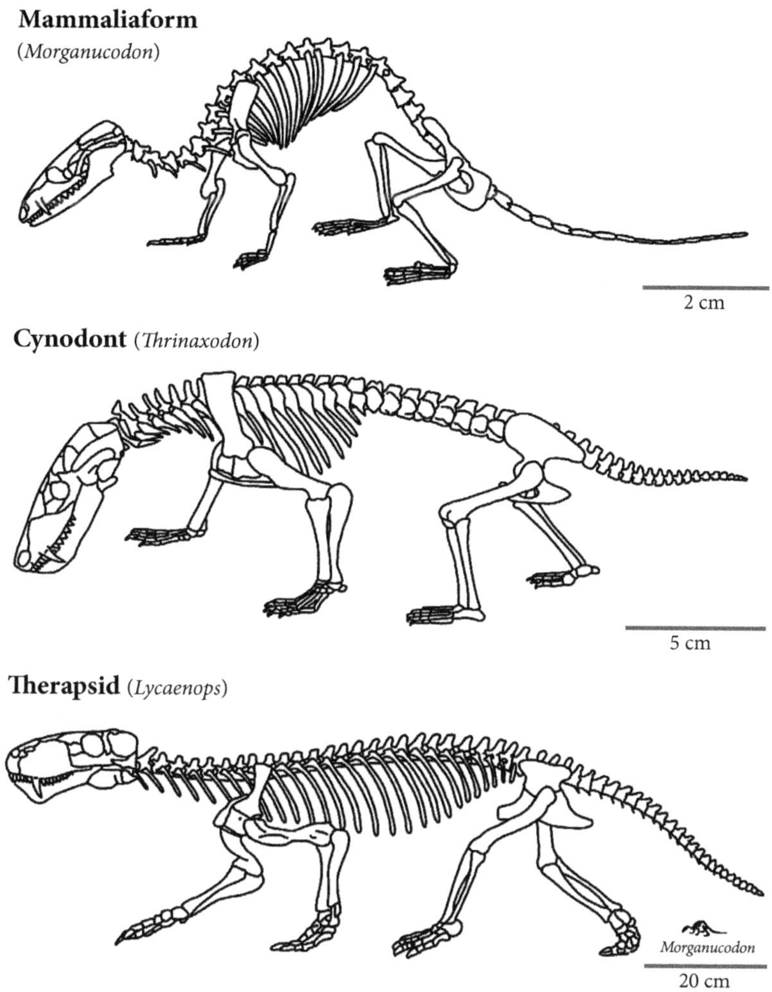

Figure 6.4 Representative stem mammals. These three skeletons are drawn to the same size but, as the scale bars reveal, most mammaliaforms were smaller than most cynodonts, which in turn were smaller than most therapsids. The small image of *Morganucodon* at the bottom right, drawn to the same scale as *Lycaenops*, drives this point home. Another important difference between therapsids and the younger stem mammals is that the latter lack lumbar ribs, which suggests the presence of a diaphragm.

Adapted from Rowe (2017), courtesy of Timothy Rowe.

6.3. Synapsid Brain-Body Scaling

The cranial endocasts of early stem mammals are generally incomplete, because the cerebral hemispheres of these animals were encased in cartilage, rather than bone. However, endocasts from a few well-preserved middle stem mammals indicate that these animals had elongate, tubular cerebral hemispheres that were taller than wide (Figure 6.5). They also had relatively large olfactory bulbs, a well-developed

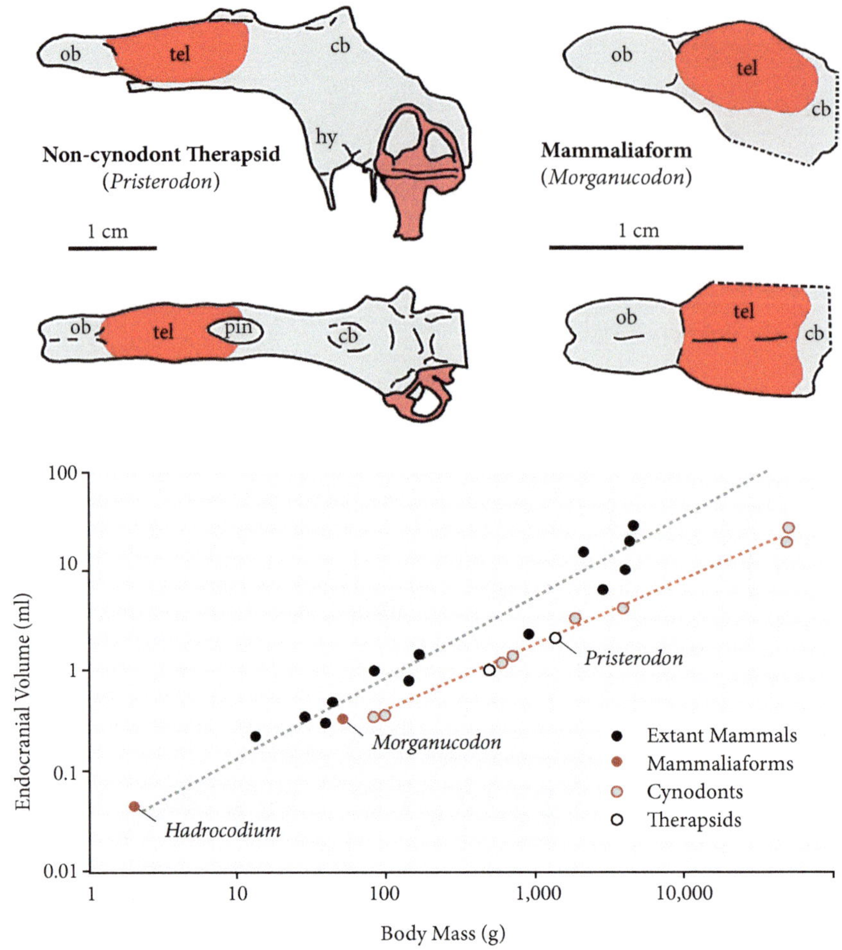

Figure 6.5 Stem mammal endocasts. Shown at the top are lateral and dorsal views of endocasts for two different stem mammals. The approximate location of the telencephalon is shown in red, the vestibular apparatus in pink. The graph compares endocast volumes and estimated body masses between middle stem mammals (non-cynodont therapsids), late stem mammals (non-mammaliaform cynodonts), the most recent stem mammals (non-mammalian mammaliaforms), and extant mammals. Abbreviations: cb – cerebellum; hy – hypothalamus; ob – olfactory bulb; pin – pineal; tel – telencephalon. Adapted from Rowe et al. (2011), Laaß (2015).

parietal opening (for the "parietal eye"; see Appendix), and a substantial cerebellum with a laterally directed parafloccular lobe. Little more can be said about the brains of middle stem mammals, because in these animals, as in non-avian sauropsids, the brain did not completely fill the endocranial cavity. However, the available endocasts suggest that, relative to body size, middle stem mammals had smaller brains than extant mammals (Figure 6.5; Laaß, 2015).

The endocasts of the most recent stem mammals, such as *Hadrocodium* and *Morganucodon*, reveal some major phylogenetic transformations. Most obviously, the cerebral hemispheres are proportionately larger than in early stem mammals, and the olfactory bulbs are relatively huge. The cerebral hemispheres are also wider than the olfactory bulbs, especially caudally. These changes in brain shape may relate to the small body size of more recent stem mammals, because brain-body ratios generally decrease with body size (aka Haller's rule), which means that at small body sizes the brain takes up a disproportionate fraction of the animal's skull. This fact, in turn, may lead to mechanical stresses during embryogenesis that modify brain shape (Striedter and Northcutt, 2006). Alternatively, brain shape may have changed in recent stem mammals because they evolved de novo, or greatly expanded, their neocortex (Rowe et al., 2011). This hypothesis remains speculative, however, because determining the boundaries between neocortex and olfactory cortex in stem mammal endocasts is essentially impossible.

To determine whether evolution changed brain size relative to body size, scientists often examine a species' "encephalization quotient," which is defined as the ratio between the species' actual brain size and the brain size one would expect it to have, given the animals' typical body size (Boddy et al., 2012). The expected brain size, in turn, is based on the best-fit line through all the data points for the relevant taxonomic group. Using this approach and extant mammals as the relevant taxonomic group, Rowe et al. (2011) determined that the encephalization quotients of recent stem mammals are intermediate between those of crown mammals and the older stem mammals. However, a plot of endocranial volume versus body size (Figure 6.5) reveals that the best-fit line for crown mammals has a steeper slope than that for stem mammals (Laaß, 2015) and that the data points for the most recent stem mammals lie roughly where those lines converge. Although this observation is consistent with the hypothesis that stem mammals became more highly encephalized through a series of incremental evolutionary steps (Rowe et al., 2011), the data are also consistent with a single evolutionary change in the brain-body scaling rule of the most recent stem mammals. That is, these animals may have increased the rate at which brain size increases with body size, so that subsequent increases in body size "automatically" led to larger and larger increases in relative brain size.

Comparing brain-body scaling between extant mammals and non-avian sauropsids reveals that the brains of the former are 5–10 times larger than those of the latter, even at small body sizes (Figure 6.6). Therefore, some additional increases in relative brain size (i.e., in the encephalization quotient) must have occurred after the first true mammals emerged. Indeed, some primates and toothed whales have brains that are at least twice as large as those of other mammals at the same body size (Figure 6.7; Montgomery et al., 2013). These evolutionary increases are widely thought to be associated with increases in behavioral complexity and flexibility—i.e., intelligence. Of course, comparing levels of intelligence across species is notoriously difficult, mainly because even objectively identical tests of intelligence may be biased for or against some species because of differences in their sensory and motor systems, motivation, and sensitivity to interactions with humans. That said, some

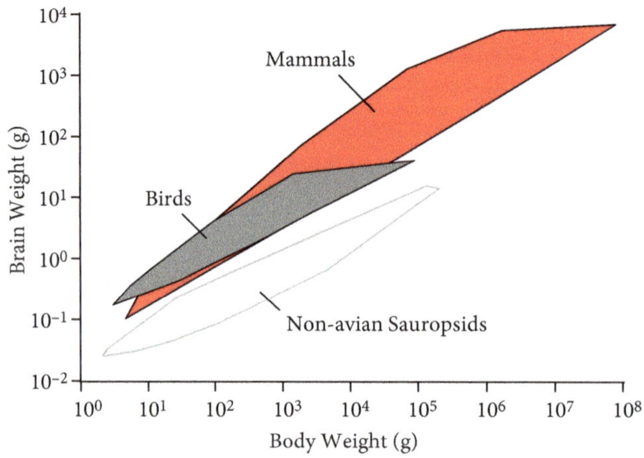

Figure 6.6 Convergent increases in relative brain size of mammals and birds. This graph depicts the minimum convex polygons enclosing all the data points for brain and body size in birds, mammals, and non-avian sauropsids (aka reptiles). It shows that the brains of extant mammals and birds are 5–10 times heavier, relative to body weight, than those of non-avian sauropsids. Given that birds are a derived lineage within reptiles (i.e., flying dinosaurs; see Chapter 5), we can conclude that relative brain size increased independently in the avian and mammalian lineages.
Adapted from Striedter (2005).

measures of intelligence are useful in comparative studies (Mackintosh et al., 1985; Emery and Clayton, 2004; Prior et al., 2008; Lefebvre, 2011).

Extant birds have increased their relative brain size just as much as mammals have, relative to the non-avian sauropsids (Figure 6.6). Relative brain size is especially high in songbirds and parrots, which are close relatives of one another (Hackett et al., 2008) and, together, account for the vast majority of birds. Moreover, these birds have a higher density of neurons, per gram of brain tissue, than other birds (Figure 6.8). Indeed, their neuronal density is even higher than that of primates, which in turn have more neurons per gram of brain tissue than other mammals, especially at larger brain sizes (Herculano-Houzel et al., 2007; Olkowicz et al., 2016). Overall, we can conclude that large brains, relatively to body size, evolved independently in birds and mammals, especially within some select lineages, and that these increases in relative brain size were augmented by convergent increases in neuron density. This conclusion, combined with the idea that neuron number correlates with computational capacity, supports the general hypothesis that primates, toothed whales, parrots, and songbirds are more intelligent than other mammals or birds. To give just one example, many species in these four groups of animals are capable of imitating sounds and using tools (Nottebohm, 1972; Striedter, 2013), two capacities that are otherwise quite rare.

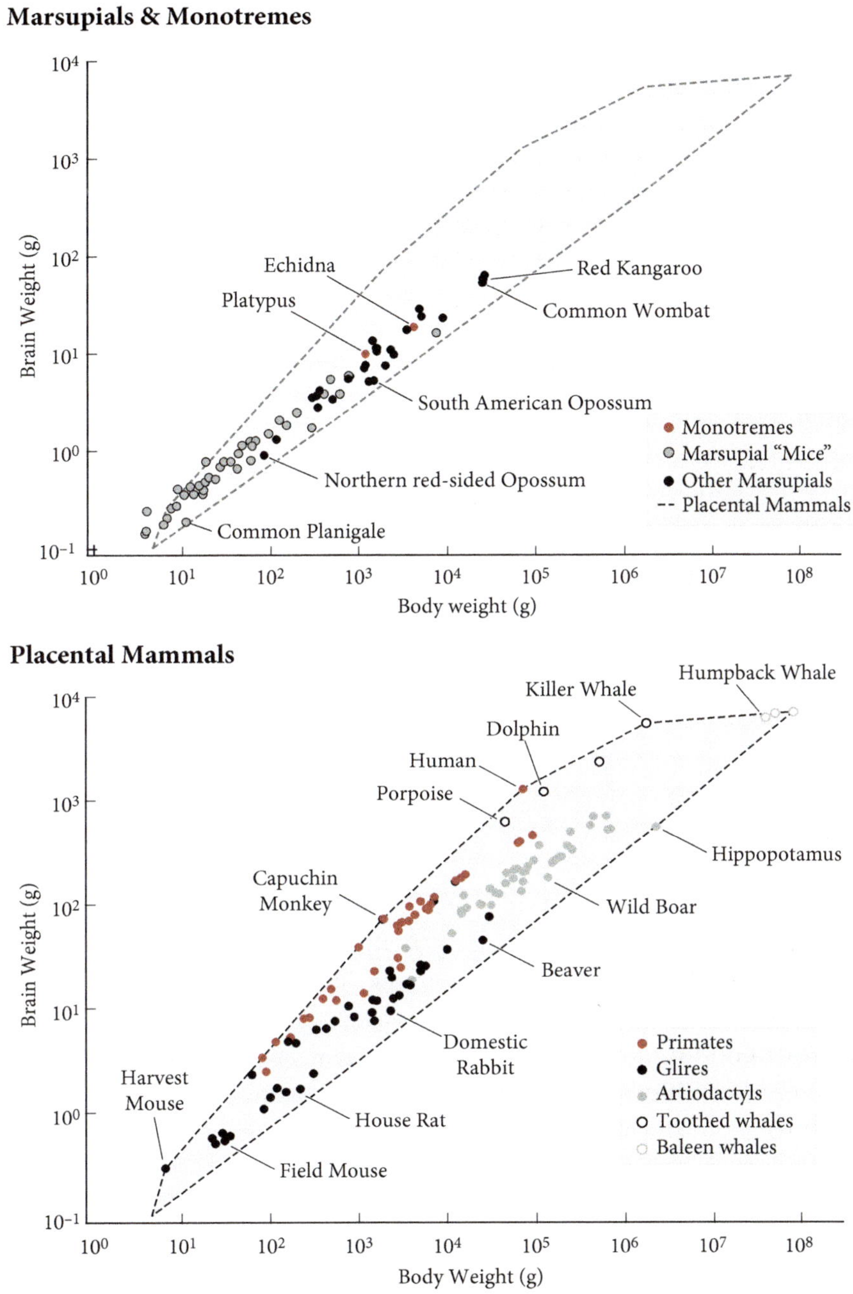

Figure 6.7 Mammalian brain-body size scaling. Monotremes have brains that are roughly as large as one would expect for placental mammals of their body size. Large marsupials have relatively small brains, relative to body size, but small marsupials, notably marsupial "mice" (Dasyuridae) have brains that are as least as large as those of small placental mammals. Among placental mammals (bottom graph), primates and toothed whales have large brains for their body size. The dashed line indicates the minimum convex polygon that encompasses all the data points for placental mammals. Data from Mangold-Wirz (1966), Möller (1973), McNab and Eisenberg (1989), Ashwell (2008), Ridgway et al. (2016).

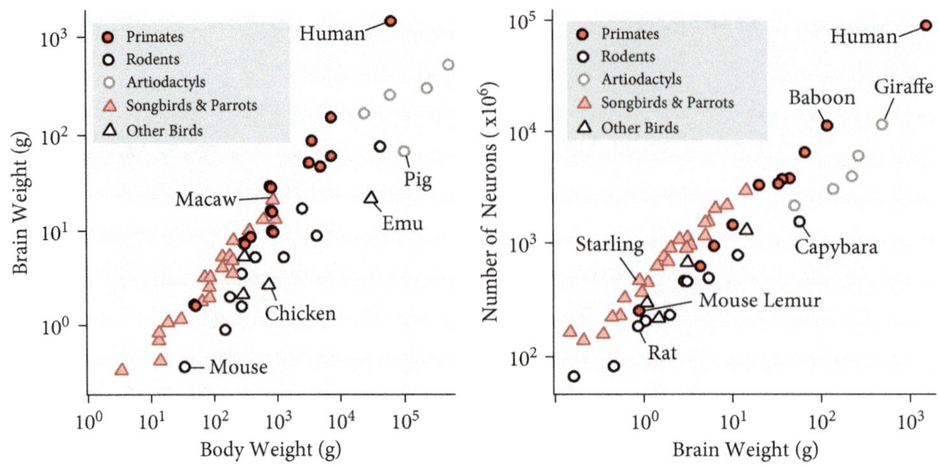

Figure 6.8 Neuron number scaling in mammals and birds. Within mammals, relative brain size is larger in primates than most other mammals. Among birds, it is the parrots and songbirds that have the largest brains, relative to body size (left). These species differences are even greater when one considers neuron numbers because, at any given brain weight, primates tend to have more neurons than other mammals (especially at larger brain sizes), and songbirds and parrots have more neurons per gram of brain tissue than other birds (right). Indeed, neuron density is higher in songbirds and parrots than in primates. Note that domestication probably reduced the relative brain size of the pig in this data set (see Kruska and Röhrs, 1974); whether a similar domestication effect exists for chickens is unclear.
Adapted from Olkowicz et al. (2016).

Although relative brain size is widely recognized as a fairly good predictor of intelligence, absolute brain size also correlates with some measures of intelligence, especially within a taxonomic group (Striedter, 2005; Marino, 2006; Deaner et al., 2007; MacLean et al., 2014; but see Benson-Amram, 2016). Thus, we can infer that early mammals, with their tiny bodies and brains, were not too smart. Moreover, the fact that brain and body size increased, at least on average, as mammals diversified, suggests that their intelligence likewise increased. This would be consistent with the view that the mammals with the largest brains, notably toothed whales, elephants, and apes, are also the most intelligent. One should note, however, that all hypotheses linking brain size to intelligence remain uncertain and controversial (e.g., Healy and Rowe, 2007; Manger et al., 2013). In any case, absolute brain size increased independently in many different mammalian lineages. Also of interest is that, within each lineage, absolute brain size correlates with the degree of neocortical folding (Figure 6.9). Most likely, this correlation results from the fact that the developing neocortex expands in surface area much more rapidly than the underlying tissue (Mota and Herculano-Houzel, 2015), which generates mechanical buckling forces that ultimately cause the neocortex to fold (Striedter et al., 2014).

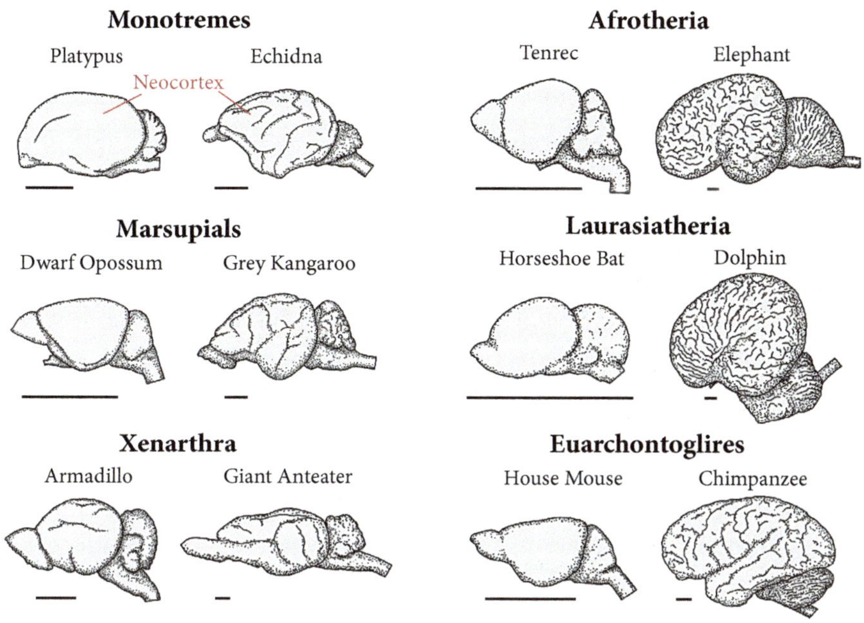

Figure 6.9 Convergent evolution of large and folded neocortices. Shown here are lateral views of some of the smallest and some of the largest brains in each of the six major mammalian lineages (see Figure 6.1). All scale bars equal 1 mm. The small brains tend to have a smooth neocortex, whereas the large brains tend to exhibit numerous cortical folds (gyri and sulci). These data support the hypothesis that mammalian neocortex tends to become folded when it becomes large, and that this process played out repeatedly, in all the major lineages.
Based on photographs at brainmuseum.org.

6.4. Paleoecology, Physiology, and Behavior

Before discussing how the nervous system was modified during synapsid phylogeny, it is useful to consider the ecological conditions in which synapsids lived and how they adapted to those conditions with changes in their general behavior and physiology. This task is complicated, because stem mammals existed for such a long time (~140 million years) before true mammals emerged (see Figure 6.3). Moreover, this period included some drastic ecological disturbances, leading to two of the planet's five principal mass extinctions. Thus, the origin of mammals is a long, complex story.

The synapsids probably diverged from the sauropsids (reptiles and birds) more than 300 mya, during the Carboniferous period. They soon diversified into several different lineages, including the various pelycosaurs and diverse therapsids. However, most of these early synapsids died out at the end of the Permian period as part of the planet's largest mass extinction (see Chapter 5). Still, a few therapsids made it through those dark days, including the cynodont lineage that ultimately led to crown mammals. The cynodonts then diversified in the Mesozoic era, which extended from 251 to 65.5 mya and includes, in order, the Triassic, Jurassic and Cretaceous periods.

Early during the Mesozoic era, the supercontinent Pangea broke up into Gondwana in the South and Laurasia in the North. Atmospheric oxygen levels were considerably lower than they are today, and carbon dioxide levels were higher (Berner, 2006). Perhaps because of all that carbon dioxide (a greenhouse gas), the planet's climate in the early Mesozoic was significantly warmer than it is today, with ocean surface temperatures reaching as high as 40°C in the tropics (Sun et al., 2012). During the Triassic period the globe's climate still exhibited large seasonal variations, but these diminished as the continents subdivided further, expanding the areas with a less variable, coastal climate. Moreover, by the end of the Mesozoic, average surface temperatures exhibited much less variation from North to South, with polar ice caps being largely nonexistent.

Terrestrial vegetation underwent great change during the Mesozoic, with cycads, conifers, and other gymnosperms dominating the land for most of this era, and flowering plants (angiosperms) starting to take hold toward its end. Animal life on land was likely dominated by insects and sauropsids, most notably the dinosaurs. Birds originated sometime in the middle Mesozoic but, like mammals, they did not diversify extensively until after this long era had closed.

6.4.1. Becoming Small and Nocturnal

The Mesozoic is often referred to as the Age of Reptiles or, more commonly, the Age of Dinosaurs. Indeed, the fossil record of Mesozoic dinosaurs includes at least 1,000 documented species. As the Mesozoic era progressed, dinosaurs and other archosaurs increased in average body size (Figure 6.10), leading to some gigantic dinosaurs that weighed as much as 50 tons. Those enormous body sizes would have helped to protect these animals from predators, notably other dinosaurs, setting up a bit of an arms race. Larger dinosaurs would also have found it easier to stay cool on hot days and warm at night, because their low body surface-to-volume ratio would have reduced the rate of heat transfer across the skin. In addition, a low surface-to-volume ratio minimizes water loss across the skin, which would have been adaptive in the hot and often arid inland areas of Mesozoic continents.

Although the dinosaurs capture the public imagination, stem mammals coexisted with them. In contrast to dinosaurs, however, Mesozoic stem mammals reduced their average body size substantially (Figure 6.10). Although this reduction is not seen in every lineage of early synapsids, it is clear that, over the long run, the smaller synapsids were more successful than their larger relatives (Sookias et al., 2012). The most likely explanation for this trend is that the early synapsids expanded into niches that were not occupied by the much larger dinosaurs, thus avoiding direct competition with them (though some smaller dinosaurs surely did enjoy eating a synapsid or two whenever they could). Specifically, it is likely that

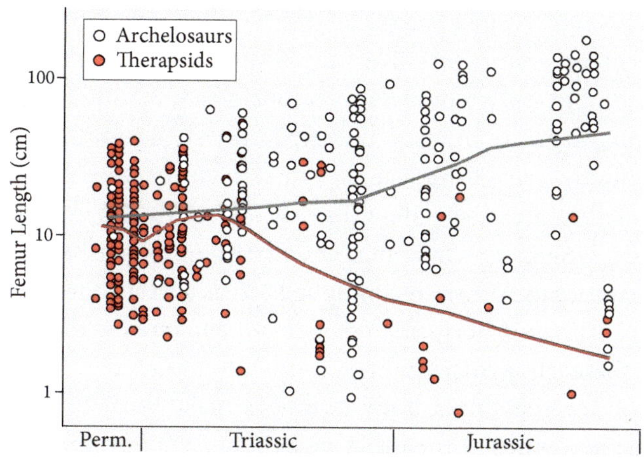

Figure 6.10 Body size divergence. Each open circle represents a fossil archelosaur (i.e., extinct turtles, crocodilians, and dinosaurs), and the red circles represent various therapsids (i.e., stem mammals; see Figure 6.3). The length of the femur, plotted on the *y*-axis, is a proxy for body size. Plotted along the *x*-axis is the midpoint of each taxon's age range. The solid lines represent local group averages and demonstrate that therapsids decreased in average body size during the late Triassic and Jurassic, while archelosaurs became larger, at least on average.
Adapted from Sookias et al. (2012)

Mesozoic synapsids became increasingly arboreal and started feeding mainly on insects, rather than plants and larger animals. In addition, they became more nocturnal, while the archosaurs were diurnal. Support for this hypothesis comes from broad comparative analyses of behavioral activity patterns (Maor et al., 2017) and mammalian visual systems, which seem to have been adapted to low light conditions (we return to this topic in Section 6.5).

A key innovation that allowed stem mammals to become nocturnal, and thus avoid competition with dinosaurs, is that they became capable of generating their own body heat. Without this capacity for endothermy, the small synapsids would have gotten too cold at night to run around and hunt for food, because both neurons and muscles work best when their operating temperature is stable and warm (e.g., Stecker and Baylor, 2009). This much is clear, but the details of how and when endothermy evolved remain the subject of heated debates.

6.4.2. The Origins of Synapsid Endothermy

Endothermic animals can elevate their body temperature beyond ambient levels by generating their own, internal heat. This ability allows them to maintain a relatively high and stable body temperature (as long as they can also avoid overheating). By this definition, birds and mammals are the only fully endothermic vertebrates (endothermy also evolved in some social insects; Heinrich, 1972). Pythons can elevate their body temperature by shivering while they are incubating eggs (Harlow and Grigg, 1984), but most non-avian sauropsids are ectothermic, which means that they can only regulate their body temperature behaviorally (e.g., by basking in the sun). Similarly, a few fishes can elevate the temperature of some body parts, most notably the brain and specific muscles, but they are generally ectothermic. Given these data, we can conclude that endothermy evolved independently in extant mammals and birds. When during synapsid or sauropsid phylogeny did endothermy emerge? The most likely answer is that it appeared gradually, in a series of steps. Let us consider the synapsids first.

A key method for generating body heat is to make the cell and mitochondrial membranes leakier, which means that energy must be expended to restore the vital ion and proton gradients across those membranes; an inevitable byproduct of this energy expenditure is heat. Of course, it is impossible to determine from the fossil record when synapsids might have evolved more leaky membranes and, thus, elevated their basal metabolic rates. However, we can note that average body temperatures and body size-corrected basal metabolic rates are higher in most placental mammals than marsupials, higher in marsupials than monotremes, and higher in monotremes than in non-avian sauropsids (Grigg et al., 2004; McNab, 2008; Lovegrove, 2012). This is one indication that endothermy did not evolve in one fell swoop.

When mammals feel cold, they can elevate their body temperature by means of several tissue-specific mechanisms. The prime example is shivering, which involves repetitive muscle contractions that generate heat as a metabolic byproduct but, otherwise, perform no useful work (Rowland et al., 2015). In addition, some mammals can uncouple muscle activation from contraction, which causes large numbers of calcium ions to flow down their concentration gradients without performing mechanical work. Restoring those calcium gradients requires the activity of enzymatic ion transporters, which generates heat (Pant et al., 2016). Finally, some mammals possess brown fat (brown adipose tissue), which expresses high levels of a protein that allows protons to bypass the adenosine triphosphate (ATP) generating mechanism in mitochondria, thereby reducing ATP generation but increasing heat production (Argyropoulos and Harper, 2002). Because monotremes and marsupials do not possess brown fat (Hayward and Lisson, 1992), we can infer that this form of thermogenesis is an innovation of placental mammals. Shivering and non-shivering muscular thermogenesis are probably more ancient capabilities, though more comparative research is needed to support this hypothesis. Similar forms of muscular thermogenesis are found also in birds (Bicudo et al., 2001), but they probably resulted from convergent evolution.

An important complement to generating body heat is the prevention of heat loss when it is cold outside. One uniquely mammalian solution to this problem is to cover the body with hair, which prevents heat loss by trapping an insulating layer of air next to the skin. Multituberculates and other extinct crown mammals clearly had hair, as did *Castorocauda lutrasimilis*, an aquatic mammaliaform from the middle Jurassic that likely weighed around 500 g (Ji et al., 2006). Thus, fur is at least 170 my old. Since hair is less likely than bone to fossilize, it is possible that fur evolved even earlier in the synapsid lineage. However, the currently available data indicate that fur evolved after stem mammals had already become relatively small, which is consistent with the idea that thermal insulation becomes increasingly important as endotherms decrease in size.

Endothermic stem mammals also retained body heat by evolving a convoluted set of mucous membranes, called the maxillary nasal turbinates (Figure 6.11), inside the nasal cavity, where they warm up the inhaled air and pull some heat (and

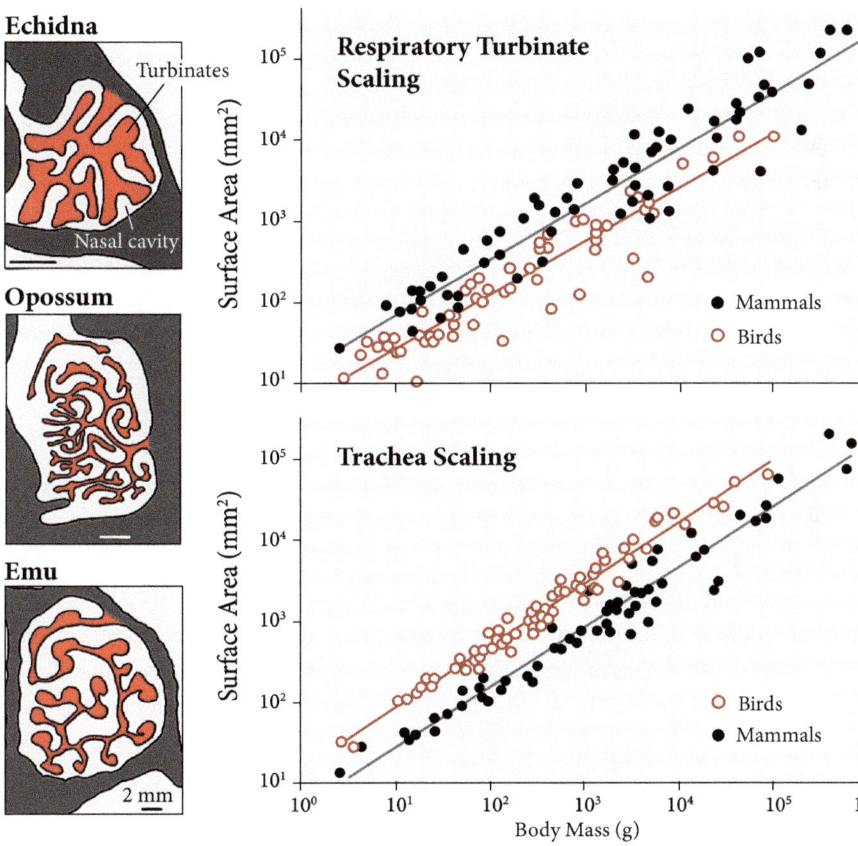

Figure 6.11 Turbinate and trachea scaling in mammals and birds. Shown on the left are transverse sections through the nasal cavity of two mammals and a bird, with the respiratory turbinates depicted in red (all scale bars equal 2 mm). Their large epithelial surfaces make it easier for warm-blooded mammals and birds to retain body heat and moisture. Intriguingly, the surface area of the respiratory turbinates is larger, relative to body mass, in mammals than in birds (top). However, birds make up for having smaller turbinates by having a longer trachea (bottom), which also helps retain heat and moisture.

Adapted from Owerkowicz et al. (2015).

moisture) back out of the exhaled air. Such nasal turbinates are not found in non-avian sauropsids, but birds have independently evolved quite similar structures (Figure 6.11). Because mucous membranes don't fossilize, we cannot be certain which stem mammals had nasal turbinates. However, the nasal turbinates are attached to the walls of the nasal cavity by bony ridges, and such ridges have been described in a few cynodonts (e.g., *Thrinaxodon*) and even some non-cynodont therapsids (Hillenius and Ruben, 2004). Moreover, in one non-cynodont therapsid, the cartilaginous components of the nasal turbinates have been exquisitely preserved (Laaß et al., 2011). Therefore, we can infer that at least some non-cynodont therapsids were probably endothermic. Since these animals were fairly large, we can further surmise that endothermy evolved before the synapsids reduced their body size. If this is true, then the body size reduction might have necessitated the evolution of fur, but not of endothermy itself.

In order to burn the energy required for endothermy, mammals rely extensively on oxidative phosphorylation, which means that they must inhale a lot of air and then distribute the oxygen efficiently. This need for oxygen must have been especially acute in the Mesozoic, when oxygen levels were relatively low. One way in which stem mammals solved this problem was to augment rib-based breathing (see Chapter 5) with the evolution of a muscular diaphragm, which likely first appeared in therapsids (Hirasawa and Kuratani, 2013). Such a diaphragm would have allowed them to inhale more air and, in contrast to their ectothermic ancestors, to breathe while on the run (Carrier, 1987). In addition, late stem mammals evolved a hard palate beneath their nasal cavity, which made it possible to breathe with a mouth full of food. A related innovation was the evolution of a four-chambered heart, which allowed freshly oxygenated blood to be separated from deoxygenated blood, thereby maximizing oxygen absorption in the lungs. Mammals also made their capillaries thinner and shrank their red blood cells (e.g., by removing their cell nuclei), which improved oxygen delivery to their tissues (Ruben, 1995; Snyder and Sheafor, 1999). Collectively, these evolutionary changes allowed endothermic mammals to burn far more energy than their ectothermic ancestors. Indeed, active mammals in the wild consume 20–30 times as much metabolic energy as active lizards of the same body size (Bennett and Ruben, 1979). Even at rest, mammalian metabolic rates are 5–10 times higher than those of non-avian sauropsids (Ruben, 1995).

Of course, burning energy requires not only oxygen, but also food. One way in which stem mammals increased their calorie intake was to feed on insects, which can be very nutritious, and to adopt the nocturnal niche, which was relatively free of other vertebrate predators. In this context, endothermy was extremely useful, as it allowed for faster and more sustained movements, especially at night. Similarly useful were an endothermy-induced enhancement of axonal conduction speed and muscle efficiency, improvements to sensory systems that work well at night, and the evolution of proportionately larger brains (see Sections 6.5 and 6.7). However, neural tissue is itself metabolically expensive to develop and operate (Isler and van

Schaik, 2006; Sobrero et al., 2011). Thus, the evolution of endothermic, nocturnal synapsids involves a causal feedback loop: enhanced neural systems and endothermy required more food intake, but they also made it possible for those animals to eat much more.

6.4.3. Endothermy in Birds

At some point during the Mesozoic, early birds or their immediate ancestors also became endothermic. Given that today's non-avian sauropsids are all ectothermic, avian endothermy surely evolved independently of its mammalian counterpart. However, it is based on a variety of very similar adaptations. For example, birds, like mammals, have a four-chambered heart, small red blood cells, and nasal turbinates (Figure 6.11; Owerkowicz et al., 2015). They also possess high metabolic rates, even higher than those of mammals at the same body size, and they shiver when they get cold. In fact, some evidence suggests that birds can, like mammals, generate muscular heat by uncoupling muscle activation from contraction (Bicudo et al., 2001).

Despite these similarities, avian endothermy differs from its mammalian analog in multiple respects, as one would expect for independently evolved characters. Most notably, birds insulate their bodies with feathers, which are modified scales (Sawyer and Knapp, 2003) and are even more effective than fur at trapping air. Birds also have a specialized system of internal air sacs that push and pull air through the lung unidirectionally. This mechanism makes avian lungs far more efficient at gas exchange than mammalian lungs, in which air flow is bidirectional (i.e., air passes in and out along the same pathway). Extant crocodilians also exhibit unidirectional airflow through the lungs (Farmer and Sanders, 2010; Butler et al., 2012), but crocodilians lack the air sacs of birds. Therefore, we can conclude that early archosaurs had already evolved more efficient lungs than other amniotes, but that lung efficiency increased further in the lineage leading to birds. A related innovation of extant birds is their unusually long trachea, which augments the nasal turbinates' ability to resorb heat and moisture from exhaled air (Figure 6.11; Owerkowicz et al., 2015).

Precisely when endothermy evolved in the avian lineage remains a subject of debate. Large archosaurs, including the large theropods ancestral to birds, probably had fairly stable body temperatures, because the low surface-to-volume ratios of their large bodies would have buffered them against fluctuations in ambient temperature. These animals would also have been capable of warming up through exercise and then storing this heat inside their bodies. However, they could not generate their own, internal body heat without moving. Therefore, true endothermy probably evolved later, as theropods became smaller (see Figure 5.10 in Chapter 5). If feathers are taken as an indicator of endothermy, then we can surmise that endothermy predates the origin of true flight, because some non-avian theropods had at least a few feathers (Chuong et al., 2003; Xu et al., 2003; Clarke and Pörtner,

2010). None of the fossil theropods exhibits the bony ridges associated with nasal turbinates in synapsids, but such ridges are also absent in extant birds, despite their large cartilaginous turbinates (see Figure 6.11). Nonetheless, it has been suggested that theropod nasal cavities were too narrow to accommodate large turbinates, implying that non-avian theropods were probably ectothermic (Ruben et al., 1998). In sum, all we can presently conclude is that sauropsid endothermy probably evolved either with the origin of birds or shortly before then.

6.4.4. Surviving an Asteroid

A 10 km wide asteroid hit the earth at the end of the Mesozoic era, 65.5 mya, and the sequelae of this impact killed off more than 75% of all land plant and animal species, as well as numerous marine invertebrates. This hypothesis is based in part on the discovery that a geological layer of clay at the Cretaceous-Paleogene (aka Cretaceous-Tertiary) boundary is enriched in the element iridium, which is common in asteroids but rare on earth (Alvarez et al., 1980). The hypothesis was later corroborated by the discovery of a huge impact crater (180 km wide and 20 km deep) centered on the town of Chicxulub in Southern Mexico (Schulte et al., 2010). Because this asteroid hit land (at least in part), it generated a vast cloud of dust and debris that must have darkened the sky for several months, chilling the planet and killing off most photosynthetic plankton, algae, and land plants. The impact seems also to have released large amounts of gypsum and other sulfur-rich compounds into the atmosphere, leading to years of sulfurous smog and acid rain.

At roughly the same time, Western India experienced an enormous series of volcanic eruptions that covered parts of the continent with lava that was 2 miles deep, creating the Deccan Traps. These eruptions would have spewed vast amounts of CO_2 into the atmosphere, leading to global warming and ocean acidification, much as analogous eruptions in Siberia had done at the end of the Permian. There have been long and acrimonious debates about the relative significance of the asteroid impact and the Deccan Trap eruptions in causing the end-Cretaceous extinction. We suspect that both cataclysms contributed substantially, though the climatic effects of the volcanic eruptions were probably much more prolonged. Intriguingly, it has been suggested that the asteroid impact may have triggered or accelerated volcanic activity around the world (Brannen, 2017).

In any case, the combined effects of these two global disasters killed off a huge number of species, ranging from the bottom of the food chain to top predators. Indeed, no tetrapods larger than 25 kg survived the end-Cretaceous extinction. The dinosaurs and pterosaurs (see Section 5.1.4) died out, but the extinction also engulfed many other groups of vertebrates, insects, and plants (Longrich et al., 2011, 2012). Still, most lineages bounced back after the catastrophe had passed. In fact, ray-finned fishes diversified rapidly after the end-Cretaceous extinction (Friedman and Sallan, 2012), as did mammals and birds. The mammals diversified at least in part by invading niches that dinosaurs had previously occupied, and birds probably benefited from the post-Mesozoic (i.e., Paleogene) diversification of insects and flowering plants. As fascinating as these late radiations are, our principal aim in the rest of this chapter is to understand the changes in the nervous system that accompanied the earlier, Mesozoic stages of synapsid evolution—the ones that set the stage for their later success. We begin with a consideration of synapsid sensory systems.

6.5. Modified Sensory Abilities

As stem mammals became increasingly nocturnal, they adapted their visual system to low light conditions and expanded several other senses, especially hearing and olfaction. They also evolved a novel kind of somatosensory system, involving hair. These non-visual senses could compensate for the limited utility of vision at night. Many other aspects of mammalian sensory systems were modified later, during the Paleogene, but we here cover only a few of these subsequent changes. In particular, we discuss some aspects of primate sensory biology.

6.5.1. Vision

Early mammals became less dependent on vision than their ancestors had been, mainly because they were active primarily at night. This shift to the nocturnal niche entailed multiple changes in the visual system. Some later groups of mammals became more diurnal again and, in association with this shift, modified their retinas for vision in a broader range of light intensities. Among other things, they evolved enhanced color vision and, in primates, a fovea.

6.5.1.1. Nocturnal Vision

Late stem mammals lost the parietal eye of their ancestors and, therefore, detected light only through their lateral eyes (Quay, 1979; Benoit et al., 2015). In nocturnal lizards, birds, and primates, the lateral eyes are relatively wide in relation to their length (i.e., the ratio of cornea diameter to axial eye length is relatively large; Heesy and Hall, 2010). This difference in eye shape is probably adaptive for nocturnal animals, because increasing cornea diameter allows more light to enter the eye, and decreasing axial length concentrates the incoming light on the retina. Importantly, the nocturnal eye shape seems to have been the primitive condition for all mammals, which is consistent with the hypothesis that early mammals were nocturnal. Similarly, most mammals have eyes that are more forward-facing than they are in lizards, snakes, or birds (Heesy and Hall, 2010). Such frontally directed eyes are good for vision in low light, as they effectively double the chance of detecting photons that come from light sources in front of the animal. They also allow for depth perception via stereoscopic vision (Pettigrew, 1986), which is consistent with the hypothesis that early mammals hunted for insects among the small branches of bushes and trees. Primates later converged their eyes even further, which is consistent with the view that they, too, hunted insects in the arboreal fine branch niche, where depth perception is extremely useful (Cartmill, 1992).

As discussed in previous chapters, most vertebrate retinas contain both rod and cone photoreceptors. The former are used for vision in low light, but saturate when it is bright. In contrast, cone photoreceptors work only when light levels are relatively high. Given these physiological observations, it is not surprising that most nocturnal vertebrates have rod-dominated retinas, though most retain at least a few cones (Jacobs, 1993). It is likely, therefore, that early mammals also evolved rod-dominated retinas when they became more nocturnal. Perhaps because of this shift, mammalian retinas tend to be simpler than those of other tetrapods. For example, mammalian retinas contain a smaller number and diversity of amacrine cells than one finds in sauropsids and frogs (Dowling, 1968; Dubin, 1970).

6.5.1.2. Color Vision and Foveae

In contrast to rod photoreceptors, cone photoreceptors come in several different types that express different types of opsin molecules, which are tuned to different wavelengths of light. By comparing levels of activity across those different cone types, animals can discriminate different colors of light. Indeed, most fishes and tetrapods possess four different cone types and excellent color vision. Mammals, in contrast, have lost some of these cone types and the corresponding opsin genes (Figure 6.12). Specifically, all mammals lack the "green-sensitive opsin," also known as Rh2, which is structurally similar to the rhodopsin expressed in rods (Okano et al., 1992). In addition, monotremes lost the ultraviolet/violet-sensitive opsin SWS1, and therians lost the blue-sensitive opsin SWS2 (Bowmaker, 2008).

The loss of these opsins and their associated cone types probably reflects relaxed selection for excellent color vision, which is consistent with the hypothesis that early mammals went through a "nocturnal bottleneck" and, therefore, emphasized low-light vision with rods over color vision with cones (Walls, 1942). It should be noted, however, that neither early monotremes nor early therians were totally color blind; they simply had to rely on just two types of opsins (rather than 3 or 4) to analyze colors, thus making some colors harder to discriminate. The color vision of early mammals also had less spatial resolution than that of their ancestors, because the retinas of early mammals contained far more rods than cones.

As some mammalian lineages later became more diurnal, they had to "reinvent" some of the functionalities that they had lost during their nocturnal bottleneck period. Specifically, they re-evolved additional cone types and improved color vision. For example, several marsupials evolved a middle wavelength (cyan) sensitive cone type that probably expresses a duplicated version of the rod opsin (Rh1) gene (though this hypothesis remains to be confirmed; Cowing et al., 2008; Ebeling et al., 2010). Old World monkeys and apes (i.e., catarrhine primates; Figure 6.13)

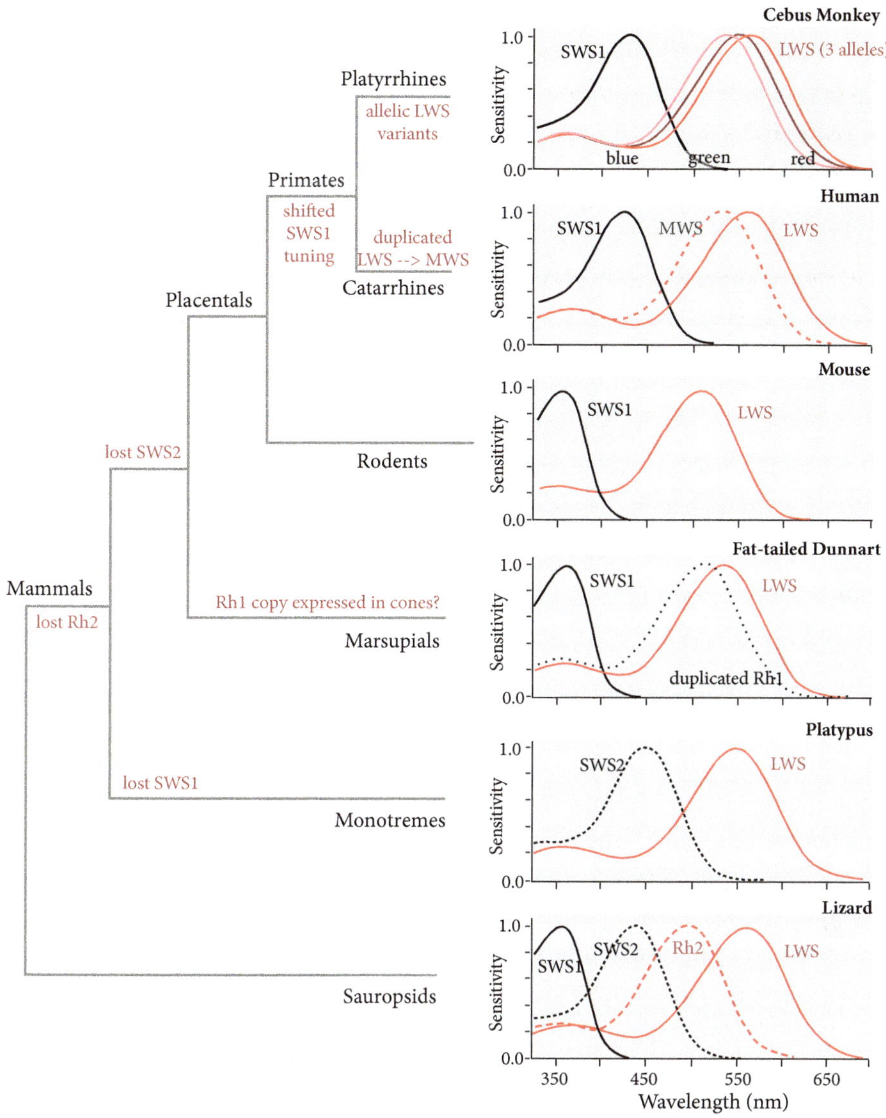

Figure 6.12 Changes in cone photopigments. The graphs indicate how sensitive the various cone photopigments (opsins) are to light of different wavelengths. The inferred evolutionary changes in the opsin repertoire and frequency tuning are indicated in red on the phylogeny.

Abbreviations: LWS – long wavelength sensitive opsins (which exhibit functionally significant allelic variation in platyrrhines); MWS – middle wavelength sensitive opsin; Rh1 + Rh2 – rhodopsins #1 and #2; SWS1 + SWS2 – short wavelength sensitive opsins #1 and #2.

Adapted from Jacobs (2008) and Davies et al. (2012).

evolved a different kind of middle wavelength (green) sensitive cone that expresses a duplicated and slightly modified version of the ancestral long wavelength (red) sensitive opsin (see Figure 6.12). This innovation allowed catarrhine primates to discriminate red from green, which most mammals have trouble with. That change,

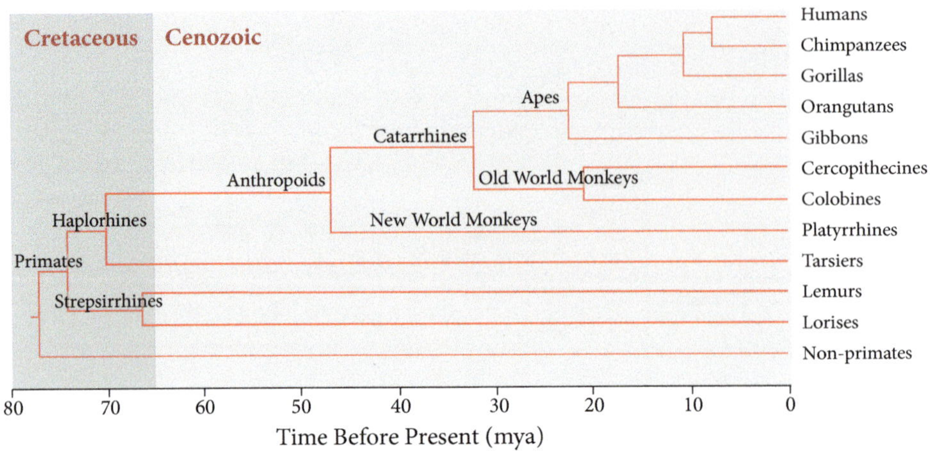

Figure 6.13 Primate time tree. The illustrated phylogeny is based on an analysis of complete mitochondrial genomes from 65 primate species, as well as 25 non-primates. Divergence times were estimated using 16 fossil calibration points (confidence intervals are not shown). This phylogeny implies that the major primate lineages originated well before the end-Cretaceous extinction, but this conclusion is controversial.
Adapted from Pozzi et al. (2014).

in turn, helped them identify ripe fruit, as well as young and nutritious leaves, which tend to be reddish where catarrhine primates evolved (Osorio and Vorobyev, 1996; Dominy et al., 2003). It may also have helped them determine whether their conspecifics looked pale or had healthy, reddish skin (Changizi et al., 2006). New World monkeys (platyrrhine primates) overcame their red-green color blindness by a third mechanism: they evolved multiple alleles of the red opsin gene, each of which is maximally sensitive to a slightly different wavelength (see Figure 6.12). Since the red opsin gene is located on the X chromosome, females that have two different alleles on their two X chromosomes can discriminate red from green. However, males and females with just one of the red opsin alleles are not so lucky (Jacobs, 2008). Importantly, all of these different ways of improving color vision beyond the ancestral mammalian condition evolved independently of one another.

While some mammals improved their color vision, others lost color vision completely. In particular, a variety of fully nocturnal mammals (e.g., owl monkeys and raccoons) lost their UV/violet-sensitive opsin SWS1; so did the cetaceans (dolphins and whales) and pinnipeds (sea lions and seals) (Peichl, 2005). Independently of one another, these lineages all accumulated mutations that rendered their SWS1 gene non-functional. These species all retain the red-sensitive cones, but having just one type of cone is thought to be insufficient for color vision. Another interesting aspect of opsin evolution is that, with the loss of the blue opsin SWS2 in early therians, the spectral tuning of SWS1 shifted upward, from ultraviolet to violet (Figure 6.12). Perhaps this upward shift filled the gap in spectral sensitivity that was created by the loss of SWS2. Shifts in the tuning of SWS1 also occurred in birds, but they evolved independently of those in mammals and likely served other functions (Hunt and Peichl, 2014).

One of the most important innovations in primate retinas is the evolution of a fovea, defined as a depression (or pit) in the central retina where photoreceptor density is very high but other retinal neurons are pushed aside. Cones, in particular, are highly concentrated in primate foveae. As we discussed in Chapter 5, most birds and many lizards have one or two such foveae per retina (see Figure 5.16), and the earliest synapsids may have had them as well. However, most mammals only have a so-called area centralis, which features an increased cell density but lacks the fovea-defining pit. The only mammals to possess a genuine fovea are the haplorhine primates (tarsiers, all monkeys, and apes; Figure 6.13); the other, extrafoveal regions of primate retinas have a relatively low density of photoreceptors, at least in comparison to non-mammalian amniotes. Several recent studies have identified the biomechanical factors that likely shape the primate fovea during development (Springer and Hendrickson, 2005; Provis et al., 2013), and these studies raise doubts about the widely held hypothesis that the displacement of retinal cell nuclei away from the foveal center is an adaptation for reduced light scattering within the fovea (Springer, 1999). However, those studies do remain consistent with the hypothesis that primate foveas evolved to facilitate high acuity vision in the central visual field. In that sense, the central foveae of sauropsids and primates are an excellent example of independent, convergent evolution.

6.5.2. Hearing

Whereas vision is severely limited at night, the sense of hearing is unaffected. Indeed, stem mammals expanded their hearing ability in several respects as they went through the "nocturnal bottleneck." For example, they evolved external ear flaps (pinnae), which help mammals localize sounds, especially when those ear flaps are movable. More importantly, stem mammals evolved a tympanic membrane (ear drum). As we discussed in Chapters 4 and 5, ear drums evolved independently in anuran amphibians, lepidosaurs, and archosaurs. They evolved a fourth time in stem mammals, though the precise timing of their origin is difficult to determine (Kitazawa et al., 2015; Maier and Ruf, 2015).

6.5.2.1. Mammalian Middle Ears

Stem mammals evolved two middle ear bones (the malleus and incus) that, in series with the stapes, transfer vibrations from the ear drum to the inner ear. Both the incus and the malleus are homologous to jaw bones in early synapsids, but their size progressively decreased until they completely detached from the lower jaw and became fully incorporated into the middle ear (Figure 6.14, top). This detachment of the middle ear bones from the lower jaw is thought to have occurred at least twice

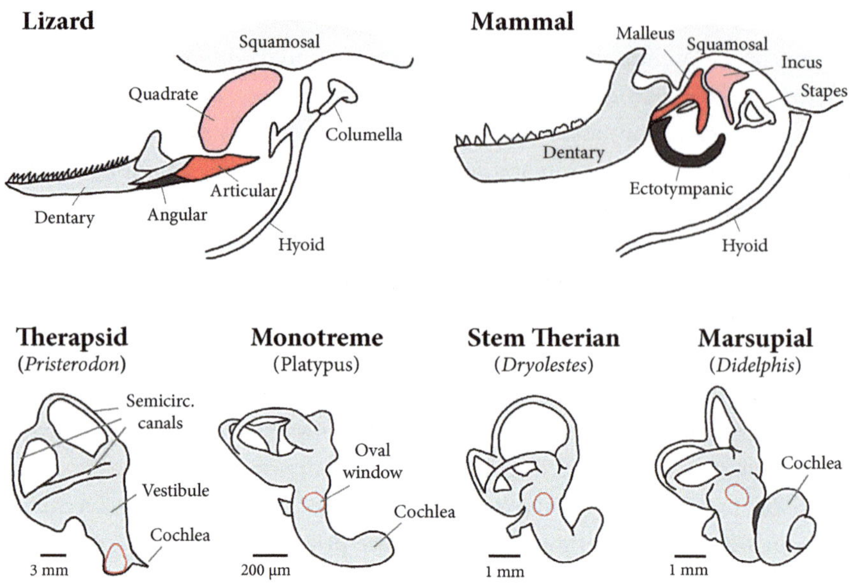

Figure 6.14 Middle and inner ear evolution. The lower jaw of mammals consists solely of the dentary bone. In contrast to the reptilian condition, the angular, articular, and quadrate bones in mammals are not part of the jaw but, instead, form two of the middle ear bones (malleus and incus) and a bone that holds the tympanum (colors and shading indicate homologies). The bottom drawings depict the vestibular apparatus and cochlea in four synapsids. These data suggest that the cochlea lengthened and curled progressively in the lineage leading to therian mammals (i.e., placental mammals and marsupials).

Adapted from Maier and Ruf (2015, with permission from John Wiley & Sons), Laaß (2016), Luo et al., 2011).

(i.e., convergently) within mammals, and may also have occurred in the lineage leading to the tiny non-mammalian mammaliaform *Hadrocodium* (Rowe, 1996; Martin and Luo, 2005; Ramírez-Chaves et al., 2016).

As the middle ear bones decreased in size and became independent of the jaw, early mammals and mammaliaforms became able to hear and chew at the same time. More importantly, they became more sensitive to high-frequency airborne sounds. These changes improved the ability of early mammals to hear the sounds made by potential prey (notably nocturnal insects), to avoid approaching predators, and to communicate effectively with one another in the dark. Improved high-frequency hearing may also have improved the ability of early mammals to localize sounds, because interaural sound level differences provide important cues to a sound's spatial origin and increase with sound frequency (see Chapter 5).

6.5.2.2. Modified Inner Ears

Evolutionary changes in the inner ear further enhanced mammalian hearing. Whereas the vestibular apparatus of mammals is similar to that of other amniotes, the auditory portion of their inner ear is very different (Figure 6.14, bottom). In early stem mammals the auditory sensory epithelium is located inside a large chamber (called the vestibule). In contrast, middle stem mammals have a short but distinct cochlear duct that probably contained the auditory sensory epithelium. This cochlear duct then lengthened as late stem mammals evolved, and it became slightly curved in at least some early mammals (Rodrigues et al., 2013). In fact, extant monotremes retain such a curved cochlea. It is only in therian mammals that the cochlea became the snail-shaped, coiled structure that is familiar to most students of biology. This coiling allowed the cochlea to become extremely long (more than 50 mm in large whales) while still fitting within the skull.

The effect of cochlea elongation and coiling on hearing remains debatable. One possibility is that elongation of the cochlea improves an animal's ability to discriminate between sounds of different frequencies because, in mammals at least, it tends to make the individual hair cells more sharply tuned to specific sound frequencies. More confusing is the link between cochlea elongation and hearing range. Although elongation of the cochlea correlates with the evolution of high-frequency hearing in stem mammals and early mammals, as well as sauropsids (see Chapter 5), cochlea elongation in therians tends to shift an animal's hearing range toward low frequencies (Manley, 2012). Humans, for example, have a very long cochlea (~ 30 mm) and can hear low frequencies better than many other mammals with shorter cochleae (Figure 6.15). This confusing pattern is related, at least in part, to the fact that very large mammals tend to have very long cochleae as well as bulky middle ear bones, which are inherently less sensitive to high sound frequencies.

The sensory epithelium inside mammalian cochleae was also modified in several substantial ways. For example, the basilar membrane has become more flexible in mammals than in lizards, allowing it to vibrate in response to sound. A similar thinning of the basilar membrane occurred convergently in birds and crocodilians (Manley, 2012). In addition, therian mammals have lost the so-called lagenar macula, a patch of hair cells near the tip of the cochlea in monotremes and many non-mammals that probably serves vestibular, rather than auditory, functions (Ladhams and Pickles, 1996).

Best studied is the fact that mammals, including monotremes, evolved two very distinct types of hair cells within the basilar papilla. The inner hair cells, located close to the central axis of the cochlear spiral, are innervated by sensory axons that transmit auditory information into the brain. Outer hair cells also receive some sensory innervation, but most of their innervation is efferent, meaning that it transmits signals emanating from the brain. This efferent innervation is thought to modulate the principal function of the outer hair cells, which is to amplify sound-induced vibrations of the membrane on which the hair cells sit (i.e., the basilar membrane). This active amplification makes the inner hair cells more responsive to sounds by a factor of approximately 100 (Ren et al., 2011). Its underlying mechanism involves the molecule prestin, which is modified in mammals and highly expressed in the side walls of the outer hair cells (Dallos and Fakler, 2002; Okoruwa et al., 2008).

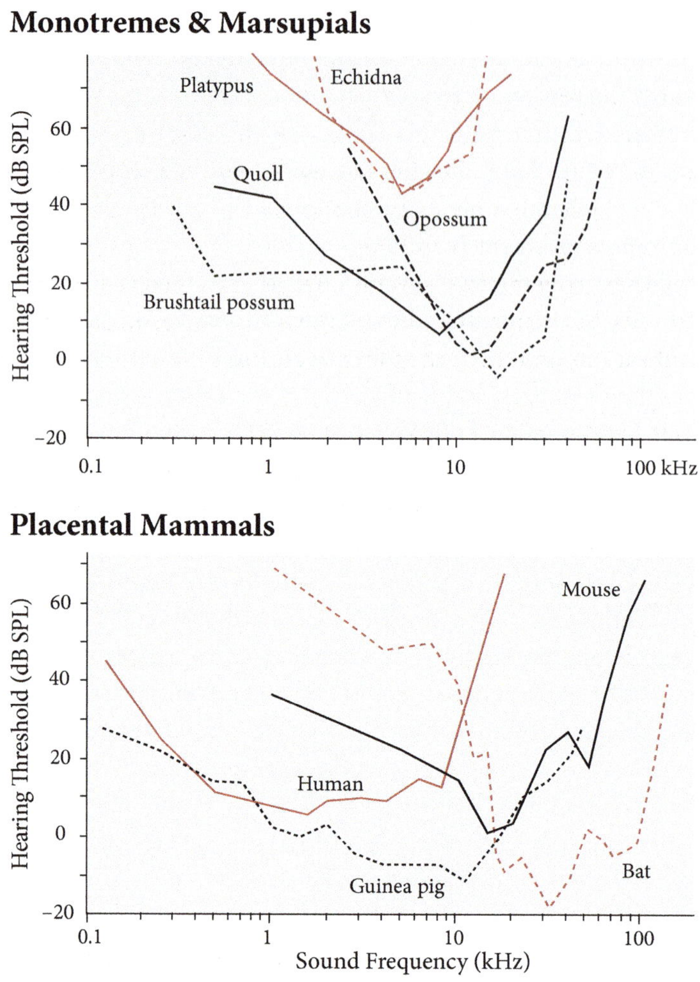

Figure 6.15 Mammalian hearing abilities. The illustrated audiograms show that the auditory system of monotremes is not as sensitive to quiet sounds as that of other mammals and is less capable of sensing frequencies above 10 kHz. The data also show that bats are unusual in their ability to hear very high frequencies (i.e., ultrasound) and that humans are better than mice and bats at hearing low-frequency sounds.
Adapted from Vater et al. (2004).

When deflection of the hair cell stereocilia depolarizes an outer hair cell, the prestin molecules change shape, causing the hair cell to change in length (Ashmore, 2008). That, in turn, exerts mechanical forces on the basilar membrane that amplify the sound-induced vibrations.

Lizards, snakes, and turtles do not have distinct inner and outer hair cells, but the hair cells of birds and crocodilians vary systematically in length, ranging from "short hair cells" on one extreme to "tall hair cells" on the other (Köppl, 2011). Remarkably, the short hair cells receive only efferent innervation and express a homolog of the mammalian prestin molecule in their cell membrane, just like mammalian outer hair cells. However, unlike mammalian outer hair cells, they do not change their length in response to membrane depolarization (Tan et al., 2011; Xia et al., 2016). Instead, they probably amplify sound-induced deflections of the hair cell stereocilia by actively moving (twitching) those stereocilia from side to side (Beurg et al., 2013). In short, the similarities between the short hair cells of birds and the outer hair cells of mammals are superficial rather than deep, suggesting that they resulted from convergent evolution.

Within the therian mammals, several lineages have evolved the ability to hear ultrasonic sounds. Most notably, ultrasonic hearing evolved in bats that produce ultrasonic vocalizations and then listen for the echoes in an effort to detect potential obstacles and prey (Madsen and Surlykke, 2013). In fact, echolocation is thought to have evolved independently in at least two different lineages of bats (Li et al., 2008). Some of these echolocating bats have elongated the portion of their cochlea that represents the ultrasonic frequencies associated with the returning echoes (Neuweiler, 1990). Moreover, all of the echolocating bats have modified their prestin molecules in similar, functionally significant ways (Liu et al., 2014). Even more remarkable is that toothed whales, which evolved ultrasonic hearing and echolocation independently of bats, evolved the same modifications of their prestin proteins (Li et al., 2010; Liu et al., 2014). It is an excellent example of convergent evolution at the molecular level.

6.5.3. Olfaction and the Vomeronasal System

Early mammals clearly had a well-developed sense of smell, but the extent to which it was enhanced during phylogeny is difficult to determine. Late stem mammals certainly had a longer snout and larger nasal cavity than earlier synapsids, but some of that increase is due to an expansion of the anterior nasal cavity, which houses the respiratory turbinates that are involved in temperature regulation (see Section 6.4.2). These animals also had olfactory turbinates, located more posteriorly, closer to the brain and away from the direct route taken by inhaled air (Crompton and Musinsky, 2015), but the boundary between these two turbinate types cannot be determined with certainty in extinct animals. Therefore, it seems likely that the size of the olfactory epithelium expanded in the lineage leading to mammals, but the degree of that expansion is unclear.

Another potential indicator of increased olfaction in stem mammals is that the olfactory bulbs of late stem mammals are very large, relative to the rest of the brain (Macrini et al., 2007). This is especially true for the most recent stem mammals *Morganucodon* and *Hadrocodium* (see Figure 6.5; Rowe et al., 2011). These data certainly suggest that olfaction became more important as stem mammals evolved, but one should note that the olfactory bulb in several basal mammalian lineages scales with negative allometry (Ribeiro et al., 2014), which means that it becomes proportionately larger as brain size decreases. To determine whether the olfactory bulbs of

Morganucodon and *Hadrocodium* really increased in size over those of their ancestors, one would want to know how it scales in earlier stem mammals and non-avian sauropsids. To our knowledge, such a comparison has not yet been performed.

Perhaps the strongest evidence for an evolutionary expansion of olfactory abilities in the synapsid lineage is that most extant mammals have a larger olfactory receptor (OR) repertoire than lizards, chickens, and fishes (Figure 6.16; Niimura, 2012). Since the platypus has a smaller OR gene repertoire than most therian mammals, we can surmise that this gene family expanded in early therians. However, this hypothesis remains tentative as long as we have no data on the OR gene repertoire in echidnas. Also puzzling is that the frog *Xenopus* has as many functional OR genes as most therians (see Figure 6.16). This large OR repertoire in *Xenopus* is probably

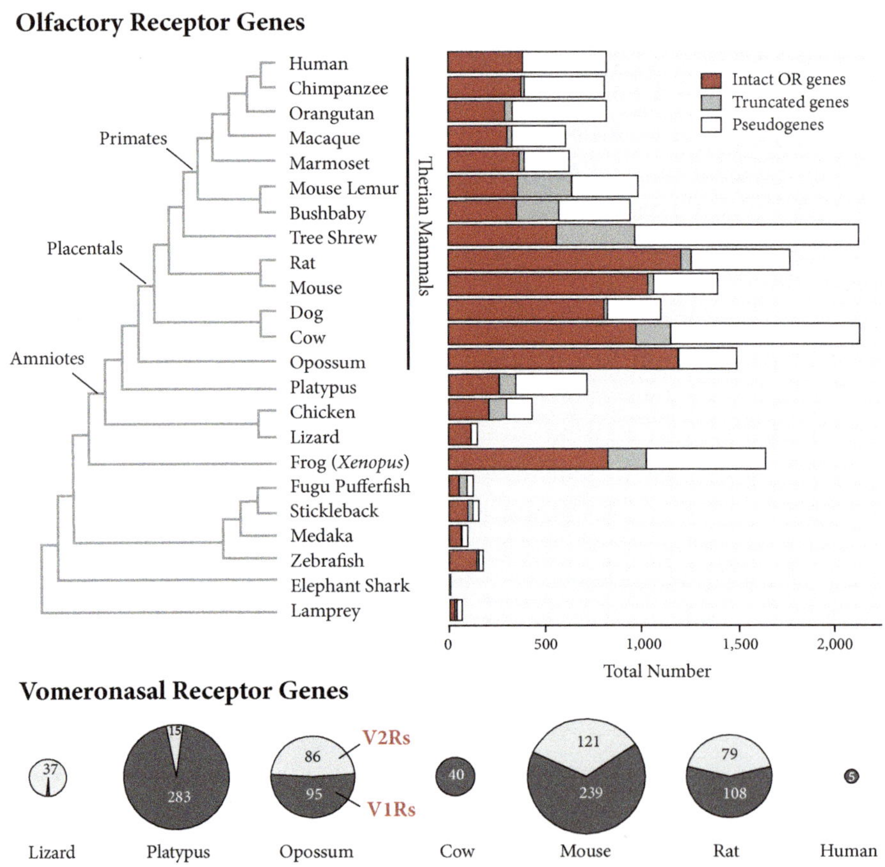

Figure 6.16 Olfactory and vomeronasal receptor diversity. The top diagram shows that the olfactory receptor gene repertoire probably increased with the origin of therian mammals and decreased again with the origin of primates. The gray rectangles represent genes that were incompletely sequenced and, therefore, may or may not be functional. The pie charts show the number of V1 and V2 vomeronasal receptor genes that have been identified in representative amniotes.

Adapted from Niimura (2012) and Brykczynska et al. (2013).

the result of convergent evolution, but this hypothesis also requires additional data. Conversely, it is clear that the number of functional OR genes has decreased in haplorhine primates, in concert with a decrease in the size of their olfactory bulbs (Gilad et al., 2003; Heritage, 2014). Even more drastic reductions in olfactory bulb and OR repertoire size occurred in the toothed whales, which have dismantled their olfactory system almost entirely (Kishida et al., 2015).

In addition to the main olfactory system, most mammals have retained the vomeronasal system of their ancestors. As we discussed briefly in Chapter 5, the vomeronasal epithelium contains receptors that belong to two different gene families: the V1 and V2 receptors. The V1 receptors are thought to bind mainly small airborne molecules, whereas the V2 receptors seem specialized for water-soluble substances, though this distinction remains debatable (Emes et al., 2004; Shi and Zhang, 2007). In any case, comparative studies have shown that lizards and snakes have numerous V2 genes but only a small handful of V1 genes (Brykczynska et al., 2013). In contrast, mammals tend to have more V1 genes than functional V2 genes (Figure 6.16). These data are consistent with early mammals having expanded their V1 receptor repertoire. The alternative hypothesis, that sauropsids lost most of their V1 receptor types, is inconsistent with the scarcity of non-functional V1 genes in lizards and snakes. Cows and dogs seem to lack functional V2 receptors entirely, but the lineage leading to opossums expanded its repertoire of V2 receptors (Young and Trask, 2007). Catarrhine primates (including humans) lack an anatomically defined vomeronasal system, but they have retained a few functional V1 receptors. Birds and toothed whales are even more extreme, having lost the anatomically defined vomeronasal system as well as all its receptors (Shi and Zhang, 2007; Kishida et al., 2015).

6.5.4. Somatosensation

In contrast to sauropsids, mammals generally lack scales. Instead, their skin tends to be covered with hair. As noted earlier, a dense layer of hair (fur) provides useful insulation against the cold. However, many mammalian hairs also have a sensory function, because they are innervated by axons that can sense when the hair is bent. Thus, hairs can provide information about direct physical contact with external objects, as well as information about air movements (Yu et al., 2016). The latter function is especially important in bats, which use wing hairs for flight control (Sterbing-D'Angelo et al., 2011). Feathers in birds provide an analogous function, sensing air flow during flight (Brown and Fedde, 1993).

In addition to their regular body hair, most therian mammals possess long and densely innervated vibrissae (aka whiskers), especially on parts of their face (Pocock, 1914). Monotremes lack these facial vibrissae, but all extant monotremes have such unusual snouts that their condition in this respect is probably derived rather than primitive (Benoit et al., 2016). Many rodents and at least some marsupials and afrotherians frequently move their facial vibrissae symmetrically and rhythmically, using specialized muscles (Mitchinson et al., 2011). This active whisking behavior seems to have evolved repeatedly within the therians, although early mammals probably had at least some limited capacity for whisker movements (Grant et al., 2013; Muchlinski et al., 2018). The functional benefits of active whisking have not been fully resolved (Mitchinson et al., 2007), but whisking clearly extends the range over which vibrissae can detect external objects (Arkley et al., 2017). More generally, it seems reasonable to speculate that facial vibrissae would have been very useful for early mammals in their presumed niche, scrambling across small branches in the dark. It would also have been adaptive for early mammals that dug burrows in the ground and had to navigate through them.

Many birds also have "whiskers" on their face, especially near the base of the beak (Cunningham et al., 2010). These avian whiskers look superficially similar to mammalian vibrissae and probably have some somatosensory function (in addition to protecting the eyes from debris during feeding or in flight), but they are modified feathers, rather than hair, and clearly evolved independently of their mammalian counterparts. In addition, several groups of birds (notably kiwis, shorebirds, ibises, ducks, and geese, as well as parrots) possess a highly sensitive tactile organ near the tip of their beak (Cunningham et al., 2003). Ibises and kiwis use this bill tip organ for "touch at a distance," sensing the movements of nearby buried prey as they probe the sand with their long bills (Cunningham et al., 2009, 2010). Shorebirds (e.g., sandpipers and snipes) probably do so as well. In contrast, ducks use their bill tip organ to help them sort prey and seeds from murky water, and parrots make use of it when they peel seeds or manipulate other food items inside their beak. Although the bill tip organs in these distantly related avian lineages are similar in cellular anatomy (e.g., containing numerous "Herbst corpuscles"; Gottschaldt, 1985), they also exhibit important differences and clearly evolved independently of one another. Their presence in kiwis, which belong to the most basal lineage of extant birds (i.e., the paleognath birds; see Figure 5.8 in Chapter 5), suggests that the bill tip organ may have been an innovation of early birds that was then lost and later re-evolved several times. This hypothesis is supported by the presence of bony pits at the tip of the beak in other paleognath birds, notably emus and ostriches, as well as some extinct paleognaths (Crole and Solely, 2017).

Aside from the axons innervating vibrissae and hair, mammals have a variety of other sensory axons in their skin. Some of them terminate as free nerve endings, while others are associated with specialized accessory cells. Sauropsids have a similarly broad array of somatosensory receptors in their skin (von Düring and Miller, 1979), but we are not aware of any comprehensive studies that have tried to homologize the various somatosensory receptor types between sauropsids and mammals. Some, such as Merkel cells and Pacinian corpuscles, seem to be broadly conserved (Berkhoudt, 1979; Toyoshima and Shimamura, 1991). However, sauropsids also have some unique receptor types, including the Herbst corpuscles mentioned earlier and the dome pressure receptors of crocodilians (Di-Poï and

Milinkovitch, 2013). Moreover, mammals clearly have their own specialized somatosensory receptors. For example, monotremes have evolved electroreceptors on their snout (or beak) that are innervated by sensory axons of the somatosensory system (Proske et al., 1998) and are very different from the electroreceptors of fishes (see Chapter 3). Considerable variation also exists in the distribution of the various somatosensory receptors across the skin. Primates, for example, have an unusually high density of specialized somatosensory receptors in their generally clawless and hairless fingertips (Soligo and Müller, 1999). This "somatosensory fovea" may help them grasp objects, such as branches or fruit, and sense their texture and elasticity (Hoffmann et al., 2004; Verendeev et al., 2015).

Finally, mammals have elaborated their sense of body position and movement (i.e., proprioception). In particular, they have evolved complex muscle spindles that provide animals with information about muscle forces and lengths (Proske and Gandevia, 2012). As noted in Chapter 4, fishes have at best a few, very simple muscle spindles. Muscle spindles are more common in some amphibians (Mandal and Anderson, 2010), and they are widespread in all amniotes. Mammalian muscle spindles are more complex than those of lizards, which contain just a single intrafusal muscle fiber (Proske, 1969). Avian muscle spindles are also more complex than their lizard homologs, but their structural complexity is different from that of their mammalian counterparts (Maier, 1992a). Overall, these data suggest that spindle complexity increased independently in birds and mammals. Also of interest is that the motor axons that innervate the muscle spindles in lizards tend to be collateral branches of the motor axons that innervate the main, extrafusal muscle fibers (Proske, 1967). In contrast, mammalian extrafusal and intrafusal fibers tend to be innervated separately by alpha and gamma motor neurons, respectively. Birds seem to be more like mammals than like lizards in this respect (Maier, 1992b), but the evidence is not definitive. If confirmed, this similarity between avian and mammalian muscle spindles would likely represent yet another example of convergent evolution between these two remarkably successful lineages.

6.6. Breathing, Chewing, and Moving Around

Over the long course of synapsid phylogeny, stem mammals significantly altered how they breathed, how they chew their food, and how they walked and ran. Next to nothing is known about the neural correlates of the changes in behavior, but they were crucial to the success of this lineage.

As noted earlier, stem mammals evolved a muscular diaphragm, which increased their breathing efficiency. Combined with the mucus membranes of the respiratory turbinates, the improved breathing helped endothermic stem mammals retain their hard-earned body heat. What we have not yet discussed is that mammals also pant, a behavior we may define as taking a series of shallow breaths in through the nose and out through the mouth (Biewener et al., 1985). Importantly, panting decreases body temperature through evaporative cooling, especially along the turbinates. This ability to shed internal heat would have been especially important for the smaller stem mammals, since their high surface-to-volume ratio would have caused them to heat up rapidly when it was hot outside. Many reptiles also pant (Tattersall et al., 2006), but their lack of nasal turbinates makes shedding heat more difficult. Avian panting appears to be at least as efficient as in mammals, especially because of evaporative cooling in the air sacs (Richards, 1970). Both mammals and birds also use their enhanced breathing abilities for another very important purpose, namely vocalization. Most non-avian sauropsids vocalize rarely or not at all (Ferrara et al., 2013), whereas most birds and many mammals do so frequently. Extensive vocalization in these lineages probably evolved in concert with their improved breathing and hearing abilities.

Stem mammals greatly enhanced their ability to process food inside the mouth, that is, to chew. With rare exceptions (Jones et al., 2012), non-mammalian tetrapods grab their food and simply gulp it down. Non-avian sauropsids sometimes seem to "chew" (Mills, 1972), but this behavior is much simpler than that of mammals. Specifically, late stem and early crown mammals evolved a more flexible jaw joint and more complex jaw muscles, with allowed the lower jaw to move more freely, including from side to side (Crompton and Parker, 1978). They also evolved a highly movable tongue and hard palate (see Section 6.4), which makes it easier to manipulate food inside the mouth. In addition, stem mammals evolved a highly differentiated set of teeth, including the premolar and molar teeth, which are perfect for crushing food and cutting it into smaller pieces (Ungar, 2010). Indeed, stem mammals are often classified according to their dentition not only because teeth predominate in the fossil record, but also because stem mammal teeth vary enormously. The main benefit of extensive chewing is that the food is cut into smaller pieces before it enters the gastrointestinal tract, thereby creating more surface area for enzyme action and speeding up digestion. This, in turn, allows for higher rates of food intake, which would have been important for later stem mammals because, as noted earlier, endothermy requires a lot of energy. Another, possibly incidental benefit of extensive chewing is that it liberates food odorants, which can enter the nasal cavity through the back of the mouth, stimulating olfactory receptors and combining with the information from taste buds to create flavor (Rowe and Shepherd, 2015).

The way synapsids stand and move across the ground also changed substantially over evolutionary time. Early stem mammals had a fully sprawling stance and gait, just like extant lizards and urodeles (see Figure 5.19 in Chapter 5). Middle stem mammals may have been capable of standing more erect at times, but they still preferred the sprawling mode of locomotion (Kemp, 1978; Blob, 2001). Even the extant monotremes extend their upper legs out sideways as they walk, though their feet are typically positioned under the body (Jenkins, 1971; Fish et al., 2001; Ashwell, 2013). Multituberculate mammals (see Figure 6.3) are likely to have had a partly sprawling stance and gait as well (Kielan-Jaworowska and Hurum, 2006). A fully erect stance and parasagittal gait, in which the legs are moved within a plane that lies parallel to the midline (sagittal) plane, seems not to have evolved until shortly before the origin of therian mammals. As mentioned in Chapter 5, the main benefits of this parasagittal gait and erect stance is that the long limb bones can act as vertical struts that bear most of the body's weight without fatiguing the limb muscles. In addition, this new form of locomotion reduces the side-to-side bending of the body that impairs the ability of lizards to breathe while they are running (see Figure 5.19 in Chapter 5). In short, parasagittal locomotion greatly increased the efficiency of terrestrial locomotion. Additional efficiencies arose in lineages that run on their toes or nails (e.g., carnivores and horses, respectively), rather than planting their whole foot on the ground.

In conjunction with evolutionary changes in locomotion, the vertebral column of synapsids became more highly differentiated (Jones et al., 2018). In basal amniotes and pelycosaurs the vertebral column is divisible into a sacral region and three more anterior divisions. Early therapsids then added a distinct pectoral region, which probably evolved in conjunction with changes in the forelimb and pectoral girdle. Therian mammals subsequently added a distinct lumbar region that lacks ribs and is associated with the hindlimb. Intriguingly, a more highly regionalized vertebral column evolved independently in the sauropsid lineage, but these animals added a distinct anterior region that was probably related to elongation of the neck.

Primates modified the ancestral mammalian pattern of locomotion by relying more on their hindlimbs for powering locomotion and using the forelimbs mainly for changing movement direction ("steering") and grasping tree branches or food (Demes et al., 1994; Schmitt, 2010). They also evolved at least partly opposable big toes and thumbs. All of these changes were associated with early primates becoming even more arboreal than their ancestors (Cartmill, 1992). Indeed, most extant primates spend much of their life in trees. Humans are the principal exception. As hominins became fully bipedal (much as birds had done much earlier), they spent more time walking on the ground and used their hands for non-locomotor activities. In association with these shifts, hominins lost the opposable big toe and perfected the opposable thumb (Andrews, 2015; Feix et al., 2015).

6.7. Brain Enlargement and Reorganization

As described in Section 6.3, synapsids evolved larger brains, relative to body size, along the lineage leading to the crown mammals. Endothermy was surely a prerequisite for this increased encephalization, but the larger brains may also have been one of the adaptations that made it possible for early mammals to obtain the nutrition required to maintain an elevated metabolic rate and, thus, achieve more stable body temperatures. One might further surmise that the ability to obtain more and better food improved further as mammals increased in absolute brain and body size. But did these behavioral and cognitive improvements result from simply scaling up the old, ancestral brains, or were they the consequence of significant anatomical and physiological innovations? To answer this question, it is important to explore how mammalian brains have changed, not just in size, but also in their organization. The most striking changes have occurred in the forebrain, but the lower regions of the brain have also undergone some interesting modifications.

6.7.1. Hindbrain Auditory Circuits

Just as the inner ear of mammals differs dramatically from that of sauropsids, so do the cochlear nuclei. Specifically, mammals have three separate cochlear nuclei, rather than two (see Chapter 5). These three cell groups are called the anteroventral, posteroventral, and dorsal cochlear nuclei. Each receives direct input from the sensory neurons that innervate the cochlea, and all of these inputs are tonotopically organized (Figure 6.17). However, the three nuclei differ in connectivity, cell morphology, and physiology (Malmierca and Merchán, 2004). Of special interest are the "spherical bushy cells" in the anteroventral cochlear nucleus, which receive auditory input through unusually large synapses. They are specialized for maintaining fine temporal information in the auditory signal (Carr and Soares, 2002).

The dorsal cochlear nucleus of mammals differs from the ventral cochlear nuclei in that it generally contains a large number of "granule cells" that resemble cerebellar granule cells in terms of structure, function, and developmental origin. These neurons receive not only auditory information, but also somatosensory information about the position of the external ears (pinnae). Integration of this ear position information with auditory signals probably improves an animal's ability to localize sounds, especially along the vertical dimension (Oertel and Young, 2004). In this context, it is intriguing that the granular layer of the dorsal cochlear nucleus is reduced in the lineage leading to humans (Figure 6.17), as well as in dolphins and whales (Moore, 1980). These changes are likely linked to reductions in the size and mobility of their external ears.

Finding homologs of the three mammalian cochlear nuclei in sauropsids is difficult, if not impossible. As reviewed in Chapter 5, sauropsids have two main cochlear nuclei, called nucleus angularis and magnocellularis. The former projects directly to the auditory midbrain, whereas the latter projects to nucleus laminaris, which encodes interaural time differences via axonal delay lines (see Figure 5.21). None of these cell groups is very similar to the dorsal cochlear nucleus of mammals. However, there is also no obvious correspondence between the two ventral cochlear nuclei of mammals and the sauropsid cochlear nuclei (Miller, 1980; Ryugo and Parks, 2003; Grothe et al., 2004). Moreover, in mammals the cochlear nuclei are densely connected with one another and contain numerous inhibitory interneurons, but this is not the case in birds. Based on these observations, we conclude that the mammalian and avian cochlear nuclei represent independent elaborations of a much simpler hindbrain auditory nucleus in their last common ancestor.

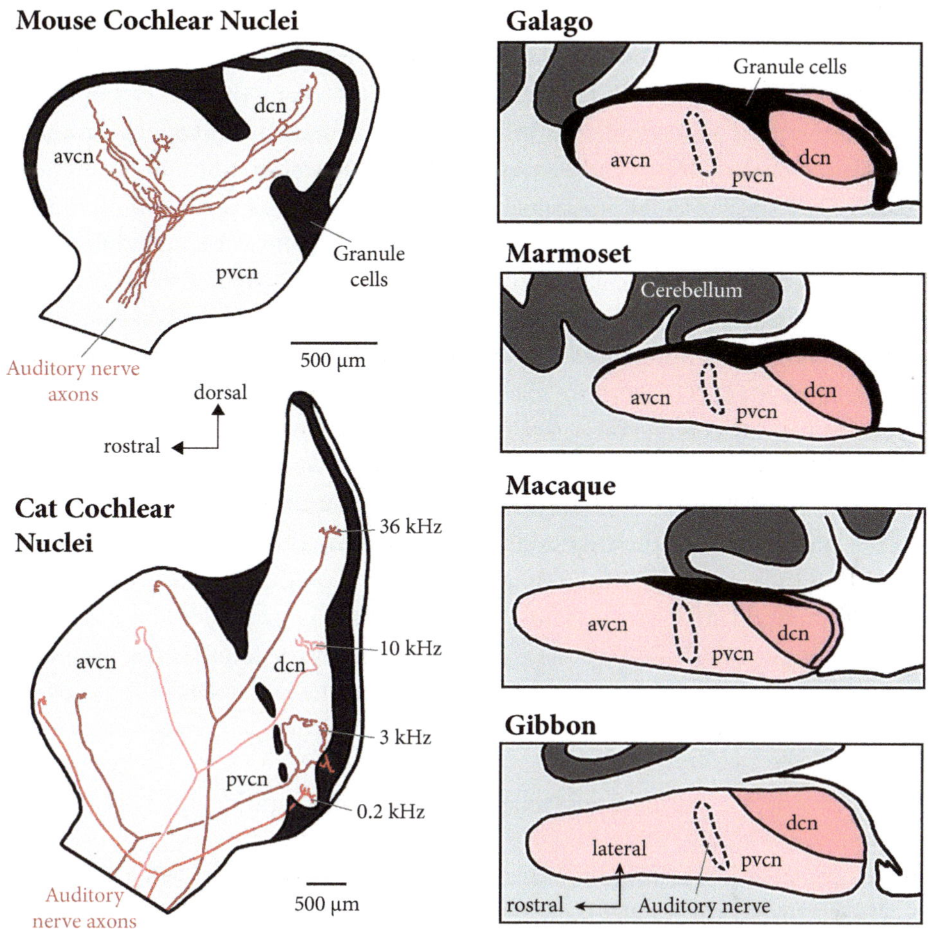

Figure 6.17 Mammalian cochlear nuclei. Mammals have three cochlear nuclei, namely the anteroventral, posteroventral, and dorsal cochlear nuclei (avcn, pvcn, and dcn). The drawings on the left show (in sagittal sections) how multiple auditory nerve axons branch shortly after entering the cochlear nucleus complex. The four illustrated axons in the cat were tuned to different sound frequencies (0.2–36 kHz). The drawings on the right depict horizontal sections through the cochlear nuclei of four primate species. Importantly, the layer of cerebellum-like granule cells that is a prominent feature of the cochlear nuclei in most mammals is reduced in monkeys and eliminated in apes (e.g., gibbons).

Adapted from Berglund and Brown (1994), Ryugo and May (1993, with permission from John Wiley & Sons), Moore (1980).

One might consider them to be "field homologs," but this interpretation tends to obscure their novelty (see Chapter 2). Just as tympanic ears and high-frequency hearing evolved independently in mammals and birds, so did their elaborate cochlear nuclei.

That said, the auditory hindbrain of birds and mammals does exhibit some fascinating similarities. Most striking is the fact that many neurons in nucleus magnocellularis of birds are very similar to the spherical bushy cells of mammals in terms of having large, round cell bodies and receiving auditory input via unusually large synapses. We consider these similarities to represent convergent evolution and to reflect functional constraints on how to maximize temporal precision in neuronal circuits (Carr and Soares, 2002). Furthermore, nucleus laminaris of birds is quite similar to the medial superior olivary nucleus (MSO) of mammals, insofar as both structures are specialized for processing interaural time differences. However, in contrast to nucleus laminaris, the MSO is located in the ventral medulla, far from the cochlear nuclei. Furthermore, it does not use delay lines to compute interaural time differences but, instead, relies on precisely timed inhibition (Myoga et al., 2014).

Finally, mammals alone possess a lateral superior olivary nucleus (LSO) in the anterior ventral medulla, close to the MSO. This LSO receives auditory input from both the left and the right cochlear nuclei, with the contralateral input being relayed through the nucleus of the trapezoid body (MNTB). This circuitry is crucial for computing interaural intensity differences, which provide another useful cue for localizing sounds. As far as we know, neither the LSO nor the MNTB has a clear homolog in sauropsids. In fact, the entire superior olivary complex (including the MNTB) appears to be a mammalian innovation (Grothe et al., 2004).

6.7.2. Cerebellum and Related Areas

The mammalian cerebellum resembles that of other amniotes in terms of internal structure (Llinás, 1969; Glickstein, 2007; Yopak et al. 2017), but it exhibits some unique features. Based on both functional and anatomical data, the mammalian cerebellum can be divided into three main longitudinal zones: a medially located vermis, an adjacent paravermis, and the cerebellar hemispheres. Birds and other sauropsids have homologs of the vermis and paravermis, but they have at best "rudimentary" cerebellar hemispheres (Goodman, 1969; Pakan et al., 2007). Sauropsids also lack the deep cerebellar nucleus associated with those hemispheres (i.e., the dentate nucleus). These species differences are interesting, because the cerebellar hemispheres in mammals receive strong projections from the neocortex via the pontine nuclei, which are another mammalian innovation (as discussed later and in Section 5.5.2).

Given that the cerebellar hemispheres in mammals are involved in motor as well as cognitive control (e.g., Schmahmann, 1991), it is reasonable to speculate that these functions were enhanced by the emergence of cerebellar hemispheres in early mammals. An interesting twist to this story is that, in contrast to the cerebellar hemispheres, the mammalian vermis and paravermis require fibroblast growth factor 8 (Fgf8) for normal development, whereas the avian cerebellum can develop independently of Fgf8 (Butts et al., 2014). These data suggest either that the avian cerebellum is homologous to the mammalian cerebellar hemispheres, or that the role of Fgf8 in the development of the cerebellar vermis and paravermis changed during amniote phylogeny. We favor the latter hypothesis.

Another striking aspect of cerebellar variation is that the cerebellum exhibits numerous transverse folds (or folia) in all adult mammals, but merely a simple bend in lizards and turtles (see Figure 5.2). Birds are similar to mammals in having a highly folded cerebellum (see Figure 5.9), but this similarity evolved independently. Most likely, the folding in both lineages arises because the cerebellar cortex expands faster in area than in thickness, which would generate buckling forces similar to those apparently responsible for folding the neocortex (Striedter et al., 2014). However, the mechanisms underlying cerebellar foliation remain poorly understood (Sudarov and Joyner, 2007). What does seem clear is that the degree of cerebellar folding increases with cerebellum surface area (Sultan and Braitenberg, 1993), while the thickness of the cerebellar cortex is much less variable.

Although the neocortex gets most of the attention in mammalian brains, the cerebellum contains 80% of all the neurons in a human brain (Azevedo et al., 2009), most of them being the small and densely packed cerebellar granule cells. As mammalian brains increased in absolute size over evolutionary time, the percentage of the brain that consists of cerebellum remained a roughly constant 12%–14%, whereas the neocortex fraction increased from less than 20% to more than 80%; at least, that is the scaling relationship observed across extant species that vary in absolute brain size (Figure 6.18; Clark et al., 2001). However, this does not mean that the cerebellum remained invariant across mammalian phylogeny. It is better to think of cerebellum volume as increasing in concert with neocortex volume, just at a lower rate. In terms of numbers of neurons, the cerebellum actually expands faster than the neocortex with increasing brain size, adding roughly four cerebellar neurons for every one neuron that is added to the neocortex, at least in rodents (Herculano-Houzel et al., 2011).

Cerebellum volume and neuron numbers in mammals scale rather predictably with increasing brain size, but some departures from the rules are obvious. For example, the cerebellum of the elephant occupies 25% of the brain's total volume and contains an astonishing 97.5% of all its neurons (~250 billion; Herculano-Houzel et al., 2014). More subtle is the fact that the cerebellum of apes is larger than one would expect, relative to the size of their neocortex. A double-logarithmic plot of cerebellum volume against neocortex volume (Figure 6.19) reveals that the regression lines for apes and monkeys have very similar slopes but different *y*-intercepts (Smaers, 2014). This kind of evolutionary change in scaling relationships is called a "grade shift" (Pagel and Harvey, 1988) and is seen in many different domains of morphological evolution. This particular case is interesting, because it is widely assumed that the complex cognitive abilities of great apes (including humans) stem largely from the fact that they increased the size of their neocortex. This is true, as we discuss shortly, but the great apes increased the size of their cerebellum even more (Barton and Venditti, 2014). In particular, they dramatically enlarged their cerebellar hemispheres, which, as noted previously,

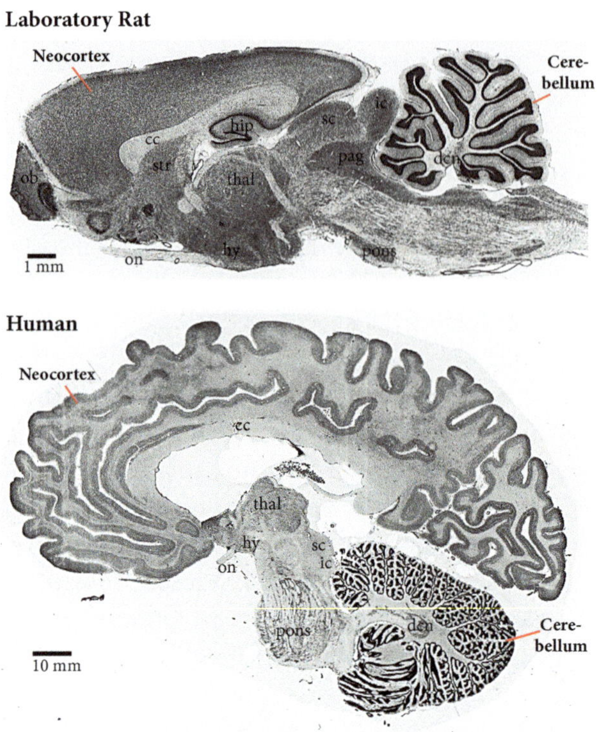

Figure 6.18 Rat brain versus human brain. These parasagittal Nissl-stained sections through the brains of a laboratory rat and a human illustrate that the neocortex is proportionately much larger, relative to the remaining brain, in humans than in rats.

Abbreviations: cc – corpus callosum; dcn – deep cerebellar nuclei; hip – hippocampus; hy – hypothalamus; ic – inferior colliculus; ob – olfactory bulb; on – optic nerve; pag – periaqueductal gray; sc – superior colliculus; thal – thalamus.

Rat brain from BrainMaps, http://brainmaps.org (retrieved on 3/15/2017); human brain section from the Yakovlev-Haleem Collection (images from zoomablebrain.bio.uci.edu).

are densely interconnected with the neocortex. In contrast, the cerebellar vermis, which is densely interconnected with the medulla and spinal cord, is relatively small in apes (MacLeod et al., 2003).

Several areas linked to the cerebellar hemispheres have also changed in size during primate phylogeny. For instance, the deep cerebellar nucleus that conveys output from the cerebellar hemispheres to other brain regions is very large and complexly folded in apes, which is why it is called the dentate (teeth-like) nucleus. In addition, the small-celled "parvocellular" red nucleus, which receives major inputs from the dentate nucleus (Figure 6.19) and projects heavily to the ipsilateral inferior olive, is very large in apes. In contrast, the "magnocellular" division of the red nucleus, which projects strongly to the contralateral spinal cord, is small in the great apes and poorly differentiated in humans (ten Donkelaar, 1988; Yamaguchi and Soto, 2006). The relative reduction of this rubrospinal tract in apes is accompanied by a corresponding increase in the extent of spinal projections from the

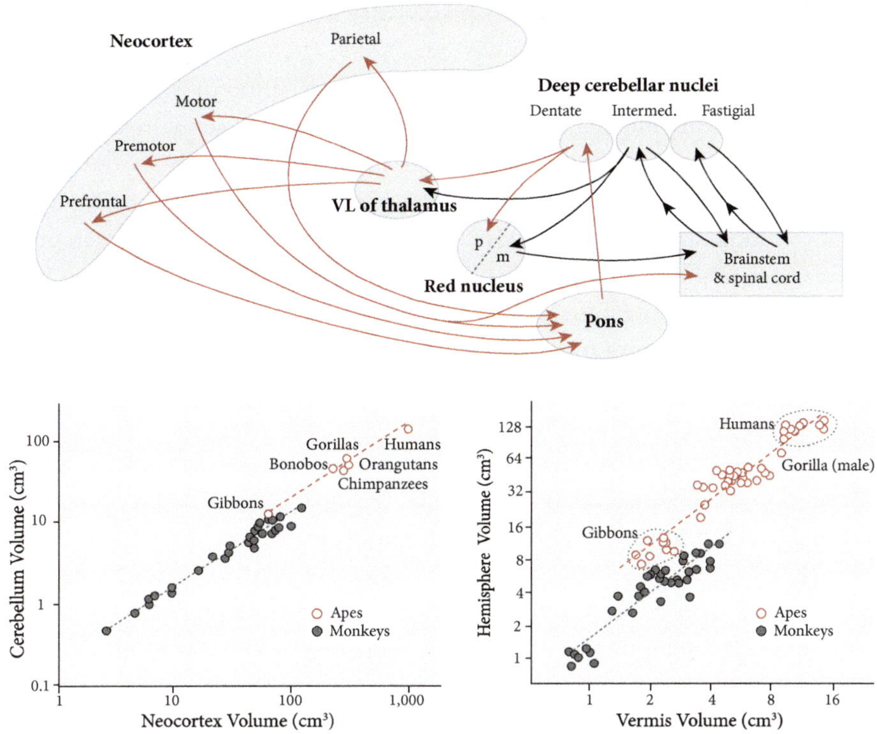

Figure 6.19 Cerebrocerebellar relationships in primates. The diagram depicts the major connections between the neocortex and the dentate nucleus, which is a deep cerebellar nucleus that gets strong inputs from the cerebellar hemispheres (not shown). The red nucleus in primates has magnocellular and parvocellular divisions (m and p, respectively). The graphs show that, relative to the neocortex, the cerebellum is larger in apes than in monkeys (left) and that, relative to the cerebellar vermis, the cerebellar hemispheres are disproportionately large in apes (right).

Adapted from Rilling (2007), MacLeod et al. (2003), Barton and Venditti (2014).

neocortex. Apparently, the corticospinal tract functionally replaced the rubrospinal tract as apes evolved.

Finally, apes have evolved a set of very large pontine nuclei that transmit information from the neocortex and the midbrain roof to the cerebellar hemispheres and dentate nucleus (Figures 6.18 and 6.19). All mammals have such pontine nuclei, but non-avian sauropsids do not (as far as we know). Birds do have pontine nuclei, but they are small and receive relatively little telencephalic input (Zeier and Karten, 1971; Freedman et al., 1975). Therefore, the available data suggest that pontine nuclei evolved independently in birds and mammals and serve at least somewhat different functions. In mammals they form part of a looping circuit between the cerebellar hemispheres and the neocortex, and this loop is thought to have both cognitive and motor functions (Strick et al., 2009). Their function in birds remains unknown.

6.7.3. A Modest Midbrain Roof

The midbrain roof in mammals is divided into the superior and inferior colliculi, which are homologous to the optic tectum and torus semicircularis of non-mammalian vertebrates, respectively. Despite the difference in names, these regions are relatively conservative in their embryonic gene expression patterns, internal organization, and main connections. They do, however, vary substantially in proportional size and adult position.

As we described in previous chapters, the optic tectum is a large, ballooning, and highly laminated structure in most non-mammalian vertebrates. It is the principal recipient of retinal projections in amphibians, reptiles, and birds, and lesions of the optic tectum impair most visually guided behaviors in these species (Vanegas, 1984). In contrast, the mammalian superior colliculus is much smaller, relative to other brain regions (see Figure 6.18). It does exhibit multiple laminae and contains several interesting cell types that are also found in sauropsids (May, 2006), but it does not form an externally visible "balloon" that covers a prominent tectal ventricle. Clearly, stem mammals reduced the size of the optic tectum homolog (at least in terms of relative volume). It is tempting to attribute this reduction to the ecological shift of stem mammals into the nocturnal niche. However, stem mammals expanded the main dorsal thalamic target of the retina, the dorsal lateral geniculate nucleus (LGNd), even as they reduced the size of their optic tectum. Thus, it is better to think of stem mammals as shifting much of their visual processing away from the midbrain and into the forebrain. While the mammalian midbrain remains heavily involved in the control of eye movements and other orienting behaviors, the forebrain in mammals became increasingly concerned with the identification of complex visual objects.

The trend of reducing the midbrain's relative contribution to visual processing continued within mammals, especially primates. Thus, the upper, visual portion of the superior colliculus is 2–3 times larger than the LGNd in hamsters and rats but only 1/8 the size of the LGNd in macaque monkeys (Schiller, 2010). Furthermore, in rabbits and rats all retinal ganglion cells project to both the superior colliculus and the LGNd via axon collaterals, whereas in macaque monkeys fewer than 10% of the retinal ganglion cells project to the superior colliculus (Perry and Cowey, 1984). Indeed, it seems that in primates the superior colliculus gets most of its visual input from the visual cortex, rather than the retina. Primates also direct much of the output from the superior colliculus toward the visual cortex, via the pulvinar nucleus of the thalamus (see Section 6.7.4). A major function of this ascending pathway is to direct spatial attention to salient visual targets, especially when no overt orienting movements are involved (Berg et al., 2017). The optic tectum also mediates covert spatial attention in non-primates (Knudsen and Schwartz, 2017), but the enormous size of the pulvinar in primates (Robinson and Petersen, 1992) suggests that this important cognitive function was elaborated during primate phylogeny.

The torus semicircularis is the principal auditory region of the midbrain in all vertebrates. In non-mammals, it begins its development caudal to the optic tectum but then expands anteriorly (Figure 6.20) and eventually comes to lie ventral to the optic tectum in transverse sections through adult brains (see Figure 4.20). In mammals, however, the homolog of the torus semicircularis expands caudally and dorsally, forming a bilateral pair of "mounds" (colliculus means "little hill" in Latin) that lies caudal (inferior in humans) to the superior colliculi (Figure 6.21). It is difficult to judge without quantitative data whether this evolutionary shift in position is associated with an evolutionary change in size, relative to the rest of the brain. However, given the expanded hearing abilities of mammals compared to non-avian sauropsids, it seems likely that early mammals expanded the proportional size of their inferior colliculi. It is certainly the case that, among extant mammals, the species with exceptional hearing abilities also have unusually large inferior colliculi (Figure 6.21).

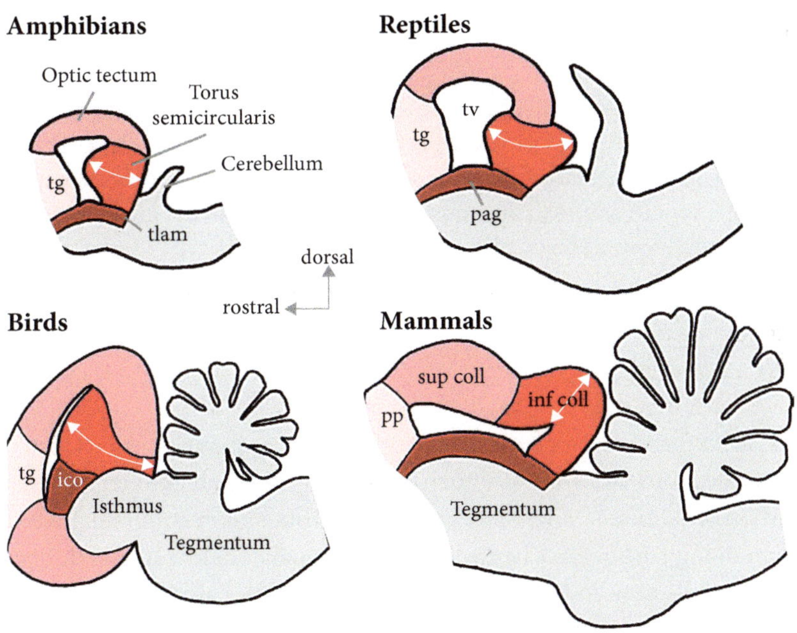

Figure 6.20 The midbrain roof of tetrapods. The torus semicircularis bulges into the tectal ventricle (tv) in amphibians, reptiles, and birds, whereas its homolog in mammals—the inferior colliculus (inf coll)—bulges outward, forming a small bump on the external surface of the brain. The mammalian homolog of the tectal gray (tg) is called the posterior pretectal nucleus (pp). The likely homolog of the periaqueductal gray (pag) is called torus pars laminaris (tlam) in amphibians and intercollicular area (ico) in birds. The white arrows indicate the axis along which the torus semicircularis (or inferior colliculus) thickens during development.

Additional abbreviation: sup coll – superior colliculus.

Adapted from Puelles et al. (1994), with permission of John Wiley & Sons.

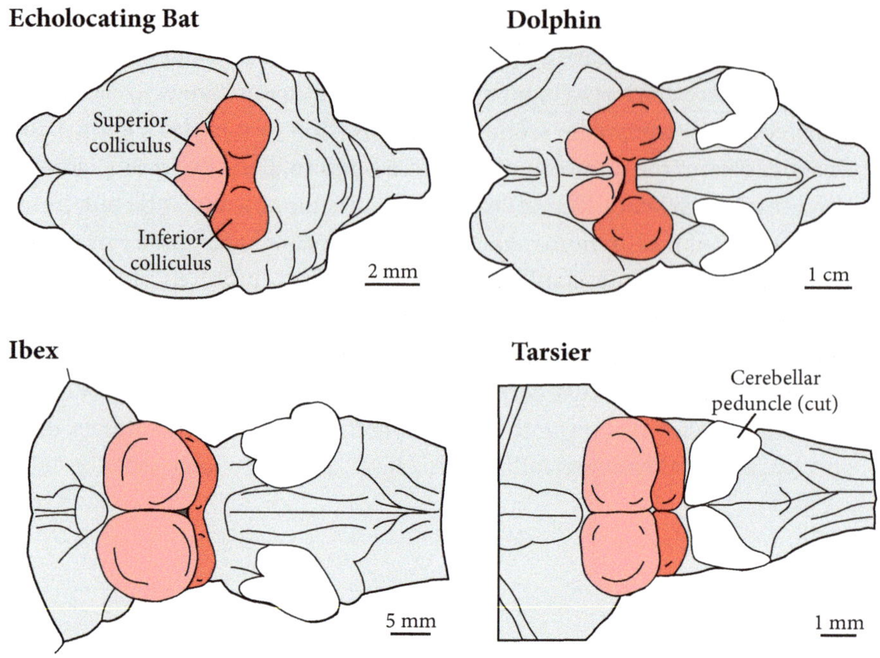

Figure 6.21 The superior and inferior colliculi of four mammals in dorsal view. In echolocating bats and bottlenose dolphins the inferior colliculi, which process mainly auditory information, are larger than the superior colliculi, which process mainly visual information. In contrast, the inferior colliculi are relatively small in mammals that are not auditory specialists, such as tarsiers and ibexes (a type of goat). In bats the two pairs of colliculi are visible in dorsal views, but in the other species they can be seen only after the overlying telencephalon and cerebellum have been removed.
Adapted from Baron et al. (1996), Longworthy (1932), Schober and Brauer (1974), Tilney (1927).

The mammalian inferior colliculus contains several subdivisions, but the ascending projections from the auditory brainstem nuclei all converge onto its central division (Malmierca and Merchán, 2004). This observation suggests that the expansion of high-frequency hearing and additional cochlear nuclei did not entail the evolution of a new and separate ascending auditory pathway through the midbrain. This observation is consistent with what happened with the evolution of tympanic ears and a novel auditory brainstem nucleus in frogs (see Chapter 4): those innovations were not accompanied by the emergence of an obviously novel auditory nucleus within the torus semicircularis. In general, these findings indicate that homologous brain areas may process information from non-homologous peripheral structures (Wilczynski, 1984). Put differently, old brain regions are able to accommodate new sensory inputs. This phylogenetic plasticity makes it easier for useful changes to evolve, increasing the brain's "evolvability" (Pigliucci, 2008). However, as we will see shortly, this conclusion does not imply that novel brain regions never arise in higher brain regions. Indeed, as mentioned in Chapter 4, even the homologies of the auditory toral nuclei remains debatable.

6.7.4. An Enlarged, More Complex Thalamus

The thalamus (aka the dorsal thalamus) is the most variable division of the diencephalon in amniotes. Most obvious is that the thalamus is much larger, relative to other diencephalic areas, in mammals and birds than in non-avian sauropsids. Along with this difference in size comes a substantial difference in complexity, as the thalamus contains a much larger number of distinct cell groups in birds and mammals than in turtles or lizards (Figure 6.22). Because of this difference in structural complexity, it is extremely difficult—indeed, often impossible—to homologize individual thalamic cell groups between the major amniote lineages. Similarities in relative position, histochemistry, and neural connections do suggest a few clear-cut homologies, but many thalamic cell groups defy the expectation of strict one-to-one

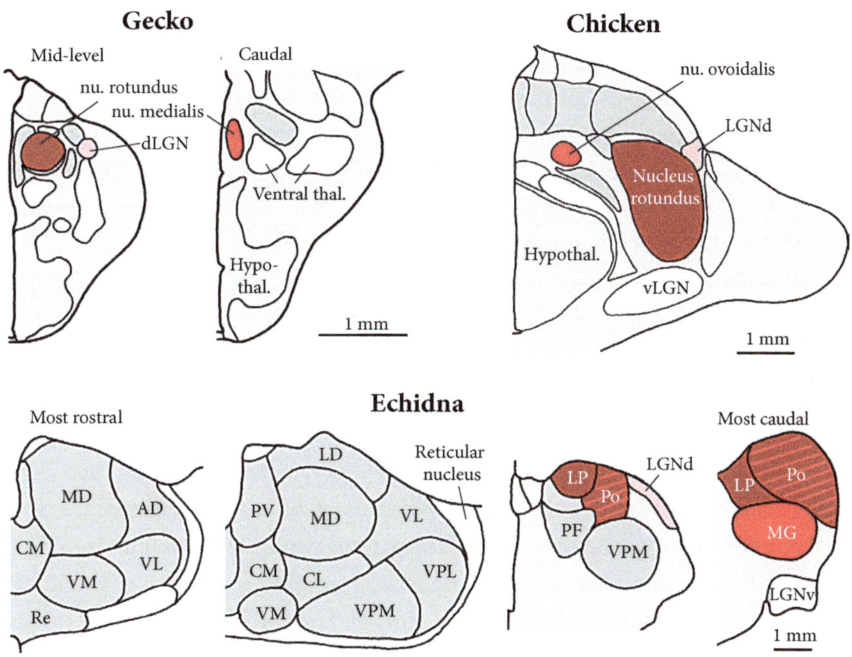

Figure 6.22 Parcellation of the thalamus in amniotes. Shown here are schematic transverse sections through the diencephalon of a Tokay gecko (at two rostrocaudal levels), a domestic chicken, and an echidna (at four levels). The nuclei of the posterior thalamic zone, which receive inputs from the midbrain roof, are shaded dark red (main visual target) or medium red (main auditory target). The posterior nucleus (Po) in mammals receives multisensory input from the midbrain roof. The nuclei of the anterior thalamic zone are shaded either gray or, in the case of the dorsal lateral geniculate nucleus (LGNd), pink. Clearly, the anterior thalamic zone is much larger and more complex in mammals and chickens than geckos. Many of the thalamic nuclei in geckos and chickens are not individually identified, and the nuclei in the echidna's thalamus are identified only by their standard abbreviations.

Adapted from Northcutt (1978), Puelles et al. (2007), Ashwell and Paxinos, (2005).

correspondences between the major amniote lineages. The problem is exacerbated by the fact that amphibians have only three major cell groups within their thalamus (see Chapter 4).

6.7.4.1. Thalamic Homologies and Novelties

As previewed briefly in Chapter 5, Ann Butler proposed an interesting solution to this dilemma (Butler, 1994, 1995). She started with the observation that the amphibian thalamus can be divided into a posterior division that receives sensory input via the midbrain roof and an anterior division that receives converging input from multiple sensory modalities, including direct projections from the retina (Neary and Northcutt, 1983). Next, Butler argued that the posterior division, which she called the collothalamus, is divisible into two principal regions: one receives its input mainly from the optic tectum (or superior colliculus), the other mainly from the torus semicircularis (or inferior colliculus). These two divisions of the posterior thalamic division are present in amphibians and, although they have been given a variety of names, they are generally thought to have strict homologs in the various amniotes (Figure 6.22). In contrast, Butler argued that the various cell groups that make up the anterior division of the thalamus in tetrapods are homologous to one another only as derivatives of a conserved embryonic field, which gives rise to a single adult cell group in amphibians (called nucleus anterior), but differentiates into multiple cell groups in amniotes.

Moreover, Butler proposed that each of the cell groups in the anterior thalamus of amniotes retains only a subset of the connections of the amphibian anterior nucleus, which is presumed to represent the primitive condition. In essence, Butler invoked Ebbesson's "parcellation theory," according to which evolution may divide a single ancestral cell group into multiple descendant nuclei that each retain only a subset of the ancestral features (Ebbesson, 1980; Striedter, 2005). Particularly important is Butler's argument that the expansion of the anterior thalamic division creates a novel cell group, called the dorsal lateral geniculate nucleus (LGNd) that retains the direct retinal input of its ancestral homolog but has lost the ancestral non-visual sensory inputs. Since a very similar retinorecipient cell group is also found in reptiles and birds, Butler argued that LGNd must have evolved in the last common ancestor of amniotes, which would make the LGNd homologous across all extant amniotes. It is important to note, however, that the LGNd is small in most amniotes, becoming large and structurally complex (laminated) only in highly visual mammals, such as squirrels, tree shrews, and primates (Kaas, 2007).

In contrast to the LGNd, the other nuclei of the anterior thalamic division are highly divergent across amniotes. Non-avian sauropsids exhibit two principal nuclei within this group (called the dorsomedial and dorsolateral anterior nuclei), but birds possess several additional cell groups that have no obvious homologs in lizards or turtles. Mammals present a different pattern entirely. In addition to the LGNd, their anterior thalamic division includes an anterior group of nuclei, the anterior intralaminar nuclei, a mediodorsal nucleus with strong connections to the prefrontal cortex, a thalamic motor nucleus, and a large somatosensory nucleus that, itself, contains multiple subdivisions (Figure 6.22). Some authors have proposed homologies for some of these mammalian cell groups in birds. For example, histochemical and connectional similarities have been used to identify putative homologs of the mammalian intralaminar and mediodorsal nuclei in birds (Veenman et al., 1997). The problem with these proposals is that these nuclei are less distinct, if not entirely lacking, in non-avian sauropsids. Therefore, it seems likely that the putative homologs in birds and mammals are, instead, the result of independent parcellation events that produced somewhat similar but not strictly homologous cell groups. The same line of reasoning suggests that mammals and birds independently evolved a motor thalamic nucleus that receives input from the cerebellum and basal ganglia (Medina et al., 1997; Wullimann, 2011).

While the anterior division of the thalamus clearly expanded in early mammals, the posterior division was probably reduced in size. The largest component of this posterior division in sauropsids is nucleus rotundus, which receives massive inputs from the optic tectum. Its likely homolog in mammals is the LP/pulvinar complex, but this complex is small and indistinct in most mammals. This is not surprising, given that early mammals reduced the size of the optic tectum (i.e., their superior colliculus): as the optic tectum shrank, its main thalamic target was likewise reduced in size. Remarkably, this correlation breaks down in primates. Even though primates have a relatively small superior colliculus, their LP/pulvinar complex is huge (Figure 6.23). In particular, the pulvinar has dramatically expanded in primates and contains at least some subdivisions that have no obvious homolog in other mammals (Kaas, 2007; Baldwin et al., 2017).

6.7.4.2. Thalamic Connections with the Telencephalon

Across all tetrapods, virtually all of the thalamic (aka dorsal thalamic) nuclei have ascending projections to the telencephalon. The nuclei of the anterior thalamic division consistently project to the medial and the dorsal pallium (assuming that amphibians have a dorsal pallium at all; see Figures 4.24 and 4.29), but the ascending projections of the posterior nuclei vary considerably among the major tetrapod lineages. In amphibians, the central nucleus, which receives strong auditory input from the midbrain roof, projects heavily to the striatum; so does the adjacent lateral nucleus, which receives most of its input from the optic tectum (Neary and Northcutt, 1983; Roth et al., 2003). In contrast, the homologs of these thalamic nuclei in amniotes project not only to the striatum but also to the overlying pallium. In mammals the nuclei of the posterior thalamic division (notably the LP/pulvinar, posterior, and posterior intralaminar regions) project mainly to parts of the neocortex, though minor projections to the pallial amygdala (and claustrum) have been observed (Doron and LeDoux, 1999; Guirado et al., 2005). In reptiles and birds, the nuclei of the posterior thalamic division project mainly to the dorsal ventricular ridge (DVR), which we discussed in Chapter 5 (see Figures 5.24–5.27). These observations have been used to the argue that the sauropsid DVR is homologous to

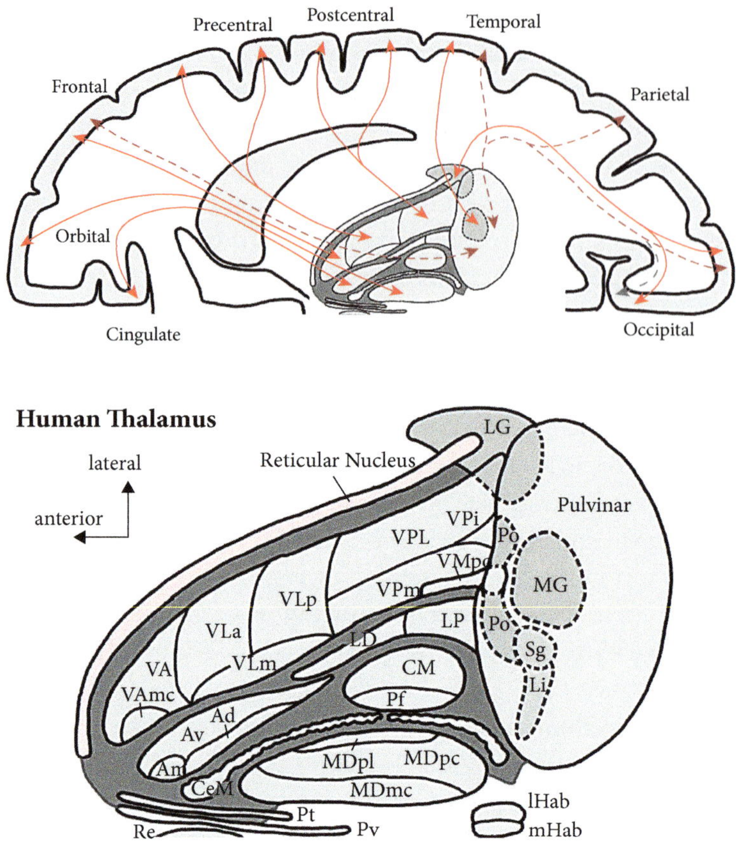

Figure 6.23 The human thalamus. Shown at the top is a schematic horizontal section through the right hemisphere of a human brain (anterior is to the left). The red arrows indicate reciprocal connections between all major cortical areas and the major thalamic cell groups (the connections of some thalamic nuclei are not shown). The dashed red arrows represent the cortical connections of the pulvinar, which are unusually widespread. The diagram on the bottom shows a close-up of the human thalamus, revealing its many subdivisions. The reticular thalamic nucleus, which is part of the prethalamus is depicted in pink.

Abbreviations: Ad – anterodorsal; Am – anteromedial; Av – anteroventral; CeM – central medial; LD – lateral dorsal; LG – dorsal lateral geniculate; lHab – lateral habenula; CM – centromedian; Li – limitans; LP – lateral posterior; MDmc – magnocellular mediodorsal; MDpc – parvocellular mediodorsal; MDpl – paralaminar mediodorsal; MG – medial geniculate; mHab – medial habenula; Pf – parafascicular; Pt – parataenial; Po – posterior; Pv – paraventricular; Re – reuniens; Sg – suprageniculate; VA – ventral anterior; VAmc – magnocellular ventral anterior; VLa – anterior ventral lateral; VLp – posterior ventral lateral; VLm – medial ventral lateral; VMpo – ventromedial posterior; VPm – ventral posteromedial; VPl – ventral posterolateral; VPi – ventral posterior inferior.

Adapted from Nieuwenhuys et al. (2008) with permission from Springer Nature.

parts of the mammalian neocortex (Karten, 1969). However, we think it is at least as likely that axons from the posterior thalamic nuclei that ancestrally terminated only in the striatum "invaded" (during phylogeny) the overlying pallial areas, which happened to be the DVR in sauropsids and primarily the neocortex in mammals. In other words, we propose that a phylogenetic "invasion" of posterior thalamic axons into the pallium occurred independently in mammals and sauropsids and targeted different, non-homologous brain regions in the two lineages. We will revisit this idea shortly.

An interesting, related question is whether the telencephalic areas that receive thalamic input reciprocate those projections. Reciprocal connections between thalamic nuclei and their telencephalic targets seem to be the rule in all mammals (Jones, 2007). Such reciprocity is also observed for the pallium-projecting nuclei of the anterior thalamic division in sauropsids, although it is frequently unclear whether the back-projections match the ascending pathways as precisely as they do in mammals (Wu and Karten, 1998; Pritz, 2015). In contrast, the projection targets of the posterior thalamic division in the DVR of sauropsids do not project back to the thalamus. Indeed, as far as we know, no parts of the sauropsid DVR project back to the nuclei of the posterior thalamic division. Therefore, we conclude that early mammals expanded the extent to which their pallium is reciprocally connected with the thalamus. According to this hypothesis, the postulated role of the mammalian thalamus in relaying information between cortical areas (Sherman and Guillery, 2006) must also be a mammalian innovation.

Compared to the palliothalamic connections, the reciprocal connections between the thalamus and the reticular nucleus of the prethalamus are more conserved across all amniotes (Díaz et al., 1994; Pritz, 2016). These reciprocal projections account for sleep-related oscillations in mammals (Huguenard and McCormick, 2007), but whether this is true also in sauropsids remains unclear. It is interesting to note, however, that at least some sleep-related oscillations have now been recorded in both lizards and birds as well as mammals (Shein-Idelson et al., 2016).

6.7.5. Striatopallidal Circuits

As discussed in previous chapters, the vertebrate telencephalon is divisible into a pallium and a subpallium. The main divisions of the subpallium, in turn, are the striatum, pallidum, septum, preoptic area, and pallial amygdala. Among these divisions, the most significant variation is found in the striatum and the pallidum, which we may collectively refer to as the striatopallidal complex. Historically, some influential neuroscientists had thought that the sauropsid dorsal ventricular ridge (DVR) was part of the striatum, but embryological data has long suggested a pallial origin (see Striedter, 1997; Puelles et al., 2000). Moreover, immunohistochemical studies revealed that only the region ventral to the DVR contains high levels of dopamine and acetylcholine, two key features of the striatum in mammals (Figure 6.24; Juorio and Vogt, 1967; Karten, 1969). Since then, many studies have revealed numerous similarities between the striatopallidal complex of birds, non-avian sauropsids, and mammals (Reiner et al., 1998; Puelles et al., 2000; Kuenzel et al., 2011). Many of these similarities are primitive; that is, they were inherited from

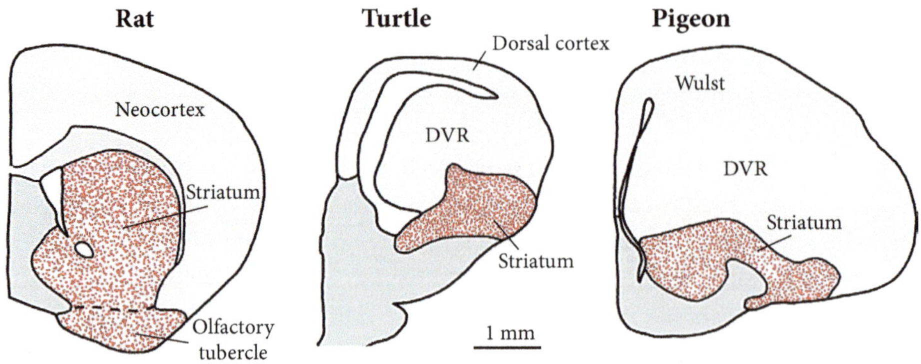

Figure 6.24 The striatum of amniotes. These schematic transverse sections through the right hemispheres of a rat, a turtle and a pigeon, depict dopamine-containing axons and terminals in red (weak dopamine labeling is not shown). Most of the dopaminergic axons terminate in the striatum, which lies ventral to the dorsal ventricular ridge (DVR) in the sauropsids.
Adapted from Smeets et al. (1987), Wynne and Güntürkün (1995), Zhou et al. (2001).

pre-amniote ancestors. Here we focus on some major evolutionary changes that occurred as amniotes evolved and, in particular, on those that characterize the synapsid lineage.

In amphibians, the striatum projects to the pallidum, but the descending projections of these two structures are otherwise quite similar (see Figure 4.28 in Chapter 4). In contrast, the striatum and the pallidum have very different connections in amniotes. Specifically, the striatum of amniotes has strong reciprocal connections with the substantia nigra, whereas the pallidum provides most of the other outputs of the striatopallidal complex. Assuming that the amphibian condition is primitive, we hypothesize that the striatum and the pallidum became more differentiated from one another with the origin of amniotes. The degree of segregation between dorsal and ventral divisions of the striatopallidal complex also appears to have increased with the origin of amniotes (Veenman et al., 1995), although even in mammals the division of the striatum into dorsal and ventral divisions represents two ends of a continuum, rather than a strict dichotomy (Voorn et al., 2004).

More importantly, amniotes added the pallium as a major source of sensory input to the striatum (Figure 6.25). In amphibians and other anamniotes the striatum receives almost all of its non-olfactory sensory input via the thalamus. Amniotes do possess such thalamostriatal pathways (e.g., from the intralaminar and midline thalamic nuclei in mammals; Giménez-Amaya et al., 1995; Smith et al., 2004), but the principal sensory thalamic nuclei in amniotes project mainly to the pallium, which then provides strong inputs to the striatum (Veenman et al., 1995). Given that the pallium in amniotes performs extensive sensory processing (more on this shortly), this evolutionary “augmentation” of the ancestral thalamostriatal

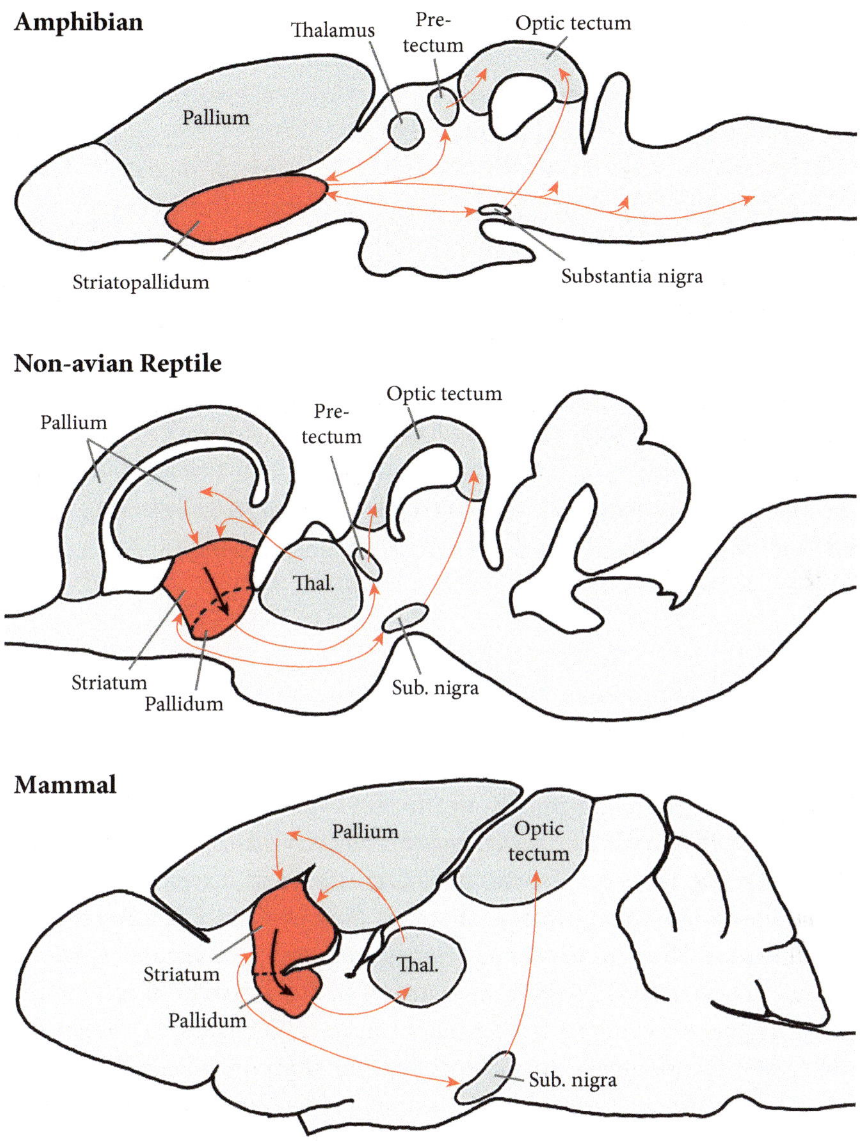

Figure 6.25 Circuits through the striatum and pallidum. These schematic sagittal sections show the principal circuits through the striatopallidal complex in representative tetrapods. The projections from the striatopallidal complex to the medulla and spinal cord were lost in amniotes, but thalamopalliostriatal pathways were added or expanded significantly. Pallidopretectotectal pathways were lost in mammals, but a strong projection from the pallidum to the thalamus appeared.
Adapted from Marin et al. (1997), Rodrigues et al. (2008), and brainmuseum.org.

pathways provides the striatum of amniotes with additional, more highly processed, and more nuanced sensory input.

The origin of amniotes was also associated with the loss of long descending projections from the pallidum to the medulla and spinal cord (Figure 6.25).

However, amniotes retain the ancestral output pathways that course through the substantia nigra and pretectum to the optic tectum, which has its own projections to the medulla and spinal cord. Presumably those tectal projections can compensate for the loss of the direct pallidal outputs.

Further changes occurred within the lineage leading to mammals. Especially important is that mammals gained a pathway from the pallidum to parts of the thalamus that project to the frontal lobe (Figure 6.25), setting up a pallidothalamocortical loop that is thought to play a major role in the control of behavior (Alexander et al., 1986; Middleton and Strick, 2002). Birds have evolved a similar pathway (Medina et al., 1997), but it is relatively small. More importantly, no such pathway has been described in any non-avian sauropsid. Therefore, it seems likely that pallidothalamopallial loops evolved independently in mammals and birds (Wullimann, 2017a). Intriguingly, the shift of pallidal output toward the thalamus in mammals was associated with the loss of pallidal outputs to the pretectum (Reiner et al., 1998). Thus, we concur with Veenman et al. (1995) that the mammalian striatopallidal complex controls behavior primarily through a pallidothalamocortical loop, whereas this function is assumed mainly by the pallidopretectotectal pathway in birds.

Mammals are also unique in having a "patch-matrix" type of organization in their striatum (Graybiel and Ragsdale, 1978; Reiner et al., 1998; Brimblecombe and Cragg, 2016) and in having distinct external and internal divisions of the pallidum, with the former projecting mainly to the subthalamic nucleus. Sauropsids and (probably) amphibians do have a subthalamic nucleus (Jiao et al., 2000; Maier et al., 2010), but its inputs do not (apparently) originate from a separate pallidal division. Sauropsids are also unusual in that they have lost the dopamine transporter gene and seem, instead, to use the norepinephrine transporter to re-uptake dopamine (Lovell et al., 2015). Within the sauropsids, birds have evolved a number of additional, apparently unique striatopallidal features. For example, songbirds have evolved a specialized "area X" inside their striatum that contains not only striatum-typical medium spiny neurons but also numerous pallidum-like neurons (Farries et al., 2005) that project to the ventral pallidum and to a specialized thalamic nucleus that is involved in song learning and control (Gale et al., 2008; Gale and Perkel, 2010). Whether these unusual features are unique to songbirds or common to all birds remains unknown.

6.7.6. Evolution of the Neocortex

The most unique component of the mammalian pallium is the neocortex. Especially distinctive is that it contains a larger number of cellular layers (laminae) than the adjacent olfactory and hippocampal cortices, which concentrate most of their neuronal cell bodies into just a single cellular layer. The region most likely to be a neocortex homolog in non-avian sauropsids, called the dorsal cortex, also features just a single cell-dense layer, sandwiched between two relatively fibrous laminae (Figure 6.26). In contrast, mammalian neocortex typically contains at least five cellular layers and a thin superficial layer consisting mainly of axons. Although some of the cellular laminae can be further subdivided in some lineages, and cetaceans lack layer 4 (Knopf et al., 2016), early mammals almost certainly had the kind of "six-layered" neocortex seen in rats and other small mammals (Brodmann, 1999; Schmolke and Künzle, 1997).

One important feature of most neocortical neurons is that their dendrites and axons tend to extend radially within the neocortex, creating a distinctly columnar organization. Furthermore, most of the inputs to the neocortex ascend radially within a neocortical column, having entered it from the underlying white matter.

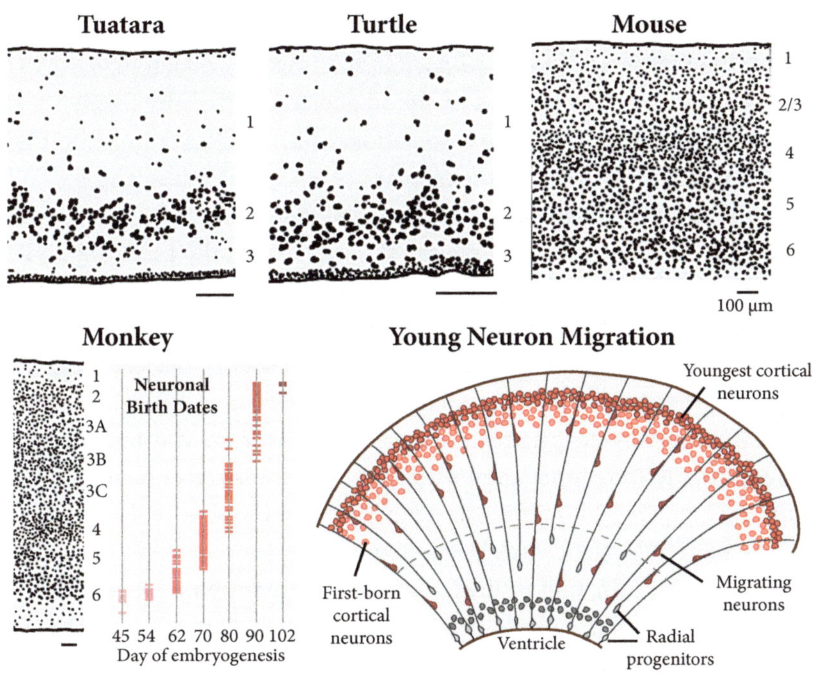

Figure 6.26 Evolution and development of the neocortex. The dorsal cortex of non-avian sauropsids (e.g., tuataras and turtles) has only three layers, with most of the neuronal cell bodies concentrated in layer 2. In contrast, mammalian neocortex contains at least six layers, some of which are further subdivided in primates. Injections of tritiated thymidine on specific days of embryonic development have been used to determine when the neurons in specific cortical layers are born (i.e., stop dividing). These data show that progressively younger (i.e., later born) neurons occupy progressively more superficial cortical layers (bottom left). This means that young neurons, which are born near the ventricular surface, must migrate past older neurons to reach the top of the cortical plate. In this migration, they crawl along the radial processes of radial progenitor cells (bottom right).

Adapted from (Naumann and Laurent, 2017), Rakic (1974), Striedter et al. (2014); monkey cortical layers after Balaram and Kaas (2014).

The internal circuitry of the neocortex varies across cortical areas, but it is generally true that the principal thalamic inputs target layer 4, which then projects mainly to layers 2 and 3, which in turn project mainly to layers 5 and 6. The latter, deep cortical layers then project out of the neocortex, notably back to the thalamus and to the striatum. Several visual, somatosensory, and motor cortical areas have additional descending projections to the optic tectum and to motor regions in the spinal cord and medulla, respectively. It is worth pointing out, however, that these long descending pathways vary considerably across the various mammalian lineages (e.g., Nudo et al., 1995).

Developmentally, a distinguishing feature of mammalian neocortex is that neurons in the upper cortical layers are born later than neurons in the deeper cortical layers (Figure 6.26), which means that young cortical neurons must always migrate past the older neurons. This orderly inside-out pattern of neural development (Angevine and Sidman, 1961; Rakic, 1974; He et al., 2015) is apparently unique to mammalian neocortex and, to some degree, mammalian olfactory cortex (Bayer, 1986). The most lateral component of the neocortex, called the insular cortex, is unusual in that it develops in close association with the claustrum, a non-cortical brain region that lies just deep to the insular cortex. Luis Puelles and his collaborators recently interpreted the claustrum and the insular cortex as comprising the entire lateral pallium, while they assigned the more ventrally positioned olfactory cortex to the ventral pallium (Watson and Puelles, 2016; Puelles et al., 2016; Puelles, 2017). This is a significant departure from the traditional tetrapartite model of pallial organization, which considered the olfactory cortex to be derived from both the lateral and the ventral pallium (see Chapter 4; Wullimann, 2017b). Regardless of the specific models, it is clear that mammalian neocortex (including insular cortex) develops adjacent to the hippocampus medially and the olfactory cortex laterally.

6.7.6.1. The Search for Neocortex Homologs

Given these data, it is difficult to find a homolog of mammalian neocortex in amphibians. The amphibian pallium does not contain any six-layered regions, and there is no indication that any parts of it develop in an inside-out manner. As we noted in Chapter 4, thalamic inputs terminate mainly in the amphibian medial pallium, which is probably homologous to the mammalian hippocampus (and, most likely, the adjacent parahippocampal and entorhinal cortices). A few thalamic axons extend into more dorsal and lateral pallial regions, but these pallial divisions also receive inputs from the olfactory bulb (see Figure 4.29), which the mammalian neocortex does not (except for a small projection to part of the entorhinal cortex; Biella and de Curtis, 2000). Long descending projections emanate exclusively from the medial and olfacto-recipient lateral pallial regions in amphibians (see Figure 4.31).

Nonetheless, a small region between the medial and lateral pallia of amphibians has traditionally been identified as a dorsal pallium and homologized to the mammalian neocortex, mainly on the basis of its relative position. There is little agreement on the precise borders of this area, but if it really is the homolog of mammalian neocortex and represents the primitive condition for amniotes, then it must have expanded enormously along the lineage leading to mammals, picking up a large number of novel features along the way. This expansion of the ancestral dorsal pallium might have been accomplished by restricting the secondary olfactory projections to the lateral and ventral divisions of the pallium (and just a small part of the medial pallium) in early amniotes. As we hypothesized in Chapter 3, such phylogenetic reductions of olfactory dominance over the pallium occurred repeatedly among anamniotes. We suspect that it happened again with the emergence of amniotes and that the restriction of the secondary olfactory projections may have been carried even further in early mammals and birds, such that their olfactory bulbs came to project exclusively to the ventral pallium (Reiner and Karten, 1985; Atoji and Wild, 2014; Puelles et al., 2016).

Assuming that the last common ancestor of all amniotes had some sort of dorsal pallium (for more on this, see Chapter 7), the most likely homolog of neocortex in non-avian sauropsids is the dorsal cortex (excluding its most caudal component, which is probably homologous to part of the mammalian hippocampus; Tosches et al., 2018). In turtles this region receives visual input from a structure that has been called the dorsal lateral geniculate nucleus (LGNd; see Figure 5.29 in Chapter 5) and projects back to it (Hall et al., 1977), supporting the homology hypothesis. However, individual axons from the LGNd in turtles pass through the dorsal cortex tangentially, having entered from its lateral edge. This trajectory is very different from that taken by thalamocortical axons in mammals, and it causes the neurons in the dorsal cortex of turtles to have much larger spatial receptive fields than one observes in mammalian primary visual cortex (V1; Mulligan and Ulinski, 1990). Moreover, in lizards the LGNd projects only to the pallial thickening, which lies lateral to the dorsal cortex (see Figure 5.29; Kenigfest et al., 1997). These observations raise doubts about whether the dorsal cortex of turtles contains a V1 homolog, but they don't necessarily call into question its homology with the neocortex in general. Similarly, the rostral portion of the dorsal cortex receives somatosensory inputs, suggesting it is homologous to mammalian somatosensory cortex (Medina, 2007), but it does not contain a homolog of mammalian auditory cortex. Overall, these observations indicate that the connections of the dorsal pallium have changed considerably as mammals, lizards, and turtles diverged from one another.

The dorsal pallium also changed drastically in the avian lineage, where it is called the Wulst (see Chapter 5). For instance, the rostral Wulst in some bird species has projections to the medulla and upper spinal cord (Wild and Williams, 2000). Because such long descending projections have not been described in non-avian sauropsids, they probably evolved convergently in mammals and birds. An even more impressive example of convergent evolution is the visual Wulst of owls, which bears a striking but deceptive similarity to V1 in mammals, both in terms of internal organization and neuronal response properties (see Figure 5.29 in Chapter 5; Pettigrew, 1979; Liu and Pettigrew, 2003).

Most comparative neurobiologists agree that the dorsal cortex and Wulst are homologous at least to part of the mammalian neocortex, but they have strident debates about the homology of the sauropsid DVR (Karten, 1969; Aboitiz, 1993; Northcutt and Kaas, 1995; Striedter, 1997: Puelles et al., 2017). The observation that the anterior DVR (ADVR) receives sensory inputs from thalamic cell groups that receive input from the midbrain roof (see Chapter 5) suggested that it may be homologous to parts of mammalian neocortex other than V1 (whose homolog was thought to be in the dorsal cortex and Wulst; Karten, 1969). The finding that thalamorecipient neurons in the ADVR project to other ADVR neurons suggested that the former neurons might be homologous to layer 4 neurons in mammalian neocortex, while the latter are homologous to layers 2 and 3 (Figure 6.27). Finally, since the higher order neurons in the ADVR project to the PDVR, which in turn projects to lower brain regions, it was proposed that the entire sensorimotor circuit through the sauropsid DVR is homologous to the canonical cortical circuit (Karten and Shimizu, 1989). This "homologous circuit" hypothesis received some interesting support from a recent comparative molecular study, which found that genes expressed in layer 4 of mammalian neocortex are also expressed in the thalamorecipient neurons of the ADVR (Figure 6.27; Dugas-Ford et al., 2012), while genes expressed in the deep cortical layers are expressed in the PDVR (called arcopallium in birds).

Although the homologous circuit hypothesis is very intriguing, it is not consistent with other data (Montiel et al., 2015). For example, the intra-DVR circuits are much more complex in birds than in most lizards and turtles, implying that at least some new cell types must have evolved in birds (or in the last common ancestor of birds and crocodilians). Another complication is that the PDVR does not project to most of the thalamic cell groups that project to the ADVR (Zeier and Karten, 1971), whereas the mammalian circuit does feature prominent feedback projections to the thalamus. The comparative molecular data are similarly problematic, because the most recent comparative analysis of cellular gene expression patterns suggests cell type homologies between the avian DVR and the mammalian neocortex that differ from earlier hypotheses (Dugas-Ford et al., 2012; Briscoe et al., 2018).

Moreover, several of the selected genes are expressed not only in the targeted cortical layers but also in the claustrum and pallial (basolateral) amygdala (Figure 6.27). Those additional expression domains are consistent with extensive comparative developmental data indicating that the sauropsid DVR is homologous to the ventral and lateral pallium of mammals, which include the claustrum, the endopiriform nucleus, and the pallial amygdala (Figure 6.28). According to this "claustroamygdaloid hypothesis," the ventral pallial division and, to a lesser extent, the lateral pallium expanded much more in sauropsids than in mammals during embryological development (see Figure 5.24; Bruce and Neary, 1995; Striedter, 1997; Medina et al., 2013; Puelles et al., 2017). As they did so, the projections of the olfactory bulb became restricted to the ventral pallium. More importantly, they

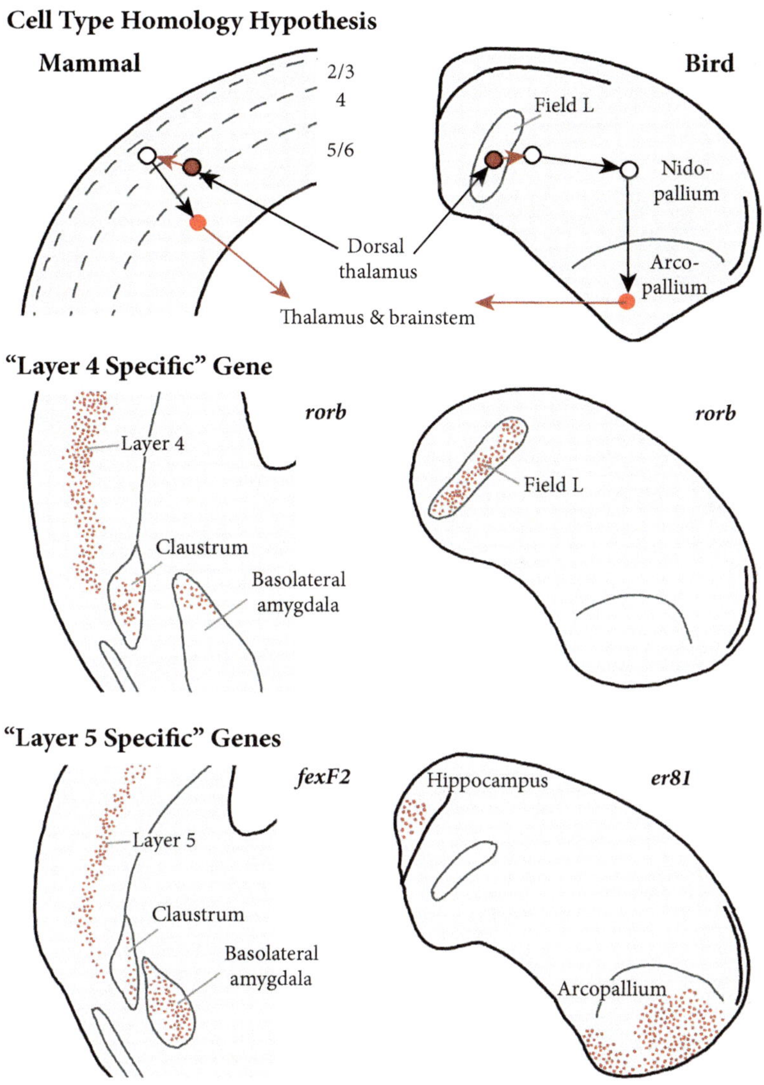

Figure 6.27 Testing cell type homologies. Harvey Karten (1969) proposed that neurons in layer 4 of mammalian neocortex are homologous to thalamorecipient neurons in the avian DVR (e.g., auditory neurons in Field L), and that neurons in layer 5 of mammalian neocortex are homologous to neurons in the avian arcopallium. Dugas-Ford et al. (2012) tested this hypothesis using genes that are differentially expressed in cortical layer 4 (e.g., *rorb*) versus layer 5 (e.g., *fezF2* and *er81*). They reported unequivocal support for Karten's hypothesis. However, as shown here, several of the "layer-specific" genes are also expressed in the claustrum and basolateral amygdala, which is more consistent with the hypothesis that the avian DVR is mainly a ventral pallial derivative.

Adapted from Dugas-Ford et al. (2012) and Dugas-Ford and Ragsdale (2015), with additional data from the Allen Brain Atlas, www.brain-map.org.

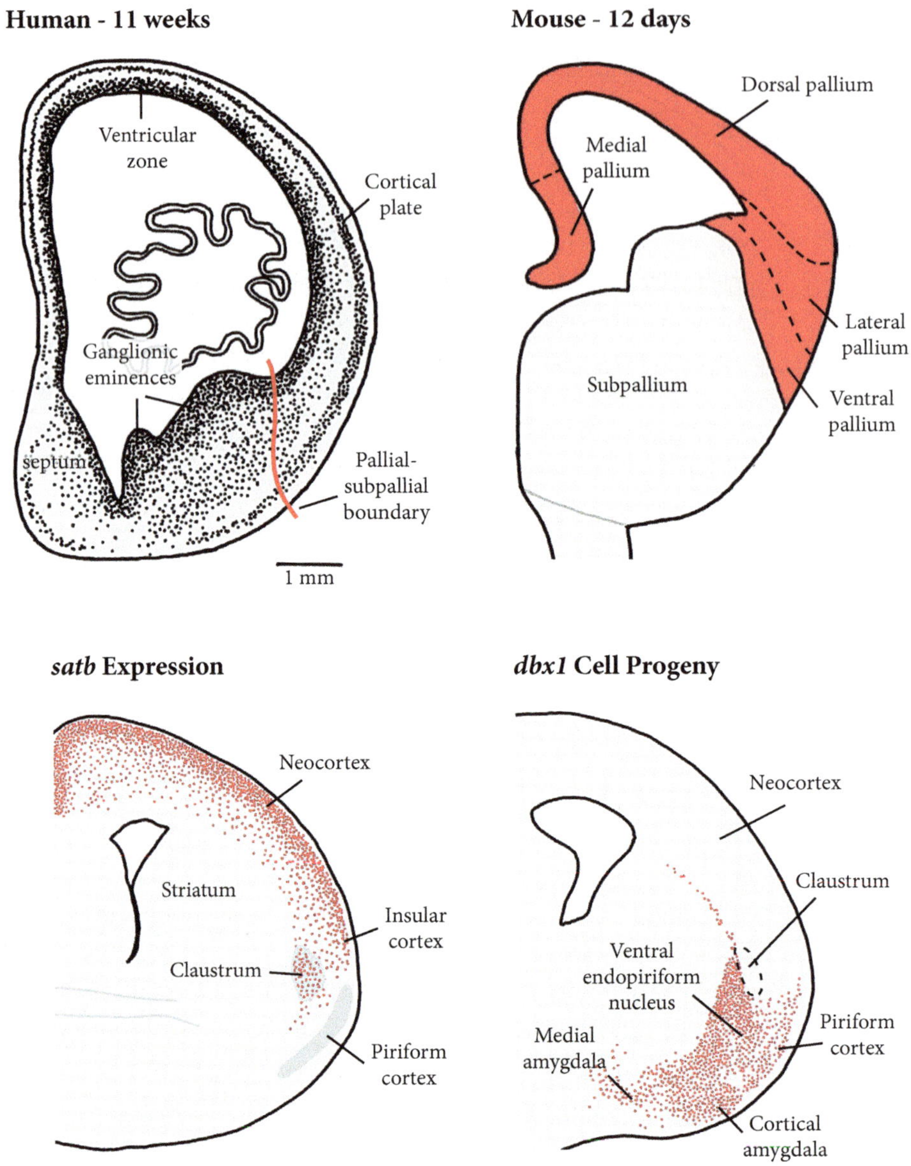

Figure 6.28 Pallial divisions in mammals. At early stages of development, the mammalian telencephalon features several intraventricular ridges (ganglionic eminences) that together with the septum form the subpallium (top left). The pallium features a ventricular zone full of dividing cells and a cortical plate replete with young neurons. The diagram at the top right depicts the four pallial divisions in red. The bottom panels show that the gene *satb* is expressed in the neocortex, insular cortex, and claustrum, thus extending to the ventral border of the lateral pallium (according to Puelles et al., 2017). In contrast, cells derived from *dbx1* expressing cells in the ventral pallium form the piriform cortex, ventral endopiriform nucleus, and pallial amygdala.
Adapted from Nieuwenhuys et al. (2008), Medina (2007), Puelles et al. (2015, 2016).

remained restricted to the brain surface, while the deeper regions elaborated their ability to process other, non-olfactory inputs (see Chapter 5).

How is one to decide between these competing hypotheses of DVR homology? Comparative neurobiologists who believe that neuronal connections and cell types are highly conserved across phylogeny tend to favor the homologous circuit hypothesis, but our broad look at brain evolution, especially forebrain evolution, suggests that neuronal connections are quite variable and that novel cell types must have evolved repeatedly. In contrast, comparative embryologists tend to be convinced that developmental origins are more broadly conserved. Of course, as we have argued in the preceding pages, evolution does sometimes change neural development in fundamental ways, leading for example to the emergence of the midbrain and telencephalon in early vertebrates (see Chapter 2). However, such changes in early brain development seem to be relatively rare (Striedter, 1999; Nieuwenhuys and Puelles, 2016). Therefore, we are more impressed by similarities in brain development than by similarities in adult connectivity. As important as these considerations are, the critical step in any homology analysis is to determine whether the observed similarities, be they connectional or developmental, can be traced back to a common evolutionary origin (see Chapter 1). For such analyses, data on non-avian sauropsids are critical. It is important to note, therefore, that recent single-cell transcriptomics data from lizards and turtles strongly support the hypothesis that the ADVR is homologous to the mammalian claustrum and pallial amygdala (Tosches et al., 2018).

One factor that further inclines us toward the claustroamygdaloid hypothesis is that we are not troubled by the implication that birds and mammals independently evolved very similar telencephalic circuitry. As we have stressed throughout this chapter, birds and mammals are convergent in numerous respects; why should the telencephalon be any different? Another important consideration is that, according to our analysis, the connections and functions of the telencephalon must have changed extensively, even if one disregards the avian data. After all, most of the connections that we associate with the key functions of mammalian neocortex are not associated with the pallium in amphibians or non-avian sauropsids. These derived mammalian features include extensive unimodal inputs from the thalamus, projections from the pallium back to the thalamus, and long descending projections to the midbrain and medulla. If these features evolved once, in the synapsid lineage, why shouldn't they be able to evolve again in the sauropsid lineage? Again, additional data on non-avian sauropsids would be very useful, but the important message is this: the connections of the telencephalon have changed dramatically across phylogeny, regardless of how one views the question of DVR homology.

6.7.6.2. Cortical Expansion and Areal Differentiation

Early crown mammals were very small in body size (Rowe et al., 2011; O'Leary et al., 2013), which means that their brains must also have been small, likely weighing less than 1g (Laaß, 2015). Furthermore, we know that proportional neocortex size increases with absolute brain size (Finlay and Darlington, 1995). This correlation allows us to infer that early mammals must have had a proportionately small neocortex, likely occupying just 10%–15% of the telencephalon's total volume (Stephan et al., 1981; Clark et al., 2001). As mentioned in Section 6.3, absolute brain and body size later increased in some members of most mammalian lineages, leading to a substantial increase in average brain size. As brain size increased, neocortical surface area expanded much more than neocortical thickness, which led to the appearance of neocortical folding when brain size increased beyond about 3 g (Striedter et al., 2014; Mota and Herculano-Houzel, 2015). Thus, the fact that echidnas have a highly folded neocortex (see Figure 6.9) does not reflect the retention of a primitive characteristic (Lewitus et al., 2014) but, instead, an increase in absolute brain size that occurred within the monotreme lineage. Of course, brain size also decreased in some relatively late branches of the mammalian phylogenetic tree, and in those species the degree of neocortical folding was probably reduced (Kelava et al., 2013).

As the neocortex expanded in surface area, the number of distinct cortical areas increased. Comparative studies of various small-brained mammals indicate that the neocortex of the earliest mammals contained primary as well as secondary visual, somatosensory, and auditory areas (Krubitzer et al., 2011; Dooley et al., 2013). Sandwiched between the principal sensory domains were small strips of additional cortex (Figure 6.29). Since marsupials reportedly lack a primary motor cortex, and the evidence for a motor cortex in monotremes is weak, we believe that early mammals did not possess a primary motor cortex (Lende, 1963; Kaas, 2011, 2017). However, these animals probably did possess a handful of small cingulate, perirhinal, and frontal cortical areas. All in all, it has been estimated that the earliest mammals had fewer than 20 neocortical areas (Kaas, 2008). In contrast, the neocortex of strepsirhine primates contains approximately 50 distinct areas (Wong and Kaas, 2010), and human neocortex is estimated to contain at least 180 cortical areas (Glasser et al., 2016). In general, the available data indicate that the number of cortical areas increased in multiple mammalian lineages. This increase in cortical area number probably reflects an increase in absolute brain size, but the neocortex of many large-brained species (e.g., whales and elephants) has not been studied in enough detail to estimate how many cortical areas they might contain (Manger et al., 2013).

Nonetheless, the available data indicate that, as brain size increases, the primary and secondary sensory cortices do not expand as much as the entire neocortex (Kaskan et al., 2005). Instead, the territory between those sensory cortices expands disproportionately (Figure 6.29), and it is there that the additional areas tend to appear in larger cortices. Historically, those new cortical areas have often been regarded as multimodal "association cortex," but most of them are linked to one main sensory modality (Kaas, 1999). Diurnal primates, for example, are thought to have upward of 30 distinct visual areas in their neocortex (Felleman and Van Essen, 1991), and bats seem to have more auditory areas than other mammals (Kössl et al., 2015). In addition, later placental mammals added a number of new premotor

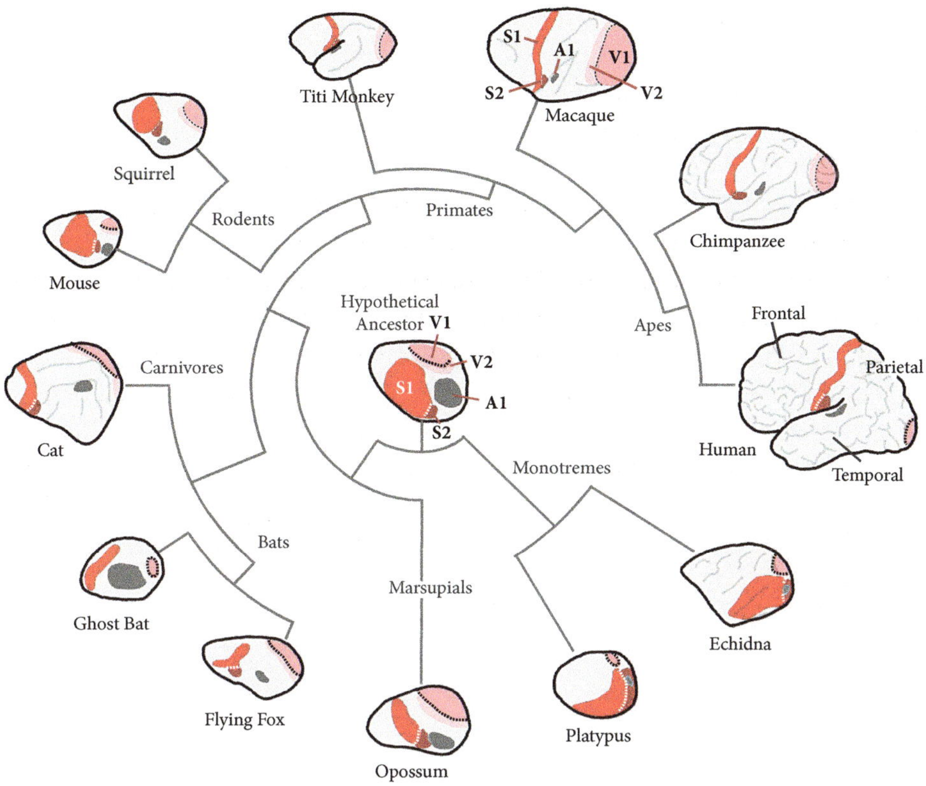

Figure 6.29 Sensory areas of mammalian neocortex. All mammals studied to date exhibit primary visual, auditory, and somatosensory areas (V1, A1, and S1). It is therefore parsimonious to assume that the last common ancestor of all mammals also shared these cortical areas (center). Early mammals probably also possessed small secondary visual and somatosensory areas (V2 and S2), as well as a handful of areas that are not so tightly linked to specific sensory modalities. Importantly, the cortical territories between the principal sensory areas expanded disproportionately in several lineages as absolute brain size increased.

Adapted from Krubitzer and Seelke (2012, courtesy of Leah Krubitzer), Krubitzer and Stolzenberg (2014), Kaas (2017).

areas. Most of these are located in the posterior portion of the frontal lobe, but new areas critical for the control of reaching movements also evolved within primate parietal cortex (Kaas, 2004). Finally, additional cortical areas evolved in the prefrontal cortex of primates (Figure 6.30), which became disproportionately large as primate brains enlarged (Preuss and Goldman-Rakic, 1991; Passingham and Smaers, 2014). Some investigators have argued that an amalgam of these prefrontal areas exists in rodents, but this kind of "field homology" argument does not do justice to their novel anatomical and physiological features (Passingham and Wise, 2012; Wise, 2017).

If additional cortical areas evolved independently in diverse lineages as those taxa expanded their neocortex, then one would expect that some of the added areas

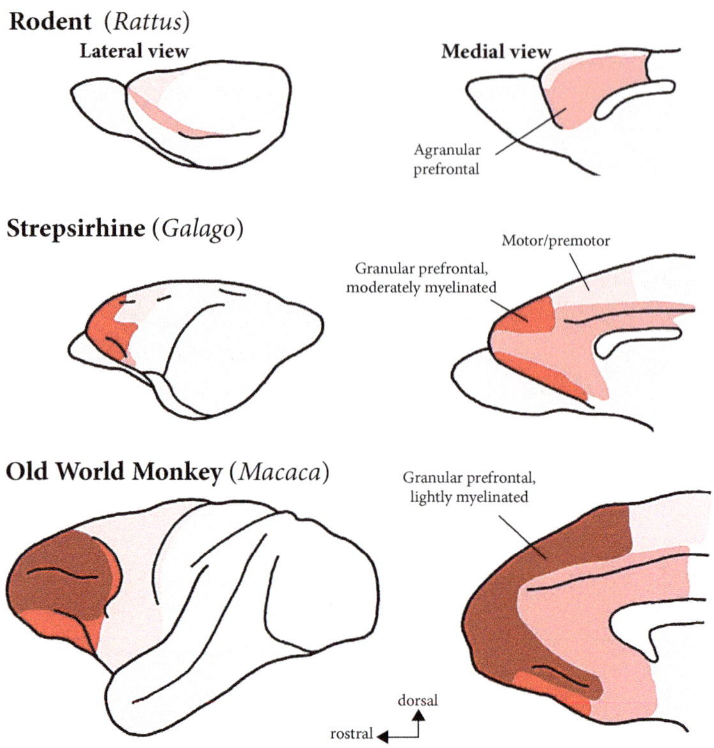

Figure 6.30 Prefrontal cortex expansion in primates. Compared to rodents, primates have a proportionately larger prefrontal cortex that includes not only agranular but also granular cortical areas. Moreover, old world monkeys have lightly myelinated granular prefrontal areas that seem to have no homologs in strepsirhine primates, such as the bushbaby *Galago*.

Adapted from Preuss and Goldman-Rakic (1991), Wu et al. (2000), and Wise (2017).

have no homologs in other lineages, even if they exhibit extensive similarities. In support of this hypothesis, it is difficult, if not impossible, to homologize some of the higher order visual areas between primates and carnivores, which expanded their cortical visual systems independently of one another (Lyon, 2007). The alternative hypothesis posits that early mammals already had more than 20 higher order visual areas, which were retained in all later mammals but have not been identified in small-brained extant mammals. In support of this alternative, recent studies have shown that rats and mice do have more visual cortical areas than previously suspected (Wang and Burkhalter, 2007). However, the spatial arrangement of the 9–10 reported visual areas in mice is different from the layout observed in primates, suggesting that some of the putative cortical areas in mice might be cortical "modules" rather than areas (see Laramé and Boire, 2015). Furthermore, comparative studies that include squirrels, which are more highly visual rodents, indicate that primates have at least a few visual cortical areas that rodents do not possess (Negwer et al., 2017). Additional studies will be needed to determine the extent to which this variation in areal differentiation varies with neocortex size.

The behavioral consequences of neocortex expansion are still uncertain (Lefebvre, 2012). Several studies have shown that proportional neocortex size correlates with social complexity, at least among the haplorhine primates (Dunbar, 1992; Barton, 1996; Shultz and Dunbar, 2007). More recently it has been shown that eating fruits, rather than leaves, is the best predictor of primate brain size, relative to body size (DeCasien et al., 2017). However, the latter study focused on relative brain size, rather than proportional neocortex size, and these two measures do not always correlate tightly. Therefore, the "social brain hypothesis" is far from dead. Nor is it logically incompatible with diet-based or other ecological hypotheses. Indeed, we suspect that many of the cognitive abilities conferred by neocortical expansion would be useful for navigating complex social relationships as well as finding ripe fruit, whose distribution in the wild varies enormously over both space and time (Murray et al., 2016). For that matter, the same cognitive capacities would probably also enhance the ability to stalk and hunt elusive prey.

Finding functional correlates for the addition of specific cortical areas is even more challenging than trying to explain evolutionary changes in the neocortex as a whole. For example, what is the benefit of primates having evolved some new granular prefrontal areas? Lesion studies indicate that these regions are involved in a wide variety of complex cognitive functions related to generation of behavioral goals (Passingham and Wise, 2012). One of those functions is the ability to solve problems by means of clever strategies, rather than slow trial-and-error learning (Murray et al., 2016).

This finding is interesting because primates, especially the large-brained ones, tend to be better than other mammals at solving problems fast. For example, when rodents are given a series of pairwise discrimination tasks, in which they must learn which member of a pair is linked to a reward, they slowly get better at learning the task (i.e., they gradually learn to pick the rewarded item faster for each new pair of objects). Macaque monkeys, in contrast, improve much more rapidly. After learning roughly 200 different object-reward associations and then being presented with a novel object pair, they pick the rewarded item on the second trial more than 70% of the time (Figure 6.31). How do they manage this feat? Most likely the monkeys learn that, if the first trial was not rewarded, they should pick the other object on the second trial; otherwise they should continue picking the object they chose on the first trial. Since learning this win-stay, lose-shift strategy seems to require intact connections between the prefrontal cortex and other cortical areas (Figure 6.31; Browning et al., 2006), it seems reasonable to speculate that the macaques excel at this task at least in part because they had evolved an expanded prefrontal cortex.

Using such data, Murray et al. (2016; see also Wise, 2017) have argued that the evolutionary addition of granular prefrontal areas, in concert with some changes in posterior parietal cortex, enabled anthropoid primates (i.e., monkeys and apes) to forage more effectively for ripe fruits distributed in patches across large territories. This increase in foraging efficiency would have been very beneficial, given that anthropoids feed on volatile resources, notably fruits, and are subject to

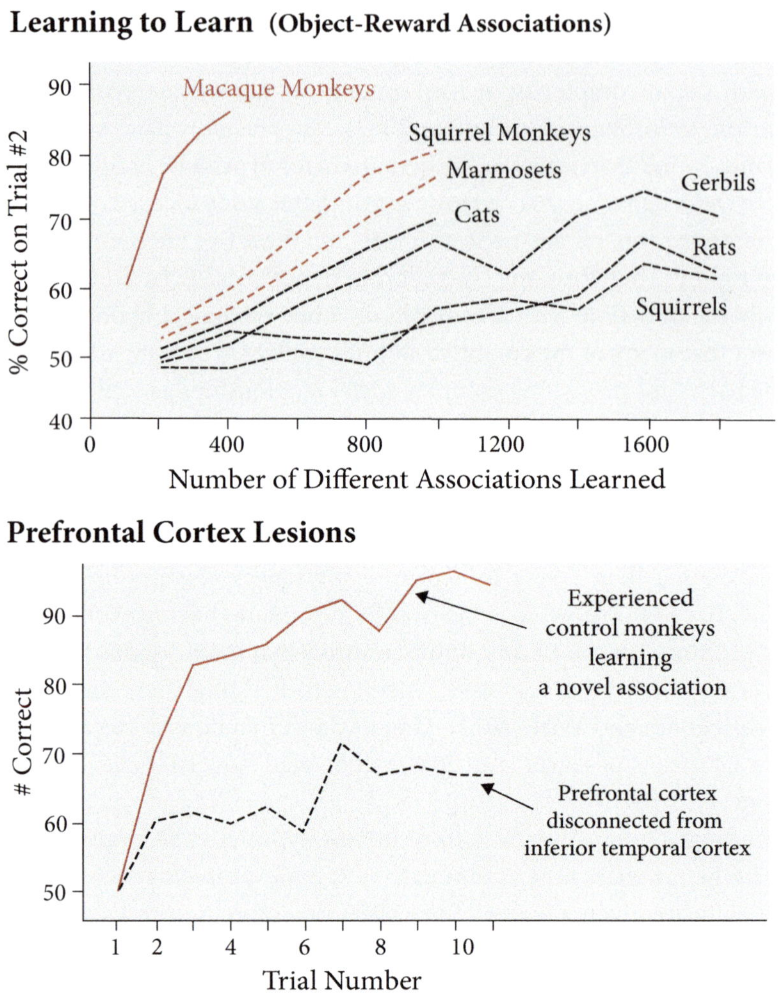

Figure 6.31 Learning set acquisition in macaques. Animals were taught a series of pairwise object discrimination tasks; once they had learned one object-reward association, they were switched to the next one (i.e., a different pair of objects). Macaque monkeys rapidly became faster at learning each new discrimination, often making the correct choice already on the second trial. They probably solved the task by adopting a win-stay, lose-shift strategy. Other mammals, including New World monkeys, took significantly longer to learn this strategy (if they did so at all). The graph on the bottom shows that macaques fail to acquire a learning set when their prefrontal cortex is disconnected from the inferior temporal lobe.

Adapted from Warren (1966), Passingham (1982), and Browning et al. (2006).

substantial predation. We suspect that more efficient foraging was not the only benefit of having an enlarged and more complex prefrontal cortex, but it may well have been the principal factor driving the evolutionary expansion of the prefrontal cortex in anthropoids.

6.7.6.3. The Corpus Callosum: An Innovation of Placental Mammals

Only placental mammals have a corpus callosum, which interconnects neocortical areas on the two sides of the brain by means of reciprocal, excitatory, and largely topographic connections (Figure 6.32; see Suárez et al., 2014). Marsupials

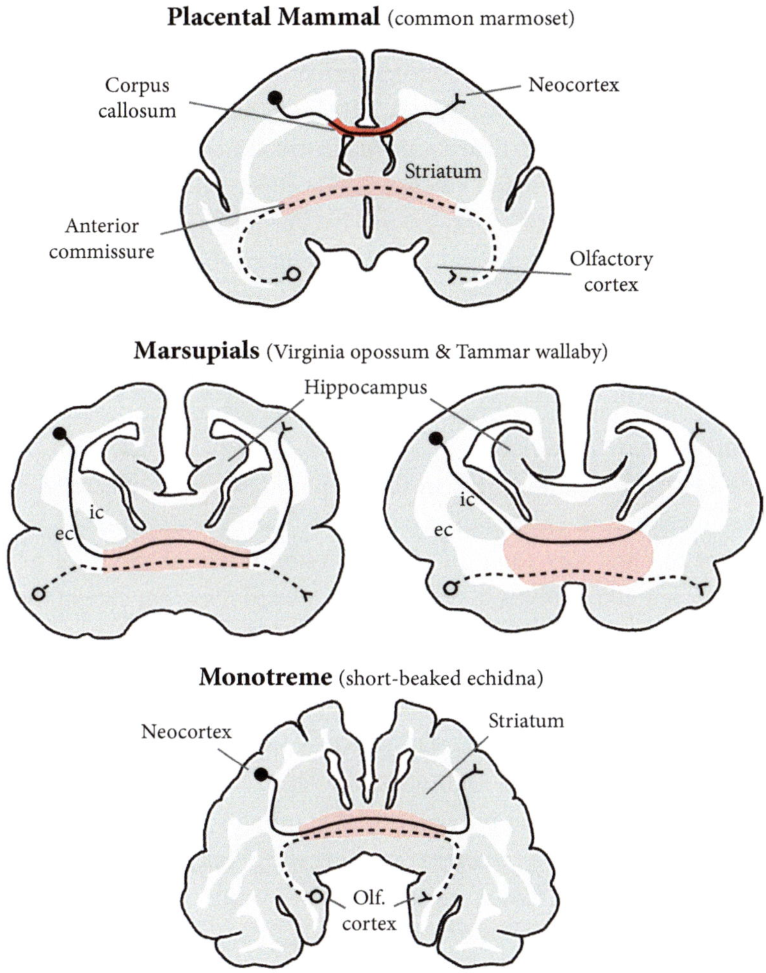

Figure 6.32 Commissural connections in mammals. Only placental mammals (top) have a corpus callosum, through which neocortical neurons project to the contralateral neocortex. Marsupials (middle) do have neocortical neurons that project to the contralateral neocortex, but the axons of these neurons pass through the anterior commissure, which in all mammals carries commissural axons from the olfactory cortex (as well as several other structures; not shown). The commissural neocortical axons take somewhat different routes in polyprotodontid marsupials (e.g., opossums; left) versus diprodontid marsupials (e.g., wallabies; right). Monotremes probably have neocortical neurons that project to the contralateral neocortex through the anterior commissure (bottom), but this hypothesis awaits experimental verification.

Abbreviations: ec – external capsule; ic – internal capsule.

Adapted from Ashwell et al. (2016), with permission from Elsevier.

do have numerous axons that interconnect the left and right neocortices, but these axons cross the midline in the anterior commissure, which is shared among most vertebrates and typically carries midline-crossing axons from the olfactory cortex, the pallial amygdala, the olfactory bulb, and various subpallial regions. The fact that neocortical axons in marsupials course within the anterior commissure, rather than a corpus callosum, explains why the anterior commissure is significantly larger in marsupials than in placental mammals, relative to overall brain size (Ashwell et al., 2016). Because the anterior commissure of monotremes is similar to that of marsupials in relative size (Ashwell et al., 2016), it is reasonable to hypothesize that monotremes also have neocortical axons that cross in the anterior commissure, but we know of no direct, published evidence to that effect. Looking beyond mammals, one finds that commissural connections between the neocortex homologs of sauropsids (i.e., the avian Wulst and the dorsal cortex of lizards and turtles) are either weak (Martínez-Garcia et al., 1990) or lacking entirely (Ulinski, 1990; Letzner et al., 2015). We interpret these data to mean that the dorsal pallium acquired reciprocal commissural connections with the origin of mammals, but that routing these axons through a corpus callosum, rather than the anterior commissure, was an innovation of placental mammals.

Why do neocortical axons cross in the corpus callosum in placental mammals but in the anterior commissure of marsupials? One potential answer is based on the idea that the first-developing ("pioneer") commissural axons of neocortical neurons in placental mammals may be guided toward the other side of the brain by axons of hippocampal neurons that project through the hippocampal commissure, which develops early enough for this idea to be feasible (Ashwell et al., 1996). In monotremes, however, the neocortical commissural axons develop long before the hippocampal commissure, making it impossible for them to use this structure as a preexisting "bridge" across the midline (Mihrshahi, 2006; Suárez, 2016). Another, complementary answer is that only placental mammals possess a corpus callosum because only they develop a transient "midline sling" (aka glial sling) that attracts the pioneer neocortical commissural axons by secreting semaphorin 3C, an guidance molecule (Niquille et al., 2009; Piper et al., 2009). Furthermore, the formation of the corpus callosum requires prior remodeling and intermingling of the glial cells that normally separate the two sides of the brain. Because this glial cell remodeling dorsal and rostral to the hippocampal commissure does not occur in marsupials (Gobius et al., 2017), it would be impossible for the neocortical axons of marsupials to cross in that location (i.e., form a corpus callosum), even if they grew toward the dorsal midline.

What benefit did early placental mammals derive from having their neocortical commissural axons cross in the newly formed corpus callosum, rather than the anterior commissure? One might answer that the corpus callosum is beneficial because it provides a shorter, faster route for axons to connect the two hemispheres (Ringo et al., 1994). This increase in interhemispheric conduction speed would facilitate the fusion of visual and motor representations across the midline, which in turn would enhance central vision and bimanual coordination (see Aboitiz and Montiel, 2003). This argument is attractive, but the stated benefits would mainly accrue to primates with frontally directed eyes and large brains, in which the speed of information transmission between distant brain regions becomes a serious constraint (Phillips et al., 2015). It is more difficult to fathom what the benefits of the corpus callosum might have been in early placental mammals, which probably weighted less than 250 g and had eyes that were directed far more laterally than they are in primates (Heesy, 2004; O'Leary, 2013). Thus, it remains unclear what, if any, benefit early placental mammals derived from their corpus callosum.

In contrast, the emergence of commissural neocortical pathways in the earliest mammals was probably adaptive because these connections (regardless of their route through the forebrain) would have allowed the left and right neocortices to work more closely together. In particular, these commissural connections facilitated the "transfer" of memories between the neocortices on both sides of the brain (Webster, 1975; van der Knaap et al., 2011). Such interhemispheric memory transfer is useful whenever important information preferentially enters one telencephalic hemisphere (e.g., through just one eye or from one-half of the visual field) but must later be used by the other hemisphere. Some non-mammals, notably pigeons, are capable of transferring some forms of memory between the two sides of their brain (Watanabe, 1985) but, as noted previously, birds lack direct commissural connections between their neocortex homologs. In addition to facilitating memory transfer, commissural neocortical projections in early mammals would have helped to synchronize activity patterns in the left and right neocortices (Aboitiz et al., 2003; for an avian solution to the bilateral synchronization problem, see Schmidt et al., 2004). Collectively, these effects would have decreased the relatively ancient tendency for the left and right hemispheres of vertebrates to specialize for different behavioral functions (Bisazza et al., 1998). The tendency toward functional lateralization later increased again, as absolute brain size increased in the various mammalian lineages (Ringo et al., 1994; Wey et al., 2013; Hänggi et al., 2014; Phillips et al., 2015).

6.7.7. The Hippocampus: Old but Modified

The hippocampus is a relatively ancient structure, with widely accepted homologs in all jawed vertebrates. Comparative neuroanatomists often refer to the hippocampus as the medial pallium, because it generally develops at the pallium's medial edge, just dorsal to the septum (Figure 6.33; see also Figure 5.24 in Chapter 5). In mammals, for instance, the hippocampus originally develops in a dorsomedial location, even though in rodents it becomes displaced into the posterior telencephalon, and in primates most of the hippocampus ends up inside the medial temporal lobe. Only in the ray-finned fishes, which exhibit telencephalic eversion rather than

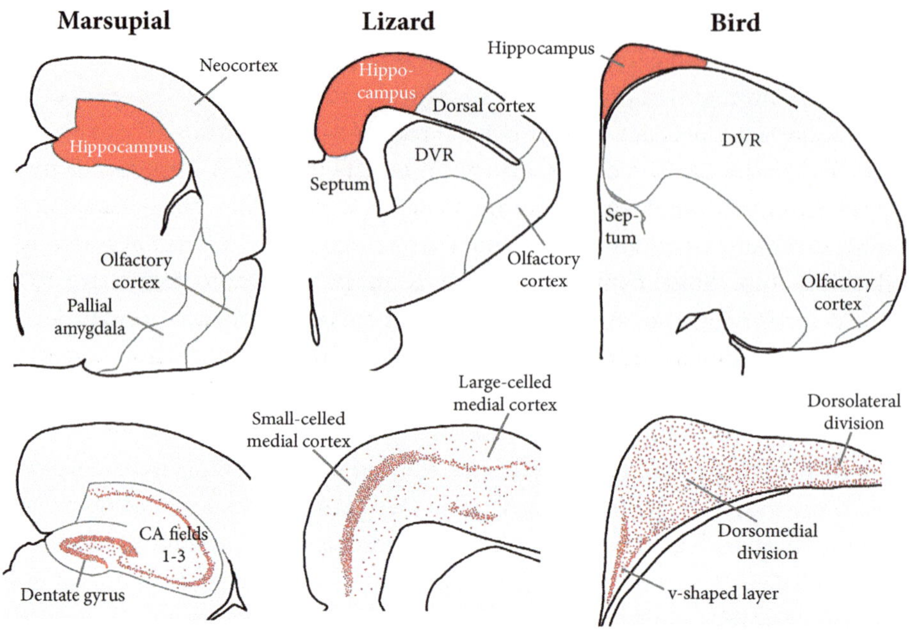

Figure 6.33 The hippocampus of amniotes. The hippocampus (shaded red) occupies a dorsomedial position in the telencephalon of most amniotes. However, the cellular architecture varies considerably between mammals, lizards, and birds (shown here are transverse sections through the hippocampus of a stripe-faced dunnart, a gecko, and a zebra finch). This variation makes it difficult to homologize the various hippocampal divisions across the major amniote lineages.
Adapted from Ashwell (2010), Striedter (2016), Karten et al. (2013).

evagination, does the hippocampus homolog come to occupy a dorsolateral position in adult brains (see Figure 3.27).

Functionally, the hippocampus clearly plays a major role in certain forms of spatial memory and navigation, and this function appears to be broadly conserved. The most detailed studies have been performed in mammals, notably rodents, and these have shown that the hippocampus is necessary for learning spatial relationships, which can then be used as a "cognitive map" to find target locations from novel starting points (O'Keefe and Nadel, 2003; Schiller et al., 2015). Many birds and non-avian sauropsids also need an intact hippocampus for this kind of spatial memory (Rodríguez et al., 2002; Bingman et al., 2017). Even amphibians and teleost fishes use their medial pallium to help them learn spatial relationships (Salas et al., 1996; Sotelo et al., 2016).

Despite this high degree of functional conservation, the cytoarchitecture of the hippocampus homologs differs dramatically between the principal amniote lineages (Figure 6.33). Whereas the mammalian hippocampus contains at least three cytoarchitecturally distinct divisions, including the dentate gyrus and several CA fields, only two divisions have been described in the hippocampus of lizards. The avian hippocampus contains three major divisions, and each of these can be subdivided further (Atoji and Wild, 2004). Given these differences, it is not surprising that attempts to homologize specific subdivisions across amniotes have been contentious.

Especially controversial is whether non-mammals have a homolog of the dentate gyrus (Papp et al., 2007; Atoji et al., 2016). A V-shaped cell layer at the ventromedial edge of the hippocampus in some birds superficially resembles the C-shaped mammalian dentate gyrus (Figure 6.33), but many birds and all non-avian sauropsids lack this feature. Some connectional data do support the homology between the avian V-shaped cell layer and the mammalian dentate gyrus, but other data lead to different conclusions (Székely, 1999; Kahn et al., 2003). The strongest support for the hypothesis that the ventromedial edge of the hippocampus in sauropsids is homologous to the mammalian dentate gyrus comes from the observation that both of these structures express *prox1* during early development (Abellán et al., 2014; Briscoe and Ragsdale, 2018; Tosches et al., 2018). However, if we suppose the dentate gyrus is homologous across amniotes, then we must also admit that the circuits running through the hippocampus must have changed considerably as amniotes diversified (Striedter, 2016).

The connections between the hippocampus and the rest of the brain were also modified considerably during phylogeny. In amphibians, the medial pallium is the principal integrative center of the pallium, receiving multimodal sensory information from the thalamus, combining it with inputs from the olfactory bulb and olfactory cortex (i.e., the lateral and ventral pallium), and ultimately conveying its outputs to a variety of subpallial targets, including the septum, striatopallidal complex, and diencephalon (see Figure 4.31 in Chapter 4). Many of these connections exist also in mammals, but they seem to be of minor functional significance compared with the connections between the hippocampus and neocortex. Specifically, the mammalian hippocampus sits atop a sensory processing hierarchy that passes through a series of neocortical areas (Felleman and Van Essen, 1991), and its projections back to the neocortex are thought to mediate its core functions of long-term memory formation and recall (McKenzie et al., 2015). Thus, information that flowed directly to and from the hippocampus in anamniotes became routed through the neocortex in mammals. Whether this evolutionary re-routing occurred with the origin of mammals or in early amniotes remains unclear, because very little is known about how the sauropsid hippocampus interacts with other brain regions (Shanahan et al., 2013). Regardless, the emergence in mammals of 20–200 distinct neocortical areas, each with its own specialized representations (Murray et al., 2016), surely had profound consequences for the functions of the hippocampus in mammals.

If the internal structure and connections of the hippocampus have changed so much over evolutionary time, how can its role in spatial memory and navigation be so well conserved? The problem with trying to answer this question is that we know almost nothing about what other functions, in addition to spatial memory, the hippocampus might perform in non-mammals. It is clearly needed for "episodic memory" in primates and rodents, and some authors have claimed that birds are capable of storing "episodic-like" memories (Clayton and Dickinson, 1998; Eichenbaum et al., 2005). However, the term "episodic-like" probably obscures potentially important species differences (Suddendorf and Corballis, 2010) and, in any case, whether birds need an intact hippocampus for their "episodic-like" memories remains unknown. Indeed, as far as we know, all research on the functions of the hippocampus in non-mammalian vertebrates has focused exclusively on spatial memory. We strongly suspect that the functions of the hippocampus changed substantially, perhaps by adding new functions, during the course of vertebrate phylogeny (e.g., Danjo et al., 2018; Omer et al., 2018), but the lack of relevant data in non-mammals makes this hypothesis speculative for now.

References

Abellán A, Desfilis E, Medina L. 2014. Combinatorial expression of *Lef1, Lhx2, Lhx5, Lhx9, Lmo3, Lmo4*, and *Prox1* helps to identify comparable subdivisions in the developing hippocampal formation of mouse and chicken. *Front Neuroanat* 8:59.

Aboitiz F. 1993. Further comments on the evolutionary origin of the mammalian brain. *Med Hypotheses* 41:409–418.

Aboitiz F, López J, Montiel J. 2003. Long distance communication in the human brain: timing constraints for inter-hemispheric synchrony and the origin of brain lateralization. *Biol Res* 36:89–99.

Aboitiz F, Montiel J. 2003. One hundred million years of interhemispheric communication: the history of the corpus callosum. *Braz J Med Biol Res* 36:409–420.

Alexander GE, DeLong MR, Strick PL. 1986. Parallel organization of functionally segregated circuits linking basal ganglia and cortex. *Annu Rev Neurosci* 9:357–381.

Alvarez LW, Alvarez W, Asaro F, Michel HV. 1980. Extraterrestrial cause for the Cretaceous-Tertiary extinction. *Science* 208:1095–1108.

Andrews P. 2015. *An Ape's View of Human Evolution*. Cambridge: Cambridge University Press.

Angevine JB, Sidman RL. 1961. Autoradiographic study of cell migration during histogenesis of cerebral cortex in the mouse. *Nature* 192:766–768.

Argyropoulos G, Harper M-E. 2002. Uncoupling proteins and thermoregulation. *J Appl Physiol* 92:2187–2198.

Arkley K, Tiktak GP, Breakell V, Prescott TJ, Grant RA. 2017. Whisker touch guides canopy exploration in a nocturnal, arboreal rodent, the Hazel dormouse (*Muscardinus avellanarius*). *J Comp Physiol A* 203:133–142.

Ashmore J. 2008. Cochlear outer hair cell motility. *Physiol Rev* 88:173–210.

Ashwell KWS. 2008. Encephalization of Australian and New Guinean marsupials. *Brain Behav Evol* 71:181–199.

Ashwell K. 2010. *The Neurobiology of Australian Marsupials*. Cambridge: Cambridge University Press.

Ashwell K. 2013. *Neurobiology of Monotremes*. Clayton South, Australia: CSIRO.

Ashwell KWS. 2016. Anterior commissure versus corpus callosum: a quantitative comparison across mammals. *Zoology* 119:126–136.

Ashwell KW, Waite PM, Marotte L. 1996. Ontogeny of the projection tracts and commissural fibres in the forebrain of the tammar wallaby (*Macropus eugenii*): timing in comparison with other mammals. *Brain Behav Evol* 47:8–22.

Ashwell KWS, Paxinos G. 2005. Cyto- and chemoarchitecture of the dorsal thalamus of the monotreme *Tachyglossus aculeatus*, the short beaked echidna. *J Chem Neuroanat* 30:161–183.

Atoji Y, Sarkar S, Wild JM. 2016. Proposed homology of the dorsomedial subdivision and V-shaped layer of the avian hippocampus to Ammon's horn and dentate gyrus, respectively. *Hippocampus* 26:1608–1617.

Atoji Y, Wild JM. 2004. Fiber connections of the hippocampal formation and septum and subdivisions of the hippocampal formation in the pigeon as revealed by tract tracing and kainic acid lesions. *J Comp Neurol* 475:426–461.

Atoji Y, Wild JM. 2014. Efferent and afferent connections of the olfactory bulb and prepiriform cortex in the pigeon (*Columba livia*). *J Comp Neurol* 522:1728–1752.

Azevedo FAC, Carvalho LRB, Grinberg LT, Farfel JM, Ferretti REL, Leite REP, Jacob Filho W, Lent R, Herculano-Houzel S. 2009. Equal numbers of neuronal and nonneuronal cells make the human brain an isometrically scaled-up primate brain. *J Comp Neurol* 513:532–541.

Balaram P, Kaas JH. 2014. Towards a unified scheme of cortical lamination for primary visual cortex across primates: insights from NeuN and VGLUT2 immunoreactivity. *Front Neuroanat* 8:81.

Baldwin MKL, Balaram P, Kaas JH. 2017. The evolution and functions of nuclei of the visual pulvinar in primates. *J Comp Neurol* 525:3207–3226.

Baron G, Stephan H, Frahm HD. 1996. *Comparative Neurobiology in Chiroptera*. Basel: Birkhauser.

Barrowclough GF, Cracraft J, Klicka J, Zink RM. 2016. How many kinds of birds are there and why does it matter? *PLoS ONE* 11:e0166307.

Barton RA. 1996. Neocortex size and behavioural ecology in primates. *Proc Roy Soc Lond B* 263:173–177.

Barton RA, Venditti C. 2014. Rapid evolution of the cerebellum in humans and other great apes. *Curr Biol* 24:2440–2444.

Bayer SA. 1986. Neurogenesis in the rat primary olfactory cortex. *Int J Dev Neurosci* 4:251–271.

Bennett AF, Ruben JA. 1979. Endothermy and activity in vertebrates. *Science* 206:649–654.

Benoit J, Abdala F, Van den Brandt MJ, Manger PR, Rubidge BS. 2015. Physiological implications of the abnormal absence of the parietal foramen in a late Permian cynodont (Therapsida). *Naturwissensch* 102:501–4.

Benoit J, Manger PR, Rubidge BS. 2016. Palaeoneurological clues to the evolution of defining mammalian soft tissue traits. *Sci Rep* 6:25604.

Benson-Amram S, Dantzer B, Stricker G, Swanson EM, Holekamp KE. 2016. Brain size predicts problem-solving ability in mammalian carnivores. *Proc Natl Acad Sci USA* 113:2532–2537.

Berg DJ, Kan JY, Marino RA, Itti L, Munoz DP, White BJ. 2017. Superior colliculus neurons encode a visual saliency map during free viewing of natural dynamic video. *Nat Commun* 8:14263.

Berglund AM, Brown MC. 1994. Central trajectories of type II spiral ganglion cells from various cochlear regions in mice. *Hear Res* 75:121–130.

Berkhoudt H. 1979. The morphology and distribution of cutaneous mechanoreceptors (Herbst and Grandry corpuscles) in bill and tongue of the mallard (*Anas platyrhynchos* L.). *Neth J Zool* 30:1–34.

Berner RA. 2006. GEOCARBSULF: A combined model for Phanerozoic atmospheric O_2 and CO_2. *Geochim Cosmochim Acta* 70:5653–5664.

Beurg M, Tan X, Fettiplace R. 2013. A prestin motor in chicken auditory hair cells: active force generation in a nonmammalian species. *Neuron* 79:69–81.

Bicudo J, Vianna CR, Chaui-Berlinck JG. 2001. Thermogenesis in birds. *Biosci Rep* 21:181–188.

Biella G, de Curtis M. 2000. Olfactory inputs activate the medial entorhinal cortex via the hippocampus. *J Neurophysiol* 83:1924–1931.

Biewener AA, Soghikian GW, Crompton AW. 1985. Regulation of respiratory airflow during panting and feeding in the dog. *Respir Physiol* 61:185–195.

Bingman VP, Rodriguez F, Salas C. 2017. The hippocampus of nonmammalian vertebrates. In: Kaas JH, Striedter GF (eds.), *Evolution of Nervous Systems*, 2nd ed. Amsterdam: Elsevier, pp. 479–489.

Bisazza A, Rogers LJ, Vallortigara G. 1998. The origins of cerebral asymmetry: a review of evidence of behavioural and brain lateralization in fishes, reptiles and amphibians. *Neurosci & Biobehav Rev* 22:411–426.

Blob R. 2001. Evolution of hindlimb posture in nonmammalian therapsids: biomechanical tests of paleontological hypotheses. *Paleobiology* 27:14–38.
Boddy AM, McGowen MR, Sherwood CC, Grossman LI, Goodman M, Wildman DE. 2012. Comparative analysis of encephalization in mammals reveals relaxed constraints on anthropoid primate and cetacean brain scaling. *J Evol Biol* 25:981–994.
Bowmaker JK. 2008. Evolution of vertebrate visual pigments. *Vision Res* 48:2022–2041.
Brannen P. 2017. *The Ends of the World*. New York: HarperCollins.
Brimblecombe KR, Cragg SJ. 2016. The striosome and matrix compartments of the striatum: a path through the labyrinth from neurochemistry toward function. *ACS Chem Neurosci* 8:235–242.
Briscoe SD, Albertin CB, Rowell JJ, Ragsdale CW. 2018. Neocortical association cell types in the forebrain of birds and alligators. *Curr Biol* 28:1–11.
Brodmann K. 1999. Brodmann's Localisation in the Cerebral Cortex. World Scientific Publishing Company.
Brown RE, Fedde MR. 1993. Airflow sensors in the avian wing. *J Exp Biol* 179:13–30.
Browning PGF, Easton A, Gaffan D. 2006. Frontal-temporal disconnection abolishes object discrimination learning set in macaque monkeys. *Cereb Cortex* 17:859–864.
Bruce LL, Neary TJ. 1995. The limbic system of tetrapods: a comparative analysis of cortical and amygdalar populations. *Brain Behav Evol* 46:224–234.
Brykczynska U, Tzika AC, Rodriguez I, Milinkovitch MC. 2013. Contrasted evolution of the vomeronasal receptor repertoires in mammals and squamate peptiles. *Genome Biol Evol* 5:389–401.
Butler, AB. 1994. The evolution of the dorsal thalamus of jawed vertebrates, including mammals: cladistic analysis and a new hypothesis. *Brain Res Rev* 19:29–65.
Butler, AB. 1995. The dorsal thalamus of jawed vertebrates: a comparative viewpoint. *Brain Behav Evol* 46:209–223.
Butler RJ, Barrett PM, Gower DJ. 2012. Reassessment of the evidence for postcranial skeletal pneumaticity in triassic archosaurs, and the early evolution of the avian respiratory system. *PLoS ONE* 7:e34094.
Butts T, Green MJ, Wingate RJT. 2014. Development of the cerebellum: simple steps to make a "little brain." *Development* 141:4031–4041.
Carr CE, Soares D. 2002. Evolutionary convergence and shared computational principles in the auditory system. *Brain Behav Evol* 59:294–311.
Carrier DR. 1987. The evolution of locomotor stamina in tetrapods: circumventing a mechanical constraint. *Paleobiology* 13:326–341.
Cartmill M. 1992. New views on primate origins. *Evol Anthropol* 1:105–111.
Changizi MA, Zhang Q, Shimojo S. 2006. Bare skin, blood and the evolution of primate colour vision. *Biol Lett* 2:217–221.
Chuong C-M, Wu P, Zhang F-C, Xu X, Yu M, Widelitz RB, Jiang T-X, Hou L. 2003. Adaptation to the sky: defining the feather with integument fossils from mesozoic China and experimental evidence from molecular laboratories. *J Exp Zool B* 298:42–56.
Clark DA, Mitra PP, Wang SS. 2001. Scalable architecture in mammalian brains. *Nature* 411:189–193.
Clarke A, Pörtner H-O. 2010. Temperature, metabolic power and the evolution of endothermy. *Biol Rev* 85:703–727.
Clayton NS, Dickinson A. 1998. Episodic-like memory during cache recovery by scrub jays. *Nature* 395:272–274.
Cowing JA, Arrese CA, Davies WL, Beazley LD, Hunt DM. 2008. Cone visual pigments in two marsupial species: the fat-tailed dunnart (*Sminthopsis crassicaudata*) and the honey possum (*Tarsipes rostratus*). *Proc Roy Soc Lond B* 275:1491–1499.
Crole MR, Soley JT. 2017. Bony pits in the ostrich (*Struthio camelus*) and emu (*Dromaius novaehollandiae*) bill tip. *Anat Rec* 300:1705–1715.
Crompton AW, Musinsky C. 2015. Evolution of the mammalian nose. In: Dial KP, Shubin N, Brainerd EL (eds.), *Great Transformations in Vertebrate Evolution*. Chicago: University of Chicago Press, pp. 189–203.

Crompton AW, Parker P. 1978. Evolution of the mammalian masticatory apparatus. *Am Sci* 66:192–201.

Cunningham SJ, Alley MR, Castro I. 2010. Facial bristle feather histology and morphology in New Zealand birds: Implications for function. *J Morphol* 272:118–128.

Cunningham SJ, Castro I, Jensen T, Potter MA. 2010. Remote touch prey-detection by Madagascar crested ibises *Lophotibis cristata urschi*. *J Avian Biol* 41:350–353.

Cunningham SJ, Castro I, Potter MA. 2009. The relative importance of olfaction and remote touch in prey detection by North Island brown kiwis. *Anim Behav* 78:899–905.

Cunningham SJ, Corfield JR, Iwaniuk AN, Castro I, Alley MR, Birkhead TR, Parsons S. 2013. The anatomy of the bill tip of kiwi and associated somatosensory regions of the brain: comparisons with shorebirds. *PLoS ONE* 8:e80036.

Dallos P, Fakler B. 2002. Prestin, a new type of motor protein. *Nat Rev Mol Cell Biol* 3:104–111.

Danjo T, Toyoizumi T, Fujisawa S. 2018. Spatial representations of self and other in the hippocampus. *Science* 359:213–218.

Davies WIL, Collin SP, Hunt DM. 2012. Molecular ecology and adaptation of visual photopigments in craniates. *Mol Ecol* 21:3121–3158.

Deaner RO, Isler K, Burkart J, Van Schaik C. 2007. Overall brain size, and not encephalization quotient, best predicts cognitive ability across non-human primates. *Brain Behav Evol* 70:115–124.

DeCasien AR, Williams SA, Higham JP. 2017. Primate brain size is predicted by diet but not sociality. *Nature Ecol Evol* 1:0112.

Demes B, Larson SG, Stern JT, Jungers WL. 1994. The kinetics of primate quadrupedalism. *J Hum Evol* 26:353–374.

Di-Poï N, Milinkovitch MC. 2013. Crocodylians evolved scattered multi-sensory micro-organs. *EvoDevo* 4:19.

Díaz C, Yanes C, Trujillo CM, Puelles L. 1994. The lacertidian reticular thalamic nucleus projects topographically upon the dorsal thalamus: experimental study in *Gallotia galloti*. *J Comp Neurol* 343:193–208.

Dominy NJ, Svenning J-C, Li W-H. 2003. Historical contingency in the evolution of primate color vision. *J Hum Evol* 44:25–45.

Dooley JC, Franca JOG, Seelke AMH, Cooke DF, Krubitzer LA. 2013. A connection to the past: *Monodelphis domestica* provides insight into the organization and connectivity of the brains of early mammals. *J Comp Neurol* 521:3877–3897.

Doron NN, LeDoux JE. 1999. Organization of projections to the lateral amygdala from auditory and visual areas of the thalamus in the rat. *J Comp Neurol* 412:383–409.

Dowling JE. 1968. Synaptic organization of the frog retina: an electron microscopic analysis comparing the retinas of frogs and primates. *Proc Roy Soc Lond B* 170:205–228.

Dubin MW. 1970. The inner plexiform layer of the vertebrate retina: a quantitative and comparative electron microscopic analysis. *J Comp Neurol* 140:479–505.

Dugas-Ford J, Ragsdale CW. 2015. Levels of homology and the problem of neocortex. *Annu Rev Neurosci* 38:351–368.

Dugas-Ford J, Rowell JJ, Ragsdale CW. 2012. Cell-type homologies and the origins of the neocortex. *Proc Natl Acad Sci USA* 109:16974–16979.

Dunbar RIM. 1992. Neocortex size as a constraint on group size in primates. *J Hum Evol* 22:469–493.

Ebbesson SO. 1980. The parcellation theory and its relation to interspecific variability in brain organization, evolutionary and ontogenetic development, and neuronal plasticity. *Cell Tissue Res* 213:179–212.

Ebeling W, Natoli RC, Hemmi JM. 2010. Diversity of color vision: not all Australian marsupials are trichromatic. *PLoS ONE* 5:e14231.

Eichenbaum H, Fortin NJ, Ergorul C, Wright SP, Agster KL. 2005. Episodic recollection in animals: "If it walks like a duck and quacks like a duck . . . ." *Learn Motiv* 36:190–207.

Emery NJ, Clayton NS. 2004. The mentality of crows: convergent evolution of intelligence in corvids and apes. *Science* 306:1903–1907.

Emes RD, Beatson SA, Ponting CP, Goodstadt L. 2004. Evolution and comparative genomics of odorant- and pheromone-associated genes in rodents. *Genome Res* 14:591–602.

Farmer CG, Sanders K. 2010. Unidirectional airflow in the lungs of alligators. *Science* 327:338–340.

Farries MA, Ding L, Perkel DJ. 2005. Evidence for "direct" and "indirect" pathways through the song system basal ganglia. *J Comp Neurol* 484:93–104.

Feix T, Kivell TL, Pouydebat E, Dollar AM. 2015. Estimating thumb-index finger precision grip and manipulation potential in extant and fossil primates. *J R Soc Interface* 12:20150176.

Felleman DJ, Van Essen DC. 1991. Distributed hierarchical processing in the primate cerebral cortex. *Cereb Cortex* 1:1–47.

Ferrara CR, Vogt RC, Sousa-Lima RS. 2013. Turtle vocalizations as the first evidence of posthatching parental care in chelonians. *J Comp Psychol* 127:24–32.

Finlay BL, Darlington RB. 1995. Linked regularities in the development and evolution of mammalian brains. *Science* 268:1578–1584.

Fish FE, Frappell PB, Baudinette RV, MacFarlane PM. 2001. Energetics of terrestrial locomotion of the platypus *Ornithorhynchus anatinus*. *J Exp Biol* 204:797–803.

Foley NM, Springer MS, Teeling EC. 2016. Mammal madness: is the mammal tree of life not yet resolved? *Philos Trans R Soc Lond B* 371:20150140–11.

Freedman SL, Feirabend HK, Vielvoye GJ, Voogd J. 1975. Re-examination of the ponto-cerebellar projection in the adult white leghorn (*Gallus domesticus*). *Acta Morphol Neerl Scand* 13:236–238.

Friedman M, Sallan LC. 2012. Five hundred million years of extinction and recovery: a Phanerozoic survey of large-scale diversity patterns in fishes. *Palaeontol* 55:707–742.

Gale SD, Perkel DJ. 2010. Anatomy of a songbird basal ganglia circuit essential for vocal learning and plasticity. *J Chem Neuroanat* 39:124–131.

Gale SD, Person AL, Perkel DJ. 2008. A novel basal ganglia pathway forms a loop linking a vocal learning circuit with its dopaminergic input. *J Comp Neurol* 508:824–839.

Gilad Y, Man O, Pääbo S, Lancet D. 2003. Human specific loss of olfactory receptor genes. *Proc Natl Acad Sci USA* 100:3324–3327.

Giménez-Amaya JM, Mcfarland NR, las Heras de S, Haber SN. 1995. Organization of thalamic projections to the ventral striatum in the primate. *J Comp Neurol* 354:127–149.

Glasser MF, Coalson TS, Robinson EC, Hacker CD, Harwell J, Yacoub E, Ugurbil K, Andersson J, Beckmann CF, Jenkinson M, Smith SM, Van Essen DC. 2016. A multi-modal parcellation of human cerebral cortex. *Nature* 536:171–178.

Glickstein M, Oberdick J, Voogd J. 2007. Evolution of the Cerebellum. In: Kaas JH (ed.), *Evolution of Nervous Systems*. San Diego, CA: Academic Press, pp. 413–442.

Gobius I, Suárez R, Morcom L, Paolino A, Edwards TJ, Kozulin P, Richards LJ. 2017. Astroglial-mediated remodeling of the interhemispheric midline during telencephalic development is exclusive to eutherian mammals. *Neural Dev* 12:9.

Goodman DC. 1969. Behavioral aspects of cerebellar stimulation and ablation in the frog and alligator and their relationship to cerebellar evolution. In: Llinás R (ed.), *Neurobiology of Cerebellar Evolution and Development*. Chicago: American Medical Association, pp. 467–479.

Gottschaldt KM. 1985. Structure and function of avian somatosensory receptors. In: King AS, McLelland J (eds.), *Form and Function in Birds*, Vol. 3. London: Academic Press, pp. 375–461.

Grant RA, Haidarliu S, Kennerley NJ, Prescott TJ. 2013. The evolution of active vibrissal sensing in mammals: evidence from vibrissal musculature and function in the marsupial opossum *Monodelphis domestica*. *J Exp Biol* 216:3483–3494.

Graybiel AM, Ragsdale CW. 1978. Histochemically distinct compartments in the striatum of human, monkeys, and cat demonstrated by acetylthiocholinesterase staining. *Proc Natl Acad Sci USA* 75:5723–5726.

Grigg GC, Beard LA, Augee ML. 2004. The evolution of endothermy and its diversity in mammals and birds. *Physiol Biochem Zool* 77:982–997.

Grothe B, Carr CE, Casseday JH, Fritzsch B, Köppl C. 2004. The evolution of central pathways and their neural processing patterns. In: Manley GA, Fay RR (eds.), *Evolution of the Vertebrate Auditory System*. New York: Springer, pp. 289–359.

Guirado S, Real MÁ, Dávila JC. 2005. The ascending tectofugal visual system in amniotes: new insights. *Brain Res Bull* 66:290–296.

Hackett SJ, Kimball RT, Reddy S, Bowie RCK, Braun EL, Braun MJ, Chojnowski JL, Cox WA, Han K-L, Harshman J, et al. 2008. A phylogenomic study of birds reveals their evolutionary history. *Science* 320:1763–1768.

Hall JA, Foster RE, Ebner FF, Hall WC. 1977. Visual cortex in a reptile, the turtle (*Pseudemys scripta* and *Chrysemys picta*). *Brain Res* 130:197–216.

Hänggi J, Fövenyi L, Liem F, Meyer M, Jäncke L. 2014. The hypothesis of neuronal interconnectivity as a function of brain size-a general organization principle of the human connectome. *Front Hum Neurosci* 8:915.

Harlow P, Grigg G. 1984. Shivering thermogenesis in a brooding diamond python, *Python spilotes spilotes*. Copeia 4:959–965.

Hayward JS, Lisson PA. 1992. Evolution of brown fat: its absence in marsupials and monotremes. *Can J Zool* 70:171–179.

He S, Li Z, Ge S, Yu Y-C, Shi S-H. 2015. Inside-out radial migration facilitates lineage-dependent neocortical microcircuit assembly. *Neuron* 86:1159–1166.

Healy SD, Rowe C. 2007. A critique of comparative studies of brain size. *Proc Roy Soc Lond B* 274:453–464.

Heesy CP. 2004. On the relationship between orbit orientation and binocular visual field overlap in mammals. *Anat Rec* 281:1104–1110.

Heesy CP, Hall MI. 2010. The nocturnal bottleneck and the evolution of mammalian vision. *Brain Behav Evol* 75:195–203.

Heinrich B. 1972. Patterns of endothermy in bumblebee queens, drones and workers. *J Comp Physiol A* 77:65–79.

Herculano-Houzel S, Avelino-de-Souza K, Neves K, Porfírio J, Messeder D, Mattos Feijó L, Maldonado J, Manger PR. 2014. The elephant brain in numbers. *Front Neuroanat* 8:46.

Herculano-Houzel S, Collins CE, Wong P, Kaas JH. 2007. Cellular scaling rules for primate brains. *Proc Natl Acad Sci USA* 104:3562–3567.

Herculano-Houzel S, Ribeiro P, Campos L, Valotta da Silva A, Torres LB, Catania KC, Kaas JH. 2011. Updated neuronal scaling rules for the brains of Glires (rodents/lagomorphs). *Brain Behav Evol* 78:302–314.

Heritage S. 2014. Modeling olfactory bulb evolution through primate phylogeny. *PLoS ONE* 9:e113904.

Hillenius WJ, Ruben JA. 2004. The evolution of endothermy in terrestrial vertebrates: Who? When? Why? *Physiol Biochem Zool* 77:1019–1042.

Hirasawa T, Kuratani S. 2013. A new scenario of the evolutionary derivation of the mammalian diaphragm from shoulder muscles. *J Anat* 222:504–517.

Hoffmann JN, Montag AG, Dominy NJ. 2004. Meissner corpuscles and somatosensory acuity: the prehensile appendages of primates and elephants. *Anat Rec* 281A:1138–1147.

Huguenard JR, McCormick DA. 2007. Thalamic synchrony and dynamic regulation of global forebrain oscillations. *Trends Neurosci* 30:350–356.

Hunt DM, Peichl L. 2014. S cones: evolution, retinal distribution, development, and spectral sensitivity. *Vis Neurosci* 31:115–138.

Isler K, van Schaik CP. 2006. Metabolic costs of brain size evolution. *Biol Lett* 2:557–560.

Jacobs GH. 1993. The distribution and nature of colour vision among the mammals. *Biol Rev* 68:413–471.

Jacobs GH. 2008. Primate color vision: a comparative perspective. *Vis Neurosci* 25:619–633.

Jenkins FA. 1971. Limb posture and locomotion in the Virginia opossum (*Didelphis marsupialis*) and in other non-cursorial mammals. *J Zool Lond* 165:303–315.

Ji Q, Luo Z-X, Yuan CX, Tabrum AR. 2006. A swimming mammaliaform from the Middle Jurassic and ecomorphological diversification of early mammals. *Science* 311:1123–1127.

Jiao Y, Medina L, Veenman CL, Toledo C, Puelles L, Reiner A. 2000. Identification of the anterior nucleus of the ansa lenticularis in birds as the homolog of the mammalian subthalamic nucleus. *J Neurosci* 20:6998–7010.

Jones EG. 2007. *The Thalamus*, 2nd ed. Cambridge: Cambridge University Press.

Jones KE, Angielczyk KD, Polly PD, Head JJ, Fernandez V, Lungmus JK, Tulga S, Pierce SE. 2018. Fossils reveal the complex evolutionary history of the mammalian regionalized spine. *Science* 361:1249–1252.

Jones MEH, O'higgins P, Fagan MJ, Evans SE, Curtis N. 2012. Shearing mechanics and the influence of a flexible symphysis during oral food processing in *Sphenodon* (Lepidosauria: Rhyncho cephalia). *Anat Rec* 295:1075–1091.

Juorio AV, Vogt M. 1967. Monoamines and their metabolites in the avian brain. *J Physiology* 189:489–518.

Kaas JH. 1999. The transformation of association cortex into sensory cortex. *Brain Res* Bull 50:425.

Kaas JH. 2004. Evolution of somatosensory and motor cortex in primates. *Anat Rec* 281A:1148–1156.

Kaas JH. 2007. The evolution of the dorsal thalamus in mammals. In: Kaas JH (ed.), *Evolution of Nervous Systems*. Amsterdam: Elsevier, pp. 499–516.

Kaas JH. 2008. The evolution of the complex sensory and motor systems of the human brain. *Brain Res Bull* 75:384–390.

Kaas JH. 2011. Reconstructing the areal organization of the neocortex of the first mammals. *Brain Behav Evol* 78:7–21.

Kaas JH. 2017. The organization of neocortex in early mammals. In: Kaas JH, Herculano-Houzel S (eds.), *Evolution of Nervous Systems*, 2nd ed. Amsterdam: Elsevier, pp. 87–101.

Kahn MC, Hough GE, Eyck Ten GR, Bingman VP. 2003. Internal connectivity of the homing pigeon (*Columba livia*) hippocampal formation: an anterograde and retrograde tracer study. *J Comp Neurol* 459:127–141.

Karten HJ. 1969. The organization of the avian telencephalon and some speculations on the phylogeny of the amniote telencephalon. *Ann N Y Acad Sci* 167:164–179.

Karten HJ, Brzozowska-Prechtl A, Lovell PV, Tang DD, Mello CV, Wang H, Mitra PP. 2013. Digital atlas of the zebra finch (*Taeniopygia guttata*) brain: A high-resolution photo atlas. *J Comp Neurol* 521:3702–3715.

Karten HJ, Shimizu T. 1989. The origins of neocortex: connections and lamination as distinct events in evolution. *J Cog Neurosci* 1:291–301.

Kaskan PM, Franco ECS, Yamada ES, De Lima Silveira LC, Darlington RB, Finlay BL. 2005. Peripheral variability and central constancy in mammalian visual system evolution. *Proc Roy Soc Lond B* 272:91–100.

Kelava I, Lewitus E, Huttner WB. 2013. The secondary loss of gyrencephaly as an example of evolutionary phenotypical reversal. *Front Neuroanat* 7:16.

Kemp TS. 1978. Stance and gait in the hindlimb of a therocephalian mammal-like reptile. *J Zool Lond* 186:143–161.

Kenigfest N, Martínez-Marcos A, Belekhova M, Font C, Lanuza E, Desfilis E, Martínez-García F. 1997. A lacertilian dorsal retinorecipient thalamus: a re-investigation in the old-would lizard *Podarcis hispanica*. *Brain Behav Evol* 50:313–334.

Kielan-Jaworowska Z, Hurum JH. 2006. Limb posture in early mammals: sprawling or parasagittal. *Acta Palaeontol Pol* 51:393–406.

Kishida T, Thewissen J, Hayakawa T, Imai H, Agata K. 2015. Aquatic adaptation and the evolution of smell and taste in whales. *Zoological Lett* 1:9.

Kitazawa T, Takechi M, Hirasawa T, Adachi N, Narboux-Nême N, Kume H, Maeda K, Hirai T, Miyagawa-Tomita S, Kurihara Y, et al. 2015. Developmental genetic bases behind the independent origin of the tympanic membrane in mammals and diapsids. *Nat Commun* 6:6853.

Knopf JP, Hof PR, Oelschläger HHA. 2016. The neocortex of Indian river dolphins (genus *Platanista*): comparative, qualitative and quantitative analysis. *Brain Behav Evol* 88:93–110.

Knudsen EI, Schwarz JS. 2017. The optic tectum: a structure evolved for stimulus selection. In: Kaas JH, Striedter GF (eds.), *Evolution of Nervous Systems*, 2nd ed. Amsterdam: Elsevier, pp. 387–408.

Köppl C. 2011. Birds—same thing, but different? Convergent evolution in the avian and mammalian auditory systems provides informative comparative models. *Hear Res* 273:65–71.

Kössl M, Hechavarria J, Voss C, Schaefer M, Vater M. 2015. Bat auditory cortex—model for general mammalian auditory computation or special design solution for active time perception? *Eur J Neurosci* 41:518–532.

Krubitzer L, Campi KL, Cooke DF. 2011. All rodents are not the same: a modern synthesis of cortical organization. *Brain Behav Evol* 78:51–93.

Krubitzer LA, Seelke A. 2012. Cortical evolution in mammals: the bane and beauty of phenotypic variability. *Proc Natl Acad Sci USA* 109:10647–10654.

Krubitzer L, Stolzenberg DS. 2014. The evolutionary masquerade: genetic and epigenetic contributions to the neocortex. *Curr Opin Neurobiol* 24:157–165.

Kruska D, Röhrs M. 1974. Comparative-quantitative investigations on brains of feral pigs from the Galapagos Islands and of European domestic pigs. *Z Anat Entwicklungsgesch* 144:61–73.

Kuenzel WJ, Medina L, Csillag A, Perkel DJ, Reiner A. 2011. The avian subpallium: new insights into structural and functional subdivisions occupying the lateral subpallial wall and their embryological origins. *Brain Res* 1424:67–101.

Laaß M. 2015. Virtual reconstruction and description of the cranial endocast of *Pristerodon mackayi* (Therapsida, Anomodontia). *J Morphol* 276:1089–1099.

Laaß M. 2016. The origins of the cochlea and impedance matching hearing in synapsids. *Acta Palaeontol Pol* 61:267–280.

Laaß M, Hampe O, Schudack M, Hoff C, Kardjilov N, Hilger A. 2011. New insights into the respiration and metabolic physiology of *Lystrosaurus*. *Acta Zool* 92:363–371.

Ladhams A, Pickles JO. 1996. Morphology of the monotreme organ of Corti and macula lagena. *J Comp Neurol* 366:335–347.

Laramée M-E, Boire D. 2015. Visual cortical areas of the mouse: comparison of parcellation and network structure with primates. *Front Neural Circuits* 8:149.

Lefebvre L. 2011. Taxonomic counts of cognition in the wild. *Biology Letters* 7:631–633.

Lefebvre L. 2012. Primate encephalization. *Prog Brain Res* 195:393–412.

Lende RA. 1963. Cerebral cortex: a sensorimotor amalgam in the Marsupiala. *Science* 141:730–732.

Letzner S, Simon A, Güntürkün O. 2015. Connectivity and neurochemistry of the commissura anterior of the pigeon (*Columba livia*). *J Comp Neurol* 524:343–361.

Lewitus E, Kelava I, Kalinka AT, Tomancak P, Huttner WB. 2014. An adaptive threshold in mammalian neocortical evolution. *Plos Biol* 12:e1002000.

Li G, Wang J, Rossiter SJ, Jones G, Cotton JA, Zhang S. 2008. The hearing gene *Prestin* reunites echolocating bats. *Proc Natl Acad Sci USA* 105:13959–13964.

Li Y, Liu Z, Shi P, Zhang J. 2010. The hearing gene *Prestin* unites echolocating bats and whales. *Curr Biol* 20:R55–R56.

Liu GB, Pettigrew JD. 2003. Orientation mosaic in barn owl's visual Wulst revealed by optical imaging: comparison with cat and monkey striate and extra-striate areas. *Brain Res* 961:153–158.

Liu Z, Qi F-Y, Zhou X, Ren H-Q, Shi P. 2014. Parallel sites implicate functional convergence of the hearing gene *Prestin* among echolocating mammals. *Mol Biol Evol* 31:2415–2424.

Llinás R (ed.). 1969. *Neurobiology of Cerebellar Evolution and Development*. Chicago: American Medical Association.

Longrich NR, Bhullar B-AS, Gauthier JA. 2012. Mass extinction of lizards and snakes at the Cretaceous-Paleogene boundary. *Proc Natl Acad Sci USA* 109:21396–21401.

Longrich NR, Tokaryk T, Field DJ. 2011. Mass extinction of birds at the Cretaceous-Paleogene (K-Pg) boundary. *Proc Natl Acad Sci USA* 108:15253–15257.

Longworthy OR. 1932. A description of the central nervous system of the porpoise (*Tursiops truncatus*). *J Comp Neurol* 54:437–499.

Lovegrove BG. 2012. The evolution of endothermy in Cenozoic mammals: a plesiomorphic-apomorphic continuum. *Biol Rev* 87:128–162.

Lovell PV, Kasimi B, Carleton J, Velho TA, Mello CV. 2015. Living without DAT: Loss and compensation of the dopamine transporter gene in sauropsids (birds and reptiles). *Sci Rep* 5:14093.

Luo Z-X. 2007. Transformation and diversification in early mammal evolution. *Nature* 450:1011–1019.
Luo Z-X, Ruf I, Schultz JA, Martin T. 2011. Fossil evidence on evolution of inner ear cochlea in Jurassic mammals. *Proc Roy Soc Lond B* 278:28–34.
Lyon DC. 2007. The evolution of visual cortex and visual systems. In: Kaas JH (ed.), *Evolution of Nervous Systems*. Amsterdam: Elsevier, pp. 267–306.
Mackintosh NJ, Wilson B, Boakes RA, Barlow HB. 1985. Differences in mechanisms of intelligence among vertebrates [and Discussion]. *Philos Trans R Soc Lond B* 308:53–65.
MacLean EL, Hare B, Nunn CL, Addessi E, Amici F, Anderson RC, Aureli F, Baker JM, Bania AE, Barnard AM, et al. 2014. The evolution of self-control. *Proc Natl Acad Sci USA* 111:E2140–E2148.
MacLeod CE, Zilles K, Schleicher A, Rilling JK, Gibson KR. 2003. Expansion of the neocerebellum in Hominoidea. *J Hum Evol* 44:401–429.
Macrini TE, Rougier GW, Rowe T. 2007. Description of a cranial endocast from the fossil mammal *Vincelestes neuquenianus* (Theriiformes) and its relevance to the evolution of endocranial characters in Therians. *Anat Rec* 290:875–892.
Madsen PT, Surlykke A. 2013. Functional convergence in bat and toothed whale biosonars. *Physiology* 28:276–283.
Maier A. 1992a. The avian muscle spindle. *Anat Embryol* 186:1–25.
Maier A. 1992b. Sensory and motor innervation of bird intrafusal muscle fibers. *Comp Biochem Physiol* 103:635–639.
Maier S, Walkowiak W, Luksch H, Endepols H. 2010. An indirect basal ganglia pathway in anuran amphibians? *J Chem Neuroanat* 40:21–35.
Maier W, Ruf I. 2015. Evolution of the mammalian middle ear: a historical review. *J Anat* 228:270–283.
Malmierca MS, Merchán MA. 2004. Auditory system. In: Paxinos G (ed.), *The Rat Nervous System*. Amsterdam: Elsevier, pp. 997–1082.
Mandal R, Anderson CW. 2010. Identification of muscle spindles in the submentalis muscle of the marine toad, *Bufo marinus* and its potential proprioceptive capacity in jaw-tongue coordination. *Anat Rec* 293:1568–1573.
Manger PR, Spocter MA, Patzke N. 2013. The evolutions of large brain size in mammals: the 'over-700-gram club quartet.' *Brain Behav Evol* 82:68–78.
Mangold-Wirz K. 1966. Cerebralisation und Ontogenesemodus bei Eutherien. *Acta Anat* 63:449–508.
Manley GA. 2012. Evolutionary paths to mammalian cochleae. *J Assoc Res Otolaryngol* 13:733–743.
Maor R, Dayan T, Ferguson-Gow H, Jones KE. 2017. Temporal niche expansion in mammals from a nocturnal ancestor after dinosaur extinction. *Nat Ecol Evol* 1:1889–1895.
Marin O, Smeets WJ, González A. 1997. Distribution of choline acetyltransferase immunoreactivity in the brain of anuran (*Rana perezi, Xenopus laevis*) and urodele (*Pleurodeles waltl*) amphibians. *J Comp Neurol* 382:499–534.
Marino L. 2006. Absolute brain size: did we throw the baby out with the bathwater? *Proc Natl Acad Sci USA* 103:13563–13564.
Martin T, Luo Z-X. 2005. Homoplasy in the mammalian ear. *Science* 307:861–862.
Martínez-García F, Amiguet M, Schwerdtfeger WK, Olucha FE, Lorente MJ. 1990. Interhemispheric connections through the pallial commissures in the brain of *Podarcis hispanica* and *Gallotia stehlinii* (Reptilia, Lacertidae). *J Morphol* 205:17–31.
May PJ. 2006. The mammalian superior colliculus: laminar structure and connections. *Prog Brain Res* 151:321–378.
McKenzie S, Keene CS, Farovik A, Bladon J, Place R, Komorowski R, Eichenbaum H. 2015. Representation of memories in the cortical-hippocampal system: results from the application of population similarity analyses. *Neurobiol Learn Mem* 134A:178–191.
McNab BK. 2008. An analysis of the factors that influence the level and scaling of mammalian BMR. *Comp Biochem Physiol A* 151:5–28.

McNab BK, Eisenberg JF. 1989. Brain size and its relation to the rate of metabolism in mammals. *Am Nat* 133:157–167.

Medina L. 2007. Do birds and reptiles possess homologues of mammalian visual, somatosensory, and motor cortices? In: Kaas JH (ed.), *Evolution of Nervous Systems.* Amsterdam: Elsevier, pp. 163–194.

Medina L, Abellán A, Desfilis E. 2013. A never-ending search for the evolutionary origin of the neocortex: rethinking the homology concept. *Brain Behav Evol* 81:150–153.

Medina L, Veenman CL, Reiner A. 1997. Evidence for a possible avian dorsal thalamic region comparable to the mammalian ventral anterior, ventral lateral, and oral ventroposterolateral nuclei. *J Comp Neurol* 384:86–108.

Middleton FA, Strick PL. 2002. Basal-ganglia "projections" to the prefrontal cortex of the primate. *Cereb Cortex* 12:926–935.

Mihrshahi R. 2006. The corpus callosum as an evolutionary innovation. *J Exp Zool* 306B:8–17.

Miller MR. 1980. The cochlear nuclei of snakes. *J Comp Neurol* 192:717–736.

Mills JR. 1972. Evolution of mastication. *Proc R Soc Med* 65:392–396.

Mitchinson B, Grant RA, Arkley K, Rankov V, Perkon I, Prescott TJ. 2011. Active vibrissal sensing in rodents and marsupials. *Philos Trans R Soc Lond B* 366:3037–3048.

Mitchinson B, Martin CJ, Grant RA, Prescott TJ. 2007. Feedback control in active sensing: rat exploratory whisking is modulated by environmental contact. *Proc Roy Soc Lond B* 274:1035–1041.

Möller H. 1973. Zur Evolutionshöhe des Marsupialgehirns. *Zool Jahrb Anat* 91:434–448.

Montgomery SH, Geisler JH, McGowen MR, Fox C, Marino L, Gatesy J. 2013. The evolutionary history of cetacean brain and body size. *Evolution* 67:3339–3353.

Montiel JF, Vasistha NA, García-Moreno F, Molnár Z. 2015. From sauropsids to mammals and back: new approaches to comparative cortical development. J *Comp Neurol* 524:630–645.

Moore JK. 1980. The primate cochlear nuclei: loss of lamination as a phylogenetic process. *J Comp Neurol* 193:609–629.

Mota B, Herculano-Houzel S. 2015. Cortical folding scales universally with surface area and thickness, not number of neurons. *Science* 349:74–77.

Muchlinski MN, Wible JR, Corfe I, Sullivan M, Grant R. 2018. Good Vibrations: the evolution of whisking in small mammals. *Anat Rec.* doi.org/10.1002/ar.23989.

Mulligan KA, Ulinski PS. 1990. Organization of geniculocortical projections in turtles: isoazimuth lamellae in the visual cortex. *J Comp Neurol* 296:531–547.

Murray E, Wise S, Graham K. 2016. *The Evolution of Memory Systems.* Oxford: Oxford University Press.

Myoga MH, Lehnert S, Leibold C, Felmy F, Grothe B. 2014. Glycinergic inhibition tunes coincidence detection in the auditory brainstem. *Nat Commun* 5:3790.

Naumann RK, Laurent G. 2017. Function and evolution of the reptilian cerebral cortex. In: Kaas JH, Striedter GF (eds). *Evolution of Nervous Systems*, 2nd ed. Amsterdam: Elsevier, pp. 491–518.

Neary TJ, Northcutt RG. 1983. Nuclear organization of the bullfrog diencephalon. *J Comp Neurol* 213:262–278.

Negwer M, Liu Y-J, Schubert D, Lyon DC. 2017. V1 connections reveal a series of elongated higher visual areas in the California ground squirrel, Otospermophilus beecheyi. *J Comp Neurol* 525:1909–1921.

Neuweiler G. 1990. Auditory adaptations for prey capture in echolocating bats. *Physiol Rev* 70:615–641.

Nieuwenhuys R, Puelles L. 2016. *Towards a New Neuromorphology.* Cham: Springer.

Nieuwenhuys R, Voogd J, van Huijzen C. 2008. *The Human Central Nervous System*, 4th ed. Berlin: Springer-Verlag.

Niimura Y. 2012. Olfactory receptor multigene family in vertebrates: from the viewpoint of evolutionary genomics. *Curr Genomics* 13:103–114.

Niquille M, Garel S, Mann F, Hornung J-P, Otsmane B, Chevalley S, Parras C, Guillemot F, Gaspar P, Yanagawa Y, Lebrand C. 2009. Transient neuronal populations are required to guide callosal axons: a role for semaphorin 3C. Plos Biol 7:e1000230.

Northcutt RG. 1978. Forebrain and midbrain organization in lizards and its phylogenetic significance. In: Greenberg NL, MacLean PD (eds.), *Behavior and Neurology of Lizards*. Bethesda, MD: NIMH, pp. 11–64.

Northcutt RG, Kaas JH. 1995. The emergence and evolution of mammalian neocortex. *Trends Neurosci* 18:373–379.

Nottebohm F. 1972. The origins of vocal learning. *Am Nat* 106:116–140.

Nowak RM. 1999. *Walker's Mammals of the World*, 6th ed. Baltimore, MD: JHU Press.

Nudo RJ, Sutherland DP, Masterton RB. 1995. Variation and evolution of mammalian corticospinal somata with special reference to primates. *J Comp Neurol* 358:181–205.

O'Keefe J, Nadel L. 2003. *The Hippocampus as a Cognitive Map*. Oxford: Oxford University Press.

O'Leary MA, Bloch JI, Flynn JJ, Gaudin TJ, Giallombardo A, Giannini NP, Goldberg SL, Kraatz BP, Luo Z-X, Meng J, Ni X, Novacek MJ, Perini FA, Randall ZS, Rougier GW, Sargis EJ, Silcox MT, Simmons NB, Spaulding M, Velazco PM, Weksler M, Wible JR, Cirranello AL. 2013. The placental mammal ancestor and the post-K-Pg radiation of placentals. *Science* 339:662–667.

Oertel D, Young ED. 2004. What's a cerebellar circuit doing in the auditory system? *Trends Neurosci* 27:104–110.

Okano T, Kojima D, Fukada Y, Shichida Y, Yoshizawa T. 1992. Primary structures of chicken cone visual pigments: vertebrate rhodopsins have evolved out of cone visual pigments. *Proc Natl Acad Sci USA* 89:5932–5936.

Okoruwa OE, Weston MD, Sanjeevi DC, Millemon AR, Fritzsch B, Hallworth R, Beisel KW. 2008. Evolutionary insights into the unique electromotility motor of mammalian outer hair cells. *Evol Dev* 10:300–315.

Olkowicz S, Kocourek M, Lučan RK, Porteš M, Fitch WT, Herculano-Houzel S, Němec P. 2016. Birds have primate-like numbers of neurons in the forebrain. *Proc Natl Acad Sci USA* 113:7255–7260.

Omer DB, Maimon SR, Las L, Ulanovsky N. 2018. Social place-cells in the bat hippocampus. *Science* 359:218–224.

Osorio D, Vorobyev M. 1996. Colour vision as an adaptation to frugivory in primates. *Proc Roy Soc Lond B* 263:593–599.

Owerkowicz T, Musinsky C, Middleton KM, Crompton AW. 2015. Respiratory turbinates and the evolution of endothermy in mammals and birds. In: Dial KP, Shubin N, Brainerd EL (eds.), *Great Transformations in Vertebrate Evolution*. Chicago: University of Chicago Press, pp. 143–165.

Pagel MD, Harvey PH. 1988. The taxon-level problem in the evolution of mammalian brain size: facts and artifacts. *Am Naturalist* 132:344–359.

Pakan JMP, Iwaniuk AN, Wylie DRW, Hawkes R, Marzban H. 2007. Purkinje cell compartmentation as revealed by Zebrin II expression in the cerebellar cortex of pigeons (*Columba livia*). *J Comp Neurol* 501:619–630.

Pant M, Bal NC, Periasamy M. 2016. Sarcolipin: a key thermogenic and metabolic regulator in skeletal muscle. *Trends Endocrinol Metab* 27:881–892.

Papp G, Witter MP, Treves A. 2007. The CA3 network as a memory store for spatial representations. *Learn Mem* 14:732–744.

Passingham RE. 1982. *The Human Primate*. Oxford: W. H. Freeman.

Passingham RE, Smaers JB. 2014. Is the prefrontal cortex especially enlarged in the human brain allometric relations and remapping factors. *Brain Behav Evol* 84:156–166.

Passingham RE, Wise SP. 2012. *The Neurobiology of the Prefrontal Cortex*. Oxford: Oxford University Press.

Peichl L. 2005. Diversity of mammalian photoreceptor properties: Adaptations to habitat and lifestyle? *Anat Rec* 287A:1001–1012.

Perry VH, Cowey A. 1984. Retinal ganglion cells that project to the superior colliculus and pretectum in the macaque monkey. *Neuroscience* 12:1125–1137.

Pettigrew JD. 1979. Binocular visual processing in the owl's telencephalon. *Proc Roy Soc Lond B* 204:435–454.

Pettigrew JD. 1986. The evolution of binocular vision. In: Pettigrew JD, Sanderson KJ, Levick WR (eds.), *Visual Neuroscience*. Cambridge: Cambridge University Press, pp. 208–222.

Phillips KA, Stimpson CD, Smaers JB, Raghanti MA, Jacobs B, Popratiloff A, Hof PR, Sherwood CC. 2015. The corpus callosum in primates: processing speed of axons and the evolution of hemispheric asymmetry. *Proc Roy Soc Lond B* 282:20151535.

Pigliucci M. 2008. Is evolvability evolvable? *Nat Rev Genet* 9:75–82.

Piper M, Plachez C, Zalucki O, Fothergill T, Goudreau G, Erzurumlu R, Gu C, Richards LJ. 2009. Neuropilin 1-Sema signaling regulates crossing of cingulate pioneering axons during development of the corpus callosum. *Cerebral Cortex* 19 Suppl 1:i11–i21.

Pocock RI. 1914. On the facial vibrissae of Mammalia. *J Zool Lond* 84:889–912.

Pozzi L, Hodgson JA, Burrell AS, Sterner KN, Raaum RL, Disotell TR. 2014. Primate phylogenetic relationships and divergence dates inferred from complete mitochondrial genomes. *Mol Phylogenet Evol* 75:165–183.

Preuss TM, Goldman-Rakic PS. 1991. Myelo- and cytoarchitecture of the granular frontal cortex and surrounding regions in the strepsirhine primate *Galago* and the anthropoid primate *Macaca*. *J Comp Neurol* 310:429–474.

Prior H, Schwarz A, Güntürkün O. 2008. Mirror-induced behavior in the magpie (*Pica pica*): evidence of self-recognition. *Plos Biol* 6:e202.

Pritz MB. 2015. Crocodilian forebrain: evolution and development. *Integr Comp Biol* 55:949–961.

Pritz MB. 2016. Thalamic reticular nucleus in *Caiman crocodilus*: forebrain connections. *Neurosci Lett* 627:65–70.

Proske U. 1967. Responses of muscle spindles in the lizard. *Nature* 213:1144–1146.

Proske U. 1969. The innervation of muscle spindles in the lizard *Tiliqua nigrolutea*. *J Anat* 105:217–230.

Proske U, Gandevia SC. 2012. The proprioceptive senses: their roles in signaling body shape, body position and movement, and muscle force. *Physiol Rev* 92:1651–1697.

Proske U, Gregory JE, Iggo A. 1998. Sensory receptors in monotremes. *Philos Trans R Soc Lond B* 353:1187–1198.

Provis JM, Dubis AM, Maddess T, Carroll J. 2013. Adaptation of the central retina for high acuity vision: cones, the fovea and the avascular zone. *Prog Ret Eye Res* 35:63–81.

Puelles L. 2017. Comments on the updated tetrapartite pallium model in the mouse and chick, featuring a homologous claustro-insular complex. *Brain Behav Evol* 90:171–189.

Puelles L, Ayad A, Alonso A, Sandoval JE, Martinez-de-la-Torre M, Medina L, Ferran JL. 2015. Selective early expression of the orphan nuclear receptor Nr4a2 identifies the claustrum homolog in the avian mesopallium: Impact on sauropsidian/mammalian pallium comparisons. *J Comp Neurol* 524:665–703.

Puelles L, Kuwana E, Puelles E, Bulfone A, Shimamura K, Keleher J, Smiga S, Rubenstein JL. 2000. Pallial and subpallial derivatives in the embryonic chick and mouse telencephalon, traced by the expression of the genes *Dlx-2, Emx-1, Nkx-2.1, Pax-6,* and *Tbr-1*. *J Comp Neurol* 424:409–438.

Puelles L, la Torre M de M, Paxinos G, Watson C, Martinez S. 2007. *The Chick Brain in Stereotaxic Coordinates*. Amsterdam: Academic Press.

Puelles L, Medina L, Borello U, Legaz I, Teissier A, Pierani A, Rubenstein JLR. 2016. Radial derivatives of the mouse ventral pallium traced with *Dbx1-LacZ* reporters. *J Chem Neuroanat* 75:2–19.

Puelles L, Robles C, Martinez-de-la-Torre M, Martínez S. 1994. New subdivision schema for the avian torus semicircularis: neurochemical maps in the chick. *J Comp Neurol* 340:98–125.

Puelles L, Sandoval JE, Ayad A, del Corral R, Alonso A, Ferran JL, Martinez-de-la-Torre M. 2017. The pallium in reptiles and birds in the light of the updated tetrapartite pallium model. In: Kaas JH, Striedter GF (eds.), *Evolution of Nervous Systems*, 2nd ed. Amsterdam: Elsevier, pp. 519–555.

Quay WB. 1979. The parietal eye-pineal complex. In: Gans C, Northcutt RG (eds.), *Biology of the Reptilia*, Vol. 9A. San Diego: Academic Press, pp. 245–406.

Rakic P. 1974. Neurons in rhesus monkey visual cortex: systematic relation between time of origin and eventual disposition. *Science* 183:425–427.

Ramírez-Chaves HE, Wroe SW, Selwood L, Hinds LA, Leigh C, Koyabu D, Kardjilov N, Weisbecker V. 2016. Mammalian development does not recapitulate suspected key transformations in the evolutionary detachment of the mammalian middle ear. *Proc Roy Soc Lond B* 283:20152606–7.

Reiner A, Karten HJ. 1985. Comparison of olfactory-bulb projections in pigeons and turtles. *Brain Behav Evol* 27:11–27.

Reiner A, Medina L, Veenman CL. 1998. Structural and functional evolution of the basal ganglia in vertebrates. *Brain Res Rev* 28:235–285.

Ren T, He W, Gillespie PG. 2011. Measurement of cochlear power gain in the sensitive gerbil ear. *Nat Commun* 2:216–7.

Ribeiro PFM, Manger PR, Catania KC, Kaas JH, Herculano-Houzel S. 2014. Greater addition of neurons to the olfactory bulb than to the cerebral cortex of eulipotyphlans but not rodents, afrotherians or primates. *Front Neuroanat* 8:23.

Richards SA. 1970. Physiology of thermal panting in birds. *Ann Biol Anim Biochim Biophys* 10:151–168.

Ridgway SH, Carlin KP, Van Alstyne KR, Hanson AC, Tarpley RJ. 2016. Comparison of dolphins' body and brain measurements with four other groups of cetaceans reveals great diversity. *Brain Behav Evol* 88:235–257.

Rilling JK. 2007. The evolution of the cerebellum in anthropoid primates. In: Kaas JH (ed.), *Evolution of Nervous Systems*. Amsterdam: Elsevier, pp. 149–156.

Ringo JL, Doty RW, Demeter S, Simard PY. 1994. Time is of the essence: a conjecture that hemispheric specialization arises from interhemispheric conduction delay. *Cereb Cortex* 4:331–343.

Robinson DL, Petersen SE. 1992. The pulvinar and visual salience. *Trends Neurosci* 15:127–132.

Rodrigues PG, Ruf I, Schultz CL. 2013. Digital reconstruction of the otic region and inner ear of the non-mammalian cynodont *Brasilitherium riograndensis* (Late Triassic, Brazil) and its relevance to the evolution of the mammalian ear. *J Mammal Evol* 20:291–307.

Rodrigues S-L, Maseko BC, Ihunwo AO, Fuxe K, Manger PR. 2008. Nuclear organization and morphology of serotonergic neurons in the brain of the Nile crocodile, *Crocodylus niloticus*. *J Chem Neuroanat* 35:133–145.

Rodríguez F, López JC, Vargas JP, Gómez Y, Broglio C, Salas C. 2002. Conservation of spatial memory function in the pallial forebrain of reptiles and ray-finned fishes. *J Neurosci* 22:2894–2903.

Roth G, Grunwald W, Dicke U. 2003. Morphology, axonal projection pattern, and responses to optic nerve stimulation of thalamic neurons in the fire-bellied toad *Bombina orientalis*. *J Comp Neurol* 461:91–110.

Rowe T. 1988. Definition, diagnosis, and origin of Mammalia. *J Vert Paleontol* 8:241–264.

Rowe T. 1996. Coevolution of the mammalian middle ear and neocortex. *Science* 273:651–654.

Rowe TB. 2017. The emergence of mammals. In: Kaas JH, Herculano-Houzel S (eds.), *Evolution of Nervous Systems*, Vol. 2, 2nd ed. Amsterdam: Elsevier, pp. 1–52.

Rowe TB, Macrini TE, Luo Z-X. 2011. Fossil evidence on origin of the mammalian brain. *Science* 332:955–957.

Rowe TB, Shepherd GM. 2015. Role of ortho-retronasal olfaction in mammalian cortical evolution. *J Comp Neurol* 524:471–495.

Rowland LA, Bal NC, Periasamy M. 2015. The role of skeletal-muscle-based thermogenic mechanisms in vertebrate endothermy. *Biol Rev* 90:1279–1297.

Ruben J. 1995. The evolution of endothermy in mammals and birds: from physiology to fossils. *Annu Rev Physiol* 57:69–95.

Ruben J, Jones T, Geist N. 1998. Respiratory physiology of the dinosaurs. *Bioessays* 20:852–859.

Ryugo DK, May SK. 1993. The projections of intracellularly labeled auditory nerve fibers to the dorsal cochlear nucleus of cats. *J Comp Neurol* 329:20–35.

Ryugo DK, Parks TN. 2003. Primary innervation of the avian and mammalian cochlear nucleus. *Brain Res Bull* 60:435–456.

Salas C, Broglio C, Rodr guez F, L pez JC, Portavella M, Torres B. 1996. Telencephalic ablation in goldfish impairs performance in a 'spatial constancy' problem but not in a cued one. *Behav Brain Res* 79:193–200.
Sawyer RH, Knapp LW. 2003. Avian skin development and the evolutionary origin of feathers. *J Exp Zool B* 298B:57–72.
Schiller D, Eichenbaum H, Buffalo EA, Davachi L, Foster DJ, Leutgeb S, Ranganath C. 2015. Memory and space: towards an understanding of the cognitive map. *J Neurosci* 35:13904–13911.
Schiller PH. 2010. The superior colliculus and visual function. In: Darian-Smith I (ed.), *Handbook of Physiology, the Nervous System, Sensory Processes*. Oxford: Oxford University Press, pp. 457–505.
Schmahmann JD. 1991. An emerging concept. The cerebellar contribution to higher function. *Arch Neurol* 48:1178–1187.
Schmidt MF, Ashmore RC, Vu ET. 2004. Bilateral control and interhemispheric coordination in the avian song motor system. *Ann N Y Acad Sci* 1016:171–186.
Schmitt D. 2010. Primate locomotor evolution: biomechanical studies of primate locomotion and their implications for understanding primate neuroethology. In: Platt ML, Ghazanfar AA (eds.), *Primate Neuroethology*. Oxford: Oxford University Press, pp. 31–63.
Schmolke C, Künzle H. 1997. On the presence of dendrite bundles in the cerebral cortex of the Madagascan lesser hedgehog tenrec and the red-eared pond turtle. *Anat Embryol* 196:195–213.
Schober W, Brauer K. 1974. Makromorphologie des Zentralnervensystems. II. Teil. Das Gehirn. In: Helmcke JG, Starck D, Wermuth H (eds.), *Handbuch der Zoologie*. Berlin: Walter de Gruyter, pp. 1–26.
Schulte P, Alegret L, Arenillas I, Arz JA, Barton PJ, Bown PR, Bralower TJ, Christeson GL, Claeys P, Cockell CS, et al. 2010. The Chicxulub asteroid impact and mass extinction at the Cretaceous-Paleogene boundary. *Science* 327:1214–1218.
Shanahan M, Bingman VP, Shimizu T, Wild M, Güntürkün O. 2013. Large-scale network organization in the avian forebrain: a connectivity matrix and theoretical analysis. *Front Comput Neurosci* 7:89.
Shein-Idelson M, Ondracek JM, Liaw H-P, Reiter S, Laurent G. 2016. Slow waves, sharp waves, ripples, and REM in sleeping dragons. *Science* 352:590–595.
Sherman SM, Guillery RW. 2006. *Exploring the Thalamus and Its Role in Cortical Function*. Cambridge, MA: MIT Press.
Shi P, Zhang J. 2007. Comparative genomic analysis identifies an evolutionary shift of vomeronasal receptor gene repertoires in the vertebrate transition from water to land. *Genome Res* 17:166–174.
Shultz S, Dunbar R. 2007. The evolution of the social brain: anthropoid primates contrast with other vertebrates. *Proc Roy Soc Lond B* 274:2429–2436.
Smaers JB. 2014. Modeling the evolution of the cerebellum: from macroevolution to function. *Prog Brain Res* 210:193–216.
Smeets WJ, Jonker AJ, Hoogland PV. 1987. Distribution of dopamine in the forebrain and midbrain of the red-eared turtle, *Pseudemys scripta elegans*, reinvestigated using antibodies against dopamine. *Brain Behav Evol* 30:121–142.
Smith Y, Raju DV, Pare J-F, Sidibe M. 2004. The thalamostriatal system: a highly specific network of the basal ganglia circuitry. *Trends Neurosci* 27:520–527.
Snyder GK, Sheafor BA. 1999. Red blood cells: centerpiece in the evolution of the vertebrate circulatory system. *Integr Comp Biol* 39:189–198.
Sobrero R, May-Collado LJ, Agnarsson I, Hernández CE. 2011. Expensive brains: "brainy" rodents have higher metabolic rate. *Front Evol Neurosci* 3:2.
Soligo C, Müller AE. 1999. Nails and claws in primate evolution. *J Hum Evol* 36:97–114.
Sookias RB, Butler RJ, Benson RBJ. 2012. Rise of dinosaurs reveals major body-size transitions are driven by passive processes of trait evolution. *Proc Roy Soc Lond B* 279:2180–2187.
Sotelo MI, Daneri MF, Bingman VP, Muzio RN. 2016. Telencephalic neuronal activation associated with spatial memory in the terrestrial toad *Rhinella arenarum*: participation of the medial pallium during navigation by geometry. *Brain Behav Evol* 88:149–160.

Springer AD. 1999. New role for the primate fovea: a retinal excavation determines photoreceptor deployment and shape. *Vis Neurosci* 16:629–636.

Springer AD, Hendrickson AE. 2005. Development of the primate area of high acuity, 3: Temporal relationships between pit formation, retinal elongation and cone packing. *Vis Neurosci* 22:171–185.

Stecker MM, Baylor K. 2009. Peripheral nerve at extreme low temperatures 1: Effects of temperature on the action potential. *Cryobiology* 59:1–11.

Stephan H, Frahm H, Baron G. 1981. New and revised data on volumes of brain structures in insectivores and primates. *Folia Primatol* 35:1–29.

Sterbing-D'Angelo S, Chadha M, Chiu C, Falk B, Xian W, Barcelo J, Zook JM, Moss CF. 2011. Bat wing sensors support flight control. *Proc Natl Acad Sci USA* 108:11291–11296.

Strick PL, Dum RP, Fiez JA. 2009. Cerebellum and nonmotor function. Annu Rev Neurosci 32:413–434.

Striedter GF. 1997. The telencephalon of tetrapods in evolution. *Brain Behav Evol* 49:179–213.

Striedter GF. 1999. Homology in the nervous system: of characters, embryology and levels of analysis. In: Bock GK, Cardew G (eds.), *Novartis Found Symp 222 – Homology*. New York: Wiley, pp. 158–172.

Striedter GF. 2005. *Principles of Brain Evolution*. Sunderland, MA: Sinauer Associates.

Striedter GF. 2013. Bird brains and tool use: beyond instrumental conditioning. *Brain Behav Evol* 82:55–67.

Striedter GF. 2016. Evolution of the hippocampus in reptiles and birds. *J Comp Neurol* 524:496–517.

Striedter GF, Northcutt RG. 2006. Head size constrains forebrain development and evolution in ray-finned fishes. *Evol Dev* 8:215–222.

Striedter GF, Srinivasan S, Monuki ES. 2014. Cortical folding: when, where, how, and why? *Annu Rev Neurosci* 38:291–307.

Suárez R. 2016. Evolution of telencephalic commissures: conservation and change of developmental systems in the origin of brain wiring novelties. In: Kaas JH, Herculano-Houzel S (eds.), *Evolution of Nervous Systems*, Vol. 2, 2nd ed. Amsterdam: Elsevier, pp. 205–223.

Suárez R, Gobius I, Gobius, Richards LJ. 2014. Evolution and development of interhemispheric connections in the vertebrate forebrain. *Front Hum Neurosci* 8:497.

Sudarov A, Joyner AL. 2007. Cerebellum morphogenesis: the foliation pattern is orchestrated by multi-cellular anchoring centers. *Neural Dev* 2:26.

Suddendorf T, Corballis MC. 2010. Behavioural evidence for mental time travel in nonhuman animals. *Behav Brain Res* 215:292–298.

Sues HD, Reisz RR. 1998. Origins and early evolution of herbivory in tetrapods. *Trends Ecol Evolut* 13:141–145.

Sultan F, Braitenberg V. 1993. Shapes and sizes of different mammalian cerebella: a study in quantitative comparative neuroanatomy. *J Hirnforsch* 34:79–92.

Sun Y, Joachimski MM, Wignall PB, Yan C, Chen Y, Jiang H, Wang L, Lai X. 2012. Lethally hot temperatures during the Early Triassic greenhouse. *Science* 338:366–370.

Székely AD. 1999. The avian hippocampal formation: subdivisions and connectivity. *Behav Brain Res* 98:219–225.

Tan X, Pecka JL, Tang J, Okoruwa OE, Zhang Q, Beisel KW, He DZZ. 2011. From zebrafish to mammal: functional evolution of prestin, the motor protein of cochlear outer hair cells. *J Neurophysiol* 105:36–44.

Tattersall GJ, Cadena V, Skinner MC. 2006. Respiratory cooling and thermoregulatory coupling in reptiles. *Respir Physiol Neurobiol* 154:302–318.

ten Donkelaar HJ. 1988. Evolution of the red nucleus and rubrospinal tract. *Behav Brain Res* 28:9–20.

Tilney F. 1927. The brain stem of *Tarsius*: a critical comparison with other primates. *J Comp Neurol* 43:371–432.

Tosches MA, Yamawaki TM, Naumann RK, Jacobi AA, Tushev G, Laurent G. 2018. Evolution of pallium, hippocampus, and cortical cell types revealed by single-cell transcriptomics in reptiles. *Science* 360:881–888.

Toyoshima K, Shimamura A. 1991. Ultrastructure of Merkel corpuscles in the tongue of the finch, *Lonchura striata*. *Cell Tissue Res* 264:427–436.

Ulinski PS. 1990. The cerebral cortex of reptiles. In: Jones EG, Peters A (eds.), *Cerebral Cortex*, Vol. 8A. Boston: Springer, pp. 139–215.

Ungar PS. 2010. *Mammal Teeth*. Baltimore, MD: Johns Hopkins University Press.

Vanegas H. 1984. *Comparative Neurology of the Optic Tectum*. New York; London: Plenum Press.

van der Knaap LJ, van der Ham IJM. 2011. How does the corpus callosum mediate interhemispheric transfer? A review. *Behav Brain Res* 223:211–221.

Vater M, Meng J, Fox RC. 2004. Hearing organ evolution and specialization: early and later mammals. In: Manley GA, Fay RR (eds.), *Evolution of the Vertebrate Auditory System*. New York: Springer, pp. 256–288.

Veenman CL, Medina L, Reiner A. 1997. Avian homologues of mammalian intralaminar, mediodorsal and midline thalamic nuclei: immunohistochemical and hodological evidence. *Brain Behav Evol* 49:78–98.

Veenman CL, Wild JM, Reiner A. 1995. Organization of the avian corticostriatal projection system: a retrograde and anterograde pathway tracing study in pigeons. *J Comp Neurol* 354:87–126.

Verendeev A, Thomas C, McFarlin SC, Hopkins WD, Phillips KA, Sherwood CC. 2015. Comparative analysis of Meissner's corpuscles in the fingertips of primates. *J Anat* 227:72–80.

von Düring M, Miller MR. 1979. In: Gans C, Northcutt RG (eds.), *Biology of the Reptilia*, Vol. 9A. San Diego: Academic Press, pp. 407–442.

Voorn P, Vanderschuren LJMJ, Groenewegen HJ, Robbins TW, Pennartz CMA. 2004. Putting a spin on the dorsal-ventral divide of the striatum. *Trends Neurosci* 27:468–474.

Walls GL. 1942. *The Vertebrate Eye and Its Adaptive Radiation*. Bloomfield Hills, MI: Cranbrook Institute of Science.

Wang Q, Burkhalter A. 2007. Area map of mouse visual cortex. *J Comp Neurol* 502:339–357.

Warren JM. 1966. Reversal learning and the formation of learning sets by cats and rhesus monkeys. *J Comp Physiol Psychol* 61:421–428.

Watanabe S. 1985. Interhemispheric transfer of visual discrimination in pigeons with supraoptic decussation (DSO) lesions before and after monocular learning. *Behav Brain Res* 17:163–170.

Watson C, Puelles L. 2016. Developmental gene expression in the mouse clarifies the organization of the claustrum and related endopiriform nuclei. *J Comp Neurol* 525:1499–1508.

Webster DM. 1975. Interhemispheric transfer in a marsupial: a behavioural investigation of interhemispheric transfer of visual information in *Trichosurus vulpecula*. PhD Thesis, University of Auckland, NZ.

Wey H-Y, Phillips KA, McKay DR, Laird AR, Kochunov P, Davis MD, Glahn DC, Blangero J, Duong TQ, Fox PT. 2014. Multi-region hemispheric specialization differentiates human from nonhuman primate brain function. *Brain Struct Funct* 219:2187–2194.

Wible JR, Rougier GW, Novacek MJ, Asher RJ. 2007. Cretaceous eutherians and Laurasian origin for placental mammals near the K/T boundary. *Nature* 447:1003–1006.

Wilczynski W. 1984. Central neural systems subserving a homoplasous periphery. *Am Zool* 24:755–763.

Wild JM, Williams MN. 2000. Rostral Wulst in passerine birds. I. Origin, course, and terminations of an avian pyramidal tract. *J Comp Neurol* 416:429–450.

Wilson DE, Reeder DM. 2005. *Mammal Species of the World*, 3rd ed. Baltimore, MD: Johns Hopkins University Press.

Wise SP. 2017. The evolution of the prefrontal cortex in early primates and anthropoids. In: Kaas JH, Krobitzer L (eds.), *Evolution of Nervous Systems*, 2nd ed. Amsterdam: Elsevier, pp. 387–422.

Wong P, Kaas JH. 2010. Architectonic subdivisions of neocortex in the galago (*Otolemur garnetti*). *Anat Rec* 293:1033–1069.

Wu CC, Karten HJ. 1998. The thalamo-hyperstriatal system is established by the time of hatching in chicks (*Gallus gallus*): a cholera toxin B subunit study. *Vis Neurosci* 15:349–358.

Wu CW, Bichot NP, Kaas JH. 2000. Converging evidence from microstimulation, architecture, and connections for multiple motor areas in the frontal and cingulate cortex of prosimian primates. *J Comp Neurol* 423:140–177.

Wullimann MF. 2011. Basal ganglia: insights into origins from lamprey brains. *Curr Biol* 21:R497–R500.

Wullimann. 2017a. Nervous system architecture in vertebrates. In: Shepherd SV (ed.), *The Wiley Handbook of Evolutionary Neuroscience*. Chichester, UK: Blackwell-Wiley, pp. 236–278.

Wullimann MF. 2017b. Should we redefine the classic lateral pallium? *J Comp Neurol* 525:1509–1513.

Wynne B, Güntürkün O. 1995. Dopaminergic innervation of the telencephalon of the pigeon (*Columba livia*): a study with antibodies against tyrosine hydroxylase and dopamine. *J Comp Neurol* 357:446–464.

Xia A, Liu X, Raphael PD, Applegate BE, Oghalai JS. 2016. Hair cell force generation does not amplify or tune vibrations within the chicken basilar papilla. *Nat Commun* 7:13133.

Xu X, Zhou Z, Wang X, Kuang X, Zhang F, Du X. 2003. Four-winged dinosaurs from China. *Nature* 421:335–340.

Yamaguchi K, Goto N. 2006. Development of the human magnocellular red nucleus: a morphological study. *Brain Dev* 28:431–435.

Yopak KE, Pakan JMP, Wylie D. 2017. The cerebellum of nonmammalian vertebrates. In: Kaas JH, Striedter GF (eds.), *Evolution of Nervous Systems*, Vol. 1, 2nd ed. Amsterdam: Elsevier, pp. 373–385.

Young JM, Trask BJ. 2007. V2R gene families degenerated in primates, dog and cow, but expanded in opossum. *Trends Genet* 23:212–215.

Yu YSW, Graff MM, Bresee CS, Man YB, Hartmann MJZ. 2016. Whiskers aid anemotaxis in rats. *Sci Adv* 2:e1600716.

Zeier H, Karten HJ. 1971. The archistriatum of the pigeon: organization of afferent and efferent connections. *Brain Res* 31:313–326.

Zhou F-M, Liang Y, Dani JA. 2001. Endogenous nicotinic cholinergic activity regulates dopamine release in the striatum. *Nat Neurosci* 4:1224–1229.

7
Synthesis
Patterns and Principles

Vertebrates originated more than 550 million years ago, in the Ediacaran Period, and then underwent a number of major transitions as they diversified. In the preceding chapters we reviewed those transitions, stressing how they were reflected in the sensory, motor, and central nervous systems of the respective animals. In the following section, we briefly review some highlights of those earlier chapters. Then we step back to look for general patterns and trends, asking if they can be explained by evolutionary rules or principles.

7.1. Major Transitions in Vertebrate Phylogeny

Many of the evolutionary changes we discussed in the preceding chapters represent what one might call "key innovations" (Hunter, 1998). However, most of the changes occurred in a piecemeal fashion over millions of years (Donoghue and Purnell, 2009), rather than suddenly, and arose not in isolation from the rest of the body, but in concert with a variety of other modifications. Still, most of the major changes we discussed are likely to have played a major role in the respective lineage's success, as reflected mainly in its longevity and in the number of species that it spawned. In the following subsections, as well as Section 7.2, we briefly reprise the principal changes in vertebrate bodies, behavior, and ecology; later in the chapter we elaborate on some of the associated changes in the brain.

7.1.1. The Origin of Vertebrates

The earliest vertebrates were inconspicuous suspension feeders that somehow made it through the sizable extinction event at the end of the Ediacaran period (see Figure 2.3 in Chapter 2). Then, during the Cambrian, early vertebrates evolved pharyngeal muscles that allowed them to pump water through their pharynx, thereby increasing the rate at which the animals could obtain food. Accompanied by the evolution of vascularized gills, the increased water flow also facilitated gas exchange, enabling early vertebrates to evolve larger bodies and become more powerful swimmers.

Brains Through Time. Georg F. Striedter and R. Glenn Northcutt, Oxford University Press (2020). © Oxford University Press.
DOI: 10.1093/oso/9780195125689.001.0001

These changes in the body of early vertebrates were accompanied by major transformations of the sense organs. Particularly striking is the transformation of a small unpaired photoreceptor organ into a pair of relatively large, image-forming eyes. They allowed early vertebrates to see potential food items, landscapes, and predators at a distance. Early vertebrates also evolved a complex vestibular apparatus, which helped them become more efficient swimmers and stabilized retinal images during locomotion, thereby minimizing vision blur. In addition, early vertebrates evolved mechanosensory and electrosensory lateral lines, taste buds, and a well-developed olfactory system that included paired olfactory bulbs. Meanwhile, the brains of early vertebrates increased in size and neuron number. Importantly, they featured a midbrain and a telencephalon. Despite a recent suggestion that amphioxus adults possess a telencephalon-like brain region (Benito-Gutiérrez et al., 2018), the preponderance of current evidence suggests that both of these brain divisions are vertebrate innovations.

Most of these transformations were linked to the evolution of two novel embryonic tissues, namely placodes and the neural crest. Dramatic changes also happened at the genetic level, as the entire genome seems to have duplicated twice during early vertebrate evolution. Collectively, this multifaceted set of evolutionary changes allowed at least some vertebrates to survive a major mass extinction at the end of the Cambrian period, which probably resulted from global cooling.

7.1.2. The Emergence of Jawed Vertebrates

Vertebrates evolved moveable jaws 450–500 million years ago and did so gradually, experimenting with various structural designs. An obvious function for jaws is to grab prey, but closing the mouth also improved the ability of early jawed vertebrates to pump water across their gills. Around the same time, jawed vertebrates evolved paired fins, which helped them steer during swimming. Collectively, these innovations made the early jawed vertebrates more efficient swimmers and fiercer predators. Many early vertebrates evolved heavy body armor to fend off predators, but those heavily armored lineages (including most placoderms) died out at the end of the Devonian (see Figure 3.8 in Chapter 3). In contrast, the lighter, more agile species made it through that period-ending extinction.

Along with becoming more efficient swimmers and hunters, early jawed vertebrates improved their ability to sense head rotations by adding a third semicircular canal. The addition of a proper cerebellum likely helped to process this vestibular information and, more generally, improve sensorimotor coordination. Large cerebellum-like structures were also used to process inputs from the lateral line sensory systems. In addition, early jawed vertebrates evolved a novel class of chemosensory receptors, the vomeronasal receptors, and expanded their olfactory receptor repertoire. These sensory modifications helped early jawed vertebrates to locate and identify potential prey at greater distances; they also helped the animals to navigate across familiar territory. Indeed, we propose to resurrect the old idea that the olfactory systems were of major importance to early vertebrates, including the early gnathostomes, and dominated telencephalic processing.

Equipped with these innovations and modifications, jawed vertebrates began to diversify in the Ordovician period and flourished in the Devonian, which is why the latter period is often called the "Age of Fishes." By the end of the Devonian, the jawed vertebrates comprised three major radiations that survived to today, namely the cartilaginous, ray-finned, and lobe-finned fishes (see Figure 3.8 in Chapter 3). The cartilaginous fishes eventually gave rise to modern sharks and rays, whereas the ray-finned fishes gave rise to teleosts, which comprise roughly half of all living vertebrate species. Both cartilaginous and ray-finned fishes underwent further modifications and innovations, including major changes to the jaws. In this book, we have barely scratched the surface of what could be said about these fascinating lineages, partly because relevant data are scarce, especially for cartilaginous fishes, but also because we decided to focus more heavily on the lobe-finned fishes and their descendants, the tetrapods. We partially compensate for this intentional neglect by emphasizing cartilaginous fishes and teleosts more heavily in later sections of the present chapter.

7.1.3. The First Terrestrial Vertebrates

When aquatic oxygen levels are as low as they were at the end of the Devonian, many aquatic vertebrates gulp air, which can carry a much higher concentration of oxygen than water. Some of these air-breathing vertebrates could pull themselves onto and across land, using modified pectoral fins. Over millions of years, the pectoral and pelvic fins of these lobe-finned, tetrapod-like fishes gradually transformed into four sturdy limbs (see Figure 4.6 in Chapter 4). The vertebral column of these animals also stiffened dorsoventrally to prevent the body from sagging and dragging on the ground. Although these changes allowed the early tetrapods to walk on land, the animals still needed to stay close to water, both for reproduction and to prevent desiccation of the adults. Aside from providing easier access to oxygen, moving onto land allowed early tetrapods to escape from large aquatic predators and to dine on the many invertebrates, especially insects, that had preceded them into terrestrial habitats.

With the invasion of land came major changes in the motor and sensory systems. For instance, early tetrapods evolved a complex set of limb muscles that required novel mechanisms of neural control. They also needed to evolve novel methods for capturing prey, such as tongue protrusion and "pouncing," because sucking prey into the mouth by expanding its cavity works well in water but not air. Eyes that are well adapted for vision under water likewise do not function well in air (see Figure 4.10 in Chapter 4). However, once the eyes have been appropriately modified, vision in air can work over much longer distances. Some of these modifications occurred before the origin of terrestrial tetrapods, in lobe-finned fishes that were still fully aquatic but spent time at the water surface, looking onto land. Early tetrapods also modified their olfactory system to detect a larger variety of airborne odorants (see Figure 4.15 in Chapter 4). In contrast, the mechanosensory and electrosensory lateral line systems are virtually useless on land, which is why the terrestrial stages of early tetrapods either lost them or buried them under an epithelium, to be "re-used" for sensing underwater vibrations when the animals return to water for breeding.

The brains of early tetrapods did not become larger or more complex with the invasion of land; if anything, they became simpler. This trend toward brain simplification was carried further in some lungfishes and urodele amphibians, both of which exhibit unusually large neurons that do not migrate far from where they were born and tend not to differentiate into discrete cell groups. Early tetrapods were probably less "paedomorphic" than modern salamanders and lepidosirenid lungfishes (see Section 4.5.1 in Chapter 4), but their brains were nonetheless relatively simple, perhaps as an adaptation to low levels of oxygen under water.

One lineage of early tetrapods probably gave rise to all modern amphibians, which then branched into urodeles, caecilians, and anurans (see Figure 4.7 in Chapter 4). Of these three groups, the anurans were the most successful, mainly because of their hopping mode of locomotion and their diverse modes of reproduction, including the evolution of a highly effective tadpole stage. Most anurans also evolved tympanic ears, which allowed them to hear high-frequency airborne sounds. In tandem with this enhanced hearing capacity, most anurans evolved more complex auditory pathways and species-specific vocalizations, used mainly in the context of reproduction. In fact, the interplay between anuran vocalizations and hearing preferences may have been a major factor promoting speciation in this clade (Wilkins et al., 2013). A separate lineage of early tetrapods gave rise to amniotes, which became more fully independent of water and, thus, came to dominate the terrestrial habitat.

7.1.4. The Origin and Diversification of Ectothermic Amniotes

Some early tetrapods evolved "amniotic eggs," which resist dehydration but still allow for high levels of gas exchange. This remarkable innovation allowed early amniotes to lay their eggs on land, albeit in moist environments, and thus become fully terrestrial. Early amniotes also minimized water loss from their own bodies by evolving lipid-covered scales. Such scales limit gas exchange across the skin, but early amniotes compensated for this constraint by evolving larger, more complex lungs, as well as movable ribs that make it possible to suck air into the lungs, rather than pushing it in. Locomotion across land was facilitated by the evolution of a lighter skeleton and longer legs (see Figure 5.11 in Chapter 5). Early tetrapods also evolved a flexible neck, which facilitated prey capture. Even more important was the evolution of "kinetic skulls" in the ancestors of modern lizards and snakes; such skulls allow the jaws to be opened extremely wide and, thus, take in much larger prey. In conjunction with all these changes in the skeletomotor system, early amniotes evolved more complex muscle spindles and topographically arranged motor neuron pools, both of which probably improved sensorimotor control.

Early amniotes evolved fully terrestrial vision and an expanded vomeronasal system. Several amniote lineages (synapsids, lepidosaurs, and archosaurs) also evolved tympanic ears and a more slender stapes, which greatly improved their ability to detect airborne sounds. Along with these changes in the sense organs came a major expansion of the sensory pathways in the brain. Especially expanded were the dorsal midbrain, the dorsal diencephalon, and the telencephalic pallium. Particularly interesting is that the ventral pallium expanded dramatically in the lineage leading to modern reptiles and birds (i.e., sauropsids), while the dorsal pallium expanded in the lineage leading to mammals (i.e., synapsids). The synapsid dorsal pallium also evolved a unique "six-layered" organization, which is why the designation "neocortex" is appropriate. As the ascending visual, auditory, and somatosensory pathways to the pallium expanded in amniotes, the olfactory projections became more restricted. In general, the telencephalon did not just expand in amniotes; it was dramatically reorganized. We come back to this issue in Section 7.5.

Amniotes flourished throughout the Permian, Jurassic and Triassic periods, all the way to the end of the Cretaceous, 66 mya (see Figure 5.1). The sauropsids in particular did very well, diversifying into many successful lineages, including the dinosaurs. Early synapsids (i.e., stem mammals) held their own during this period (see Figure 6.3), but the crown therians (placental mammals and marsupials) did not become highly successful until all of the large dinosaurs died out during the end-Cretaceous mass extinction. Some sauropsids did make it through this terrible period, when a massive asteroid struck the planet, but the surviving sauropsids never again reached the spectacular body sizes of the extinct dinosaurs. Instead, the remaining dinosaurs gave rise to birds, most of which were rather small. Indeed, no land animals greater than about 25 kg survived the end-Cretaceous extinction, presumably because the terrestrial food chains collapsed. This is a good reminder that, in the face of a mass extinction, having a large brain and well-developed sensorimotor systems may not suffice to keep the species alive.

7.1.5. The Rise of Endothermic Amniotes

Both birds and mammals evolved the ability to generate internal body heat, which makes their body temperature less dependent on the external environment. This remarkable achievement involved many related innovations, such as the evolution of respiratory turbinates and body hair or feathers to reduce heat loss. Birds and mammals also increased their basal metabolic rate and evolved a diaphragm or (in archosaurs) diaphragm-like muscles to boost their respiratory volume. Importantly, all of these changes occurred independently, and thus convergently, in mammals and birds. Indeed, the evolution of these two lineages features a great number of additional convergences. For example, the ancestors of both birds and mammals dramatically reduced their body size (see Figures 5.10 and 6.4 in Chapters 5 and 6]), which allowed them to feed mainly on insects and may have been a major reason why they survived the end-Cretaceous extinction. They also evolved an erect stance and parasagittal (rather than sprawling) gait. Moreover, birds and mammals both elongated their cochleae, allowing for better high-frequency hearing and improved frequency discrimination. Finally, birds and mammals both evolved much larger brains, relative to body size; we will return to this subject shortly.

Although birds and mammals evolved a large number of convergent similarities, they also diverged in various respects. Especially important is that early mammals became nocturnal, whereas birds remained in the diurnal niche. In association with this difference, early mammals evolved highly light-sensitive, rod-dominated retinas and lost one or more of the cone opsins used for color vision. Perhaps to compensate for the reduced importance of vision in their nocturnal niche, early mammals expanded their olfactory and vomeronasal receptor repertoires and evolved the ability to "sniff" for odorants. In contrast, birds have a relatively small (though still useful!) olfactory system and have lost their vomeronasal system entirely. An obvious divergence in the brains of birds and mammals is that the former possess a large optic tectum, while mammals have greatly reduced their homolog of this structure, the superior colliculus (though it remains an important structure). Within the telencephalon, birds continued to expand their ventral pallium and extended this expansion to the lateral pallium (see Figure 5.24 in Chapter 5). Mammals, in contrast, continued to expand their dorsal pallium, especially during later stages of mammalian phylogeny, when bodies and brains increased in absolute size. Both groups retained their medial pallium, which is called the hippocampus in birds and mammals and the medial cortex in other amniotes.

Both birds and mammals have been extremely successful during the last 50 million years (see Figure 5.14). Between them, they account for almost half of all vertebrate species, with teleosts representing most of the rest. Why were these two lineages so successful on land? Mammals did well by invading the nocturnal niche, whereas birds took to the air, evading their flightless predators and in pursuit of novel foods. What both lineages shared is that they became endothermic, which allowed them to feed even when environmental temperatures were relatively cold. Moreover, their high metabolic rate allowed them to support large brains, facilitating both food finding and pursuit, as well as various intraspecific interactions that probably increased their speciation rates.

Building and maintaining large brains and sense organs requires a great deal of metabolic energy, which means that birds and mammals have little choice but to consume a large number of calories. This constraint could become problematic if there were another mass extinction that disrupts mammalian and avian food supplies (some say that such an extinction has already begun). In times of global ecological disturbances, highly specialized and energy-demanding species often do less well than species that are more frugal and less specialized. Of course, the outcomes of mass extinctions are hard to predict. As the preceding pages have hopefully made clear, vertebrate phylogeny comprises a long series of unique historical events. That said, one can certainly step back and look for large-scale patterns and trends in vertebrate phylogeny, which is what we shall now attempt.

7.2. General Patterns of Evolutionary Change

When we step back and look at all of the evolutionary changes mentioned in this book, we can discern some overarching patterns, trends, and principles. In the following sections, we first review some very general patterns and trends, as well as their implications. We defer a discussion of changes in brain size, brain complexity, and neuronal circuits to Section 7.3.

7.2.1. Changes in Taxonomic Diversity, Body Size, and Complexity

As one looks across the broad swath of vertebrate phylogeny, it is obvious that there have been some marked shifts in the diversity of major lineages. Placoderms, for example, were very successful until the end of the Devonian period, when they were essentially wiped out. Similarly, non-avian dinosaurs dominated the Mesozoic era but died out at the end of the Cretaceous period. These lineage-ending mass extinctions left large holes in the globe's ecology, but those lacunae were quickly (by geological standards) filled through the diversification of other lineages. Thus, the disappearance of placoderms was accompanied by major radiations of the cartilaginous and bony fishes (Figure 7.1), and the extinction of the dinosaurs was soon followed by a great diversification of mammals and birds (see Figure 5.14 in Chapter 5).

This is not to say that bony and cartilaginous fishes "functionally replaced" all of the placoderms, or that birds and mammals filled all the niches previously occupied by dinosaurs. Nor would it be accurate to say that the extinguished lineages had been in direct competition with the more successful ones. However, the extinctions of major radiations did, at some key moments in vertebrate phylogeny, open up new sets of ecological possibilities, into which the surviving lineages opportunistically diversified. In some cases the new species may have filled empty niches, but they also created a large number of novel ones. Indeed, the overall diversity of vertebrates increased exponentially over the last 500 million years, periodic mass extinctions notwithstanding (see Figure 5.14 in Chapter 5). This enormous increase in diversity implies a corresponding increase in ecological complexity.

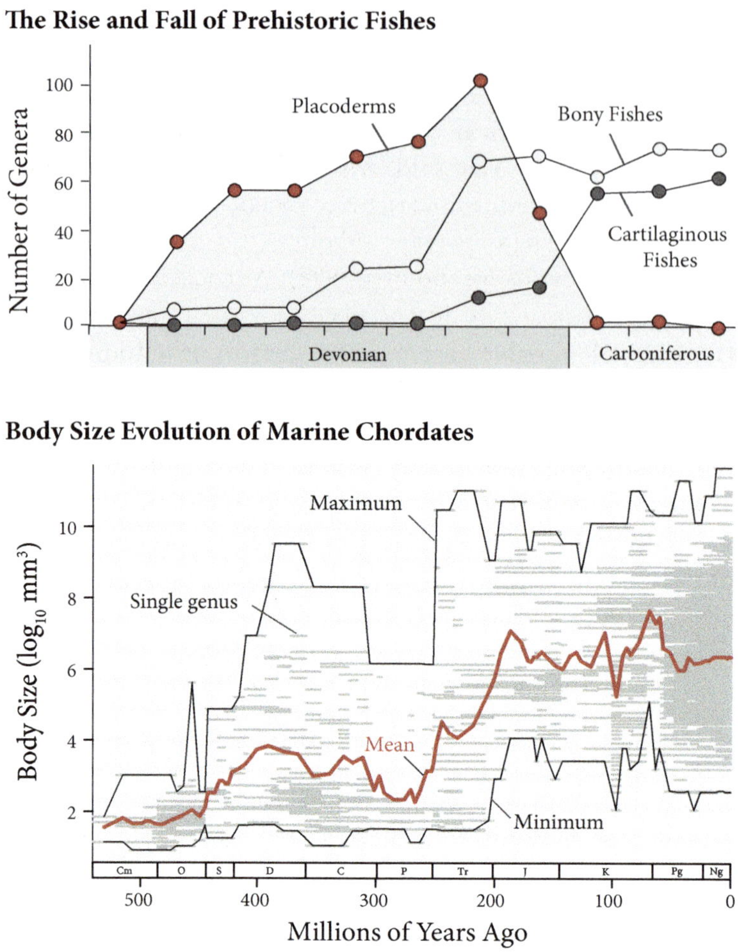

Figure 7.1 Historical trends in vertebrate phylogeny. As shown in the top graph, ray-finned fishes and bony fishes diversified shortly before and after placoderms became extinct (adapted from Carr, 1995). The bottom graph shows how mean, minimum, and maximum body sizes changed across vertebrate phylogeny; it is based on data from 1,182 genera, which are represented by the gray horizontal bars (Heim et al., 2015). Geological periods are indicated by their abbreviations along the *x*-axis of the bottom graph (please refer to Figure 1.3 for their full names).
Courtesy of Noel A. Heim.

Even as different vertebrate lineages pursued their own, idiosyncratic paths, they tended to increase in body size. The immediate ancestors of chordates were probably quite small, but vertebrate body sizes have generally increased by several orders of magnitude since the Cambrian period, albeit independently in different lineages. This trend exists not only for maximum and average body size, but also for minimum body size (Figure 7.1). An analysis of more than 17,000 marine animal genera (including chordates but also several additional phyla) revealed that this evolutionary increase in body size, often referred to as "Cope's rule" (Stanley,

1973), was a general phenomenon and exceeded what one would expect from a "random walk" model of body size evolution (Heim et al., 2015). In other words, the trend was caused at least in part by selection for increased body size. Of course, as we noted in Chapters 5 and 6, the ancestors of birds and mammals decreased their body size and, thus, are clear exceptions to Cope's rule. However, even in these clades, maximum body size increased again as the respective lineages diversified.

In tandem with the general increase in body size, vertebrates became increasingly complex, at least on average (Bonner, 1988). We concede that there is no ideal, objective way to quantify complexity and that *bigger* doesn't necessarily mean *more complex* (Ruse, 1993). However, over the long run, vertebrates certainly seem to have increased in the "number of distinguishable kinds of components" (Bullock, 2002) that their bodies contain. In particular, vertebrate jaws, limbs, and brains have become more complex, at least on average, since the Devonian. There have been some reversals in these trends, notably the simplification of the jaw in stem mammals, the loss of limbs in many squamates (as well as caecilian amphibians and cetaceans), and neuronal simplification in lepidosirenid lungfishes and urodele amphibians. Still, an overarching trend toward increased complexity is apparent, at least to us.

7.2.2. Evolutionary Divergence and Convergence

As Darwin recognized, evolution by natural selection causes species to diverge from one another in both structure and function. Even random walk models of evolution generate divergence between species, though this divergence increases only with the square root of phylogenetic distance (Letten and Cornwell, 2014). It is not surprising, therefore, that vertebrates have diverged drastically. They may be built according to some common "vertebrate body plan" (Haeckel, 1866), but the implementation of that plan varies enormously across the various vertebrate lineages. No manta ray will be mistaken for a hummingbird! Even at the molecular level, where conservation is often said to be highest, divergence has been substantial (Striedter, 2019). Aside from individual genes diverging in their DNA sequence, vertebrate evolution has featured duplications of the entire genome, dramatic expansions of select gene families, losses of some genes and gene families, as well as major changes in gene regulation (e.g., Lee et al., 2007; Cotney et al, 2013; Vierstra et al., 2014).

Given that evolutionary divergence is to be expected, it is surprising that strikingly similar features sometimes pop up in distantly related lineages. Two of the best known instances of such convergent evolution are the evolution of flight in birds, pterosaurs, and bats, and the evolution of very similar body shapes in dolphins, lamnid sharks, tunas, and Jurassic ichthyosaurs (Shadwick, 2005; Lingham-Soliar, 2016). Although such convergences have long been recognized, they used to be regarded as relatively rare. This has begun to change, largely because comparative molecular data have frequently revealed that many superficially similar species, once thought to be close relatives, are only distantly related to one another (Conway Morris, 2003; Losos, 2017). Moreover, many interesting features have now been mapped across a wide diversity of species, including the basal lineages that are so crucial to reconstructing the phylogenetic history of individual traits. For example, mapping the presence of limbs onto squamate phylogeny reveals that limbs were lost in more than 20 different lineages, not just in snakes (Wiens et al., 2006). Another good example is viviparity, the birthing of live young, which evolved independently in mammals, coelacanths, several groups of cartilaginous fishes, some teleosts, and over 100 times among lizards and snakes (Dulvy and Reynolds, 1997; Wourms, 1981; Blackburn, 2006). Even some placoderms had intra-uterine embryos with umbilical cords (Long et al., 2008). A clear example of convergence in the neural realm is the evolution of retinal foveae in primates and various sauropsids (see Section 6.5.1 in Chapter 6).

The simplest explanation for the prevalence of evolutionary convergence is that, for many biological problems, there are just a limited number of excellent solutions. For example, the "thunniform" body shape of tunas, dolphins, and ichthyosaurs is ideal for maximizing swimming speed in the open ocean. Similarly, the nasal turbinates of birds and mammals are an ideal solution for the problem of retaining body heat. Another interesting example is that many feeding- and respiration-related features in amphibian tadpoles evolved independently in several lineages, presumably because there is only a limited number of ways to build tadpoles that are well-adapted for scraping food off the substrate (Roelants et al., 2011). Given the existence of such "design rules," it is not surprising that different species encountering similar problems, either at different locations on the globe or in different geological periods, have sometimes evolved similar solutions.

A different kind of potential explanation posits that convergence is based on the "deep homology" of underlying genes (Shubin et al., 2009). Indeed, it does appear that convergence is more frequent between species that are closely related—and therefore likely to share more genes—than among distant relatives (e.g., Blackburn, 2006; Ord and Summers, 2015). However, most convergent similarities probably involve at least some lineage-specific genes, as well as genes that were uniquely co-opted into the feature's development in one or both of the convergent lineages (Wagner, 2007). For example, complex image-forming eyes evolved independently in vertebrates and cephalopods (e.g., squid), but most of the genes that are differentially expressed in squid eyes appear to be unique to cephalopods (Yoshida et al., 2015). Moreover, more than half of the genes expressed in the eyes of human fetuses are not expressed in cephalopod eyes, even though they are deeply conserved (as genes).

In general, we think it is best not to conflate the homology of genes with the homology or non-homology of higher-level characters (see Chapter 2). Thus, convergent evolution can be studied at several different levels of biological organization, including that of molecules (Liu et al., 2014a, b), and explanations of convergence can likewise be sought at (and across) multiple levels of analysis (e.g., van Dyke et al., 2014). In any case, the study of convergent evolution is one of the best windows we have into the rules by which evolution operates. With that in mind, let us now discuss the patterns and trends that have characterized the evolution of vertebrate brains. Specifically, let us discuss evolutionary changes in brain size, in the size of individual brain regions, in the fundamental "Bauplan" of vertebrate brains, and in some key neuronal circuits.

7.3. Trends in the Evolution of Brain Size

Comparative neurobiologists have amassed more data on vertebrate brain and body sizes than they have collected for any other neural trait. These data are usually log-transformed and then subjected to some sort of regression analysis that allows investigators to calculate for each species how much its brain size deviates from the best-fit regression line for the larger taxonomic group. Species with negative deviations (i.e., negative residuals) can be interpreted as having brains that are smaller than one would expect for a species of their lineage and body size. In contrast, species with positive residuals are said to have larger than expected brains, relative to body size. Importantly, this variation in relative brain size (aka encephalization) can be mapped onto the relevant phylogenies. Parsimony analyses (see Chapter 1) can then be used to determine which lineages exhibited decreases in relative brain size and which became more highly encephalized. The following sections summarize some of those results.

7.3.1. Independent Increases in Relative Brain Size

A comparative analysis of brain and body sizes across the vertebrates reveals that relative brain size increased numerous times, in many different lineages. First, consider the cyclostomes (Figure 7.2). Hagfishes clearly have larger brains, relative to body size, than lampreys do. Since hagfishes also have more complex brains than either lampreys or the invertebrate chordates (see Chapter 2, Figure 2.22), we can infer that relative brain size probably increased substantially in the lineage leading to modern hagfishes. Relative brain size also increased at least once within lampreys, within the genus *Lampetra,* and it probably decreased in the genus *Ichthyomyzon*.

Even more dramatic variation is observed among the cartilaginous fishes (Figure 7.3). In particular, manta rays and some of their close relatives are much more encephalized than other rays and skates. Similarly, hammerhead sharks and a few other galeomorph sharks have much larger brains, relative to body size, than squalomorph sharks and holocephalans. Based on this distribution, we conclude that relative brain size increased at least three times within the cartilaginous fishes. Among the ray-finned fishes, relative brain size changed even more frequently.

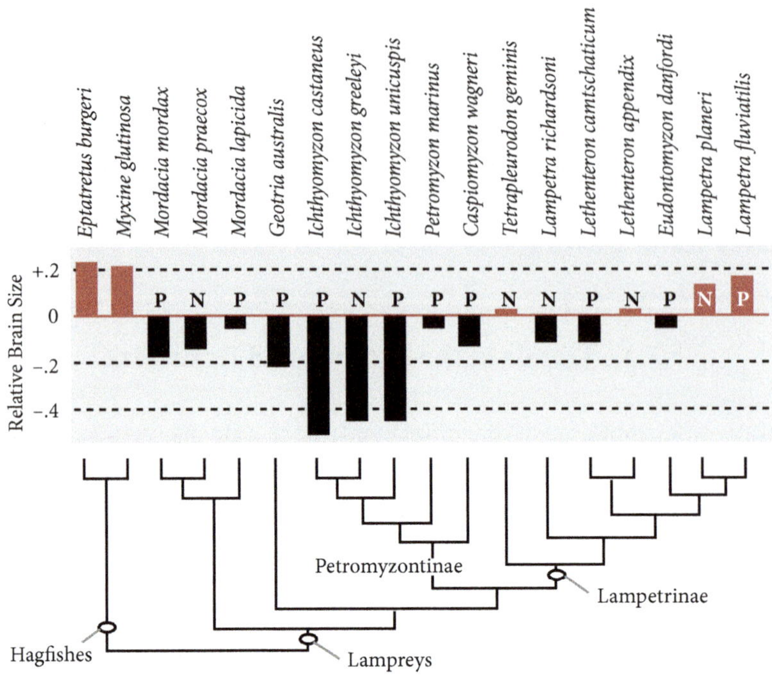

Figure 7.2 Brain-body scaling in cyclostomes. Brain and body sizes for 16 lamprey and two hagfish species were regressed against one another and used to calculate residuals, which represent "relative brain size" (aka degree of encephalization) for each species. Given the phylogenetic relationships of these species, we can infer that hagfishes are more highly encephalized than lampreys (positive residuals are shown in red), that members of the genus *Ichthyomyzon* have relatively small brains relative to body size, and that members of the genus *Lampetra* are more highly encephalized than the other lampreys in the dataset. Statistical speaking, lamprey species that parasitize other fishes (P) tend to be less encephalized than non-parasitic species (N; Salas et al., 2017), but close relatives tend to have similar relative brain sizes even if they differ in feeding mode.
Based on data in Salas (2016).

The available data suggest at least 10 independent increases in relative brain size across all marine teleosts (Figure 7.4), as well as a handful of substantial decreases in encephalization.

Moving from fishes to tetrapods, we again find multiple changes in relative brain size. Urodeles probably became less encephalized than their ancestors (see Striedter, 2005), perhaps in conjunction with the secondary simplification of their brains (paedomorphosis; see Chapter 4). However, two major increases in relative brain size occurred with the origins of birds and mammals. Within each of these lineages, additional increases in relative brain size are evident. As noted in Chapter 5, parrots and songbirds have relatively large brains, even for birds (see Figure 5.20; Olkowicz et al., 2016). Relatively large brains are also found in owls

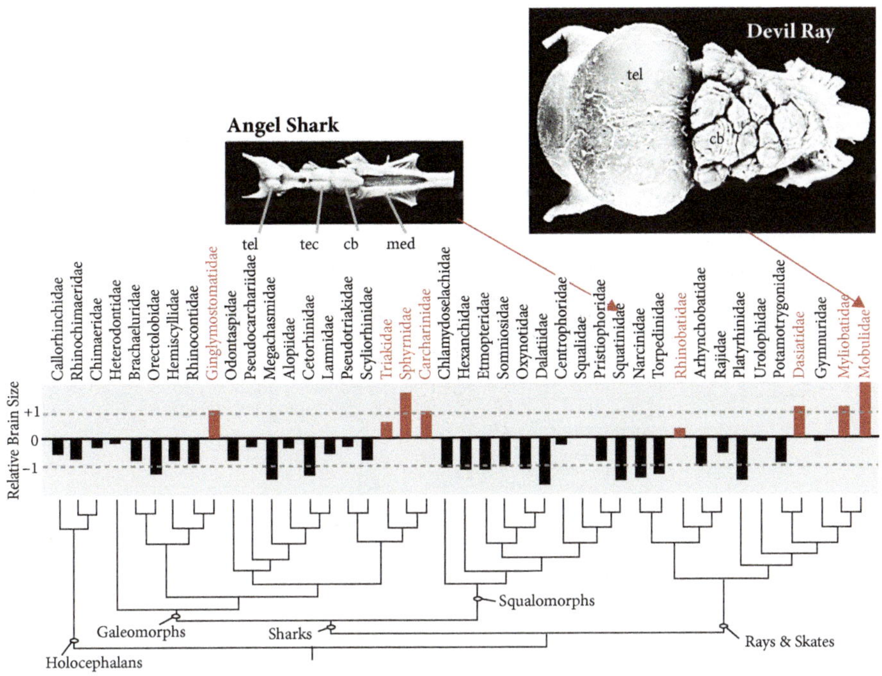

Figure 7.3 Brain-body scaling in cartilaginous fishes. Relative brain size in this diagram represents the residuals calculated from a regression of brain size versus body size. When relative brain size (aka degree of encephalization) is plotted against phylogeny, it becomes apparent that relative brain size increased at least three times within the cartilaginous fishes (positive residuals are shown in red). Some decreases in relative brain size also occurred, but a parsimony analysis suggests that the increases were more dramatic. Especially encephalized are the devil and manta rays (Mobulidae and Myliobatidae), the hammerhead, requiem and hound sharks (Sphyrnidae, Carcharinidae, and Triakidae), and the carpet and nurse sharks (Ginglymostomatidae). In general, the highly encephalized species tend to have a massive telencephalon and a very large, complexly folded cerebellum.
Based on data in Yopak (2012). Photographs from Northcutt (1985), with permission from John Wiley & Sons.

and their relatives, as well as pelicans and hornbills (Figure 7.5; Sayol et al., 2016b; Fristoe et al., 2017). Among mammals, primates are highly encephalized, and so are toothed whales (Figure 7.6; Striedter, 2005; Boddy et al., 2012; Herculano-Houzel, 2012). Importantly, all of these increases in relative brain size occurred independently of one another. In contrast, decreases in relative brain size were relatively rare, at least within the amniotes.

In theory, evolutionary increases in relative brain size could stem from decreases in body size, rather than increases in absolute brain size. However, in most vertebrate lineages, absolute body weight increased more frequently than it decreased (see Figure 7.1), implying that "encephalization by dwarfism" or by reducing skeletal density was not a major factor in the evolution of vertebrate brain size. One

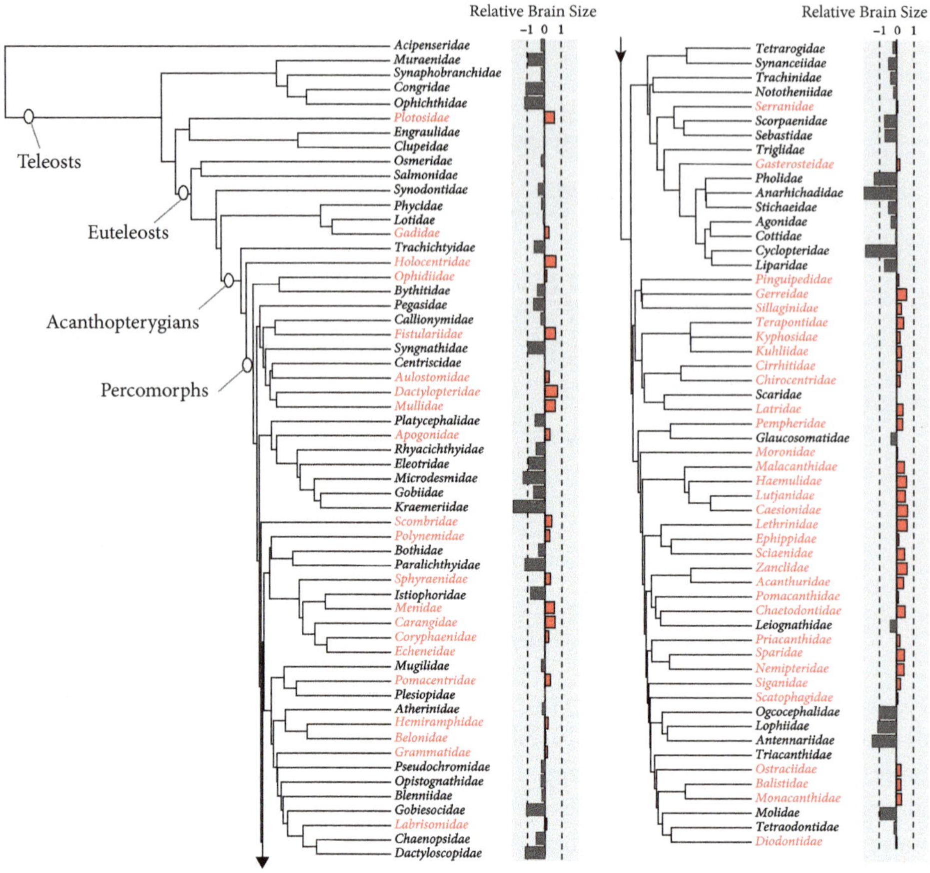

Figure 7.4 Brain-body scaling in marine ray-finned fishes. Relative brain sizes were calculated as residuals from a log-log regression of brain mass on body mass and then averaged for each family.

The raw data were obtained from http:/www.fishbase.org, peer-reviewed papers, and new collections (see Iglesias et al., 2015). The time-calibrated phylogeny was generated by Rabosky et al. (2013). Courtesy of Dan Warren, Alex Dornburg, and Teresa L. Iglesias.

might also argue that the number of neurons in a brain is a better measure for comparative analyses than brain volume (Herculano-Houzel, 2016). Indeed, it has been shown that neuron density is unusually high in large-brained songbirds and parrots, relative to other birds (Olkowicz et al., 2016), and in large primates, relative to other mammals of similar body size (Herculano-Houzel et al., 2007). However, neuron numbers and densities are not yet available for many lineages, including all anamniotes. Therefore, for the time being, we are constrained to use brain size for broad comparative analyses. Based on these surveys, we conclude that vertebrate phylogeny involved far more increases than decreases in relative and absolute brain size.

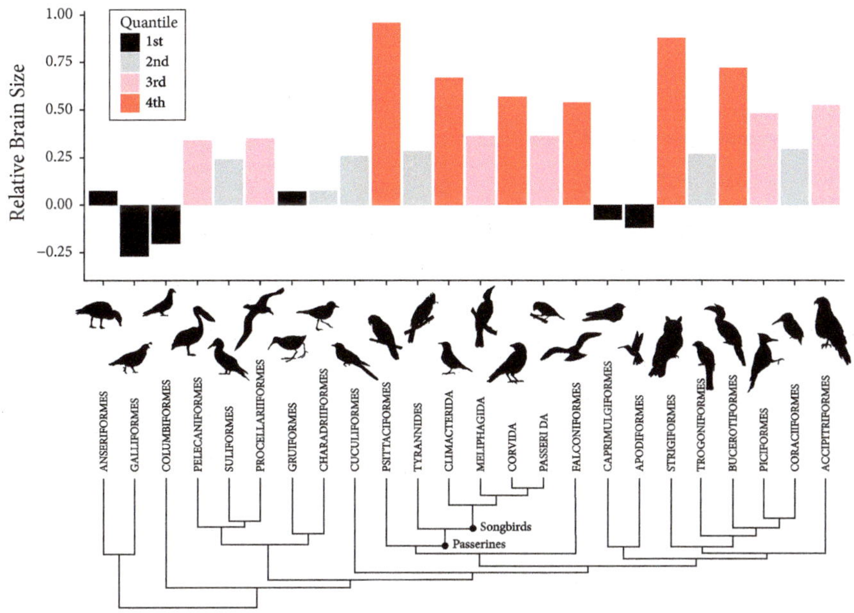

Figure 7.5 Relative brain size evolution in the major avian lineages. The diagram represents the mean residuals of a phylogenetically corrected log-log regression between brain size and body size across multiple orders of birds (infraorders for passerines). A phylogenetic analysis of these data reveals that relative brain size increased several times independently, both within songbirds (e.g., corvids) and in some other avian orders, notably parrots and owls (Psittaciformes and Strigiformes, respectively). The most basal avian lineages tend to have relatively small brains, but Caprimulgiformes and Apodiformes (e.g., hummingbirds) have evolved relatively small brains secondarily.
Based on data in Sayol et al. (2018); courtesy of Ferran Sayol.

7.3.2. Functional Correlates of Evolutionary Changes in Brain Size

Many studies have sought to explain evolutionary changes in brain size by examining how those changes correlate with variation in life history, behavioral complexity, or cognitive prowess. Since this literature is very large, a few examples must suffice.

Among lampreys, parasitic species tend to have somewhat smaller brains, at least on average, than species that do not feed on the blood of other fishes (Figure 7.2; Salas et al., 2017). Within cartilaginous fishes, relative brain size is highest in species that live either on complex reefs or in the open ocean; in contrast, deep-sea bottom-dwelling sharks tend to be poorly encephalized (Yopak, 2012). In teleosts as well, deep-sea species tend to have relatively small brains (Iglesias et al., 2015), while fast open ocean predators and reef-dwelling species tend to be highly encephalized (based on an informal analysis of Figure 7.4). In birds, relative brain size correlates

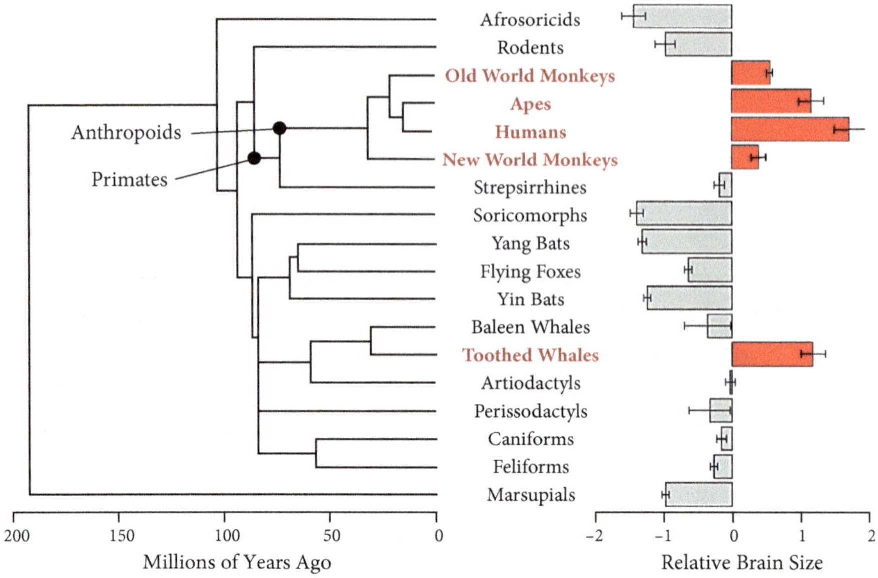

Figure 7.6 Relative brain size evolution in mammals. Shown here is a timetree of mammals together with the results of a phylogenetically corrected log-log regression of brain size against body size for a sample of 1,003 species (some minor lineages were omitted from the diagram). The data show that relative brain size increased repeatedly within the primate lineages, especially in anthropoid primates. Relative brain size increased independently in the toothed whales.
Courtesy of Jeroen Smaers and Daphne Hudson.

most strongly with cognition-related parameters, such as living in highly variable environments, invading new habitats, inventing novel feeding methods, and general problem-solving (e.g., Sol et al., 2005; Overington et al., 2009; Sayol et al., 2016a; b). Among mammals, brain size has been found to correlate strongly with diet and social complexity, as well as diverse measures related to problem-solving ability (van Dongen, 1998; Lefebvre et al., 2004; Dunbar and Shultz, 2007; MacLean et al., 2014; Benson-Amram, 2016; DeCasien et al., 2017).

These and other efforts to identify functional correlates of relative brain size have yielded an abundance of correlations but little consistency. Some correlations hold across multiple taxonomic groups, but many are specific to just a few taxa, in part because the animals live in very different environments. Even within a lineage, it is not uncommon for different studies to report correlations with different behavioral or ecological parameters. This is not surprising, because neural functions that are useful in one behavioral context might well prove useful in other contexts. It has long been argued, for example, that cognitive capacities that can be used to manipulate objects might also be used to manipulate conspecifics (Humphrey, 1976). Thus, links between relative brain size and ecological parameters do not preclude causal links to social behavior; they are not incompatible hypotheses (Parker, 2015). Nonetheless, it leaves investigators with no simple, overarching answer to the question of why so many lineages have increased their degree of encephalization. This, of course, may be expected if different lineages increased their relative brain size through different mechanisms or for different reasons.

A related issue is that the evolutionary benefits of increased relative brain size, whatever they may be, must be "paid for." Having a large brain requires a high metabolic rate (Mink et al., 1981), which in turn requires a high rate of calorie intake (Fonseca-Azevedo and Herculano-Houzel, 2012). Alternatively or in addition, increased encephalization may require a sizable reduction in other metabolically expensive organs, such as the gut (Aiello and Wheeler, 1995; Tsuboi et al., 2015; Liao et al., 2016). Large brains also require a great deal of energy to develop, which probably explains why many large-brained species leave a relatively small number of offspring per year but invest heavily in their development, either by evolving large eggs or through extensive parental care and long maturation periods (Barton and Capellini, 2011; Isler, 2011; Tsuboi et al., 2015).

Although these findings demonstrate the existence of general principles that govern evolutionary changes in brain size, many of those rules vary across lineages; they are not like the universal principles that scientists (inspired by physics) often strive for (Smith and Morowitz, 1982). For example, relative brain size is not inversely correlated with gut size in birds or non-primate mammals (Jones and MacLarnon, 2004; Isler and Van Schaik, 2006; Navarrete et al., 2011). Those species found other strategies for saving energy. We are left with the relatively simple (but also vague) global rule that the benefits of having a large brain must, somehow, outweigh the metabolic costs. An important aim for future research would be to understand why different lineages adopted different strategies for offsetting the costs of building and maintaining large brains.

Yet another complication in the search for functional correlates of brain size variation is that some behavioral or cognitive measures correlate more strongly with absolute brain size than relative brain size. Across a broad sample of non-human primates, for example, cognitive ability, as assessed by a variety of complementary measures, correlates more robustly with absolute brain size than relative brain size (Deaner et al., 2007). Similarly, absolute brain size was the best predictor of performance on two tasks that require "self-control" across a broad range of species, including some birds as well as various mammals (MacLean et al., 2014). One may argue that some species do poorly on these tasks for reasons other than cognitive capacity (Jelbert et al., 2016), and that quantifying cognitive abilities is notoriously difficult (Macphail, 1982). Still, it is difficult to resist the notion that some aspects of "intelligence" are linked to changes in brain size. In any case, our main point here is that absolute brain size deserves at least as much attention as relative brain size when one is looking for functional correlates of evolutionary variation in brain size, whatever behaviors or cognitive capacities one is examining (Striedter, 2005).

Likewise important is the rapidly increasing data set on neuron numbers in the brains of diverse vertebrates (e.g., Herculano-Houzel et al., 2014). If neurons are the brain's main computational units, then cognitive capacity should scale more tightly with neuron number than with absolute brain size (Herculano-Houzel, 2011). Such a perspective would explain why chimpanzees seem more intelligent than cows, for example: their brains are roughly equal in absolute size but, given the neuronal scaling rules in their respective orders, chimpanzee brains almost certainly contain far more neurons (Kazu et al., 2014). However, to date there have been no statistical tests of the hypothesis that neuron number correlates with cognitive ability. Moreover, one should not neglect the possibility that the size and complexity of individual neurons is an important additional determinant of brain function. A few studies have compared the structure of specific neuron types across species and tried to correlate this variation with cognitive traits (Elston et al., 2005, 2011; Hakeem et al., 2009; Raghanti et al., 2015), but disentangling variation in neuron structure from variations in neuron number, brain size, or other morphological variables, is bound to be difficult.

For now, the main insight garnered from comparing neuron numbers across species is that neuron densities vary among lineages, especially for larger brain sizes. Given this variation, it is not surprising that efforts to discern the functional correlates of variation in brain size have yielded a large and complex array of answers (e.g., Lefebvre, 2012). One potential route forward would be to focus more explicitly on the evolutionary changes in the scaling rules and brain-behavior correlations, asking when and why the correlations changed.

7.4. The Evolution of Brain Region Size

Given the challenges of correlating total brain size with behavioral or ecological parameters, several authors have advocated shifting research attention to the size of specific brain regions, rather than the entire brain (Healy and Rowe, 2007). This research strategy works well for many sensory and motor areas, for which structure-function relationships are fairly obvious. For example, the large size of the vagal lobes in goldfish (see Figure 3.6 in Chapter 3) correlates nicely with the well-developed gustatory sense of these fishes. Similarly, it is easy to understand why stargazers, a group of teleosts that modified some of their eye muscles into electric organs, possess extremely large oculomotor nuclei (Ariëns Kappers, 1941). Additional examples are found in the preceding pages, and many more could be adduced (e.g., Corfield et al., 2015; Wylie et al., 2015).

It is more difficult to correlate the size of "higher brain regions" with specific behaviors or ecological parameters. Part of the problem is that those regions may contribute to multiple behaviors or cognitive processes. In addition, little experimental work has been conducted on higher level brain regions in non-mammalian vertebrates. For example, the anterior portion of the cerebellum has long been known to be enormously enlarged in mormyrid electric fishes (see Figure 3.18 in Chapter 3; Zhang et al., 2011), but the functional correlates of this hypertrophy remain unclear. Its enormous size in mormyrids probably relates to the electrosensory ability of these fishes, but another, distantly related group of electrosensory teleosts, the gymnotoids, perform similar behaviors without the benefit of such a large cerebellum (Bell and Szabo, 1986; Carr and Maler, 1986). Similarly, it is not clear what the various large-brained cartilaginous fishes are doing with their massive cerebellums (Figure 7.3; Figure 3.2 in Chapter 3). We suspect that the cerebellum performs a highly conserved computational function, such as "adaptive feed-forward control"; see Bastian, 2006), but the cerebellum interacts with many different neural circuits, suggesting that its core function can enhance many different behaviors or cognitive processes. If this is true, then one would not expect to find a single, broadly conserved correlation between cerebellum size and a specific behavior or ecological parameter.

Because the telencephalon is the most variable brain region in vertebrates, especially in terms of size, many researchers have sought to determine the functional correlates of that variation. In mammals and birds, telencephalon volume tends to correlate with higher cognitive functions, such as learning, memory, problem-solving, and "general intelligence." Best studied is variation in the size of the hippocampus, which is involved in spatial learning and memory across a wide variety of vertebrates (Bingman et al., 2017). In birds, its proportional size (i.e., its size relative to other brain regions) has been reported to correlate positively with the tendency to store and retrieve food, a behavior that requires good navigational memory (Krebs et al., 1989; Kamil et al., 2001; Ward et al., 2012), as well as memory for what was stored and when (Clayton and Dickinson, 1998). However, the correlation between avian hippocampus size and food caching has been questioned (see Lucas et al., 2004; Pravosudov and Roth, 2013), and some studies indicate that the number of neurons in the hippocampus may be a stronger correlate of food storing than hippocampus volume (Gould et al., 2013). Moving beyond the hippocampus, the proportional size of the non-sensory, associative components of the DVR in birds (see Chapter 5) correlates with the ability to solve problems in novel ways (Sayol et al., 2016a). In primates, too, the proportional size of the "executive brain" (neocortex plus striatum) correlates with innovation rate, tool use, and the ability to learn from conspecifics (Reader and Laland, 2002).

All of these "neuroecological" findings have attracted skeptics (e.g., Bolhuis and Macphail, 2001), but they probably do capture some real structure-function relationships, especially when those correlations are backed up by lesion studies or neurophysiological research. Even so, it is important to note that most of the reported correlations hold only for some lineages. Even for the hippocampus, robust correlations between its size and spatial memory have been reported only for birds and among humans (at least for hippocampal subregions; Maguire et al., 2006). This lack of generality probably stems mainly from the fact that brain regions may vary among species in internal structure or external connections, even if they are homologous (Striedter, 2002). We review some evidence to this effect in the following section.

7.4.1. Changes in Brain Region Complexity and Connections

Brain regions that became enlarged during phylogeny often exhibit an increased number of subdivisions. For example, the pallium of teleosts with a very large telencephalon contains more than 15 distinct subdivisions, whereas only five or so can be identified in zebrafish or other teleosts with a small telencephalon (Figure 7.7). Even more impressive is that hypertrophied brain regions often assume a highly laminar architecture. The large vagal lobes of goldfish, for example, contain 15 distinct laminae (see Figure 3.6 in Chapter 3), even though their smaller homologs in other species exhibit no such laminae. Highly laminar architectures also evolved in

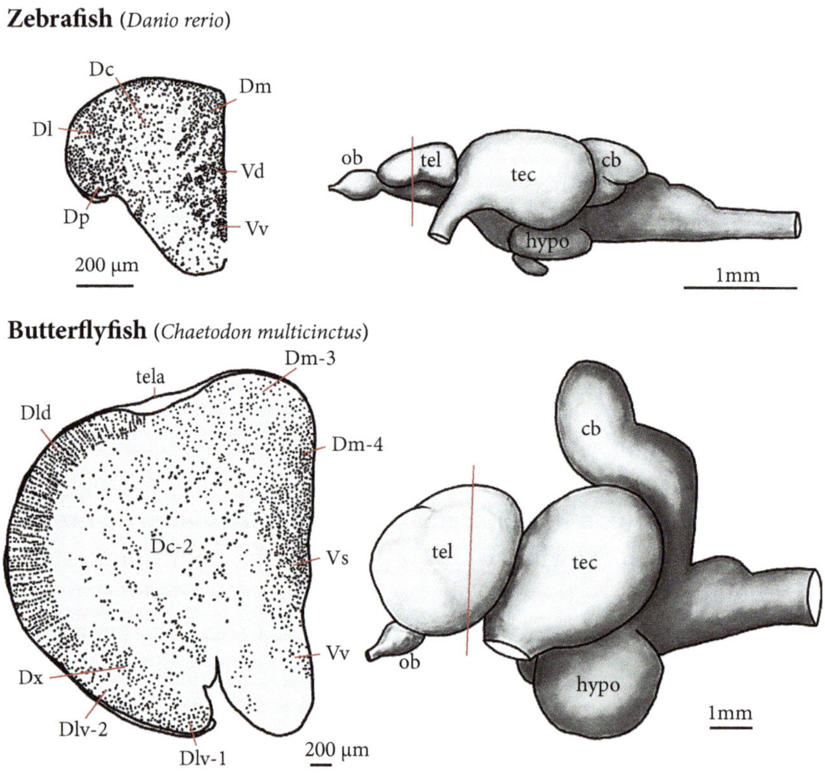

Figure 7.7 Variation in pallial complexity among teleosts. Illustrated at the top is a lateral view of the brain of a zebrafish (family Cyprinidae), together with a transverse section through one half of its telencephalon (at the level indicated by the red line). These small teleosts (which are not represented in Figure 7.4 because most cyprinids live in freshwater) have a relatively small and simple telencephalon. In contrast, the butterflyfish (family Chaetodonidae) has a very large and highly differentiated telencephalon. Especially enlarged is its pallium, which is divisible into at least 15 subdivisions, including Dm1–4, Dld, Dlv1–3, Dlp, Dc1–2, Dx, Dp, and Dd. By comparison, zebrafish have a relatively small and simple pallium. Areas Vv, Vs, and Vd are divisions of the teleost subpallium.

Adapted from Wullimann et al. (1996) and Dewan and Tricas (2014).

the hagfish telencephalon (see Figure 2.22 in Chapter 2), the torus semicircularis of gymnotoid electric fishes (see Figure 3.20 in Chapter 3), the optic tectum of teleosts and birds (see Figures 3.19 and 5.22 in Chapters 3 and 5]), the avian DVR (see Figure 5.28 in Chapter 5) and the mammalian neocortex, to name just five of the more striking examples. Presumably, these evolutionary transitions to a laminar architecture enhance the structure's computational functions, even if this is difficult to demonstrate (Guy and Staiger, 2017). In particular, lamination would be expected to minimize connection lengths, boost metabolic efficiency, and synchronize synaptic activity.

Homologous brain regions may also gain or lose connections with other cell groups, which would tend to alter the range of behaviors to which these regions contribute. Such changes in connectivity are difficult to identify when homology hypotheses are based primarily on similarities in connections (Karten and Shimizu, 1989), but they come into focus sharply when developmental criteria are emphasized (Puelles et al., 2007; Nieuwenhuys and Puelles, 2016). Even among closely related species, changes in the connectivity of clearly homologous brain regions can be significant (Striedter, 1992). For example, the set of neocortical areas that projects to the superior colliculus differs between primates, squirrels, and murine rodents (Baldwin et al., 2018). A more dramatic example is the phylogenetic restriction of olfactory projections to the telencephalic pallium (see Chapter 3). As the olfactory bulb projections became more restricted in the lineages leading to teleosts, amniotes, sharks, and rays, the vacated territories became targets for non-olfactory sensory projections from the diencephalon, notably the thalamus or the posterior tuberculum (see Section 7.5.2). These new ascending connections presumably provided the telencephalon with new kinds of information that could be used in novel ways. In short, the changes in connectivity must have modified the ancestral structure-function relationships.

These complications might suggest that correlating brain region sizes with behavioral or cognitive parameters is largely futile, and that the effort going into such research would be better spent on experimental analyses of neuronal mechanisms (Bolhuis and Macphail, 2001). However, the discovery of unexpected cross-species structure-function correlations will generate novel hypotheses about a brain region's function, which can then be tested experimentally. For example, the discovery that a specific pretectal nucleus (nucleus lentiformis mesencephali) is hypertrophied in hummingbirds, relative to other birds, suggested that it may be involved in the unique ability of hummingbirds to hover in the air for extended periods (Iwaniuk and Wylie, 2007). This hypothesis was recently confirmed by neurophysiological research showing that, compared to other birds, the neurons of this nucleus in hummingbirds exhibit unique specializations for hovering and rapid flight (Gaede et al., 2016). Even when comparative studies fail to show the expected structure-function correlations, they facilitate progress by indicating that one's original ideas were wrong or that the region's structure and functions changed across phylogeny. Our survey indicates that such changes were more common than neuroscientists have generally assumed.

7.4.2. Mosaic and Concerted Patterns of Brain Evolution

Beyond attempting to correlate the size of individual brain regions with functional parameters, we can ask whether those sizes covary with each other or with total brain size. Do brain regions vary in size independently of one another, exemplifying "mosaic evolution" (Barton and Harvey, 2000)? Or do they vary in concert with one another, which means that we should be able to predict the size of any one region from the size of other regions or total brain size (Finlay and Darlington, 1995)? Moreover, if the various brain regions were subject to "concerted evolution" (Striedter, 2005), do they all enlarge or shrink at the same rate as total brain size increases or decreases? Finally, if their rates of change are unequal, which regions tend to enlarge the most as brain size increases, and which become disproportionately small? These questions have been debated at length (Finlay et al., 2001), and evidence for both mosaic and concerted brain evolution has been adduced. Indeed, these two modes of brain evolution are not mutually exclusive. We here remark only on some of the more interesting aspects of this ongoing debate.

Even a cursory glance at the variation covered in this book suffices to show that large vertebrate brains are not just scaled-up versions of small vertebrate brains. Even at the same brain size, proportional brain region sizes vary considerably across the major vertebrate lineages. For example, the optic tectum is significantly smaller, relative to other brain regions, in mammals than in other amniotes and jawed fishes. Similarly, the telencephalon of an average mammal is roughly four times larger than that of cartilaginous fishes, relative to the size of the medulla, while the mammalian cerebellum is twice the size (Yopak et al., 2010).

Even within each major lineage, brain region proportions vary substantially. Among teleosts, for example, some families evolved large brains mainly by increasing the size of their telencephalon, while others did so mainly by increasing the size of their optic tectum (Figure 7.8). Furthermore, the evolutionary regression or expansion of a specific sensory system generally results in a dramatic reduction or expansion, respectively, of the central neural systems associated with those systems. For instance, the primary visual cortex of blind mole rats is dramatically reduced in size (though still present and responsive to auditory stimuli; Bronchti et al., 2002). Conversely, expansion of the gustatory sense in various teleosts was accompanied by enlargement of the hindbrain gustatory areas, creating prominent vagal or facial lobes. Importantly, the expansion of these gustatory lobes appears to be independent of absolute body size and to have occurred independently in diverse lineages (York et al., 2018). These and many other, similar observations support the idea that brain regions can change in size independently of one another, which is to say that they evolve mosaically. This conclusion is also consistent with

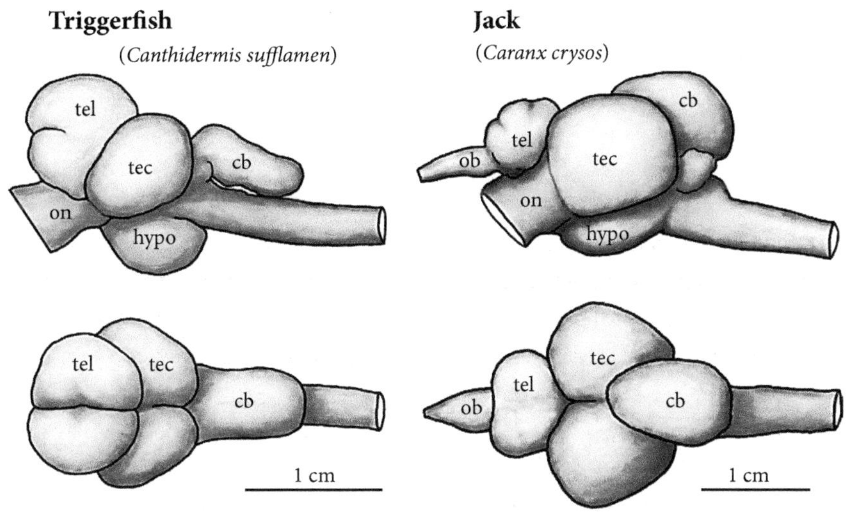

Figure 7.8 Differential enlargement of major brain regions in teleosts. Shown on the left is the brain of an ocean triggerfish (family Balistidae) from lateral (top) and dorsal (bottom) perspectives. These fishes, which hunt for prey near the ocean surface, have a relatively large brain (see Figure 7.4) and a proportionately larger telencephalon (tel) than most other teleosts. Shown on the right is the brain of a blue runner, which is an open ocean predator of the family Carangidae. These teleosts have similarly large brains, relative to body size (see Figure 7.4), but their telencephalon is proportionately small. Instead, they have disproportionately enlarged their optic tectum (tec).
Other abbreviations: cb – cerebellum; hypo – hypothalamus; ob – olfactory bulb; on – optic nerve.
Adapted from photographs in Schroeder (1980).

the generally accepted view that brain regions differ in function and, thus, with the various neuroecological correlations we reviewed earlier in this section.

However, when one compares brain region proportions across species that differ widely in absolute brain size, a pattern of concerted evolution does emerge. In a broad range of mammals, each major brain division scales predictably with absolute brain size (Finlay and Darlington, 1995; see also Hofman, 1989). Finlay and Darlington also noted that different brain regions scale with different slopes, leading to predictable changes in brain region proportions as one goes from small mammalian brains to larger ones (Figure 7.9). Since the neocortex has the highest slope, it becomes disproportionately large as brain size increases. Across primates, bats, and diverse other mammals, absolute brain size accounts for roughly 96% of the observed variation in individual brain region size, at least when one considers high-level brain regions, such as the entire neocortex. Similar findings have been reported for cartilaginous fishes (Figure 7.9), which suggests that the rules underlying concerted brain evolution are broadly conserved (Yopak et al., 2010). One should note, however, that in one large group of teleosts (cichlids), absolute brain size accounts for only 86% of the variance in major brain region volume (Gonzalez-Voyer et al., 2009) and that no comparable analysis has been performed in sauropsids.

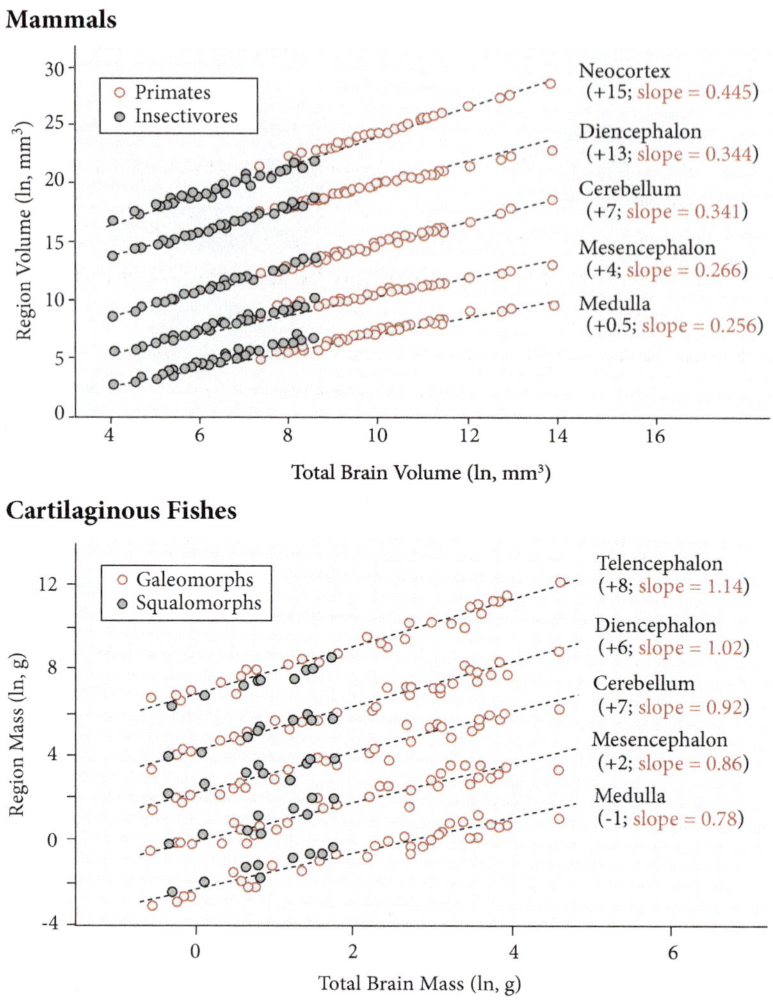

Figure 7.9 Concerted brain region evolution. Brain region sizes are here plotted against total brain size (volume or mass) in double-logarithmic coordinates for a large sample of mammals (top) and cartilaginous fishes (bottom). Although both graphs reveal some differences between the illustrated lineages, total brain size is an excellent predictor of brain region size. Importantly, the slopes of the scaling relationships vary for the different brain regions. To reveal these differences between the major brain regions, data points for each region were plotted with different offsets along the *y*-axis. Adapted from Finlay and Darlington (1995) and Yopak et al. (2010).

Therefore, it seems best to conclude only that some degree of concerted brain evolution probably exists in all vertebrates, but that the tightness of the relevant scaling rules, as well as their slopes and intercepts, varies between lineages. Moreover, even when a major brain region evolved concertedly, its various subdivisions may vary in ways that defy the applicable scaling rules (Smaers et al., 2017).

How can we explain the existence of concerted brain evolution? Finlay and Darlington (1995) noted that the slopes of the regional scaling rules correlate positively with the peak of neurogenesis in those regions, which led them to propose that evolution stretches or compresses global neurogenetic schedules to generate variation in brain size (see also Striedter, 2005). This idea has been regarded skeptically, in part because we do not yet know what kind of developmental mechanism could generate such global changes in neurogenetic schedules. In addition, it seems hard to believe that the size of a specific brain region might be the result of selection acting on the size of some other brain region or on total brain size. Thus, the suggestion that "big isocortices may be spandrels—byproducts of structural constraints for which some use is found later" has been met with disbelief by various authors (see commentary in Finlay et al., 2001). We share some of that skepticism, especially since neural tissue is so expensive metabolically. However, we also suspect that biologists tend to overestimate the degree to which natural selection can target specific traits without affecting other aspects of the organism. Just as pleiotropy can be a powerful constraint at the level of genes (Paaby and Rockman, 2013), developmental rules may limit what phenotypes can be produced, at least until evolution finds some way to change the rules.

A potential alternative explanation for the correlation between brain region scaling and neurogenesis timing is that it might have resulted from the fact that delaying neurogenesis is an ideal way to enlarge a brain region without causing concomitant decreases in the size of other, adjacent brain regions. In other words, changes in brain region size might be driving changes in neurogenesis timing, rather than being controlled by them. This hypothesis cannot explain why the most enlarged regions in vertebrate brains tend to be located in the brain's most rostral and dorsal (alar) regions, where neurogenesis tends to occur later than in the more caudal and basal regions (Finlay et al., 1998). However, delaying neurogenesis also delays structural and functional maturation. Therefore, one could argue that the brain's rostral and dorsal regions are most readily expanded in evolution because delaying their maturation is not detrimental to hatchlings or newborns; in contrast, neurons in the brain's basal regions might be needed for basic functions early in development. The enlargement and prolonged neurogenesis of (basal) motor nuclei innervating the electric organs of some fishes (Figure 7.10; Fox and Richardson, 1982) do not fit this hypothesis, but those electric organs are not used until relatively late in life. More problematic is that this hypothesis cannot explain why the various major brain divisions scale so predictably across a wide range of absolute brain sizes. However, it might help to explain evolutionary changes in the scaling rules.

Another potential explanation for the correlated evolution of diverse brain regions is that co-evolving structures may be components of a coherent functional circuit or system (Barton and Harvey, 2000; Whiting and Barton, 2003; Montgomery et al., 2016). It is fairly obvious, for example, that fishes, reptiles, and birds with large retinas tend also to have a large optic tectum, which is the retina's largest target in non-mammalian vertebrates (see Chapter 6). Several other structures intimately connected with the optic tectum also correlate in size with it, at least in birds (Gutiérrez-Ibáñez et al., 2014). However, even in the latter study, 80% of

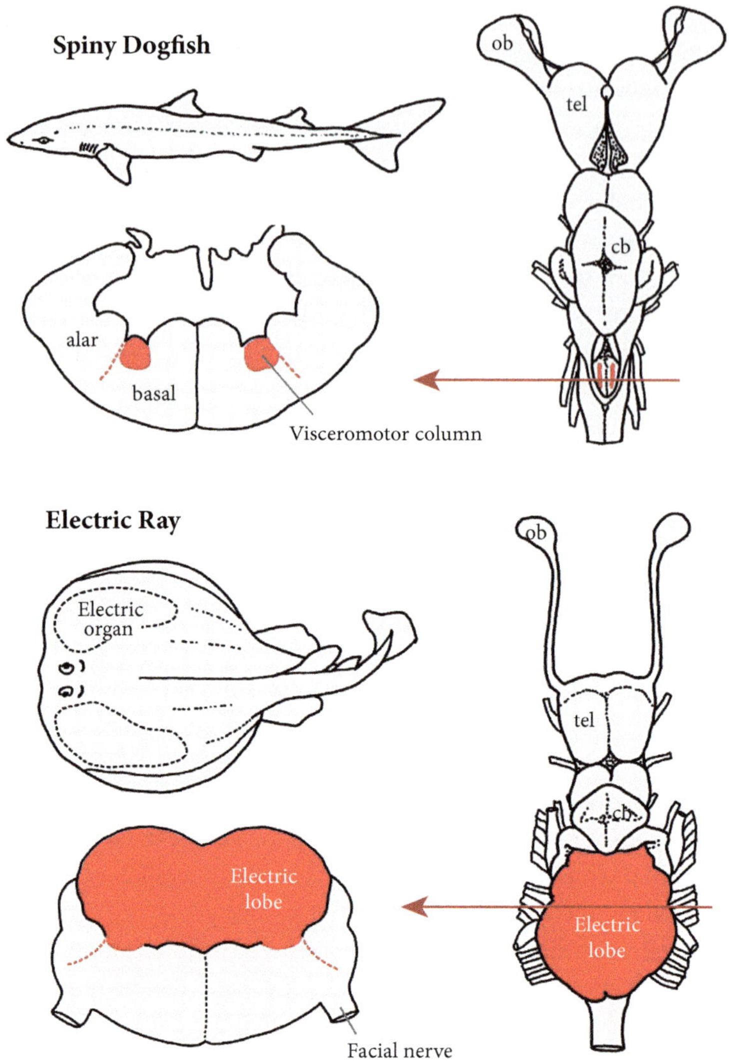

Figure 7.10 Hypertrophy of electromotor nuclei in the hindbrain of electric rays. Shown at the top is a transverse section through the hindbrain of a spiny dogfish (*Squalus acanthias*), with the lateral column of motor neurons (aka the visceromotor column) highlighted in red. By comparison, the homologous region in electric rays (of the genus *Torpedo*) is enormously enlarged (bottom). This region contains motor neurons that innervate bilateral electric organs, which represent modified pectoral muscles and are innervated by the facial, glossopharyngeal, and vagal nerves. The available data suggest that neurogenesis in these "electric lobes," which are part of the hindbrain's basal plate, extends past hatching (Fox and Richardson, 1982).
Adapted from Nieuwenhuys (2011b).

the observed variation in brain region size can be accounted for by the variation in total brain size. Moreover, some components of the avian visual system do not vary in concert with the rest of the system. These deviations might reflect functional specializations and differences in connectivity, but those hypotheses have not been tested yet. We conclude, therefore, that interconnected brain regions may vary concertedly in size, but those patterns need not be independent of the kind of concerted evolution that Finlay and Darlington have suggested. More importantly, comparative data on connectivity are so scarce that testing hypotheses about neural circuit evolution is currently possible in just a few systems (Striedter and Northcutt, 1989). Most studies simply assume that connections are conserved across the species being examined.

Having reviewed the concerted-versus-mosaic brain evolution debate, we feel compelled to reiterate the key discovery of Finlay and her collaborators, which was that the cellular mechanisms underlying brain development are broadly conserved, and that those conserved "rules" of brain development have adult structural and functional implications (Finlay et al., 2001; Cahalane et al., 2014). Specifically, the conservation of neurogenesis timing caused the neocortex and cerebellum to become disproportionately large in any lineage that increased its absolute brain size dramatically (at least among mammals and cartilaginous fishes). The functional implications of these changes remain difficult to specify, but both neocortical and cerebellar circuits can be deployed in a variety of functional contexts and tend to scale up gracefully (i.e., maintain or enhance their functional contributions as they increase in size; Finlay et al., 2011). Therefore, the changes in the brain that are produced by tweaking the parameters of the conserved developmental rules tend to be adaptive. In contrast, any hypothetical scaling rules that result in the disproportionate enlargement of specific motor or endocrine nuclei, for example, would be much less useful, at least for the majority of species. Given these considerations, it is reasonable to hypothesize that the conserved developmental rules themselves were "selected for" during phylogeny because they generated well-functioning adult nervous systems (Finlay, 2016). More generally, this extended evo-devo perspective reveals that evolution selects not only adult phenotypes but also the developmental mechanisms that give rise to them (e.g., Dyer et al., 2009). Of course, as we review in the following section, this does not mean that the rules of brain development were never modified.

7.5. Changes in the Basic Plan of Vertebrate Brains

Most neuroscientists, including most comparative neurobiologists, share a deep conviction that all vertebrate brains are built according to a common plan, sometimes referred to as the "vertebrate brain Bauplan" ("Bauplan" is German for "construction plan"). Vertebrate brains may vary in total size or in the size of their subdivisions but, according to the currently dominant view, they all share the same fundamental set of brain regions, whose connections and functions are highly conserved. As Butler and Hodos wrote in their textbook on comparative neuroanatomy, right after noting that different species may be specialized for different ecological niches: "All vertebrate central nervous systems share a common organizational scheme so that someone who is familiar with the brain of any vertebrate will also be on familiar ground when first encountering the brain of any other species" (Butler and Hodos, 2005, p. xvi).

There are, however, two competing schools of thought about what constitutes the conserved elements of the vertebrate brain Bauplan. According to the "New Neuromorphology" (Nieuwenhuys and Puelles, 2015), the conserved elements are a patchwork of "fundamental morphological units" that can be recognized by the patterns of genes that those units express during early stages of development (see Section 1.3.2). In contrast, the competing view emphasizes the conservation of cell types that retain their connectivity, as well as their "molecular fingerprints," across phylogeny (Karten, 2015; Briscoe et al., 2018). Since both schools of thought reach fundamentally different conclusions about the homology of the neocortex and a few other brain regions (e.g., Puelles et al., 2007), they cannot both be correct. Based on the information we reviewed in the preceding pages, we adopt a compromise. As outlined in the following pages, we argue that the brain Bauplan was modified in a few critical respects during vertebrate phylogeny and that some major neuronal pathways were likewise modified.

7.5.1. Adding Divisions to the Brain Bauplan

Comparative studies of gene expression have led to major advances (e.g., Medina et al., 2011; Puelles et al., 2016; Sugahara et al., 2016), but the developmental "genoarchitectonic" approach does have a few limitations. As mentioned in Chapter 1, homologous brain regions may, at least in theory, derive from nonhomologous embryonic precursors. We do not currently have enough lineage tracing data to know how often such evolutionary shifts in embryonic origin occur, but we suspect that they are relatively rare. More concerning is that gene expression patterns are probably more variable, across embryonic stages as well as species, than the published literature suggests. For example, a systematic examination of embryonic expression patterns for 1,103 genes between zebrafish and tunicates revealed far more divergence than expected (Sobral et al., 2009). Specifically, the expression patterns for homologous tissues in two different species (larval zebrafish and tunicates) were more divergent than the expression patterns for different tissues within each species. In a hierarchical clustering analysis, "tissues tended to group together by species rather than according to their homology" (Sobral et al., 2009). We do not yet have similarly comprehensive, objective data for embryonic nervous systems but, in the meantime, caution seems warranted.

The very idea of a vertebrate brain Bauplan suggests that this plan originated with vertebrates and, thus, must have evolved from a more widely conserved template. As we reviewed in Chapter 2, there is little to no evidence that invertebrate chordates possess a midbrain or cerebellum-like structures, and the evidence for a telencephalon in amphioxus is currently quite tentative. If those fundamental brain divisions were added to the more ancient "chordate brain Bauplan," why should we assume that the brain Bauplan remained immutable after the origin of vertebrates? As summarized in the following two sections, we submit that the brain Bauplan was augmented during vertebrate phylogeny by the addition of at least two fundamental units, namely a proper cerebellum and a dorsal pallium. Surprisingly, the latter division appears to have arisen independently in multiple vertebrate lineages.

7.5.1.1. Evolution of a Proper Cerebellum

Hagfishes have neither a proper cerebellum nor cerebellum-like structures (Ronan and Northcutt, 1998), which generally consist of tightly packed granule cells with long "parallel fiber" axons that contact the dendrites of other neurons, such as those in the octavolateralis area of the hindbrain (see Figure 3.17 in Chapter 3). Lampreys likewise lack a proper cerebellum (the cells that are sometimes mentioned as possible Purkinje cells in lampreys are more likely to be hindbrain motor neurons; Wicht, 1996), but they do have some cerebellum-like structures in their dorsal medulla. In contrast, all jawed vertebrates possess both cerebellum-like structures and a proper cerebellum, which we here define as containing both granule and Purkinje cells (note that teleosts do not possess deep cerebellar nuclei, though they do possess homologous neurons; Finger, 1978). Given these data, we infer that cerebellum-like structures evolved with the origin of vertebrates but were lost in hagfishes, and that a proper cerebellum is an innovation of jawed vertebrates.

Does the evolution of a proper cerebellum represent an elaboration of an ancestral cerebellum-like region, or the addition of a new division to the vertebrate brain Bauplan? We argue in favor of the latter position, because cerebellar Purkinje cells develop from a proliferative region, generally called the cerebellar ventricular zone (Figure 7.11), that lies immediately rostral to the rhombic lip, which generates the granule cells of the cerebellum (as well as several other brain regions; see Wullimann et al., 2011; Hashimoto and Hibi, 2012; Marzban et al., 2015). Since the presence of Purkinje cells is a defining feature of a proper cerebellum, these data suggest that the evolution of a proper cerebellum was associated with the emergence of a novel proliferative zone just anterior to the more broadly conserved rhombic lip. Given that the fundamental divisions of the vertebrate brain Bauplan are typically defined on the basis of their association with a unique proliferative zone (Nieuwenhuys and Puelles, 2015), it seems reasonable to propose that the evolution of a proper cerebellum in jawed vertebrates represents the addition of a novel fundamental division to the vertebrate brain Bauplan (see also Montgomery et al., 2012). Alternatively, if one is willing to accept that homologous neurons may develop from non-homologous precursor regions (see Chapter 1) and change their

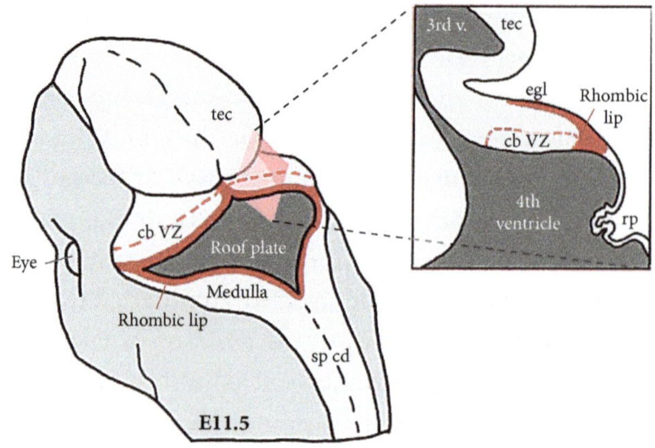

Figure 7.11 Developmental precursors of the cerebellum and cerebellum-like structures. Shown at the left is a posterolateral view of a mouse embryo at 11.5 days of embryogenesis, dissected so that the brain is visible. The drawing on the right represents a parasagittal section through the embryonic brain at the location indicated by the pink quadrangle. The rhombic lip, which gives rise to granule cells in the cerebellum and in cerebellum-like structures, is colored dark red. Immediately rostral to the cerebellar portion of the rhombic lip lies the cerebellar ventricular zone (cb VZ), which gives rise to cerebellar Purkinje cells and other inhibitory neurons of the cerebellum proper.

Additional abbreviations: 3rd v. – third ventricle; rp – roof plate; egl – external granular layer (covering the embryonic cerebellum); sp cd – spinal cord; tec – optic tectum.

Adapted from Hashimoto and Hibi (2012), with permission from John Wiley & Sons.

neurotransmitters, one could propose that Purkinje cells are homologous to the principal cells of cerebellum-like structures (see Figure 3.17 in Chapter 3).

7.5.1.2. Evolution of a Dorsal Pallium

The mammalian neocortex develops from a proliferative zone that comparative neuroembryologists refer to as the dorsal pallium. This zone is sandwiched between the medial pallium, which gives rise to the hippocampus, and the lateral pallium, which is traditionally thought to develop into the olfactory cortex. The region ventral to the lateral pallium is called the ventral pallium (Puelles et al., 2000) and is traditionally thought to give rise to the ventral part of the olfactory cortex, the pallial amygdala and the endopiriform portion of the claustrum. A more recent proposal is that the olfactory cortex develops entirely from the ventral pallium, while the lateral pallium gives rise to the dorsal claustrum and the insular cortex (Puelles et al., 2016, 2017). These ideas remain controversial (e.g., see Wullimann, 2016), but regardless of the interpretation one prefers, most comparative neurobiologists today believe that all four pallial zones (ventral, lateral, dorsal, and medial) are present in all jawed vertebrates (see Kaas, 2017), which makes their adult derivatives homologous (at least as "field homologs"; see Chapter 1). Most important for our present discussion is that virtually all comparative neurobiologists for the last 50 years have recognized a dorsal pallium in all major vertebrate lineages. Much to our surprise, we have come to doubt this view.

Our concerns are illustrated most clearly by what we see in ray-finned fishes. Teleosts, which form the largest radiation of extant ray-finned fishes, tend to have a large pallium with numerous divisions (see Figure 7.7). These pallial divisions were wisely given neutral, topological names, such as area dorsalis medialis, lateralis, posterior, etc. (abbreviated Dm, Dl, Dp, and so forth), at least in part because their homologies to pallial divisions in other vertebrates were uncertain at the time (Nieuwenhuys, 1963; Northcutt and Braford, 1980). However, most researchers have come to view the area dorsalis dorsalis (Dd) as the most likely homolog of the mammalian dorsal pallium, mainly on the basis of its obviously dorsal position. A serious problem with this view is that Dd is difficult to recognize in many teleosts and, according to some authors, absent entirely in zebrafish (Mueller et al., 2011). This observation led to the hypothesis that area dorsalis centralis (Dc) is an additional, if not the principal, component of the dorsal pallium in teleosts. This hypothesis is largely based on the observation that at least part of Dc develops from a unique proliferative zone in the dorsal region of the embryonic pallium (Mueller et al., 2011). Based on this evidence, one might conclude that Dd and/or a large part of Dc are homologous to the dorsal pallium of other vertebrates.

The principal problem with this modified homology hypothesis is that neither Dd nor Dc are evident in the pallium of *Polypterus*, which represents the most basal ray-finned fish lineage (Figure 7.12). Indeed, Rudolf Nieuwenhuys, who created the pallial terminology for teleosts (Nieuwenhuys, 1963), originally described the *Polypterus* pallium as a uniform structure with no subdivisions (Nieuwenhuys, 1969). Later studies used connectional and immunohistochemical data to describe three or four divisions within the pallium of *Polypterus* and the closely related reedfish *Erpetoichthys calabaricus* (see Nieuwenhuys, 1998). However, we strongly suspect that the pallium of these fishes consists of just two major divisions, each of which is divisible into two parts (Holmes and Northcutt, 2003). These two major divisions are most likely homologous to the medial and ventrolateral (i.e., ventral plus lateral) pallium of other vertebrates. Importantly, the pallium of *Polypterus* does not contain any neurons that seem comparable to the large cells found in the Dc of teleosts. Nor does it seem to contain any neurons that project to the optic tectum, which is the most characteristic connection of dorsal Dc (Holmes and Northcutt, 2003, and unpublished observations). In short, we find no convincing evidence for either Dd or Dc in *Polypterus*.

The next most basal group of ray-finned fishes, the sturgeons and their relatives, have a pallium that is more complex than that of *Polypterus* but still simpler than the pallium of teleosts (Figure 7.12). Northcutt and Braford (1980) identified a "Dd + Dl" region in this lineage, but they did not recognize Dd as a distinct area. Sturgeons do have a few large neurons in the central region of their pallium, which Northcutt and Braford (1980) named Dc, but this region does not seem to be associated with a unique proliferative zone, and its cells appear to lack descending projections to

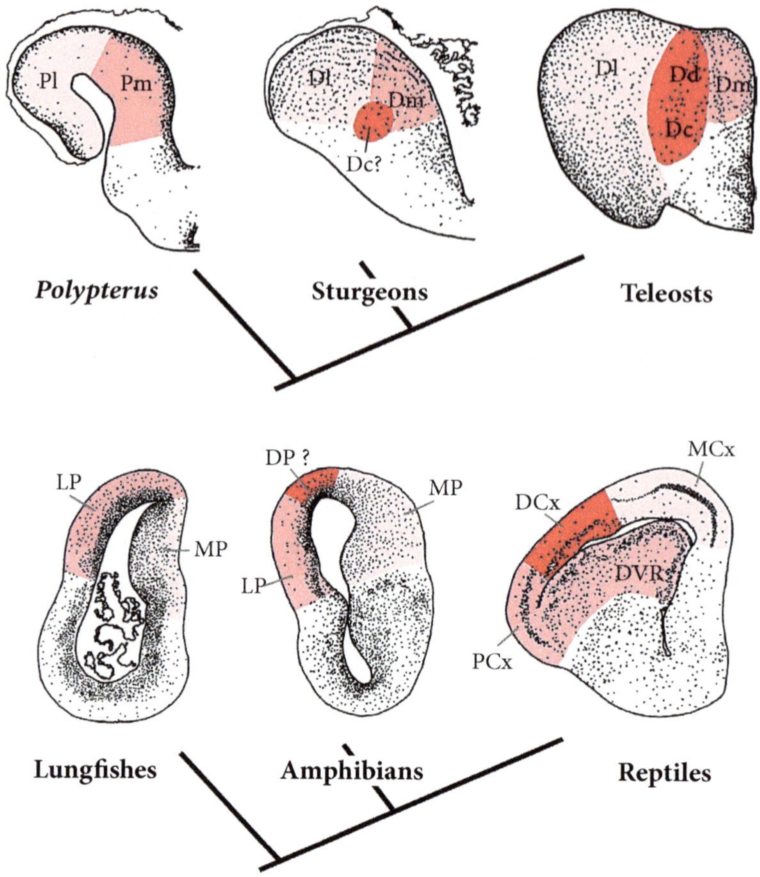

Figure 7.12 Emergence of a dorsal pallium in two separate lineages. Areas dorsalis dorsalis (Dd) and dorsalis centralis (Dc) have been regarded as a dorsal pallium in teleosts, but Dd cannot be identified in other ray-finned fishes (notably sturgeons and *Polypterus*) and whether the Dc of sturgeons is homologous to the Dc of teleosts remains doubtful. Similarly, the dorsal cortex (DCx) of non-avian sauropsids is widely regarded as a dorsal pallium, but it is difficult or impossible to identify in amphibians or lungfishes. Accordingly, we hypothesize that a dorsal pallium arose at least twice independently, namely in the lineage leading to teleosts, and in the lineage leading to amniotes.

Additional abbreviations: Dl – area dorsalis lateralis; Dm – area dorsalis medialis; DP – dorsal pallium candidate; DVR – dorsal ventricular ridge; LP – lateral pallium; MCx – medial cortex; MP – medial pallium; PCx – piriform cortex.

Adapted from Northcutt and Braford (1980); Northcutt and Davis (1983); Northcutt (2008).

the optic tectum (unpublished observations by RGN). We doubt that it is homologous to the portion of Dc that Mueller et al. (2011) regarded as the dorsal pallium in teleosts. Even more closely related to teleosts than sturgeons are the holostean fishes, *Amia* and gars. Northcutt and Braford (1980) reported that these animals lack an obvious Dc but have a large Dd, though they acknowledge that the latter region might be part of Dl. Altogether, these data suggest that Dd and Dc were not present in ancestral ray-finned fishes but, instead, evolved gradually in the lineage leading to teleosts. If this is true, then Dd and Dc cannot be homologous to the mammalian dorsal pallium. Instead, parsimony prompts us to infer that the "dorsal pallia" of amniotes and teleosts evolved independently of one another.

Did a dorsal pallium emerge with the origin of amniotes, or did the earliest tetrapods already have a dorsal pallium? The generally accepted view is that all living amniotes do have a dorsal pallium, called the dorsal cortex, Wulst, or neocortex (see Chapters 5 and 6). However, gene expression data obtained by Desfilis et al. (2017) suggest that most of the dorsal cortex in lizards is probably part of the medial pallium, leaving only a small rostral portion of the lizard pallium to serve as the likely homolog of the avian Wulst and mammalian neocortex. If this hypothesis is correct, then the dorsal pallium of early amniotes was probably quite small.

Even more doubtful is whether amphibians and lungfishes, the closest living relatives of amniotes, possess a dorsal pallium. As we reviewed in Chapter 4, the region that has been called the dorsal pallium in amphibians is so difficult to delineate that authors disagree considerably about its boundaries; its connections also overlap substantially with those of the adjacent areas. Therefore, this area is probably best regarded as a transition zone between the medial and lateral pallia (Figure 7.12). Similarly, the region that has been called the dorsal pallium in lungfishes is very small and poorly differentiated from the adjacent lateral pallium (González and Northcutt, 2009). If one abandons the assumption that amphibians and lungfishes *should* have a dorsal pallium, then the evidence for its existence in these species looks weak. Therefore, we hypothesize that a definite (albeit small) dorsal pallium arose with the origin of amniotes.

Similarly, one may ask when in phylogeny the dorsal pallium of teleosts arose. Because the most basal ray-finned fishes (e.g., *Polypterus*) have no distinct dorsal pallium, we hypothesize that this pallial division arose early in the ray-finned fish lineage. However, it is possible that basal ray-finned fishes eliminated their dorsal pallium. This alternative hypothesis would be supported if the vertebrate lineages basal to the ray-finned fishes possess a dorsal pallium. However, as we discussed in Chapter 3, the pallium of cyclostomes is very poorly understood. Lampreys are sometimes said to have a dorsal pallium, because they have a pallial region with descending projections to the optic tectum and reticular formation (Ocaña et al., 2015). However, such long descending pathways are not found in the pallium of basal ray-finned fishes, lungfishes, amphibians, or non-avian sauropsids, which means that they probably evolved independently in lampreys, teleosts, mammals, and birds. If lampreys do have a dorsal pallium, it is probably small and indistinct (see Figure 2.21 in Chapter 2). The situation in hagfishes is even more obscure. Thus, the data on cyclostomes do not provide convincing evidence that a proper dorsal pallium arose with the origin of vertebrates.

Cartilaginous fishes present more of a challenge to our hypothesis, because sharks, skates, and rays do seem to have a well-developed dorsal pallium. This region does not receive direct olfactory bulb projections (see Figure 3.29 in

Chapter 3) and, instead, receives non-olfactory inputs from the diencephalon (see Figure 3.31 in Chapter 3). These features were long considered diagnostic of a dorsal pallium. However, as we reviewed in Chapter 3, the available data indicate that the olfactory bulbs projected throughout the pallium in early vertebrates and later restricted their projections independently in several vertebrate lineages, including the cartilaginous fishes (Figure 7.13). Furthermore, most of the diencephalic inputs to the putative dorsal pallium in cartilaginous fishes probably arise from the posterior tuberculum rather than the thalamus, which provides the main input to the dorsal pallium in amniotes (see Figure 3.31 in Chapter 3). Thus, the connectional data are not as decisive as they at first appear. Similarly, some gene expression data in shark embryos have been used to argue for the presence of a dorsal pallium (Rodríguez-Moldes et al., 2017), but the similarities to other vertebrates are less than compelling.

Given these data, we propose that the "dorsal pallium" of elasmobranchs (sharks, skates, and rays) evolved independently of the dorsal pallium in amniotes and, as mentioned earlier, the "dorsal pallium" of teleosts. Although these "dorsal pallia" all occupy the same topological positon within the telencephalon and derive from very similar embryonic precursor regions, they are not homologous to one another, because they had multiple evolutionary origins (see Chapter 1, Section 1.2.3). To test

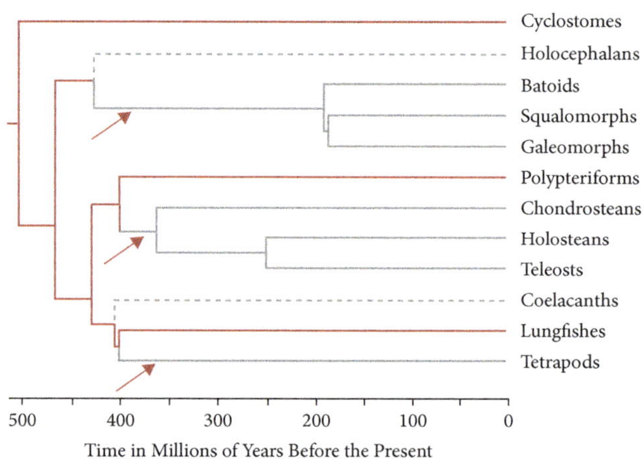

Figure 7.13 Phylogenetic restriction of olfactory bulb projections. Axon tracing studies have revealed that the olfactory bulbs project throughout most of the pallium in lampreys and hagfishes (i.e., cyclostomes), in basal ray-finned fishes (i.e., *Polypterus*), and in lungfishes (see Chapter 3 for more details). Given the phylogenetic relationships of these three lineages, we hypothesize that widespread olfactory bulb projections were the primitive condition for vertebrates (indicated by red lines), and that these projections later became more restricted in three separate lineages (gray lines). To test this hypothesis one would like to have data on the olfactory projections in holocephalans and coelacanths (dashed lines).

our hypothesis, it would be good to have more data on the pallium of cyclostomes. At least as useful would be experimental data on the pallium of ratfishes (the only extant holocephalans), which comprise the most basal lineage of cartilaginous fishes (Figure 7.13). Published descriptions indicate that holocephalans have a relatively large lateral pallium, but the boundaries of their medial and dorsal pallial divisions are much less clear. If we compare the pallium of holocephalans to that of amphibians and lungfishes, rather than other cartilaginous fishes, then it appears that much of what Smeets (1990) and Smeets et al. (2011) identified as the dorsal pallium in holocephalans may actually be part of the medial pallium (Figure 7.14). Part of it may also be a deep component of the lateral pallium. Immunohistochemical and gene expression data may help to resolve some of these issues, but what would be most helpful are data on the olfactory bulb and diencephalic projections in holocephalans. Our hypothesis would be strengthened considerably if most of the pallium in holocephalans were shown to receive direct inputs from the olfactory bulbs and minimal thalamic inputs.

The idea of adding a new pallial division during phylogeny goes against the widely held belief that truly novel parts cannot appear during phylogeny, because it is impossible for something to come from nothing (see Chapter 2). However, we are not the first to suggest that vertebrates may vary in number of pallial divisions they possess (e.g., Desfilis et al., 2017; Yamamoto et al., 2017; Ruiz-Reig et al.,

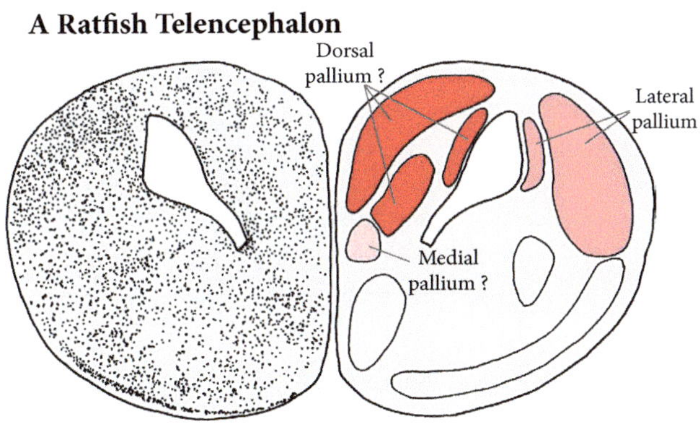

Figure 7.14 The telencephalon of a holocephalan cartilaginous fish. Shown here is a schematic transverse section through the telencephalon of the spotted ratfish (*Hydrolagus collei*). The left side presents stained cell bodies, while the right side depicts some cell group boundaries as sketched by Smeets (1990). Smeets homologized the three cell groups shaded dark red as likely homologs of the dorsal pallium in sharks and rays, and he recognized only a very small medial pallium. Alternatively, one might hypothesize that these animals have a very large medial pallium (including Smeets's putative dorsal pallium) and no dorsal pallium. To test these hypotheses, one would need experimental data, which are currently lacking.
Adapted from Smeets (1990) and Smeets et al. (2011).

2018). Moreover, it is not difficult to imagine an evolutionary–developmental scenario that could result in the *de novo* formation of a dorsal pallium. As illustrated in Figure 7.15, we already know that the medial and lateral edges of the embryonic pallium, called the hem and anti-hem, respectively, secrete diffusible signals that can influence neuronal cell fates (Assimacopoulos et al., 2003; Mangale et al., 2008). We speculate that this embryonic pallium in primitive vertebrates was so small that the signals from the hem and anti-hem met near the dorsal midline and, between them, covered the entire pallium at meaningful concentrations. Next, we propose that the embryonic pallium in the derived condition expanded tangentially. This pallial expansion would have created an area in the middle of the pallium where cells receive signals neither from the hem, nor from the anti-hem. One would expect such an area to exhibit some modified molecular interactions and, thus, some novel structural traits. The new dorsal pallium might well contain cell types that are homologous to those of other pallial regions, since homologous cell types may be located in non-homologous brain regions (see Chapter 1, Section 1.3.3), but some of its cell types may be genuinely new. We have no direct support for or against these hypotheses and here present them merely as a thought experiment. However, it is interesting to note that the early embryonic pallium is surprisingly small in basal ray-finned fishes (Nieuwenhuys, 2011a) and holocephalans (Holmgren, 1922).

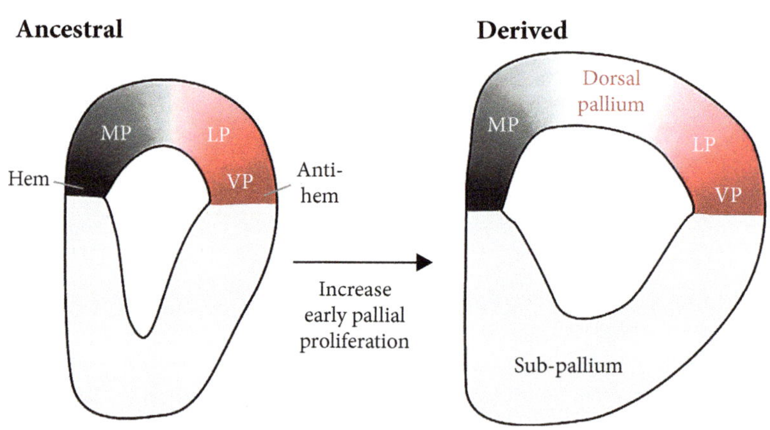

Figure 7.15 An evo-devo model of dorsal pallium origins. The hem and anti-hem in young vertebrate embryos secrete diffusible factors (gray and red shading, respectively) that influence the fate of pallial neurons. We hypothesize that these factors collectively reach the entire pallium in the primitive vertebrate condition (left) and induce the formation of the medial and ventrolateral (i.e., ventral plus lateral) pallial divisions, respectively. In the derived condition (right), the embryonic pallium has undergone more extensive proliferation and is, therefore, expanded. As a result, the diffusible factors from the hem and anti-hem fail to reach the middle of the pallium, which consequently adopts a novel fate, namely that of a dorsal pallium.
Abbreviations: MP – medial pallium; LP – lateral pallium; VP – ventral pallium.

7.5.2. Evolutionary Changes in Brain Circuitry

Neurobiologists have long been interested in the evolution of neural circuits, but they had little relevant data until a panoply of axon tracing techniques became available in the second half of the 20th century (Lanciego and Wouterlood, 2011). With these new methods, researchers soon discovered that some major neuronal pathways are more similar across the major vertebrate lineages than earlier scientists had thought (e.g., Northcutt, 1981). This unexpected discovery explains why connectional similarities became widely accepted as a good "putative homology criterion" (Striedter, 1999). After all, if neuronal connections tend to be conserved, then homologous cell groups should have similar connections. Although it is true that connectional similarities are often useful in identifying (or ruling out) homologies, the available evidence indicates that neuronal pathways did change as vertebrates diversified, and sometimes did so substantially. Evidence for this malleability has been reviewed in many of the preceding pages. Here we summarize only a few of the most fascinating examples.

As noted previously, the phylogenetic restriction of olfactory bulb projections to the telencephalon (Figure 7.13) must have altered telencephalic function profoundly, especially since the lost input was replaced by non-olfactory sensory inputs ascending from the diencephalon. Importantly, those ascending inputs originate from the thalamus (aka dorsal thalamus) in amniotes, but from the posterior tuberculum in ray-finned fishes (Figure 7.16; see also Figure 3.23 in Chapter 3). The latter region is small in many vertebrates and not typically regarded as a sensory region, but it is a major target of both olfactory and gustatory projections in many fishes. In teleosts it also conveys auditory, lateral line, and visual information. Thus, the evidence is pretty clear that teleosts and amniotes independently evolved multisensory ascending pathways to telencephalic regions that no longer receive direct olfactory inputs from the olfactory bulbs. As we discussed earlier in this chapter, the data on cartilaginous fishes are less definitive. Sharks and rays clearly possess pallial regions that receive non-olfactory sensory information, and most investigators have assumed that this information comes from the thalamus. However, at least some of this information may instead derive from the posterior tuberculum (see Figure 3.31 in Chapter 3). Therefore, we cannot assume that non-olfactory sensory information *must* be relayed by the thalamus, just because this is what happens in amniotes.

Although mammals and sauropsids (reptiles and birds) possess non-olfactory sensory projections from the thalamus to the telencephalic pallium, the principal targets of these projections are different in the two lineages (see Chapters 5 and 6). In mammals the sensory nuclei of the thalamus target mainly the dorsal pallium (i.e., the neocortex), whereas in sauropsids the sensory thalamic regions project primarily to the ventral pallium (i.e., the DVR). As far as we're concerned, these targets are not homologous to one another. Therefore, one would expect their subsequent projections and pathways to differ in a number of respects. However, the

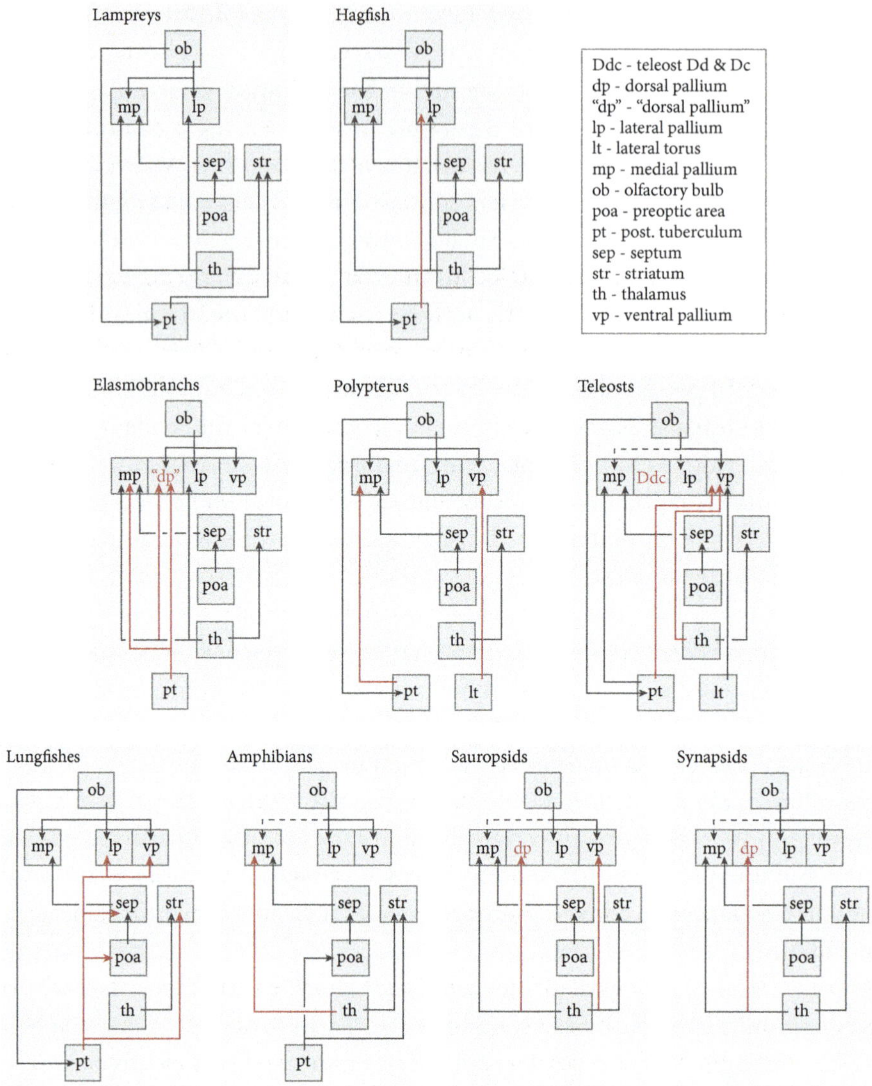

Figure 7.16 Variation in telencephalic pathways. This schematic diagram illustrates the major pathways to the telencephalon in nine vertebrate groups. Red arrows indicate pathways that are thought to be derived for the lineage, and dashed arrows indicate pathways that are present but dramatically reduced. Although it is difficult to reconstruct the history of some pathways, it is clear that the sensory pathways to the telencephalon have changed substantially during the course of vertebrate phylogeny. The independently evolved dorsal pallial divisions in cartilaginous fishes ("dp"), teleosts (Ddc), and amniotes (dp) are shown in red font.

circuits through the avian DVR are surprisingly similar to circuits through mammalian neocortex (e.g., Karten and Shimizu, 1989; Wang et al., 2010). One can use this similarity to argue that the DVR is actually homologous to part of the mammalian neocortex, but this hypothesis seems unlikely in light of the developmental evidence. Moreover, the referenced studies explicitly highlight the similarities and downplay or ignore the differences in connectivity. They also tend to disregard non-avian sauropsids, which generally have a simpler DVR than birds (see Chapter 5). In light of these issues, we suspect that most of the circuit similarities between the avian DVR and the mammalian neocortex are the result of convergent evolution, which is rampant between mammals and birds in any case (Striedter and Northcutt, 2017).

The circuits of the pallium in teleosts are even more divergent, with the non-olfactory sensory pathways targeting both the medial and the lateral pallial homologs (Dl and Dm; Northcutt, 2006; Yamamoto and Ito, 2008). These pallial regions have descending projections to the hypothalamus and diencephalon; they also have projections to area dorsalis centrals (Dc), which has even longer descending projections. This pattern of connectivity has been described as being similar to the mammalian intracortical pathway and to the circuits coursing through the avian DVR (Ito and Yamamoto, 2009). Again, however, such comparisons ignore numerous differences (e.g., see Murakami et al., 1983). Crucially, we already know that many of the cited connections are not present in basal ray-finned fishes or, for that matter, amphibians. Therefore, the similarities probably represent another instance of convergent evolution. Whether the circuits through the pallium of teleosts are "functionally equivalent" (Ito and Yamamoto, 2009) to those of birds or mammals is unknown, since we know virtually nothing about how those circuits actually function. If structurally and functionally equivalent circuit motifs did evolve independently in teleosts, mammals, and birds (Shanahan, 2013; van den Heuvel et al., 2016), that would be fascinating, because it would be indicative of some very general rules of neural circuit design (e.g., Mengistu et al., 2016).

As the dorsal and ventral pallial divisions expanded in early mammals and sauropsids, respectively, the roles of the medial pallium and striatum were modified. In amphibians the medial pallium is the largest division of the pallium, receives most of the ascending sensory input, and originates most of the pallium's descending projections (see Figure 4.31 in Chapter 4). By contrast, the medial pallium of most amniotes (i.e., the hippocampus) receives most of its sensory inputs from other pallial regions, and targets mainly other telencephalic regions (Striedter, 2016). Because of these changes in connectivity, the functions of the medial pallium in amniotes are enhanced by the innovative, highly specialized representations that it receives from the dorsal and ventral pallium (in mammals and sauropsids, respectively). Similarly, the striatum in amniotes receives not only the relatively ancient thalamic inputs, but also novel inputs, carrying more specialized representations, from the dorsal and ventral pallia. These evolutionary shifts in function may have been more quantitative than qualitative, as most connections of the medial pallium and striatum are qualitatively conserved across most vertebrates. However, quantitative changes in connection strength may still modify the role a region plays within the brain's overall circuitry. Moreover, we feel compelled to stress, again, that the existing studies may have focused on the similarities more than the differences (e.g., see Figure 6.27).

The telencephalon receives the lion's share of attention in comparative brain research, but the optic tectum is at least as important in most non-mammalian vertebrates. It receives sensory input of multiple modalities, is beautifully laminated in most sauropsids and teleosts, and projects to a wide variety of motor regions in the midbrain tegmentum and medulla. It clearly is the major integrative sensorimotor region in the vast majority of non-mammalian vertebrates. In mammals, however, the tectum's size and role were greatly diminished, especially in the lineage leading to humans. Our optic tectum (i.e. our superior colliculus) still plays a major role in the control of eye and head movements, as well as spatial attention (Knudsen and Schwarz, 2016), but its role in stimulus identification was probably reduced as those functions were "shifted" into the neocortex (Aboitiz, 1993; Striedter, 2002). This "corticalization" of functions in large mammalian brains has long been suspected on the basis of comparative brain lesion work (Ferrier, 1876; James, 1890), but the underlying mechanisms have rarely been explored. The corticalization of motor functions is probably related to the emergence or expansion of projections from the dorsal pallium to lower motor neurons (Nudo and Frost, 2007). Regarding the corticalization of sensory functions, we know that primates drastically reduced the fraction of retinal ganglion cells that project to the superior colliculus (Perry and Cowey, 1984) and greatly enlarged their pulvinar, which plays a major role in some of the functions that the optic tectum ancestrally performed (Shipp, 2004; Kaas and Lyon, 2007). However, the mechanisms of corticalization probably also involved some qualitative losses and gains of neural connections (Striedter, 2005; Herculano-Houzel et al., 2015). For example, part of the pulvinar in anthropoid primates has lost the phylogenetically ancient input from the superior colliculus and, instead, receives visual input directly from the retina (Baldwin et al., 2018).

The cerebellum proper (see Section 7.5.1) is one of the most highly conserved regions in gnathostome brains. Its internal organization is remarkably similar across species, though the shape of the Purkinje cell dendrites exhibits some interesting variation (Meek and Nieuwenhuys, 1991), the neurons of the deep nuclei are part of the cerebellar cortex in teleosts (Murakami and Morita, 1987), and novel types of interneurons were added during mammalian phylogeny (Yopak et al., 2017). Although the principal inputs and outputs of the cerebellum are widely thought to be likewise conserved, substantial variation does exist. For example, the cerebellum in birds and diverse other vertebrates receives strong input from a pretectal nucleus that processes information about optic flow (i.e., the perceived "flow" of the visual world around an organism as it moves through its environment), but this projection was apparently lost in mammals (Pakan and Wylie, 2006). As if to compensate for this absence, the cerebellum in mammals receives major inputs from the neocortex via the pontine nuclei and projects back to the neocortex via the thalamus. Because these pathways are not found in non-avian sauropsids or anamniotes, they probably originated in early mammals (though more research on this topic would be invaluable). Thus, even if the cerebellum's internal structure and function are highly conserved, the set of circuits that the cerebellum can modulate has likely changed during phylogeny.

All of this variation in neural connectivity is likely just the tip of the iceberg, since comparative neuroanatomists have historically been more interested in finding species similarities than differences, if only because species differences require more complex explanations and because it's easier to find expected patterns than novel ones. Hopefully the future will bring more extensive efforts to demonstrate both species differences and similarities in neuronal circuits. Research on basal lineages, such as holocephalans, coelacanths, lungfishes, monotremes, tuataras, and ratites, will be especially helpful. Although researchers often like to compare distantly related species, e.g., mammals versus birds, comparisons among closely related species are often less controversial and may, nonetheless, reveal some fascinating variation in neuronal circuitry (Northcutt and Wullimann, 1988; Striedter, 1992).

7.6. Conclusion: Natural History through Time

The Nobel Prize–winning biologist Francois Jacob once compared evolution to a "tinkerer," who does not "produce novelties from scratch" but, instead, works with old, highly conserved parts (Jacob, 1977). As Jacob pointed out, this analogy works best at the molecular level, where the majority of genes have homologs in distantly related species. However, the notion of evolution as a tinkerer has taken hold also at higher levels of analysis, exemplified by the idea that all vertebrate brains are built from a conserved set of "building blocks" (e.g., cell types or brain regions). As we just discussed, there is considerable merit in this idea, but Jacob's proposal has sometimes been carried too far. It seems to us that contemporary neurobiologists often interpret "tinkering" as merely varying some marginal, superficial details, while the fundamental aspects of nervous system structure and function remain conserved. Essentially, they regard the variation across species as "noise." This is clearly not what Jacob had in mind, as he stressed the ability of evolution to combine old components in novel ways. As he put it: "Novelties come from previously unseen association of old material." (p. 1163). Thus, new spatiotemporal expression patterns of "old genes" can lead to morphological innovations (see Figure 7.15). For example, Jacob himself considered the neocortex to be uniquely mammalian (Jacob, 1977).

A second substantial problem with the current state of evolutionary neurobiology is that it focuses primarily on structure, often ignoring function. This attitude is understandable, since comparative functional studies must control for extra variables, such as species differences in sensory or motor capacities, as well as physiology. However, most of the variation in brain anatomy presumably has some functional correlates, and even the conserved features and general principles were probably conserved because they were useful. As we reviewed in Sections 7.3 and 7.4, many comparative neurobiologists do try to get at those functions by correlating variation in brain structure against variation in species ecology and behavior. However, those attempts are often limited by insufficient knowledge of both the structural variation and the behavioral biology. Furthermore, the correlative studies tend to emphasize the behaviors of today's species, neglecting the fact that many of the relevant changes occurred long ago, under different ecological conditions. We hope that this book's emphasis on paleoecology will help to fill that gap or, at least, will show that reconstructing ancient environments and selective pressures is possible. One can, for example, reconstruct the kinds of conditions in which mammals first evolved, and this helps us to understand many evolutionary changes in their bodies and brains (see Chapter 6). In general, we hope to promote what Daniel Lehrman (1971) called the "natural history orientation." As he saw it, this perspective or stance emphasizes "questions arising from the natural life of a particular species" rather than "questions applied to an arbitrarily selected species from a generalized theoretical framework" (p. 464).

Our interests in natural history and neurobiological species differences are intertwined, because a consideration of multiple species in their natural environments naturally leads one to contemplate behavioral differences; and the recognition of behavioral differences, in turn, prompts one to wonder how the nervous systems of those species diverged. This interest in "species idiosyncrasies" notwithstanding, we remain equally interested in general rules and principles that help connect our observations across species, be they observations of similarities or differences. In short, we deem it essential to "consider simultaneously similarities and differences between species so that each illuminates the other" (Lehrman, 1971, p. 467).

In closing, we acknowledge that our ability to document evolutionary changes in brain anatomy far outpaces our understanding of how those structural changes have altered brain function and behavior. It is relatively easy to explain how the evolution of a fovea or an optic tectum might have benefited now-extinct vertebrates, but explaining evolutionary changes at higher levels of the brain, especially within the telencephalon, will require a much deeper understanding of how brains work. In our view, this limitation does not obviate the need to discover how brain anatomy has changed during phylogeny. To the contrary, it underscores the importance of comparative neuroanatomy. Future neuroscientists have no greater achievement to anticipate than the synthesis of evolutionary neuroanatomy with the physiological and behavioral changes that neural innovations engendered. When that day comes, we have no doubt that it will involve an integrated theory encompassing anatomy, physiology, and evolution: structure, function, and history. Developmental neurobiology will nicely complement this synthesis by revealing the proximate mechanisms of evolutionary divergence. As we consider the panoply of innovations in neural systems, understanding the ecological circumstances in which they emerged is certain to inform our understanding of how their homologs function today. In this sense, understanding brain evolution may well expedite the "deeper understanding of how brains work" that neuroscience strives to achieve.

References

Aboitiz F. 1993. Further comments on the evolutionary origin of the mammalian brain. *Med Hypotheses* 41:409–418.

Aiello LC, Wheeler P. 1995. The expensive-tissue hypothesis: the brain and the digestive system in human and primate evolution. *Curr Anthropol* 36:199–221.

Ariëns Kappers CU. 1941. The oculomotor nucleus of the electric stargazers: *Astroscopus guttatus* and *Astroscopus y-graecum*. *Proc Roy Netherlands Acad Arts Sci* 44:369–377.

Assimacopoulos S, Grove EA, Ragsdale CW. 2003. Identification of a Pax6-dependent epidermal growth factor family signaling source at the lateral edge of the embryonic cerebral cortex. *J Neurosci* 23:6399–6403.

Baldwin MKL, Young NA, Matrov D, Kaas JH. 2018. Cortical projections to the superior colliculus in gray squirrels (*Sciurus Carolinensis*). *Eur J Neurosci* 49:1008–1023.

Barton RA, Capellini I. 2011. Maternal investment, life histories, and the costs of brain growth in mammals. *Proc Natl Acad Sci USA* 108:6169–6174.

Barton RA, Harvey PH. 2000. Mosaic evolution of brain structure in mammals. *Nature* 405:1055–1058.

Bastian AJ. 2006. Learning to predict the future: the cerebellum adapts feedforward movement control. *Curr Opin Neurobiol* 16:645–649.

Bell CC, Szabo T. 1986. Electroreception in mormyrid fish: central anatomy. In: Bullock TH, Heiligenberg W (eds.), *Electroreception*. New York: Springer, pp. 375–421.

Benito-Gutiérrez E, Stemmer M, Rohr SD, Schuhmacher LN, Tang J, Marconi A, Jékely G, Arendt D. 2018. Patterning of a telencephalon-like region in the adult brain of amphioxus. *bioRxiv* 307629, doi: 10.1101/307629.

Benson-Amram S, Dantzer B, Stricker G, Swanson EM, Holekamp KE. 2016. Brain size predicts problem-solving ability in mammalian carnivores. *Proc Natl Acad Sci USA* 113:2532–2537.

Bingman VP, Rodriguez F, Salas C. 2017. The hippocampus of nonmammalian vertebrates. In: Kaas JH, Striedter GF (eds.), *Evolution of Nervous Systems*, 2nd ed. Amsterdam: Elsevier, pp. 479–489.

Blackburn DG. 2006. Squamate reptiles as model organisms for the evolution of viviparity. *Herpetological Monographs* 20:131.

Boddy AM, McGowen MR, Sherwood CC, Grossman LI, Goodman M, Wildman DE. 2012. Comparative analysis of encephalization in mammals reveals relaxed constraints on anthropoid primate and cetacean brain scaling. *J Evol Biol* 25:981–994.

Bolhuis JJ, Macphail EM. 2001. A critique of the neuroecology of learning and memory. *Trends Cogn Sci* 5:426–433.

Bonner JT. 1988. *The Evolution of Complexity by Means of Natural Selection*. Princeton, NJ: Princeton University Press.

Briscoe SD, Albertin CB, Rowell JJ, Ragsdale CW. 2018. Neocortical association cell types in the forebrain of birds and alligators. *Curr Biol* 28:1–11.

Bronchti G, Heil P, Sadka R, Hess A, Scheich H, Wollberg Z. 2002. Auditory activation of "visual" cortical areas in the blind mole rat (*Spalax ehrenbergi*). *Eur J Neurosci* 16:311–329.

Bullock TH. 2002. Grades in neural complexity: how large is the span? *Integr Comp Biol* 42:757–761.

Butler AB, Hodos W. 2005. *Comparative Vertebrate Neuroanatomy*. Hoboken, NJ: Wiley.

Cahalane DJ, Charvet CJ, Finlay BL. 2014. Modeling local and cross-species neuron number variations in the cerebral cortex as arising from a common mechanism. *Proc Natl Acad Sci USA* 111:17642–17647.

Carr C, Maler L. 1986. Electroreception in gymnotiform fish: central anatomy and physiology. In: Bullock TH, Heiligenberg W (eds.), *Electroreception*. New York: Wiley, pp. 319–373.

Carr RK. 1995. Placoderm systematics diversity, and evolution. *Bull Mus Natl Hist Nat Paris* 17:85–125.

Clayton NS, Dickinson A. 1998. Episodic-like memory during cache recovery by scrub jays. *Nature* 395:272–274.

Conway Morris S. 2003. *Life's Solution*. Cambridge: Cambridge University Press.

Corfield JR, Price K, Iwaniuk AN, Gutiérrez-Ibáñez C, Birkhead T, Wylie DR. 2015. Diversity in olfactory bulb size in birds reflects allometry, ecology, and phylogeny. *Front Neuroanat* 9:102.

Cotney J, Leng J, Yin J, Reilly SK, DeMare LE, Emera D, Ayoub AE, Rakic P, Noonan JP. 2013. The evolution of lineage-specific regulatory activities in the human embryonic limb. *Cell* 154:185–196.

Deaner RO, Isler K, Burkart J, Van Schaik C. 2007. Overall brain size, and not encephalization quotient, best predicts cognitive ability across non-human primates. *Brain Behav Evol* 70:115–124.

DeCasien AR, Williams SA, Higham JP. 2017. Primate brain size is predicted by diet but not sociality. *Nat Ecol Evol* 1:0112.

Desfilis E, Abellán A, Sentandreu V, Medina L. 2017. Expression of regulatory genes in the embryonic brain of a lizard and implications for understanding pallial organization and evolution. *J Comp Neurol* 526:166–202.

Dewan AK, Tricas TC. 2014. Cytoarchitecture of the telencephalon in the coral reef multiband butterflyfish (*Chaetodon multicinctus*: perciformes). *Brain Behav Evol* 84:31–50.

Donoghue PCJ, Purnell MA. 2009. The evolutionary emergence of vertebrates from among their spineless relatives. *Evo Edu Outreach* 2:204–212.

Dulvy NK, Reynolds JD. 1997. Evolutionary transitions among egg-laying, live-bearing and maternal inputs in sharks and rays. *Proc Roy Soc Lond B* 264:1309–1315.

Dunbar RIM, Shultz S. 2007. Evolution in the social brain. *Science* 317:1344–1347.

Dyer MA, Martins R, da Silva Filho M, Muniz JAPC, Silveira LCL, Cepko CL, Finlay BL. 2009. Developmental sources of conservation and variation in the evolution of the primate eye. *Proc Natl Acad Sci USA* 106:8963–8968.

Elston GN, Benavides-Piccione R, Elston A, Manger PR, Defelipe J. 2011. Pyramidal cells in prefrontal cortex of primates: marked differences in neuronal structure among species. *Front Neuroanat* 5:2.

Elston GN, Benavides-Piccione R, Elston A, Zietsch B, Defelipe J, Manger P, Casagrande V, Kaas JH. 2005. Specializations of the granular prefrontal cortex of primates: Implications for cognitive processing. *Anat Rec* 288A:26–35.

Ferrier D. 1876. *The Functions of the Brain*. London: Smith, Elder.

Finger TE. 1978. Efferent neurons of the teleost cerebellum. *Brain Res* 153:608–614.

Finlay BL. 2016. Principles of network architecture emerging from comparisons of the cerebral cortex in large and small brains. *Plos Biol* 14:e1002556.

Finlay BL, Darlington RB. 1995. Linked regularities in the development and evolution of mammalian brains. *Science* 268:1578–1584.

Finlay BL, Darlington RB, Nicastro N. 2001. Developmental structure in brain evolution. *Behav Brain Sci* 24:263–308.

Finlay BL, Hersman MN, Darlington RB. 1998. Patterns of vertebrate neurogenesis and the paths of vertebrate evolution. *Brain Behav Evol* 52:232–242.

Finlay BL, Hinz F, Darlington RB. 2011. Mapping behavioural evolution onto brain evolution: the strategic roles of conserved organization in individuals and species. *Philos Trans R Soc Lond B, Biol Sci* 366:2111–2123.

Fonseca-Azevedo K, Herculano-Houzel S. 2012. Metabolic constraint imposes tradeoff between body size and number of brain neurons in human evolution. *Proc Natl Acad Sci USA* 109:18571–18576.

Fox GQ, Richardson GP. 1982. The developmental morphology of *Torpedo marmorata*: electric lobe-electromotoneuron proliferation and cell death. *J Comp Neurol* 207:183–190.

Fristoe TS, Iwaniuk AN, Botero CA. 2017. Big brains stabilize populations and facilitate colonization of variable habitats in birds. *Nat Ecol Evol* 1:1706–1715.

Gaede AH, Goller B, Lam JPM, Wylie DR, Altshuler DL. 2017. Neurons responsive to global visual motion have unique tuning properties in hummingbirds. *Curr Biol* 27:279–285.

González A, Northcutt RG. 2009. An immunohistochemical approach to lungfish telencephalic organization. *Brain Behav Evol* 74:43–55.

González-Voyer A, Winberg S, Kolm N. 2009. Brain structure evolution in a basal vertebrate clade: evidence from phylogenetic comparative analysis of cichlid fishes. *BMC Evol Biol* 9:238–13.

Gould KL, Gilbertson KE, Hrvol AJ, Nelson JC, Seyfer AL, Brantner RM, Kamil AC. 2013. Differences in relative hippocampus volume and number of hippocampus neurons among five corvid species. *Brain Behav Evol* 81:56–70.

Gutiérrez-Ibáñez C, Iwaniuk AN, Moore BA, Fernández-Juricic E, Corfield JR, Krilow JM, Kolominsky J, Wylie DR. 2014. Mosaic and concerted evolution in the visual system of birds. *PLoS ONE* 9:e90102.

Guy J, Staiger JF. 2017. The functioning of a cortex without layers. *Front Neuroanat* 11:54.

Haeckel E. 1866. Generelle Morphologie der Organismen: Allgemeine Grundzüge der organischen Formen-Wissenschaft, mechanisch begründet durch die von Charles Darwin reformierte Descendenz-Theorie. Berlin.

Hakeem AY, Sherwood CC, Bonar CJ, Butti C, Hof PR, Allman JM. 2009. Von Economo neurons in the elephant brain. *Anat Rec* 292:242–248.

Hashimoto M, Hibi M. 2012. Development and evolution of cerebellar neural circuits. *Dev Growth Differ* 54:373–389.

Healy SD, Rowe C. 2007. A critique of comparative studies of brain size. *Proc Roy Soc Lond B* 274:453–464.

Heim NA, Knope ML, Schaal EK, Wang SC, Payne JL. 2015. Cope's rule in the evolution of marine animals. *Science* 347:867–870.

Herculano-Houzel S. 2011. Brains matter, bodies maybe not: the case for examining neuron numbers irrespective of body size. *Ann N Y Acad Sci* 1225:191–199.

Herculano-Houzel S. 2012. Neuronal scaling rules for primate brains: the primate advantage. *Prog Brain Res* 195:325–340.

Herculano-Houzel S. 2016. *The Human Advantage*. Cambridge, MA: MIT Press.

Herculano-Houzel S, Avelino-de-Souza K, Neves K, Porfírio J, Messeder D, Mattos Feijó L, Maldonado J, Manger PR. 2014. The elephant brain in numbers. *Front Neuroanat* 8:46.

Herculano-Houzel S, Collins CE, Wong P, Kaas JH. 2007. Cellular scaling rules for primate brains. *Proc Natl Acad Sci USA* 104:3562–3567.

Herculano-Houzel S, Kaas JH, de Oliveira-Souza R. 2015. Corticalization of motor control in humans is a consequence of brain scaling in primate evolution. *J Comp Neurol* 524:448–455.

Hofman MA. 1989. On the evolution and geometry of the brain in mammals. *Prog Neurobiol* 32:137–158.

Holmes PH, Northcutt RG. 2003. Connections of the pallial telencephalon in the Senegal bichir, *Polypterus*. *Brain Behav Evol* 61:113–147.

Holmgren N. 1922. Points of view concerning forebrain morphology in lower vertebrates. *J Comp Neurol* 34:391–459.

Humphrey N. 1976. The social function of intellect. In: Bateson PPG, Hinde RA (eds.), *Growing Points in Ecology*. Cambridge: Cambridge University Press, pp. 303–317.

Hunter JP. 1998. Key innovations and the ecology of macroevolution. *Trends Ecol* Evolut 13:31–36.

Iglesias TL, Dornburg A, Brandley MC, Alfaro ME, Warren DL. 2015. Life in the unthinking depths: energetic constraints on encephalization in marine fishes. *J Evol Biol* 28:1080–1090.

Isler K. 2011. Energetic trade-offs between brain size and offspring production: marsupials confirm a general mammalian pattern. *Bioessays* 33:173–179.

Isler K, Van Schaik C. 2006. Costs of encephalization: the energy trade-off hypothesis tested on birds. *J Hum Evol* 51:228–243.

Ito H, Yamamoto N. 2009. Non-laminar cerebral cortex in teleost fishes? *Biol Lett* 5:117–121.

Iwaniuk AN, Wylie DRW. 2007. Neural specialization for hovering in hummingbirds: hypertrophy of the pretectal nucleus Lentiformis mesencephali. *J Comp Neurol* 500:211–221.

Jacob F. 1977. Evolution and tinkering. *Science* 196:1161–1166.

James W. 1890. *The Principles of Psychology*. New York: Holt.

Jelbert SA, Taylor AH, Gray RD. 2016. Does absolute brain size really predict self-control? Hand-tracking training improves performance on the A-not-B task. *Biol Lett* 12:20150871.

Jones KE, MacLarnon AM. 2004. Affording larger brains: testing hypotheses of mammalian brain evolution on bats. *Am Nat* 164:E20–31.

Kaas JH (ed.). 2017. *Evolution of Nervous Systems*. Amsterdam: Elsevier.

Kaas JH, Lyon DC. 2007. Pulvinar contributions to the dorsal and ventral streams of visual processing in primates. *Brain Res Rev* 55:285–296.

Kamil AC, Goodyear AJ, Cheng K. 2001. The use of landmarks by Clark's nutcrackers: first tests of a new model. *J Navig* 54:429–435.

Karten HJ. 2015. Vertebrate brains and evolutionary connectomics: on the origins of the mammalian "neocortex." *Philos Trans R Soc Lond B* 370:20150060.

Karten HJ, Shimizu T. 1989. The origins of neocortex: connections and lamination as distinct events in evolution. *J Cog Neurosci* 1:291–301.

Kazu RS, Maldonado J, Mota B, Manger PR, Herculano-Houzel S. 2014. Cellular scaling rules for the brain of Artiodactyla include a highly folded cortex with few neurons. *Front Neuroanat* 8:128.

Knudsen EI, Schwarz JS. 2016. The optic tectum: a structure evolved for stimulus selection. In: Kaas JH, Striedter GF (eds.), *Evolution of Nervous Systems*, 2nd ed. Amsterdam: Elsevier, pp. 387–408.

Krebs JR, Sherry DF, Healy SD, Perry VH, Vaccarino AL. 1989. Hippocampal specialization of food-storing birds. *Proc Natl Acad Sci USA* 86:1388–1392.

Lanciego JL, Wouterlood FG. 2011. A half century of experimental neuroanatomical tracing. *J Chem Neuroanat* 42:157–183.

Lee C-T, Risom T, Strauss WM. 2007. Evolutionary conservation of microRNA regulatory circuits: an examination of microRNA gene complexity and conserved microRNA-target interactions through metazoan phylogeny. *DNA Cell Biol* 26:209–218.

Lefebvre L. 2012. Primate encephalization. *Prog Brain Res* 195:393–412.

Lefebvre L, Reader SM, Sol D. 2004. Brains, innovations and evolution in birds and primates. *Brain Behav Evol* 63:233–246.

Lehrman DS. 1971. Behavioral science, engineering and poetry. In: Tobach E (ed.), *The Biopsychology of Development*. New York: Academic Press, pp. 459–471.

Letten AD, Cornwell WK. 2014. Trees, branches and (square) roots: why evolutionary relatedness is not linearly related to functional distance. *Methods Ecol Evol* 6:439–444.

Liao WB, Lou SL, Zeng Y, Kotrschal A. 2016. Large brains, small guts: the expensive tissue hypothesis supported within anurans. *Am Nat* 188:693–700.

Lingham-Soliar T. 2016. Convergence in thunniform anatomy in lamnid sharks and jurassic ichthyosaurs. *Integr Comp Biol* 56:1323–1336.

Liu Z, Qi F-Y, Zhou X, Ren H-Q, Shi P. 2014a. Parallel sites implicate functional convergence of the hearing gene prestin among echolocating mammals. *Mol Biol Evol* 31:2415–2424.

Liu Z, Wang W, Zhang T-Z, Li G-H, He K, Huang J-F, Jiang X-L, Murphy RW, Shi P. 2014b. Repeated functional convergent effects of NaV1.7 on acid insensitivity in hibernating mammals. *Proc Roy Soc Lond B* 281:2013–2950.

Long JA, Trinajstic K, Young GC, Senden T. 2008. Live birth in the Devonian period. *Nature* 453:650–652.

Losos JB. 2017. *Improbable Destinies*. London: Penguin.

Lucas JR, Brodin A, De Kort SR, Clayton NS. 2004. Does hippocampal size correlate with the degree of caching specialization? *Proc Roy Soc Lond B* 271:2423–2429.

MacLean EL, Hare B, Nunn CL, Addessi E, Amici F, Anderson RC, Aureli F, Baker JM, Bania AE, Barnard AM, et al. 2014. The evolution of self-control. *Proc Natl Acad Sci USA* 111:E2140–E2148.

Macphail EM. 1982. *Brain and Intelligence in Vertebrates*. New York: Oxford University Press.

Maguire EA, Woollett K, Spiers HJ. 2006. London taxi drivers and bus drivers: a structural MRI and neuropsychological analysis. *Hippocampus* 16:1091–1101.

Mangale VS, Hirokawa KE, Satyaki PRV, Gokulchandran N, Chikbire S, Subramanian L, Shetty AS, Martynoga B, Paul J, Mai MV, Li Y, Flanagan LA, Tole S, Monuki ES. 2008. Lhx2 selector activity specifies cortical identity and suppresses hippocampal organizer fate. *Science* 319:304–309.

Marzban H, Del Bigio MR, Alizadeh J, Ghavami S, Zachariah RM, Rastegar M. 2015. Cellular commitment in the developing cerebellum. *Front Cell Neurosci* 8:450.

Medina L, Bupesh M, Abellán A. 2011. Contribution of genoarchitecture to understanding forebrain evolution and development, with particular emphasis on the amygdala. *Brain Behav Evol* 78:216–236.

Meek J, Nieuwenhuys R. 1991. Palisade pattern of mormyrid Purkinje cells: a correlated light and electron microscopic study. *J Comp Neurol* 306:156–192.

Mengistu H, Huizinga J, Mouret J-B, Clune J. 2016. The evolutionary origins of hierarchy. *PLoS Comput Biol* 12:e1004829.

Mink JW, Blumenschine RJ, Adams DB. 1981. Ratio of central nervous system to body metabolism in vertebrates: its constancy and functional basis. *Am J Physiol* 241:R203–12.

Montgomery JC, Bodznick D, Yopak KE. 2012. The cerebellum and cerebellum-like structures of cartilaginous fishes. *Brain Behav Evol* 80:152–165.

Montgomery SH, Mundy NI, Barton RA. 2016. Brain evolution and development: adaptation, allometry and constraint. *Proc Roy Soc Lond B* 283:20160433.

Mueller T, Dong Z, Berberoglu MA, Guo S. 2011. The dorsal pallium in zebrafish, *Danio rerio* (Cyprinidae, Teleostei). *Brain Res* 1381:95–105.

Murakami T, Morita Y. 1987. Morphology and distribution of the projection neurons in the cerebellum in a teleost, *Sebastiscus marmoratus*. *J Comp Neurol* 256:607–623.

Murakami T, Morita Y, Ito H. 1983. Extrinsic and intrinsic fiber connections of the telencephalon in a teleost, *Sebastiscus marmoratus*. *J Comp Neurol* 216:115–131.

Navarrete A, van Schaik CP, Isler K. 2011. Energetics and the evolution of human brain size. *Nature* 480:91–93.

Nieuwenhuys R. 1963. The comparative anatomy of the actinopterygian forebrain. J Hirnforsch 7:171–192.

Nieuwenhuys R. 1969. A survey of the structure of the forebrain in higher bony fishes (Osteichthyes). *Ann NY Acad Sci* 167:31–64.

Nieuwenhuys R. 1998. Brachiopterygian fishes. In: Nieuwenhuys R, ten Donkelaar HJ, Nicholson C (eds.), *The Central Nervous System of Vertebrates*. Berlin; Heidelberg: Springer, pp. 655–699.

Nieuwenhuys R. 2011a. The development and general morphology of the telencephalon of actinopterygian fishes: synopsis, documentation and commentary. *Brain Struct Funct* 215:141–157.

Nieuwenhuys R. 2011b. The structural, functional, and molecular organization of the brainstem. *Front Neuroanat* 5:1–17.

Nieuwenhuys R, Puelles L. 2015. *Towards a New Neuromorphology*. Berlin: Springer.

Nieuwenhuys R, Puelles L. 2016. *Towards a New Neuromorphology*. Cham: Springer.

Northcutt RG. 1981. Evolution of the telencephalon in nonmammals. *Annu Rev Neurosci* 4:301–350.

Northcutt RG. 1985. Brain phylogeny: speculations on pattern and cause. In: Cohen MJ, Strumwasser F (eds.), *Comparative Neurobiology: Modes of Communication in the Nervous System*. New York: Wiley, pp. 351–378.

Northcutt RG. 2006. Connections of the lateral and medial divisions of the goldfish telencephalic pallium. *J Comp Neurol* 494:903–943.

Northcutt RG. 2008. Forebrain evolution in bony fishes. *Brain Res Bull* 75:191–205.

Northcutt RG, Braford MR. 1980. New observations on the organization and evolution of the telencephalon of actinopterygian fishes. In: Ebbesson, SOE (ed.), *Comparative Neurology of the Telencephalon*. Boston: Springer, pp. 41–98.

Northcutt RG, Davis RE. 1983. Telencephalic organization in ray-finned fishes. In: Davis RE, Northcutt RG (eds.), *Fish Neurobiology*, Vol. 2, pp. 203–236.

Northcutt RG, Wullimann MF. 1988. The visual system in teleost fishes: morphological patterns and trends. In: Atema J, Fay RR, Popper AN, Tavolga WN (eds.), *Sensory Biology of Aquatic Animals*. New York: Springer, pp. 515–552.

Nudo RJ, Frost SB. 2007. The evolution of motor cortex and motor systems. In: Kaas JH (ed.), *Evolution of Nervous Systems*. Amsterdam: Springer, pp. 373–395.

Ocaña FM, Suryanarayana SM, Saitoh K, Kardamakis AA, Capantini L, Robertson B, Grillner S. 2015. The lamprey pallium provides a blueprint of the mammalian motor projections from cortex. *Curr Biol* 25:413–423.

Olkowicz S, Kocourek M, Lučan RK, Porteš M, Fitch WT, Herculano-Houzel S, Němec P. 2016. Birds have primate-like numbers of neurons in the forebrain. *Proc Natl Acad Sci USA* 113:7255–7260.

Ord TJ, Summers TC. 2015. Repeated evolution and the impact of evolutionary history on adaptation. *BMC Evol Biol* 15:137.

Overington SE, Morand-Ferron J, Boogert NJ, Lefebvre L. 2009. Technical innovations drive the relationship between innovativeness and residual brain size in birds. *Anim Behav* 78:1001–1010.

Paaby AB, Rockman MV. 2013. The many faces of pleiotropy. *Trends Genet* 29:66–73.

Pakan JMP, Wylie DRW. 2006. Two optic flow pathways from the pretectal nucleus lentiformis mesencephali to the cerebellum in pigeons (*Columba livia*). *J Comp Neurol* 499:732–744.

Parker ST. 2015. Re-evaluating the extractive foraging hypothesis. *New Ideas Psychol* 37:1–12.

Perry VH, Cowey A. 1984. Retinal ganglion cells that project to the superior colliculus and pretectum in the macaque monkey. *Neurosci* 12:1125–1137.

Pravosudov VV, Roth TC II. 2013. Cognitive ecology of food hoarding: the evolution of spatial memory and the hippocampus. *Annu Rev Ecol Evol Syst* 44:173–193.

Puelles L, Kuwana E, Puelles E, Bulfone A, Shimamura K, Keleher J, Smiga S, Rubenstein JL. 2000. Pallial and subpallial derivatives in the embryonic chick and mouse telencephalon, traced by the expression of the genes *Dlx-2, Emx-1, Nkx-2.1, Pax-6*, and *Tbr-1*. *J Comp Neurol* 424:409–438.

Puelles L, la Torre M de M, Paxinos G, Watson C, Martinez S. 2007. *The Chick Brain in Stereotaxic Coordinates: An Atlas Featuring Neuromeric Subdivisions and Mammalian Homologies*. Amsterdam: Academic Press.

Puelles L, Medina L, Borello U, Legaz I, Teissier A, Pierani A, Rubenstein JLR. 2016. Radial derivatives of the mouse ventral pallium traced with Dbx1-LacZ reporters. *J Chem Neuroanat* 75:2–19.

Puelles L, Sandoval JE, Ayad A, del Corral R, Alonso A, Ferran JL, Martinez-de-la-Torre M. 2017. The pallium in reptiles and birds in the light of the updated tetrapartite pallium model. In: Kaas JH, Striedter GF (eds.), *Evolution of Nervous Systems*, 2nd ed. Amsterdam: Elsevier, pp. 519–555.

Rabosky DL, Santini F, Eastman J, Smith SA, Sidlauskas B, Chang J, Alfaro ME. 2013. Rates of speciation and morphological evolution are correlated across the largest vertebrate radiation. *Nat Commun* 4:1–8.

Raghanti MA, Spurlock LB, Treichler FR, Weigel SE, Stimmelmayr R, Butti C, Thewissen JGMH, Hof PR. 2015. An analysis of von Economo neurons in the cerebral cortex of cetaceans, artiodactyls, and perissodactyls. *Brain Struct Funct* 220:2303–2314.

Reader SM, Laland KN. 2002. Social intelligence, innovation, and enhanced brain size in primates. *Proc Natl Acad Sci USA* 99:4436–4441.

Rodríguez-Moldes I, Santos-Durán GN, Pose-Méndez S, Quintana-Urzainqui I, Candal E. 2017. The brains of cartilaginous fishes. In: Kaas JH, Striedter GF (eds.), *Evolution of Nervous Systems*, 2nd ed. Amsterdam: Elsevier, pp. 77–97.

Roelants K, Haas A, Bossuyt F. 2011. Anuran radiations and the evolution of tadpole morphospace. *Proc Natl Acad Sci USA* 108:8731–8736.

Ronan M, Northcutt RG. 1998. The central nervous system of hagfishes. In: Jørgensen JM, Lomholt JP, Weber RE, Malte H (eds.), *The Biology of Hagfishes*. Dordrecht: Springer Netherlands, pp. 452–479.

Ruiz-Reig N. 2018. The caudo-ventral pallium is a novel pallial domain expressing Gdf10 and generating Ebf3-positive neurons of the medial amygdala. *Brain Struct Funct* 223:3279–3295.

Ruse M. 1993. Evolution and progress. *Trends Ecol Evolut* 8:55–59.

Salas C. 2016. The central nervous system of agnathans. PhD dissertation, University of Western Australia.

Salas CA, Yopak KE, Lisney TJ, Potter IC, Collin SP. 2017. The central nervous system of jawless vertebrates: encephalization in lampreys and hagfishes. *Brain Behav Evol* 89:33–47.

Sayol F, Downing PA, Iwaniuk AN, Maspons J, Sol D. 2018. Predictable evolution towards larger brains in birds colonizing oceanic islands. *Nat Comm* 9:2820.

Sayol F, Lefebvre L, Sol D. 2016a. Relative brain size and its relation with the associative pallium in birds. *Brain Behav Evol* 87:69–77.

Sayol F, Maspons J, Lapiedra O, Iwaniuk AN, Székely T, Sol D. 2016b. Environmental variation and the evolution of large brains in birds. *Nat Commun* 7:13971.

Schroeder DM. 1980. The telencephalon of teleosts. In: Ebesson SOE (ed.), *Comparative Neurology of the Telencephalon*. Boston, MA: Springer US, pp. 99–115.

Shadwick R. 2005. How tunas and lamnid sharks swim: an evolutionary convergence. *Am Sci* 93:524–531.

Shanahan M. 2013. Large-scale network organization in the avian forebrain: a connectivity matrix and theoretical analysis. *Front Comput Neurosci* 7:89.

Shipp S. 2004. The brain circuitry of attention. *Trends Cogn Sci* 8:223–230.

Shubin N, Tabin C, Carroll S. 2009. Deep homology and the origins of evolutionary novelty. *Nature* 457:818–823.

Smaers JB, Gómez-Robles A, Parks AN, Sherwood CC. 2017. Exceptional evolutionary expansion of prefrontal cortex in great apes and humans. *Curr Biol* 27:714–720.

Smeets WJAJ. 1990. The telencephalon of cartilaginous fishes. In: Jones EG, Peters A (eds.), *Comparative Structure and Evolution of Cerebral Cortex*, Part I, Vol. 8A, *Cerebral Cortex*. Boston, MA: Springer US, pp. 3–30.

Smeets W, Nieuwenhuys R, Roberts BL. 2011. *The Central Nervous System of Cartilaginous Fishes*. Berlin; Heidelberg: Springer.

Smith TF, Morowitz HJ. 1982. Between history and physics. *J Mol Evol* 18:265–282.

Sobral D, Tassy O, Lemaire P. 2009. Highly divergent gene expression programs can lead to similar chordate larval body plans. *Curr Biol* 19:2014–2019.

Sol D, Duncan RP, Blackburn TM, Cassey P, Lefebvre L. 2005. Big brains, enhanced cognition, and response of birds to novel environments. *Proc Natl Acad Sci USA* 102:5460–5465.

Stanley SM. 1973. An explanation for Cope's rule. *Evolution* 27:1–26.

Striedter GF. 1992. Phylogenetic changes in the connections of the lateral preglomerular nucleus in ostariophysan teleosts: a pluralistic view of brain evolution. *Brain Behav Evol* 39:329–357.

Striedter GF. 1999. Homology in the nervous system: of characters, embryology and levels of analysis. *Novartis Found Symp* 222:158–72.

Striedter GF. 2002. Brain homology and function: an uneasy alliance. *Brain Res Bull* 57:239–242.

Striedter GF. 2005. *Principles of Brain Evolution*. Sunderland, MA: Sinauer Associates.

Striedter GF. 2013. Bird brains and tool use: beyond instrumental conditioning. *Brain Behav Evol* 82:55–67.

Striedter GF. 2016. Evolution of the hippocampus in reptiles and birds. *J Comp Neurol* 524:496–517.

Striedter GF. 2019. Variation across species and levels: implications for model species research. *Brain Behav Evol* 93:57–69.

Striedter GF, Northcutt RG. 1989. Two distinct visual pathways through the superficial pretectum in a percomorph teleost. *J Comp Neurol* 283:342–354.

Striedter GF, Northcutt RG. 2017. Rampant convergence between mammals and birds in body and brain. *Brain Behav Evol* 89:293.

Sugahara F, Pascual-Anaya J, Oisi Y, Kuraku S, Aota S-I, Adachi N, Takagi W, Hirai T, Sato N, Murakami Y, Kuratani S. 2016. Evidence from cyclostomes for complex regionalization of the ancestral vertebrate brain. *Nature* 531:97–100.

Tsuboi M, Husby A, Kotrschal A, Hayward A, Buechel SD, Zidar J, Løvlie H, Kolm N. 2015. Comparative support for the expensive tissue hypothesis: big brains are correlated with smaller gut and greater parental investment in Lake Tanganyika cichlids. *Evolution* 69:190–200.

van den Heuvel MP, Bullmore ET, Sporns O. 2016. Comparative connectomics. *Trends Cogn Sci* 20:345–361.

van Dongen PAM. 1998. Brain size in vertebrates. In: Nieuwenhuys R, ten Donkelaar HJ, Nicholson C (eds.), *The Central Nervous System of Vertebrates*. Berlin; Heidelberg: Springer, pp. 2099–2134.

Van Dyke JU, Brandley MC, Thompson MB. 2014. The evolution of viviparity: molecular and genomic data from squamate reptiles advance understanding of live birth in amniotes. *Reproduction* 147:R15–26.

Vierstra J, Rynes E, Sandstrom R, Zhang M, Canfield T, Hansen RS, Stehling-Sun S, Sabo PJ, Byron R, Humbert R, et al. 2014. Mouse regulatory DNA landscapes reveal global principles of cis-regulatory evolution. *Science* 346:1007–1012.

Wagner GP. 2007. The developmental genetics of homology. *Nat Rev Genet* 8:473–479.

Wang Y, Brzozowska-Prechtl A, Karten HJ. 2010. Laminar and columnar auditory cortex in avian brain. *Proc Natl Acad Sci USA* 107:12676–12681.

Ward BJ, Day LB, Wilkening SR, Wylie DR, Saucier DM, Iwaniuk AN. 2012. Hummingbirds have a greatly enlarged hippocampal formation. *Biol Lett* 8:657–659.

Whiting BA, Barton RA. 2003. The evolution of the cortico-cerebellar complex in primates: anatomical connections predict patterns of correlated evolution. *J Hum Evol* 44:3–10.

Wicht H. 1996. The brains of lampreys and hagfishes: characteristics, characters, and comparisons. *Brain Behav Evol* 48:248–261.

Wiens JJ, Brandley MC, Reeder TW. 2006. Why does a trait evolve multiple times within a clade? Repeated evolution of snakelike body form in squamate reptiles. *Evolution* 60:123–141.

Wilkins MR, Seddon N, Safran RJ. 2013. Evolutionary divergence in acoustic signals: causes and consequences. *Trends Ecol Evolut* 28:156–166.

Wourms JP. 1981. Viviparity: the maternal-fetal relationship in fishes. *Am Zool* 21:473–515.

Wullimann MF. 2016. Should we redefine the classic lateral pallium? *J Comp Neurol* 525:1509–1513.

Wullimann MF, Mueller T, Distel M, Babaryka A. 2011. The long adventurous journey of rhombic lip cells in jawed vertebrates: a comparative developmental analysis. *Front Neuroanat* 5:27.

Wullimann MF, Rupp B, Reichert H. 1996. *Neuroanatomy of the Zebrafish Brain*. Basel: Birkhauser.

Wylie DR, Gutiérrez-Ibáñez C, Iwaniuk AN. 2015. Integrating brain, behavior, and phylogeny to understand the evolution of sensory systems in birds. *Front Neurosci* 9:281.

Yamamoto K, Bloch S, Vernier P. 2017. New perspective on the regionalization of the anterior forebrain in Osteichthyes. *Dev Growth Differ* 59:175–187.

Yamamoto N, Ito H. 2008. Visual, lateral line, and auditory ascending pathways to the dorsal telencephalic area through the rostrolateral region of the lateral preglomerular nucleus in cyprinids. *J Comp Neurol* 508:615–647.

Yopak KE. 2012. Neuroecology of cartilaginous fishes: the functional implications of brain scaling. *J Fish Biol* 80:1968–2023.

Yopak KE, Lisney TJ, Darlington RB, Collin SP, Montgomery JC, Finlay BL. 2010. A conserved pattern of brain scaling from sharks to primates. *Proc Natl Acad Sci USA* 107:12946–12951.

Yopak KE, Pakan JMP, Wylie D. 2017. The cerebellum of nonmammalian vertebrates. In: Kaas JH, Striedter GF (eds.), *Evolution of Nervous Systems*, 2nd ed. Amsterdam: Elsevier, pp. 373–385.

York RA, Byrne A, Abdhilleh K, Patil C, Streelman JT, Finger TE, Fernald RD. 2018. Behavioral evolution drives hindbrain diversification among Lake Malawi cichlid fish. BioRxiv:467282.

Yoshida MA, Ogura A, Ikeo K, Shigeno S, Moritaki T, Winters GC, Kohn AB, Moroz LL. 2015. Molecular evidence for convergence and parallelism in evolution of complex brains of cephalopod molluscs: insights from visual systems. *Integr Comp Biol* 55:1070–1083.

Zhang Y, Shi Z, Magnus G, Meek J, Han VZ, Qiao JT. 2011. Functional circuitry of a unique cerebellar specialization: the valvula cerebelli of a mormyrid fish. *Neuroscience* 182:11–31.

APPENDIX

Evolution of the Cranial Nerves

The cranial nerves transmit action potentials from the brain to the muscles and glands of the head and neck, and they convey sensory information from the head (as well as some parts of the trunk) back to the brain. Most neurobiologists at some point find it necessary to learn the names and numbers (usually written in Roman numerals) associated with each of the 12 cranial nerves in human brains, as well as some of their functions (Table A.1). Memorizing this information can be challenging, which is why students have invented numerous mnemonics to help them (e.g., "On old Olympus's towering top a Finn and German viewed some hops" for the names of the cranial nerves in numerical sequence, and "Some say marry money but my brother says big brains matter more" to indicate whether a nerve is sensory, motor, or both). Complex as this information may be, the cranial nerves become even more complex when one looks beyond humans and other mammals to non-mammalian vertebrates. This broader perspective reveals that many vertebrates have more than 12 cranial nerves and that their organization varies substantially among the major lineages of vertebrates and their close relatives.

To illustrate the problem, consider the nerves in the head of amphioxus, the most intensely studied invertebrate chordate (see Chapter 2). Although the caudal limits of the head and brain are difficult to determine in amphioxus, it is quite obvious that the nerves in the head of amphioxus are remarkably similar to those in the rest of the body (Figure A.1). Collectively, they form a long series of repeating elements, each associated with a specific segment of the body musculature (i.e., a myotome). Each body segment features a dorsal nerve, carrying mainly sensory axons (as well as a few motor axons), and a ventral nerve that consists of modified muscle fibers that contact the floor of the central nervous system (see Chapter 2, Figure 2.18; Fritsch and Northcutt, 1993; Wicht and Lacalli, 2005). Only the two most rostral nerves depart from this pattern by extending rostrally away from the brain and carrying the axons of specialized mechanosensory cells at the very tip of the head (Lacalli, 2004). In contrast, the brain of a typical shark exhibits a great variety of cranial nerves that look quite different from those in the trunk (Figure A.1; Norris and Hughes, 1920). Moreover, the cranial nerves in sharks do not exhibit an obvious segmental pattern, at least compared to the clear segmentation of the spinal nerves. Even if we consider only the nerve roots (bottom of Figure A.1), the spatial pattern in the head is quite complex. Clearly, some of the cranial nerves in sharks (or other fishes) do not have obvious homologs in mammals or other amniotes. How can one account for this variation?

A.1. The Segmental Paradigm

The traditional approach to handling the complexity of vertebrate cranial nerves has been to postulate that the head, just like the trunk, can be divided into a series of segments, even if those segments are not obvious. According to this head segmentation paradigm (see Onai et al., 2014), the cranial nerves of vertebrates are "serial homologs" (i.e., corresponding elements in different segments of the body) of vertebrate spinal nerves (Figure A.2; Goodrich, 1918).

Given this framework, one can ask for each cranial nerve which head segment it serves (Figure A.2) and whether it corresponds to a dorsal spinal nerve or a ventral spinal nerve (carrying mainly sensory and motor axons, respectively). Since most proponents of the head segmentation framework recognize eight segments within vertebrate heads (Figure A.2), the paradigm predicts the existence of at least 16 cranial nerves on each side of the brain. In addition, it is generally acknowledged that vertebrates have several "head-specific" sense organs that require their own cranial nerves. These include the olfactory and optic nerves, as well as the nerves innervating the ear and, in anamniotes, the various lateral line receptors (see Chapter 3). Yet another complication is

Table A.1 Names and Innervation Targets of the 12 Principal Cranial Nerves in Mammals

Number	Nerve Name	Innervation
I	Olfactory	Olfactory epithelium
II	Optic	Retina
III	Oculomotor	Internal and external eye muscles
IV	Trochlear	External eye muscles
V	Trigeminal	Jaw muscles; touch to face and snout; tear glands
VI	Abducens	External eye muscles
VII	Facial	Taste buds; facial muscles; salivary and tear glands
VIII	Vestibulocochlear	Cochlea; vestibular apparatus
IX	Glossopharyngeal	Taste buds; pharynx; salivary glands
X	Vagus	Taste buds; viscera; pharynx; larynx
XI	Spinal accessory	Neck and shoulder muscles
XII	Hypoglossal	Tongue muscles

that vertebrates posses not only striated (somatic) muscle but also smooth muscle, which exhibits a distinct pattern of innervation, both in the trunk and in the head.

Considering all these complexities, it is not surprising that segmental models of cranial nerve organization postulate as many as four distinct cranial nerves per head segment (Butler and Hodos, 2005). An obvious problem with such proposals is that the number of actually observed cranial nerves is substantially lower than 32 (i.e., fewer than 4 x 8). To account for this discrepancy, one must allow for many of the postulated nerves to be either missing or combined into "mixed nerves." Indeed, the notion of cranial nerve mixing and disappearance is critical to all segmental models of cranial nerve organization (e.g., Johnston, 1905), although authors tend to disagree on which nerves were combined and which were lost in the various vertebrate lineages.

Despite these problems, the head segmentation paradigm has historically derived support from several embryological observations. One important discovery was that the hindbrain of embryonic vertebrates is divisible into seven or eight discrete segments, called rhombomeres, and that similar segments can also be identified in the midbrain and forebrain (Wilkinson et al., 1989; Lee et al., 1993; Kiecker and Lumsden, 2005). Another influential observation was that the series of mesodermal somites, which gives rise to skeletal muscles in the trunk, extends into the head up to the otic capsule (i.e., the developing ear). Even the mesoderm rostral to the otic capsule (the preotic mesoderm) exhibits some incomplete segments, generally referred to as somitomeres (Figure A.2; Meier, 1981). Then there is the obvious segmentation of the mesoderm in the pharyngeal arches, which are closely related to the more dorsally located somitomeres (Noden, 1983) and clearly subdivided by the series of pharyngeal pouches. Finally, the placodes and neural crest cells that give rise to most of the sensory axons passing through the cranial nerves (see Chapter 2) are divided into discrete patches and streams of migrating cells. Collectively, these data show that the head of embryonic vertebrates contains many different subregions, some of which form periodically repeating elements. But do they form a single series of head segments that is serially homologous to the series of segments in the trunk? Recent studies suggest that they do not.

Shigeru Kuratani and his collaborators have accumulated an impressive body of evidence indicating that the preotic mesoderm in early vertebrates formed a continuous sheet, rather than discrete segments. The somitomeres mentioned in the preceding paragraph may exist in amniotes, but they lack clear homologs in lampreys or sharks (Kuratani et al., 1999; Kuratani, 2008a). The preotic mesoderm of lampreys does show some bulges and indentations, as well as molecular regionalization (Suzuki et al., 2016), but it does not appear to be intrinsically segmented. The mesoderm that gives rise to the pharyngeal arches also seems not to be intrinsically segmented

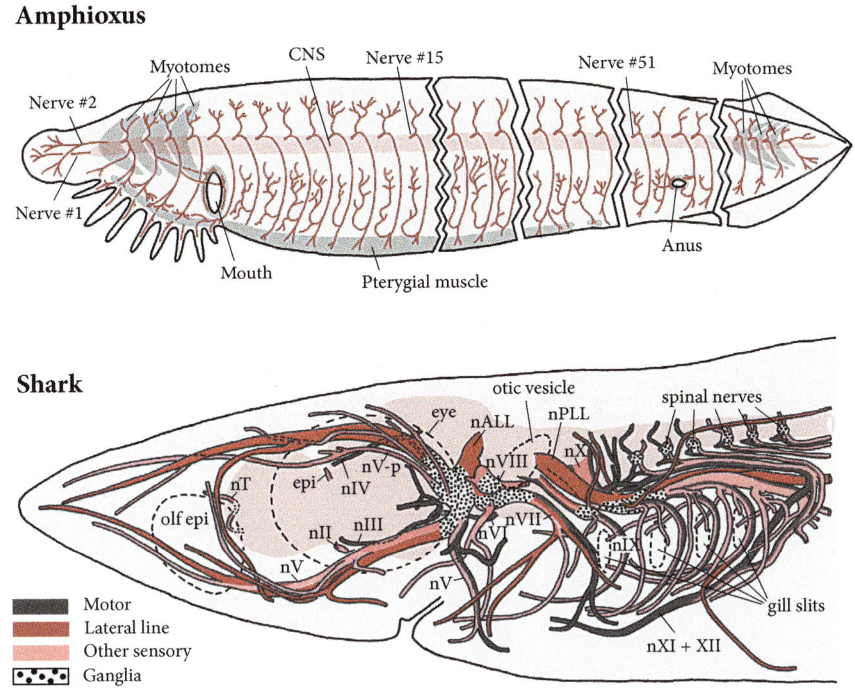

Figure A.1 The relative complexity of vertebrate cranial nerves. Shown at the top are the nerves in the trunk and head of amphioxus, an invertebrate chordate. Except for the two most rostral nerves (#s 1 and 2), the nerves in the head of amphioxus are very similar to those in the trunk. Each of these nerves issues from the central nervous system (CNS) in betwen two myotomes, which are the principal muscles of the trunk (only the most rostral and most caudal myotomes are depicted. The bottom diagram illustrates the cranial nerves of a shark (*Squalus acanthias*). They clearly differ from the spinal nerves and are much more complex than the nerves in the head of amphioxus. The nerves are labeled with Roman numerals; for the corresponding proper names, see Table A.1.

Additional abbreviations: epi – epiphyseal nerve; nT – terminal nerve; nALL – anterior lateral line nerve; nPLL – posterior lateral line nerve; nV-p – profudus component of the trigeminal nerve; olf epi – olfactory epithelium.

Adapted from Wicht and Lacalli (2005, © Canadian Science Publishing or its licensors) and Norris and Hughes (1920, with permission from John Wiley & Sons).

but, instead, becomes divided into several subregions through interactions with the underlying endoderm and otic capsule (Figure A.3; Kuratani et al., 1999). Most interesting is that the position of several hindbrain nerves (V, VII, and IX) is governed not by the mesoderm but by specific hindbrain rhombomeres and their associated neural crest cells (Kuratani and Eichele, 1993). In contrast, the location of all spinal nerves is specified by the trunk mesoderm, because the neural crest cells and motor axons comprising those nerves can only migrate through the anterior portion of each somite (Keynes and Stern, 1985; Bronner-Fraser and Stern, 1991; Keynes et al., 1991). The finding that the position of some cranial nerves is regulated by hindbrain rhombomeres does support the notion that these nerves are segmentally organized, just like the hindbrain. However, it is not consistent with the idea that the cranial nerves are serially homologous to spinal nerves. The gene networks that specify muscle development also differ substantially between head and trunk mesoderm (Sambasivan et al., 2011; Adachi et al., 2012).

Goodrich's Model of Head Segmentation

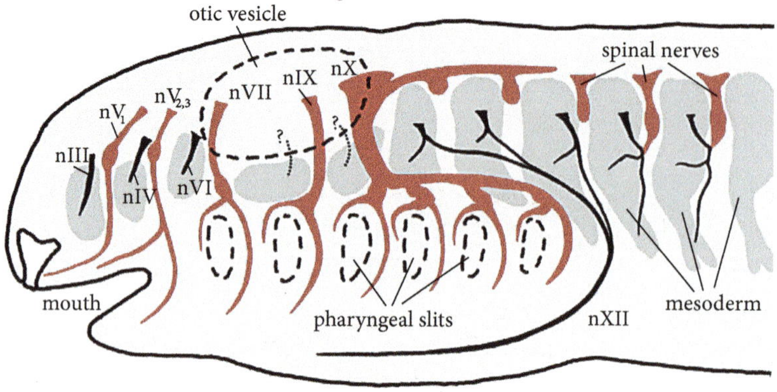

The Four-Nerves-Per-Segment Model

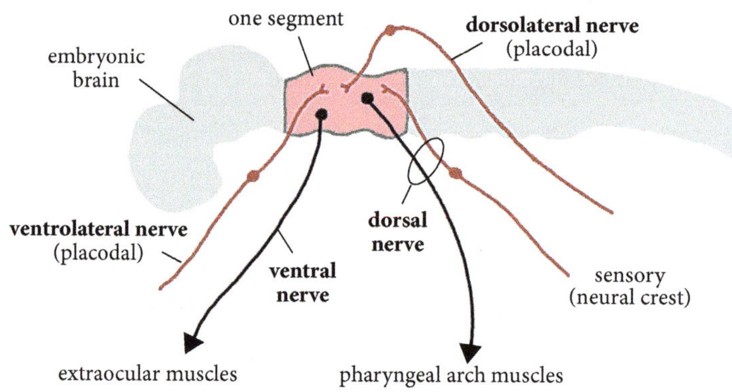

Figure A.2 Segmental models of cranial nerve organization. Goodrich (1918) accepted the idea that vertebrate heads contain a series of segments that are serially homologous to those of the trunk. Moreover, he proposed that each head segment contains dorsal and ventral nerves, which are serially homologous to the dorsal and ventral roots of vertebrate spinal nerves. Thus, cranial nerve III and the profundus branch of nV (nV_1) belong to the 1st head segment, nIV and the remaining components of nV are part of the second head segment, and so forth (see Table A.1 for the proper names of the cranial nerves). According to this scheme, the ventral nerves associated with the 4th and 5th head segments must be vestigial (indicated by question marks). Missing from Goodrich's model are the cranial nerves that develop from placodes (notably the lateral line and epibranchial placodes). To accomodate these nerves, Ann Butler and others have proposed that each "theoretical head segment" actually contains four cranial nerves (per side), as shown in the bottom diagram. To make this proposal consistent with biological reality, one must also suppose that not all of these nerves are present in all head segments and that some of them have coalesced to form "mixed nerves."

Adapted from Goodrich (1918), Northcutt (1993), Butler and Hodos (2005).

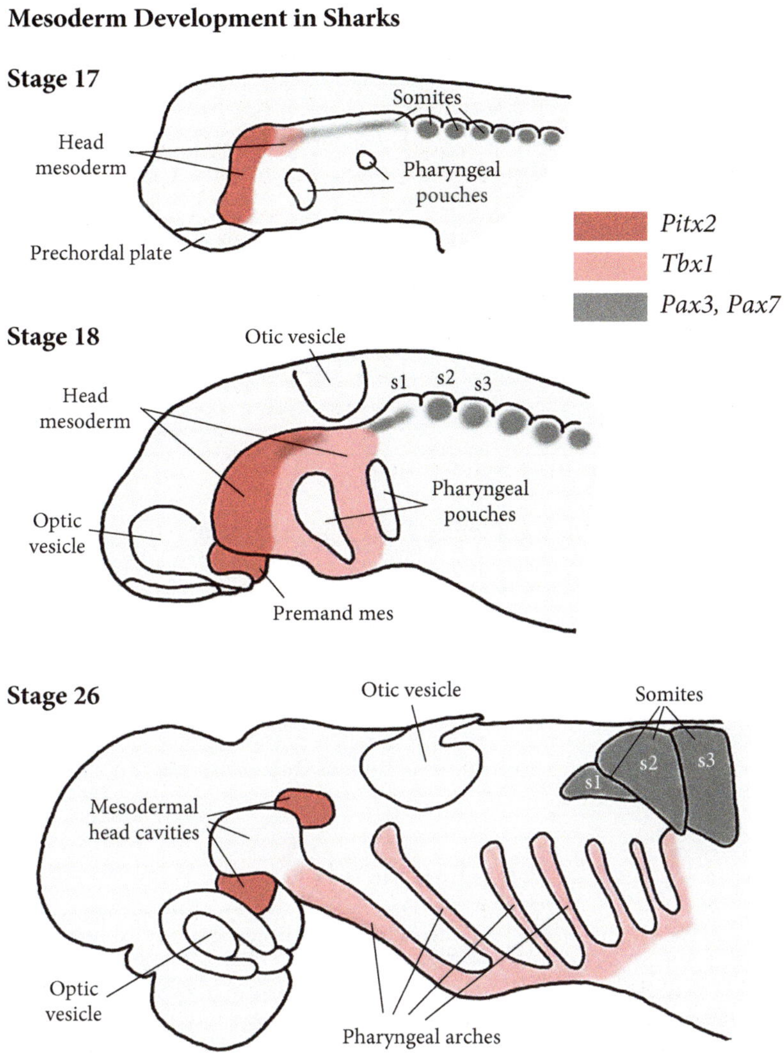

Figure A.3 Genetic parcellation of the mesoderm. Adachi et al. (2012) examined the expression of several genes known to be important for mesoderm development in embryos of the cloudy catshark (*Sciliorhinus torazame*). They found that *Pitx2* is selectively expressed in the dorsal (paraxial) portion of the head mesoderm, whereas *Tbx1* expression characterizes the pharygeal mesoderm. Neither of these tissues express high levels of *Pax3* or *Pax7*, which are characteristic of somites in the trunk and postotic region of the head. The *Pitx2* expressing region eventually develops into three "head cavities" that give rise to the extraocular muscles (among other derivatives). Although these cavities are segment-like, they are specified by a different set of genes than the segmentally arranged somites and, thus, are unlikely to be serial homologs of them.

Adapted from Adachi et al. (2012), with permission from John Wiley & Sons.

Table A.2 The Six Main Groups of Cranial Nerves in Vertebrates

Series	Nerve # or Name	Defining Feature
Olfactory Group	I, vomeronasal nerve, terminal nerve (n0)	Develop from, or in close association with, the olfactory placode
CNS Tracts	II, epiphyseal and parietal "nerves"	Interconnect regions of the CNS
Oculomotor Series	III, IV, VI	Innervate external eye muscles
Branchiomeric Series	V, VII, IX, X	Innervate pharyngeal arch derivatives (jaws, gills, heart)
Octavolateral Series	VIII, six lateral line nerves (AD, AV, OT, M, ST, Po)	Derived from octavolateral placodes
Occipital Group	XI, XII	Rostral spinal nerves that pass through the cranium

A.2. Six Groups of Cranial Nerves

Given the various difficulties with the segmental model of cranial nerve organization, we see no point in discussing it further, especially as it has been extensively reviewed elsewhere (e.g., Northcutt, 1993; Butler and Hodos, 2005; Kuratani, 2008a). Instead, it is more fruitful to think of vertebrate cranial nerves as forming several groups or "series" of nerves (Table A.2) that differ from each other not only in location and function, but also, and critically, in their mode of development (Liem et al., 2001). Although the nerves in some of these groups form serially repeating elements, collectively they do not represent a continuous segmental series.

A.2.1. The Olfactory Group

The olfactory nerve (cranial nerve I) connects the olfactory epithelium to the olfactory bulb. When the distance between these two structures is very short (e.g., in most mammals), the axons of the olfactory sensory neurons form multiple axon bundles, but in species where the olfactory epithelium lies far rostral to the brain (e.g., frogs), those fascicles coalesce to form a single nerve (Daston et al., 1990). In species that have a distinct vomeronasal epithelium, the vomeronasal axons travel separately from the olfactory nerve for at least part of their journey to the accessory olfactory bulb, but the vomeronasal nerve is entirely separate from the olfactory nerve only in lizards and snakes (Parsons, 1967). The neurons that constitute both of these nerves have their cell bodies in the sensory epithelia and develop from the olfactory placode (which apparently contains some neural crest derivatives; Whitlock and Westerfield, 2000; Forni and Wray, 2012).

In addition to the olfactory and vomeronasal nerves, many anamniotes have a terminal nerve (see Pinkus, 1895; Wirsig-Wiechmann et al., 2002; Bartheld, 2004). Because this nerve was identified after the canonical 12 cranial nerves had already been named, it is sometimes called the supernumerary nerve or nerve number zero (see Vilensky, 2014). The cells that form the terminal nerve have been reported to arise from the olfactory placode or from neural crest cells closely associated with it (Schwanzel-Fukuda and Pfaff, 1990; Whitlock et al., 2003). However, a recent study in zebrafish traces their origin to the anterior preplacodal ectoderm and argues against a contribution from the neural crest (Aguillon et al., 2018).

The cell bodies of terminal nerve neurons tend to migrate centrally during development, so that many of them end up scattered along the olfactory nerve, in a distinct terminal nerve ganglion, or inside the brain, with substantial variation across species. The peripheral processes of these neurons extend into the olfactory epithelium and (in mammals) the nasal septum, where they presumably serve some still obscure sensory function and may modulate the responses of other sensory neurons to odorants (Eisthen et al., 2000). Their central processes project to a variety of different regions in the brain, especially the preoptic area and hypothalamus (Demski and Northcutt, 1983). In teleosts, a substantial number of terminal nerve axons project to the retina, forming what is sometimes called the olfactoretinal pathway (Münz et al., 1982). These axons tend to bypass the olfactory bulbs, which raises difficult questions about the extent to which these terminal nerve fibers overlap with the "extrabulbar olfactory projections" (i.e., axons that originate in the olfactory epithelium but project to brain regions other than the olfactory bulb) that have been reported in some vertebrates (Bartheld, 2004; Eisthen and Polese, 2007; D'aniello et al., 2015).

One of the most interesting features of the terminal nerve is that many of its neurons express gonadotropin releasing hormone (GnRH; LHRH in mammals). These data suggest that the terminal nerve plays some role in reproductive behavior, and some experimental evidence supports this hypothesis (Demski, 1984; Schwanzel-Fukuda and Pfaff, 1990). However, other terminal nerve neurons express other peptides (e.g., FRMFamide), and the functional significance of this heterogeneity remains unclear. At least some components of the terminal nerve can be identified in most vertebrates, including lampreys and lungfishes (Bartheld, 2004; see Pombal and Megías, 2018). However, the terminal nerve is poorly developed in adult humans and is missing entirely in adult bats (Brown, 1980). Adult toothed whales lack olfactory and vomeronasal systems, but they do have a terminal nerve, suggesting that this nerve in these species may perform non-sensory functions, such as local regulation of blood flow (Buhl and Oelschläger, 1986).

A.2.2. Central Nervous System Tracts

Because the retina develops as an evagination of the embryonic forebrain, the optic nerve (nII) is not really a cranial nerve at all, but a tract that connects two different components of the central nervous system. Compared to other axon tracts, the optic nerve contains a large amount of connective tissue interspersed among its axon fascicles, which may facilitate nerve flexion during eye movements (Jeffery et al., 1995). The unusual pleated form of the optic nerve in many teleosts (resembling a folded ribbon; Scholes, 1991) may further accommodate bending of the optic nerve, but it may also facilitate the channeling of light through the top of the head and out the eye (aka optic nerve-transmitted eyeshine; Fritsch et al., 2017). Although all vertebrates possess an optic nerve, it is quite small in "blind" species whose retina never fully forms or partially degenerates (Besharse and Brandon, 1974; Berti et al., 2001).

The axons of the optic nerve project mainly to contralateral brain regions, but substantial ipsilateral projections are present in many vertebrate lineages. The presence of these ipsilateral projections is associated with frontally directed eyes in some vertebrate lineages (notably primates and carnivores), where they may improve depth perception, as well as vision in dim light. In addition, ipsilateral retinal projections may facilitate forelimb-eye coordination when manipulating objects in the contralateral visual field (Larsson, 2011). However, much of the variation in the presence and extent of ipsilateral retinal projections remains difficult to explain and may well be random (Ward et al., 1995).

The pineal (or epiphysial) and parapineal (aka parietal) nerves are also tracts rather than nerves, because the pineal and parapineal/parietal organs develop as dorsally directed evaginations of the diencephalic midline roof (Concha and Wilson, 2001). Lampreys possess both pineal and parapineal organs (Figure A.4), which contain photosensory cells as well as neurons (Cole and Youson, 1982). Some extinct ostracoderms and lizards also had "four eyes" (i.e., the two lateral eyes, as well as photosensitive pineal and parapineal organs), as evidenced by the presence of two openings along the dorsal midline of the skull (Edinger, 1956; Smith et al., 2018). In extant jawed vertebrates, however, one median eye is much better developed than the other (Figure A.4; Oksche, 1984; Ekström and Meissl, 2003). Teleosts, for example, tend to have a large pineal eye and a diminutive parapineal. In contrast, many lizards (and the tuatara) have a well-developed parapineal organ, which is generally called the parietal eye (Quay, 1979). Frogs have a "frontal organ" that penetrates through the skull and is often considered homologous to lizard parietal eyes (Adler, 1976); alternatively, it may be an elaborated dorsal component of the pineal organ. In mammals, the parapineal is rudimentary and the pineal organ has lost its ability to sense light directly.

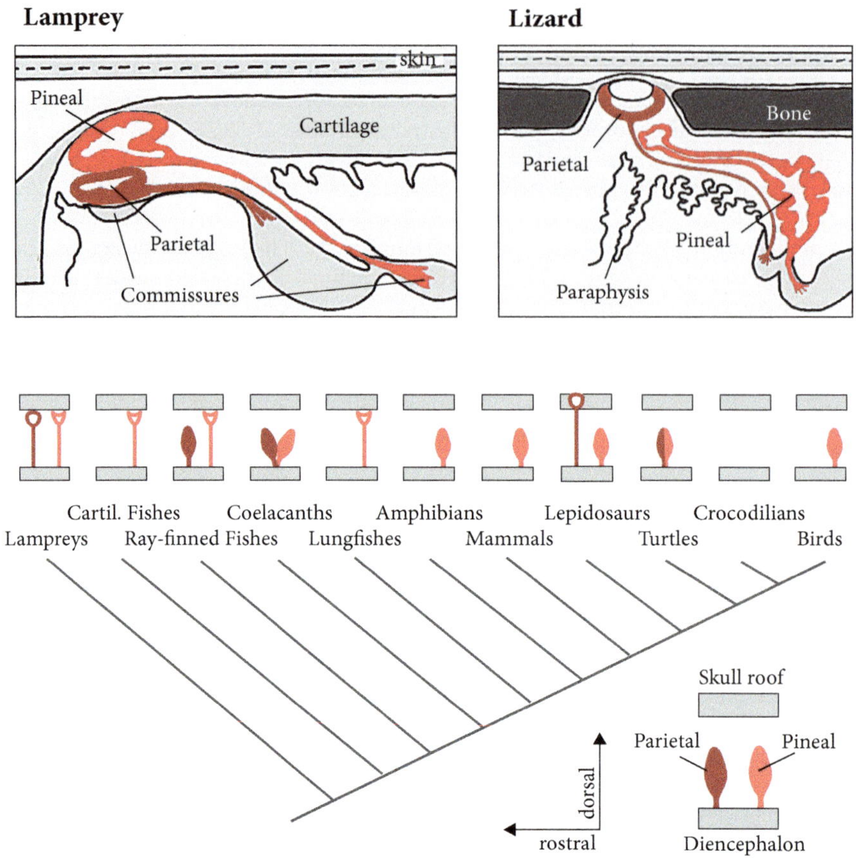

Figure A.4 Parietal and pineal organs. Shown at the top are the pineal and parapineal (aka parietal) organs of a lamprey and a lizard in mid-sagittal views (rostral to the left). The cladogram at the bottom presents the phylogenetic distribution of various parietal and pineal morphologies in extant vertebrates schematically. "Parietal eyes" are present in lampreys and some lepidosaurs (i.e., tuataras and various lizards); in the latter, they penetrate through the overlying bone. The illustrated amphibian pattern represents the condition in urodeles. Anuran amphibians have a "frontal organ" (not shown) that looks like a "parietal eye" but may instead be part of the pineal.
Adapted from Edinger (1956), Smith et al. (2018, with permission from Elsevier).

The neurons of the pineal and parapineal organs receive synaptic input from the photosensitive cells (when present) and project to several brain regions, including the preoptic area, the hypothalamus, and the habenula (Korf and Wagner, 1981; Ekström and van Veen, 1983; Puzdrowski and Northcutt, 1989). Intriguingly, the habenular projections are frequently asymmetrical, with the parapineal organ projecting preferentially to the left habenula (see Concha and Wilson, 2001). This observation suggests that the pineal and parapineal organs might have originated as two lateral eyes that rotated by 90 degrees to become midline structures, with the parapineal coming to lie rostral to the pineal (Concha and Wilson, 2001); however, this hypothesis remains highly speculative. In addition to synapsing on neurons that then project to the brain, most photosensitive cells in the pineal organ synthesize and release the sleep-regulating hormone melatonin into their immediate environment (Ekström and Meissl, 2003). The pineal organs of mammals and snakes have lost their light-sensing ability (Quay, 1979) but still release the sleep-regulating hormone melatonin, which is why they are generally referred to as pineal *glands*.

A.2.3. The Oculomotor Series

Muscles that move the eyeball in its socket (i.e., the extraocular muscles) are found in all vertebrates except hagfishes, which have degenerate eyes. Based on data from lampreys and sharks (Suzuki et al., 2016), we know that these muscles develop from the *Pitx2*-expressing dorsal portion of the head mesoderm (see Figure A.3). The most rostrally developing eye muscles end up being innervated by the oculomotor nerve (nIII); those that develop more caudally are innervated by the trochlear nerve (nIV); and the embryologically most caudal muscles are innervated by the abducent nerve (nIV; *abducent* is an adjective, *abducens* a noun). Although lampreys and jawed vertebrates all have these three cranial nerves and six main extraocular muscles, homologizing these muscles across the major vertebrate lineages is surprisingly difficult (Suzuki et al., 2016). Indeed, the extraocular muscles have undergone some substantial changes during vertebrate phylogeny (Figure A.5).

Assuming the lamprey condition is primitive, Fritzsch et al. (1990) proposed that the dorsal rectus muscle of early vertebrates split into two separate muscles in the lineage leading to elasmobranchs, whereas the rostral rectus underwent a similar division in the lineage leading to bony fishes. Moreover, the caudal rectus of early vertebrates (and retained in lampreys) is thought to

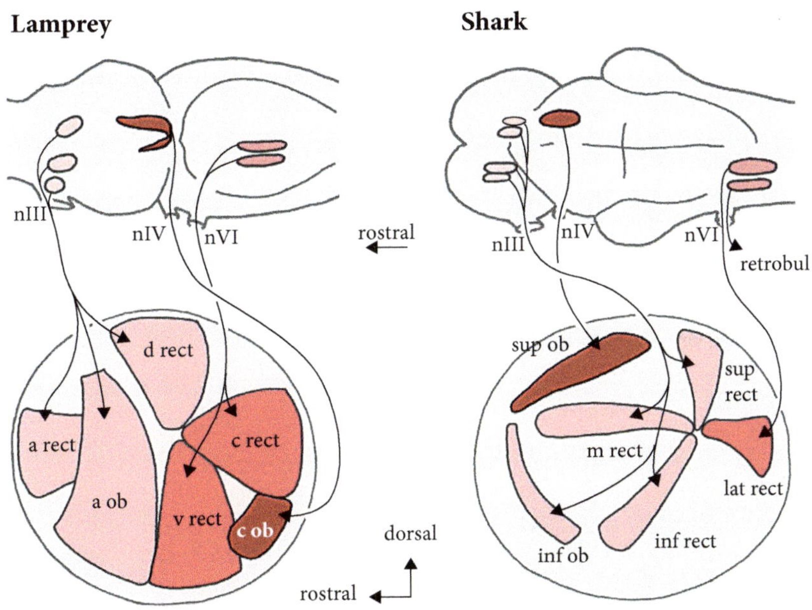

Figure A.5 Extraocular muscles and their innervation. Shown at the top are dorsal views of the midbrain and hindbrain in a lamprey and a shark, indicating the positions of the motor neurons giving rise to the oculomotor (nIII), trochlear (nIV), and abducent nerves (nVI). Shown at the bottom are the extraocular muscles, shaded according to their innervation pattern (which also reflects a common embryonic origin). Lampreys have three extraocular muscles that are innervated by the oculomotor nerve (nIII), whereas sharks (and other gnathostomes) have four, suggesting that one of the lamprey muscles split during phylogeny. The caudal rectus (c rect) muscle of lampreys is thought to be homologous to the retrobulbar muscle (retrobul) of gnathostomes, which retracts the eyeball deeper into the head.

Other abbreviations: a rect – anterior rectus; a ob – anterior oblique; c ob – caudal oblique; d rect – dorsal rectus; inf ob – inferior oblique; inf rect – inferior rectus; lat rect – lateral rectus; m rect – medial rectus; sup ob – superior oblique; sup rect – superior rectus; v rect – ventral rectus.

Adapted from Fritzsch et al. (1990) and Suzuki et al. (2016).

have evolved into the *retractor bulbi*, a rarely mentioned but important extraocular muscle that pulls the eyeball toward the interior of the head. Contraction of this seventh extraocular muscle in gnathostomes protects the eye by closing the nictitating membrane in species that possess this "third eyelid" (i.e., most jawed vertebrates but not primates). The *retractor bulbi* also helps to constrict the pharynx, thereby helping to push food into the esophagus and air into the lungs (e.g., in frogs). In coelacanths the *retractor bulbi* is represented by the basicranial muscle, which attaches to the front of the upper jaw and the ventral portion of the back of the skull. Although this muscle is innervated by the abducent nerve (Bemis and Northcutt, 1991), it has clearly lost its ancestral oculomotor function. Instead, the basicranial muscle of coelacanths is thought to increase the force of the animal's bite by rotating the rostral portion of the skull downward around the intracranial joint (Dutel et al., 2015). The intracranial joint, in turn, was an innovation of early lobe-finned fishes that was lost independently in lungfishes and tetrapods.

In addition to receiving motor innervation, the extraocular muscles send proprioceptive sensory information back to the brain, notably from muscle spindles. The sensory axons innervating these muscle spindles course through part of the trigeminal nerve (nV; see Figure A.1) and have their cell bodies in the trigeminal nerve's semilunar ganglion (Manni et al., 1970). Although only some mammals are known to have muscle spindles in their extraocular muscles, non-mammalian eye muscles do contain some other types of sensory nerve endings (Maier et al., 1974). Whether these unencapsulated receptors are also innervated by the trigeminal nerve remains unknown.

Aside from the extraocular muscles, which consist of striated muscle fibers, vertebrate eyes contain some smooth muscles, notably the pupillary constrictors and dilators, as well as ciliary muscles that can modify the shape or position of the lens. In most vertebrates, these muscles receive both sympathetic and parasympathetic innervation (McDougal and Gamlin, 2015). The parasympathetic axons travel through spinal nerves and synapse on neurons in the superior cervical ganglion, whose axons reach the intrinsic eye muscles through the trigeminal nerve. The sympathetic innervation comes from neurons in the Edinger-Westphal nucleus, which is closely associated with the oculomotor nuclei in amniotes (Kozicz et al., 2011), and passes through the oculomotor nerve with a relay in the ciliary ganglion. Although some elements of this autonomic innervation of the eye are broadly conserved, there are numerous variations among the major vertebrate lineages (Neuhuber and Schrödl, 2011). It is worth noting, for example, that the pupillary muscles in many sauropsids contain a substantial number of striated muscle fibers (Douglas, 2018).

A.2.4. The Branchiomeric Series

The trigeminal, facial, glossopharyngeal, and vagal nerves (nV, nVII, nIX and nX) are called the branchiomeric nerves (Figure A.6), because they innervate the branchial arches (*branchia* is the Greek word for "gills") and form a segmental series (*merism* in biology refers to serially repeating parts). The branchial arches (aka pharyngeal arches) separated the gill slits from one another in early vertebrates, and most of them still do so in fishes and aquatic amphibians. Developmentally, the striated muscles of the branchial arches arise from the *Tbx1*-positive, lateral portion of the head mesoderm (see Figure A.3), which become segmented by a series of endodermal pouches that break through the overlying mesoderm and ectoderm. However, the segmentation of the branchiomeric nerves is not driven by the segmentation of the mesoderm, but by the segmentation of the hindbrain into rhombomeres (see Figure 1.16 in Chapter 1). In particular, cranial nerves V, VII, and IX are attached to rhombomeres 2, 4, and 6, at least during early development (the position of their adult roots varies somewhat across vertebrate lineages; Kuratani and Horigome, 2000). Cranial nerve X is not associated with a specific rhombomere but, instead, forms through the coalescence of multiple nerve roots in the most caudal hindbrain (see Northcutt and Brändle, 1995).

The branchiomeric nerves also carry sensory information, not just from the striated muscles that they innervate (e.g., from muscle spindles) but also from touch-, temperature-, and pain-sensitive nerve endings in the head and neck, as well as oxygen or carbon dioxide sensors

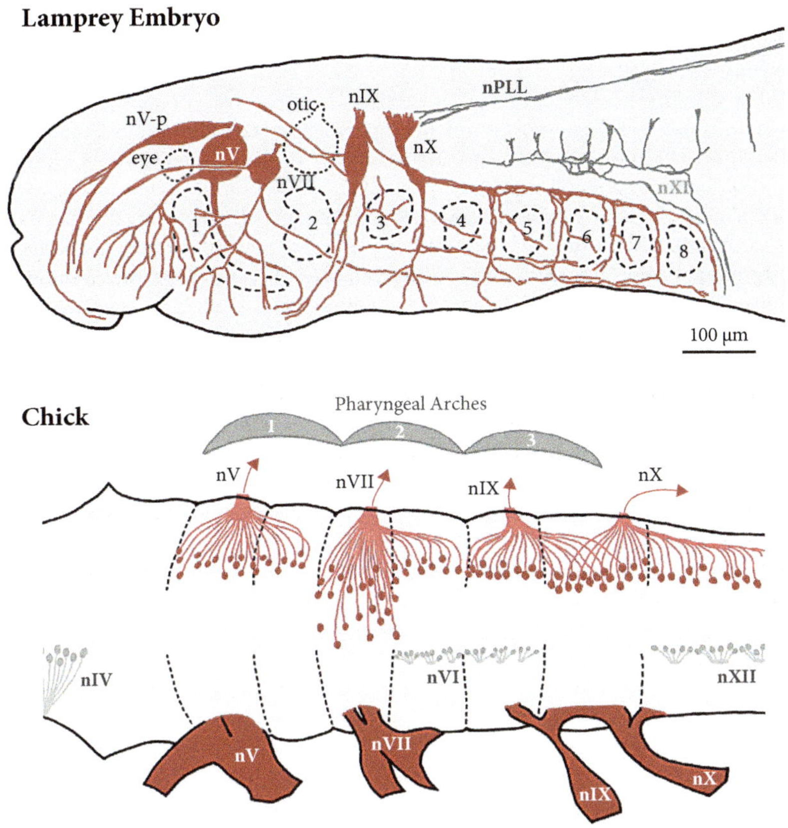

Figure A.6 The branchiomeric nerves. Shown at the top is a lateral view of a lamprey embryo in which the branchiomeric nerves (red) form a clear rostrocaudal series and the profundus nerve (nV-p) is separate from the rest of the trigeminal nerve (they fuse later). The nerves associated with the extraocular eye muscles do not develop until much later in lamprey development. Shown at the bottom is a schematic dorsal view of a chick's hindbrain. The top half shows the cell bodies and axons of the branchiomeric motor neurons (red). The bottom half shows the branchiomeric nerve roots, as well as the motor neurons of the trochlear, abducent, and hypoglossal nerves (nIV, nVI, and nXII; shown in gray). The dashed black lines represent rhombomere boundaries. The first three branchiomeric nerves (nV, nVII, and nIX) exit the brain at rhombomeres 2, 4, and 6, respectively.

Additional abbreviations: 1–8 – pharyngeal arches; nPLL – posterior lateral line nerve; otic – otic capsule. Adapted from Kuratani et al. (1997) and Kiecker and Lumsden (2005, with permission from Springer Nature).

in some blood vessels (Butler et al., 1977). Cranial nerves VII, IX, and X also innervate taste buds. Although the developmental origin of taste buds remains a matter of debate, the sensory neurons that innervate taste buds clearly derive from a series of placodes that lie dorsal to the branchial arches, which is why they are called the epibranchial placodes (Harlow and Barlow, 2007). The axons of the gustatory neurons do not form separate cranial nerves; instead, they enter the brain through three of the branchiomeric nerves (all except for nV). However, the cell bodies of the placodally derived gustatory neurons tend to be located farther away from the brain than those of the neural crest-derived somatosensory and proprioceptive neurons, forming separate ganglia (at least in amniotes). Several of the branchiomeric nerves also contain autonomic sensory and motor axons that innervate smooth muscles throughout the body, the heart, and diverse glands.

Further complicating any attempts to fully understand the branchiomeric nerves is that they underwent substantial changes during vertebrate phylogeny. We discuss these changes in Section A.3. For now, we provide merely brief summaries of their most general features (for more detailed accounts, see Brodal, 1967; Butler, 2002).

A.2.4.1. Trigeminal Nerve

The trigeminal nerve (nV) derives its name from the fact that it includes three major branches (*trigemini* means "triplets" in Latin). One of these, the ophthalmic branch, is a separate "profundal" nerve in some anamniotes, especially during embryogenesis (Northcutt and Brändle, 1995; Piotrowski and Northcutt, 1996; Kuratani et al., 1997, 2000). This branch of the trigeminal nerve is purely sensory, conveying touch and pain information from the dorsal snout, as well as the cornea. The second major branch of the trigeminal nerve, the maxillary branch, is both sensory and motor in cyclostomes (Oisi et al., 2013), but purely sensory in jawed vertebrates, innervating for example the nasal sinuses and upper teeth. The third, mandibular branch of the trigeminal nerve includes both sensory and motor components. In cyclostomes, it innervates the muscles of the rasping tongue (see Chapter 2, Section 2.2.1) and a pharyngeal valve called the velum; in jawed vertebrates, it innervates the muscles of the jaw. This observation is consistent with the idea that the trigeminal nerve innervates the derivatives of the most rostral branchial arch, which in gnathostomes develops into the principal jaw elements (see Chapter 3, Figure 3.14). In mammals, the mandibular branch also innervates the tensor tympani, one of two small muscles attached to one of the middle ear bones that is derived from one of the ancestral jaw bones (see Chapter 6, Figure 6.14).

The sensory neurons of the trigeminal nerve are all derived from neural crest precursors, except for those of the ophthalmic branch (or profundus nerve), which derive from a distinctive profundal placode (Northcutt and Brändle, 1995). The cell bodies of the various trigeminal sensory neurons lie in one or more large trigeminal ganglia (depending on the species and stage of development), but the neurons that innervate muscle spindles of the jaw have their cell bodies in the midbrain, which is why they are called mesencephalic trigeminal neurons (see Chapter 3, Figure 3.15).

A.2.4.2. Facial Nerve

The facial nerve (nVII) is the third nerve in the branchiomeric series. It innervates muscles that are derived from the second pharyngeal arch, which is called the hyoid arch. In cyclostomes, these muscles are associated with the most rostral gill slit (Guimond et al., 2003). In gnathostomes, they include muscles that suspend the jaw from the braincase and facilitate jaw opening. The fact that some of the accessory jaw bones were incorporated into the middle ear of amniotes (see Figure 6.14 in Chapter 6), explains why the facial nerve in sauropsids and mammals innervates the stapedius muscle of the middle ear (Counter et al., 1981). In mammals, the facial nerve also innervates a wide variety of facial muscles, including those that lower the eyelids, move the lips and cheeks, and furrow the brow. Besides innervating all these striated muscles, the facial nerve in terrestrial tetrapods provides parasympathetic innervation to the tear glands and two of the salivary glands.

Like the other branchiomeric nerves, the facial nerve contains not only motor axons but also a variety of sensory nerve fibers. For example, it includes sensory axons that innervate muscle spindles. In addition, the facial nerve carries somatosensory information from some portions of the face and scalp, although in amniotes this function is performed primarily by the trigeminal nerve. In contrast to the trigeminal nerve, the facial nerve also includes neurons that are derived from the most rostral epibranchial placode and innervate taste buds. In all vertebrates some of these taste buds lie inside the pharynx. Some fishes have taste buds on their external body surface (e.g., on their barbels and pectoral fins), and some (e.g., catfishes) have them on their entire body surface. These external taste buds are always innervated by a branch of the facial nerve

(Northcutt, 2004); those on the trunk are innervated by the facial nerve's recurrent branch. In mammals, the facial nerve innervates taste buds mainly on the rostral two-thirds of the tongue, but birds do not have taste buds on the rostral portion of the tongue. Instead, their facial nerve innervates taste buds on the rostral interior portion of the lower beak (Ganchrow et al., 1986). Finally, the facial nerve in most anamniotes contains numerous sensory axons associated with the lateral line system, which we discuss in Section A.2.5.

A.2.4.3. Glossopharyngeal Nerve

Compared to the other branchiomeric nerves, the glossopharyngeal nerve (nIX) is relatively simple. In anamniotes it innervates the muscles of the third gill arch and conveys sensory information back from them, including information from taste buds. In amniotes, the third pharyngeal arch gives rise to only one minor muscle (the stylopharyngeus). However, the glossopharyngeal nerve of amniotes does innervate the smooth muscle of the parotid salivary gland, and it conveys sensory information from taste buds on the back of the tongue, some parts of the pharynx, and the ear drum. In addition, the glossopharyngeal carries information from oxygen sensors in the carotid sinus and carotid body of amniotes. In anamniotes, the (presumably) homologous receptors are more widely distributed (e.g., on the gills) and innervated by the trigeminal and facial nerves, as well as the glossopharyngeal nerve (Milsom and Burleson, 2007).

A.2.4.4. Vagal Nerve

The vagal nerve (nX) is the most caudal of the branchiomeric nerves. In aquatic anamniotes, it innervates the most caudal set of gill muscles. The fact that the number of gill slits varies from seven in cyclostomes and some cartilaginous fishes to just three in bony fishes probably explains why the vagal nerve attaches to the caudal brainstem though a variable number of distinct roots. However, those roots fuse into a single nerve trunk and at least two ganglia before they branch again on their way to the periphery. In addition to innervating the most caudal gill muscles, the vagus innervates muscles in the roof of the pharynx, which are especially well developed in teleosts that use them to sort edible items from debris (see Figure 3.6; Finger, 2009). In tetrapods, the caudal pharyngeal arches give rise mainly to the laryngeal muscles, which are innervated by a special (recurrent) branch of the vagus.

The vagal nerve also has a branch that provides parasympathetic innervation to the heart. This pathway is common to all vertebrates, except for hagfishes, which lack cardiac innervation entirely (Greene, 1902). In lampreys the vagus does innervate the heart, but its axons target only the heart's arterial pole (Higashiyama et al., 2016). Moreover, vagal activation speeds up the cardiac rhythm in lampreys, which is the opposite of what happens in other vertebrates (Taylor et al., 1999). In addition to the cardiac muscle, the vagus in all vertebrates innervates the smooth muscles of the esophagus, gut, lungs, and other internal organs. The same branch also conveys chemosensory and mechanosensory information back to the brain from the internal organs. The vagus carries some gustatory information from the pharynx, but the vagal contribution to the sense of taste is relatively small in amniotes. Finally, parts of the vagal nerve complex fuse with nerves that innervate neuromasts of the lateral line system (see Section A.2.5).

A.2.5. The Octavolateral Series

In addition to the olfactory and epibranchial placodes, most aquatic anamniotes have a series of six or seven octavolateral placodes that give rise to the sensory cells of the lateral line system and inner ear, as well as to the neurons innervating those sensors (Northcutt, 1992; Northcutt et al., 1994). These placodes form a rostrocaudal series (Figure A.7), but they do not align in any simple way with the series of segments in the brain or with the branchial arches. The middle placode in the octavolateral series is called the octaval placode; it gives rise to the hair cells of the inner ear and to its innervation, the octaval nerve (nVIII). The other octavolateral placodes generate the mechanosensory and electrosensory lateral line sensors and nerves.

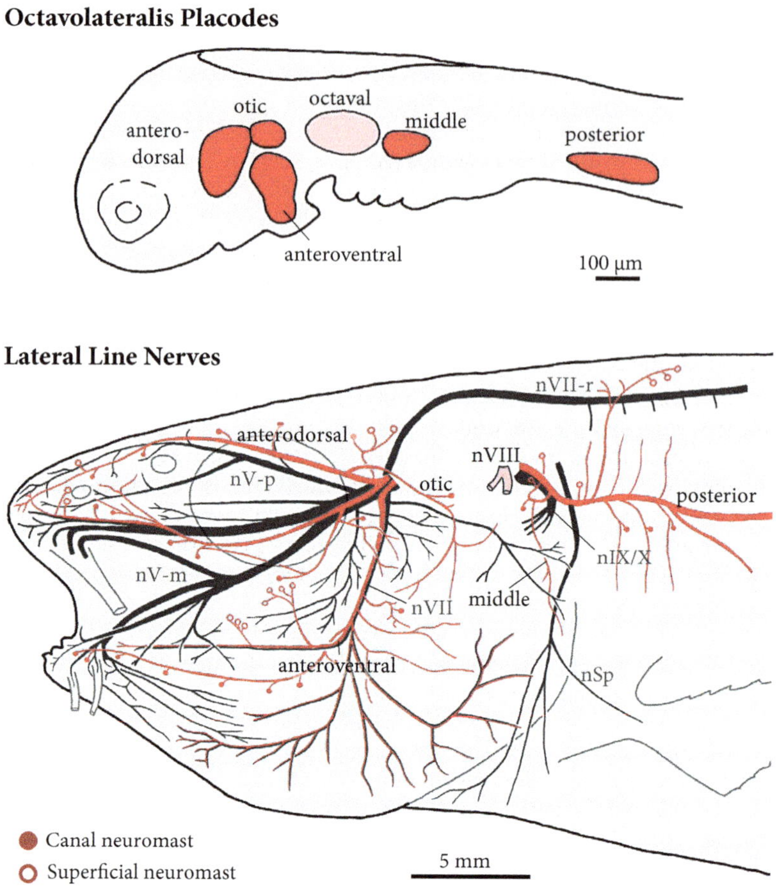

Figure A.7 Octavolateralis placodes and lateral line nerves. Shown at the top is a lateral view of a catfish embryo, highlighting its octaval placode, its three preotic placodes, and two postotic placodes (note that catfishes lack the superatemporal postotic placode found in most other fishes and amphibians). The posterior placode is in the process of migrating down the embryo's trunk. Shown at the bottom is a lateral view of the head of a juvenile catfish, depicting its lateral line nerves (red) and their relationship to some of the other cranial nerves (black). Each of the major lateral line nerves (anterodorsal, anteroventral, otic, middle, and posterior) is derived from a different placode. As their axons course toward the brain, they join either other lateral line nerves or one of the branchiomeric nerves (nV, nVII, nIX, nX), but they maintain their separate identities (see Figure A.8).

Additional abbreviations: nSP – first spinal nerve; nIX/X – complex of nIX and nX; nV-m – mandibular branch of nV; nV-p – profudus branch of nV; nVII-r – recurrent ramus of nVII.

Adapted from Northcutt (2003) and Northcutt et al. (2000).

A.2.5.1. Octaval Nerve

The eighth cranial nerve (nVIII), is often called the vestibulocochlear nerve, but this name is not appropriate for the many vertebrates that do not have a cochlea (see Chapters 4 and 5, Figures 4.14 and 5.17). Instead, it is better to call it the octaval nerve. It comprises mainly the axons of the sensory neurons that innervate the hair cells of the vestibular and auditory epithelia in the inner ear.

In most vertebrates, including cyclostomes (Fritzsch et al., 1989), the octaval nerve also contains some axons that have their cell bodies in the hindbrain and provide the hair cells of the inner ear with efferent innervation. Although these efferent axons course through the octaval nerve, they are thought to be the axons of modified facial motor neurons that were phylogenetically "rerouted" to the inner ear (Roberts and Meredith, 1992; Fritzsch and Elliott, 2017). When activated, these efferent axons tend to inhibit the hair cells, which may protect them from sensory overload during self-initiated movements or vocalizations (Sienknecht et al., 2014). In mammals and birds, these efferent axons target primarily the outer hair cells of the cochlea, which respond by changing their shape (Ashmore, 2008). However, even mammals and birds retain some efferent axons that target inner hair cells, as well as vestibular hair cells. The function of these "enigmatic efferents" (Roberts and Meredith, 1992) remains poorly understood.

A.2.5.2. Lateral Line Nerves

Besides the octaval placode, most fishes and amphibians possess six lateral line placodes (Figure A.7). Present-day amphibians lack one of the preotic placodes (i.e., rostral to the octaval placode), and catfishes lack one of the postotic ones (Figure A.7), but a comparative analysis suggests that early vertebrates had three preotic lateral line placodes and three postotic ones. Collectively, these placodes give rise to all of the hair cells of the lateral line system, which aggregate into neuromasts (see Chapter 2, Figure 2.15). In addition, some of the preotic placodes generate electroreceptors (Northcutt, 1992), which tend to develop at the edges of the placodes and are innervated by the same lateral line nerves as the neuromasts (Northcutt et al., 1994, 1995; Baker et al., 2013).

Aquatic anamniotes possess five or six separate lateral line nerves, one for each placode (Cole, 1897; Northcutt, 1989, 1992). These nerves are separate from each other and the other cranial nerves during embryonic development (Northcutt and Brändle, 1995). However, as development proceeds, some of the lateral line nerves coalesce with one another or with one of the branchiomeric nerves, which means that not all of them enter the brain as distinct roots. For example, the anterodorsal and otic lateral line nerves in catfishes fuse and enter the brain jointly (Figure A.7). Similarly, the anteroventral lateral line nerve combines with part of the facial nerve long before the two enter the brain. Despite this merging of the nerves, their axons do not intermingle; the branchiomeric and lateral line nerves also tend to maintain separate ganglia (Figure A.8).

Despite this physical segregation, many older studies did not recognize the lateral line nerves as distinct nerves. Instead, lateral line axons that enter the brain with one of the branchiomeric nerves were identified as a special functional component of the branchiomeric nerves; thus, they were not given a separate name. For example, in the original illustration of a shark's cranial nerves shown in Figure A.1, the two main sets of lateral line axons entering the brain were identified as the "lateral line roots of the facial and vagal nerves" (Norris and Hughes, 1920). We re-labeled these roots in the figure as the anterior and posterior lateral line nerves (nALL and nPLL), because they are fully separate from the facial and vagal nerves. This nomenclature is found in many papers on the lateral line nerves, with some recognizing an additional middle lateral line nerve (Boord and Campbell, 1977; Puzdrowski, 1989; Song and Northcutt, 1991; Pombal and Megías, 2018). Although the terms remain useful, the embryology suggests that both the anterior and posterior lateral line nerves are actually composites of several cranial nerves, which deserve their own names (e.g., Northcutt and Bemis, 1993; Piotrowski and Northcutt, 1996).

Although the lateral line nerves consist primarily of sensory axons, they do contain a few of the "enigmatic efferents" mentioned in the previous section. Some of these efferent neurons project to hair cells in both the inner ear and the lateral line (Hellmann and Fritzsch, 1996).

A.2.6. The Occipital Group

The two most caudal cranial nerves are the spinal accessory and hypoglossal nerves (Figure A.9). They are motor nerves that exit from the most rostral spinal cord, but then turn rostrally and course inside the head for at least part of their path. It is probably best to think of them as rostral spinal nerves, but they are traditionally included among the cranial nerves because they course at least partly within the skull. Either way, the spinal accessory and hypoglossal nerves are located

Preotic Nerves and Ganglia

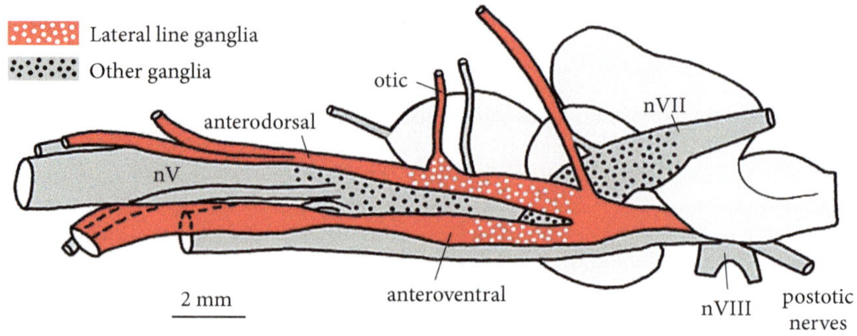

Postotic Nerves and Ganglia

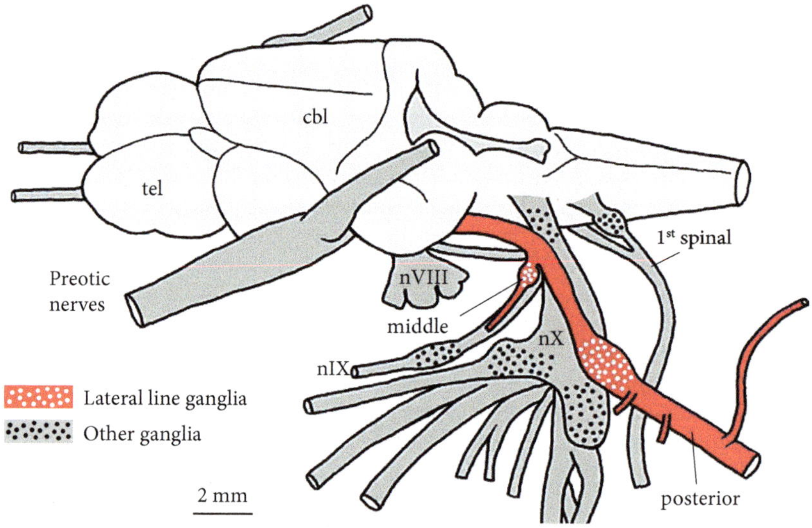

Figure A.8 Lateral line nerves and ganglia. The top diagram represents a lateral view of a catfish brain, highlighting the preotic lateral line nerves and ganglia (see Figure A7) in relation to other cranial nerves. The bottom diagram does the same for the postotic lateral line nerves and ganglia (shown from a dorsolateral perspective). Although the lateral line nerves are closely associated with some of the other nerves, they have separate ganglia, and their axons do not intermingle with those of other cranial nerves.

Adapted from Northcutt et al. (2000).

in the transition zone between the head and neck, which is itself variable across species (since the neck is a tetrapod innovation; see Chapter 4, Section 4.1.4). Both nerves innervate muscles that are derived from the most rostral postotic somites (Noden, 1983; Tada and Kuratani, 2015). Thus, they are clearly distinct from the branchiomeric nerves.

A.2.6.1. Spinal Accessory Nerve

Like the vagal nerve, the spinal accessory nerve (nXI) has multiple roots that fuse to form a common trunk. Its cell bodies are also located more dorsally than those of the motor neurons in the spinal cord's ventral horn (Tada and Kuratani, 2015). However, unlike the

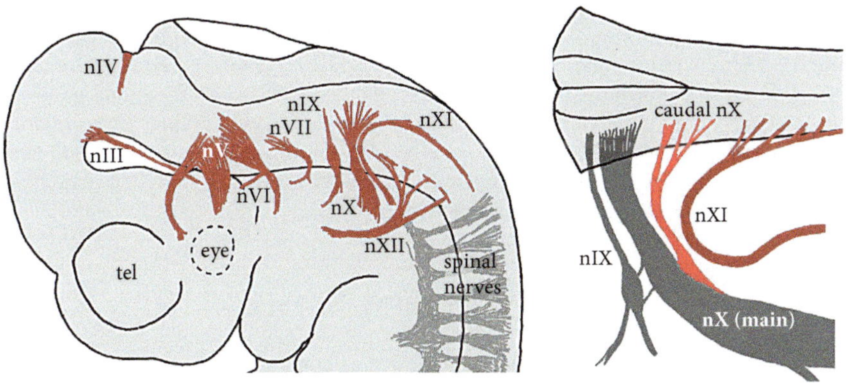

Figure A.9 The spinal accessory and hypoglossal nerves. Shown on the left is a lateral view of a mouse at 10.5 days of embryogenesis. The drawing reveals the complex spatial relatioships between the vagal, spinal accessory, and hypoglossal nerves (nX, nXI, and nXII, respectively). The diagram on the right is a dorsal view of the hindbrain (rostral to the left). It shows that nXI is distinct from the caudal portion of the vagus nerve (caudal nX), even though the latter is sometimes considered to be part of nXI.
Adapted from Lachman et al. (2002), Campos et al. (2011), Tada and Kuratani (2015).

motor neurons of the vagal nerve, the neurons of the spinal accessory nerve lie caudal to the hindbrain. Some caudal rootlets of the vagus are sometimes misidentified as the cranial root of the spinal accessory nerve (Figure A.9), but these two cranial nerves target very different muscles (Lachman et al., 2002; Campos et al., 2011). Instead of innervating branchiomeric muscles, the spinal accessory nerve innervates epibranchial and hypobranchial muscles, which lie dorsal and ventral to the pharynx, respectively (Adachi et al., 2018). In jawed fishes the spinal accessory nerve (sometimes called the occipital nerve or nerves) innervates an additional muscle, namely the cucullaris muscle, which evolved into the principal neck muscles of tetrapods. A putative homolog of the spinal accessory nerve has been identified in lampreys (Tada and Kuratani, 2015), but these animals lack a homolog of the cucullaris muscle.

A.2.6.2. Hypoglossal Nerve

The hypoglossal nerve (nXII) courses through the skull for part of its path, but it clearly consists of several fused spinal nerves, which exit the spinal cord ventrally (Figure A.9; Tada and Kuratani, 2015). Like the spinal accessory nerve, the hypoglossal nerve innervates part of the hypobranchial musculature. In tetrapods it innervates mainly the muscles of the tongue, which develop from the anterior portion of the hypobranchial muscle precursor region. The posterior hypobranchial precursor region develops into the infrahyoid muscles of the neck, which are innervated by part of the hypoglossal nerve in birds. In mammals, these muscles are innervated by rostral cervical spinal nerves (the *ansa cervicalis*), suggesting that these nerves may be homologous to part of the hypoglossal nerve in non-mammals. In birds, part of the hypoglossal nerve also innervates the syrinx, the avian vocal organ.

Although the hypoglossal nerve is primarily a motor nerve, it does carry sensory axons coming from the tongue, at least in some species. These hypoglossal afferent neurons generally have their cell bodies in the ganglia of the adjacent vagal or spinal nerves. Like the hair cell efferents we mentioned in Section A.2.5, they may have been "re-routed" during phylogeny (Anderson and Nishikawa, 1997).

A.3. Evolutionary Transformations of the Cranial Nerves

As we discussed in Section A.1, many researchers have wondered whether the cranial nerves represent a rostral continuation of the spinal nerves and are, therefore, serially homologous to them. We no longer support this serial homology hypothesis (except for the spinal accessory and hypoglossal nerves) and, therefore, reject the segmental paradigm of cranial nerve organization (see Section A.1). In the present section, we ask a different question: Do the cranial nerves of vertebrates have homologs in the invertebrate chordates? When did they originate, and how were they modified at key junctures in vertebrate phylogeny?

A.3.1. The Origin of Vertebrates

According to the "new head hypothesis" (Gans and Northcutt, 1983; Northcutt and Gans, 1983), the origin of vertebrates was marked by several key innovations that, collectively, gave vertebrates a very different type of head (see Chapter 2). These innovations include the appearance of neural crest cells that migrate extensively and give rise to several different types of cells, including neurons. In addition, only vertebrates have placodes that give rise to a variety of sensory cells and the neurons that innervate them. A third key innovation of early vertebrates was the development of branchiomeric muscles (although amphioxus larvae have similar muscles that degenerate at metamorphosis; Yasui et al., 2013) and a chambered heart, both of which develop from head mesoderm and neural crest. Indeed, cephalic mesoderm is a vertebrate innovation in its own right.

Some authors have argued that some or all of these "vertebrate novelties" have precursors in the invertebrate chordates and are, therefore, not really "new." For example, Diogo et al. (2015) argued that "the 'new' head arose instead by elaboration and modification of existing tissues, cell populations and gene networks through evolutionary 'tinkering.'" (p. 470). We agree that most building blocks of the "new head" had some sort of precursors in pre-vertebrate ancestors, but these preexisting components were combined in novel ways to form new structures with novel functions, which is precisely how Francois Jacob conceived of "evolutionary tinkering" in his influential paper (Jacob, 1977). In any case, there can be little doubt that the innovations associated with the "new head" of vertebrates allowed these animals to adopt a far more active way of life than that exhibited by their filter-feeding ancestors (see Chapter 2).

How did the evolutionary changes in the head's sensory systems and musculature affect the cranial nerves? Assuming that the nerves of amphioxus are representative of those in the immediate ancestors of vertebrates, this question amounts to asking how the cranial nerves of vertebrates differ from the cephalic nerves of amphioxus. The short answer is that the changes were profound.

With regard to the sensory axons and nerves, the dorsal nerves in the head of amphioxus probably carry mechano- and chemosensory information from axon-bearing sensory cells in the pharynx and skin (Figure A.10; Lacalli, 2004). However, amphioxus does not have any of the neural crest-derived ganglia associated with the somatosensory axons in vertebrate cranial nerves. Nor does amphioxus possess any of the nerves in the olfactory and octavolateralis groups, since these nerves are derived from placodes that amphioxus lacks. The absence of epibranchial placodes in amphioxus further implies that these animals lack homologs of the gustatory axons that course through most of the branchiomeric nerves. One may argue that amphioxus does possess an optic nerve (or tract) that connects the median eye to the brain, but amphioxus certainly does not have the paired optic nerves typical of vertebrates. Whether amphioxus has epiphyseal or parietal nerves remains uncertain, since it is unclear whether they possess a homolog of the vertebrate pineal complex (see Chapter 2). Thus, we can conclude that virtually all of the sensory nerves in the head of vertebrates emerged with the origin of vertebrates.

Motor commands emerge from the central nervous system of amphioxus through ventral nerves that consist of thin muscle fiber processes that extend up to the central nervous system, where they receive synaptic input from motor neurons (Figure A.10; see also Figure 2.18 in Chapter 2). Whether this arrangement is a derived feature of amphioxus remains unclear, but either amphioxus or early vertebrates must have reconfigured their ventral motor nerves

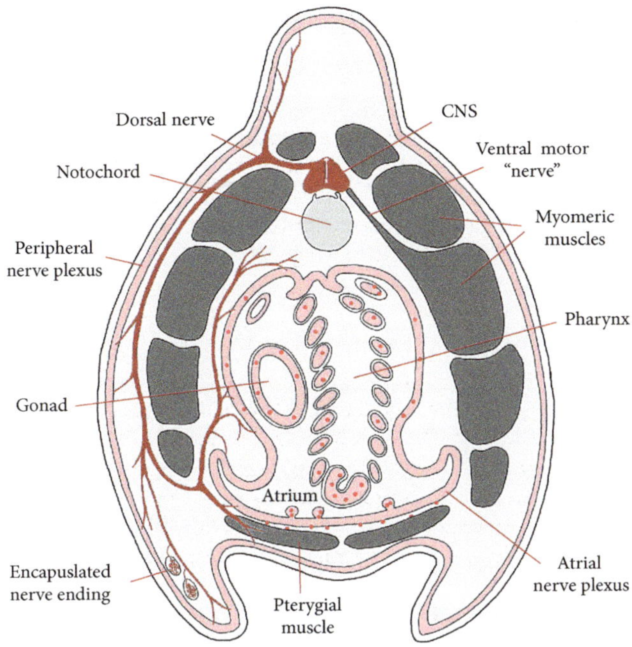

Figure A.10 The peripheral nervous system of amphioxus. This schematic transverse section through the body of amphioxus near the caudal end of the pharynx (see Figure A.1) highlights the peripheral distribution of one dorsal nerve (dark red). This nerve innervates encapsulated nerve endings as well as networks of peripheral neurons (i.e., plexuses; shaded pink, with red dots indicating cell bodies) in the skin and the wall of the atrium, which surrounds the pharynx. The ventral "nerves" of amphioxus consist of slender processes of mymoeric muscles (dark gray) that extend to the ventrolateral surface of the central nervous system (CNS), where they receive synaptic input from motor axons. However, some motor axons exit the CNS through the dorsal nerves; they innervate primarily the pterygial muscle, which lies ventral to the pharynx and atrium (see also Figure A.1).
Adapted from Wicht and Lacalli (2005 © Canadian Science Publishing or its licensors).

substantially. Some motor axons do exit the central nervous system of amphioxus, but they do so through the dorsal nerves. These motor axons innervate various internal organs (e.g., the gonads) and the pterygial muscle, which lies ventral to the pharynx (Figures A.1 and A.10; Wicht and Lacalli, 2005). The dorsally exiting motor axons of amphioxus have sometimes been homologized to the motor components of the branchiomeric nerves (Gans, 1989), but it is far from certain that the pterygial muscle of amphioxus is homologous to the branchial muscles of vertebrates. Alternatively, it might be homologous to the hypobranchial muscles of vertebrates, which develop from rostral somites and migrate around the pharynx to their adult position ventral to the pharynx (Adachi et al., 2018). Far more certain is that amphioxus lacks homologs of the vertebrate extraocular muscles, which means that the nerves of the oculomotor series are yet another vertebrate innovation (Figure A.11).

In contrast to all vertebrates, amphioxus does not possess smooth muscle fibers or glands. It also lacks the sympathetic and parasympathetic ganglia that characterize the autonomic nervous system of jawed vertebrates. Amphioxus does, however, have an extensive atrial nervous system, which innervates the pharynx and several internal organs (Figure A.9). This innervation may be homologous to parts of the autonomic nervous system of vertebrates, especially since the

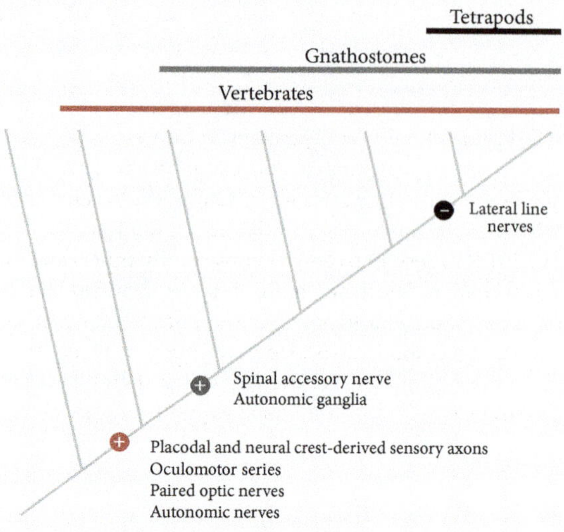

Figure A.11 Major transformations in cranial nerve organization. Most cranial nerves emerged with the origin of vertebrates; this includes all of the placode-derived nerves and the neural crest-derived components of the branchiomeric nerves. Well-developed spinal accessory nerves and autonomic ganglia emerged with the origin of jawed vertebrates (gnathostomes). After that, the main trunks of the cranial nerves remained relatively invariant, until the lateral line nerves were lost in the tetrapod lineage.

atrial nervous system includes a complex network of peripheral neurons and ganglia (Wicht and Lacalli, 2005). However, other data indicate that lampreys and hagfishes have at best some evolutionary precursors of autonomic ganglia (e.g., Johnels, 1956; Häming et al., 2011). Therefore, we conclude that autonomic ganglia are an innovation not of vertebrates, but of jawed vertebrates (Fritzsch et al., 2017).

A.3.2. The Origin of Gnathostomes

Aside from changes in the autonomic nervous system, the origin of jawed vertebrates brought with it several changes in the skeletal motor system. However, these changes were relatively minor with regard to the cranial nerves.

The details of how gnathostome jaws evolved remain debatable (e.g., Shigetani et al., 2005; Mallatt, 2008), but they clearly involved substantial transformations of the first two pharyngeal arches. The muscles of these two arches are innervated by the trigeminal and facial nerves in both cyclostomes and gnathostomes, but homologizing these muscles and the associated peripheral branches of the nerves is very difficult, if not impossible. The principal reason for this difficulty is that cyclostomes independently evolved a complex feeding apparatus of their own (see Chapter 2, Section 2.2.1) that contains numerous muscles without obvious homologs in gnathostomes (Yalden, 1985; Ziermann et al., 2014). Although the trigeminal nerve consists of three major branches in both cyclostomes and gnathostomes, its middle branch carries motor axons only in cyclostomes, raising serious questions about its homology to the middle (maxillary) branch of the trigeminal nerve in gnathostomes (Oisi et al., 2013).

The operation of vertebrate jaws involves not only modified branchial muscles, but also hypobranchial muscles, which, as their name suggests, lie below the pharynx (at least in part; Adachi et al., 2018). Their function in cyclostomes remains obscure, but in gnathostomes they attach to the lower jaw and the pectoral girdle such that their contraction opens the jaw. Despite this major transformation, the hypobranchial muscles in gnathostomes are innervated by the hypoglossal nerve (nXII), just like their homologs in cyclostomes.

Another innovation of early gnathostomes is the cucullaris muscle, which originally connected the back of the skull to the pectoral girdle and, in tetrapods, became transformed into the trapezius and sternocleidomastoideus muscles of the neck (Trinajstic et al., 2013; Kuratani, 2013; Diogo and Ziermann, 2014). These muscles are consistently innervated by the spinal accessory nerve (nXI). Although cyclostomes lack obvious homologs of the cucullaris muscle and the spinal accessory nerve (Kuratani, 2008b), some potential homologs of these structures have been identified in lampreys (Tada and Kuratani, 2015). However, because these structures in lampreys lack several important developmental and molecular features of their gnathostome counterparts, it is reasonable to classify the cucullaris muscle and nXI as gnathostome innovations (Figure A.11).

A.3.3. The Origin of Tetrapods

The emergence of fully terrestrial tetrapods was accompanied by several substantial changes in their sensory systems. Chief among them was the loss of the lateral line system. In contrast to teleosts, which lost only the electrosensory portion of their lateral line system (see Chapter 3), terrestrial tetrapods lost both the mechanosensory and electrosensory components of the lateral line. Accordingly, they also lost all of the nerves associated with this sensory system. A second major change was that terrestrial tetrapods sequestered their vomeronasal receptors into a separate epithelium that projects to a separate target in the brain, namely the accessory olfactory bulb. As a result of these changes, some tetrapods evolved a distinct vomeronasal nerve (Parsons, 1967). In early tetrapods, however, the vomeronasal axons probably joined with the olfactory axons on their path into the brain. A third major innovation was the evolution of tympanic ears, which proceeded independently in frogs, in mammals, and at least once within sauropsids. This expansion of hearing capacities was accompanied by an expansion of the octaval nerve, but the auditory axons do not form a separate cranial nerve.

On the motor side of the ledger, adult terrestrial tetrapods lost their gills, which means that the branchiomeric nerves VII, IX, and X lost most of their motor functions. In addition, terrestrial tetrapods evolved the ability to walk. This change in locomotor behavior involved changes in spinal circuits and spinal nerves, rather than cranial nerves, and shall not concern us here. Early tetrapods also evolved a distinct and mobile neck. As mentioned in the previous section, this transformation involved the evolution of more complex neck muscles, but those muscles remained innervated by the spinal accessory nerve (nXI). The fourth set of motor innovations in early tetrapods related to the evolution of a muscular tongue (Iwasaki, 2002). Most fishes rely on suction feeding to capture prey and transport food through the oral cavity, but suction feeding does not work in air. Tetrapods solved this problem, in part, by evolving a fleshy tongue that could be extended to target prey (Deban et al., 2007) and used to manipulate food within the oral cavity. The muscles of the tongue evolved from the hypobranchial musculature of pre-tetrapod fishes and, accordingly, are innervated by the same cranial nerve, namely the hypoglossal nerve (nXII). Thus, despite occasional assertions to the contrary (e.g., Székely and Matesz, 1993), the enormous changes in the skeletomotor system of early tetrapods were not accompanied by the evolution of new cranial nerves; nor did they prompt the loss of any motor nerves. Most of the changes happened in the periphery, among the muscles and bones, and in the central mechanisms of neural control (Matesz et al., 2014).

Two additional innovations of terrestrial tetrapods are tear and salivary glands. The former are essential for life on land, as they prevent the eyes from drying out. The latter are useful not only for digestion but also for lubricating the oral cavity and GI tract so that ingested food does not get stuck. The parasympathetic axons innervating these glands travel through the facial, trigeminal, and glossopharyngeal nerves. Thus, again, major changes in the periphery did not entail substantial modifications of the preexisting cranial nerves.

A.4. Conclusions

As this overview has shown, the cranial nerves vary dramatically between vertebrates and their invertebrate relatives, and they exhibit substantial variation even within the vertebrates. This variation is not captured by the simplified schema of "the twelve cranial nerves" that most students of neuroanatomy are asked to learn (see Table A.1). Nor is it emphasized in the segmental paradigm of cranial nerve organization (see Figure A.2), which focuses on the serial repetition of a conserved pattern, rather than variation. In fact, the segmental paradigm neglects fundamental differences between cranial and spinal nerves, as well as differences between the head of vertebrates and that of the invertebrate chordates. Therefore, we have here taken a different approach, which is to divide the cranial nerves into six major groups according to their shared features (Table A.2) and then to ask when in phylogeny those nerves originated and how they were subsequently modified.

That said, it is important to note that, after the origin of vertebrates, the cranial nerves underwent remarkably little change. The loss of the lateral line nerves in terrestrial tetrapods was a major transformation but, beyond that, most of the variation is limited to the peripheral branches of the established cranial nerves. Some axons were phylogenetically "re-routed" and some of the cranial nerves vary in the precise location where they enter or exit the brain, but the major trunks and ganglia of the cranial nerves are highly conserved. This phylogenetic stability probably reflects the fact that the major trunks of the cranial nerves are established early in development, following broadly conserved molecular cues. Later developing axons, which tend to be more variable across phylogeny, tend to follow those early paths and, thus, join the previously established nerves. As a result, major transformations in the sensory or motor periphery, or within the brain, are often not accompanied by major changes in the cranial nerves. Still, the existing variation is substantial enough to create serious problems for anyone who tries to summarize the organization of the cranial nerves in terms of any simple, general scheme. As for the nervous system generally, the "vertebrate Bauplan" has undergone substantial remodeling.

References

Adachi N, Pascual-Anaya J, Hirai T, Higuchi S, Kuratani S. 2018. Development of hypobranchial muscles with special reference to the evolution of the vertebrate neck. *Zool Lett* 4:5.

Adachi N, Takechi M, Hirai T, Kuratani S. 2012. Development of the head and trunk mesoderm in the dogfish, *Scyliorhinus torazame*. II. Comparison of gene expression between the head mesoderm and somites with reference to the origin of the vertebrate head. *Evol Dev* 14:257–276.

Adler K. 1976. Extraocular photoreception in amphibians. *Photochem Photobiol* 23:275–298.

Aguillon R, Batut J, Subramanian A, Madelaine R, Dufourcq P, Schilling TF, Blader P. 2018. Cell-type heterogeneity in the early zebrafish olfactory epithelium is generated from progenitors within preplacodal ectoderm. *eLife* 7:e32041.

Anderson CW, Nishikawa KC. 1997. The functional anatomy and evolution of hypoglossal afferents in the leopard frog, *Rana pipiens*. *Brain Res* 771:285–291.

Ashmore J. 2008. Cochlear outer hair cell motility. *Physiol Rev* 88:173–210.

Baker CVH, Modrell MS, Gillis JA. 2013. The evolution and development of vertebrate lateral line electroreceptors. *J Exp Biol* 216:2515–2522.

Bartheld von CS. 2004. The terminal nerve and its relation with extrabulbar "olfactory" projections: lessons from lampreys and lungfishes. *Microsc Res Tech* 65:13–24.

Bemis WE, Northcutt RG. 1991. Innervation of the basicranial muscle of *Latimeria chalumnae*. *Environ Biol Fish* 32:147–158.

Berti R, Durand JP, Becchi S, Brizzi R, Keller N, Ruffat G. 2001. Eye degeneration in the blind cave-dwelling fish *Phreatichthys andruzzii*. *Can J Zool* 79:1278–1285.

Besharse JC, Brandon RA. 1974. Postembryonic eye degeneration in the troglobitic salamander *Typhlotriton spelaeus*. *J Morphol* 144:381–405.

Boord RL, Campbell CBG. 1977. Structural and functional organization of the lateral line system of sharks. *Am Zool* 17:431–441.

Brodal A. 1967. *The Cranial Nerves: Anatomy and Anatomo-Clinical Correlations*. Oxford: Blackwell Scientific.

Bronner-Fraser M, Stern C. 1991. Effects of mesodermal tissues on avian neural crest cell migration. *Dev Biol* 143:213–217.

Brown JW. 1980. Developmental history of nervus terminalis in embryos of insectivorous bats. *Anat Rec* 196:23–35.

Buhl EH, Oelschläger HA. 1986. Ontogenetic development of the nervus terminalis in toothed whales: evidence for its non-olfactory nature. *Anat Embryol* 173:285–294.

Butler AB. 2002. Cranial Nerves. In: Ramachandran V (ed.), *Encyclopedia of the Human Brain*, Vol. 2. Amsterdam: Elsevier, pp. 65–81.

Butler AB, Hodos W. 2005. *Comparative Vertebrate Neuroanatomy*. Hoboken, NJ: John Wiley & Sons.

Butler PJ, Taylor EW, Short S. 1977. The effect of sectioning cranial nerves V, VII, IXand X on the cardiac response of the dogfish *Scyliorhinus canicula* to environmental hypoxia. *J Exp Biol* 69:233–245.

Campos D, Rieger A, Mohr H, Morphol JEJ, 2011. 2011. Anatomy and evolution of accessory nerve: cranial or spinal origins? A review. *J Morph Sci* 28:222–227.

Cole FJ. 1897. On the cranial nerves of *Chimæra monstrosa* (Linn. 1754); with a discussion of the lateral line system, and of the morphology of the chorda tympani. *Trans Roy Soc Edinburgh* 38:631–680.

Cole WC, Youson JH. 1982. Morphology of the pineal complex of the anadromous sea lamprey, *Petromyzon marinus* L. *Am J Anat* 165:131–163.

Concha ML, Wilson SW. 2001. Asymmetry in the epithalamus of vertebrates. *J Anat* 199:63–84.

Counter SA, Borg E, Lännergren J. 1981. Basic contraction properties of the avian stapedius muscle. *Acta Physiol Scand* 111:105–108.

D'aniello B, Luongo L, Rastogi RK, Di Meglio M, Pinelli C. 2015. Tract-tracing study of the extrabulbar olfactory projections in the brain of some teleosts. *Microsc Res Tech* 78:268–276.

Daston MM, Adamek GD, Gesteland RC. 1990. Ultrastructural organization of receptor cell axons in frog olfactory nerve. *Brain Res* 537:69–75.

Deban SM, O'Reilly JC, Dicke U, van Leeuwen JL. 2007. Extremely high-power tongue projection in plethodontid salamanders. *J Exp Biol* 210:655–667.

Demski LS. 1984. The evolution of neuroanatomical substrates of reproductive behavior: Sex steroid and LHRH-specific pathways including the terminal nerve. *Am Zool* 24:809–830.

Demski LS, Northcutt RG. 1983. The terminal nerve: a new chemosensory system in vertebrates? *Science* 220:435–437.

Diogo R, Kelly RG, Christiaen L, Levine M, Ziermann JM, Molnar JL, Noden DM, Tzahor E. 2015. A new heart for a new head in vertebrate cardiopharyngeal evolution. *Nature* 520:466–473.

Diogo R, Ziermann JM. 2014. Muscles of chondrichthyan paired appendages: comparison with osteichthyans, deconstruction of the fore-hindlimb serial homology dogma, and new insights on the evolution of the vertebrate neck. *Anat Rec* 298:513–530.

Douglas RH. 2018. The pupillary light responses of animals; a review of their distribution, dynamics, mechanisms and functions. *Prog Ret Eye Res*. 10.1016/j.preteyeres.2018.04.005.

Dutel H, Herbin M, Clément G, Herrel A. 2015. Bite force in the extant coelacanth *Latimeria*: The role of the intracranial joint and the basicranial muscle. *Curr Biol* 25:1228–1233.

Edinger T. 1956. Paired pineal organs. *Prog Neurobiol* 2:121–129.

Eisthen HL, Delay RJ, Wirsig-Wiechmann CR, Dionne VE. 2000. Neuromodulatory effects of gonadotropin releasing hormone on olfactory receptor neurons. *J Neurosci* 20:3947–3955.

Eisthen HL, Polese G. 2007. Evolution of vertebrate olfactory subsystems. In: Kaas JH (ed.), *Evolution of Nervous Systems*, Vol. 2. Amsterdam: Elsevier, pp. 355–406.

Ekström P, Meissl H. 2003. Evolution of photosensory pineal organs in new light: the fate of neuroendocrine photoreceptors. *Philos Trans R Soc Lond B* 358:1679–1700.

Ekström P, van Veen T. 1983. Central connections of the pineal organ in the three-spined stickleback, *Gasterosteus aculeatus* L. (Teleostei). *Cell Tissue Res* 232:141–155.

Finger TE. 2009. Evolution of gustatory reflex systems in the brainstems of fishes. *Integr Zool* 4:53–63.

Forni PE, Wray S. 2012. Neural crest and olfactory system: new prospective. *Mol Neurobiol* 46:349–360.

Fritsch R, Ullmann JFP, Bitton P-P, Collin SP, Michiels NK. 2017. Optic-nerve-transmitted eyeshine, a new type of light emission from fish eyes. *Frontiers Zool* 14:14.

Fritzsch B, Dubuc R, Ohta Y, Grillner S. 1989. Efferents to the labyrinth of the river lamprey (*Lampetra fluviatilis*) as revealed with retrograde tracing techniques. *Neurosci Lett* 96:241–246.

Fritzsch B, Elliott KL. 2017. Evolution and development of the inner ear efferent system: Transforming a motor neuron population to connect to the most unusual motor protein via ancient nicotinic receptors. *Front Cell Neurosci* 11:181–189.

Fritzsch B, Elliott KL, Glover JC. 2017. Gaskell revisited: new insights into spinal autonomics necessitate a revised motor neuron nomenclature. *Cell Tissue Res* 370:195–209.

Fritzsch B, Northcutt RG. 1993. Cranial and spinal nerve organization in amphioxus and lampreys: evidence for an ancestral craniate pattern. *Acta Anat (Basel)* 148:96–109.

Fritzsch B, Sonntag R, Dubuc R, Ohta Y, Grillner S. 1990. Organization of the six motor nuclei innervating the ocular muscles in lamprey. *J Comp Neurol* 294:491–506.

Ganchrow JR, Ganchrow D, Oppenheimer M. 1986. Chorda tympani innervation of anterior mandibular taste buds in the chicken (*Gallus gallus domesticus*). *Anat Rec* 216:434–439.

Gans C. 1989. Stages in the origin of vertebrates: analysis by means of scenarios. *Biol Rev* 64:221–268.

Gans C, Northcutt RG. 1983. Neural crest and the origin of vertebrates: a new head. *Science* 220:268–273.

Goodrich ES. 1918. On the development of the segments of the head in *Scyllium*. *Q J Microsc Soc* 63:1–30.

Greene CW. 1902. Contributions to the physiology of the California hagfish, *Polistotrema stouti*. II. The absence of regulative nerves for the systemic heart. *Am J Physiol* 6:318–324.

Guimond JC, Auclair F, Lund JP, Dubuc R. 2003. Anatomical and physiological study of respiratory motor innervation in lampreys. *Neurosci* 122:259–266.

Häming D, Simoes-Costa M, Uy B, Valencia J, Sauka-Spengler T, Bronner-Fraser M. 2011. Expression of sympathetic nervous system genes in lamprey suggests their recruitment for specification of a new vertebrate feature. *PLoS ONE* 6:e26543–8.

Harlow DE, Barlow LA. 2007. Embryonic origin of gustatory cranial sensory neurons. *Dev Biol* 310:317–328.

Hellmann B, Fritzsch B. 1996. Neuroanatomical and histochemical evidence for the presence of common lateral line and inner ear efferents and of efferents to the basilar papilla in a frog, *Xenopus laevis*. *Brain Behav Evol* 47:185–194.

Higashiyama H, Hirasawa T, Oisi Y, Sugahara F, Hyodo S, Kanai Y, Kuratani S. 2016. On the vagal cardiac nerves, with special reference to the early evolution of the head-trunk interface. *J Morphol* 277:1146–1158.

Iwasaki S-I. 2002. Evolution of the structure and function of the vertebrate tongue. *J Anat* 201:1–13.

Jacob F. 1977. Evolution and tinkering. *Science* 196:1161–1166.

Jeffery G, Evans A, Albon J, Duance V, Neal J, Dawidek G. 1995. The human optic nerve: fascicular organisation and connective tissue types along the extra-fascicular matrix. *Anat Embryol* 191:491–502.

Johnels AG. 1956. On the peripheral autonomic nervous system of the trunk region of *Lampetra planeri*. *Acta Zool* 37:251–286.

Johnston JB. 1905. The morphology of the vertebrate head from the viewpoint of the functional divisions of the nervous system. *J Comp Neurol Psychol* 15:175–275.

Keynes RJ, Jaques KF, Cook GM. 1991. Axon repulsion during peripheral nerve segmentation. *Dev Suppl Suppl* 2:131–139.

Keynes RJ, Stern CD. 1985. Segmentation and neural development in vertebrates. *Trends Neurosci* 8:220–223.

Kiecker C, Lumsden A. 2005. Compartments and their boundaries in vertebrate brain development. *Nat Rev Neurosci* 6:553–564.

Korf HW, Wagner U. 1981. Nervous connections of the parietal eye in adult *Lacerta* s. *sicula* Rafinesque as demonstrated by anterograde and retrograde transport of horseradish peroxidase. *Cell Tissue Res* 219:567–583.

Kozicz T, Bittencourt JC, May PJ, Reiner A, Gamlin PDR, Palkovits M, Horn AKE, Toledo CAB, Ryabinin AE. 2011. The Edinger-Westphal nucleus: a historical, structural, and functional perspective on a dichotomous terminology. *J Comp Neurol* 519:1413–1434.

Kuratani S. 2008a. Is the vertebrate head segmented? Evolutionary and developmental considerations. *Integr Comp Biol* 48:647–657.

Kuratani S. 2008b. Evolutionary developmental studies of cyclostomes and the origin of the vertebrate neck. *Dev Growth Differ* 50:S189–S194.

Kuratani S. 2013. Evolution. A muscular perspective on vertebrate evolution. *Science* 341:139–140.

Kuratani SC, Eichele G. 1993. Rhombomere transplantation repatterns the segmental organization of cranial nerves and reveals cell-autonomous expression of a homeodomain protein. *Development* 117:105–117.

Kuratani S, Horigome N. 2000. Developmental morphology of branchiomeric nerves in a cat shark, *Scyliorhinus torazame*, with special reference to rhombomeres, cephalic mesoderm, and distribution patterns of cephalic crest cells. *Zool Sci* 17:893–909.

Kuratani S, Horigome N, Hirano S. 1999. Developmental morphology of the head mesoderm and reevaluation of segmental theories of the vertebrate head: evidence from embryos of an agnathan vertebrate, *Lampetra japonica*. *Dev Biol* 210:381–400.

Kuratani S, Nobusada Y, Saito H, Shigetani Y. 2000. Morphological characteristics of the developing cranial nerves and mesodermal head cavities in sturgeon embryos from early pharyngula to late larval stages. *Zool Sci* 17:911–933.

Kuratani S, Ueki T, Aizawa S, Hirano S. 1997. Peripheral development of cranial nerves in a cyclostome, *Lampetra japonica*: morphological distribution of nerve branches and the vertebrate body plan. *J Comp Neurol* 384:483–500.

Lacalli TC. 2004. Sensory systems in amphioxus: a window on the ancestral chordate condition. *Brain Behav Evol* 64:148–162.

Lachman N, Acland RD, Rosse C. 2002. Anatomical evidence for the absence of a morphologically distinct cranial root of the accessory nerve in man. *Clin Anat* 15:4–10.

Larsson M. 2011. Binocular vision and ipsilateral retinal projections in relation to eye and forelimb coordination. *Brain Behav Evol* 77:219–230.

Lee RK, Eaton RC, Zottoli SJ. 1993. Segmental arrangement of reticulospinal neurons in the goldfish hindbrain. *J Comp Neurol* 329:539–556.

Liem KF, Bemis WE, Walker WF, Grande L. 2001. *Functional Anatomy of the Vertebrates*, 3rd ed. Belmont, CA: Brooks/Cole.

Maier A, DeSantis M, Eldred E. 1974. The occurrence of muscle spindles in extraocular muscles of various vertebrates. *J Morphol* 143:397–408.

Mallatt J. 2008. The origin of the vertebrate jaw: neoclassical ideas versus newer, development-based ideas. *Zool Sci* 25:990–998.

Manni E, Desole C, Palmieri G. 1970. On whether eye muscle spindles are innervated by ganglion cells located along the oculomotor nerves. *Exp Neurol* 28:333–343.

Matesz K, Kecskes S, Bácskai T, Rácz E, Birinyi A. 2014. Brainstem circuits underlying the prey-catching behavior of the frog. *Brain Behav Evol* 83:104–111.

McDougal DH, Gamlin PD. 2015. Autonomic control of the eye. *Compr Physiol* 5:439–473.

Meier S. 1981. Development of the chick embryo mesoblast: morphogenesis of the prechordal plate and cranial segments. *Dev Biol* 83:49–61.

Milsom WK, Burleson ML. 2007. Peripheral arterial chemoreceptors and the evolution of the carotid body. *Respir Physiol Neurobiol* 157:4–11.

Münz H, Claas B, Stumpf WE, Jennes L. 1982. Centrifugal innervation of the retina by luteinizing hormone releasing hormone (LHRH)-immunoreactive telencephalic neurons in teleostean fishes. *Cell Tissue Res* 222:313–323.

Neuhuber W, Schrödl F. 2011. Autonomic control of the eye and the iris. *Auton Neurosci: Basic & Clin* 165:67–79.

Noden DM. 1983. The embryonic origins of avian cephalic and cervical muscles and associated connective tissues. *Am J Anat* 168:257–276.

Norris HW, Hughes SP. 1920. The cranial, occipital, and anterior spinal nerves of the dogfish, *Squalus acanthias*. *J Comp Neurol* 31:293–402.

Northcutt RG. 1989. The phylogenetic distribution and innervation of craniate mechanoreceptive lateral lines. In: Coombs S, Görner P, Münz H (eds.), *The Mechanosensory Lateral Line: Neurobiology and Evolution*. New York: Springer-Verlag, pp. 17–78.

Northcutt RG. 1992. The phylogeny of octavolateralis ontogenies: a reaffirmation of Garstang's phylogenetic hypothesis. In: Webster DB, Fay RR (eds.), *The Evolutionary Biology of Hearing*. New York: Springer, pp. 21–47.

Northcutt G. 1993. A reassessment of Goodrich's model of cranial nerve phylogeny. *Acta Anat* 148:71–80.

Northcutt RG. 2003. Development of the lateral line system in the channel catfish. In: Browman HI, Skiftesvik AB (eds.), *The Big Fish Bang. Proc 26th Annual Larval Fish Conference*. Bergen, NO: Institute of Marine Research, pp. 137–159.

Northcutt RG. 2004. Taste buds: development and evolution. *Brain Behav Evol* 64:198–206.

Northcutt RG, Bemis WE. 1993. Cranial nerves of the coelacanth, *Latimeria chalumnae* [Osteichthyes: Sarcopterygii: Actinistia], and comparisons with other craniata. *Brain Behav Evol* 42 Suppl 1:1–76.

Northcutt RG, Brändle K. 1995. Development of branchiomeric and lateral line nerves in the axolotl. *J Comp Neurol* 355:427–454.

Northcutt RG, Brändle K, Fritzsch B. 1995. Electroreceptors and mechanosensory lateral line organs arise from single placodes in axolotls. *Dev Biol* 168:358–373.

Northcutt RG, Catania KC, Criley BB. 1994. Development of lateral line organs in the axolotl. *J Comp Neurol* 340:480–514.

Northcutt RG, Gans C. 1983. The genesis of neural crest and epidermal placodes: a reinterpretation of vertebrate origins. *Quart Rev Biol* 58:1–28.

Northcutt RG, Holmes PH, Albert JS. 2000. Distribution and innervation of lateral line organs in the channel catfish. *J Comp Neurol* 421:570–592.

Oisi Y, Ota KG, Kuraku S, Fujimoto S, Kuratani S. 2013. Craniofacial development of hagfishes and the evolution of vertebrates. *Nature* 493:175–180.

Oksche A. 1984. Evolution of the pineal complex: correlation of structure and function. *Ophthalmic Res* 16:88–95.

Onai T, Irie N, Kuratani S. 2014. The evolutionary origin of the vertebrate body plan: the problem of head segmentation. *Annu Rev Genom Human Genet* 15:443–459.

Parsons TS. 1967. Evolution of the nasal structure in the lower tetrapods. *Am Zool* 7:397–413.

Pinkus F. 1895. Die Hirnnerven des *Protopterus annectens*. *Morphol Arb* 4:275–346.

Piotrowski T, Northcutt RG. 1996. The cranial nerves of the Senegal bichir, *Polypterus senegalus* [Osteichthyes: Actinopterygii: Cladistia]. *Brain Behav Evol* 47:79–102.

Pombal MA, Megías M. 2018. Development and functional organization of the cranial nerves in lampreys. *Anat Rec*. doi:10.1002/ar.23821

Puzdrowski RL. 1989. Peripheral distribution and central projections of the lateral-line nerves in goldfish, *Carassius auratus*. *Brain Behav Evol* 34:110–131.

Puzdrowski RL, Northcutt RG. 1989. Central projections of the pineal complex in the silver lamprey *Ichthyomyzon unicuspis*. *Cell Tissue Res* 255:269–274.

Quay WB. 1979. The parietal eye-pineal complex. In: Gans C, Northcutt RG, Ulinsk P (eds.), *Biology of the Reptilia*, Vol. 9A. London: Academic Press, pp. 245–406.

Roberts BL, Meredith GE. 1992. The efferent innervation of the ear: variations on an enigma. In: Wegster DB, Fay RR (eds.), *The Evolutionary Biology of Hearing*. New York: Springer, pp. 185–210.

Sambasivan R, Kuratani S, Tajbakhsh S. 2011. An eye on the head: the development and evolution of craniofacial muscles. *Development* 138:2401–2415.

Scholes J. 1991. The design of the optic nerve in fish. *Vis Neurosci* 7:129–139.

Schwanzel-Fukuda M, Pfaff DW. 1990. The migration of luteinizing hormone-releasing hormone (LHRH) neurons from the medial olfactory placode into the medial basal forebrain. *Experientia* 46:956–962.

Shigetani Y, Sugahara F, Kuratani S. 2005. A new evolutionary scenario for the vertebrate jaw. *Bioessays* 27:331–338.

Sienknecht UJ, Köppl C, Fritzsch B. 2014. Evolution and development of hair cell polarity and efferent function in the inner ear. *Brain Behav Evol* 83:150–161.

Smith KT, Bhullar B-AS, Köhler G, Habersetzer J. 2018. The only known jawed vertebrate with four eyes and the bauplan of the pineal complex. *Curr Biol* 28:1101–1107.

Song JK, Northcutt RG. 1991. Morphology, distribution and innervation of the lateral-line receptors of the Florida gar, *Lepisosteus platyrhincus*. *Brain Behav Evol* 37:10–37.

Suzuki DG, Fukumoto Y, Yoshimura M, Yamazaki Y, Kosaka J, Kuratani S, Wada H. 2016. Comparative morphology and development of extra-ocular muscles in the lamprey and gnathostomes reveal the ancestral state and developmental patterns of the vertebrate head. *Zool Lett* 2:10.

Székely G, Matesz C. 1993. The hypoglossal nucleus: the appearance of the muscular tongue. *Adv Anat Embryol Cell Biol* 128:7–11.

Tada MN, Kuratani S. 2015. Evolutionary and developmental understanding of the spinal accessory nerve. *Zool Lett* 1:4.

Taylor EW, Jordan D, Coote JH. 1999. Central control of the cardiovascular and respiratory systems and their interactions in vertebrates. *Physiol Rev* 79:855–916.

Trinajstic K, Sanchez S, Dupret V, Tafforeau P, Long J, Young G, Senden T, Boisvert C, Power N, Ahlberg PE. 2013. Fossil musculature of the most primitive jawed vertebrates. *Science* 341:160–164.

Vilensky JA. 2014. The neglected cranial nerve: nervus terminalis (cranial nerve N). *Clin Anat* 27:46–53.

Ward R, Repérant J, Hergueta S, Miceli D, Lemire M. 1995. Ipsilateral visual projections in non-eutherian species: random variation in the central nervous system? *Brain Res Rev* 20:155–170.

Whitlock KE, Westerfield M. 2000. The olfactory placodes of the zebrafish form by convergence of cellular fields at the edge of the neural plate. *Development* 127:3645–3653.

Whitlock KE, Wolf CD, Boyce ML. 2003. Gonadotropin-releasing hormone (GnRH) cells arise from cranial neural crest and adenohypophyseal regions of the neural plate in the zebrafish, *Danio rerio*. *Dev Biol* 257:140–152.

Wicht H, Lacalli TC. 2005. The nervous system of amphioxus: structure, development, and evolutionary significance. *Can J Zool* 83:122–150.

Wilkinson DG, Bhatt S, Cook M, Boncinelli E, Krumlauf R. 1989. Segmental expression of Hox-2 homoeobox-containing genes in the developing mouse hindbrain. *Nature* 341:405–409.

Wirsig-Wiechmann CR, Houck LD, Feldhoff PW, Feldhoff RC. 2002. Pheromonal activation of vomeronasal neurons in plethodontid salamanders. *Brain Res* 952:335–344.

Yalden DW. 1985. Feeding mechanisms as evidence for cyclostome monophyly. *Zool J Linn Soc* 84:291–300.

Yasui K, Kaji T, Morov AR, Yonemura S. 2013. Development of oral and branchial muscles in lancelet larvae of *Branchiostoma japonicum*. *J Morphol* 275:465–477.

Ziermann JM, Miyashita T, Diogo R. 2014. Cephalic muscles of Cyclostomes (hagfishes and lampreys) and Chondrichthyes (sharks, rays and holocephalans): comparative anatomy and early evolution of the vertebrate head muscles. *Zool J Linn Soc* 172:771–802.

Index

For the benefit of digital users, indexed terms that span two pages (e.g., 52–53) may, on occasion, appear on only one of those pages.

Tables and figures are indicated by *t* and *f* following the page number